Handbook of Nanotechnology in Nutraceuticals

Nanotechnology has been emerging as an important tool in the nutraceutical and food industries to improve the overall quality of life. Nanotechnology has established a new horizon by bestowing modified properties on nanomaterials and applying them to the production of nanoformulations, nutritional supplements, and the food industry. The ***Handbook of Nanotechnology in Nutraceuticals*** highlights the impact of nanotechnology on the food industries.

The book focuses on the application of nanotechnology in nutraceuticals and the food industry to improve the overall quality of life. The book also addresses some important applications of nano-nutraceuticals in the treatment of different diseases, such as oxidative stress, cancer, neurodegenerative disorders, cardiovascular diseases, and so on.

Features

- Presents a scientometric approach to analyze the emergence of nano-nutraceuticals in cancer prevention and treatment
- Examines various strategies employed to prepare nanocarrier systems, such as nanoparticles, nanostructure lipids, phospholipid-based nanocarriers, polysaccharide-based nanostructures, and metal nanoparticles
- Discusses various regulatory issues related to nanotechnology and their application in different fields

This book is a valuable reference for nanotechnologists, scientists, and researchers working in the field of food technology, food science, pharmaceuticals, and nutraceuticals.

Handbook of Nanotechnology in Nutraceuticals

Edited by
Shakeel Ahmed
Tanima Bhattacharya
Annu
Akbar Ali

CRC Press is an imprint of the
Taylor & Francis Group, an **informa** business

First edition published 2023
by CRC Press
6000 Broken Sound Parkway NW, Suite 300, Boca Raton, FL 33487-2742

and by CRC Press
4 Park Square, Milton Park, Abingdon, Oxon, OX14 4RN
CRC Press is an imprint of Taylor & Francis Group, LLC

Library of Congress Cataloging-in-Publication Data

Names: Ahmed, Shakeel, 1989- editor. Title: Handbook of nanotechnology in nutraceuticals / edited by Shakeel Ahmed, Tanima Bhattacharya, Annu, Akbar Ali. Description: First edition. | Boca Raton : CRC Press, 2023. | Includes bibliographical references and index.
Identifiers: LCCN 2022020034 (print) | LCCN 2022020035 (ebook) | ISBN 9781032127910 (hardback) | ISBN 9781032155678 (paperback) | ISBN 9781003244721 (ebook)
Subjects: LCSH: Functional foods--Handbooks, manuals, etc. | Nanotechnology. Classification: LCC QP144.F85 H358 2023 (print) | LCC QP144.F85 (ebook) | DDC 572/.33--dc23/eng/20220720
LC record available at https://lccn.loc.gov/2022020034
LC ebook record available at https://lccn.loc.gov/2022020035

ISBN: 978-1-032-12791-0 (hbk)
ISBN: 978-1-032-15567-8 (pbk)
ISBN: 978-1-003-24472-1 (ebk)

DOI: 10.1201/9781003244721

Typeset in Times
by Deanta Global Publishing Services, Chennai, India

Contents

Preface vii
Editors ix
Contributors xi

Chapter 1 Nanoceuticals: Mystifying Composites at the Interface of Nutrition, Medicine, and Technology 1

Niloy Chatterjee and Pubali Dhar

Chapter 2 Scope of Nanotechnology in Nutraceuticals: Analysis, Challenges and Opportunities 51

Gurleen Kaur, Rajinder Kaur, and Sukhminderjit Kaur

Chapter 3 Systematic Review Analysis of Emerging Nano-Nutraceuticals in Cancer Prevention and Treatment Using a Scientometric Approach 63

Madhulika Bhati, Ranjana Aggarwal, Swathi Sreekuttan, Zahra Shekason, Asha Anish Madhavan, and Reshmi S. Nair

Chapter 4 Strategic Design of a Nanocarrier System for Nutraceuticals 79

Chenmala Karthika, Mehrukh Zehravi, Rokeya Akter, and Md. Habibur Rahman

Chapter 5 Nanostructured Lipids as a Bioactive Compound Carrier 97

Peeyush Kaushik, Deepali Tomar, Anjoo Kamboj, Hitesh Malhotra, and Rupesh K. Gautam

Chapter 6 Role of β-Glucans in Dyslipidemia and Obesity 123

Hitesh Malhotra, Peeyush Kaushik, Anjoo Kamboj, and Rupesh K. Gautam

Chapter 7 Nanotechnology-Based Delivery of Non-Opioid Therapy for Opioid Addiction 149

Mohd Adzim Khalili Rohin, Atif Amin Baig, Norhaslinda Ridzwan, and Norhayati Abd Hadi

Chapter 8 Phospholipid-Based Nanoplatforms: Evolving as Promising Carriers for Therapeutic Intervention 183

Amrita Chakraborty and Pubali Dhar

Chapter 9 Investigating the Potential of Multifunctional Nanoparticle-Based Nutraceuticals in Targeted Therapeutics 223

Giselle Amanda Borges e Soares, Ali H. Hamzah, and Tanima Bhattacharya

Chapter 10 Polysaccharide-Based Nanostructures as Nutraceutical Carriers 263

V.B. Poornima, Richa Katiyar, Shivali Banerjee, Antonio Patti, and Amit Arora

Chapter 11 Metal Nanoparticles in Encapsulation and Delivery Systems of Food Ingredients and Nutraceuticals 301

H.C. Ananda Murthy, Gezahegn Tadesse Ayanie, Tegene Desalegn Zeleke, Yilkal Dessie Sintayehu, and C.R. Ravikumar

Chapter 12 Applications of Nanotechnology-Based Approaches for Targeted Delivery of Nutraceuticals 329

Hitesh Chopra, Shivani Sharma, Saba Yousaf, Rahat Naseer, Shakeel Ahmed, and Atif Amin Baig

Chapter 13 Application of Nano-Nutraceuticals in Medicines 347

Kanika Sharma, Ahmed Salim, Shreyas R. Murthy, Gunjan Sharma, and Rupesh K. Gautam

Chapter 14 Nano-Nutraceuticals and Oxidative Stress 377

Rahat Naseer, Sadia Nawaz, Muzna Munir, Hitesh Chopra, Uday Younis Hussein Abdullah, Shakeel Ahmed, and Atif Amin Baig

Chapter 15 Nano-Nutraceuticals for the Treatment of Cancer 399

Saba Yousaf, Fouzia Qamar, Zirwah Tahir, Zoya Faisal, Muniba Khaliq, Aiza Talat, Umar Bacha, Aaiza Naveed, Hitesh Chopra, Uday Younis Hussein Abdullah, Shakeel Ahmed, and Atif Amin Baig

Chapter 16 Nano-Nutraceuticals in Neurodegenerative Disorders 417

Wardah Ali, Zirwah Tahir, Uday Younis Hussein Abdullah, Shakeel Ahmed, and Atif Amin Baig

Chapter 17 Use of Nano-Nutraceuticals as Anti-Inflammatory Tools in Cardiovascular Disease 431

Rajat Goyal, Anjali Saharan, and Rupesh K. Gautam

Chapter 18 Nanotechnology and Regulatory Issues 441

Abhijit Gupta, Saurabh Gupta, and Gunjan Mukherjee

Index 461

Preface

The rise in microbial resistance has shifted people's preference away from artificial medications and food supplements towards natural ones, particularly health-protecting foods. The resulting products, which serve as nutraceuticals, represent a new perspective in the food industries. The term *nutraceuticals* is a combination of *nutrition* and *pharmaceuticals*. Nutraceuticals are defined as natural substances that are foods or functional foods or parts of food containing bioactive phytochemicals that deliver therapeutic or health benefits, and which can also be used as a means of disease prevention and treatment. Similarly, it is determined by the implications that can hold a critical quality of essential nutrients such as vitamins, carbohydrates, minerals, proteins, lipids, and so on. Further, the scope of nutrition has been expanded to disease risk failure. The success of nutraceuticals can be attributed to their ability to provide desirable therapeutic benefits while minimizing the side effects associated with the use of pharmaceutical substances in the prevention and treatment of various diseases. Nutraceuticals have been proven to protect against a number of diseases, including cancer, diabetes, and neurodegenerative disorders.

However, the poor solubility of most phytochemicals in gastrointestinal fluid results in incomplete absorption and finally, no biological activity. Also, the resultant low bioavailability is a major concern and a subject of investigation. In order to address this issue, the principles of nanotechnology could provide a new way for efficient delivery of these nutraceuticals with the aim of enhancing their biological activity.

Nanotechnology has been emerging as an important tool in the nutraceutical and food industries to improve the overall quality of life. Nanotechnology has established a new horizon by bestowing modified properties on nanomaterials and applying them to the production of nanoformulations, nutritional supplements, and the food industry. The application of nanotechnology can increase solubility, improve quality, enhance lifetime, advance bioavailability, enable controlled release, and protect the overall stability of micronutrients. For the efficient delivery of encapsulated nutraceuticals, a number of formulation techniques have been developed, including nanoemulsions, micelles, nanocapsules, nanoparticles, nanocrystals, and so on. Besides improving the bioavailability of the nutraceuticals and their therapeutic efficacy, such nanoformulations provide targeted delivery and sustained release.

The book *Handbook of Nanotechnology in Nutraceuticals* is a comprehensive description and review of recent advances and applications of nanotechnology in nutraceuticals. The book contains a total of 18 chapters. Chapters 1 and 2 mainly focus on the applications of nanotechnology in the field of nutraceuticals. The challenges and opportunities of nanoceuticals in future market scenarios are analyzed. Chapter 3 presents a scientometric approach to analyze the emergence of nano-nutraceuticals in cancer prevention and treatment. Chapters 4 to 11 analyze various strategies employed to prepare nanocarrier systems, such as nanoparticles, nanostructure lipids, phospholipid-based nanocarriers, polysaccharide-based nanostructures, metal nanoparticles, and so on, for nutraceuticals and therapeutic applications. Chapters 12 to 17 present the applications of nano-nutraceuticals in different fields, such as medicine, oxidative stress, cancer treatment, neurodegenerative disorders, cardiovascular diseases, and so forth. Finally, Chapter 18 discusses the various regulatory issues related to nanotechnology and their application in different fields.

Therefore, this book will serve as an important reference for nanotechnologists, scientists, and researchers working in the fields of food technology, food science, pharmaceuticals, and nutraceuticals.

Finally, the editors would like to extend their sincere gratitude to all the contributors for providing help, support, and their contribution to accomplish this mission.

Shakeel Ahmed
Tanima Bhattacharya
Annu
Akbar Ali
Jammu and Kashmir, India

Editors

Dr. Shakeel Ahmed is an Assistant Professor of Chemistry in the Higher Education Department, Government of Jammu and Kashmir, India. He obtained a first degree in general science from the Government Postgraduate College Rajouri (University of Jammu) followed by a master's degree and doctoral degree in chemistry from Jamia Millia Islamia, a central university, New Delhi. He has published several research publications in the areas of green nanomaterials and biopolymers for various applications, including biomedical, packaging, and water treatment. He has published more than 20 books in the areas of nanomaterials and green materials with publishers of international repute. He is the recipient of the Young Scientist Award and Professor of the Year Award—2020. His name has been listed among the top 2% of scientists in the world in the area of polymer chemistry.

Dr. Tanima Bhattacharya is a formulation scientist, who completed a doctoral degree in food processing and nutrition science from the Indian Institute of Engineering Science and Technology, Shibpur, West Bengal, India, and gained overseas postdoctoral experience from the College of Chemistry and Chemical Engineering, Hubei University of China. She is interested in the fabrication of biocompatible nanostructures and studying their properties and applications in the areas of food science and technology and biomedical sciences. She has published several research articles in internationally reputed journals. She is the recipient of a number of awards from the Oil Technology Association India, the Council of Scientific and Industrial Research–Indian Institute of Chemical Technology, American Oil Chemist Society, and West Bengal State Council (Science Congress) for contribution to research.

Dr. Annu is an Assistant Professor in the Department of Chemistry in Lingaya's Vidyapeeth, Haryana, India. She is an Honorary Assistant Professor in the Department of Science and Engineering at the Novel Global Community Education Foundation, Australia. She worked as a Project Assistant at the Indian Institute of Technology Delhi in 2020. She obtained her graduation and postgraduation degrees in chemistry from the University of Allahabad and a doctoral degree in chemistry from Jamia Millia Islamia, a central university, New Delhi. She has published several scientific research articles in international peer-reviewed journals and many book chapters with publishers of international repute. Her research interests include the fabrication and modification of sustainable bionanocomposites, biomaterials, green synthesis of nanoparticles, hybrid nanomaterials, and modification of biopolymers and composite nanomaterials for their applications in the biomedical field, food packaging, nutraceuticals, textiles, and environmental sustainability.

Dr. Akbar Ali is an Assistant Professor at the Department of Chemistry, Kargil Campus, University of Ladakh, Union Territory of Ladakh, India. He completed his bachelor's degree in science (BSc) at Islamia College of Science and Commerce, University of Kashmir. He did his master's in chemistry in 2013 at Jamia Millia Islamia, a central university, New Delhi, India. He obtained his doctorate degree (PhD) in chemistry at the same institution in 2020. His areas of research include polymer chemistry, specializing in biological macromolecules, functionalization of natural polymers, polymer–iodine complexes, hydrogel chemistry, and green chemistry. He has published several research papers and book chapters in different internationally reputed journals in the area of polymers. He is a regular member of the Asian Polymer Association, India. He has attended and presented his research work at more than 20 national and international conferences.

Contributors

Uday Younis Hussein Abdullah
Faculty of Medicine
University Sultan Zainal Abidin
Kuala Terengganu, Malaysia

Ranjana Aggarwal
CSIR-National Institute of Science Communication and Policy Research (CSIR-NIScPR)
New Delhi, India

Shakeel Ahmed
Department of Chemistry
Government Degree College Mendhar
Jammu and Kashmir, India

Rokeya Akter
Department of Global Medical Science
Yonsei University
Gangwon-do, South Korea

Wardah Ali
Government College University
Lahore, Pakistan

Annu
Department of Chemistry Department of Applied Sciences
Galgotias College of Engineering and Technology
Greater Noida, U.P., India

Amit Arora
IITB-Monash Research Academy
Indian Institute of Technology Bombay
Bioprocessing Laboratory
Centre for Technology Alternatives for Rural Areas (CTARA)
Maharashtra, India

Gezahegn Tadesse Ayanie
Department of Applied Chemistry
School of Applied Natural Science
Adama Science and Technology University
Adama, Ethiopia

Umar Bacha
School of Health Sciences
University of Management and Technology Pakistan
Johar Town, Lahore, Pakistan

Atif Amin Baig
Faculty of Medicine
Sultan Zainal Abidin University (UniSZA), Medical Campus
Terengganu, Malaysia

Shivali Banerjee
Bioprocessing Laboratory
Centre for Technology Alternatives for Rural Areas (CTARA)
Maharashtra, India

Madhulika Bhati
CSIR-National Institute of Science Communication and Policy Research (CSIR-NIScPR)
New Delhi, India

Tanima Bhattacharya
Innovation, Incubation and Industry Laboratory Techno India
NJR Institute of Technology
Udaipur, Rajasthan, India

Amrita Chakraborty
Laboratory of Food Science and Technology Food and Nutrition Division
University of Calcutta
Kolkata, West Bengal, India

Niloy Chatterjee
Centre for Research in Nanoscience & Nanotechnology
University of Calcutta
Kolkata, West Bengal, India

Hitesh Chopra
Chitkara College of Pharmacy
Chitkara University
Rajpura, Punjab, India

Pubali Dhar
Laboratory of Food Science and Technology, Food and Nutrition Division
University of Calcutta
Kolkata, West Bengal, India

Zoya Faisal
School of Health Sciences
University of Management and Technology Pakistan
Johar Town, Lahore, Pakistan

Rupesh K. Gautam
Department of Pharmacology
MM School of Pharmacy
Maharishi Markandeshwar University
Sadopur-Ambala, Haryana, India

Abhijit Gupta
University Institute of Biotechnology
Chandigarh University
Mohali, Punjab, India

Saurabh Gupta
Department of Microbiology
Mata Gujri College
Fatehgarh, Punjab, India

Rajat Goyal
MM School of Pharmacy
Maharishi Markandeshwar University
Sadopur-Ambala, Haryana, India

Norhayati Abd Hadi
School of Nutrition & Dietetics
Faculty of Health Sciences
Gong Badak Campus
Sultan Zainal Abidin University (UniSZA)
Terengganu, Malaysia

Ali H. Hamzah
Department of Medicinal and Biological Chemistry
University of Toledo
Toledo, Ohio, USA

Anjoo Kamboj
Chandigarh College of Pharmacy
Mohali, Punjab, India

Chenmala Karthika
Department of Pharmaceutics
JSS College of Pharmacy
JSS Academy of Higher Education and Research
Mysore, Karnataka, India

Richa Katiyar
Bioprocessing Laboratory
Centre for Technology Alternatives for Rural Areas (CTARA)
Maharashtra, India

Gurleen Kaur
Department of Biotechnology
Chandigarh University
Mohali, Punjab, India

Rajinder Kaur
Department of Biotechnology
Chandigarh University
Mohali, Punjab, India

Sukhminderjit Kaur
Department of Biotechnology
Chandigarh University
Mohali, Punjab, India

Peeyush Kaushik
Guru Gobind Singh College of Pharmacy
Yamunanagar, Haryana, India

Muniba Khaliq
Department of Food Science and Human Nutrition
Faculty of Bio-Sciences
University of Veterinary and Animal Sciences
Lahore, Pakistan

Asha Anish Madhavan
Amity University
Dubai, UAE

Hitesh Malhotra
Guru Gobind Singh College of Pharmacy
Yamunanagar, Haryana, India

Gunjan Mukherjee
University Institute of Biotechnology
Chandigarh University
Mohali, Punjab, India

Muzna Munir
Riphah International University
Lahore, Pakistan

H.C. Ananda Murthy
Department of Applied Chemistry
School of Applied Natural Science
Adama Science and Technology University
Adama, Ethiopia

Shreyas R. Murthy
Department of Pharmaceutical Science
Massachusetts College of Pharmacy and Health Sciences
Boston, Massachusetts, USA

Reshmi S. Nair
Amity University
Dubai, UAE

Rahat Naseer
University of Veterinary and Animal Sciences (UVAS)
Lahore, Pakistan

Aaiza Naveed
Fatima Jinnah Medical University
Lahore, Pakistan

Sadia Nawaz
University of Veterinary and Animal Sciences (UVAS)
Lahore, Pakistan

Antonio Patti
School of Chemistry
Monash University
Clayton, Victoria, Australia

V.B. Poornima
IITB-Monash Research Academy
Indian Institute of Technology Bombay
Bioprocessing Laboratory Centre for Technology Alternatives for Rural Areas (CTARA)
Mumbai, Maharashtra, India

Fouzia Qamar
Biology Department
Lahore Garrison University
Lahore, Pakistan

Md. Habibur Rahman
Department of Pharmacy
Southeast University
Banani, Dhaka, Bangladesh

C.R. Ravikumar
Research Centre
Department of Chemistry
East West Institute of Technology
Visvesvaraya Technological University
Bangalore, India

Norhaslinda Ridzwan
School of Nutrition & Dietetics
Faculty of Health Sciences
Gong Badak Campus
Sultan Zainal Abidin University (UniSZA)
Terengganu, Malaysia

Mohd Adzim Khalili Rohin
School of Nutrition & Dietetics
Faculty of Health Sciences
Gong Badak Campus
Sultan Zainal Abidin University (UniSZA)
Terengganu, Malaysia

Anjali Saharan
MM School of Pharmacy
Maharishi Markandeshwar University
Sadopur-Ambala, Haryana, India

Ahmed Salim
Department of Pharmacology
School of Pharmaceutical Education and Research
Jamia Hamdard University
New Delhi, India

Gunjan Sharma
Department of Pharmacology
Delhi Pharmaceutical Sciences and Research University
New Delhi, India

Kanika Sharma
Department of Pharmaceutical Science
Massachusetts College of Pharmacy and Health Sciences
Boston, Massachusetts, USA

Shivani Sharma
Institute of Pharmaceutical Science
Bhaddal, Punjab, India

Zahra Shekason
Amity University
Dubai, UAE

Yilkal Dessie Sintayehu
Department of Applied Chemistry
School of Applied Natural Science
Adama Science and Technology University
Adama, Ethiopia

Giselle Amanda Borges E. Soares
Department of Medicinal and Biological Chemistry
University of Toledo
Toledo, Ohio, USA

Swathi Sreekuttan
Amity University
Dubai, UAE

Zirwah Tahir
Institute of Orthopedics and Rehabilitation
Lahore, Pakistan

Aiza Talat
School of Health Sciences
University of Management and Technology Pakistan
Johar Town, Lahore, Pakistan

Deepali Tomar
Geeta Institute of Pharmacy
Naultha, Panipat, Haryana, India

Saba Yousaf
School of Health Sciences
University of Management and Technology Pakistan
Johar Town, Lahore, Pakistan

Mehrukh Zehravi
Department of Clinical Pharmacy Girls Section
Prince Sattam Bin Abdul Aziz University
Alkharj, Saudi Arabia

Tegene Desalegn Zeleke
Department of Applied Chemistry
School of Applied Natural Science
Adama Science and Technology University
Adama, Ethiopia

1 Nanoceuticals
Mystifying Composites at the Interface of Nutrition, Medicine, and Technology

*Niloy Chatterjee and Pubali Dhar**

CONTENTS

Abbreviations 1
1.1 Introduction 2
1.2 Why Nutraceuticals? 3
1.3 Nutraceuticals to Nanoceuticals 6
1.4 Quantification of Nanoparticles and Nanomaterials in Food and Other Related Matrices 8
1.5 Nanoformulations or Nanostructures for Delivery of Nutraceuticals 12
1.5.1 Nanoparticles 12
1.5.2 Solid-Lipid Nanoparticles 14
1.5.3 Niosomes 14
1.5.4 Nanospheres 15
1.5.5 Nanoliposomes 15
1.5.6 Nanofibers 16
1.5.7 Nanoemulsion 17
1.5.8 Nanocapsules 18
1.5.9 Carbon Nanotubes 19
1.6 Bioavailability 19
1.6.1 Solubilization 24
1.6.2 Absorption 25
1.6.3 Metabolism, Escape, and Passage to Blood Circulation 26
1.7 Disadvantages/Pitfalls of Use of Nanoceuticals 26
1.8 Regulation of Nanoceuticals and Nanoproducts 27
1.9 Market Review 32
1.10 Conclusion 33
1.11 Future Prospects 34
References 35

ABBREVIATIONS

IOS International Organization for Standardization
SLN Solid-Lipid Nanoparticle
NP Nanoparticle
NM Nanomaterial
OECD Organisation for Economic Co-operation and Development

* Corresponding author: pubalighshdhar23@gmail.com or pubalighoshdhar@yahoo.co.in

DOI: 10.1201/9781003244721-1

AOCS American Oil Chemists' Society
TiO_2 Titanium Dioxide

1.1 INTRODUCTION

The concept of nutraceuticals was introduced and originated by Stephen L. DeFelice (1989), founding father and chairperson of the Modern Medicine Foundation (New York), an American association which advocates medicinal healthiness, combining nutrition and pharmaceuticals (Andlauer and Furst 2002; Kalra 2003). Nutraceuticals are characterized as diet or food sources offering a plethora of health and therapeutic benefits, such as avoidance and treatment of ailments (Shinde et al. 2014). Nutraceuticals include a variety of foodstuffs, such as extracted nutrients, nutritional complements, fortified/enriched foods, natural and functional foods, and therapeutic foods. They are therefore more accurately described as ingredients of a diet or food that appear to offer medicinal or health benefits as well as counteracting and curing diseases (El Sohaimy 2012). The term is comprehensively applied to products including selected nutrients and components, prebiotics, dietary fiber, probiotics, antioxidants, polyunsaturated fatty acids (PUFAs), various other food products and herbal supplements, specialized foods and processed foodstuffs such as cereal grains, soups, and also drinks (Das et al. 2012). The alchemists and Casimir Funk were among the biggest contributors who endorsed and encouraged the contemporary nutraceuticals paradigm by proposing the innovative notions of "elixir vitae" and "vitamins", respectively (Ronis et al. 2018). Since prehistoric times, our ancestors have invariably used food to treat infections, favoring the treatment of many bodily complaints with the assistance of a meticulous food regime. Traditionally, any particular foods or diets that can avert or treat disease are described as drugs by the United States Food and Drug Administration (USFDA) (Ruchi 2017; Wildman 2016). In recent years, people have sought to enhance their standard of living by altering their diet and swapping conventional medication for alternative organic and natural compounds, which are highly potent and have no side effects. The nutraceutical market and related R&D are therefore flourishing (Srivastava et al. 2015). The poor bioabsorption of nutraceuticals is a major question in this field that needs to be addressed. Generally, they are excreted from the body without offering any restorative advantage or therapeutic value (El Sohaimy 2012). Nanotechnology has the potential to enhance not only bio-absorption and retention but also bioavailability and sustained release of nutrients and health supplements (He and Hwang 2016; Dasgupta et al. 2015).

Nanotechnology is characterized as the design strategy, amalgamation, and exploitation of constituents of nanoscale dimensions, which gives rise to explicit physical, chemical, mechanical, electrical, and also biological properties (Jeevanandam et al. 2018) not encountered in macro or bulk counterparts. "Nano" originates from Greek and signifies "dwarf". A nanometer (nm) is a very small-scale dimension of length and mathematically can be defined as one billionth of a meter (10^{-9} meters). The field of nanoscience was previously assumed to be homogeneous and of great importance only in the domains of elementary sciences and engineering, but its diversity and multifaceted potential are being recognized and currently expanding and including the domains of biology, food and nutrition, dietetics, nutraceuticals, and biotechnology. Even though nanotransformation does not modify the biochemical skeleton of the nutrient, it has a great effect on its functioning, potency, and efficacy. The distinctive properties of nanomaterials are attributable to their tiny dimensions, typically about 1–100 nm (though there is still a lack of in-depth consensus on the concept of nanomaterials), the high surface area to volume ratio, and the high reactivity, with resulting higher absorption and communication with biological barriers and surfaces (Stone et al. 2017; Nel et al. 2013). Fabricated nanoparticles and nanomaterials, including nanoceuticals, have huge consequences for the global economy, trade and commerce, and also human livelihoods, and are used in a wide assortment of daily items, such as feedstuffs, edibles, veterinary drugs, and biocides, and in other fields, such agribusiness, cleansing, or water decontamination strategies (Colvin 2003). Nanoscience is evolving as one of the most dynamic, encouraging and fast-growing areas of R&D and innovation, with extensive prospects of ameliorating health and quality of life as well as personal satisfaction.

Extensive benefactions of preferment in nanoscale phenomena to boost human verdure have given rise to further anticipation, resulting in the development of novel biomaterials, apparatus, approaches, and tools for biological revelation and advancement (Wordpress.com). The introduction of nano-based procedures to strengthen the distribution of nutraceuticals is another recent application of nutritive linctus. Many issues associated with the impaired physiochemical profile of nutraceuticals, primarily related to their uptake and inferior water solubility, can be addressed by modern nanotechnology (Asghar et al. 2018).

Significant nanotechnology applications in the food science and nutritional disciplines relate to the search, evolution, and design of "new food constituents" having functional and bioactive functions possessing enhanced thermal stability, aqueous solubility, oral bioavailability, sensorial characteristics, and also physiological as well as biological efficiency (Gaibort et al. 2016). Since the first highlighted study on food nanotechnology, which was featured in 2003, there has been a great deal of curiosity and mystery surrounding this groundbreaking field of study. Several more prominent studies and reports have been published recently, not only to realize novel possibilities of food nanotechnology but also to unearth unexplored fields and areas of this technology in the domains of food security, food hygiene, formulation of new food products using nanotechnological advancements or attributes, and controlled, tailored structure and delivery systems for the supply of bioactive compounds and functional foods for health improvement and therapeutic applications (Thiruvengadam et al. 2018; Vandermoere et al. 2011).

The last few years have seen an upsurge in studies as well as experiments focusing on the exploitation of nanotechnology in nutraceutical assimilation, metabolic processes, and excretion, utilizing various *in vitro, in situ*, and also *in vivo* models. This chapter emphasizes the physiological assimilation, formation, and breakdown as well as the biological fate of nanoceuticals (<1000 nm), different technologies and methodologies adapted for their nanofabrication, their pros and cons as well as the present market scenario, and will be valuable for all investigators, scientists, and entrepreneurs who are interested in using food-grade or edible nanoparticle delivery systems or nanoforms of various bioactive nutrients and need background information. Furthermore, how these nanoceuticals are changing the face of human nutrition in the current scenario and potential research problems and future exploration challenges are described in this context. To our knowledge, detailed technology advancements and other facets of areas associated with nanotechnology in nutraceuticals or functional foods have not been extensively covered elsewhere, which underlies the novelty of this chapter.

1.2 WHY NUTRACEUTICALS?

Different interventions are being used to battle ailments and heal illnesses and prospective health threats. Along with pharmaceutical avenues, dietary approaches are also observed to be appropriate for the avoidance of various ailments (Santini et al. 2017). An ingredient or constituent that is only responsible for the growth and sustenance of healthy cells, organs, and tissues and performs no additional beneficial roles in physiological systems can be referred to as a food ingredient. On the contrary, any nutrient or food component that has a beneficial effect on biological or physiological processes, either directly or indirectly, can be classified in the domain of pharmaceuticals or medicines (Pandey et al. 2013; Abuajah et al. 2015). In a general sense, the term "nutraceuticals" can refer to any bioactive component or mixture present in herbal or various food products. Nutraceuticals generally include phytochemicals, bioactive ingredients, functional compounds, etc. Bioactive compounds are supplementary nutritious elements that typically exist in trivial amounts in foods. Manipulating various types of foods/nutraceuticals can be effective in relieving and managing various maladies (Keshwani et al. 2015). The conception of nutraceuticals is not exclusively novel but was previously unexplored, and the domain has progressed dramatically in the last few decades. Some of the well-known examples include β-carotene present in carrots, β-glucan from barley and oats, folic acid in broccoli and green leafy vegetables, lycopene in watermelons and tomato, glucosamine in shellfish, ω-3 fatty acid from fish oil, isoflavones in soybeans, conjugated linoleic acid from soybean oils and mushrooms, and probiotic bacilli in yogurt and curd (Tank Dharti et al. 2010; Shekhar et al. 2014). Miscellaneous bioactive nutraceuticals are found in our daily food regime and diet;

they are highly useful therapeutically and are assumed to be very important, playing a pivotal role in the maintenance of our wellbeing. Carotenoids, flavonoids and their glycosides, polyphenols, vitamins C and E, fatty acids, and so on, are well-known antioxidants that scavenge free radicals and protect from oxidative damage and also inflammatory responses; furthermore, they are highly advantageous as prophylactic food ingredients preventing different infectious diseases (Aronson 2017). Then again, fruit juice has been documented as a rich source of several antioxidants, functional components, and phytonutrients, as well as a cornucopia of vitamins, amino acids, and carbohydrates, along with minerals (Lawless et al. 2012; Khezri et al. 2016). Different fruit juices of natural origin, such as orange, apple, mango, grape, etc., are of immense nutritive value and pharmacologically significant not only for behavioral and physiological health but also in the avoidance of various chronic and acute diseases, such as cardiovascular diseases, Alzheimer's disease, cancer, chronic obstructive pulmonary disease, diabetes, cystic fibrosis, and arthritis (Ho et al. 2020; Hyson 2015). Regular consumption of green vegetables and fresh fruits in the day-to-day food regime is essential not only from a nutritional perspective but also because of their medicinal and curative possessions (Dolkar et al. 2017; Singh and Devi 2015). A short representation of various advantages offered by nutraceuticals is shown in Figure 1.1, although various other rewards are also provided by these magical dietary components (Nasri et al. 2014; Kumar and Kumar 2015; Dudeja and Gupta 2017; Souyoul et al. 2018; Srivastava et al. 2015; Williamson et al. 2020).

The utility of nutraceuticals for the mitigation and treatment of different diseases is credited to their capability to offer prudent therapeutic sustenance along with minimal side effects. Globally, researchers have recognized their inherent potential to shield against a wide range of diseases, comprising cancer, pneumonia, diabetes, asthma, and cardiovascular diseases, along with neurodegenerative conditions like Parkinson's syndrome (Daliu et al. 2019; Mecocci et al. 2014; Chanda et al. 2019). Nanotechnology can be used for nutritional medicine with better absorption and biodynamics (Kumar and Smita 2017). Driven by an increase in the market for innovative food products, fortified healthy foodstuffs, and the simultaneous analysis of healthy food ingredients, which was previously impossible, the market for such functional foods and related components is predicted to reach about $140 billion by 2020 (Intelligence 2016).

The habit of using different functional foods and nutraceuticals is growing due to the ease of obtaining nutraceuticals without any involvement of doctors or medical workers, minimal side

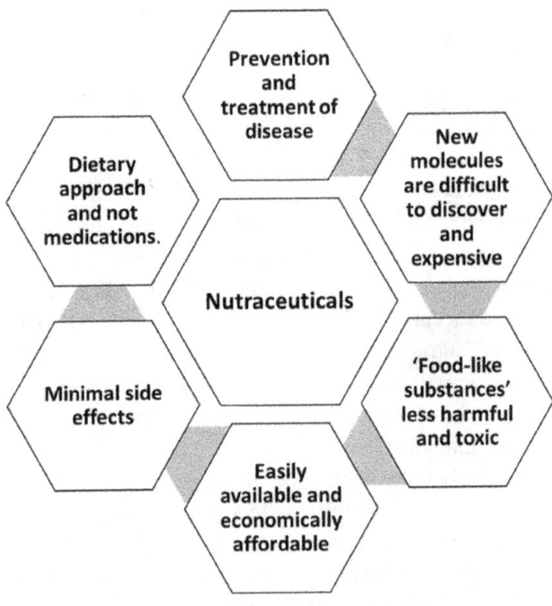

FIGURE 1.1 Various compensations offered by nutraceuticals in this new age.

effects, high potency, and great health benefits, unlike drugs. Various natural and herbal products are becoming popular, and consumer acceptance is very high, since conventional medicines struggle to combat certain diseases, and natural products are also devoid of side effects (Every-Palmer and Howick 2014; Cheuka et al. 2017). But, there are also some limitations associated with the use of these products. One of the foremost issues faced by nutraceuticals is lack of control of product quality. Nutraceuticals are generally regarded as food and not medicines, so they are controlled under the legislation for food and not drugs, which lacks the specific requisites for medications and other such related molecules. Issues include contamination with toxins, adulteration, risk to consumers, difficulties in carrying out adequate research to demonstrate potential health benefits, requirements for purity and dosage, lack of scientific basis, cost-effectiveness, interaction with compounds during uptake, etc. (Fibigr et al. 2018; Lockwood 2011; Santini et al. 2018; Keservani et al. 2014; Orhan et al. 2016; Koch et al. 2014; Gil et al. 2016; Nounou et al. 2018; Andrew and Izzo 2017; Télessy 2019; Sattigere et al. 2020; Gupta et al. 2018). Besides these, nutraceuticals also face numerous problems during handling, storage, delivery, and processing, as depicted in Figure 1.2. The diagrammatic depiction of different problems linked with nutraceuticals at different steps, from extraction or production to the stage of delivery, are clearly shown in this figure (Kessel 2011; Galanakis 2013; Lai et al. 2017; Joana Gil-Chávez et al. 2013; Datta 2017; Rangan et al. 2016; Buteyn et al. 2019).

For several decades, the food industry has been using different scientific methods and advances without recognizing them as nanoscience or technological applications, such as when fat droplets

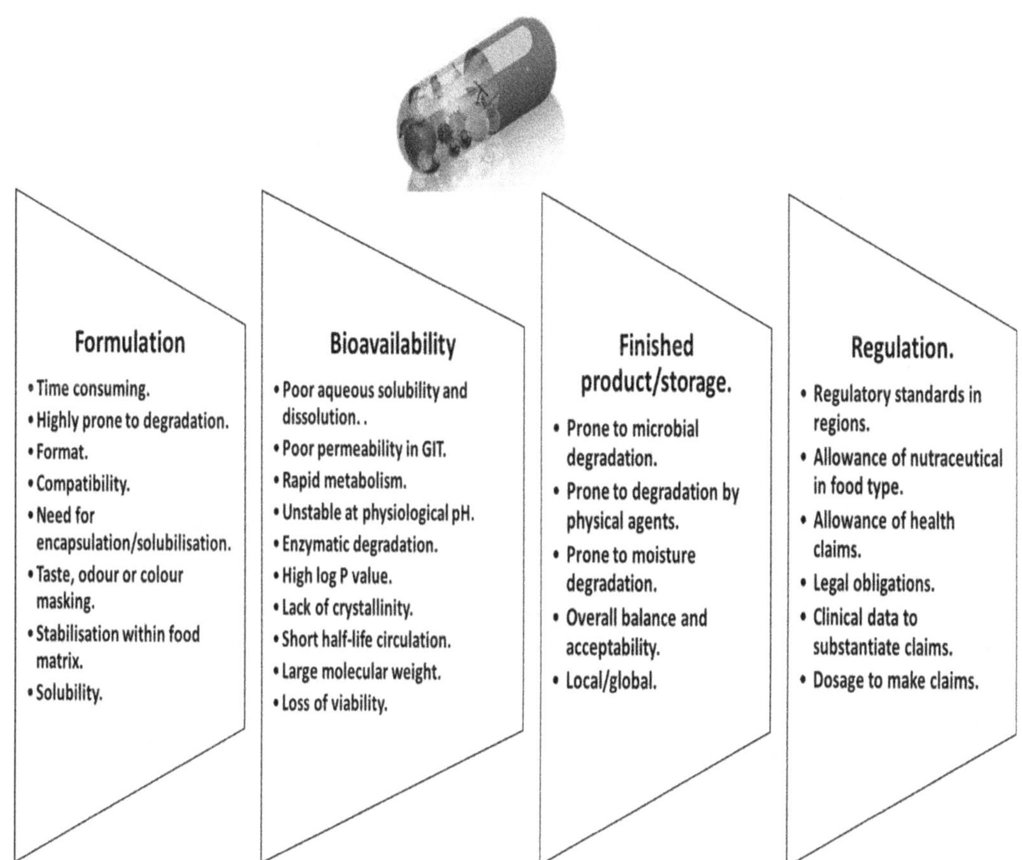

Formulation
• Time consuming.
• Highly prone to degradation.
• Format.
• Compatibility.
• Need for encapsulation/solubilisation.
• Taste, odour or colour masking.
• Stabilisation within food matrix.
• Solubility.
Bioavailability
• Poor aqueous solubility and dissolution..
• Poor permeability in GIT.
• Rapid metabolism.
• Unstable at physiological pH.
• Enzymatic degradation.
• High log P value.
• Lack of crystallinity.
• Short half-life circulation.
• Large molecular weight.
• Loss of viability.
Finished product/storage.
• Prone to microbial degradation.
• Prone to degradation by physical agents.
• Prone to moisture degradation.
• Overall balance and acceptability.
• Local/global.
Regulation.
• Regulatory standards in regions.
• Allowance of nutraceutical in food type.
• Allowance of health claims.
• Legal obligations.
• Clinical data to substantiate claims.
• Dosage to make claims.

FIGURE 1.2 Various hurdles confronted in nutraceutical technology at various stages from manufacturing to marketing.

are emulsified by an underlying dilution mechanism using nanosized droplets of water to produce healthy products, such as mayonnaise with low fat content. This method retains the flavor and other sensory characteristics of the high-fat variety but provides healthier alternatives to customers (Li et al. 2020). Water-insoluble ingredients or additives can be easily dissolved in various food products by utilizing nanotechnology. The use of nanomaterials may provide a selective advantage, and no further fat or surfactants may be required, as the same consistency can be achieved using low amounts. Developments in uniformity, palate, savor, and surface characteristics are easily attained by "nanotexturing" foodstuffs by the application of various stable emulsions, nanostructures, and nanoformulations (Mousa and Abd El-Hady 2018; Smolkova et al. 2015; Chau 2015).

1.3 NUTRACEUTICALS TO NANOCEUTICALS

The provision of wholesome foods that are cost-effective, advantageous, tempting, and convenient for consumers is a key aim for food manufacturers (Khan et al. 2013; Bayram and Gökırmaklı 2020). The established effectiveness of a functional food attributed to various desired therapeutic effects increases its desirability as a supplement or adjunct to nutritional supplements (Kearney 2010; McCarron 2016). Researchers have recognized and established numerous nutraceuticals since the interest in nutraceutical components and functional foods emerged over two decades ago. However, unexpectedly few nutraceuticals have been successfully integrated as ingredients in functional foods, amid substantial research and development efforts in this domain. There are various scientific, commercial, and bureaucratic obstacles that still need to be overcome if functional foods are to prosper in the marketplace (Asioli et al. 2017).

The fantastic field of "nanotechnology" wields mysterious powers to invigorate every domain with its magic touch and is gradually conquering the food business sector with its incredible potential, especially nutraceuticals and other such bioactive food components (Chaudhry et al. 2017). The concept of "nanoceuticals" is becoming prevalent, and commercial dairy/food additives comprising nanoparticles or nutrients in nano form are nowadays available (Paul and Dewangan 2015). The conversion of "nutra to nano" can be foreseen as a rational development of scientific knowledge and technology-driven advancement and is greatly anticipated to improve the capture of various components and their transport to the desired site, intensify the essence of flavors, augment the bioavailability of phytochemicals, introduce antimicrobial nanomaterials or composites into the food matrix, and enhance the shelf life or storage period of foods (Usmani et al. 2019; Fathi et al. 2018). The origins of this change are shown in detail in Figure 1.3 in the different stages of product development (Kakkar et al. 2016; Malik et al. 2021; Shastry and Rai 2018; Bernela et al. 2018). In addition, nanoencapsulation of nutraceuticals creates a defensive shield against degradation and increases their potency, culminating in a sustained release profile of the ingredient and enhanced aqueous stability (Anandharamakrishnan 2014; Chellaram et al. 2014). The development of nano-nutraceuticals with improved stability in the production chain will be of great importance to food companies and industries looking to optimize nutrient content and will thus eventually be advantageous to customers (Dasgupta et al. 2015). Nanoscience is rapidly establishing itself as one of the most fascinating and dynamic emerging disciplines of study, with a clear goal of improving healthcare and human wellbeing (Javeri 2016). Nanoceuticals involve the manipulation and control of attributes of food substances at the component level, resulting in an increase in the ingestion of supplements and nutrients of nanoscale dimensions, not only boosting the nutritive content of the foods but also increasing the overall stability and bioavailability of the products. Nanotechnology is quickly becoming one of the most fascinating and rapidly increasing areas of study and development, with the goal of improving health and wellbeing (Dasgupta et al. 2015).

Nanoencapsulation methods can easily incorporate supplements for optimal delivery to target areas (Assadpour and Mahdi Jafari 2019; Fathi et al. 2014). The function of different foods as well as dietary products has developed radically in recent years from being merely a vital source of nourishment and growth to contributing to human health and wellbeing and protection from various diseases (McClements 2020). When encapsulated in nanocarriers, nutrients may be rendered resistant

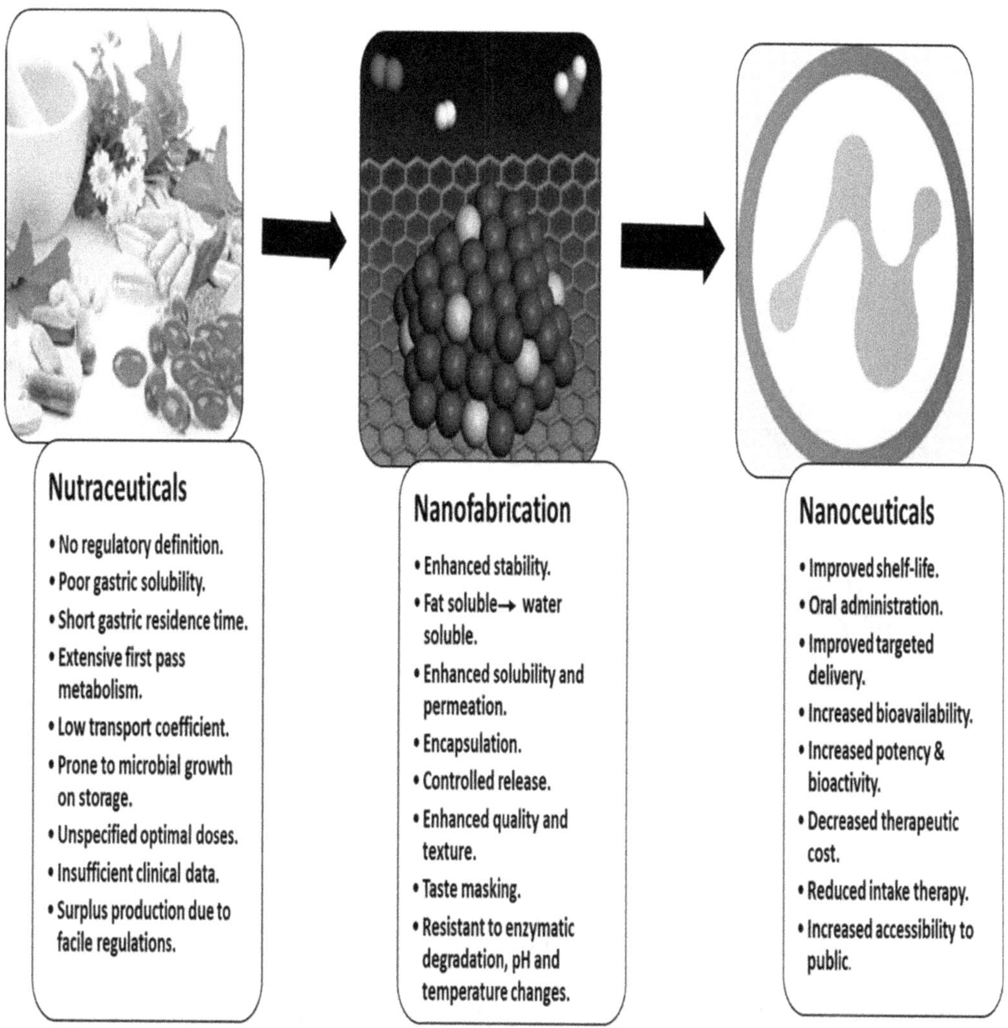

Nutraceuticals
• No regulatory definition.
• Poor gastric solubility.
• Short gastric residence time.
• Extensive first pass metabolism.
• Low transport coefficient.
• Prone to microbial growth on storage.
• Unspecified optimal doses.
• Insufficient clinical data.
• Surplus production due to facile regulations.
Nanofabrication
• Enhanced stability.
• Fat soluble→ water soluble.
• Enhanced solubility and permeation.
• Encapsulation.
• Controlled release.
• Enhanced quality and texture.
• Taste masking.
• Resistant to enzymatic degradation, pH and temperature changes.
Nanoceuticals
• Improved shelf-life.
• Oral administration.
• Improved targeted delivery.
• Increased bioavailability.
• Increased potency & bioactivity.
• Decreased therapeutic cost.
• Reduced intake therapy.
• Increased accessibility to public.

FIGURE 1.3 General pathway of transformation from nutraceutical to nanoceutical.

to severe denaturing chemicals, proteases, and other enzymes present in the environment. They confer enhanced stability against fluctuations of physical conditions like temperature, pH, heat, etc., when encapsulated in certain tailor-made nanocarriers that can now be formulated. Nanoceuticals are termed "healthier foods" because of their greater bioavailability, effectiveness, and superiority at the same price. As a result, smaller amounts are required, lowering the cost of nutritional constituents in terms of individual food or quantity and thereby, indirectly benefitting more individuals quantitatively as well as qualitatively, possibly facilitating the alleviation of deficiencies of nutrients as well as food resources in underdeveloped countries (Thulasi et al. 2013).

Various nutraceuticals, such as ω-3 fatty acids, lycopene, vitamins such as D_2, B_{12}, C, and E, and some helpful probiotic bacterial species, have had commercial success in this area and are widely used. Vitamin E, coenzyme Q, and beta-carotene have all been shown in studies to have therapeutic potential (Mishra et al. 2019). In the market, a variety of such marvelous ingredients are increasingly gaining relevance in nano form, with some examples shown in Figure 1.4. The available literature implies that there are a number of unanswered questions about nanoceuticals, including potential human health hazards associated with their use, bioaccumulation potential, overall effect on the ecosystem, etc. (Cushen et al. 2012). Nanomaterials have potential relevance in every sphere

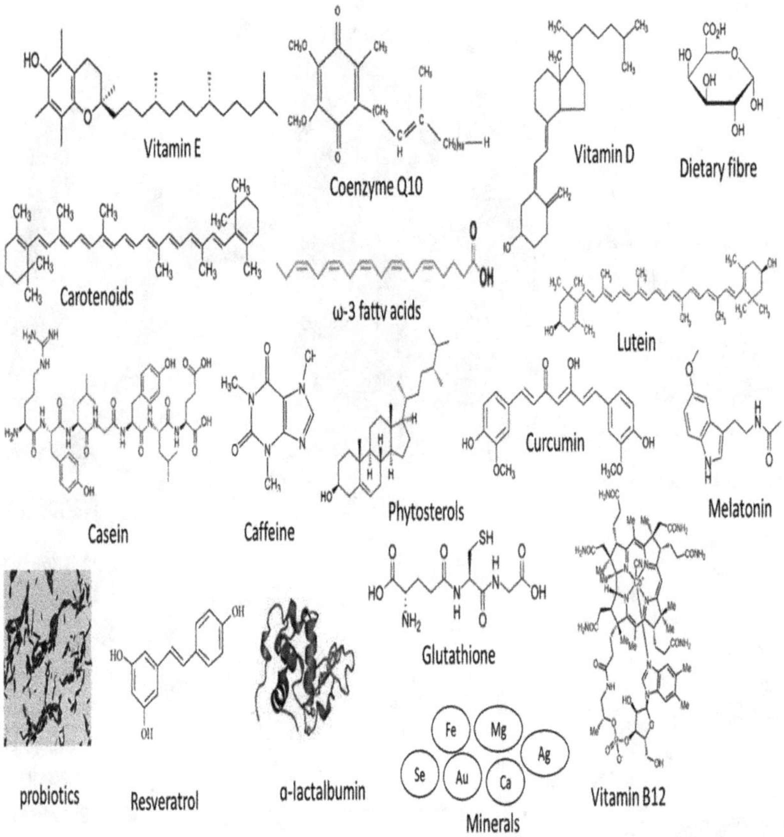

FIGURE 1.4 Some of the nutrients that are widely available in nanoparticulate form in the market.

of life; currently, they are being developed with caution, and progress may be impeded by a dearth of governance and associated hazards (Kirdar 2015).

In the health food market, progress in improving safety and wellbeing is achieved through the application of nanomaterials, nanopolymers, and nanoparticles as well as nanosized supplements and nutraceuticals (McClements and Xiao 2017). Cutting-edge technology currently uses nanodelivery systems or matrices as well as nanocarriers to capture bioactive ingredients and transport them through the active endocytic barrier, thereby boosting overall ingestion and metabolism in the digestive tract, conferring precision targeting and a manifold increase in bioactivity in the systemic circulation (Momin et al. 2013). Cellulose, wheat, micronized starch, rice flour, and a variety of spices and herbs have long been produced in Asian countries and are highly beneficial for medical and gastronomic purposes. Carbon nanoparticles have recently been discovered in meals containing caramelized sugar, such as corn flakes and bread (Sk et al. 2012). Manufacturers of nanoceuticals generally claim that (a) the absorption of nutrients will be complete and uniform, (b) hydrophobic nutrients will be transformed into hydrophilic counterparts, leading to increased absorption, (c) the beneficial effect will be enhanced with lower doses, (d) the threat of grave or deleterious effects will be reduced, and (e) supplement permeability across physiological and biological barriers will be increased (Paul and Dewangan 2015).

1.4 QUANTIFICATION OF NANOPARTICLES AND NANOMATERIALS IN FOOD AND OTHER RELATED MATRICES

Based on their chemical nature, nanomaterials in nutraceutical preparations can be organic or inorganic. Natural/organic nanomaterials can act either as the active constituent (e.g. self-assembled ω-3 unsaturated fats, curcumin, polyphenols, etc.) or as frameworks for the transport of active constituents (e.g. phospholipids, protein, carbohydrates, etc.). Inorganic nanomaterials can also be the active components themselves, as in the case of nanosilver and zinc oxide, or can be matrices or carriers for nutrients (such as nanotubes, nanospheres, etc.). The equivalent physicochemical properties making these new materials promising for some application fields may influence human wellbeing. To forestall health hazards related to the use of nanomaterials, it is critical to assess their vulnerability and risk (Handford et al. 2014; Kulkarni et al. 2016). In nutraceuticals, nanomaterials are utilized to improve the bioavailability of functional constituents. In this regard, consumers are exposed to nanomaterials by the oral route, and their effect on human wellbeing should be carefully assessed (Kermanizadeh et al. 2016).

The unique qualities of nanomaterials that contribute to their high efficacy necessitate a fresh approach to completely comprehend their behavior and fate as well as any negative consequences associated with them. Various physicochemical endpoints have been defined within this framework (Rasmussen et al. 2018), mostly referring to composition, state of dispersion, aggregation, size and size distribution, surface area, and porosity. Many projects are presently underway to determine whether existing characterization methods can be adapted and used for nanomaterials or whether new, more appropriate approaches must be devised (Lin et al. 2014; Laborda et al. 2016). The processing of nanomaterials in complex matrices for functional food and nutraceutical applications may result in alteration to their physicochemical properties (Contado 2015). These changes can also happen during digestion, which has a big impact on the stability and bioavailability of active nanoingredients (DeLoid et al. 2017). As a result, it's important to characterize both pure nanomaterials and nanomaterials in complex matrices, including food, biological tissues, and fluids (i.e. secondary characterization), in order to better understand their efficacy and fate over their entire life cycle. The best method for physicochemical characterization of nanomaterials is determined by their type and the measuring environment (McClements et al. 2016) (Table 1.1).

Although the aforementioned techniques can be used to measure nanoparticles in their pure form, measuring and quantifying nanoparticles within biological cells and tissues, food, and additional biotic or food-associated milieus poses significant problems and challenges due to the lack of appropriate equipment and measurement strategies (Picó 2016; Singh et al. 2014). Different foods inherently contain significant quantities of nanoparticles or nanomaterials naturally without any processing or treatment, such as silica, proteins, titanium dioxide, iron, and various micronutrients, which makes the detection of added nanoparticles challenging because it eliminates approaches like elemental mapping where background levels are already high. The elemental mapping techniques are generally used in case of nanoparticles like silver and gold, which are not found naturally in foods or related matrices and surfaces. If we focus on quantifying the surface volume or area of a given mass of nanoparticles instead of qualitative detection strategies, this is also hindered by several hurdles, as contemporary technologies, such as adsorption of nitrogen, are unable to distinguish between surface area and porosity. Another issue to ponder is the rapid aggregation potential of nanomaterials, which happens naturally, being driven by entropy, in biological environments, causing agglomeration of small particles into large clumps and leading to a progressive rise in size with time. Because of this aggregation, measuring surface area or particle number is nearly impossible in biological or food system–associated matrices (Stamm et al. 2012).

Despite the success that has been achieved in enhancing the therapeutic efficacy as well as the bioavailability of nanoprocessed food components and nutrients using various *in vitro, in situ* and *in vivo* models, investigations and experimentations, further animal research, and human studies are needed to conclude their overall effectiveness. Furthermore, the threat of toxicity of the precipitating nanomaterials as a side effect of better bioavailability must be taken into account and assessed, as they pose a high risk not only in biological systems but also to the environment. It is essential to determine susceptibility and risk in order to avoid health problems associated with the use of nanomaterials. For nutraceuticals, nanomaterials are often used to enhance the bioavailability of active ingredients. Consumers are also orally exposed to nanomaterials, and their effects on human health must be carefully assessed. In 2011, the Technical Committee of the European Food Safety Authority (EFSA) and in 2012, various Organisation

TABLE 1.1
Various Physicochemical Endpoints for Nanomaterial Characterization in Foods and Instruments Used in Characterization

Physicochemical endpoints	Description	Currently available methods for characterization	
		Organic nanomaterials	**Inorganic nanomaterials**
Chemical composition/ Identity	Purity, nature, formulants, impurities, impurities/ coatings/encapsulating materials	Gas chromatography (GC), high-pressure liquid chromatography (HPLC), either standalone or combined with mass spectrometry (MS), capillary electrophoresis	Energy dispersive X-ray spectroscopy (EDX), atomic absorption spectroscopy (AAS), optical emission spectroscopy (OES), inductively coupled plasma-mass spectrometry (ICP-MS)
Size of particle	Principal size of particles along with their overall range, overall distribution of size, and similar parameters, also secondary (agglomerates) if present	Micrography: Atomic force microscopy (AFM), transmission electron microscope (TEM), scanning electron microscope (SEM); separative methods: chromatography and flow separation such as hydrodynamic chromatography and field flow fractionation (FFF), respectively; centrifugation methods; light dispersing methods (static and dynamic light scattering and also multi-angle light scattering)	
Physical form and morphology	Physical form and crystalline shape	Microscopy (AFM, SEM, TEM), X-ray diffraction (XRD)	
Particle and mass concentration	Concentration expressed as no. of particles per mass (dry powder) or particle mass per volume (dispersion)	Particle light scattering, microscopy, single particle ICP-MS Mass: Similar to chemical composition methods, gravimetric methods, centrifugal separation	
Specific area of surface		Brunauer–Emmett–Teller (BET) analysis (N_2 analysis)	
Surface chemistry	Biochemical/Chemical modifications that add new functionality or modify surface reactiveness	Any appropriate biochemical methods suitable for particular criteria reported in this chapter	
Stability or charge at the surface	Zeta potential	Electrophoretic methods	
Redox potential		Potentiometric methods	
Solubility and partition properties	In suitable solvents and their distribution between organic and aqueous solutions	Available standard methods (OECD, ISO, AOCS)	
Viscosity, density, pour density, dustiness		Available standard methods (OECD, ISO, AOCS)	
Chemical reactivity and photocatalytic efficacy		Kinetic capacities of the compound, catalyzed and/or biochemical responses	

Sources: From Lin, P.C. et al., *Biotechnology Advances*, 32, 4, 711–726, 2014; Loeschner, K. et al., *Analytical and Bioanalytical Chemistry*, 405, 25, 8185–8195, 2013; Helsper, J.P. et al., *Analytical and Bioanalytical Chemistry*, 405, 4, 1181–1189, 2013; Kumar, P.S. et al., Characterization techniques for nanomaterials. In *Nanomaterials for solar cell applications*, Elsevier, 2019.

for Economic Co-operation and Development (OECD) countries presented a systematic methodology for evaluating the probable hazards resulting from nanotechnologies applied to the feed and food chain as a whole (Gruere 2012; EFSA Scientific Committee 2011).

Physicochemical transitions of nanoceuticals subjected to the digestive cycle lead to their breakdown and release of active compounds and constituents. Because these transformations are explicitly associated with the unique physicochemical properties of those nanoformulations, a case-by-case strategy is necessary to determine their efficacy and safety. Nevertheless, the explosive growth of nanotechnologies in the food sector, such as nutraceuticals, demands quick, economical, and robust methods for gauging their effectiveness and safety (Szakal et al. 2014b). Numerous integrative approaches for the production of new, effective, and secure nutraceutical formulations are therefore important in this new era.

Because direct measurements are not always possible, and matrix/nanomaterial separation is required, the sample preparation technique is a critical challenge in characterization and detection of nanomaterials in complicated food matrices (Hu et al. 2016). Acid digestion has been used for total concentration elemental analysis (Zhang et al. 2012; Geiss et al. 2013), although it is incompatible with organic nanomaterials such as liposomes and micelles. Gray et al. (2013) proposed alkaline digestion with tetramethylammonium hydroxide (TMAH) and enzymatic digestion (Dan et al. 2015; Peters et al. 2014) as viable approaches for dissolving or modifying nanomaterials in the food matrix. Though the examples in this literature are not technically related to the nutraceuticals business, extraction technologies could allow the non-destructive transfer of nanomaterials from food into simpler matrices (Singh et al. 2014). Centrifugation and filtering can help separate nanomaterials from liquid samples, but there are certain drawbacks, such as particle retention on the filter membrane (Ladner et al. 2012) and changes to nanomaterial properties. There is currently no approach for characterizing, detecting, and quantifying nanomaterials in all dietary matrices. The contrast between inorganic and organic nanomaterials is the most important factor to consider. Conventional approaches such as inductively coupled plasma optical emission spectroscopy (ICP-OES), atomic absorption spectrometry (AAS), energy dispersive X-ray spectroscopy (EDX) (Golobič et al. 2012), or even inductively coupled plasma mass spectrometry (ICP-MS) (Lin et al. 2011), and innovative methods such as surface plasmon resonance (SPR) and particle induced X-ray emission (PIXE) (Raz et al. 2012; Lozano et al. 2013), and single-particle inductively coupled plasma mass spectrometry, a relatively new ICP-MS approach (SP-ICP-MS) (Zhang et al. 2012; Mitrano et al. 2012a,b; Peters et al. 2015; Tuoriniemi et al. 2012), represent a potential tool for inorganic nanomaterial characterization in complex matrices. SP-ICP-MS was found to be capable of distinguishing between nanoparticles and dissolved ions of the same element, providing information on both number and mass concentration at the same time. Moreover, SP-ICP-MS can determine not only the shape of the nanomaterial analyte but also the overall size distribution by taking into account some assumptions such as particle density (Singh et al. 2014), although more research is needed (studying the symmetry of the peaks obtained from SP-ICP-MS data). Indeed, SP-ICP-MS necessitates the use of highly diluted liquid samples, with possible nanomaterial dissolution or modification due to sample preparation. The sensitivity of SP-ICP-MS largely depends on particle composition, relative abundance of the particular isotope, and also the number of ions dissolved in the whole system in the background as well as the actual sample, and it is not sensitive for particles smaller than 15–20 nm (Szakal et al. 2014a). It is frequently essential to combine separation methods such as high-pressure liquid chromatography (HPLC) or field-flow fractionation (FFF) with fluorescence-based detection systems or UV-visible (UV-Vis) spectroscopic arrangements to tackle this problem (Szakal et al. 2014a; Wagner et al. 2015). Separative techniques, when combined with appropriate detection devices, can determine the size and volume as well as the overall size intensity of inorganic nanomaterials (Baer et al. 2010; Geiss et al. 2013; Veverka et al. 2014). Labeling with fluorescent tags or radiolabels is another option. Carbon-based nanomaterials are more difficult to detect due to their compositional closeness to food matrices, and only a few specialized approaches for chemical analysis are readily available. Analytical electron microscopy, Fourier transform infrared spectroscopy (FTIR), Auger electron spectroscopy (AES), X-ray photoelectron spectroscopy (XPS), Chemical force microscopy, and also secondary ion mass spectrometry (SIMS) are commonly applied for studying the surface characteristics of nanomaterials (Veverka et al. 2014; Dudkiewicz et al. 2011). Although these and other approaches are costly and demand a high level of knowledge, the chemical information they provide is not available through other methods (Szakal et al. 2014b). Nanomaterials cannot be described solely on the basis of compositional data. Imaging analysis methods can also provide important information on physical form, morphology, and particle size distribution. SEM and TEM (scanning and transmission electron microscopy, respectively) are the most often used techniques (Wang et al. 2015; Hu et al. 2002; Hinkley et al. 2015). Techniques working on scattering of light, such as dynamic and static light scattering (DLS/SLS) can also be very beneficial to determine the overall particle size distribution of nanomaterials in liquid samples. These techniques are fast, robust, sensitive, and low-cost methods for determining the average size, size distribution, size intensity, and also polydispersity of the nanomaterials during fabrication (Onoue et al. 2015; Hu et al. 2002; Venturini et al. 2014; Peters et al. 2012) as well as life cycle (such as in the digestive process) (Ge et al. 2015; Madureira et al. 2015). Another significant feature of dispersions, such as nutraceutical formulations, is stability, which may be easily assessed using laser Doppler microelectrophoresis over time in terms of zeta potential (Peters et al. 2012; Hu et al. 2002; Schulzova et al. 2012) At present, no unique or validated analytical techniques are available for characterizing nanomaterials in complex matrices such as functional foods, so a multi-approach is strongly recommended to accurately collect the essential physicochemical parameters required to assess the behavior and fate of nanomaterials from product synthesis to the food chain.

1.5 NANOFORMULATIONS OR NANOSTRUCTURES FOR DELIVERY OF NUTRACEUTICALS

The science of nanotechnology is currently being used for generating a plethora of novel nutraceutical oral administration solutions with increased bioavailability. Nano- and microencapsulation have gained popularity in recent years for medicinal, nourishment, and sustenance uses (Suganya and Anuradha 2017). These micro- and nanoencapsulation strategies are very useful in food applications for several reasons: (1) the selective ingredient or component can be protected from untimely degradation (throughout fabrication and storage as well as processing) or unwanted interactions with the environment; (2) bad or foul smell, taste, and other sensory characteristics can be masked; (3) the processing of the components can be indirectly facilitated by improving their dispersibility and solubility; (4) the release can be prolonged or regulated as per need or location; and/or (5) the overall bioavailability can be greatly improved, especially via the oral route (Gunasekaran and Ko 2014; Kwak 2014; Quintanilla-Carvajal et al. 2010). Edible-grade nanomaterials similar to food can be prepared from a variety of ingredients as well as biomolecules (carbohydrates, proteins, polyphenols, lipids, minerals, vitamins, etc.) utilizing a variety of techniques (ultrasonication, homogenization by high speed/pressure, antisolvent precipitation, maceration/grinding, coacervation, and spontaneous emulsification) (McClements 2014). Various nanocarriers (liposomes, nanoparticles, nanoemulsions, nanocapsules, nanodispersions, and polymeric micelles) possess varied functional features and physicochemical assets resulting from the diversity of building blocks and production processes. The physical structure, inherent composition, overall size, surface charge, physical state, and digestibility of nanoparticles, for example, can all differ to a great extent. Figure 1.5 depicts the most commonly utilized nanocarriers for nutraceutical delivery (Assadpour and Mahdi Jafari 2019; Rezaei et al. 2019; Katouzian et al. 2017; Dima et al. 2020).

1.5.1 NANOPARTICLES

Nutraceuticals are generally entrapped in nanoparticles by mixing an aqueous solution of the nutraceutical along with a co-solvent and also a hydrophobic polymer to water (in the presence or

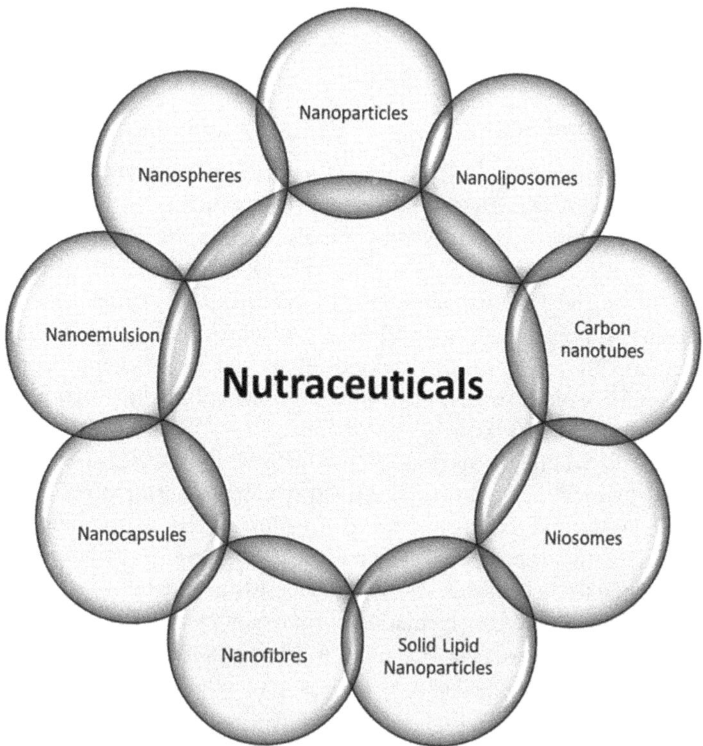

FIGURE 1.5 Various nanocarriers or delivery systems used for the delivery of nanoceuticals.

absence of surfactant). As a result of mixing of these components, the polymeric material forms a matrix or carrier, which finally aggregates, leading to condensation, and the nutraceutical is trapped inside the polymeric core (Semyonov et al. 2014). The main features for nanoparticle synthesis are utilization of substances such as proteins, polysaccharides, and synthetic polymers (Abd El-Salam and El-Shibiny 2012; Zimet et al. 2011). The polymers employed for nanoparticle preparation can be natural or synthetic. The natural polymers most frequently used in the preparation of polymeric nanoparticles are gelatin, chitosan, albumin, and sodium alginate, while the synthetic alternatives include polyanhydrides, polyglycolides (PGA), poly-aminoacids, polylactides (PLA), polyorthoesters, poly (lactide co-glycolides) (PLGA), polycyanoacrylates, polyglutamic acid, polycaprolactone, polymalic acid, etc. The Food and Drug Administration of the United States has approved poly (lactic-co-glycolic acid), a widely available polymeric material, for therapeutic use due to its biodegradability and excellent biocompatibility (Yu and Huang 2013). Colloidal particles are typically larger than micelles, ranging in size between 100 and 200 nm and with higher polydispersity. The medications are dispersed, encapsulated, enclosed, chemically bonded, or bound to the polymer matrix of these particles (Letchford and Burt 2007).

Nanoparticles possess several advantages, such as lower dosage, superior bioavailability, better shelf life, mechanical and thermal stability against pH changes, controlled release formulations, and site-specific targeting, and also can be employed for administration through various routes (Dadwal et al. 2018). One crucial point when formulating nanoparticles is that the diminutive size range and large surface area can lead to particle agglomeration, which makes the physical handling of nanoparticles complex in both fluid and dehydrated forms. Moreover, partial drug release and burst liberation is another problem associated with small particle size and large surface area. Thus, to overcome such problems and concerns about surface alteration, drug release inhibition approaches and release control parameters have to be considered and should be completely understood before nanoparticles can be used clinically or made commercially available (Anselmo and Mitragotri. 2016; Schütz et al. 2013).

1.5.2 Solid-Lipid Nanoparticles

These formulations were initiated in 1991 as exceptional delivery systems for real time–based polymer and lipid-based colloidal delivery services. The lipids, which are solid in nature, are added to these formulations, and if required, surfactants are also incorporated. The mean particle size of solid-lipid nanoparticles (SLNs) ranges from 40 to 1000 nm. Lipids used in the preparation of SLNs are durable at room as well as body temperature. Triglycerides, waxes, and hard fats are a few lipid materials, while lecithin, polysorbate 80, and polyvinyl alcohol are used as surfactants in SLN preparation (Mitri et al. 2011). The various methods employed for the preparation of SLNs include ultrasonication, solvent injection, solvent diffusion, supercritical fluid extraction of emulsions, and spray-drying (Madureira et al. 2015).

They possess several advantages, such as good biocompatibility, increased stability, protection against physical and chemical degradation, sustained release and drug targeting of active ingredients, possibility of oral, dermal, and parenteral drug administration, and decreased potential for acute and chronic toxicity (Mitri et al. 2011; Chen et al. 2013). Upon formulation of SLNs, polymorphic transitions occur due to their instability, as they possess a tendency to form aggregates. These polymorphic transitions escort the transformation of lipid nanoparticles (from spheroid to disk-like), thus causing a palpable boost in particle face quarter. Hydrophobic attraction occurs between non-polar regions on particle surfaces, contributing to aggregation and gelation, and thus limiting their relevance. To overcome this difficulty, high levels of emulsifier are added, which forms a protective shell around the lipid nanoparticles (Jenning and Gohla 2001).

1.5.3 Niosomes

Hydrophilic and lipophilic compounds that are used in excess in pharmaceuticals or as nutraceuticals can be entrapped into a double-layer structure of non-ionic surfactants in water, forming self-agglomeration vesicles called niosomes (Pando et al. 2015). Unlike liposomes, which are prepared by phospholipid-based artificial vesicles, niosomes are prepared from non-ionic surfactants; they are less expensive and are stable towards oxidative degradation. Polyoxyethylene alcohol, polyoxyethylene glycol alkyl ethers, alkyl ethoxylate, and alkyl phenol ethoxylate are a few non-ionic surfactants employed in niosome preparation. The various methods that can be employed to prepare niosomes include trans-membrane pH gradient, sonication, microfluidization, dried-reconstituted vesicles, and the bubble method (Rentel et al. 1999; Fraile et al. 2015).

One of the most widely used natural polyphenolic compounds is resveratrol, a nutraceutical with several benefits, such as cancer medication, reducing cardiovascular risk, and anti-inflammatory activity. One of the major limitations encountered with resveratrol is its photosensitive nature, as this compound occurs in cis and trans forms, with only trans-resveratrol having the health benefits. Exposure to light causes conversion of the trans form of resveratrol to the cis form, which is inactive. Thus, niosomes containing resveratrol were prepared using the thin film hydration method. Upon characterization, the niosomal formulation exhibited high resveratrol entrapment efficiency and good stability (Pando et al. 2013).

Niosomes are chemically stable, osmotically active, and highly durable, with long storage periods; they possess hydrophilic heads, and thus, their surface can be modified; they have high compatibility with natural components and fewer side effects due to their non-ionic nature; they are biodegradable and non-immunogenic; formulations can be loaded with lipotropic and hydrophilic compounds; they can advance the action of nutraceuticals by shielding them from the biological environment, with consequential superior accessibility, and limit the bioactive moiety to the destined cell; formulations are based on aqueous media, which tends to enhance accessibility for patients, and in contrast to phospholipids, the management of surfactants do not necessitate any exceptional safety measures (Moghassemi and Hadjizadeh 2014).

1.5.4 Nanospheres

Nanospheres are a homogeneous system of nanoparticles ranging in diameter from 10 to 200 nm. Their physical nature can be either crystalline or amorphous, where the bioactive moiety is entrapped in the polymer matrix. Normally, the method employed for the formation of nanospheres depends on the nature of the polymer: emulsion polymerization, solvent evaporation, and nanoprecipitation (Vilela et al. 2012). The pharmaceutical industries based on dietary supplements and nutraceuticals are now developing a new delivery system using nanospheres, which are nanosized particles from bioactive natural non-hazardous phospholipids and other similar constituents present in the human cellular system (Tan et al. 2012).

The rewards of such delivery systems are enhancement of biological and physicochemical stability, including solubility, shelf life, etc., efficient transport in the cellular microenvironment, greater release and targeted delivery of the active component inside the cell, appreciably greater potency and bioavailability, indirectly leading to greater biological action, site-specific targeting, improved half-life in the body, negligible side effects, and superior therapeutic value. These beneficial features have led to extensive utilization of nanospheres around the globe. Enzyme immobilization is a classic technique for improving the stability of enzymes, but in the present scenario, encapsulation is used. Nanospheres are employed for the encapsulation of enzymes via biomimetic silicification involving α-chymotrypsin and a fungal protease from *Aspergillus oryzae*. Encapsulations appreciably augment the susceptibility and stability of bioactive compounds against enzymatic, thermal, and various physicochemical environmental factors, paving the way for efficient transport as well as prolonged functionality of the bioactive compounds (Madadlou et al. 2010).

1.5.5 Nanoliposomes

Liposomes are spherical vesicles having a diameter in the range from 20 nm to several micrometers. Liposomes are made up of double-layered lipid moieties, hydrophobic in nature, surrounding aqueous hydrophilic layers with nutrient, nutraceutical, or other such bioactive components. A glycerol backbone containing sterols (e.g., cholesterol) or phospholipids, either natural (e.g., lecithin), or manmade (e.g., distearoyl phosphatidyl ethanolamine) in nature, can be used to make them. Unilamellar liposomes of nanodimensions ranging in size between 20 and 1000 mm are of particular interest. For the fabrication of liposomes, a high concentration of surfactant is generally required, typically around 20% by weight or greater. At such large amounts, there is a high risk of toxicity. So, various novel components or techniques are being sought, and this is an active area of research for pharmaceutical and edible applications. As a result, there is growing curiosity about developing new techniques of encapsulation (Jeetah et al. 2014; Akbarzadeh et al. 2013). Nanoliposomes are formed by the essential hydrophilic–hydrophobic interface, which connects phospholipids and water molecules (Aadinath et al. 2016). Aggregation of vesicles, which turn out to be microsized particles, upon storage is commonly observed when the nanometric size is formulated. However, these are supposed to have enough stability to maintain their size, and such vesicles preserve their nanoscale dimensions from production to consumption of the formulation. Passive or active targeting mechanisms are employed for targeted therapy via nanoliposomes. The various methods employed include extrusion, sonication, microfluidization, and various heating techniques (Xia et al. 2014). For the quality evaluation of formulated products, new approaches are being applied. Various characteristics are being studied to confirm their quality, such as overall distribution of size, stability by surface charge or zeta potential evaluation, encapsulation efficacy within the liposomal structure, lamellar nature of the structure, etc. (Gulseren and Corredig 2013; Shin et al. 2015).

They increase the bioavailability of the active ingredient, improve controlled release, enable precision targeting, and demonstrate high therapeutic activity at the target site; they can be applied for pulmonary therapies and as gene delivery vectors for stabilization of sensitive biomaterials, and they reduce toxicity and amplify the resourceful allocation of bioactive resources. They exhibit a longer residence time in the bloodstream because they are not recognized by macrophages and also prevent unwanted interactions with other molecules (Mohammadi et al. 2021; Lu et al. 2014).

The astonishing benefits of nanoliposomes have led scientists around the globe to further investigate the potential of these formulation for bioactive compounds. Carotenoids have antioxidant properties and are therefore generally added to foods, but there seems to be a major problem of sensitivity of this molecule to light, heat, and oxygen. To overcome this problem, β-carotene-loaded nanoliposomes were prepared via an ethanol injection method. This formulation has been successfully developed to deliver hydrophobic compounds (de Freitas Zompero et al. 2015).

1.5.6 Nanofibers

Nanofibers are the most exploited novel nanocarrier systems for delivery, as they can be formulated in a desirable shape using various synthetic and natural polymers. Nanofibers have specific physicochemical properties such as bulky surface–mass proportion, better-quality spontaneous performance, and high porosity. In the present scenario, nanofibers have found wide applications in drug delivery, tissue engineering, medical implants, environmental engineering and biotechnology (via membranes and filters), and energy (via solar cells and fuel cells). One use of nanofibers involves application as delivery systems for the transport of antimicrobial agents, enzymes, medicines, coloring and flavor agents, antioxidants, and other functional compounds (Ohgo et al. 2003; Avci et al. 2018). Nanofibers are employed in the controlled release of a bioactive moiety, such that a defined mixture can be stabilized and concealed in these formulations for controlled discharge in suitable surroundings. The microporous structure of nanofibers favors cell proliferation, cell adhesion, cellular rearrangement, and differentiation, all extremely preferred assets in the field of tissue engineering and related applications. Several techniques such as self-assembly, electrospinning, and phase separation are employed for their synthesis. Among these, electrospinning is mostly used, as this process does not lead to deterioration of the active compound (Cheng et al. 2014; Garcia-Moreno et al. 2016). Nanofibers are prepared from both natural and synthetic polymers; some examples of natural polymers are hyaluronic acid, collagen, gelatin, elastin, chitosan, silk, and wheat protein. Examples of synthetic polymers include PLA, poly (ethylene terephthalate), PLGA, and poly (ethylene glycol).

Electrospun nanofibers can be used for the delivery of nutrients so as to guard the bioactive moiety during storage and as a carrier for relocating the compounds to the necessary position in the system. The bioactive compounds are protected from pH, heat, and moisture so that they are released under appropriate conditions in the targeted location (Avci et al. 2013; Fung et al. 2011; Rezaei et al. 2016). For the aforementioned reasons, various studies have been performed using nanofibers as delivery systems for nutraceuticals. Bioactive compounds can be easily incorporated in nanofibers utilizing the electrospinning method. One natural aromatic plant extract is eugenol, which has antibacterial action but seems to have limited use due to its low stability and degradation via environmental exposure. Thus, a non-toxic oligosaccharide, cyclodextrin, is used to encapsulate eugenol. To enhance the heat resilience, robustness, and gradual discharge of eugenol in the fiber matrix, electrospun polyvinyl alcohol (PVA) nanofibers incorporating eugenol and cyclodextrin were created (Kayaci et al. 2013).

The major limitation of the use of nanofibers is the sudden discharge of the bioactive compound before they reach their target. Also, nanofibers seem to interfere with the normal gastrointestinal tract (GIT) function, as they are immediately absorbed prior to arriving at the target site. Furthermore, elevated doses of synthetic surfactants also lead to adverse effects (Montaño-Leyva et al. 2011).

1.5.7 Nanoemulsion

Nanoemulsions can be defined as lipid-based structures that can improve the solubilization of water-insoluble substances in an indirect way. Nanoemulsions are well-defined formulations of lipid droplets in the 10–200 nm range. They are regarded as key vehicles for delivering bioactive substances within physiological or biological systems, being administered by various routes. They can be also readily used along with various food matrices or systems and are currently being widely used in such scenarios. Besides their popularity in the domains of food sciences, they are also attracting a lot of attention as well as curiosity in the areas of pharmaceutics (McClements 2013a,b; McClements and Jafari 2018; Solans et al. 2005). They are isotropic in nature, kinetically stable, and thermodynamically unstable, comprising a suitable emulsifier that stabilizes the droplets of oil containing the hydrophobic medication in the surrounding water system. These carrier systems are physically homogeneous, having hydrodynamic diameters within the range of 50–1000 nm. In terms of liquid consistency, they range from clear transparent solutions to opaque. Small droplets generally are very clear in nature with a bluish hue, while droplets having a diameter around 500 nm are whitish and milky. Certain hydrophobic ingredients, such as lutein and curcumin, have already been encapsulated using these systems to increase human absorption. Previous nanoemulsion formulations using oil phase (Zambrano-Zaragoza et al. 2011; Lane et al. 2014) were limited to the use of oils from fish and from other animal sources, which are unsuitable and a matter of concern among vegans and vegetarians. So, the emulsion industry shifted to plant oils and similar alternatives. Emulsions are formed by mixing two solutions of different polarity and applying high energy, whereby minute droplets of one phase are generated within the other solution or phase. In this way, one liquid becomes encapsulated in the other, forming either water-in-oil or oil-in-water systems. The droplets formed will tend to fuse or aggregate together by Oswald ripening at rest until the two phases separate out and become completely unmixed. Surfactants or surface active agents that adsorb at the boundary of the phases and hinder the fusion of colliding droplets can be used to stabilize emulsions. This separation occurs in case of nanoemulsions, but very slowly. Surface active molecules called surfactants, which are generally amphiphilic in nature, are attached to the oil–water interface, thereby maintaining the hydrophilicity-lipophilicity balance (HLB), resulting in long-term stabilization of the whole system. Nowadays, emulsions of another type, so-called Pickering emulsions, are also being formed, in which solid surfactant molecules are adsorbed at the interface, with several novel applications (Taylor 1995). Emulsions, especially oil-in-water (O/W), are highly promising and beneficial carrier systems due to their unique features and widespread applications in the food, therapeutics, petroleum, chemical, and food sectors. The fact that the interaction between water and oil moieties is highly unfavorable entropically indicates that such biphasic systems are thermodynamically unstable. As a result, decreasing the droplet size of the inner phase and gradually increasing the oil density along with viscosity of the aqueous phase might improve emulsion stability (Shakeel and Ramadan 2010). Nanoemulsions have distinct physicochemical features, such as optical transparency, bulk viscosity, minuscule size, huge surface area, and physical stability, in comparison to typical microemulsion systems due to their tiny droplet sizes (Peng et al. 2010). Because of their outstanding interfacial diffusivity and biodegradability, most research until now has focused on the use of surfactants, synthetic in nature and with very low molecular weight (e.g., Spans and Tweens), as opposed to big biopolymers like polysaccharides and proteins (Ghosh et al. 2013). Concerns have been raised regarding the use of synthetic emulsifiers in food systems, as there is a big question mark regarding their toxicity, safety, and absorption in the physiological system, which greatly hinders their usage in the food sector. Nanodispersions may be useful carrier systems for incorporating bioactive ingredients as well as peptides and various other nutraceuticals in food matrices, due to which they can be effectively utilized by the human system as nutraceutical and functional agents, allowing them to exhibit their bioactivities more efficiently *in vivo* (Adjonu et al. 2014). A wide variety of lipoproteins serve as key transporters of cholesterol through the bloodstream. Large lipoprotein particles transfer dietary cholesterol from the intestine to the liver. The liver secretes cholesterol-containing very-low-density lipoproteins, which are partially transformed into low-density lipoprotein (LDL) by lipoprotein lipase. LDL cholesterol is the predominant form in which cholesterol is transported throughout the tissues of the body after its exit from the liver, while high-density lipoprotein (HDL) is its counterpart by which cholesterol from different tissues of the body is transported to the liver (Pan et al. 2004). The moieties of cholesterol are generally used as scaffolding and commonly utilized to deliver anticancer medicines selectively to tumor tissues. Physical or biological targeting strategies are used in this method. In a technique known as ultrasound-aided emulsification, low-frequency ultrasound waves are applied to disperse one liquid phase into another phase. Under the influence of a process known as cavitation, this approach makes homogenization and dispersion much easier. The overall size and poly-disperseness of the core droplets in the nanoemulsion matrix are influenced by several factors, such as power, time, frequency, and the aqueous phase and oil phase used in the process. Hydrostatic pressure, gas concentration, and temperature all have an impact on the process (Kaltsa et al. 2013; Khadem and Sheibat-Othman 2019). Droplet disruption may be the final result of the cumulative influence of these small implosion events. According to recent studies and research, in emulsification formulation and progression, sonication has developed as a more economical yet more potent technology than homogenization for obtaining smaller and more uniform sizes of core droplets and less polydispersity. Nanostructured lipid carriers (NLCs) became predominant in the scientific scenario after 1999 as a gradual development in the domain of nanotechnology, being a favorable approach for encapsulation as successors of SLNs (Pardeike et al. 2009; Aditya et al. 2014). These are submicron-sized constituent particles made up of a firm matrix, usually a mixture of oils and solid lipids, and an outer layer stabilized by a surfactant. They are generally formed of generally recognized as safe (GRAS) status ingredients and are very benign to use. They are usually the carrier systems of choice in the case of encapsulating lipophilic ingredients, as they possess various benefits such as biodegradability, safety, and non-toxicity for human use (Müller et al. 2000). A variety of aspects, with the most important parameter being the oil type, influence the physicochemical properties of NLCs (Badea et al. 2015).

1.5.8 Nanocapsules

Nanocapsules are nanoparticles that have a globular void shape, with dimensions of about 0.2 μm, entrapping a bioactive compound. These delivery systems are also referred to as hollow polymer nanostructures. The distinctiveness of this delivery system is mainly due to its specific core and shell (Esmaeili and Ebrahimzadeh 2015). Interfacial polymerization of monomers and preformed polymer methods are used for formulating nanocapsules.

Techniques for polymerization of monomeric compounds are basically employed for proteins and peptides but not for bioactive compounds due to the presence of residual solvents that are proven to be toxic. The problem of organic solvent associated with this technique is surmounted by preformed polymerization, which involves the usage of natural and synthetic polymers. The polymers used for the preparation of nanocapsules are selected to increase curative benefits and reduce side effects, with some of the important polymers including poly-ε-caprolactone and polysaccharide gums (Dos Santos et al. 2016). Nanocapsules can load and encapsulate vaccines, drugs, diagnostics, bioactive compounds, and nutraceuticals, which can be administered by various routes, such as oral, topical, parenteral, or pulmonary routes. As a bioactive compound is embedded in the lipid core, it offers several advantages for delivering nutraceuticals: (a) tissue irritation is reduced at the deposition site; (b) the bioactive compound is protected from the external medium; and (c) drug encapsulation efficiency is enhanced. In one of the studies, curcumin and olive oil were enclosed in nanocapsules using a human serum albumin shell. These nanocapsules appreciably improved the entrapment efficiency, as observed using a photodegradation study (Jain et al. 2018).

1.5.9 Carbon Nanotubes

This delivery system is comprised entirely of carbon particles, set through a sequence of the simplest aromatic structure hooked on a tubiform configuration. The two most commonly used carbon nanoforms are single-walled and multi-walled nanotubes (SWNTs/MWNTs). SWNTs are made of a single barrel-sheet-shaped grapheme, whereas MWNTs consist of numerous coordinated graphene sheets. The dimensions of SWNTs are 0.02 × 0.002 μm, while MWNTs have dimensions of around 1 × 0.05 μm. Carbon nanotubes (CNTs) possess featured properties that include extremely low mass and a sequenced arrangement with immense automatic, voltaic, thermal, and surface area. These characteristics of the delivery system have been lately utilized in biomedical fields too (Khodakovskaya et al. 2012). CNTs have been used for their exceptional sorption capacity and high stability in extraction procedures. Moreover, CNTs have been found to present an effective extraction scheme, as they have a cave-like structure and can interact with a wide range of foreign molecules. CNTs have been utilized as solid-phase adsorbents for mining polyphenols from honey samples. This system demonstrates promising convenience for the scrutiny of honey and was used for improving the strength of polyphenols in honey extracts (Smith et al. 2014) (Table 1.2).

1.6 BIOAVAILABILITY

The formulation and fabrication of novel food ingredients or components in nanostructured forms and precise and highly potent carrier structures for nutraceuticals and nutritional complements has been a key thrust area of current nanotechnology applications in food (Jafari and McClements 2017). Many entrepreneurs, industries as well as start-ups, along with government sectors are trying to invest in this booming sector. The concept of bioavailability represents the percentage of the actual dosage available in the active circulation within the physiological system and so is actually responsible for the active effect. In real scenarios, most often, it is described as the percentage of the actual oral or supplied dose that enters blood system and performs the biological activity in the body (Martin 2005). Conversely, absorption, or uptake, applies to the percentage of the actual given dosage that is captured by the barrier of the human intestine. Given the association, these two concepts are very important for the pharmaceutical as well as the food sector, since these doses actually determine the amount that must be supplied within the biological system and the actual amount that performs the functionality in the system. Most often, the uptake dose is not fully bioavailable due to physiological barriers, absorption, various physicochemical factors within the body-system, efflux transporters, and also first-pass metabolism (Harmsen 2007). There has to be a clear understanding of the various biological processes controlling bioavailability and also uptake in order to develop active nanoparticulate delivery systems for not only nutrients but also nutraceuticals and other active food components (McClements et al. 2015). Bioavailability is affected by two main clusters of factors: (1) development elements (particle size, excipients, and physical state – crystalline or amorphous) and (2) physiological elements (gastric emptying rate, pH of lumen and small intestine, and intestinal wall changes). One of the most active and burgeoning domains of contemporary research in the food and nutrition industry is improvement of the bioavailability of nutrients and nutraceuticals. In their endeavor, the industries are trying to formulate nanocarriers that possess enhanced time in the systemic circulation as well as in the GIT. For this approach, they are trying to coat the surface of nutrients using various biomolecules such as polysaccharides, proteins, etc. In order to augment the bioavailability of nutraceutical compounds, the food industry is presently trying to increase the circulation as well as the retention period of such traditional carriers within the GIT, in nano form in particular by coating their outer surfaces with protein, thereby greatly altering their binding characteristics in the cellular microenvironment and their functions in the GIT, since certain biomolecules are actually capable of binding to certain specific sites bearing

TABLE 1.2
Different Nanocarriers Used in Nutraceutical Transport and Their Advantages

	Nutraceutical	Method/Carrier	Results	Reference
Nanoparticles	α-Lactalbumin	Desolvation process derived from coacervation method	Controlled drug delivery achieved	Mehravar et al. (2009)
	Elshotzia splendens	Ionic gelation	Improved antioxidant action	Lee et al. (2010)
	Vitamin D_3	Oleoyl alginate ester	Sustained rate in GIF	Li et al. (2011)
	Selenite	Chitosan	Low toxicity and improved antioxidant property	Luo et al. (2010)
	Whey protein	Ethyl hexanoate and propylene glycol	Improved retention time	Giroux and Britten (2011)
	Quercetin from *Albizia chinensis*	Solvent evaporation	Improved solubility, stability, and permeability	Kumari et al. (2011)
	Epigallocatechin gallate (EGCG)	β-Lactoglobulin	Antioxidant activity	Li et al. (2012)
	Vitamin D	Zein	Hydrophobic nutrient transport	Luo et al. (2012)
	Vitamin C	Chitosan	Stable medium to incorporate and transport vitamins	de Britto et al. (2012)
	EGCG	Caseino-phospho peptide and chitosan	Enhanced antioxidant activity	Hu et al. (2013)
	3,30-Diindolylmethane and indole-3-carbinol	Zein and zein/ Carboxymethylchitosan	Controlled release and enhanced stability against severe conditions	Luo et al. (2013)
	Soy protein	Ionic gelation method	Controlled release of hydrophobic nutraceuticals	Teng et al. (2013)
	Soy protein	Folic acid	Enhanced delivery of anticancer drugs	Teng et al. (2013)
	β-Lactoglobulin	Carbodiimide-catalyzed approach	Controlled release, enhanced absorption of nutraceuticals, desired solubility	Teng et al. (2013)
	α-Lactalbumin	Partial hydrolysis of α-lactalbumin by glu-c v8	Sensitivity of substances to surroundings was reduced	Balandrán-Quintana et al. (2013)
	Vitamin D	Soybean β-conglycinin	Enhances hydrophobic drug delivery	Levinson et al. (2014)
	Fish oil	Zein via alcohol evaporation	Enhanced bioavailability	Soltani and Madadlou (2015)

(Continued)

TABLE 1.2 (CONTINUED)
Different Nanocarriers Used in Nutraceutical Transport and Their Advantages

	Nutraceutical	Method/Carrier	Results	Reference
	β-Lactoglobulin	Low antisolvent content before mild evaporation	Controlled release	Teng et al. (2014)
	Genistein	Nucleation complexation	Enhanced bioavailability	Semyonov et al. (2014)
	Caffeine	Desolvation with ethanol	Stability enhancement to digestion	Bagheri et al. (2014)
	Rosmarinic acid	Hot melt ultrasonication method	Stability of compound	Campos et al. (2014)
	Whey protein isolate, beet pectin	Protein–polysaccharide mixtures are thermally treated to achieve protein nanoparticle formation	Protection against chemical degradation	Arroyo-Maya and McClements (2014)
	Cocoa	Nanoprecipitation	Antiradical activity of compound preserved	Quiroz-Reyes et al. (2014)
	Trans-resveratrol	Cetyl palmitate and Tween 60	Enhanced bioavailability	Neves et al. (2016)
	White tea extract	Nanoprecipitation	Controlled release and antioxidant activity	Sanna et al. (2015)
	γ-Oryzanol	Nanoprecipitation and solvent evaporation	Better encapsulation	Ghaderi et al. (2014)
	β-Lactoglobulin	Internal gelation	High cellular uptake and cytotoxicity	Ha et al. (2015)
Solid-Lipid Nanoparticles	Cruciferin	Cold gelation	Increased stability	Abkari and Wu (2016)
	Quercetin	Poly d,l-lactide	Plasma membrane permeable, loading efficient, cost-effective, suitable for oral administration	Kumari et al. (2012)
	Lutein	Cetyl palmitate, carnauba wax, glyceryl tripalmitate, capryl glycoside, caprylic/capric triglyceride, and Tween 80	Decrement of solubility across dermis	Mitri et al. (2011)
	Caffeic acid	Chitosan, alginate, and pectin	Improved bioavailability	Fathi et al. (2013)
	Rosmarinic acid	Carnauba wax and Tween 80	Protection, stability improvement, and increased bioavailability	Madureira et al. (2015)
Niosomes	Ovalbumin	Sucrose esters, cholesterol, and diacetyl phosphate	Oral vaccine delivery systems	Rentel et al. (1999)
	Ellagic acid	Span 60, Tween 60, PEG 400, propylene glycol, and methanol	Enhanced biopharmaceutical properties, solubility, and permeability	Junyaprasert et al. (2012)

(Continued)

TABLE 1.2 (CONTINUED)
Different Nanocarriers Used in Nutraceutical Transport and Their Advantages

	Nutraceutical	Method/Carrier	Results	Reference
	Resveratrol	Mechanical agitation, sonication using Span 80, Span 60, and cholesterol	Photosensitive protection, stereoisomer intactness	Pando et al. (2013)
	Lactic acid	Sodium dodecyl sulfate	Lactic acid extraction in aqueous phases	Fraile et al. (2015)
	α-Tocopherol	Hydration of thin films using sorbitan monostearate and cholesterol	Stability against adverse environmental and gastrointestinal conditions	Abaee and Madadlou (2016)
	Chrysin and luteolin	Span 60	Enhanced bioavailability	Myung et al. (2016)
Nanospheres	Whey protein α-lactalbumin	V8 protease enzyme	Site-specific delivery	Esmaielzadeh et al. (2011)
	Waxy maize starch	Nanoprecipitation (acetic anhydride, phthalic anhydride, fluorescein isothiocyanate, dibutyltin dilaureate)	Pickering emulsion stabilization using starch-based nanospheres	Tan et al. (2012)
	Gallic, caffeic acids, catechin, and rutin	Cetyltetramethylammonium chloride and hydrogen tetrachloroaureate	Enhanced bioavailability	Vilela et al. (2012)
Nanoliposomes	Curcumin	Click chemistry produced curcumin–phospholipid conjugates	Enhanced affinity for amyloid protein	Mourtas et al. (2011)
	Crocin	Dehydration and rehydration method	Enhancement of cytotoxic effects	Mousavi et al. (2011)
	Tea polyphenol	High-pressure homogenization	More efficient catechin incorporation in milk phospholipid bilayer than in soy phospholipid bilayer	Gulseren and Corredig (2013)
	Vitamin D	Thin film hydration and sonication using lecithin and cholesterol	Enhancement of physical properties and storage stability	Mohammadi et al. (2014)
	Curcumin	Ethanol injection method	Enhanced bioavailability, enhanced shelf life as well as encapsulation efficiency and high mucoadhesive property	Shin et al. (2013)
	Vitamin C	Double emulsion technique using vibrant high-pressure microfluidization	Provide long-term storage	Yang et al. (2013)
	Allicin	Reverse phase evaporation	Shelf-life increase, protection against unfavorable conditions, and weakening of offensive odor	Lu et al. (2014)
	β-Lactoglobulin	Reverse phase evaporation	Gastrointestinal stability	Ma et al. (2014)

(Continued)

TABLE 1.2 (CONTINUED)
Different Nanocarriers Used in Nutraceutical Transport and Their Advantages

	Nutraceutical	Method/Carrier	Results	Reference
	Vitamin E	Chitosan	Thermostability and economic feasibility	Xia et al. (2014)
	Zataria multiflora	Thin film evaporation, ethanol injection, and sonication	Enhancement of physical properties and storage stability	Khatibi et al. (2014)
	Vitamin A palmitate	Thin film hydration sonication	Better encapsulation	Pezeshky et al. (2016)
Nanofibers	*Bombyx mori* and *Samia cynthia ricini*	Electrospinning technique using hexafluoroacetone	Nanoscale fibers obtained from polymer solution	Ohgo et al. (2003)
	Lactobacillus acidophilus	Electrospinning technique using PVA	Thermal protection of protection in heat-processed foods	Fung et al. (2011)
	Durum wheat straw	Electrospinning method using trifluoroacetic acid	Alternative raw material for production of cellulose nanofibers as renewable source	Montaño-Leyva et al. (2011)
	Lawsonia inermis	Electrospinning technique using poly ethylene oxide and PVA	Antibacterial action	Avci et al. (2013)
	Aloe vera rind	Chemo-mechanical method	Cellulose nanofibrils were obtained	Cheng et al. (2014)
	Fish oil	Electrospinning using PVA	Enhancement of bioactivity	Garcia Moreno et al. (2016)
Nanoemulsions	Pea protein	Water–pea protein–sunflower oil mixture	Enhanced bioavailability	Donsi et al. (2010)
	Capsaicin	Tween 80, Alginate, sodium salt, chitosan	Improvement of stability	Choi et al. (2011)
	Curcumin	Purity gum 136 ultra	Enhanced bioavailability	Abbas et al. (2015)
	Canthaxanthin ketocarotenoid	Sunflower oil, sorbitan monododecanoate, polyoxyethylene, sorbitan monooleate, and potassium sorbate	Lowering of surface tension	Taghi Gharibzahedi et al. (2015)
Carbon nanotubes	Curcumin	Poly-D-lysine	Entrapment of hydrophobic bioactive compound	Sadeghi et al. (2013)
	Tobacco	Chemical vapor deposition	Regulation of cell division	Khodakovskaya et al. (2012)
	Quercetin and rutin	SWCNTs, SWCNT-OH, SWCNT-COOH	Antioxidants with lower solubility are transported with improved scavenging rate and specific targeting	Nichita and Stamatin (2013)
Nanocapsules	Vitamin E	Nanoprecipitation using polycaprolactone	Enhancement of physical or chemical stability	Khayata et al. (2012)
	Crataegus azorolus	Polymer deposition solvent evaporation	Enhanced therapeutic action	Esmaeili et al. (2013)

(Continued)

TABLE 1.2 (CONTINUED)
Different Nanocarriers Used in Nutraceutical Transport and Their Advantages

Nutraceutical	Method/Carrier	Results	Reference
Casein	Rapeseed and soybean oil	Stabilization of hydrophobic interaction	Ghasemi and Abbasi (2014)
Egg albumin and soya protein	Emulsion preparation	Controlled release delivery system obtained	Gupta and Ghosh (2014)
Lycopene	Dextran and chitosan as matrix	Protection against moisture and thermal degradation	Perez-Masia et al. (2015)
Aloe vera	Emulsion and solvent penetration using PEG–polybutylene adipate–polyethylene glycol polymer and olive oil	Enhancement of medicinal effects	Esmaeili and Ebrahimzadeh (2015)
α-Tocopherol	Emulsification-diffusion	Enhancement of therapeutic actions	Galindo-Pérez et al. (2015)
Eugenol	Solvent displacement using polycaprolactone	Enhanced therapeutic potential in periodontal infections	Pramod et al. (2016)

receptor oligosaccharides found on the epithelial cell surface (Rein et al. 2013; Pathakoti et al. 2017). Experiments have shown that lectin can covalently bind to carriers made of poly (methylvinyl ether)/maleic anhydride, which increases their retention time, decreasing their rate of elimination in the GIT. Again, food nanomaterials coated with bovine serum albumin can strongly adhere to stomach mucosa receptors and retain their bioactivity for longer periods of time. Surfaces enriched with specific biomolecules may have further properties that may mediate absorption of food entities by a specific population of cells. From the standpoint of food scientists, bioavailability is cumulatively determined by assimilation, accumulation, digestion, and elimination; however, in the case of food formulations, the key emphasis is on assimilation and digestion. Oral nutraceuticals must enter through three routes to reach bloodstream or circulatory system and are therefore transformed by various mechanisms in the physiological system, as depicted in Figure 1.6 (McClements et al. 2015; Espín et al. 2007; AlAli et al. 2021; McClements 2019).

1.6.1 Solubilization

Food nutrients and nutraceuticals are dissipated in the small intestine, specifically the lumen part. Their overall solubility depends on the surface functional group or chemical moiety present in the native chemical or on the surface of any carrier (Wooster et al. 2017). They can be either hydrophobic or hydrophilic in nature. The hydrophilic ingredients dissolve in the aqueous environment of the lumen, while the hydrophobic ones are dissolved by various micelles that form in the lumen (Corte-Real and Bohn 2018). The micelles are formed of phospholipids as well as endogenous bile salts and exogenous emulsifiers from foods (Singh et al. 2015). As the micelles possess a hydrophobic/lipophilic core, they can readily dissolve compounds that are insoluble in water. Lipids or fats present in different food matrices can greatly contribute to the development of such micelles, which are usually generated after lipase digestion and are generally amphiphilic in nature. For example, phospholipids are generally digested into lysophospholipids and free fatty acids, while triglycerides yield free fatty acids, monoglycerides, and diglycerides (Huang et al. 2010).

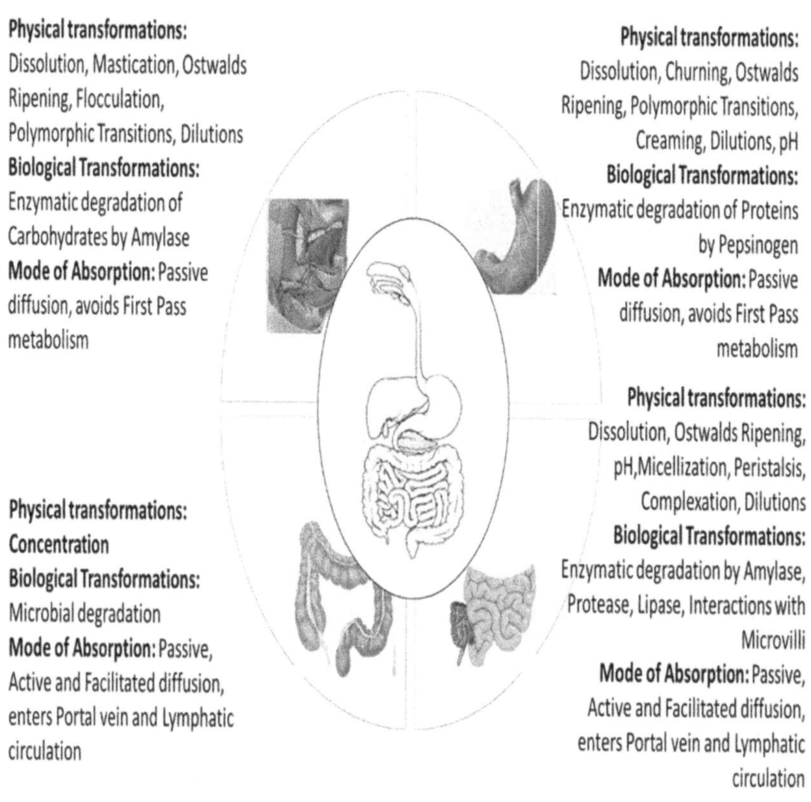

FIGURE 1.6 Various transformations of nanoceuticals occurring at various steps in the digestive system.

1.6.2 Absorption

Various nutrients, nutraceuticals, or nutrient-like ingredients that are solubilized are generally captured by the small intestine, mainly through various epithelial pathways. Various organs such as the pharynx, mouth, esophagus, stomach, large intestine (including the colon), and small intestine (duodenum, jejunum, and ileum) constitute the digestive system (Kiela and Ghishan 2016). The nutraceuticals and nutrients are mainly absorbed in the small intestine, mostly by enterocytes (epithelial absorption cells) found in the jejunum, especially the lumen (Wapnir and Teichberg 2002). The molecularly distributed nutraceuticals are transported through the lumen of the small intestine by its intrinsic aqueous solution. Several mechanisms aid their movement through epithelial layers (Porter and Charman 2001). They may diffuse with the aid of thermal movements and also the concentration gradient present in the epithelial layer. (1) Hydrophilic compounds appear to disperse by a paracellular diffusion mechanism. Although there are tight junctions, and the inter-epithelial gaps are only a few angstroms wide, hydrophilic components can still diffuse through tight junctions or inter-epithelial gaps sandwiched between neighboring cells. Nanomaterials with size >20 nm can distort these occluding junctions between adjacent cells of the epithelium in a reversible manner with liberation of various bioactive compounds into the blood circulation. Several components, such as chitosan and polyacrylate as well as their thiolated derivatives, are well-known disruptors of tight junctions and are generally used for the fabrication of edible or food-grade nanomaterials, which indirectly enhances their paracellular transportation mechanism. (2) Another route is transcellular diffusion, in which hydrophilic as well as lipophilic ingredients are entrapped or distributed in micellar structures, which aids their spread through the phospholipid membrane and also the epithelial cell cytosol. (3) The last mechanism includes mediated transport, in which various receptors carry the bioactive components and help them cross the epithelium. This transport may be driven by a positive concentration gradient or by utilization of energy (i.e., facilitated or active transport, respectively), but no metabolic energy source is generally required. Some nutrients are transported back to the lumen from epithelial layers by various cell surface proteins, for instance various transporters and P-glycoproteins (Xiaohui et al. 2007; Bakke et al. 2010).

1.6.3 Metabolism, Escape, and Passage to Blood Circulation

The nutraceuticals that are absorbed can enter the systemic circulation by three routes: directly passing into the lymph of the intestine, directly moving into the blood circulation, or crossing the liver through the portal vein and finally, joining the circulatory system (Faulks and Southon 2008). The two different types of components enter circulation via different mechanisms. The hydrophilic or polar ingredients primarily disseminate through the blood vessels and are transferred to the liver through the veins, primarily the hepatic portal vein. Metabolism (first-pass hepatic metabolism) can then occur in the liver. On the other hand, highly non-polar lipophilic compounds are too large to enter blood vessels, so they are introduced in the chylomicrons, which travel directly into the systemic circulation through lymphatic transport, avoiding the metabolism of the first step (Xu et al. 2020). The metabolism is actually a necessary physiological alteration, mainly through degradation of components by various processes. The metabolic changes may happen to these nutraceuticals either in the liver (hepatic metabolism) or in these digestive tract/ epithelium (colonic metabolism) during their transport into the systemic circulation after absorption and digestion (Jones et al. 2019). Besides, nano-nutrients may often escape endosomal or lysosomal engulfment, leading to their direct transport and transfer to the Golgi complex as well as the endoplasmic reticulum. Such nanoparticles can also undergo exocytosis through the membrane (Gonçalves et al. 2018) (Figure 1.7).

1.7 DISADVANTAGES/PITFALLS OF USE OF NANOCEUTICALS

While there is a continuous development of nanoparticles in the field of drug and nutraceutical transport, diagnostics, therapeutics, and engineering technology, many concerns have arisen about their adverse effects on the atmosphere and environmental as well as biological systems, and hence, their fate must be properly evaluated (Bundschuh et al. 2018; Jamuna and Ravishankar 2014). When various nanoparticles are exposed to biological molecules or macromolecules, this leads to bio-corona, exhibiting far-reaching effects in the physiological environment. This bio-corona is very important in the cellular microenvironment and organelles, which could lead to undesirable outcomes for cells (Shannahan 2017). Different nanoparticles can cause immunotoxicity, cell death, cytotoxicity, and genotoxicity, and also generate free radicals, leading to oxidative stress, as well as anomalous expression of various genes associated with epigenetic processes or proteome-stimulated drastic variations (Bahadur et al. 2016; Jeevanandam et al. 2018). Due to the effects of such extracellular components, the DNA sequence as a whole may not change, but genetic alterations can result from the epigenetic background, which may indirectly lead to gene expression. This can occur via various mechanisms such as histone modifications (chromatin remodeling leading to methylation, acetylation, phosphorylation, ATP-ribosylation, and ubiquitination (Patil-Rajpathak and Patil 2020). In various regulatory landscapes of gene as well as post-transcriptional modification, DNA methylation induced by microRNA is a very important event, triggering a wide multitude of cascades (Pogribna and Hammons 2021; Wong et al. 2017). It has been also shown that lifestyle and environmental factors may cause epigenetic modifications. The mechanism of trans-generational inheritance may often shift the place as well as the timing of epigenetic modifications. Nanoparticles such as TiO_2 are widely used as anticaking agents in the food industry as well as for packaging material to increase food shelf life and prevent microbial contamination. A wide variety of *in vitro* studies with these nanoparticles indicated augmented intensities of DNA damage and also increased levels of DNA lesions due to oxidation after their exposure to living systems (Magdolenova et al. 2014).

Another issue of concern is the environmental fate of nanoceuticals or various nanomaterials used in this procedure. They can cause environmental hazards, affecting living organisms as well

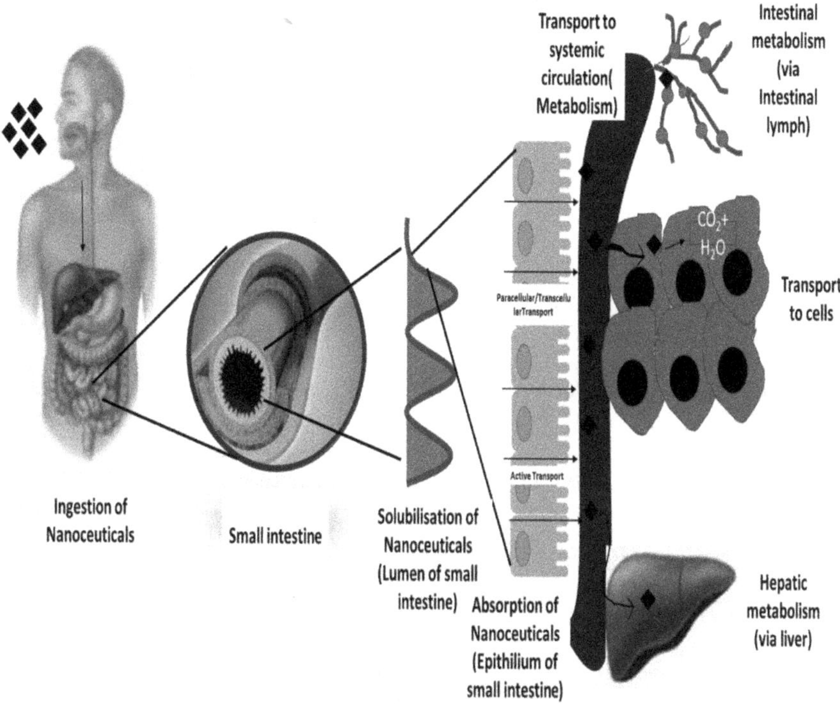

FIGURE 1.7 Bioavailability of nutraceuticals in the human body. (From Faulks and Southon 2008; Sensoy 2014; Maznah et al. 2005; Sharif and Khalid 2018.)

as the ecosystem. Tangible footsteps are the need of the hour and must be taken by different regulatory authorities of international repute to evaluate the human and environmental effects of nanotechnology by performing studies involving the assessment of hazardous effects (Bundschuh et al. 2018). Such experiments may involve how the fate of nanomaterials changes in the overall life cycle or environment as an approach involved during the production, development, delivery, usage, and probably, recovery of nanoceuticals (Table 1.3).

1.8 REGULATION OF NANOCEUTICALS AND NANOPRODUCTS

The marketability of a particular food component or merchandise as a dietetic complement or fortified product involves a regulatory process, as the label of medicinal properties associated with any such product may add a constraint to comply with the criteria governing drugs or medications (Anadón et al. 2021; Bianco and Sipp 2014). Such products are regulated with regard to side effects, overall effectiveness, correct dosage, consistency of analysis procedures, and marketing authorization techniques. For numerous years, various nutritional supplements were treated as food under the aegis of the FDA to guarantee that they would be equally healthy, nutritious, and safe, and that their branding was correct and not misleading (Darrow et al. 2020; Foote and Berlin 2004). The Food Safety and Quality Act, passed by the Government of India in 2006, incorporated and streamlined a number of rules governing nutraceuticals, food supplements, food ingredients, and comestibles (Gulati et al. 2021). The Food Safety and Standards Authority (FSSA) is established under the Act. In 2007, the FDA established a nanotechnology technical committee to address healthcare and compliance issues related to the rising use of nanomaterials in diet, supplements, and cosmetic products (Hamburg 2012). A preliminary guideline statement on nanomaterials was issued by the FDA in April 2012, which comprised approval policy guidelines, the first on foodstuff and another on cosmetic products. Despite the fact that neither statement specifically mentions

TABLE 1.3
Various Formulated Nanoceuticals and Their Enhanced Therapeutic Efficiency

Nutraceuticals	Nanoceuticals	Therapeutic effects	Reference
Curcumin	*Polymeric nanoparticles encapsulating curcumin evaluated against pancreatic tumor cells*	a. Apoptosis induction b. Inhibition of nuclear factor kappa B (NFκB) c. Depletion of pro-inflammatory cytokines like IL-8, TNFα, IL-6, etc.	Bisht et al. (2007)
	Biocompatible heat-reactive polymeric nanoparticles loaded with curcumin	Toxicity toward PC3 cancer cell line via apoptosis	Rejinold et al. (2011)
	Curcumin-loaded chitosan nanoparticles	a. Cytotoxic activity against cancerous cells b. Non-toxic to normal cells	Anitha et al. (2012)
	Curcumin-loaded nanoparticles against prostate cancer	a. Good internalization of nanoparticles and better tumor deterioration in xenograft mice b. Reduction in expression of nuclear β-catenin and androgen receptors c. Regression in STAT3 and AKT phosphorylation d. Restriction of Mcl-1 and Bcl-xL proteins e. PARP cleavage f. Oncogenic miR21 depletion and increase in miR-205 levels	Yallapu et al. (2014)
	Curcumin-loaded hydrogel nanoparticle–derived clusters against lung cancer cell line (A-549)	Increment in apoptosis rates	Teong et al. (2015)
	Curcumin nanoparticles	a. Specificity toward PC3 cancer cells as compared with normal HEK cells b. Hemolysis % for both (nanocurcumin and native) was similar	Adahoun et al. (2017)
Resveratrol	*Resveratrol nanoparticles*	Better active targeting drug delivery system due to lack of surface amines	Nair et al. (2010)
	Resveratrol-loaded gelatin nanoparticles	a. Upregulation in cellular toxicity against NCI-H460 cells b. DNA damage and generation of reactive oxygen species (ROS) c. Enhanced bioavailability and longer half-life	Karthikeyan et al. (2013)
	Resveratrol–bovine serum albumin nanoparticles	Apoptosis in ovary cancer cells via caspase cascade	Guo et al. (2010)
	Colloidal mesoporous silica nanoparticles of resveratrol	a. Superior cytotoxicity to native resveratrol via PARP and cIAP1 pathways b. Inhibition of lipopolysaccharide-induced NF-κB activation in macrophages	Summerlin et al. (2016)
	SLNs of resveratrol modified by adding apo lipoprotein E	Better cellular uptake of resveratrol	Neves et al. (2016)

(Continued)

TABLE 1.3 (CONTINUED)
Various Formulated Nanoceuticals and Their Enhanced Therapeutic Efficiency

Nutraceuticals	Nanoceuticals	Therapeutic effects	Reference
Green tea polyphenols	*PLGA nanoparticles of tea polyphenols*	Enhanced protection of mouse skin tissue against DNA damage	Srivastava (2013)
	Self-assembled micellar nanocomplexes of green tea catechins and herceptin	a. Adjuvant efficacy leading to better tumor specificity and antiproliferative activity b. Increased blood half-life	Chung et al. (2014)
	Gold-conjugated green tea nanoparticles	a. Upregulation in anticancer activity and protection of liver tissues b. Apoptosis indicated via increase in Bax levels, depletion in Bcl2 and Caspase 3 levels c. Antioxidant activity and better cellular uptake	Mukherjee et al. (2015)
	Green tea polyphenols	a. Increased specific *in vitro* cytotoxicity in bladder cancer cells as compared with normal Vero cells b. Induction of apoptosis via mitochondrial pathway c. Downregulation of tumor cells when given orally to C3H/HeN mice	Granja et al. (2016)
	EGCG–gold nanoparticles	a. Increased apoptosis in B16F10 murine melanoma cells via intrinsic pathway b. Enhanced biocompatibility and less injury to erythrophils c. Depletion in tumor growth *in vivo*	Granja et al. (2016)
	EGCG-conjugated gold nanoparticles	a. Higher stability and perseverance of antioxidant profile b. Apoptosis in neuroblastoma cells	Granja et al. (2016)
Quercetin	*Protein–quercetin bioactive nanoparticles*	Preservation of antioxidant potential by encapsulation	Fang et al. (2011)
	Quercetin-embedded nanoparticles	Sustained release and enhanced anticancer efficacy	Pandey et al. (2015)
	Quercetin-loaded PLGA nanoparticles	a. Targeted apoptosis in HepG2 cell line b. Efficiently repressed tumor growth in solid tumor–bearing mouse model	Guan et al. (2016)
Ferulic acid	*Ferulic acid (FA) nanoparticles*	a. Enhanced anticancer effect via increase in ROS generation b. Upregulation in mitochondrial polarization, DNA injury, and TBARS	Merlin et al. (2012)
	FA-loaded chitosan nanoparticles	Characteristic morphological changes of apoptosis in ME-180 cells	Panwar et al. (2016)
	FA–inorganic nanohybrids	Two times higher anticancer activity	Kim et al. (2013)
Lycopene	*Encapsulated lycopene against leukemia cell line (K562)*	a. Restriction in cellular growth and enhancement in apoptosis b. Inhibition of telomerase activity	Gharib and Faezizadeh (2014)
	Nanoemulsion delivering gold nanoparticles and lycopene	Likely combined effects on HT-29 cell line	Huang et al. (2015)

(Continued)

TABLE 1.3 (CONTINUED)
Various Formulated Nanoceuticals and Their Enhanced Therapeutic Efficiency

Nutraceuticals	Nanoceuticals	Therapeutic effects	Reference
Piperine	*Piperine-loaded PEG–PLGA nanoparticles against breast cancer cells*	Adjuvant therapeutic effect with Paclitaxel along with conjugated aptamer nanoparticles Noteworthy decrease in required paclitaxel dosages, thus making nanoscale piperine a safe adjuvant in cancer chemotherapy	Pachauri et al. (2015)
	Rapamycin- and piperine-loaded polymeric nanoparticles	Better absorption profile with improved bioavailability of rapamycin in combination with piperine as chemo-sensitizer	Katiyar et al. (2016)
	Piperine and piperlongumine-containing nanoparticles	a. Decreased *in vitro* migration and incursion of triple-negative breast cancer cells b. Bioactivity of phytochemicals was uncompromised	Ghassemi-Rad and Hoskin (2015)
Capsaicin	*Capsaicin-loaded trimethyl chitosan–based nanoparticles*	a. Increase in Bax levels and depletion in BCL-2 as well as MDR-1 genes in human HepG2 hepatocarcinoma cells b . Blank trimethyl-chitosan NPs had a little anti-apoptotic activity of their own too	Elkholi et al. (2014)
	Capsaicin-loaded nanoparticles	a. Increased uptake in human glioblastoma U251 cells b. Enhanced anticancer potential due to their ability to cross the blood–brain barrier	Jiang et al. (2015)
Propolis	*Propolis nanofood*	a. Easy dispersion in aqueous media b. Identical *in vitro* toxicity to that of free propolis against a battery of human pancreatic cancer cells	Kim et al. (2008)
	Stingless bee propolis capped silver nanoparticles	a. Having enhanced anticancer potential against human lung cancer cell line (A-549) b. Of greater compatibility, less costly, and eco-friendly	Kothai and Jayanthi (2014)
Eugenol	*Eugenol-loaded nanoemulsions*	a. Superior anti-proliferative profile against colon cancer b. ROS production leading to apoptosis	Majeed et al. (2014)
Gingerol	*Ginger-derived nanoparticles*	a. Protective effect toward liver against alcohol-induced liver damage b. Activation of antioxidant genes, which further decrease ROS generation due to nuclear factor erythroid 2-related factor-2 activation	Zhuang et al. (2015)
Diosgenin	*Self-assembled nanoparticles of diosgenin*	Optimum size, higher drug loading capacity, slow release of the drugs, and higher combined efficiency	Li et al. (2015)

(*Continued*)

TABLE 1.3 (CONTINUED)
Various Formulated Nanoceuticals and Their Enhanced Therapeutic Efficiency

Nutraceuticals	Nanoceuticals	Therapeutic effects	Reference
Phycocyanin	*Silver nanoparticles fabricated utilizing phycocyanin*	a. Noticeable growth inhibition of Gram +ve and Gram −ve bacteria b. Promising cytotoxic effect against breast cancer (MCF-7) cell line as well as in ascites tumor–possessing mice	El-Naggar et al. (2017)
	Selenium nanoparticles designed with phycocyanin	Protectivity by mitigating apoptotic effects of palmitic acid in pancreatic β cells by hindering mitochondria depolarization	Liu et al. (2018)
	Phycocyanin nanoformulation delivering paclitaxel	Better cellular uptake led to enhanced efficiency of drug against U87MG cell line	Agrawal et al. (2017)
	Phycocyanin as a nanocargo for doxorubicin	a. Stimulation of ROS-intervening apoptosis b. Controlling drug resistance	Huang et al. (2017)

dietary supplements, the FDA considers them to be a food segment, and the prospective food guidance is expected to have a substantial impact on the nutritional supplements business (Soltani and Pouypouy 2019). Nanoceuticals, like all other nutritional supplements, are unlicensed and thus, can be marketed with little to no evidence of wellbeing or efficiency. Since the FDA appears to overlook and govern products rather than approaches or techniques, it is unable to tackle the entire spectrum of issues and possible dangers associated with nanomaterial formulations. In addition to their direct consequences for health, the various fabrication technologies involve the discharge of wastes into the environment throughout their production, development, and use (Wacker et al. 2016). Several nano-related foodstuffs and related supplementary items are expected to fall within the Department of Combination Products' authority, since they need to be formulated or fabricated using different ingredients. However, according to the Institute of Food Science and Technology (IFST), nanomaterials will be regarded as highly hazardous till further research demonstrates the opposite. The FDA reorganized its operations as a consequence, and the Food Additive Health Office was founded to supervise and regulate the endorsement of nutritional supplements, nutritional components, nutraceuticals, and food or color additives (Buzby 2010). It may be exciting to increase the digestibility of nutrients and phytochemicals, but it can also be harmful.

Nanotechnology law is advancing in everyday items, with firms located in the United States having most products, while companies from Asia and Europe come in second and third place, respectively. A number of manufacturers are readily using the term "nanotechnology" and highlighting its minuscule dimensionality and its associated properties to showcase their products in some Asian countries, claiming that small ingredients can be readily taken up with novel functions, despite the fact that most ingredients have not had their size and function claims proved with rigorous scientific evidence (Rauscher et al. 2014). Some current regulations, such as the Occupational Safety and Health Act, the Toxic Substances Control Act (TSCA), the Food, Drugs and Cosmetic Act, and various new ecological legislations, can help deliver a legal foundation for regulation of nanoproducts in some nations (e.g., the United States) (Bobst and Kraska 2018). The Codex Alimentarius Commission, a joint Food and Agriculture Organization/World Health Organization (FAO-WHO) intergovernmental agency, promotes the synchronization of all food safety and standards studies conducted by worldwide governmental and non-governmental organizations. The Codex Alimentarius Commission, which is engaged in the establishment of global standards for food safety, considers the usage of nanomaterials as well as other nanoscale advancements in agriculture and food security in order to prevent legislative vulnerabilities (Tritscher et al. 2013). There is no global regulation of nanotechnology or related goods at this time. Only a few government bodies or groups from around the world have produced rules and legislation to recognize and oversee nanotechnology use. A variety of professional bodies have indeed expressed their views on the uses of such nanomaterials and their impact on human health and quality of life, comprising the EU Scientific Committee on Consumer Products, the U.K. Committees on Toxicity, the U.K. Royal Society and Royal Academy of Engineering (Royal Society and Royal Academy of Engineering), Scientific Committee on Emerging and Newly Found Health Risks of the European Commission, the German Federal Institute for Risk Assessment, Mutagenicity and Carcinogenicity of Chemicals in Food, the U.S. FDA, Consumer Products and the Environment, the Dutch Institute of Food Safety, Wageningen UR (RIKILT), and Center for substances and integrated risk assessment (RIVM), the U.S. National Institute of Occupational Health (NIOSH), and the National Institute of Public Health and the Environment.

1.9 MARKET REVIEW

In recent decades, nanotechnology has prepared the path for a billion-dollar worldwide economy (Rai and Rai 2017). Several big food companies are at the frontier of food science and are working to improve nanoceutical marketing. Although most nanoscale materials are manufactured and used in the United States, the European Union also possesses about a 30% share of nanoproducts commercially and is progressing at a relatively fast pace. The United Kingdom accounts for about 33% of the market in the European Union. In the European Union as well as the United Kingdom, a wide assortment of customer merchandise incorporating nanomaterials is currently present (Lee and Jose 2008). The entire worldwide market size and various companies active in the field of nano-nutraceuticals and similar food products are estimated to be diversified given the lack of accurate information due to marketing and ecologic sensitivities, according to a survey on the fabrication and usage of nanomaterials in these, in the United Kingdom conducted by the Central Science Laboratory in New York (Chaudhry et al. 2008; Chaudhry and Castle 2011). The rapid development of current nanoceuticals contributes to the overall satisfaction of customers and also desires to keep up with emerging developments to enhance their wellbeing. It is also generally expected that there will be an enormous rise in the number of businesses joining the nanotechnology industry in the near future. In terms of the agriculture and food sectors, the European Union is the global leader, followed by America, Japan, and China (Kakkar et al. 2018). The United States has the largest nutraceutical industry based on ethnic and legislative effects. People from the United States primarily consume nutritional supplements, and their laws and regulations, especially with new product promotion, are lenient. The Emerging Nanotechnologies (PEN) project of the Woodrow Wilson International Center for Scholars compiles and releases an online resource of commercially available nanotechnology-based consumer items globally, and 54 ingredients or products, accounting for 57.3% of the items listed, have indeed been classified as nutritional supplements (Currall et al. 2006).

The market appeal of nanoceuticals is evident from society's reaction to the products. A variety of factors, such as perceived risks and benefits, consistency, product reliability, general insights, value, beliefs, and ethnic standards, play a significant role in embracing new food technologies in the eyes of consumers (Kozinets et al. 2017). It has additionally been seen that the nanoceutical market is directed by clients and that organizations should cautiously think about their interest in the positioning of their products. Purchasers are more unwilling to purchase food products that utilize nanomaterials as ingredients, for example nanoceuticals, than to utilize products prepared using nanotechnology. A few sociologists accept that shoppers will generally keep up with the convenience of the food and that their astonishing increments with the diminished vicinity of the food to the various foods, and diets as well. It is therefore important for administrative organizations to safeguard the trust of clients and the public authority when performing and supporting risk evaluation studies for nanoceuticals (Buzby 2010).

Practical food varieties at present contribute the biggest portion of the nano market of India, followed by dietetic complements. Lifestyle-related sicknesses are common in metropolitan as well as provincial and rural parts of India, so nanonutrients will remain popular in the near future. The globalization of the nutraceutical and utilitarian food ventures presents significant difficulties to partners, not least the administrative heterogeneity between nations associated with the commercial center. At a time when an enormous section of the Indian populace of 1.3 billion has a place in the working class, their attention to overall wellbeing and excellence in health has invigorated the development of the nutraceuticals business (Keservani et al. 2010). According to the Indian Nutraceuticals Market research, an increasing proportion of individuals are consuming nutritious products such as nutrients, beverages, and foodstuffs that are intended to improve their health, avoid sickness, and provide an increment of sharpness and endurance (Keservani et al. 2014). The business is expected to grow by around 30% and hit USD 6.5 billion by 2021–2022. Albeit the utilization of these items has been acknowledged by individuals, they are as yet in the undeveloped stage, all things considered. The main interest driving nutraceuticals in India is expanding obesity levels as individuals lead more stationary lives and eat larger amounts of cheap food, pushing them to attempt to battle this by taking enhancements and other nutraceuticals (Yadav et al. 2017).

Such minuscule particles are of questionable benefit and are present in considerable amounts in food supplements; they do not come under FDA regulation and cannot be investigated or tested such things until they are available in the market. The FDA would have some control, however, if the definition were refreshed. The unavoidable issue is whether nanoceuticals ought to be designated as novel elements. As indicated by the FDA, another medication has been synthetically modified and has not been recently sold. In the case of nanoceuticals, the substances are not synthetically modified; they are reduced to a minute or nano scale. Various supplement makers and a few specialists contend that nanoceuticals are not novel items according to the FDA (2014) and ought to be grouped under novel dietary constituents, since they possess unique capabilities or properties whenever they are decreased to nano scale dimensions, despite the fact that they are not synthetically modified. This is an issue that the FDA has not yet resolved in the current scenario. There is no doubt that nanotechnology items convey unmatched advantages in the fields of individual consideration and nourishment. Regardless, there are still countless unanswered worries about the reliability of this new innovation. Likewise, the most hopeful supporters of nanotechnology creation acknowledge that there may be hidden dangers associated with the boundless utilization of custom-made nano-items.

1.10 CONCLUSION

Rather than relying on synthetic medications for the treatment of catastrophic diseases, the world is shifting towards natural bioactive components that provide high nutritional efficacy with negligible toxicity. In order to prevent the spread of diseases and maintain health, new food products that show good penetration into body cells without any ill-effects are now being continuously explored in order to increase the quality and standard of life (Yeung et al. 2018). With the advancement of science and the passage of time, nanoceuticals seem to be a new and futuristic therapy of choice, particularly in the field of medicine as well as food supplements. Today, nanotechnology is known to be an important resource for human care in terms of diagnosis, prevention, and treatment. Nanotechnology-mediated delivery of natural products and food supplements has been shown to be highly successful, since nanoparticles seldom present any risk to human cells. Furthermore, because these carrier systems are biodegradable, they provide complete biocompatibility with the human body and provide a spectacular route for nutraceutical delivery (Chen and Hu 2020). Nanotechnology has the ability to enhance the overall quality of food by converting it into more delicious, safer, and more nutritive versions. Most nutraceutical technologies are still at the elementary level, but with time and technical know-how, these strategies can unfold their marvelous abilities and also show promise for generating high-value nutraceuticals. The advancement of nanotechnology-based nano-nutrition technologies can also increase the income of farmers and small businessmen, vendors, etc., and help in diversifying agroindustry besides contributing to awareness and invention (Laux et al. 2018).

Only a few nano-foodstuffs have currently progressed to the commercial stage, mostly within nations outside the European Union, with the United States being the forerunner in terms of invention, followed by Japan and China. Formation of novel emulsions, encapsulation of food ingredients, invention of food contact materials, and development of sensors to detect food contamination and adulteration are the priority zones in the food industry where nanotechnology has the potential to be used, and several advanced technologies are already in operation (He and Hwang 2016).

The transformation of nutraceuticals from macro to nano form can be seen as the cornerstone leading to the next producer/developer–customer association (Jha et al. 2011). Nanoceuticals endowed with their magical properties as a consequence of their superior bioavailability provide unparalleled assets to the nutraceutical per se that are not present in their ordinary counterparts. The gradual development has been the foremost contributor and driving force for the approval of regulators as well as officials, who have provided a green light for large-scale production of such products on the consumer market (Ashraf et al. 2021). Regulatory agencies will take practical measures to create uniform global panels that will standardize procedures and programs, unfold novel guidelines, and introduce changes to standardize the overall safety evaluation process that administers the endorsement of nanoceutical products and controls merchandise entering the market (Zanella et al. 2015; Ghoshal et al. 2018). However, the advantage over risk ratio of these nanoceuticals must also be supported by customers in order to effectively complete the marketing cycle. The goal should indeed be not to weigh merits and drawbacks only but also to make optimal use of tools and technologies while avoiding their exploitation (Johnston et al. 2020). Holistic collaboration between regulatory agencies, producers, and consumers will also aid in the long-term environmental and socio-economic expansion of this cutting-edge technology, revolutionizing the nutraceutical sector.

1.11 FUTURE PROSPECTS

It seems that nanotechnology for the delivery of nutraceuticals will be applied worldwide in the future. There will be many new innovations of nano-based nutraceutical technologies, especially for novel patents on a number of applications. The trend will be the development of new eco-favorable systems with increased quality, safety, stability, and efficiency of nutraceuticals. Also, the application of nanotechnology will be the tool for quality control and management of nutraceutical products. This will be an area of great improvement in nutraceutical medicine. It is also generally expected that there will be an exponential spike in companies joining the nanotechnology industry in the immediate future. As a result, many food companies and industries are leading global nutrition research and trying to be committed to boosting the nanoceuticals market. Despite current trends and future challenges with nutraceuticals, they are intended to provide an equivalent alternative to medicines with a phenomenal impact on improving the quality of life.

While South Asian nations currently have only a burgeoning consumer segment for nutraceuticals and food supplements, India is expected to be one of the world's largest emerging markets for nutraceuticals in the future. In India, the consumption of functional foods is expected to rise over the next five years, with functional foods and beverages obtaining a higher market share than dietary supplements (Schultz and Barclay 2009). Third-world countries like India will retain a moderate business or trade in the sector of nutraceuticals and similar such bioactive components due to the fact that huge amounts of medicinal as well as nutritive components or ingredients are being produced by the large pharmaceutical companies of the country. Moreover, recently, various start-ups and companies have also started investing in phytochemicals and related compounds, as India has a rich flora, which can be a game-changer for the economy of the country. The disbursement and spread of these novel and new age food systems to remote areas and greater accessibility to common people will play a critical role in this sector's future commercial success.

REFERENCES

Aadinath, W., Ghosh, T., & Anandharamakrishnan, C. (2016). Multimodal magnetic nano-carriers for cancer treatment: Challenges and advancements. *Journal of Magnetism and Magnetic Materials, 401*, 1159–1172.

Abaee, A., & Madadlou, A. (2016). Niosome-loaded cold-set whey protein hydrogels. *Food Chemistry, 196*, 106–113.

Abbas, S., Karangwa, E., Bashari, M., Hayat, K., Hong, X., Sharif, H. R., & Zhang, X. (2015). Fabrication of polymeric nanocapsules from curcumin-loaded nanoemulsion templates by self-assembly. *Ultrasonics Sonochemistry, 23*, 81–92.

Abd El-Salam, M. H., & El-Shibiny, S. (2012). Formation and potential uses of milk proteins as nano delivery vehicles for nutraceuticals: A review. *International Journal of Dairy Technology, 65*(1), 13–21.

Abuajah, C. I., Ogbonna, A. C., & Osuji, C. M. (2015). Functional components and medicinal properties of food: A review. *Journal of Food Science and Technology, 52*(5), 2522–2529.

Adahoun, M. A. A., Al-Akhras, M. A. H., Jaafar, M. S., & Bououdina, M. (2017). Enhanced anti-cancer and antimicrobial activities of curcumin nanoparticles. *Artificial Cells, Nanomedicine, and Biotechnology, 45*(1), 98–107.

Aditya, N. P., Macedo, A. S., Doktorovova, S., Souto, E. B., Kim, S., Chang, P. S., & Ko, S. (2014). Development and evaluation of lipid nanocarriers for quercetin delivery: A comparative study of solid lipid nanoparticles (SLN), nanostructured lipid carriers (NLC), and lipid nanoemulsions (LNE). *LWT-Food Science and Technology, 59*(1), 115–121.

Adjonu, R., Doran, G., Torley, P., & Agboola, S. (2014). Whey protein peptides as components of nanoemulsions: A review of emulsifying and biological functionalities. *Journal of Food Engineering, 122*, 15–27.

Agrawal, M., Yadav, S. K., Agrawal, S. K., & Karmakar, S. (2017). Nutraceutical phycocyanin nanoformulation for efficient drug delivery of paclitaxel in human glioblastoma U87MG cell line. *Journal of Nanoparticle Research, 19*(8), 1–9.

Akbari, A., & Wu, J. (2016). Cruciferin nanoparticles: Preparation, characterization and their potential application in delivery of bioactive compounds. *Food Hydrocolloids, 54*, 107–118.

Akbarzadeh, A., Rezaei-Sadabady, R., Davaran, S., Joo, S. W., Zarghami, N., Hanifehpour, Y & Nejati-Koshki, K. (2013). Liposome: Classification, preparation, and applications. *Nanoscale Research Letters, 8*(1), 1–9.

AlAli, M., Alqubaisy, M., Aljaafari, M. N., AlAli, A. O., Baqais, L., Molouki, A., ... & Lim, S. H. E. (2021). Nutraceuticals: Transformation of conventional foods into health promoters/disease preventers and safety considerations. *Molecules, 26*(9), 2540.

Anadón, A., Ares, I., Martínez-Larrañaga, M. R., & Martínez, M. A. (2021). Evaluation and regulation of food supplements: European perspective. In *Nutraceuticals* (pp. 1241–1271). Academic Press.

Anandharamakrishnan, C. (2014). *Techniques for nanoencapsulation of food ingredients* (Vol. 8, pp. 65–67). Springer.

Andlauer, W., & Fürst, P. (2002). Nutraceuticals: A piece of history, present status and outlook. *Food Research International, 35*(2–3), 171–176.

Andrew, R., & Izzo, A. A. (2017). Principles of pharmacological research of nutraceuticals. *British Journal of Pharmacology, 174*(11), 1177.

Anitha, A., Maya, S., Deepa, N., Chennazhi, K. P., Nair, S. V., & Jayakumar, R. (2012). Curcumin-loaded N, O-carboxymethyl chitosan nanoparticles for cancer drug delivery. *Journal of Biomaterials Science, Polymer Edition, 23*(11), 1381–1400.

Anselmo, A. C., & Mitragotri, S. (2016). Nanoparticles in the clinic. *Bioengineering & Translational Medicine, 1*(1), 10–29.

Aronson, J. K. (2017). Defining 'nutraceuticals': Neither nutritious nor pharmaceutical. *British Journal of Clinical Pharmacology, 83*(1), 8–19.

Arroyo-Maya, I. J., & McClements, D. J. (2015). Biopolymer nanoparticles as potential delivery systems for anthocyanins: Fabrication and properties. *Food Research International, 69*, 1–8.

Asghar, A., Randhawa, M. A., Masood, M. M., Abdullah, M., & Irshad, M. A. (2018). Nutraceutical formulation strategies to enhance the bioavailability and efficiency: An overview. *Role of Materials Science in Food Bioengineering*, 329–352.

Ashraf, S. A., Siddiqui, A. J., Elkhalifa, A. E. O., Khan, M. I., Patel, M., Alreshidi, M., ... & Adnan, M. (2021). Innovations in nanoscience for the sustainable development of food and agriculture with implications on health and environment. *The Science of the Total Environment, 768*, 144990.

Asioli, D., Aschemann-Witzel, J., Caputo, V., Vecchio, R., Annunziata, A., Næs, T., & Varela, P. (2017). Making sense of the "clean label" trends: A review of consumer food choice behavior and discussion of industry implications. *Food Research International*, *99*, 58–71.

Assadpour, E., & Mahdi Jafari, S. (2019). A systematic review on nanoencapsulation of food bioactive ingredients and nutraceuticals by various nanocarriers. *Critical Reviews in Food Science and Nutrition*, *59*(19), 3129–3151.

Avci, H., Ghorbanpoor, H., & Nurbas, M. (2018). Preparation of *Origanum minutiflorum* oil-loaded core–shell structured chitosan nanofibers with tunable properties. *Polymer Bulletin*, *75*(9), 4129–4144.

Avci, H. Ü. S. E. Y. İ. N., Monticello, R., & Kotek, R. (2013). Preparation of antibacterial PVA and PEO nanofibers containing *Lawsonia inermis* (henna) leaf extracts. *Journal of Biomaterials Science, Polymer Edition*, *24*(16), 1815–1830.

Badea, G., Lăcătuşu, I., Badea, N., Ott, C., & Meghea, A. (2015). Use of various vegetable oils in designing photoprotective nanostructured formulations for UV protection and antioxidant activity. *Industrial Crops and Products*, *67*, 18–24.

Baer, D. R., Gaspar, D. J., Nachimuthu, P., Techane, S. D., & Castner, D. G. (2010). Application of surface chemical analysis tools for characterization of nanoparticles. *Analytical and Bioanalytical Chemistry*, *396*(3), 983–1002.

Bagheri, L., Madadlou, A., Yarmand, M., & Mousavi, M. E. (2014). Spray-dried alginate microparticles carrying caffeine-loaded and potentially bioactive nanoparticles. *Food Research International*, *62*, 1113–1119.

Bahadar, H., Maqbool, F., Niaz, K., & Abdollahi, M. (2016). Toxicity of Nanoparticles and an Overview of Current Experimental Models. *Iranian Biomedical Journal*, *20*(1), 1–11.

Bakke, A. M., Glover, C., & Krogdahl, Å. (2010). Feeding, digestion and absorption of nutrients. In *Fish physiology* (Vol. 30, pp. 57–110). Academic Press.

Balandrán-Quintana, R. R., Valdéz-Covarrubias, M. A., Mendoza-Wilson, A. M., & Sotelo-Mundo, R. R. (2013). α-Lactalbumin hydrolysate spontaneously produces disk-shaped nanoparticles. *International Dairy Journal*, *32*(2), 133–135.

Bayram, M., & Gökırmaklı, Ç. (2020). The future of food and nanotechnology. In *The future of food* (pp. 19–35). Cambridge Scholars Publishing.

Bernela, M., Kaur, P., Ahuja, M., & Thakur, R. (2018). Nano-based delivery system for nutraceuticals: The potential future. In *Advances in animal biotechnology and its applications* (pp. 103–117). Springer.

Bianco, P., & Sipp, D. (2014). Regulation: Sell help not hope. *Nature*, *510*(7505), 336–337.

Bobst, S., & Kraska, R. C. (2018). Industrial chemicals regulation of new and existing chemicals (the toxic substances control act and similar worldwide chemical control laws). In *Regulatory toxicology* (pp. 219–252). CRC Press.

Bundschuh, M., Filser, J., Lüderwald, S., McKee, M. S., Metreveli, G., Schaumann, G. E., ... & Wagner, S. (2018). Nanoparticles in the environment: Where do we come from, where do we go to? *Environmental Sciences Europe*, *30*(1), 1–17.

Buteyn, N., Oh, Y. I., Knott, J., Bokach, P., Konyndyk, J., Tenney, J., ... & Koetje, D. (2019). Exploring nutraceuticals to enhance scientific literacy: Aligning with vision and change. *American Biology Teacher*, *81*(3), 176–185.

Buzby, J. C. (2010). Nanotechnology for food applications: More questions than answers. *Journal of Consumer Affairs*, *44*(3), 528–545.

Campos, D. A., Madureira, A. R., Gomes, A. M., Sarmento, B., & Pintado, M. M. (2014). Optimization of the production of solid Witepsol nanoparticles loaded with rosmarinic acid. *Colloids and Surfaces B: Biointerfaces*, *115*, 109–117.

Chanda, S., Tiwari, R. K., Kumar, A., & Singh, K. (2019). Nutraceuticals inspiring the current therapy for lifestyle diseases. *Advances in Pharmacological Sciences*, *2019*, 6908716.

Chau, C. F. (2015). An introduction to food nanotechnology 34. In *Handbook of food chemistry* (vol. 2015, pp. 1087–1101).

Chaudhry, Q., & Castle, L. (2011). Food applications of nanotechnologies: An overview of opportunities and challenges for developing countries. *Trends in Food Science & Technology*, *22*(11), 595–603.

Chaudhry, Q., Castle, L., & Watkins, R. (Eds.). (2017). Nanotechnologies in food: What, why and how? *Nanotechnologies in food* (pp.1–19). Royal Society of Chemistry.

Chaudhry, Q., Scotter, M., Blackburn, J., Ross, B., Boxall, A., Castle, L., ... & Watkins, R. (2008). Applications and implications of nanotechnologies for the food sector. *Food Additives and Contaminants*, *25*(3), 241–258.

Chellaram, C., Murugaboopathi, G., John, A. A., Sivakumar, R., Ganesan, S., Krithika, S., & Priya, G. (2014). Significance of nanotechnology in food industry. *APCBEE Procedia*, *8*, 109–113.
Chen, C., Fan, T., Jin, Y., Zhou, Z., Yang, Y., Zhu, X., ... & Huang, Y. (2013). Orally delivered salmon calcitonin-loaded solid lipid nanoparticles prepared by micelle–double emulsion method via the combined use of different solid lipids. *Nanomedicine*, *8*(7), 1085–1100.
Chen, J., & Hu, L. (2020). Nanoscale delivery system for nutraceuticals: Preparation, application, characterization, safety, and future trends. *Food Engineering Reviews*, *12*(1), 14–31.
Cheng, J. P., Wang, B. B., Zhao, M. G., Liu, F., & Zhang, X. B. (2014). Nickel-doped tin oxide hollow nanofibers prepared by electrospinning for acetone sensing. *Sensors and Actuators B: Chemical*, *190*, 78–85.
Cheng, S., Panthapulakkal, S., Sain, M., & Asiri, A. (2014). *Aloe vera* rind cellulose nanofibers-reinforced films. *Journal of Applied Polymer Science*, *15*(131).
Cheuka, P. M., Mayoka, G., Mutai, P., & Chibale, K. (2017). The role of natural products in drug discovery and development against neglected tropical diseases. *Molecules*, *22*(1), 58–99.
Choi, A. J., Kim, C. J., Cho, Y. J., Hwang, J. K., & Kim, C. T. (2011). Characterization of capsaicin-loaded nanoemulsions stabilized with alginate and chitosan by self-assembly. *Food and Bioprocess Technology*, *4*(6), 1119–1126.
Chung, J. E., Tan, S., Gao, S. J., Yongvongsoontorn, N., Kim, S. H., Lee, J. H., ... & Ying, J. Y. (2014). Self-assembled micellar nanocomplexes comprising green tea catechin derivatives and protein drugs for cancer therapy. *Nature Nanotechnology*, *9*(11), 907–912.
Colvin, V. L. (2003). The potential environmental impact of engineered nanomaterials. *Nature Biotechnology*, *21*(10), 1166–1170.
Contado, C. (2015). Nanomaterials in consumer products: A challenging analytical problem. *Frontiers in Chemistry*, *3*, 48–67.
Corte-Real, J., & Bohn, T. (2018). Interaction of divalent minerals with liposoluble nutrients and phytochemicals during digestion and influences on their bioavailability—A review. *Food Chemistry*, *252*, 285–293.
Currall, S. C., King, E. B., Lane, N., Madera, J., & Turner, S. (2006). What drives public acceptance of nanotechnology? *Nature Nanotechnology*, *1*(3), 153–155.
Cushen, M., Kerry, J., Morris, M., Cruz-Romero, M., & Cummins, E. (2012). Nanotechnologies in the food industry: Recent developments, risks and regulation. *Trends in Food Science & Technology*, *24*(1), 30–46.
Dadwal, A., Baldi, A., & Kumar Narang, R. (2018). Nanoparticles as carriers for drug delivery in cancer. *Artificial Cells, Nanomedicine, and Biotechnology*, *46*(sup2), 295–305.
Daliu, P., Santini, A., & Novellino, E. (2019). From pharmaceuticals to nutraceuticals: Bridging disease prevention and management. *Expert Review of Clinical Pharmacology*, *12*(1), 1–7.
Dan, Y., Zhang, W., Xue, R., Ma, X., Stephan, C., & Shi, H. (2015). Characterization of gold nanoparticle uptake by tomato plants using enzymatic extraction followed by single-particle inductively coupled plasma–mass spectrometry analysis. *Environmental Science & Technology*, *49*(5), 3007–3014.
Darrow, J. J., Avorn, J., & Kesselheim, A. S. (2020). FDA approval and regulation of pharmaceuticals, 1983–2018. *JAMA*, *323*(2), 164–176.
Das, L., Bhaumik, E., Raychaudhuri, U., & Chakraborty, R. (2012). Role of nutraceuticals in human health. *Journal of Food Science and Technology*, *49*(2), 173–183.
Dasgupta, N., Ranjan, S., Mundekkad, D., Ramalingam, C., Shanker, R., & Kumar, A. (2015). Nanotechnology in agro-food: From field to plate. *Food Research International*, *69*, 381–400.
Datta, S. (2017). Sourcing, supply chain, and manufacturing of nutraceutical and functional foods. In *Developing new functional food and nutraceutical products* (pp. 179–193). Academic Press.
de Britto, D., de Moura, M. R., Aouada, F. A., Mattoso, L. H., & Assis, O. B. (2012). N, N, N-trimethyl chitosan nanoparticles as a vitamin carrier system. *Food Hydrocolloids*, *27*(2), 487–493.
de Freitas Zômpero, R. H., López-Rubio, A., de Pinho, S. C., Lagaron, J. M., & de la Torre, L. G. (2015). Hybrid encapsulation structures based on β-carotene-loaded nanoliposomes within electrospun fibers. *Colloids and Surfaces B: Biointerfaces*, *134*, 475–482.
DeLoid, G. M., Wang, Y., Kapronezai, K., Lorente, L. R., Zhang, R., Pyrgiotakis, G., ... & Demokritou, P. (2017). An integrated methodology for assessing the impact of food matrix and gastrointestinal effects on the biokinetics and cellular toxicity of ingested engineered nanomaterials. *Particle and Fibre Toxicology*, *14*(1), 1–17.
Dima, C., Assadpour, E., Dima, S., & Jafari, S. M. (2020). Bioactive-loaded nanocarriers for functional foods: From designing to bioavailability. *Current Opinion in Food Science*, *33*, 21–29.

Dolkar, D., Bakshi, P., Wali, V. K., Sharma, V., & Shah, R. A. (2017). Fruits as nutraceuticals. *Eco Env Cons, 23*, S113–S118.

Donsì, F., Senatore, B., Huang, Q., & Ferrari, G. (2010). Development of novel pea protein-based nanoemulsions for delivery of nutraceuticals. *Journal of Agricultural and Food Chemistry, 58*(19), 10653–10660.

Dos Santos, P. P., Flores, S. H., de Oliveira Rios, A., & Chiste, R. C. (2016). Biodegradable polymers as wall materials to the synthesis of bioactive compound nanocapsules. *Trends in Food Science & Technology, 53*, 23–33.

Dudeja, P., & Gupta, R. K. (2017). Nutraceuticals. In *Food safety in the 21st century* (pp. 491–496). Academic Press.

Dudkiewicz, A., Tiede, K., Loeschner, K., Jensen, L. H. S., Jensen, E., Wierzbicki, R., & Molhave, K. (2011). Characterization of nanomaterials in food by electron microscopy. *TrAC Trends in Analytical Chemistry, 30*(1), 28–43.

EFSA Scientific Committee. (2011). Guidance on the risk assessment of the application of nanoscience and nanotechnologies in the food and feed chain. *EFSA Journal, 9*(5), 2140–2175.

El Sohaimy, S. A. (2012). Functional foods and nutraceuticals-modern approach to food science. *World Applied Sciences Journal, 20*(5), 691–708.

El-Naggar, N. E. A., Hussein, M. H., & El-Sawah, A. A. (2017). Bio-fabrication of silver nanoparticles by phycocyanin, characterization, in vitro anticancer activity against breast cancer cell line and *in vivo* cytotxicity. *Scientific Reports, 7*(1), 1–20.

Elkholi, I. E., Hazem, N. M., ElKashef, W. F., Sobh, M. A., Shaalan, D., Sobh, M., & El-Sherbiny, I. M. (2014). Evaluation of anti-cancer potential of capsaicin-loaded trimethyl chitosan-based nanoparticles in HepG2 hepatocarcinoma cells. *Journal of Nanoscience and Nanotechnology, 5*(240), 2.

Esmaeili, A., & Ebrahimzadeh, M. (2015). Preparation of polyamide nanocapsules of *Aloe vera* L. delivery with *in vivo* studies. *AAPS PharmSciTech, 16*(2), 242–249.

Esmaeili, A., Rahnamoun, S., & Sharifnia, F. (2013). Effect of O/W process parameters on *Crataegus azarolus* L nanocapsule properties. *Journal of Nanobiotechnology, 11*(1), 1–9.

Esmaeilzadeh, P., Fakhroueian, Z., & Miran Beigi, A. A. (2011). Synthesis of biopolymeric α-lactalbumin protein nanoparticles and nanospheres as green nanofluids using in drug delivery and food technology. *Journal of Nano Research, 16*, 89–96. Trans Tech Publications Ltd.

Espín, J. C., García-Conesa, M. T., & Tomás-Barberán, F. A. (2007). Nutraceuticals: Facts and fiction. *Phytochemistry, 68*(22–24), 2986–3008.

Every-Palmer, S., & Howick, J. (2014). How evidence-based medicine is failing due to biased trials and selective publication. *Journal of Evaluation in Clinical Practice, 20*(6), 908–914.

Fang, R., Jing, H., Chai, Z., Zhao, G., Stoll, S., Ren, F., ... & Leng, X. (2011). Design and characterization of protein-quercetin bioactive nanoparticles. *Journal of Nanobiotechnology, 9*(1), 1–14.

Fathi, M., Donsi, F., & McClements, D. J. (2018). Protein-based delivery systems for the nanoencapsulation of food ingredients. *Comprehensive Reviews in Food Science and Food Safety, 17*(4), 920–936.

Fathi, M., Martin, A., & McClements, D. J. (2014). Nanoencapsulation of food ingredients using carbohydrate-based delivery systems. *Trends in Food Science & Technology, 39*(1), 18–39.

Fathi, M., Mirlohi, M., Varshosaz, J., & Madani, G. (2013). Novel caffeic acid nanocarrier: Production, characterization, and release modeling. *Journal of Nanomaterials, 2013*, 114–122.

Faulks, R. M., & Southon, S. (2008). Assessing the bioavailability of nutraceuticals. *Delivery and Controlled Release of Bioactives in Foods and Nutraceuticals*, 3–25.

Fibigr, J., Šatínský, D., & Solich, P. (2018). Current trends in the analysis and quality control of food supplements based on plant extracts. *Analytica Chimica Acta, 1036*, 1–15.

Foote, S. B., & Berlin, R. J. (2004). Can regulation be as innovative as science and technology-the FDA's regulation of combination products. *Minn. JL Sci. & Tech., 6*, 619–644.

Fraile, R., Geanta, R. M., Escudero, I., Benito, J. M., & Ruiz, M. O. (2015). Formulation of Span 80 niosomes modified with SDS for lactic acid entrapment. *Desalination and Water Treatment, 56*(13), 3463–3475.

Fung, W. Y., Yuen, K. H., & Liong, M. T. (2011). Agrowaste-based nanofibers as a probiotic encapsulant: Fabrication and characterization. *Journal of Agricultural and Food Chemistry, 59*(15), 8140–8147.

Gaibort, J., Sen, S., & Pathak, Y. V. (2016). Consumer acceptance of nanotechnology-based foods and food innovations. In *Nanotechnology in nutraceuticals* (pp. 367–378). CRC Press.

Galanakis, C. M. (2013). Emerging technologies for the production of nutraceuticals from agricultural by-products: A viewpoint of opportunities and challenges. *Food and Bioproducts Processing, 91*(4), 575–579.

Galindo-Pérez, M. J., Quintanar-Guerrero, D., Mercado-Silva, E., Real-Sandoval, S. A., & Zambrano-Zaragoza, M. L. (2015). The effects of tocopherol nanocapsules/xanthan gum coatings on the preservation of fresh-cut apples: Evaluation of phenol metabolism. *Food and Bioprocess Technology, 8*(8), 1791–1799.

García-Moreno, P. J., Stephansen, K., van der Kruijs, J., Guadix, A., Guadix, E. M., Chronakis, I. S., & Jacobsen, C. (2016). Encapsulation of fish oil in nanofibers by emulsion electrospinning: Physical characterization and oxidative stability. *Journal of Food Engineering*, *183*, 39–49.

Ge, W., Li, D., Chen, M., Wang, X., Liu, S., & Sun, R. (2015). Characterization and antioxidant activity of β-carotene loaded chitosan-graft-poly (lactide) nanomicelles. *Carbohydrate Polymers*, *117*, 169–176.

Geiss, O., Cascio, C., Gilliland, D., Franchini, F., & Barrero-Moreno, J. (2013). Size and mass determination of silver nanoparticles in an aqueous matrix using asymmetric flow field flow fractionation coupled to inductively coupled plasma mass spectrometer and ultraviolet–visible detectors. *Journal of Chromatography A*, *1321*, 100–108.

Ghaderi, S., Ghanbarzadeh, S., & Hamishehkar, H. (2014). Evaluation of different methods to produce nanoparticle containing gammaoryzanol for potential use in food fortification. *Pharmaceutical Sciences*, *20*(4), 130–134.

Gharib, A., & Faezizadeh, Z. (2014). In vitro anti-telomerase activity of novel lycopene-loaded nanospheres in the human leukemia cell line K562. *Pharmacognosy Magazine*, *10*(Suppl 1), S157–S163.

Ghasemi, S., & Abbasi, S. (2014). Formation of natural casein micelle nanocapsule by means of pH changes and ultrasound. *Food Hydrocolloids*, *42*, 42–47.

Ghassemi-Rad, J., & Hoskin, D. W. (2015). *Nanoparticle-encapsulated piperine and piperlongumine inhibit breast cancer cell growth and metastatic activity* (pp. 3684–3684).

Ghosh, V., Saranya, S., Mukherjee, A., & Chandrasekaran, N. (2013). Cinnamon oil nanoemulsion formulation by ultrasonic emulsification: Investigation of its bactericidal activity. *Journal of Nanoscience and Nanotechnology*, *13*(1), 114–122.

Ghoshal, G., Jain, A., & Katare, O. P. (2018). Harnessing nanotechnology using nutraceuticals for cancer therapeutics and intervention. *NanoNutraceuticals*, 57–70.

Gil, F., Hernández, A. F., & Martín-Domingo, M. C. (2016). Toxic contamination of nutraceuticals and food ingredients. In *Nutraceuticals* (pp. 825–837). Academic Press.

Giroux, H. J., & Britten, M. (2011). Encapsulation of hydrophobic aroma in whey protein nanoparticles. *Journal of Microencapsulation*, *28*(5), 337–343.

Golobič, M., Jemec, A., Drobne, D., Romih, T., Kasemets, K., & Kahru, A. (2012). Upon exposure to Cu nanoparticles, accumulation of copper in the isopod *Porcellio scaber* is due to the dissolved Cu ions inside the digestive tract. *Environmental Science & Technology*, *46*(21), 12112–12119.

Gonçalves, R. F., Martins, J. T., Duarte, C. M., Vicente, A. A., & Pinheiro, A. C. (2018). Advances in nutraceutical delivery systems: From formulation design for bioavailability enhancement to efficacy and safety evaluation. *Trends in Food Science & Technology*, *78*, 270–291.

Granja, A., Pinheiro, M., & Reis, S. (2016). Epigallocatechin gallate nanodelivery systems for cancer therapy. *Nutrients*, *8*(5), 496–517.

Gray, E. P., Coleman, J. G., Bednar, A. J., Kennedy, A. J., Ranville, J. F., & Higgins, C. P. (2013). Extraction and analysis of silver and gold nanoparticles from biological tissues using single particle inductively coupled plasma mass spectrometry. *Environmental Science & Technology*, *47*(24), 14315–14323.

Gruère, G. P. (2012). Implications of nanotechnology growth in food and agriculture in OECD countries. *Food Policy*, *37*(2), 191–198.

Guan, X., Gao, M., Xu, H., Zhang, C., Liu, H., Lv, L., ... & Tian, Y. (2016). Quercetin-loaded poly (lactic-co-glycolic acid)-d-α-tocopheryl polyethylene glycol 1000 succinate nanoparticles for the targeted treatment of liver cancer. *Drug Delivery*, *23*(9), 3307–3318.

Gulati, K., Thokchom, S. K., Joshi, J. C., & Ray, A. (2021). Regulatory guidelines for nutraceuticals in India: An overview. *Nutraceuticals*, 1273–1280.

Gülseren, I., & Corredig, M. (2013). Storage stability and physical characteristics of tea-polyphenol-bearing nanoliposomes prepared with milk fat globule membrane phospholipids. *Journal of Agricultural and Food Chemistry*, *61*(13), 3242–3251.

Gunasekaran, S., & Ko, S. (2014). Rationales of nano-and microencapsulation for food ingredients. *Nano-and Microencapsulation for Foods*, 43–64.

Guo, L., Peng, Y., Yao, J., Sui, L., Gu, A., & Wang, J. (2010). Anticancer activity and molecular mechanism of resveratrol–Bovine serum albumin nanoparticles on subcutaneously implanted human primary ovarian carcinoma cells in Nude mice. *Cancer Biotherapy & Radiopharmaceuticals*, *25*(4), 471–477.

Gupta, R. C., Srivastava, A., & Lall, R. (2018). Toxicity potential of nutraceuticals. In *Computational toxicology* (pp. 367–394). Humana Press.

Gupta, S. S., & Ghosh, M. (2014). Preparation and characterisation of protein based nanocapsules of bioactive lipids. *Journal of Food Engineering*, *121*, 64–72.

Ha, H. K., Kim, J. W., Lee, M. R., Jun, W., & Lee, W. J. (2015). Cellular uptake and cytotoxicity of β-lactoglobulin nanoparticles: The effects of particle size and surface charge. *Asian-Australasian Journal of Animal Sciences*, *28*(3), 420–427.

Hamburg, M. A. (2012). FDA's approach to regulation of products of nanotechnology. *Science*, *336*(6079), 299–300.

Handford, C. E., Dean, M., Henchion, M., Spence, M., Elliott, C. T., & Campbell, K. (2014). Implications of nanotechnology for the agri-food industry: Opportunities, benefits and risks. *Trends in Food Science & Technology*, *40*(2), 226–241.

Harmsen, J. (2007). Measuring bioavailability: From a scientific approach to standard methods. *Journal of Environmental Quality*, *36*(5), 1420–1428.

He, X., & Hwang, H. M. (2016). Nanotechnology in food science: Functionality, applicability, and safety assessment. *Journal of Food and Drug Analysis*, *24*(4), 671–681.

Helsper, J. P., Peters, R. J., Brouwer, L., & Weigel, S. (2013). Characterisation and quantification of liposome-type nanoparticles in a beverage matrix using hydrodynamic chromatography and MALDI–TOF mass spectrometry. *Analytical and Bioanalytical Chemistry*, *405*(4), 1181–1189.

Hinkley, G. K., Carpinone, P., Munson, J. W., Powers, K. W., & Roberts, S. M. (2015). Oral absorption of PEG-coated versus uncoated gold nanospheres: Does agglomeration matter? *Particle and Fibre Toxicology*, *12*(1), 1–13.

Ho, K. K., Ferruzzi, M. G., & Wightman, J. D. (2020). Potential health benefits of (poly) phenols derived from fruit and 100% fruit juice. *Nutrition Reviews*, *78*(2), 145–174.

Hu, B., Ting, Y., Zeng, X., & Huang, Q. (2013). Bioactive peptides/chitosan nanoparticles enhance cellular antioxidant activity of (–)-epigallocatechin-3-gallate. *Journal of Agricultural and Food Chemistry*, *61*(4), 875–881.

Hu, X., Li, D., Gao, Y., Mu, L., & Zhou, Q. (2016). Knowledge gaps between nanotoxicological research and nanomaterial safety. *Environment International*, *94*, 8–23.

Hu, Y., Jiang, X., Ding, Y., Ge, H., Yuan, Y., & Yang, C. (2002). Synthesis and characterization of chitosan–poly (acrylic acid) nanoparticles. *Biomaterials*, *23*(15), 3193–3201.

Huang, Q., Yu, H., & Ru, Q. (2010). Bioavailability and delivery of nutraceuticals using nanotechnology. *Journal of Food Science*, *75*(1), R50–R57.

Huang, R. F., Wei, Y. J., Inbaraj, B. S., & Chen, B. H. (2015). Inhibition of colon cancer cell growth by nanoemulsion carrying gold nanoparticles and lycopene. *International Journal of Nanomedicine*, *10*, 2823–2846.

Huang, Y., He, L., Song, Z., Chan, L., He, J., Huang, W., ... & Chen, T. (2017). Phycocyanin-based nanocarrier as a new nanoplatform for efficient overcoming of cancer drug resistance. *Journal of Materials Chemistry B*, *5*(18), 3300–3314.

Hyson, D. A. (2015). A review and critical analysis of the scientific literature related to 100% fruit juice and human health. *Advances in Nutrition*, *6*(1), 37–51.

Intelligence, M. (2016). *Global nutraceuticals market-growth, trends and forecasts (2016–2021)*. Research and Markets.

Jafari, S. M., & McClements, D. J. (2017). Nanotechnology approaches for increasing nutrient bioavailability. *Advances in Food and Nutrition Research* (vol. 81, pp. 1–30). Academic Press.

Jain, S., Yadav, V., Jain, A. K., & Yadav, A. K. (2018). Nutraceuticals: A revolutionary approach for nano drug delivery. In *Nano nutraceuticals* (pp. 1–22). CRC Press.

Jamuna, B. A., & Ravishankar, R. V. (2014). Environmental risk, human health, and toxic effects of nanoparticles. *Nanomaterials for Environmental Protection*, 523–535.

Javeri, I. (2016). Application of "nano" nutraceuticals in medicine. In *Nutraceuticals* (pp. 189–192). Academic Press.

Jeetah, R., Bhaw-Luximon, A., & Jhurry, D. (2014). Polymeric nanomicelles for sustained delivery of anti-cancer drugs. *Mutation Research/Fundamental and Molecular Mechanisms of Mutagenesis*, *768*, 47–59.

Jeevanandam, J., Barhoum, A., Chan, Y. S., Dufresne, A., & Danquah, M. K. (2018). Review on nanoparticles and nanostructured materials: History, sources, toxicity and regulations. *Beilstein journal of Nanotechnology*, *9*(1), 1050–1074.

Jenning, V., & Gohla, S. H. (2001). Encapsulation of retinoids in solid lipid nanoparticles (SLN). *Journal of Microencapsulation*, *18*(2), 149–158.

Jha, Z., Behar, N., Sharma, S. N., Chandel, G., Sharma, D. K., & Pandey, M. P. (2011). Nanotechnology: Prospects of agricultural advancement. *Nano Vision*, *1*(2), 88–100.

Jiang, Z., Wang, X., Zhang, Y., Zhao, P., Luo, Z., & Li, J. (2015). Effect of capsaicin-loading nanoparticles on gliomas. *Journal of Nanoscience and Nanotechnology*, *15*(12), 9834–9839.

Joana Gil-Chávez, G., Villa, J. A., Fernando Ayala-Zavala, J., Basilio Heredia, J., Sepulveda, D., Yahia, E. M., & González-Aguilar, G. A. (2013). Technologies for extraction and production of bioactive compounds to be used as nutraceuticals and food ingredients: An overview. *Comprehensive Reviews in Food Science and Food Safety*, *12*(1), 5–23.

Johnston, L. J., Gonzalez-Rojano, N., Wilkinson, K. J., & Xing, B. (2020). Key challenges for evaluation of the safety of engineered nanomaterials. *NanoImpact*, 100219.

Jones, D., Caballero, S., & Davidov-Pardo, G. (2019). Bioavailability of nanotechnology-based bioactives and nutraceuticals. In *Advances in food and nutrition research* (Vol. 88, pp. 235–273). Academic Press.

Junyaprasert, V. B., Singhsa, P., Suksiriworapong, J., & Chantasart, D. (2012). Physicochemical properties and skin permeation of Span 60/Tween 60 niosomes of ellagic acid. *International Journal of Pharmaceutics*, *423*(2), 303–311.

Kakkar, V., Kumar, M., & Saini, K. (2018). Nanoceuticals governance and market review. *Environmental Chemistry Letters*, *16*(4), 1293–1300.

Kakkar, V., Modgill, N., & Kumar, M. (2016). From nutraceuticals to nanoceuticals. In *Nanoscience in food and agriculture* (Vol. 3, pp. 183–198). Springer.

Kalra, E. K. (2003). Nutraceutical-definition and introduction. *AAPS Pharmsci*, *5*(3), 27–28.

Kaltsa, O., Michon, C., Yanniotis, S., & Mandala, I. (2013). Ultrasonic energy input influence on the production of sub-micron o/w emulsions containing whey protein and common stabilizers. *Ultrasonics Sonochemistry*, *20*(3), 881–891.

Karthikeyan, S., Prasad, N. R., Ganamani, A., & Balamurugan, E. (2013). Anticancer activity of resveratrol-loaded gelatin nanoparticles on NCI-H460 non-small cell lung cancer cells. *Biomedicine & Preventive Nutrition*, *3*(1), 64–73.

Katiyar, S. S., Muntimadugu, E., Rafeeqi, T. A., Domb, A. J., & Khan, W. (2016). Co-delivery of rapamycin-and piperine-loaded polymeric nanoparticles for breast cancer treatment. *Drug Delivery*, *23*(7), 2608–2616.

Katouzian, I., Esfanjani, A. F., Jafari, S. M., & Akhavan, S. (2017). Formulation and application of a new generation of lipid nano-carriers for the food bioactive ingredients. *Trends in Food Science & Technology*, *68*, 14–25.

Kayaci, F., Ertas, Y., & Uyar, T. (2013). Enhanced thermal stability of eugenol by cyclodextrin inclusion complex encapsulated in electrospun polymeric nanofibers. *Journal of Agricultural and Food Chemistry*, *61*(34), 8156–8165.

Kearney, J. (2010). Food consumption trends and drivers. *Philosophical Transactions of the Royal Society B: Biological Sciences*, *365*(1554), 2793–2807.

Kermanizadeh, A., Gosens, I., MacCalman, L., Johnston, H., Danielsen, P. H., Jacobsen, N. R., ... & Wallin, H. (2016). A multilaboratory toxicological assessment of a panel of 10 engineered nanomaterials to human health—ENPRA project—the highlights, limitations, and current and future challenges. *Journal of Toxicology and Environmental Health, Part B*, *19*(1), 1–28.

Keservani, R. K., Kesharwani, R. K., Vyas, N., Jain, S., Raghuvanshi, R., & Sharma, A. K. (2010). Nutraceutical and functional food as future food: A review. *Der Pharmacia Lettre*, *2*(1), 106–116.

Keservani, R. K., Sharma, A. K., Ahmad, F., & Baig, M. E. (2014). Nutraceutical and functional food regulations in India. In *Nutraceutical and functional food regulations in the United States and around the world* (pp. 327–342). Academic Press.

Keshwani, A., Malhotra, B., & Kharkwal, H. (2015). Nutraceutical: A drug, dietary supplement and food ingredient. *Current Pharmacogenomics and Personalized Medicine (Formerly Current Pharmacogenomics)*, *13*(1), 14–22.

Kessel, M. (2011). The problems with today's pharmaceutical business—an outsider's view. *Nature Biotechnology*, *29*(1), 27–33.

Khadem, B., & Sheibat-Othman, N. (2019). Theoretical and experimental investigations of double emulsion preparation by ultrasonication. *Industrial & Engineering Chemistry Research*, *58*(19), 8220–8230.

Khan, R. S., Grigor, J., Winger, R., & Win, A. (2013). Functional food product development–Opportunities and challenges for food manufacturers. *Trends in Food Science & Technology*, *30*(1), 27–37.

Khatibi, S. A., Misaghi, A., Moosavy, M. H., Amoabediny, G., & Basti, A. A. (2014). Effect of preparation methods on the properties of *Zataria multiflora* Boiss. essential oil loaded nanoliposomes: Characterization of size, encapsulation efficiency and stability. *Pharmaceutical Sciences*, *20*(4), 141–148.

Khayata, N., Abdelwahed, W., Chehna, M. F., Charcosset, C., & Fessi, H. (2012). Preparation of vitamin E loaded nanocapsules by the nanoprecipitation method: From laboratory scale to large scale using a membrane contactor. *International Journal of Pharmaceutics*, *423*(2), 419–427.

Khezri, S., Dehghan, P., Mahmoudi, R., & Jafarlou, M. (2016). Fig juice fermented with lactic acid bacteria as a nutraceutical product. *Pharmaceutical Sciences*, *22*(4), 260–266.

Khodakovskaya, M. V., De Silva, K., Biris, A. S., Dervishi, E., & Villagarcia, H. (2012). Carbon nanotubes induce growth enhancement of tobacco cells. *ACS Nano*, *6*(3), 2128–2135.

Kiela, P. R., & Ghishan, F. K. (2016). Physiology of intestinal absorption and secretion. *Best Practice & Research Clinical Gastroenterology*, *30*(2), 145–159.

Kim, D. M., Lee, G. D., Aum, S. H., & Kim, H. J. (2008). Preparation of propolis nanofood and application to human cancer. *Biological and Pharmaceutical Bulletin*, *31*(9), 1704–1710.

Kim, H. J., Ryu, K., Kang, J. H., Choi, A. J., Kim, T. I., & Oh, J. M. (2013). Anticancer activity of ferulic acid-inorganic nanohybrids synthesized via two different hybridization routes, reconstruction and exfoliation-reassembly. *Scientific World Journal*, *2013*, 421967–421967.

Kirdar, S. S. (2015). Current and future applications of nanotechnology in the food industry. In Conference paper, ISITES2015, Valencia, Spain.

Koch, A., Brandenburger, S., Türpe, S., & Birringer, M. (2014). The need for a legal distinction of nutraceuticals. *Food and Nutrition*, *5*, 905–913.

Kothai, S., & Jayanthi, B. (2014). Anti cancer activity of silver nano particles bio-synthesized using stingless bee propolis (*Tetragonula iridipennis*) of Tamilnadu. *Asian Journal of Biomedical and Pharmaceutical Sciences*, *4*(40), 30–37.

Kozinets, R., Patterson, A., & Ashman, R. (2017). Networks of desire: How technology increases our passion to consume. *Journal of Consumer Research*, *43*(5), 659–682.

Kulkarni, A. S., Ghugre, P. S., & Udipi, S. A. (2016). Applications of nanotechnology in nutrition: Potential and safety issues. In *Novel Approaches of Nanotechnology in Food* (pp. 509–554). Academic Press.

Kumar, B., & Smita, K. (2017). Scope of nanotechnology in nutraceuticals. In *Nanotechnology Applications in Food* (pp. 43–63). Academic Press.

Kumar, K., & Kumar, S. (2015). Role of nutraceuticals in health and disease prevention: A review. *South Asian J Food Technol Environ*, *1*, 116–121.

Kumar, P. S., Pavithra, K. G., & Naushad, M. (2019). Characterization techniques for nanomaterials. In *Nanomaterials for solar cell applications* (pp. 97–124). Elsevier.

Kumari, A., Kumar, V., & Yadav, S. K. (2012). Plant extract synthesized PLA nanoparticles for controlled and sustained release of quercetin: A green approach. *Plos One*, *7*(7), e41230.

Kumari, A., Yadav, S. K., Pakade, Y. B., Kumar, V., Singh, B., Chaudhary, A., & Yadav, S. C. (2011). Nanoencapsulation and characterization of *Albizia chinensis* isolated antioxidant quercitrin on PLA nanoparticles. *Colloids and Surfaces B: Biointerfaces*, *82*(1), 224–232.

Kwak, H. S. (2014). Overview of nano- and microencapsulation for foods. *Nano- and Microencapsulation for Foods*, 1–14.

Laborda, F., Bolea, E., Cepriá, G., Gómez, M. T., Jiménez, M. S., Pérez-Arantegui, J., & Castillo, J. R. (2016). Detection, characterization and quantification of inorganic engineered nanomaterials: A review of techniques and methodological approaches for the analysis of complex samples. *Analytica Chimica Acta*, *904*, 10–32.

Ladner, D. A., Steele, M., Weir, A., Hristovski, K., & Westerhoff, P. (2012). Functionalized nanoparticle interactions with polymeric membranes. *Journal of Hazardous Materials*, *211*, 288–295.

Lai, W. T., Khong, N. M., Lim, S. S., Hee, Y. Y., Sim, B. I., Lau, K. Y., & Lai, O. M. (2017). A review: Modified agricultural by-products for the development and fortification of food products and nutraceuticals. *Trends in Food Science & Technology*, *59*, 148–160.

Lane, K. E., Li, W., Smith, C., & Derbyshire, E. (2014). The bioavailability of an omega-3-rich algal oil is improved by nanoemulsion technology using yogurt as a food vehicle. *International Journal of Food Science & Technology*, *49*(5), 1264–1271.

Laux, P., Tentschert, J., Riebeling, C., Braeuning, A., Creutzenberg, O., Epp, A., & Jakubowski, N. (2018). Nanomaterials: Certain aspects of application, risk assessment and risk communication. *Archives of Toxicology*, *92*(1), 121–141.

Lawless, L. J., Threlfall, R. T., Howard, L. R., & Meullenet, J. F. (2012). Sensory, compositional, and color properties of nutraceutical-rich juice blends. *American Journal of Enology and Viticulture*, *63*(4), 529–537.

Lee, J. S., Kim, G. H., & Lee, H. G. (2010). Characteristics and antioxidant activity of *Elsholtzia splendens* extract-loaded nanoparticles. *Journal of Agricultural and Food Chemistry*, *58*(6), 3316–3321.

Lee, R., & Jose, P. D. (2008). Self-interest, self-restraint and corporate responsibility for nanotechnologies: Emerging dilemmas for modern managers. *Technology Analysis & Strategic Management*, *20*(1), 113–125.

Letchford, K., & Burt, H. (2007). A review of the formation and classification of amphiphilic block copolymer nanoparticulate structures: Micelles, nanospheres, nanocapsules and polymersomes. *European Journal of Pharmaceutics and Biopharmaceutics*, *65*(3), 259–269.

Levinson, Y., Israeli-Lev, G., & Livney, Y. D. (2014). Soybean β-conglycinin nanoparticles for delivery of hydrophobic nutraceuticals. *Food Biophysics*, *9*(4), 332–340.

Li, A., Gong, T., Hou, Y., Yang, X., & Guo, Y. (2020). Alginate-stabilized thixotropic emulsion gels and their applications in fabrication of low-fat mayonnaise alternatives. *International Journal of Biological Macromolecules*, *146*, 821–831.

Li, B., Du, W., Jin, J., & Du, Q. (2012). Preservation of (–)-epigallocatechin-3-gallate antioxidant properties loaded in heat treated β-lactoglobulin nanoparticles. *Journal of Agricultural and Food Chemistry*, *60*(13), 3477–3484.

Li, Q., Liu, C. G., Huang, Z. H., & Xue, F. F. (2011). Preparation and characterization of nanoparticles based on hydrophobic alginate derivative as carriers for sustained release of vitamin D3. *Journal of Agricultural and Food Chemistry*, *59*(5), 1962–1967.

Li, Y., Wang, X., Cheng, S., Du, J., Deng, Z., Zhang, Y., ... & Ling, C. (2015). Diosgenin induces G2/M cell cycle arrest and apoptosis in human hepatocellular carcinoma cells. *Oncology Reports*, *33*(2), 693–698.

Lin, P. C., Lin, S., Wang, P. C., & Sridhar, R. (2014). Techniques for physicochemical characterization of nanomaterials. *Biotechnology Advances*, *32*(4), 711–726.

Lin, Q. B., Li, B., Song, H., & Wu, H. J. (2011). Determination of silver in nano-plastic food packaging by microwave digestion coupled with inductively coupled plasma atomic emission spectrometry or inductively coupled plasma mass spectrometry. *Food Additives & Contaminants: Part A*, *28*(8), 1123–1128.

Liu, Z., Fu, X., Huang, W., Li, C., Wang, X., & Huang, B. (2018). Photodynamic effect and mechanism study of selenium-enriched phycocyanin from *Spirulina platensis* against liver tumours. *Journal of Photochemistry and Photobiology B: Biology*, *180*, 89–97.

Lockwood, G. B. (2011). The quality of commercially available nutraceutical supplements and food sources. *Journal of Pharmacy and Pharmacology*, *63*(1), 3–10.

Loeschner, K., Navratilova, J., Købler, C., Mølhave, K., Wagner, S., von der Kammer, F., & Larsen, E. H. (2013). Detection and characterization of silver nanoparticles in chicken meat by asymmetric flow field flow fractionation with detection by conventional or single particle ICP-MS. *Analytical and Bioanalytical Chemistry*, *405*(25), 8185–8195.

Lozano, O., Toussaint, O., Dogne, J. M., & Lucas, S. (2013, April). The use of PIXE for engineered nanomaterials quantification in complex matrices. In *Journal of Physics: Conference Series*, *429*(1), 012010. IOP Publishing.

Lu, Q., Lu, P. M., Piao, J. H., Xu, X. L., Chen, J., Zhu, L., & Jiang, J. G. (2014). Preparation and physicochemical characteristics of an allicin nanoliposome and its release behavior. *LWT-Food Science and Technology*, *57*(2), 686–695.

Luo, Y., Teng, Z., & Wang, Q. (2012). Development of zein nanoparticles coated with carboxymethyl chitosan for encapsulation and controlled release of vitamin D3. *Journal of Agricultural and Food Chemistry*, *60*(3), 836–843.

Luo, Y., Wang, T. T., Teng, Z., Chen, P., Sun, J., & Wang, Q. (2013). Encapsulation of indole-3-carbinol and 3, 3′-diindolylmethane in zein/carboxymethyl chitosan nanoparticles with controlled release property and improved stability. *Food Chemistry*, *139*(1–4), 224–230.

Luo, Y., Zhang, B., Cheng, W. H., & Wang, Q. (2010). Preparation, characterization and evaluation of selenite-loaded chitosan/TPP nanoparticles with or without zein coating. *Carbohydrate Polymers*, *82*(3), 942–951.

Ma, J., Guan, R., Chen, X., Wang, Y., Hao, Y., Ye, X., & Liu, M. (2014). Response surface methodology for the optimization of beta-lactoglobulin nano-liposomes. *Food & Function*, *5*(4), 748–754.

Madadlou, A., Iacopino, D., Sheehan, D., Emam-Djomeh, Z., & Mousavi, M. E. (2010). Enhanced thermal and ultrasonic stability of a fungal protease encapsulated within biomimetically generated silicate nanospheres. *Biochimica et Biophysica Acta (BBA)-General Subjects*, *1800*(4), 459–465.

Madureira, A. R., Campos, D. A., Fonte, P., Nunes, S., Reis, F., Gomes, A. M., ... & Pintado, M. M. (2015). Characterization of solid lipid nanoparticles produced with carnauba wax for rosmarinic acid oral delivery. *RSC Advances*, *5*(29), 22665–22673.

Madureira, A. R., Pereira, A., Castro, P. M., & Pintado, M. (2015). Production of antimicrobial chitosan nanoparticles against food pathogens. *Journal of Food Engineering*, *167*, 210–216.

Magdolenova, Z., Collins, A., Kumar, A., Dhawan, A., Stone, V., & Dusinska, M. (2014). Mechanisms of genotoxicity. A review of in vitro and *in vivo* studies with engineered nanoparticles. *Nanotoxicology*, *8*(3), 233–278.

Majeed, H., Antoniou, J., & Fang, Z. (2014). Apoptotic effects of eugenol-loaded nanoemulsions in human colon and liver cancer cell lines. *Asian Pacific Journal of Cancer Prevention*, *15*(21), 9159–9164.

Malik, R., Rathi, J., Manchanda, D., Makhija, M., Kushwaha, D., Katiyar, P., ... & Purohit, D. (2021). Nanoceuticals as an emerging field: Current status and future prospective. *Current Nutrition & Food Science*, *17*(7), 679–689.

Martin, Y. C. (2005). A bioavailability score. *Journal of Medicinal Chemistry*, *48*(9), 3164–3170.

Maznah, I., Loh, S. P., & Waffaa, M. H. (2005). Bioavailability studies of nutraceuticals. *Malays. J. Med. Health Sci*, *1*, 1–12.

McCarron, E. (2016). *Nanotechnology and food: Investigating consumers' acceptance of foods produced using nanotechnology* (Doctoral dissertation, Dublin Business School).

McClements, D. J. (2013a). Edible lipid nanoparticles: Digestion, absorption, and potential toxicity. *Progress in Lipid Research*, *52*(4), 409–423.

McClements, D. J. (2013b). Nanoemulsion-based oral delivery systems for lipophilic bioactive components: Nutraceuticals and pharmaceuticals. *Therapeutic Delivery*, *4*(7), 841–857.

McClements, D. J. (2014). Emulsion-based delivery systems. *Nanoparticle-and microparticle-based delivery systems: Encapsulation, protection and release of active compounds.* (pp. 191–264) CRC press.

McClements, D. J. (2019). Nutraceuticals: Superfoods or Superfads? In *Future foods* (pp. 167–201). Copernicus.

McClements, D. J. (2020). Nano-enabled personalized nutrition: Developing multicomponent-bioactive colloidal delivery systems. *Advances in Colloid and Interface Science*, *282*, Doi: 102211-102211.

McClements, D. J., & Jafari, S. M. (2018). General aspects of nanoemulsions and their formulation. In *Nanoemulsions* (pp. 3–20). Academic Press.

McClements, D. J., & Xiao, H. (2017). Is nano safe in foods? Establishing the factors impacting the gastrointestinal fate and toxicity of organic and inorganic food-grade nanoparticles. *NPJ Science of Food*, *1*(1), 1–13.

McClements, D. J., DeLoid, G., Pyrgiotakis, G., Shatkin, J. A., Xiao, H., & Demokritou, P. (2016). The role of the food matrix and gastrointestinal tract in the assessment of biological properties of ingested engineered nanomaterials (iENMs): State of the science and knowledge gaps. *NanoImpact*, *3*, 47–57.

McClements, D. J., Li, F., & Xiao, H. (2015). The nutraceutical bioavailability classification scheme: Classifying nutraceuticals according to factors limiting their oral bioavailability. *Annual Review of Food Science and Technology*, *6*, 299–327.

Mecocci, P., Tinarelli, C., Schulz, R. J., & Polidori, M. C. (2014). Nutraceuticals in cognitive impairment and Alzheimer's disease. *Frontiers in Pharmacology*, *5*, 147.

Mehravar, R., Jahanshahi, M., & Saghatoleslami, N. (2009). Production of biological nanoparticles from α-lactalbumin for drug delivery and food science application. *African Journal of Biotechnology*, *8*(24), 6822–6827.

Merlin, J. J., Prasad, N. R., & Shibli, S. M. A. (2012). Ferulic acid loaded poly-d, l-lactide-co-glycolide nanoparticles: Systematic study of particle size, drug encapsulation efficiency and anticancer effect in non-small cell lung carcinoma cell line in vitro. *Biomedicine & Preventive Nutrition*, *2*(1), 69–76.

Mishra, S. P., Padhiary, A. K., Nandi, A., & Pattnaik, A. (2019). Review on role of nano-micro nutrients in vegetable crops. *Int. J. Curr. Microbiol. App. Sci*, *8*(10), 277–282.

Mitrano, D. M., Lesher, E. K., Bednar, A., Monserud, J., Higgins, C. P., & Ranville, J. F. (2012a). Detecting nanoparticulate silver using single-particle inductively coupled plasma–mass spectrometry. *Environmental Toxicology and Chemistry*, *31*(1), 115–121.

Mitrano, D. M., Barber, A., Bednar, A., Westerhoff, P., Higgins, C. P., & Ranville, J. F. (2012b). Silver nanoparticle characterization using single particle ICP-MS (SP-ICP-MS) and asymmetrical flow field flow fractionation ICP-MS (AF4-ICP-MS). *Journal of Analytical Atomic Spectrometry*, *27*(7), 1131–1142.

Mitri, K., Shegokar, R., Gohla, S., Anselmi, C., & Müller, R. H. (2011). Lipid nanocarriers for dermal delivery of lutein: Preparation, characterization, stability and performance. *International Journal of Pharmaceutics*, *414*(1–2), 267–275.

Moghassemi, S., & Hadjizadeh, A. (2014). Nano-niosomes as nanoscale drug delivery systems: An illustrated review. *Journal of Controlled Release*, *185*, 22–36.

Mohammadi, A., Jafari, S. M., Mahoonak, A. S., & Ghorbani, M. (2021). Liposomal/nanoliposomal encapsulation of food-relevant enzymes and their application in the food industry. *Food and Bioprocess Technology*, *14*(1), 23–38.

Mohammadi, M., Ghanbarzadeh, B., & Hamishehkar, H. (2014). Formulation of nanoliposomal vitamin D3 for potential application in beverage fortification. *Advanced Pharmaceutical Bulletin*, *4*(Suppl 2), 569–575.

Momin, J. K., Jayakumar, C., & Prajapati, J. B. (2013). Potential of nanotechnology in functional foods. *Emirates Journal of Food and Agriculture*, *25*(1), 10–19.

Montaño-Leyva, B., Rodriguez-Felix, F., Torres-Chávez, P., Ramirez-Wong, B., López-Cervantes, J., & Sanchez-Machado, D. (2011). Preparation and characterization of durum wheat (*Triticum durum*) straw cellulose nanofibers by electrospinning. *Journal of Agricultural and Food Chemistry, 59*(3), 870–875.

Mourtas, S., Canovi, M., Zona, C., Aurilia, D., Niarakis, A., La Ferla, B., ... & Antimisiaris, S. G. (2011). Curcumin-decorated nanoliposomes with very high affinity for amyloid-β1-42 peptide. *Biomaterials, 32*(6), 1635–1645.

Mousa, R. M., & Abd El-Hady, D. (2018). Nano-food technology and nutrition. In *Environmental nanotechnology* (pp. 59–74). Springer.

Mousavi, S. H., Moallem, S. A., Mehri, S., Shahsavand, S., Nassirli, H., & Malaekeh-Nikouei, B. (2011). Improvement of cytotoxic and apoptogenic properties of crocin in cancer cell lines by its nanoliposomal form. *Pharmaceutical Biology, 49*(10), 1039–1045.

Mukherjee, S., Ghosh, S., Das, D. K., Chakraborty, P., Choudhury, S., Gupta, P., ... & Chattopadhyay, S. (2015). Gold-conjugated green tea nanoparticles for enhanced anti-tumor activities and hepatoprotection—Synthesis, characterization and *in vitro* evaluation. *Journal of Nutritional Biochemistry, 26*(11), 1283–1297.

Müller, R. H., Mäder, K., & Gohla, S. (2000). Solid lipid nanoparticles (SLN) for controlled drug delivery–a review of the state of the art. *European Journal of Pharmaceutics and Biopharmaceutics, 50*(1), 161–177.

Myung, Y., Yeom, S., & Han, S. (2016). A niosomal bilayer of sorbitan monostearate in complex with flavones: A molecular dynamics simulation study. *Journal of Liposome Research, 26*(4), 336–344.

Nair, H. B., Sung, B., Yadav, V. R., Kannappan, R., Chaturvedi, M. M., & Aggarwal, B. B. (2010). Delivery of antiinflammatory nutraceuticals by nanoparticles for the prevention and treatment of cancer. *Biochemical Pharmacology, 80*(12), 1833–1843.

Nasri, H., Baradaran, A., Shirzad, H., & Rafieian-Kopaei, M. (2014). New concepts in nutraceuticals as alternative for pharmaceuticals. *International Journal of Preventive Medicine, 5*(12), 1487–1499.

Nel, A., Xia, T., Meng, H., Wang, X., Lin, S., Ji, Z., & Zhang, H. (2013). Nanomaterial toxicity testing in the 21st century: Use of a predictive toxicological approach and high-throughput screening. *Accounts of Chemical Research, 46*(3), 607–621.

Neves, A. R., Martins, S., Segundo, M. A., & Reis, S. (2016). Nanoscale delivery of resveratrol towards enhancement of supplements and nutraceuticals. *Nutrients, 8*(3), 131–131.

Neves, A. R., Queiroz, J. F., & Reis, S. (2016). Brain-targeted delivery of resveratrol using solid lipid nanoparticles functionalized with apolipoprotein E. *Journal of Nanobiotechnology, 14*(1), 1–11.

Nichita, C., & Stamatin, I. (2013). The antioxidant activity of the biohybrides based on carboxylated/hydroxylated carbon nanotubes-flavonoid compounds. *Digest Journal of Nanomaterials & Biostructures (DJNB), 8*(1), 445–455.

Nounou, M. I., Ko, Y., Helal, N. A., & Boltz, J. F. (2018). Adulteration and counterfeiting of online nutraceutical formulations in the United States: Time for intervention? *Journal of Dietary Supplements, 15*(5), 789–804.

Ohgo, K., Zhao, C., Kobayashi, M., & Asakura, T. (2003). Preparation of non-woven nanofibers of Bombyx mori silk, *Samia cynthia* ricini silk and recombinant hybrid silk with electrospinning method. *Polymer, 44*(3), 841–846.

Onoue, S., Uchida, A., Nakamura, T., Kuriyama, K., Hatanaka, J., Tanaka, T., ... & Yamada, S. (2015). Self-nanoemulsifying particles of coenzyme Q10 with improved nutraceutical potential. *PharmaNutrition, 3*(4), 153–159.

Orhan, I. E., Senol, F. S., Skalicka-Wozniak, K., Georgiev, M., & Sener, B. (2016). Adulteration and safety issues in nutraceuticals and dietary supplements: Innocent or risky. *Nutraceuticals, Nanotechnology in the Agri-Food Industry; Grumezescu, AM, Ed*, 153–182.

Pachauri, M., Gupta, E. D., & Ghosh, P. C. (2015). Piperine loaded PEG-PLGA nanoparticles: Preparation, characterization and targeted delivery for adjuvant breast cancer chemotherapy. *Journal of Drug Delivery Science and Technology, 29*, 269–282.

Pan, G., Øie, S., & Lu, D. R. (2004). Uptake of the carborane derivative of cholesteryl ester by glioma cancer cells is mediated through LDL receptors. *Pharmaceutical Research, 21*(7), 1257–1262.

Pandey, M. M., Rastogi, S., & Rawat, A. K. S. (2013). Indian traditional Ayurvedic system of medicine and nutritional supplementation. *Evidence-Based Complementary and Alternative Medicine, 2013*, 1–12.

Pandey, S. K., Patel, D. K., Thakur, R., Mishra, D. P., Maiti, P., & Haldar, C. (2015). Anti-cancer evaluation of quercetin embedded PLA nanoparticles synthesized by emulsified nanoprecipitation. *International Journal of Biological Macromolecules, 75*, 521–529.

Pando, D., Beltrán, M., Gerone, I., Matos, M., & Pazos, C. (2015). Resveratrol entrapped niosomes as yoghurt additive. *Food Chemistry, 170*, 281–287.

Pando, D., Gutiérrez, G., Coca, J., & Pazos, C. (2013). Preparation and characterization of niosomes containing resveratrol. *Journal of Food Engineering, 117*(2), 227–234.

Panwar, R., Sharma, A. K., Kaloti, M., Dutt, D., & Pruthi, V. (2016). Characterization and anticancer potential of ferulic acid-loaded chitosan nanoparticles against ME-180 human cervical cancer cell lines. *Applied Nanoscience, 6*(6), 803–813.

Pardeike, J., Hommoss, A., & Müller, R. H. (2009). Lipid nanoparticles (SLN, NLC) in cosmetic and pharmaceutical dermal products. *International Journal of Pharmaceutics, 366*(1–2), 170–184.

Pathakoti, K., Manubolu, M., & Hwang, H. M. (2017). Nanostructures: Current uses and future applications in food science. *Journal of Food and Drug Analysis, 25*(2), 245–253.

Patil-Rajpathak, Y., & Patil, N. (2020). Epigenetic toxicity of nanoparticles. In *Advances in Bioengineering* (pp. 129–138). Springer.

Paul, S. D., & Dewangan, D. (2015). Nanotechnology and neutraceuticals. *Int J Nanomater Nanotechnol Nanomed, 1*, 30–33.

Peng, L. C., Liu, C. H., Kwan, C. C., & Huang, K. F. (2010). Optimization of water-in-oil nanoemulsions by mixed surfactants. *Colloids and Surfaces A: Physicochemical and Engineering Aspects, 370*(1–3), 136–142.

Pérez-Masiá, R., Lagaron, J. M., & Lopez-Rubio, A. (2015). Morphology and stability of edible lycopene-containing micro-and nanocapsules produced through electrospraying and spray drying. *Food and Bioprocess Technology, 8*(2), 459–470.

Peters, R., Herrera-Rivera, Z., Undas, A., van der Lee, M., Marvin, H., Bouwmeester, H., & Weigel, S. (2015). Single particle ICP-MS combined with a data evaluation tool as a routine technique for the analysis of nanoparticles in complex matrices. *Journal of Analytical Atomic Spectrometry, 30*(6), 1274–1285.

Peters, R., Kramer, E., Oomen, A. G., Herrera Rivera, Z. E., Oegema, G., Tromp, P. C., ... & Peijnenburg, A. A. (2012). Presence of nano-sized silica during in vitro digestion of foods containing silica as a food additive. *ACS Nano, 6*(3), 2441–2451.

Peters, R. J., Rivera, Z. H., van Bemmel, G., Marvin, H. J., Weigel, S., & Bouwmeester, H. (2014). Development and validation of single particle ICP-MS for sizing and quantitative determination of nano-silver in chicken meat. *Analytical and Bioanalytical Chemistry, 406*(16), 3875–3885.

Pezeshky, A., Ghanbarzadeh, B., Hamishehkar, H., Moghadam, M., & Babazadeh, A. (2016). Vitamin A palmitate-bearing nanoliposomes: Preparation and characterization. *Food Bioscience, 13*, 49–55.

Picó, Y. (2016). Challenges in the determination of engineered nanomaterials in foods. *TrAC Trends in Analytical Chemistry, 84*, 149–159.

Pogribna, M., & Hammons, G. (2021). Epigenetic effects of nanomaterials and nanoparticles. *Journal of Nanobiotechnology, 19*(1), 1–18.

Porter, C. J., & Charman, W. N. (2001). Intestinal lymphatic drug transport: An update. *Advanced Drug Delivery Reviews, 50*(1–2), 61–80.

Pramod, K., Aji Alex, M. R., Singh, M., Dang, S., Ansari, S. H., & Ali, J. (2016). Eugenol nanocapsule for enhanced therapeutic activity against periodontal infections. *Journal of Drug Targeting, 24*(1), 24–33.

Quintanilla-Carvajal, M. X., Camacho-Díaz, B. H., Meraz-Torres, L. S., Chanona-Pérez, J. J., Alamilla-Beltrán, L., Jimenéz-Aparicio, A., & Gutiérrez-López, G. F. (2010). Nanoencapsulation: A new trend in food engineering processing. *Food Engineering Reviews, 2*(1), 39–50.

Quiroz-Reyes, C. N., Ronquillo-de Jesús, E., Duran-Caballero, N. E., & Aguilar-Méndez, M. Á. (2014). Development and characterization of gelatin nanoparticles loaded with a cocoa-derived polyphenolic extract. *Fruits, 69*(6), 481–489.

Rai, S., & Rai, A. (2017). Nanotechnology: The secret of fifth industrial revolution and the future of next generation. *Jurnal Nasional, 7*(2), 61–66.

Rangan, A., Manjula, M. V., & Satyanarayana, K. G. (2016). Trends and methods for nanobased delivery for nutraceuticals. In *Emulsions* (pp. 573–609). Academic Press.

Rasmussen, K., Rauscher, H., Mech, A., Sintes, J. R., Gilliland, D., González, M., ... & Bleeker, E. A. (2018). Physico-chemical properties of manufactured nanomaterials-Characterisation and relevant methods. An outlook based on the OECD Testing Programme. *Regulatory Toxicology and Pharmacology, 92*, 8–28.

Rauscher, H., Roebben, G., Amenta, V., Boix, S. A., Calzolai, L., Emons, H., ... & Stamm, H. (2014). Towards a Review of the EC Recommendation for a Definition of the Term "Nanomaterial". In *Science and Policy Report by the Joint Research Centre of the European Commission*. European Union.

Raz, S. R., Leontaridou, M., Bremer, M. G., Peters, R., & Weigel, S. (2012). Development of surface plasmon resonance-based sensor for detection of silver nanoparticles in food and the environment. *Analytical and Bioanalytical Chemistry, 403*(10), 2843–2850.

Rein, M. J., Renouf, M., Cruz-Hernandez, C., Actis-Goretta, L., Thakkar, S. K., & da Silva Pinto, M. (2013). Bioavailability of bioactive food compounds: A challenging journey to bioefficacy. *British Journal of Clinical Pharmacology, 75*(3), 588–602.

Rejinold, N. S., Muthunarayanan, M., Divyarani, V. V., Sreerekha, P. R., Chennazhi, K. P., Nair, S. V., ... & Jayakumar, R. (2011). Curcumin-loaded biocompatible thermoresponsive polymeric nanoparticles for cancer drug delivery. *Journal of Colloid and Interface Science, 360*(1), 39–51.

Rentel, C. O., Bouwstra, J. A., Naisbett, B., & Junginger, H. E. (1999). Niosomes as a novel peroral vaccine delivery system. *International Journal of Pharmaceutics, 186*(2), 161–167.

Rezaei, A., Fathi, M., & Jafari, S. M. (2019). Nanoencapsulation of hydrophobic and low-soluble food bioactive compounds within different nanocarriers. *Food Hydrocolloids, 88*, 146–162.

Rezaei, B., Ghani, M., Shoushtari, A. M., & Rabiee, M. (2016). Electrochemical biosensors based on nanofibres for cardiac biomarker detection: A comprehensive review. *Biosensors and Bioelectronics, 78*, 513–523.

Ronis, M. J., Pedersen, K. B., & Watt, J. (2018). Adverse effects of nutraceuticals and dietary supplements. *Annual Review of Pharmacology and Toxicology, 58*, 583–601.

Ruchi, S. (2017). Role of nutraceuticals in health care: A review. *International Journal of Green Pharmacy (IJGP), 11*(03), S385–S394.

Sadeghi, R., Kalbasi, A., Emam-jomeh, Z., Razavi, S. H., Kokini, J., & Moosavi-Movahedi, A. A. (2013). Biocompatible nanotubes as potential carrier for curcumin as a model bioactive compound. *Journal of Nanoparticle Research, 15*(11), 1–11.

Sanna, V., Chamcheu, J. C., Pala, N., Mukhtar, H., Sechi, M., & Siddiqui, I. A. (2015). Nanoencapsulation of Natural Triterpenoid Celastrol for Prostate Cancer Treatment. *International Journal of Nanomedicine, 10*, 6835–6846.

Santini, A., Cammarata, S. M., Capone, G., Ianaro, A., Tenore, G. C., Pani, L., & Novellino, E. (2018). Nutraceuticals: Opening the debate for a regulatory framework. *British Journal of Clinical Pharmacology, 84*(4), 659–672.

Santini, A., Tenore, G. C., & Novellino, E. (2017). Nutraceuticals: A paradigm of proactive medicine. *European Journal of Pharmaceutical Sciences, 96*, 53–61.

Sattigere, V. D., Ramesh Kumar, P., & Prakash, V. (2020). Science-based regulatory approach for safe nutraceuticals. *Journal of the Science of Food and Agriculture, 100*(14), 5079–5082.

Schultz, W. B., & Barclay, L. (2009). *A hard pill to swallow: Barriers to effective FDA regulation of nanotechnology-based dietary supplements.* Woodrow Wilson International Center for Scholars, Project on Emerging Nanotechnoloties, Pew Charitable Trusts.

Schulzova, V., Krtkova, V., Tomaniova, M. Hajslova, J. (2012) *Using of ELSD for determination of organic engineered nanoparticles.* NANOCON.

Schütz, C. A., Juillerat-Jeanneret, L., Mueller, H., Lynch, I., & Riediker, M. (2013). Therapeutic nanoparticles in clinics and under clinical evaluation. *Nanomedicine, 8*(3), 449–467.

Semyonov, D., Ramon, O., Shoham, Y., & Shimoni, E. (2014). Enzymatically synthesized dextran nanoparticles and their use as carriers for nutraceuticals. *Food & Function, 5*(10), 2463–2474.

Sensoy, I. (2014). A review on the relationship between food structure, processing, and bioavailability. *Critical Reviews in Food Science and Nutrition, 54*(7), 902–909.

Shakeel, F., & Ramadan, W. (2010). Transdermal delivery of anticancer drug caffeine from water-in-oil nanoemulsions. *Colloids and Surfaces B: Biointerfaces, 75*(1), 356–362.

Shannahan, J. (2017). The biocorona: A challenge for the biomedical application of nanoparticles. *Nanotechnology Reviews, 6*(4), 345–353.

Sharif, M. K., & Khalid, R. (2018). Nutraceuticals: Myths versus realities. In *Therapeutic foods* (pp. 3–21). Academic Press.

Shastry, R. P., & Rai, V. R. (2018). Market potential of food nanotechnology innovations. In *Nanotechnology Applications in the Food Industry* (pp. 59–72). CRC Press.

Shekhar, V., Jha, A. K., & Dangi, J. S. (2014, January). Nutraceuticals: A re-emerging health aid. In Proceedings of the International Conference on Food, Biological and Medical Sciences (FBMS-2014) (pp. 28–29), Bangkok, Thailand.

Shin, G. H., Chung, S. K., Kim, J. T., Joung, H. J., & Park, H. J. (2013). Preparation of chitosan-coated nanoliposomes for improving the mucoadhesive property of curcumin using the ethanol injection method. *Journal of Agricultural and Food Chemistry, 61*(46), 11119–11126.

Shin, G. H., Kim, J. T., & Park, H. J. (2015). Recent developments in nanoformulations of lipophilic functional foods. *Trends in Food Science & Technology, 46*(1), 144–157.

Shinde, N., Bangar, B., Deshmukh, S., & Kumbhar, P. (2014). Nutraceuticals: A review on current status. *Research Journal of Pharmacy and Technology, 7*(1), 110–113.

Singh, G., Stephan, C., Westerhoff, P., Carlander, D., & Duncan, T. V. (2014). Measurement methods to detect, characterize, and quantify engineered nanomaterials in foods. *Comprehensive Reviews in Food Science and Food Safety, 13*(4), 693–704.

Singh, H., Ye, A., & Ferrua, M. J. (2015). Aspects of food structures in the digestive tract. *Current Opinion in Food Science, 3*, 85–93.

Singh, S. A. U. R. A. B. H., & Devi, M. B. (2015). Vegetables as a potential source of nutraceuticals and phytochemicals: A review. *Int J Med Pharm Sci, 5*, 1–14.

Sk, M. P., Jaiswal, A., Paul, A., Ghosh, S. S., & Chattopadhyay, A. (2012). Presence of amorphous carbon nanoparticles in food caramels. *Scientific Reports, 2*(1), 1–5.

Smith, S. C., Ahmed, F., Gutierrez, K. M., & Rodrigues, D. F. (2014). A comparative study of lysozyme adsorption with graphene, graphene oxide, and single-walled carbon nanotubes: Potential environmental applications. *Chemical Engineering Journal, 240*, 147–154.

Smolkova, B., El Yamani, N., Collins, A. R., Gutleb, A. C., & Dusinska, M. (2015). Nanoparticles in food. Epigenetic changes induced by nanomaterials and possible impact on health. *Food and Chemical Toxicology, 77*, 64–73.

Solans, C., Izquierdo, P., Nolla, J., Azemar, N., & Garcia-Celma, M. J. (2005). Nano-emulsions. *Current Opinion in Colloid & Interface Science, 10*(3–4), 102–110.

Soltani, A. M., & Pouypouy, H. (2019). Standardization and regulations of nanotechnology and recent government policies across the world on nanomaterials. In *Advances in phytonanotechnology* (pp. 419–446). Academic Press.

Soltani, S., & Madadlou, A. (2015). Gelation characteristics of the sugar beet pectin solution charged with fish oil-loaded zein nanoparticles. *Food Hydrocolloids, 43*, 664–669.

Souyoul, S. A., Saussy, K. P., & Lupo, M. P. (2018). Nutraceuticals: A review. *Dermatology and Therapy, 8*(1), 5–16.

Srivastava, S., Sharma, P. K., & Kumara, S. (2015). Nutraceuticals: A review. *J. Chronother. Drug Deliv, 6*, 1–10.

Srivastava, A. K. (2013). Synthesis of PLGA nanoparticles of tea polyphenols and their strong *in vivo* protective effect against chemically induced DNA damage. *International Journal of Nanomedicine, 8*, 1451–1462.

Srivastava, S., Sharma, P. K., & Kumara, S. (2015). Nutraceuticals: A review. *J. Chronother. Drug Deliv, 6*, 1–10.

Stamm, H., Gibson, N., & Anklam, E. (2012). Detection of nanomaterials in food and consumer products: Bridging the gap from legislation to enforcement. *Food Additives & Contaminants: Part A, 29*(8), 1175–1182.

Stone, V., Miller, M. R., Clift, M. J., Elder, A., Mills, N. L., Møller, P., & Kuhlbusch, T. A. (2017). Nanomaterials versus ambient ultrafine particles: An opportunity to exchange toxicology knowledge. *Environmental Health Perspectives, 125*(10), 106002.

Suganya, V., & Anuradha, V. (2017). Microencapsulation and nanoencapsulation: A review. *International Journal of Pharmaceutical and Clinical Research, 9*(3), 233–239.

Summerlin, N., Qu, Z., Pujara, N., Sheng, Y., Jambhrunkar, S., McGuckin, M., & Popat, A. (2016). Colloidal mesoporous silica nanoparticles enhance the biological activity of resveratrol. *Colloids and Surfaces B: Biointerfaces, 144*, 1–7.

Szakal, C., Tsytsikova, L., Carlander, D., & Duncan, T. V. (2014a). Measurement methods for the oral uptake of engineered nanomaterials from human dietary sources: Summary and outlook. *Comprehensive Reviews in Food Science and Food Safety, 13*(4), 669–678.

Szakal, C., Roberts, S. M., Westerhoff, P., Bartholomaeus, A., Buck, N., Illuminato, I., & Rogers, M. (2014b). Measurement of nanomaterials in foods: Integrative consideration of challenges and future prospects. *ACS Nano, 8*(4), 3128–3135.

Taghi Gharibzahedi, S. M., Razavi, S. H., & Mousavi, M. (2015). Optimal development of a new stable nutraceutical nanoemulsion based on the inclusion complex of 2-hydroxypropyl-β-cyclodextrin with canthaxanthin accumulated by *Dietzia natronolimnaea* HS-1 using ultrasound-assisted emulsification. *Journal of Dispersion Science and Technology, 36*(5), 614–625.

Tan, Y., Xu, K., Liu, C., Li, Y., Lu, C., & Wang, P. (2012). Fabrication of starch-based nanospheres to stabilize pickering emulsion. *Carbohydrate Polymers, 88*(4), 1358–1363.

Tank Dharti, S., Gandhi, S., & Shah, M. (2010). Nutraceuticals-portmanteau of science and nature. *Int J Pharm Sci Rev Res, 5*(3), 33–38.

Taylor, P. (1995). Ostwald ripening in emulsions. *Colloids and Surfaces A: Physicochemical and Engineering Aspects, 99*(2–3), 175–185.

Télessy, I. G. (2019). Nutraceuticals. In *The Role of Functional Food Security in Global Health* (pp. 409–421). Academic Press.

Teng, Z., Luo, Y., & Wang, Q. (2013a). Carboxymethyl chitosan–soy protein complex nanoparticles for the encapsulation and controlled release of vitamin D3. *Food Chemistry, 141*(1), 524–532.

Teng, Z., Li, Y., Luo, Y., Zhang, B., & Wang, Q. (2013b). Cationic β-lactoglobulin nanoparticles as a bioavailability enhancer: Protein characterization and particle formation. *Biomacromolecules, 14*(8), 2848–2856.

Teng, Z., Li, Y., Niu, Y., Xu, Y., Yu, L., & Wang, Q. (2014). Cationic β-lactoglobulin nanoparticles as a bioavailability enhancer: Comparison between ethylenediamine and polyethyleneimine as cationizers. *Food Chemistry, 159*, 333–342.

Teng, Z., Luo, Y., Wang, T., Zhang, B., & Wang, Q. (2013). Development and application of nanoparticles synthesized with folic acid conjugated soy protein. *Journal of Agricultural and Food Chemistry, 61*(10), 2556–2564.

Teong, B., Lin, C. Y., Chang, S. J., Niu, G. C., Yao, C. H., Chen, I. F., & Kuo, S. M. (2015). Enhanced anticancer activity by curcumin-loaded hydrogel nanoparticle derived aggregates on A549 lung adenocarcinoma cells. *Journal of Materials Science. Materials in Medicine, 26*(1), 5357–5357.

Thiruvengadam, M., Rajakumar, G., & Chung, I. M. (2018). Nanotechnology: Current uses and future applications in the food industry. *3 Biotech, 8*(1), 1–13.

Thulasi, A., Rajendran, D., Jash, S., Selvaraju, S., Jose, V. L., Velusamy, S., & Mathivanan, S. (2013). Nanobiotechnology in animal nutrition. In Sampath, K. T. & Ghosh, J. (Eds.), *Animal Nutrition and Reproductive Physiology (Recent Concepts)* (1st ed., pp. 499–516).

Tritscher, A., Miyagishima, K., Nishida, C., & Branca, F. (2013). Ensuring food safety and nutrition security to protect consumer health: 50 years of the Codex Alimentarius Commission. *Bulletin of the World Health Organization, 91*, 468–468.

Tuoriniemi, J., Cornelis, G., & Hassellöv, M. (2012). Size discrimination and detection capabilities of single-particle ICPMS for environmental analysis of silver nanoparticles. *Analytical Chemistry, 84*(9), 3965–3972.

Usmani, S., Arif, M., & Hasan, S. M. (2019). Therapeutic potential of metalloherbal nanoceuticals: Current status and future perspectives. *Nutraceuticals and Natural Product Derivatives: Disease Prevention & Drug Discovery*, 279–303.

Vandermoere, F., Blanchemanche, S., Bieberstein, A., Marette, S., & Roosen, J. (2011). The public understanding of nanotechnology in the food domain: The hidden role of views on science, technology, and nature. *Public Understanding of Science, 20*(2), 195–206.

Venturini, M., Mazzitelli, S., Micetic, I., Benini, C., Fabbri, J., Mucelli, S. P., ... & Nastruzzi, C. (2014). Analysis of operating conditions influencing the morphology and in vitro behaviour of chitosan coated liposomes. *Journal of Nanomedicine & Nanotechnology, 5*(4), 1–8.

Veverka, M., Dubaj, T., Gallovič, J., Jorík, V., Veverková, E., Mičušík, M., & Šimon, P. (2014). Beta-glucan complexes with selected nutraceuticals: Synthesis, characterization, and stability. *Journal of Functional Foods, 8*, 309–318.

Vilela, D., González, M. C., & Escarpa, A. (2012). Gold-nanosphere formation using food sample endogenous polyphenols for in-vitro assessment of antioxidant capacity. *Analytical and Bioanalytical Chemistry, 404*(2), 341–349.

Wacker, M. G., Proykova, A., & Santos, G. M. L. (2016). Dealing with nanosafety around the globe: Regulation vs. innovation. *International Journal of Pharmaceutics, 509*(1–2), 95–106.

Wagner, S., Legros, S., Löschner, K., Liu, J., Navratilova, J., Grombe, R., & Hofmann, T. (2015). First steps towards a generic sample preparation scheme for inorganic engineered nanoparticles in a complex matrix for detection, characterization, and quantification by asymmetric flow-field flow fractionation coupled to multi-angle light scattering and ICP-MS. *Journal of Analytical Atomic Spectrometry, 30*(6), 1286–1296.

Wang, Z., Neves, M. A., Isoda, H., & Nakajima, M. (2015). Preparation and characterization of micro/nano-emulsions containing functional food components. *Japan Journal of Food Engineering, 16*(4), 263–276.

Wapnir, R. A., & Teichberg, S. (2002). Regulation mechanisms of intestinal secretion: Implications in nutrient absorption. *Journal of Nutritional Biochemistry, 13*(4), 190–199.

Wildman, R. E. (Ed.). (2016). Nutraceuticals and functional foods. *Handbook of nutraceuticals and functional foods* (pp.1–22) CRC Press.

Williamson, E. M., Liu, X., & Izzo, A. A. (2020). Trends in use, pharmacology, and clinical applications of emerging herbal nutraceuticals. *British Journal of Pharmacology, 177*(6), 1227–1240.

Wong, B. S. E., Hu, Q., & Baeg, G. H. (2017). Epigenetic modulations in nanoparticle-mediated toxicity. *Food and Chemical Toxicology, 109*, 746–752.

Wooster, T. J., Moore, S. C., Chen, W., Andrews, H., Addepalli, R., Seymour, R. B., & Osborne, S. A. (2017). Biological fate of food nanoemulsions and the nutrients they carry–internalisation, transport and cytotoxicity of edible nanoemulsions in Caco-2 intestinal cells. *RSC Advances, 7*(64), 40053–40066.

Xia, S., Tan, C., Xue, J., Lou, X., Zhang, X., & Feng, B. (2014). Chitosan/tripolyphosphate-nanoliposomes core-shell nanocomplexes as vitamin E carriers: Shelf-life and thermal properties. *International Journal of Food Science & Technology, 49*(5), 1367–1374.

Xiaohui, L., Haiyan, S., Guanghua, P., & Shenghua, Z. (2007). Caco-2 cell model and development of its application to absorption of food nutrients. *Academic Periodical of Farm Products Processing, 2*.

Xu, A. X., West, E. A., & Rogers, M. A. (2020). Encapsulation of nutraceuticals. In *Nutraceuticals and Human Health* (pp. 79–104).

Yadav, J., Tripathy, S., Dahiya, M., & Dureja, H. (2017). Emerging nutraceutical regulations in India. *Applied Clinical Research, Clinical Trials and Regulatory Affairs, 4*(2), 91–98.

Yallapu, M. M., Khan, S., Maher, D. M., Ebeling, M. C., Sundram, V., Chauhan, N., ... & Chauhan, S. C. (2014). Anti-cancer activity of curcumin loaded nanoparticles in prostate cancer. *Biomaterials, 35*(30), 8635–8648.

Yang, S., Liu, C., Liu, W., Yu, H., Zheng, H., Zhou, W., & Hu, Y. (2013). Preparation and characterization of nanoliposomes entrapping medium-chain fatty acids and vitamin C by lyophilization. *International Journal of Molecular Sciences, 14*(10), 19763–19773.

Yeung, A. W. K., Mocan, A., & Atanasov, A. G. (2018). Let food be thy medicine and medicine be thy food: A bibliometric analysis of the most cited papers focusing on nutraceuticals and functional foods. *Food Chemistry, 269*, 455–465.

Yu, H., & Huang, Q. (2013). Bioavailability and delivery of nutraceuticals and functional foods using nanotechnology. *Bio-Nanotechnology: A Revolution in Food, Biomedical and Health Sciences*, 593–604.

Zambrano-Zaragoza, M. L., Mercado-Silva, E., Gutiérrez-Cortez, E., Castaño-Tostado, E., & Quintanar-Guerrero, D. (2011). Optimization of nanocapsules preparation by the emulsion–diffusion method for food applications. *LWT-Food Science and Technology, 44*(6), 1362–1368.

Zanella, M., Ciappellano, S. G., Venturini, M., Tedesco, E., Manodori, L., & Benetti, F. (2015). utraceuticals and nanotechnology. *Dietary Ingredients & Supplements, 26*(4), 26–31.

Zhang, Z., Kong, F., Vardhanabhuti, B., Mustapha, A., & Lin, M. (2012). Detection of engineered silver nanoparticle contamination in pears. *Journal of Agricultural and Food Chemistry, 60*(43), 10762–10767.

Zhuang, X., Deng, Z. B., Mu, J., Zhang, L., Yan, J., Miller, D., ... & Zhang, H. G. (2015). Ginger-derived nanoparticles protect against alcohol-induced liver damage. *Journal of Extracellular Vesicles, 4*, 28713–28713.

Zimet, P., Rosenberg, D., & Livney, Y. D. (2011). Re-assembled casein micelles and casein nanoparticles as nano-vehicles for ω-3 polyunsaturated fatty acids. *Food Hydrocolloids, 25*(5), 1270–1276.

2 Scope of Nanotechnology in Nutraceuticals

Analysis, Challenges and Opportunities

*Gurleen Kaur, Rajinder Kaur, and Sukhminderjit Kaur**

CONTENTS

2.1 Introduction 51
2.2 Nutraceuticals and Their Nanoencapsulation 53
 2.2.1 Resveratrol 53
 2.2.2 Curcumin 54
 2.2.3 Lycopene 54
 2.2.4 β-Carotene 55
 2.2.5 Vitamin C 55
 2.2.6 Lutein 56
2.3 Challenges Associated with Nanoencapsulation of Nutraceuticals 57
2.4 Opportunities for Nanoencapsulation in the Field of Nutraceuticals 59
2.5 Conclusion 60
References 60

2.1 INTRODUCTION

The term *nutraceutical* is a fusion of "nutrition" and "pharmaceutical", invented in the late 1900s by Dr. Stephen De Felice, which specifies health-boosting nutritive and beneficial components. extracted from or present as constituents of natural food. that support physiological or therapeutic assistance beyond fundamental nutritional requirements. Nutraceuticals comprise a great assortment of components, from antioxidants, biologically active peptides, lipids, amino acids, vitamins, phenolic compounds, minerals and carotenoids to plant metabolites (Leena et al., 2020). Food and medicines are both important aspects of human existence, as food contributes towards basic wellbeing, nutrition and growth, whereas medicines are taken to treat diseases. The purpose of consuming edible material and medicines is completely different, because neither can fulfil the function of the other; however, nutraceuticals act as both food and medicine in one. The understanding of nutraceuticals is not totally new. Traditional civilizations preferred their conventional remedial systems, which are currently considered as alternative or complementary medicine. Alternative medicine ranges from the consumption of herbs and shrubs having therapeutic potential to the utilization of food as a curative agent, commonly referred to as nutritional therapy (Maurya et al., 2021). Nutraceuticals work as analeptic agents in the prevention, cure and reduction of risk associated with different diseases like obesity, cardiac disease, diabetes and cancer. Nutraceuticals

* Corresponding author: sukhminderjit.uibt@cumail.in

DOI: 10.1201/9781003244721-2

function through different methods that are useful for the cure of cancer, including triggering the host immune response, programmed cell death, arrest of the cell cycle, eradication of expanded cells by cytostatic reagents, and destruction of noxious substances from the active elements of chemotherapeutic drugs (Ghani et al., 2019). It is known that individuals who consume high concentrations of saturated fats are often affected by diabetes. Diabetes can be inhibited by the prohibition of saturated and trans fats. R-lipoic acid is a natural antioxidant coenzyme that functions as a nutraceutical by reducing the level of lactate and pyruvate in the serum of individuals affected by diabetes mellitus. Several nutraceuticals, such as gamma oryzanol, which are obtained from rice bran, are acknowledged to prevent the absorption of cholesterol from the gut and augment the secretion of cholesterol. Bergamot is a nutraceutical derived from polyphenol, which is recognized to reduce cholesterol by depressing the levels of diverse receptors, such as oxidized low density lipoprotein and malondialdehyde receptors as well as phosphorylated protein kinase B receptor (Ghani et al., 2019). Nutraceuticals possess diversified healing and remedial aspects, such as antibacterial, antimicrobial, anti-inflammatory, antioxidant, antithyroid, antitumor and cardiac disease prevention properties. However, the effectiveness of nutraceutical agents is limited due to weak stability, degradation, inadequate water solubility, lack of bioavailability, inappropriate interactions with the gastrointestinal system, reduced penetration of epithelial cells, and low solubility in intestinal fluids. Furthermore, nutraceuticals show hydrophobic behavior and are sensitive to oxygen, light, pH and heat, which influences the efficient inclusion of health-stimulating nutraceutical components (Leena et al., 2020).

To fulfill the growing worldwide requirements for nutraceuticals, it is important to design an effective delivery procedure that is competent to protect nutraceuticals from unwanted deterioration and increases their potency and bioavailability (Figure 2.1). Encapsulation approaches at the nano and micro levels have been taken into consideration to overcome the issues associated with effective delivery and stability of nutraceuticals. The concern related to the solubility and absorption of nutraceuticals can be addressed by the approach of nanotechnology. The surface area of a specific component increases with a reduction in size, therefore assisting biological processes (Prabu et al., 2012). The inclusion of nutraceutical components into nanocarriers or nanodelivery mechanisms is frequently used in different areas of interest. There are several nanotechnology-based encapsulation mechanisms, comprising solid lipid nanoparticles, nanoemulsions, polymeric nanoparticles,

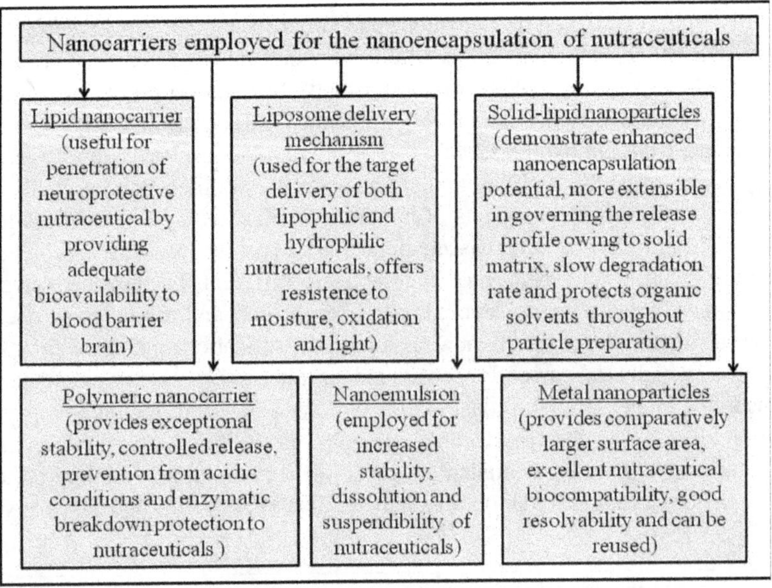

FIGURE 2.1 Nanocarriers used for the encapsulation of nutraceuticals.

nanosuspensions, liposomes and nanostructure lipid carriers (Sharma et al., 2021). Submicron emulsions are a subcategory of nanoparticles that aid in enhancing bioavailability and dissolution and increasing half-life with less abdominal irritation. Lipid nanocarriers provide improved transit of neuroprotective nutraceuticals for better bioavailability to the blood–brain barrier. Inorganic nanoparticles and dendrimers are the class of nanocarriers employed for increasing the distribution and bioavailability of anticancer bioactive agents. A liposomal delivery mechanism is used for the target delivery of both lipophilic and hydrophilic nutraceuticals and provides resistance to moisture, oxidation and light (Figure 2.1). In comparison to lipid nanocarriers, micelles (polymeric nanocarriers) offer exceptional stability, controlled release, protection from severe acidic conditions and enzymatic breakdown protection to nutraceuticals (Sharma et al., 2021). The advantages and novelty of the use of nanotechnology for nutraceutical-based industries have revolutionized human life and have encouraged and generated researchers' interest in nutraceuticals. This chapter elucidates the scope of nanotechnology in nutraceuticals, analysis, and associated challenges and opportunities.

2.2 NUTRACEUTICALS AND THEIR NANOENCAPSULATION

2.2.1 Resveratrol

Resveratrol is a nutraceutical that is derived from phenylalanine and is found in many types of herbs and shrubs, especially red grapes, where it accounts for approximately 0.1 mg/g of resveratrol in their rhytidome. Resveratrol has been acknowledged for its positive influence on curing different diseases by demonstrating neuroprotective, antioxidant, antimicrobial, cardioprotective, anti-inflammatory, anticancer and antiviral characteristics. The anticancer activity ascribed to resveratrol is achieved through various cell signaling pathways. Melanoma is a skin cancer condition wherein pigment-producing cells become cancerous and develop a greater propensity to metastasis and an increased proliferative index, which causes skin cancer–associated death (Carletto et al., 2016). Natural or synthetic chemotherapeutic agents have been utilized with immunotherapy to restrict the progress of skin cancer, but resistance occurs, which generates a challenging situation for melanoma treatment. In comparison to synthetic chemotherapeutic agents, the nutraceutical resveratrol assists in suppressing cancer cell growth, proliferation and cell death. However, as well as the therapeutic properties of this nutraceutical, certain drawbacks are associated with it, such as low solubility, high sensitivity to light, low bioavailability and rapid oxidation. To overcome these issues, a controlled resveratrol delivery system has been used. Carletto et al. (2016) worked on inhibition of murine melanoma tumor growth (B16F10 cell line) using resveratrol-loaded nanocapsules. Poly (ε-caprolactone) (PCL) nanoparticles were used for the preparation of resveratrol-loaded nanocapsules, wherein interfacial deposition of polymer assisted in the formation of PCL nanocapsules. As a result, resveratrol nanocapsules demonstrated immense drug loading potential, prevented metastasis, reduced the extent of tumor, decreased the viability of the B16F10 cell line, prevented pulmonary hemorrhage and enhanced the necrotic region (Carletto et al., 2016). This study indicated that with the integration of a nanotechnology approach, the nutraceutical resveratrol worked as a potent candidate for skin cancer treatment. In another literature study, zein was used as an encapsulating agent for the nanoencapsulation of the nutraceutical resveratrol via an electrospraying mechanism. The oral route of administration of resveratrol in humans would be expected to result in higher absorption, but $<1\%$ bioavailability has been reported, because it is metabolized in the liver, and only a low concentration of resveratrol reaches the expected target tissues and the systematic circulation. The researchers evaluated that encapsulation of resveratrol using zein nanoparticles showed excellent absorptivity and biological accessibility in a dynamic gastrointestinal system. The biological activity of the nutraceutical was strengthened, as the zein matrix was competent to retain the antioxidant potential of the nutraceutical (Jayan et al., 2019). Consequently, zein nanoparticles assist in the oral consumption of resveratrol in a tablet formulation or by integration into a convenient food system.

2.2.2 Curcumin

A natural phenol component obtained from *Curcuma longa* L. rhizome that is broadly utilized as a conventional herbal remedy for different diseases is curcumin. The pharmacological and biological characteristics associated with curcumin are attributed to its phenol and methoxy groups (Li et al., 2019). Curcumin is widely acknowledged for its different biological properties, such as anti-inflammatory, antioxidant, antiviral, antidiabetic, antitumor, cholesterol-lowering, antimicrobial and anti-mutagenic aspects. There are some limitations related to the use of curcumin, such as rapid excretion, low chemical stability leading to a reduced shelf life, accelerated metabolism, and poor absorption causing poor bioavailability (Rafiee et al., 2019). The therapeutic importance of curcumin is limited because it has a high decomposition rate at basic and neutral pH, lower disintegration (approximately lesser than or equal to 0.125 mg/L in aqueous solution) and high photochemical degradation. Even for optimum pharmacological influence, an oral dose of curcumin of greater than 8 g per day is recommended; therefore, the enhancement of bioavailability becomes an important concern. Novel drug carriers for diagnostic and therapeutic objectives can be developed by employing nanoparticles to increase the bioavailability and accumulation of curcumin at the targeted site of delivery. Different nanocarriers have been employed for the targeted delivery of curcumin, such as nanogel, liposomes, polymeric nanoparticles and micelles. In a literature study, Li et al., (2019) performed the nanoencapsulation of curcumin to enhance its bioavailability and examined the hepatoprotective effect of curcumin against CCl_4 (tetrachloromethane)-induced acute hepatic injury via intraperitoneal injection in mice. Tetrachloromethane is recognized as a liver toxin extensively utilized to induce hepatic injury in laboratory animals. A thin film dispersion technique was used for the synthesis of curcumin-loaded nanoliposomes. The three enzymes alanine transaminase (ALT), alkaline phosphatase (ALP) and aspartate transaminase (AST) indicate liver injury. The researchers reported that the levels of ALT, ALP and AST were greatly reduced by the application of curcumin-loaded nanoliposomes, indicating the importance of the nutraceutical curcumin in preventing liver injury. Greater penetration of the cell membrane by curcumin, greater stability, excellent encapsulation and adequate delivery of curcumin to the target site were observed. Curcumin nanoliposomes also illustrated potential resistance against lipid peroxidation, which is attributed to the increase in bioavailability and extended circulation time of curcumin nanoparticles in blood (Li et al., 2019). In another literature study, Grama et al. (2013) reported on the effectiveness of curcumin nanoparticles in delaying cataracts in diabetic rats. Curcumin nanoparticles were synthesized using an emulsion diffusion evaporation method, and poly(lactide-co-glycolic) acid was used as an encapsulating agent. Nanocurcumin played an eminent role in contributing to delaying the progression of diabetic cataracts in rats, which is attributed to the potential activity of curcumin in disrupting the biological processes of disease promotion, such as oxidative stress, solubilization of protein and glycation of protein. It was also reported that the administration of nanocurcumin through the oral route provided potential benefits. The improved action of nanocurcumin was ascribed to its enhanced bioavailability (Grama et al., 2013). Thus, curcumin-loaded nanoparticles act as potential candidates against diabetic cataracts in rats.

2.2.3 Lycopene

Lycopene belongs to the family of carotenoids and the group of carotenes, hydrocarbons that do not possess functional groups. Lycopene has lipophilic properties and is obtained from microbes and plants. Lycopene is endowed with very high antioxidant activity, which is attributed to its ability to interrupt the series of peroxyl radical reactions during cellular oxidation and its free radical scavenging activity. Due to its antioxidant potency and singlet oxygen scavenging ability, lycopene is utilized as a lipophilic component in the food industry. Lycopene has also been acknowledged for preventing the growth of several kinds of cancer, such as blood cancer, lung cancer, prostate cancer, breast cancer and colon cancer. The other health benefit associated with lycopene is decreased risk of cardiac disease, stroke, diabetes and obesity. Lycopene has been widely used in the pharmaceutical industry and in functional foods. However, low absorption and release of lycopene from food materials, low bioavailability, physiochemical instability and poor solubility are some of the challenging issues associated with the use of lycopene for different purposes (Ashraf et al., 2020). Lycopene is stable in nature, but during storage and processing, degradation has been observed. Within humans as well as in food, the stability and solubility of lycopene can be negatively influenced due to certain factors such as light, temperature, oxygen and time duration. Therefore, encapsulation of lycopene is the progressive approach to achieve improved bioavailability, stability, solubility and resistance to severe conditions. Different nanocarriers used for the encapsulation of lycopene are lipids, biopolymer and protein-based nanocarriers, nanoliposomes, microemulsions, nanoemulsions and nanohydrogels (Ashraf et al., 2020). In a literature study, researchers worked on the encapsulation of lycopene that was obtained from tomato peel using gelatin nanofibers through the process of electrospinning. It was reported that there was an increment in the thermal stability of lycopene due to the electrospinning mechanism. Excellent retention and antioxidant activity were demonstrated by the nutraceutical lycopene when examined for approximately 14 days of storage conditions. It was also observed that lycopene showed a 10-fold increment in its antioxidant activity due to the nanoencapsulation procedure as the molecular size decreased to nanometric size. The gelatin nanofibers also assisted in improving the water solubility of lycopene (Horuz & Belibağlı, 2018). Hence, the electrospinning procedure for nanoencapsulation of lycopene proved to be beneficial in terms of increasing stability, water solubility and antioxidant activity, which makes lycopene a promising candidate to be used in food processing.

2.2.4 β-Carotene

β-Carotene, one of the natural pigments extracted from several vegetables and fruits, is well recognized as an excellent source of vitamin A and a potential nutraceutical having the ability to dissolve in lipid, fats, non-polar solvents and oil. This nutraceutical has gained much attention as a dynamic bioactive ingredient for incorporation in several edible formulations due to its favorable functional biological activities, such as antioxidant, anticancer and anti-inflammatory properties (Gul et al., 2015). β-Carotene aids in the prevention of several degenerative diseases: cataracts, rheumatoid arthritis, and immune, cancer, cardiac and neurodegenerative diseases. However, the incorporation of β-carotene is difficult to accomplish because of poor durability, water insolubility, and susceptibility to pH, light and temperature, which further causes lower bioavailability and accessibility. The process of encapsulation of such lipophilic components in a specific matrix aids in protecting the β-carotene from unfavorable environmental conditions and maintaining its nutraceutical value. Nanoemulsions are lipid-based delivery systems that have been employed to improve the bioavailability of nutraceuticals. Mahalakshmi et al. (2020 used zein as an encapsulating agent for β-carotene. Glycerol acted as a stabilizing agent. An electrospinning and a spray drying technique were selected for the nanoencapsulation and microencapsulation of β-carotene, respectively. Researchers investigated the comparative outcomes of the technique of encapsulation, particle size, and influence on stability, solubility and release behavior. The electrospinning process demonstrated 81% nanoencapsulation efficiency. In comparison to microencapsulation, nanoencapsulates demonstrated better permeability and superior dissolution properties because of increased surface area and decreased particle size. Moreover, the β-carotene nanoencapsulates were assessed in an *in vitro* stimulated gastrointestinal environment; improved release of β-carotene was observed, which accounts for better bioavailability (Mahalakshmi et al., 2020). This study elucidates the importance of nanoencapsulation over microencapsulation for better bioavailability, stability, solubility (Figure 2.2) and permeability of β-carotene.

2.2.5 Vitamin C

Vitamin C, or ascorbic acid is a water-soluble, essential vitamin that performs necessary metabolic and physiological functions in human beings but can only be obtained through the diet.

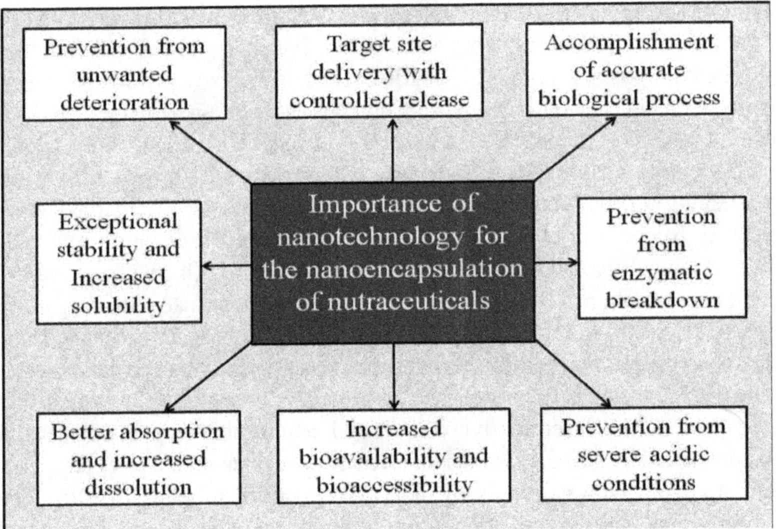

FIGURE 2.2 The important aspects associated with the application of nanotechnology for the nanoencapsulation of nutraceuticals.

Vitamin C is an essential antioxidant, competent to neutralize oxidative stress via an electron transfer or donation process; it reduces unstable nitrogen, oxygen and sulfur radicals and is capable of preventing lipid peroxidation. Vitamin C has been reported to support the absorption of folic acid, calcium and iron, therefore providing protection against allergic reactions, whereas immune suppression is caused by a reduced intracellular concentration of vitamin C. Interferon production, immunoglobulin synthesis and suppression of malignant tumor factor (interleukin-18) are beneficial characteristics associated with vitamin C; therefore, it is advised to take this nutraceutical during stress and infection. Another significant aspect is that the secretion of bile in the gallbladder is stimulated by vitamin C, which further assists in the elimination of steroid hormones (Sheraz et al., 2015). Vitamin C is also acknowledged to perform important functions in brain. Vitamin C works as an enzymatic cofactor for the synthesis of peptide hormones, tyrosine, carnitine and collagen as well as inducing the formation of myelin and the maturation of neurons. Specifically, the lack of vitamin C can give rise to neurodegenerative diseases and psychiatric disorders (Kocot et al., 2017). Apart from the different beneficial aspects related to the use of vitamin C, some challenges are also associated with it, such as difficulty with stability, target site delivery, and its degradation in aqueous medium, in the presence of metal ions and oxygen, and at higher pH (Carita et al., 2020). Encapsulation is a necessary approach whereby a carrier agent is used for the encapsulation of vitamin C with the objective to overcome issues related to its utilization. The different techniques used for the nanoencapsulation of vitamin C are spray drying, emulsification, spray chilling and complex coacervation. The different nanocarriers that assist in encapsulation of vitamin C are chitosan, poly lactic acid, starch, gelatin, casein, poly(lactide-co-glycoside) and silica (Comunian et al., 2020). In a recent study, researchers evaluated comparative studies on the oral bioavailability of liposomal and non-liposomal vitamin C. The liposomal ascorbic acid demonstrated 1.7 times higher bioavailability than the non-liposomal vitamin C. Hence, it can be concluded that nanoencapsulation approaches provide better support for the improved bioavailability of vitamin C.

2.2.6 Lutein

Lutein is considered one of the eminent functional components and is principally obtained from marigold flowers and green vegetables. Lutein is regarded as a potential antioxidant because it supports providing protection to organs and cells from the destruction caused in the body by oxygen free radicals as well as preventing blindness or impaired vision induced by macular degeneration (Liu et al., 2021). Within infant brains, lutein is the preeminent carotenoid on the grounds of neural development, whereas in adults, lutein plays a significant role in preventing oxidative stress in age-associated cognitive decline. It has been reported that intake of a lutein supplement for 12 months significantly increases concentration, interpretation skills and memory in humans. Furthermore, the consumption of this nutraceutical helps to decrease triglyceride concentration, liver cholesterol and serum cholesterol (Kishimoto et al., 2017). Recently, the food industry has been distinguished by advances in food processing technologies, with much emphasis on health-promoting components, and this has resulted in greater utilization of nutraceuticals. The fundamental aspect that affects the concentration of lutein absorbed from the diet is food processing. Lutein is susceptible to heat and light, and its bioavailability is greatly affected during processing, digestion and absorption. The bioavailability of nutraceuticals is essentially a function of absorption, transformation and bioaccessibility. An emerging discipline that aims to increase the bioaccessibility of nutraceuticals is nanotechnology. Nanoformulations, both solid and liquid, are employed for the encapsulation of lutein. Solid formulations comprise polymeric nanoparticles, nanocrystals and lipid nanoparticles, while liquid systems comprise nanoemulsions, nanoliposomes and nanopolymersomes (Bhat et al., 2020). In a literature study, researchers worked on the nanoencapsulation of lutein using zein to increase stability, water solubility, bioactivity and protection from oxidative stress. A solvent diffusion method was used for the preparation of lutein-loaded zein nanoparticles and zein-derived peptide nanoparticles. Zein-derived peptide nanoparticles showed better entrapment efficiency, stability, solubility and physiochemical attributes in comparison to lutein-loaded zein nanoparticles. Enhanced stability of lutein in stimulated intestinal fluid and gastric fluid was reported in the case of zein-derived peptide nanoparticles. Consequently, a zein-derived peptide nanocarrier was best suited for the nanoencapsulation of lutein for increased bioavailability, solubility and stability (Jiao et al., 2018) (Table 2.1).

2.3 CHALLENGES ASSOCIATED WITH NANOENCAPSULATION OF NUTRACEUTICALS

Nanoencapsulation techniques offer a better tool to protect nutraceuticals against undesirable degradation and increase the bioavailability of the nutraceutical compounds (Davidov-Pardo & McClements, 2014). For the efficient delivery of nutraceuticals, it is crucial to design an effective delivery system that can enhance the solubility, stability and bioavailability of the nutraceutical (Figure 2.3). Different methods such as solid lipid nanoparticles (SLNs), nanostructured lipid carriers, electrospinning, electrospraying, coacervation, nanospray drying, liposomes, nanoemulsions and nanogels have been used for encapsulation of nutraceuticals (Jafari & McClements, 2017; Jones et al., 2019). Apart from the economic feasibility of the above-mentioned methods of nanoencapsulation, the choice of encapsulating material is most important in deciding the stability of the product. For each nutraceutical compound, different excipient compounds need to be chosen depending upon the type of bioactive molecule. When choosing the wall material, physiochemical properties of the wall material along with biodegradability and toxicity should be taken into account. Certain nutraceuticals that show low solubility should be encapsulated in water-soluble excipients like proteins and emulsions to enhance their solubility (McClements et al., 2015). As low permeability of active molecules into the cells often limits their bioavailability, the excipient material should be chosen so as to enhance the permeability of nutraceuticals/active molecules into epithelial cells. Nutraceuticals that are degraded by metabolic processes in the gastrointestinal tract or are susceptible to enzyme degradation should be encapsulated in an excipient material that protects them from undesirable degradation. Also, the excipient material should be chemically stable to environmental stresses in the gastrointestinal tract while preserving its functional characteristics (McClements, 2012).

TABLE 2.1
Different Nutraceuticals, Their Encapsulating Agents, Techniques and Function

Nutraceutical	Nanoencapsulating agent	Nanoencapsulation technique	Function	Reference
Resveratrol	Poly (ε-caprolactone)	Interfacial deposition	Inhibition of murine melanoma tumor growth (B16F10 cell line)	(Carletto et al., 2016)
Resveratrol	Zein	Electrospraying	Enhanced biological accessibility and absorptivity in gastrointestinal system	(Jayan et al., 2019)
Resveratrol	Starch		Formation of novel functional snack with antioxidant, antidiabetic and antiobesity aspects	(Ahmad and Gani, 2021)
Resveratrol	Chitosan and γ-poly (glutamic acid)	Ionic gelation	Increment in stability	(Chung et al., 2020)
Curcumin	Soy protein	pH-driven method	Higher encapsulation efficiency and stability, acts as potent delivery system for hydrophobic nutraceuticals	(Li et al., 2021)
Curcumin	Soybean phosphatidylcholine	Thin film dispersion	Hepatoprotective property of curcumin against CCl_4-induced acute hepatic injury in mice	(Li et al., 2019)
Curcumin	Poly(lactide-co-glycolic) acid	Emulsion diffusion evaporation method	Enhanced bioavailability and effective against diabetic cataracts in rats	(Grama et al., 2013)
Lycopene	Gelatin	Electrospinning	Increased thermal stability, excellent retention, enhanced water solubility	(Horuz & Belibağlı, 2018)
Lycopene	Octenyl succinate anhydride modified starch	High-pressure homogenization	Enhanced stability of lycopene in oil-in-water emulsion, works as competent delivery system in functional food	(Li et al., 2018)
β-carotene	Zein	Electrospinning	Better permeability and superior dissolution property, improved release of β-carotene, better bioavailability	(Mahalakshmi et al., 2020)
β-carotene	Whey protein isolate	Homogenization evaporation method	Beneficial for release to intestine, reduction of peroxyl radical oxidation in Caco-2 cells	(Yi et al., 2015)
Eugenol	Chitosan	Emulsion ionic gelation mechanism	Enhanced antioxidant feature, increased efficiency of eugenol assisted in control of lipid peroxidation and selenite-induced antioxidant depletion	(Anand et al., 2021)
Vitamin C	Sodium ascorbate and phospholipids	Thin film evaporation method	Increased oral bioavailability	(Gopi & Balakrishnan, 2020)

(Continued)

TABLE 2.1 (CONTINUED)
Different Nutraceuticals, Their Encapsulating Agents, Techniques and Function

Nutraceutical	Nanoencapsulating agent	Nanoencapsulation technique	Function	Reference
Vitamin D_3	Zein	Nanoprecipitation method	Fortification of *Acca sellowiana* with vitamin D_3 showed suitability, increased physicochemical availability and stability, higher encapsulation efficiency, suitable shelf life and 80% release in stimulated gastrointestinal conditions	(de Melo et al., 2021)
Vitamin A (retinyl acetate)	Gelucire and stearic acid	Organic solvent–free sonication method	Improved oral bioaccessibility, protection against stomach digestion and storage degradation	(Resende et al., 2020)
Lutein	Zein	Solvent diffusion method	Better entrapment efficiency, increased stability, water solubility, and bioactivity in stimulated intestinal fluid and gastric fluid, and protection from oxidative stress	(Jiao et al., 2018)
Lutein	Chitosan oleic acid sodium alginate nanocarrier	Ionic gelation	Greater thermal stability, aqueous solubility, target site delivery and improved bioavailability, acts as therapeutic agent against retinopathy and macular degeneration	(Toragall et al., 2020)

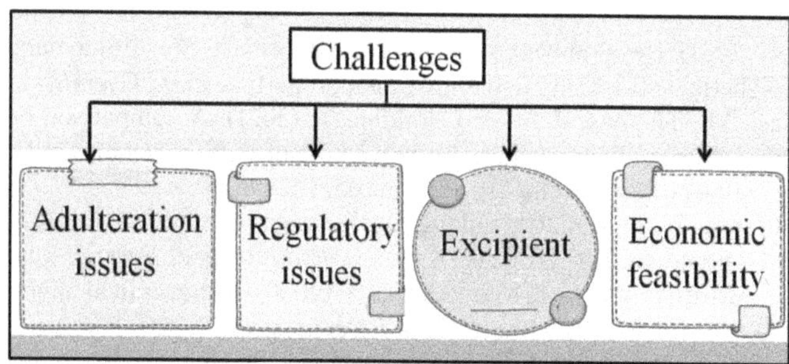

FIGURE 2.3 Challenges associated with nanoencapsulation of nutraceuticals.

Another major challenge to nanoencapsulated nutraceuticals is the adulteration issue. Certain nanoencapsulated nutraceuticals have been found to be unintentionally contaminated with heavy metals, pollens, dust and pesticides (Harris et al., 2011). Intentional adulteration with synthetic compounds for economic benefit also poses a serious challenge (Villani et al., 2015).

2.4 OPPORTUNITIES FOR NANOENCAPSULATION IN THE FIELD OF NUTRACEUTICALS

The science of nutraceuticals is currently the most fascinating area of research. Many commercial nanoencapsulated nutraceutical consortia are available in the market.

Fumed silica, nano clay, nano silver and titanium nitride are the nanomaterials that have been used for nearly 300 food products listed by the Centre for Food Safety. Chinese nano tea, nano gold and nano silver have been used in the nanoencapsulation of carotenoids and vitamins. Lipid-based nanostructures (liposomes), SLNs, nanoemulsions and self-emulsifying systems have great potential for preserving the aromas and flavors of nutraceuticals.

2.5 CONCLUSION

Nanotechnology is acknowledged as the leading approach and has revolutionized different disciplines associated with food, medicine and nutraceuticals. The benefit of nanotechnology in nutrition and food is the construction of functional foods such as nutraceuticals with better bioaccessibility, solubility, stability and bioavailability. Nutraceuticals comprise a great diversity of components, from lipids, antioxidants, biologically active peptides, amino acids, vitamins, phenolic compounds and minerals to plant metabolites and carotenoids. Nutraceuticals have proved to be beneficial in the prevention and cure of different diseases, such as cancer, lung disease, cardiovascular disease and degenerative disease, and therefore, the use of nutraceuticals has received much interest and attention from researchers. Various challenges are associated with the utilization of nutraceuticals, such as low bioavailability, low stability, solubility, unwanted deterioration and low absorption, uncontrolled release, and issues associated with target site delivery. Encapsulation at the nano and micro levels helps to overcome such problems. Various nanoencapsulation techniques employed for the nanoencapsulation of nutraceuticals include nanoemulsification, spray drying, electrospinning, solvent diffusion and nanoprecipitation. Different nanocarriers used include polymeric nanoparticles, lipid nanocarriers, SLNs and metal particles. Many researchers have utilized nanotechnology for the nanoencapsulation of nutraceuticals by taking into account the different characteristics associated with the use of nutraceuticals, such as antimicrobial, antioxidant, antitumor, antidiabetic, antithyroid and anti-inflammatory properties, and prevention of cardiovascular disease. Therefore, in the future, nanotechnology or nanocarriers will have greater commercial importance than in the past.

REFERENCES

Ahmad, M., & Gani, A. (2021). Development of novel functional snacks containing nano-encapsulated resveratrol with anti-diabetic, anti-obesity and antioxidant properties. *Food Chemistry, 352*, 129323.

Anand, T., Anbukkarasi, M., Thomas, P. A., & Geraldine, P. (2021). A comparison between plain eugenol and eugenol-loaded chitosan nanoparticles for prevention of in vitro selenite-induced cataractogenesis. *Journal of Drug Delivery Science and Technology, 65*, 102696.

Ashraf, W., Latif, A., Lianfu, Z., Jian, Z., Chenqiang, W., Rehman, A., ... & Karim, A. (2020). Technological advancement in the processing of lycopene: A review. *Food Reviews International, 38 (5),* 1–27.

Bhat, I., Yathisha, U. G., Karunasagar, I., & Mamatha, B. S. (2020). Nutraceutical approach to enhance lutein bioavailability via nanodelivery systems. *Nutrition Reviews, 78*(9), 709–724.

Carita, A. C., Fonseca-Santos, B., Shultz, J. D., Michniak-Kohn, B., Chorilli, M., & Leonardi, G. R. (2020). Vitamin C: One compound, several uses. Advances for delivery, efficiency and stability. *Nanomedicine: Nanotechnology, Biology and Medicine, 24*, 102117.

Carletto, B., Berton, J., Ferreira, T. N., Dalmolin, L. F., Paludo, K. S., Mainardes, R. M., ... & Favero, G. M. (2016). Resveratrol-loaded nanocapsules inhibit murine melanoma tumor growth. *Colloids and Surfaces B: Biointerfaces, 144*, 65–72.

Chung, J. H., Lee, J. S., & Lee, H. G. (2020). Resveratrol-loaded chitosan–γ-poly (glutamic acid) nanoparticles: Optimization, solubility, UV stability, and cellular antioxidant activity. *Colloids and Surfaces B: Biointerfaces, 186*, 110702.

Comunian, T., Babazadeh, A., Rehman, A., Shaddel, R., Akbari-Alavijeh, S., Boostani, S., & Jafari, S. M. (2020). Protection and controlled release of vitamin C by different micro/nanocarriers. *Critical Reviews in Food Science and Nutrition, 62(12)*, 3301–3322.

Davidov-Pardo, G., McClements, D. J., 2014. Resveratrol encapsulation: Designing delivery systems to overcome solubility, stability and bioavailability issues. *Trends in Food Science and Technology, 38*(2), 88e103. https://doi.org/10.1016/j.tifs.2014.05.003.

de Melo, A. P. Z., da Rosa, C. G., Noronha, C. M., Machado, M. H., Sganzerla, W. G., da Cunha Bellinati, N. V., ... & Barreto, P. L. M. (2021). Nanoencapsulation of vitamin D3 and fortification in an experimental jelly model of *Acca sellowiana*: Bioaccessibility in a simulated gastrointestinal system. *LWT, 145*, 111287.

Ghani, U., Bukhari, S. S. H., Ullah, S., Rafeeq, H., Saeed, M. M., Amjad, A., ... & Chand, S. C. (2019). A review on nutraceuticals as a therapeutic agent. *International Journal of Biosciences, 15*(5), 326–340.

Gopi, S., & Balakrishnan, P. (2020). Evaluation and clinical comparison studies on liposomal and non-liposomal ascorbic acid (vitamin C) and their enhanced bioavailability. *Journal of Liposome Research, 31*(4), 356–364.

Grama, C. N., Suryanarayana, P., Patil, M. A., Raghu, G., Balakrishna, N., Kumar, M. R., & Reddy, G. B. (2013). Efficacy of biodegradable curcumin nanoparticles in delaying cataract in diabetic rat model. *PLoS One, 8*(10), e78217.

Gul, K., Tak, A., Singh, A. K., Singh, P., Yousuf, B., & Wani, A. A. (2015). Chemistry, encapsulation, and health benefits of β-carotene-A review. *Cogent Food & Agriculture, 1*(1), 1018696.

Harris, E. S. J., Cao, S., Littlefield, B. A., Craycroft, J. A., Scholten, R., Liu, Y., Kaptchuk, T., Fu, Y., Wang, W., Liu, Y., Chen, H., Zhao, Z., Clardy, J., Woolf, A. D., & Eisenberg, D. M. (2011). Heavy metal and pesticide content in commonly prescribed individual raw Chinese herbal medicines. *Science of the Total Environment, 409*, 4297–4305.

Horuz, T. İ., & Belibağlı, K. B. (2018). Nanoencapsulation by electrospinning to improve stability and water solubility of carotenoids extracted from tomato peels. *Food Chemistry, 268*, 86–93.

Jafari, S. M., McClements, D. J. (2017). Nanotechnology approaches for increasing nutrient bioavailability. In *Advances in food and nutrition research*, first ed., vol. 81. Elsevier Inc. https://doi.org/10.1016/bs.afnr.2016.12.008.

Jayan, H., Leena, M. M., Sundari, S. S., Moses, J. A., & Anandharamakrishnan, C. (2019). Improvement of bioavailability for resveratrol through encapsulation in zein using electrospraying technique. *Journal of Functional Foods, 57*, 417–424.

Jiao, Y., Zheng, X., Chang, Y., Li, D., Sun, X., & Liu, X. (2018). Zein-derived peptides as nanocarriers to increase the water solubility and stability of lutein. *Food & Function, 9*(1), 117–123.

Jones, D., Caballero, S., & Davidov-Pardo, G. (2019). Bioavailability of nanotechnology-based bioactives and nutraceuticals. In *Food applications of nanotechnology*, first ed. Elsevier Inc. https://doi.org/10.1016/bs.afnr. 2019.02.014.

Kishimoto, Y., Taguchi, C., Saita, E., Suzuki-Sugihara, N., Nishiyama, H., Wang, W., ... & Kondo, K. (2017). Additional consumption of one egg per day increases serum lutein plus zeaxanthin concentration and lowers oxidized low-density lipoprotein in moderately hypercholesterolemic males. *Food Research International, 99*, 944–949.

Kocot, J., Luchowska-Kocot, D., Kiełczykowska, M., Musik, I., & Kurzepa, J. (2017). Does vitamin C influence neurodegenerative diseases and psychiatric disorders? *Nutrients, 9*(7), 659.

Leena, M. M., Mahalakshmi, L., Moses, J. A., & Anandharamakrishnan, C. (2020). Nanoencapsulation of nutraceutical ingredients. In *Biopolymer-based formulations* (pp. 311–352). Elsevier.

Li, D., Li, L., Xiao, N., Li, M., & Xie, X. (2018). *Physical properties of oil-in-water nanoemulsions stabilized by OSA-modified starch for the encapsulation of lycopene. Colloids and Surfaces A: Physicochemical and Engineering Aspects, 552*, 59–66.

Li, H., Zhang, X., Zhao, C., Zhang, H., Chi, Y., Wang, L., ... & Zhang, X. (2021). Entrapment of curcumin in soy protein isolate using the pH-driven method: Nanoencapsulation and formation mechanism. *LWT, 153*, 112480.

Li, J., Niu, R., Dong, L., Gao, L., Zhang, J., Zheng, Y., ... & Li, K. (2019). Nanoencapsulation of curcumin and its protective effects against CCl4-induced hepatotoxicity in mice. *Journal of Nanomaterials, 2019*.

Liu, M., Wang, F., Pu, C., Tang, W., & Sun, Q. (2021). Nanoencapsulation of lutein within lipid-based delivery systems: Characterization and comparison of zein peptide stabilized nano-emulsion, solid lipid nanoparticle, and nano-structured lipid carrier. *Food Chemistry, 358*, 129840.

Mahalakshmi, L., Leena, M. M., Moses, J. A., & Anandharamakrishnan, C. (2020). Micro-and nano-encapsulation of β-carotene in zein protein: Size-dependent release and absorption behavior. *Food & Function, 11*(2), 1647–1660.

Maurya, A. P., Chauhan, J., Yadav, D. K., Gangwar, R., & Maurya, V. K. (2021). Nutraceuticals and their impact on human health. In *Preparation of phytopharmaceuticals for the management of disorders* (pp. 229–254). Academic Press.

McClements, D. J., 2012. Nanoemulsions versus microemulsions: Terminology, differences, and similarities. *Soft Matter, 8*(6), 1719e1729. https://doi.org/10.1039/c2sm06903b.

McClements, D. J., Li, F., & Xiao, H., 2015. The nutraceutical bioavailability classification scheme: Classifying nutraceuticals according to factors limiting their oral bioavailability. *Annual Review of Food Science and Technology*, *6*(1), 299e327. https://doi.org/10.1146/annurev-food-032814-014043.

Prabu, S. L., SuriyaPrakash, T. N. K., Dinesh, K., Suresh, K., & Ragavendran, T. (2012). Nutraceuticals: A review. *Elixir Pharmacy*, *46*, 8372–8377.

Rafiee, Z., Nejatian, M., Daeihamed, M., & Jafari, S. M. (2019). Application of different nanocarriers for encapsulation of curcumin. *Critical Reviews in Food Science and Nutrition*, *59*(21), 3468–3497.

Resende, D., Lima, S. A. C., & Reis, S. (2020). Nanoencapsulation approaches for oral delivery of vitamin A. *Colloids and Surfaces B: Biointerfaces*, *193*, 111121.

Sharma, S., Sudan, P., Arora, V., Goswami, M., & Dorac, C. P. (2021). Nanotechnological advances for nutraceutical delivery. In *Nanotechnology* (pp. 149–182). Jenny Stanford Publishing.

Sheraz, M. A., Khan, M. F., Ahmed, S. O. F. I. A., Kazi, S. H., & Ahmad, I. Q. B. A. L. (2015). Stability and stabilization of ascorbic acid. *Household and Personal Care Today*, *10*, 22–25.

Toragall, V., Jayapala, N., & Vallikannan, B. (2020). Chitosan-oleic acid-sodium alginate a hybrid nanocarrier as an efficient delivery system for enhancement of lutein stability and bioavailability. *International Journal of Biological Macromolecules*, *150*, 578–594.

Villani, T. S., Reichert, W., Ferruzzi, M. G., Pasinetti, G. M., Simon, J. E., Wu, Q. (2015). Chemical investigation of commercial grape seed derived products to assess quality and detect adulteration. *Food Chemistry*, *170*, 271–280.

Yi, J., Lam, T. I., Yokoyama, W., Cheng, L. W., & Zhong, F. (2015). Beta-carotene encapsulated in food protein nanoparticles reduces peroxyl radical oxidation in Caco-2 cells. *Food Hydrocolloids*, *43*, 31–40.

3 Systematic Review Analysis of Emerging Nano-Nutraceuticals in Cancer Prevention and Treatment Using a Scientometric Approach

Madhulika Bhati[*]*, Ranjana Aggarwal, Swathi Sreekuttan, Zahra Shekason, Asha Anish Madhavan, and Reshmi S. Nair*

CONTENTS

3.1 Introduction 63
3.2 Methodology 64
 3.2.1 Publication Analysis 64
 3.2.2 Co-Occurrence Analyses Using Authors' Keywords 65
 3.2.3 International Co-Authorship Analysis 67
3.3 Research Focus 68
 3.3.1 Nano-Nutraceutical Research Trends in the United States 68
 3.3.2 Nano-Nutraceutical Research Trends in China 70
 3.3.3 Nano-Nutraceutical Research in India 71
3.4 Conclusions 75
References 75

3.1 INTRODUCTION

Cancer is an ever-increasing illness, affecting and causing mortality in more than half of the population. In 2020, the World Health Organization (WHO) (WHO 2020) reported that in 112 out of 183 countries, the first/second cause of death before the age of 70 years is cancer, and it ranks third or fourth in the remaining countries. The GLOBCAN report 2020 synopsis (Ferlay et al. 2021) indicated that breast cancer remained the leading cause of cancer death, followed by prostate and lung cancer. However, in terms of mortality, liver cancer caused most deaths, followed by lung and ovarian cancer (Figure 3.1).

Conventional drug administration in cancer is dependent on a passive diffusion through the body, during which the drug may undergo a series of biological degradations; this, in turn, reduces the chemical's efficiency and bioavailability in the blood serum, thereby yielding disappointing outcomes, such as ineffective results or wastage of a great deal of material. In order to overcome these obstacles, the recent outlook has been to focus upon nanodelivery-based systems or nanocarriers, which would aid in targeted delivery, controlled transport and optimal release of the bioactive molecules. Even though this does help surpass the previous limitations of free-form drug delivery,

[*] Corresponding author: madhulikabhati@niscpr.res.in

DOI: 10.1201/9781003244721-3

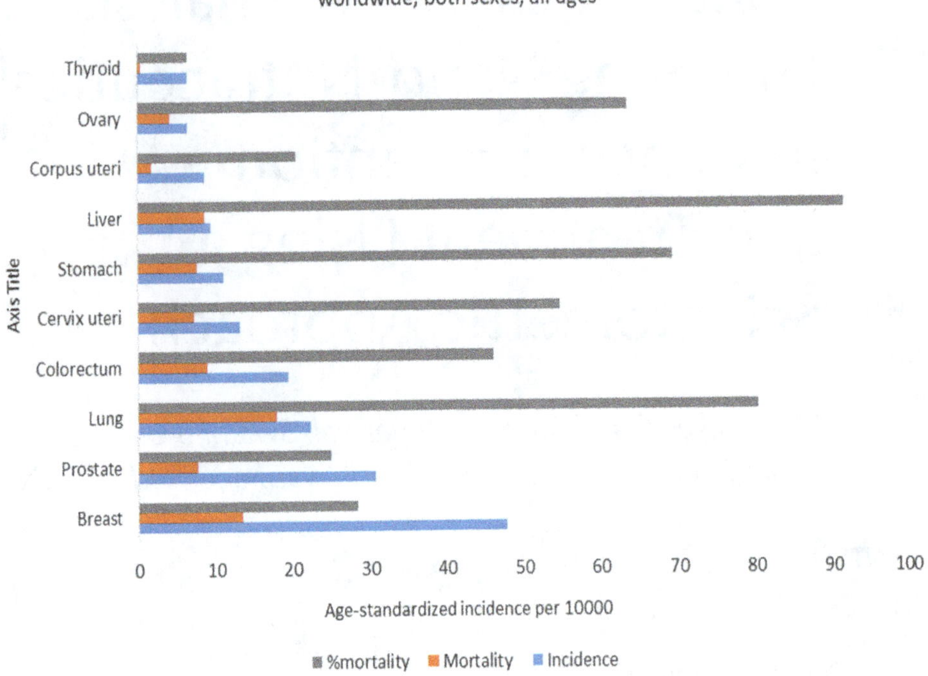

FIGURE 3.1 Estimated age-standardized incidence and mortality rate (world).

it poses its own limitations in terms of toxicity and bioaccumulation. These have recently been addressed by integrating nanotechnology with nutraceuticals, providing a promising solution for several kinds of cancer treatment.

Nutraceuticals are these bioactive functional food ingredients that provide nutrition along with therapeutic effects. They supply us with essential vitamins and minerals and help prevent diseases. They are generally classified on the basis of availability in food, chemical nature and mechanism of drug. But, they are inefficient *in vivo* due to their low water solubility, oral bioavailability and thermal stability and their premature biodegradability. Nanotechnology can be used to alter the physical and chemical properties of nutraceuticals to design better functional products. It can also be used to encapsulate these nutraceuticals to develop an efficient targeted delivery system that enables a controlled release of these nutrients, called nano-nutraceuticals (Senapati et al. 2018).

3.2 METHODOLOGY

The bibliographic database Web of Science was used to analyze the extracted publication dataset. The keywords Nano* AND nutraceuticals AND Cancer were used for the time span of 2011–2020. Ninety-two documents were retrieved. The VoS viewer visualization tool was further used to analyze the social and conceptual relationships by creating a science map. The analysis was done using organization, country, citations and co-word analysis as units of analysis (Van and Wattmann 2010).

3.2.1 Publication Analysis

The authors conducted a country-wise and year-wise analysis of publications (Figures 3.2 and 3.3) and the most prominent keywords, and co-word analysis was performed to provide the conceptual

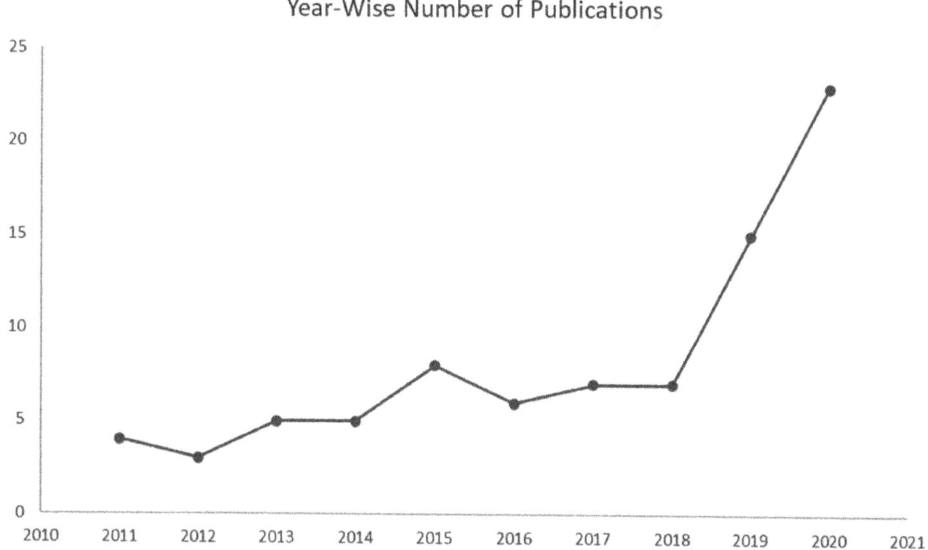

FIGURE 3.2 Year-wise publications.

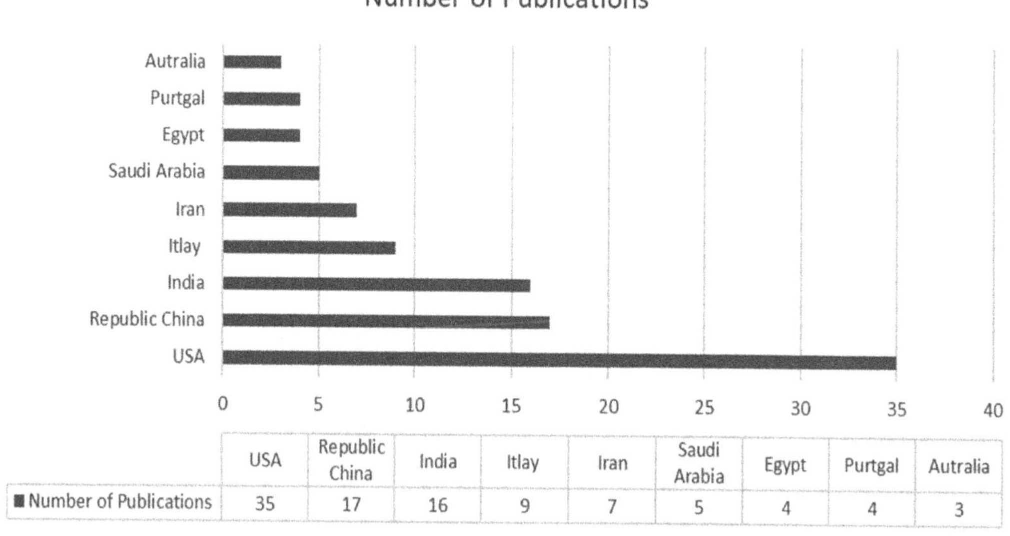

	USA	Republic China	India	Itlay	Iran	Saudi Arabia	Egypt	Purtgal	Autralia
■ Number of Publications	35	17	16	9	7	5	4	4	3

FIGURE 3.3 Country-wise publications.

and social structure of the nano-nutraceutical research (Figures 3.4 and 3.5). Exponential growth has been seen from 2011 to 2020, and the United States, China and India are the top three publishing countries.

3.2.2 Co-Occurrence Analyses Using Authors' Keywords

The authors used a total of 352 keywords in 92 papers. However, setting the minimum occurrence to 2 results in 50 keywords. A scientific map is then created based on the link strength.

Analysis of the science map (Figure 3.4) indicates that is mainly concerned with enhancing the bioavailability, controlled release and bioaccessibility of drugs via nanoencapsulation,

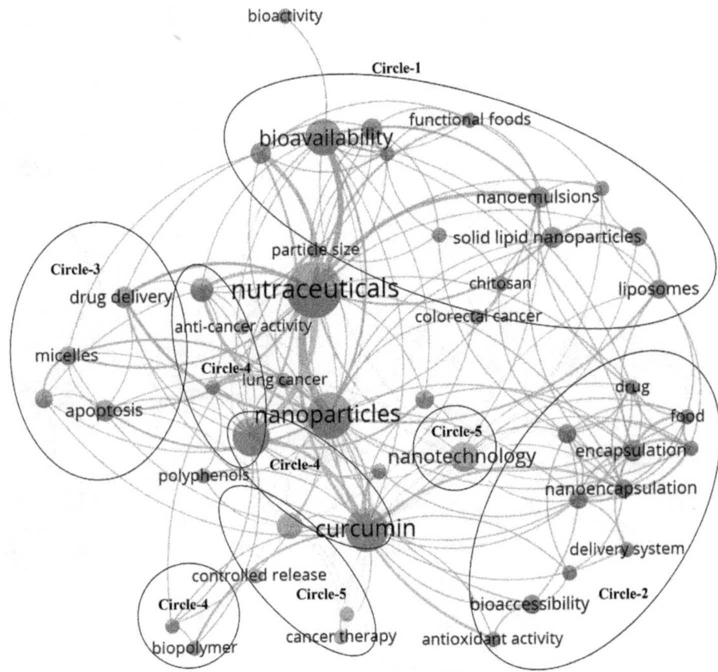

FIGURE 3.4 Science map using minimum occurrence (threshold) at 2.

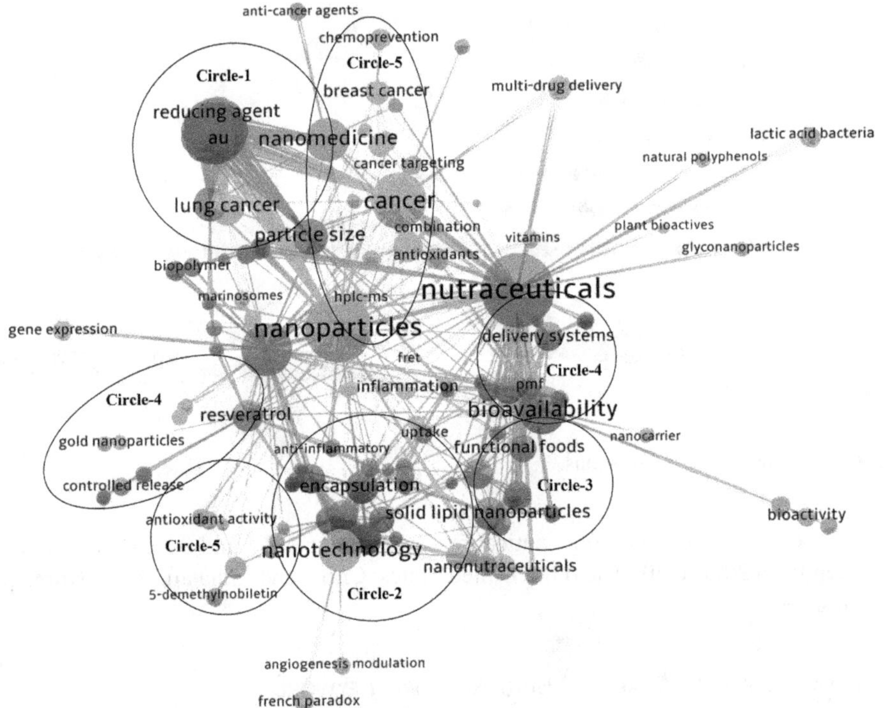

FIGURE 3.5 Science map created using minimum occurrence (threshold) at 1.

nanoemulsion, and other drug delivery systems. The primary research focus is on polyphenols, solid lipid nanoparticles, curcumin, biopolymers, micelles, liposomes and chitosan for colorectal and lung cancer therapies. Relatedness among the keywords was further explored in Figure 3.4. Overall, it is categorized into five clusters. The red cluster (or the area under Circle-1) clusters mainly show strong links between the keywords, such as solid lipid nanoparticles, liposomes, chitosan, bioavailability and colorectal cancer. One green cluster (dark; or the area under Circle-2) mainly focuses on nanoencapsulation, drug delivery system and antioxidant properties. Other green clusters (parrot green; or the area under Circle-3) indicate strong linkage between research on lung cancer, micelles, apoptosis, etc. The blue cluster (or the area under Circle-4) mainly emphasizes the role of curcumin in cancer treatment/therapy. Purple clusters (or the area under Circle-5) show linkage revealing the controlled release of bipolymer in cancer therapy.

However, when the same map was created for 352 words (Figure 3.5), some new keywords appeared, including breast cancer, resveratrol, natural polyphenol, gold nanoparticles, chemopreventive agents, biopolymer. The emergence of other words is illustrated in Figure 3.5.

3.2.3 International Co-Authorship Analysis

Figure 3.6 indicates that the United States is in the lead with the most documents, followed by China and India. In terms of international collaboration, the United States and China have a strong linkage and have published several documents authored by both countries. Sweden and Australia are among other prominent collaborators with China. The United States also shows collaboration with Japan. India has more collaborations with nations like South Korea, Saudi Arabia, Egypt and Germany. Germany has more collaboration with countries like Spain, Portugal, Iran and Pakistan.

A comprehensive review of nano-nutraceutical research is presented for the three leading countries: the United States, China and India. As shown in Figure 3.6, there is a strong correlation between the United States and China. By applying the filter "countries", 35 documents were extracted for the United States and 17 and 16 papers for China and India, respectively. However, overlapping exists, as a number of papers have been published by two countries together. These papers were categorized according to the nationalities of the first authors.

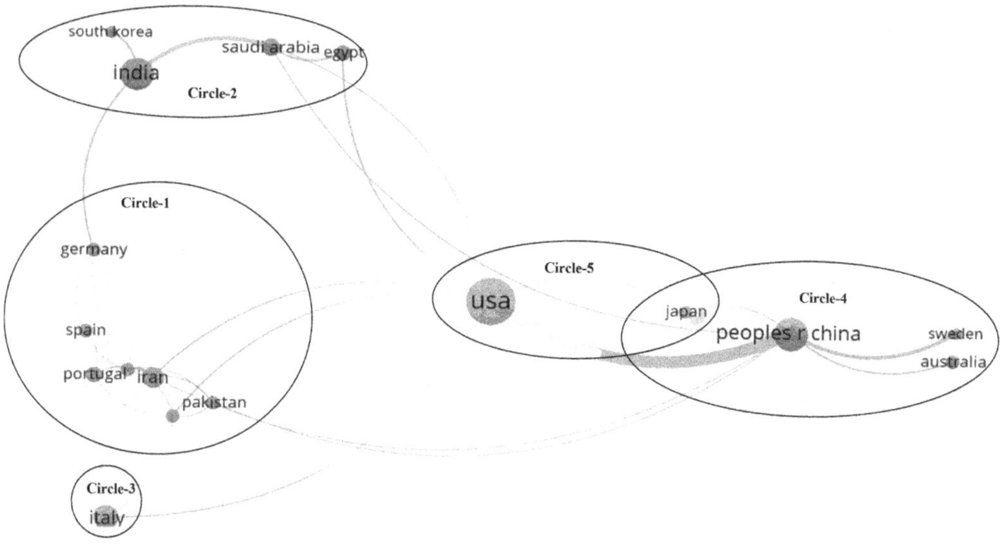

FIGURE 3.6 International co-authorship pattern in nano-nutraceutical research.

3.3 RESEARCH FOCUS

3.3.1 Nano-Nutraceutical Research Trends in the United States

The United States has published 35 documents (10 documents in collaboration with China), comprising articles as well as reviews. The authors have summarized all the review papers and their review span and focus in Table 3.1. Ten review papers have been published since 2011. The research focus is illustrated in Table 3.1. The articles will be comprehensively discussed later.

The United States has published several articles on nano-nutraceuticals, which comprise mainly preparation of nanoemulsions and nanoformulation of several bioactive compounds, such as quercetin (QC); curcumin, a yellow-colored polyphenol; and flavonoids like 5-demethylsuberosin (5DT) and myricetins. Research mainly focuses on the enhancement/increase in three important factors: solubilization, permeation and metabolism of different flavonoids and polyphenols using nanoformulation, nanoemulsion, combination strategy, Pickering emulsion agents and spontaneous micelle formation methods.

QC nanocrystals were prepared by (Lu et al. 2017). QC is a great anticarcinogen, and the crystals were prepared using a milling, spray and freeze-drying process, resulting in an increase in the solubility and dissolution rate without comprising the chemical properties of the QC. Kharat and McClements (2019) gave a successful demonstration of a recently emerging colloidal delivery system using curcumin, a natural polyphenol kunpresents in plant rhizomes.

Polymethoxyflavones (PMFs) include 5-demethyltangeretin (5DT), a hydrophobic, non-polar compound isolated from Chenpi (aged orange peel), which exhibits potent anticancer activities. It has a tendency to precipitate and sediment, resulting in low oral bioavailability and a low absorbance rate through the gastrointestinal tract (GIT) (Li et al. 2012). Both of these characteristics were improved using hydrophilic poly vinyl alcohol (PVA) and hydroxypropyl methylcellulose (HPMC) and with medium-chain triglycerides (MCT) as the oil phase. A significant increase in encapsulation efficacy was also observed. In further studies, bioavailability has been shown to have an inverse relationship with the size of the lipid droplets (Zheng et al. 2014). Using a spontaneous micelle formation method, Wang and Sukumar (2020) synthesized curcumin micelles using polysorbate 80 as surfactant and applying a cloud point cooling method, which resulted in a significant increase in the solubility of curcumin micelles compared with curcumin.

Several mechanisms, like colloidal micelles, nanoparticles, hollow microcapsules, layer-by-layer nanovesicles, nanotubes, nanogels and nanoparticles synthesized from nutraceuticals, are used to enhance characteristics of curcumin like oral bioavailability and absorption rate. A proof of concept was established to enhance the bioavailability of curcumin molecules in liposomes (Kunwar et al. 2006). Poly-D-lysine (PDL) and bovine serum albumin (BSA) were used to form bionanotubes (BNTs) as curcumin carriers through layer-by-layer deposition (Sadeghi et al. 2013). The stabilization of curcumin nanoemulsions with modified lecithin (phosphatidyl choline) was also investigated (Agame et al. 2020). A highly significant reduction in tumorigenesis index and tumor area (92% and 90%, respectively) was observed.

The myricetin group of flavonoids are known for their therapeutic potential as anti-tumor agents. Solid lipid nanoparticles (SLNs) were produced using these flavonoids. A 20-fold reduction in dosage requirement was observed compared with the free drug. Moreover, the inhibition of tumor growth rate increased to 69.3%. Nafee et al. (2020) investigated Gelucire-based SLNs comprising myricetin, Gelucire and the cell membrane component. These showed better availability and cellular internalization, resulting in better pulmonary delivery.

Resveratrol (trans-3,5,4′-trihydroxystilbene), a phenolic compound, has great therapeutic potential to treat breast, pancreatic and other cancers, and is found predominantly in red grape skin and red wine. The trans form of this compound is used on gold nanoparticles, and conjugated resveratrol-gold nanoparticles were formed using gum Arabic (GA) as a protein matrix support as well as to enhance the loading efficiency and stability (Thipe et al.2019). It was also concluded in other research that fucoxanthin has potential anticancer therapeutic efficacy when used in combination with conventional cancer treatment by reducing drug resistance (Wang et al. 2016).

TABLE 3.1
Summary of Review Papers' Research Focus: Summary Table for the United States

S. no.	Title	Reference	Total citations	Span of study	Outcome
1	Adult-onset brain tumors and neurodegeneration: Are polyphenols protective?	(Squillaro et al. 2018)	23	1999–2017	The review mainly focuses on the utilization of nutritional polyphenolic compounds such as flavonoids and non-flavonoid compounds as treatment of brain-related disorders and neurodegenerative disorders (NDs) associated with aging and tumors such as adult-onset brain tumors by limiting oxidative stress and inflammatory signaling in addition to increasing activity in protective signaling pathway.
2	Bioavailability and delivery of nutraceuticals using nanotechnology	(Huang et al. 2010)	394	1968–2009	The review article covers the enhancement of oral bioavailability and bioaccessibility by the micellar encapsulation system and nanoemulsion-based delivery systems of different phytochemicals. The review comprehensively discussed the anti-inflammatory and anticancer characteristics of the agents.
3	Emerging nano-pharmaceuticals and nano-nutraceuticals in cancer management	(Salama et al. 2020)	7	2000–2020	The papers mainly comprise the nanoformulation of natural bioactive compounds or natural extracts with a promising role in cancer treatment by inhibiting the target cell growth and changing characteristics such as pharmacokinetic and pharmacodynamic properties.
4	Curcumin as tyrosine kinase inhibitor in cancer treatment	(Golonko et al. 2019)	21	1994–2019	Review article articulate that curcumin antioxidant activity and potential cancer inhibitor. Several biochemical properties has been discussed in the paper and their responsible factor for these unique characteristics.
5	Delivery of anti-inflammatory nutraceuticals by nanoparticles for the prevention and treatment of cancer	(Nair et al. 2010)	156	1987–2010	Nutraceuticals, polyphenols, flavonoids and other nanoforms of these particles as chemopreventive agents are discussed in detail.
6	Designing food structure and composition to enhance nutraceutical bioactivity to support cancer inhibition	(McClements and Xiao 2017)	35	2001–2017	The paper discusses the role of nanoemulsions, liposomes, biopolymer nanoparticles and microgels in enhancing the therapeutic capability of nutraceuticals.
7	Liposomal nano-capsules in food science and agriculture	(Taylor et al. 2005)	331	1978–2012	This paper mainly discussed the role of liposomes as drug delivery vehicles.
8	Nano-enabled personalized nutrition: Developing multicomponent-bioactive colloidal delivery systems	(McClements 2020)	7	2008–2020	Colloidal delivery system and combination therapy as a promising strategy is discussed.

(Continued)

TABLE 3.1 (CONTINUED)
Summary of Review Papers' Research Focus: Summary Table for the United States

S. no.	Title	Reference	Total citations	Span of study	Outcome
9	Recent advances in the combination delivery of drug for leukemia and other cancers	(Al-Attar and Madihally 2020)	4	1983–2019	Detailed discussion of the potential of combination therapy using more than one chemopreventive agent, and their novel properties, like the ability for precise targeting, minimal dosage requirement, and the half-life increase of the drug in the circulation system.
10	Recent advances of curcumin and its analogues in breast cancer prevention and treatment	(Mock et al. 2015)	45	1995–2013	Curcumin is a potential candidate to prevent cancer, and the paper discusses the recent trend to develop analogs free from the limitations of curcumin.

3.3.2 Nano-Nutraceutical Research Trends in China

The research focus in China was mainly to evaluate the potential of various nutraceuticals, such as flavonoids, polyphenols, algal polysaccharides, corn hydrolysate, curcumin, and PMFs like 5DT. Molecular imaging and combination therapy with chemopreventive agents are among the prominent research topics being investigated by various stakeholders (Table 3.2)

The anticancer potential of combined ingredients from natural compounds, tea polyphenol and curcumin, on SLNs was investigated. A significant increase in inhibitory effects on liver cancer cells and carcinoma cells was observed (Li et al. 2017). The authors reported that the combination of these compounds resulted in improved solubility and chemical stability.

Molecular imprinting technology was used for the preparation of imprinted-like biopolymeric micelles (IBMs) for efficient delivery of curcumin. Encapsulation of curcumin with IBMs resulted

TABLE 3.2
Summary of Review Papers' Research Focus: Summary Table of China

S.no.	Title	Author	Total citations	Synopsis
1	Nano-complexation of proteins with curcumin: From interaction to nanoencapsulation (a review)	(Tang 2020)	8	The process of nanocomplexation, using biocompatible proteins as delivery agents for curcumin, is reviewed comprehensively.
2	Anticancerous potential of polyphenol-loaded polymeric nanotherapeutics	(Ernest et al. 2018	8	The article summarizes the pros and cons of polymeric delivery systems derived from plant polyphenols and their efficacy in cancer therapy and reveals the future research direction.
3	Nano-enabled personalized nutrition: Developing multicomponent-bioactive colloidal delivery systems	(McClements 2020)	7	This review highlights the significance of colloidal delivery systems in the development of personalized food products and also effective delivery systems.

in superior inhibition of tumor cells, enhanced oral bioavailability and a magnificent increase in solubility by a factor of 50,000 (Zhang et al. 2014)

Polymethoxyflavones (PMFs) are a group of hydrophobic, non-polar compounds isolated from Chenpi (aged orange peel), which exhibit potent anticancer activities. They have a tendency to precipitate and sediment, resulting in low oral bioavailability and a low absorbance rate through the GIT. Ning et al. (2019) improved both these characteristics using PVA and HPMC with MCT as the oil phase. A significant increase in encapsulation efficacy was also observed. Bioavailability has also been shown to have an inverse relationship with the size of the lipid droplets (Zheng et al. 2014). A spontaneous micelle formation method was used to synthesize curcumin micelles using polysorbate 80 as surfactant and applying a cloud point cooling method, resulting in a significant increase in the solubility of curcumin micelles compared with curcumin (Wang and Sukumar 2020).

Wang et al. (2016) shows that the curcumin nanoparticles (Cur NPs) stabilized by corn protein hydrolysate (CPH) exhibited better colloidal stability bioaccessibility and antioxidant activity compared to sodium caseinate (NaCas). Besides this, to improve the anticancer efficacy, a novel approach to the fabrication micellar curcumin from solution of free curcumin and food additive polysorbate 80 using slow cooling below its cloud point method (Wang et al. 2020). Rapeseed protein was used to fabricate a nanogel delivery system for hydrophobic curcumin (Wang et al. 2019).

The anticancer potential of algal polysaccharides was also explored due to their capability to secrete short-chain fatty acids (SCFAs) in the gut and to inhibit the proliferation of cancer cells and induce cell apoptosis (Ouyang et al. 2021).

3.3.3 Nano-Nutraceutical Research in India

A group of Indian researchers first delved into the anticancer, antimicrobial, anti-inflammatory, antioxidant, antidiabetic and cardioprotective effects of tea and the various constituents responsible for these effects (Rana et al. 2021). Later in the review, they discussed the use of nanoparticulate carriers such as nanoemulsions, micelles and liposomes for the encapsulation and targeted delivery of catechins, albumins and epigallocatechin gallate in order to overcome the shortcomings of free-form delivery, improve cell permeability and inhibit the chemicals' digestive degradation. They concluded that the use of a polymeric matrix for the encapsulation of the phytochemicals in tea is necessary for the development of nutraceuticals from tea.

A study by Chakraborty and Dhar (2017) also discussed this latest development in using proteins as nanovehicular carriers of bioactives such as drugs, nutraceuticals, enzymes, hormones, genes, etc. The overwhelming preference for these nanocarriers can be attributed to their tunability and modifiability due to their ability to attach surface ligands for enhanced permeation and retention (EPR) effect, their biodegradability from complex proteins to non-toxic amino acids, and their biocompatibility, feasibility and effectiveness in targeted delivery.

The review discusses the contemporary literature available on protein nanocarriers for, chemotherapeutics with a pre-emptive focus upon the modes of absorption in the GIT. Furthermore, the review then elaborates upon the various forms of protein-based nanocarriers: their source materials, fabrication, characterization, application, advantages and limitations. The review also provides a comparative analysis of *in vivo* and *in vitro* studies of the various different types of protein nanocarriers while highlighting the variation of the release pattern and molecular mechanism of the encapsulated molecule with respect to different protein types and various nanostructures. The review focuses upon the following nanocarriers and their exploited properties:

1. *Milk proteins*: Due to their excellent gelling effect and high nutritional value
   a. Casein: As nanomicelles for entrapment of hydrophobic molecules
   b. Whey proteins: As nanoemulsions for their controlled and long-term release of bioactive molecules

   c. Silk proteins: As nanoparticles due to their tunability with ligands and chemical conjugation with drugs
2. Collagen as nanohydrogels due to their improved biocompatibility
3. Gelatin nanoparticles for their versatility in drug delivery
4. Albumin–human serum albumin (HSA) and BSA as nanospheres for their excellence in targeted delivery of entrapped molecules

The article concludes by establishing the significance of understanding the molecular-level interaction between the carrier protein and the core element in order to develop an ideal protein nanocarrier that would be the safest and most effective mode of therapeutic vector in the pharmaceutical industry for several treatments, including for cancer, in the near future.

With the recent boom in demand for highly nutritionally specific products, research into hydrocolloids has seen an overwhelming surge. As the name suggests, these are particles formed from highly hydrophilic molecules, such as polysaccharides and proteins, in the form of a colloidal dispersion, which can be efficiently categorized based upon their source of origin as well as the wide range of physicochemical, structural and metabolic functions they serve.

Conventionally, the application of hydrocolloids had been restricted to their incorporation into food or pharmaceutical formulations to enhance the overall nutritional, tribological, rheological and sensorial characteristics of the system, but in recent years, its functional benefits against diabetes, hypertension, cancer and dyslipidemia, as well as its role in maintaining a healthy gut biome, and therapeutic, antimicrobial and antioxidant applications have been explored.

The latest developments are also evaluated in the field of hydrocolloids beyond functional nutraceuticals into drug delivery systems, exploiting their versatile biological and chemical properties to effectively compete with other synthetic controlled release excipients (Manzoor et al. 2020). Hydrocolloids offer a significant advantage over conventional nanocarriers like polymeric micelles and nanoparticles, as they avoid the complications of potential toxicity while also tailoring in nutritional health benefits. They further this with the application of nanotechnology to incorporate quantum size effects and surface enhancements to improve the therapeutic effects and targeted delivery of the colloidal carriers. This works hand in hand with the advancements on the nanocarrier front, as it allows the development of novel food grade vectors with increased solubility, bioavailability and permeability, and lowered degradation in the GIT. The paper encourages future applications in designing and developing nutraceutical carriers for the delivery of subsequent functional food structures. Overall, it provides a comprehensive discussion focus on their application as controlled release excipients for nutraceuticals or drug release systems.

Another study focuses upon the epigenetic mode of treatment for cancers with the help of flavonoids. This falls into the domain of nutri-epigenomics, which is a currently booming field that deals with the various mechanisms through which highly nutritive foods affect gene regulation and expression. Since cancer is caused by abnormal cellular physiology (Bunkar et al. 2019), such as genetic, transcriptomic, proteomic, metabolomic and epigenetic mutations in a cell that cause it to become malignant, a great approach to therapeutics for the treatment of malignant tumor cells as well as a precautionary measure for normal cells would be to bolster cell fortifications against genetic and epigenetic mutations, preferably with a minimally invasive and non-toxic process.

The existing literature on the subject suggests that the mitochondria play a crucial role in regulating epigenetic alterations, especially the ones responsible for cancer, by managing vital enzymes, their concentrations and cofactors, and controlling the transcription of various signaling molecules. Certain concentrations of specific polyphenolic compounds have shown significant promise in the development of new nutri-epigenomic tools to manipulate reversible epigenetic alterations, which would be effective in cancer prevention, as epigenetic changes are the principal drivers of malignancy. In order to address this, the paper suggests the use of flavonoids, since they have shown noteworthy potential to modify the epigenome through a network of interrelated anti-redox processes. Unfortunately, their limited solubility, rapid metabolism and inadequate absorption in the GIT, complexation with other bioactive molecules, and an unfortunate pharmacokinetic profile, make them unsuitable for clinical usage. Therefore, in this context, nanocarriers encapsulating bioactive flavonoids demonstrate considerable potential for enhanced transport, delivery and controlled targeting, as well as improved efficiency and pharmacokinetic parameters.

The various nanovehicular approaches reviewed in this article broadly include polymeric nanoparticles (PNPs) and micelles, SLNs, nanostructured lipid carriers. liquid crystals, liposomes, molecular inclusion complexes and microemulsions. Overall, the paper suggests alluring options using nanovectors for the safe, effective, controlled and long-term release of flavonoids for the management of epigenetic factors for the regulation and potential treatment of cancer.

A novel nanocarrier formulation for the transport and delivery of a phytochemical – grape seed extract (GSE) – was developed for the treatment of cancer. The GSE was prepared by the nanoprecipitation method and then encapsulated into a polymeric nanoparticle to form "NanoGSE" particles of size ~100 nm (Narayanan et al. 2010). The particles were then chemically modified to incorporate folic acid ligands on the surface that would help the molecular receptor to target overexpressed folate receptors on malignant cells. The nanophytodrugs produced demonstrated colloidal stability at physiological pH, which would indicate promising *in vivo* results, as the particles would be functional in the blood serum. The *in vitro* application was analyzed by the researchers by investigating the uptake of rhodamine-123-loaded NanoGSE by tumor cells with the aid of fluorescence microscopy, later confirmed with cytometry. The study concluded with promising results, as the nanoformulation depicted an almost threefold reduction in the IC_{50} values compared with conventional free-form delivery of GSE. Additionally, it was noted that the nanophytochemical substantially increased the bioavailability of the nutraceutical in cancerous cells with little to no effect on normal cells due to the EPR effect. It further substantially increased the apoptotic index of tumor cells, which was also quantified by fluorescent microscopy. Overall, the study provides a foundation for the use of nanocarriers for the delivery of nutraceutical therapeutics for efficient and targeted delivery for the management of cancer – this novel approach is termed green chemotherapy.

Rahim et al. (2018) synthesized gold nanoparticles from biological precursors from tomato juice. Various concentrations of the tomato extract were tested as reducing agents for the metal precursor tetrachloroauric acid, and the resultant nanoparticles were noted to range in size by ~10 nm. These particles were then characterized with the aid of UV-visible spectroscopy, which yielded an absorption peak at 522 nm, dynamic light scattering (DLS), which enabled the approximation of the average particle size at 10.86 ± 0.6 nm, zeta potential, transmission electron microscopy (TEM) imaging analysis and Fourier transform infrared (FTIR) spectroscopy. The stable gold nanoparticles (GNPs) were then explored for their potential application in the treatment of lung cancer of the cell line A549 as well as cervical cancer. They yielded IC_{50} values of 0.286 and 0.200 mM, respectively. This was suggested to be due to two main mechanisms: inducing cell apoptosis and producing reactive oxygen species, thereby ultimately inhibiting the growth of tumor cells. T-GNPs were found to be highly effective by virtue of their size, metallic property and capping molecules. Finally, the study discusses the prospects and implications of using nutraceuticals such as tomato juice to biosynthesize nano-nutratherapeutics for the treatment of diseases such as cancer.

Epidemiologic studies and literature have established that diet plays a crucial role in the development and proliferation of malignant cancer cells. It has also been suggested that nutraceuticals provide a promising potential in the treatment of these cancers. The aglycone of hesperidin (HRN), a glycoside found in citrus fruits, is one such example of a prospective nutraceutical molecule that has shown anticancer properties by reducing tumor mass and volume as well as the onset of chemically induced cancer in lab rats. However, studies also demonstrate the poor bioavailability of the drug due to its degradation upon oral administration and its poor solubility in the blood serum, thereby reducing its therapeutic benefit.

One study suggests the development of nanocrystalline solid dispersions (NSDs) of HRN formulations to increase the bioavailability, stability, shelf life, pharmacokinetic properties and thereby, the industrial application of HRN (Sheokand et al. 2019). The facile synthesis of these nanocrystalline formulations was done by a simple single-step process and efficacy of NSDs optimized using design of experiments (DoE) and multivariate analysis (MVA) tools. The particles were then characterized using differential scanning calorimetry (DSC) for their size, which was noted to average around 558.2 ± 68.1 nm, and crystallinity, and powder X-ray diffraction (XRD). Furthermore, the researchers compared the bioavailability of the NSD of HRN with a free-form dispersion of HRN against cancer cells upon oral administration in female rats and noted a 3.3 to 2.1-fold improvement in the blood concentrations (~70% release in 30 min) and a 4-fold decrease in the T_{max} in the optimized nanodispersions. Therefore, the study concludes the superior efficacy of the NSD formulation in cancer therapeutics.

Bansode et al. (2019) studied the difference between nano-nutraceuticals and their bulk counterparts in terms of their antioxidant, anticancer and anti-angiogenic activity. The structurally diverse bioactive compounds studied were quercetin, curcumin, anthraquinone, flavone, xanthone and vanillin. Nanoparticles of these nutraceuticals were synthesized via the reprecipitation method and characterized for their particle size and stability. *In vivo* and *in vitro* studies indicated excellent inhibition properties of nano-nutraceuticals in comparison to their bulk compounds and signify their potential for commercial usage.

Arora and Jaglan (2016) provided a detailed review of lipid-based, inorganic-based and polymer-based nanocarriers. They also focused on two of the current Food and Drug Administration (FDA)-approved and commercially available drugs involving nanocarrier systems (Abraxane and Doxil). However, this methodology still requires extensive research and clinical studies to prove fruitful for oncology.

Shende and Mallick (2020) also emphasized that nanocarriers modify the pharmacokinetics and pharmacodynamics of the nutraceuticals. Overall, these pharmaceutical nanoformulations can be used to significantly reduce or eliminate these diseases but require detailed toxicology tests. Commercial success of nano-nutraceuticals can be attained by abiding by global safety rules and regulations and gaining FDA approval.

Leena et al. (2020) studied the co-delivery of nutraceuticals, called synergism, which is being examined for variations and improvements in activity and efficiency. The synergistic effect enhances the overall delivery, release and bioavailability of the drugs. Existing nutraceutical combinations have proven their potency for antimicrobial, anti-inflammatory, antidiabetic, anticancer, anti-obesity and cardioprotective applications. Yet, this synergetic effect needs to be thoroughly studied, and the combinations should undergo multiple clinical trials to prove their efficacy. The different possible combinations and the issue of how variation in dosage affects different patients need to be addressed. Adequate positive data can render this useful for a multitude of conditions.

Thipparaboina et al. (2015) studied multi-drug therapy, the utilization of more than one drug, which is a budding area of research in the medical field. It is proven to be fruitful to treat a range of diseases such as HIV, diabetes, cardiovascular diseases and cancer. The therapy consists of embedding drugs with similar or varying therapeutic mechanisms in polymeric micelles. Polymer-based micelles are carriers constituting self-assembled copolymers and can be used for site-specific delivery of drugs or nutraceuticals. Clinical trials are still in process testing combinations between drugs, nutraceuticals, genes and small interfering RNA (siRNA). The combination of drugs and nutraceuticals, termed "phantraceuticals", can possibly improve drug bioavailability and lessen toxicity in the body. Some of the common nutraceuticals used are curcumin, chrysin, quercetin, thymoquinone and caffeic acid. Certain drawbacks associated with this method are the complicated synthesis of these polymeric micelles, lack of control of drug release and potential toxicity to the liver. The future success of this methodology depends on clinical trials and its approval by governmental organizations.

Yadav et al. (2014) investigated polyphenols derived from tea, namely catechin (CAT) and epicatechin (ECAT), which are being studied for their anticancerous nature. The normal bioavailability of these polyphenols is very low, and hence, they were encapsulated in BSA nanoparticles. The synthesized particles were characterized via TEM and atomic force microscopy (AFM), and their sizes are in the range of 50 nm. These drugs were evaluated for their release rate, and the *in vitro* studies show a controlled release. The temperature stability was also examined and showed how this encapsulated drug exhibits more stability at higher temperatures for more time. Their anticancerous activity was checked via DPPH assay and displayed improved efficacy. They have also proven their scavenging and inhibition activity against A549 cells. Results indicate the potential usage of these encapsulated nutraceuticals in pharmaceutical formulations.

Lydia et al. (2020) research about biosynthesizes gold (Au) nanoparticles (AuNPs)-based functional yoghurt having anticancer attributes. Characterization of the green synthesized gold nanoparticles was done using various instruments such as FTIR, UV-visible spectrophotometer, XRD, SEM, energy dispersive X-ray (EDX), zeta analyzer and DLS. They were rectangular-shaped particles with UV peak at 525 nm, and the FTIR spectra had the characteristic peaks of seed oil. Particle size was less than 100 nm, and they showed a potential of +34 mV. The nanoparticle-enhanced yoghurt was tested against colon and lung cancer cells, and results showed an excellent anticancerous nature. Hence, gold nanoparticle functionalized yoghurt is a prospective formulation for the medical field.

3.4 CONCLUSIONS

Cancer is an ever-increasing illness affecting and causing mortality in more than half of the population. Research trends using nano-nutraceuticals in developed and developing nations were studied using a bibliographic database and a visualization tool. Nano-nutraceuticals are playing a pivotal role in reducing tumor incidence and tumor growth. Worldwide, researchers are mainly focused on enhancing bioavailability, absorbance rate, bioaccessibility, controlled release and reduction in dosage requirement through the development of several kinds of nanocarriers/drug delivery systems, such as nanoemulsions, nanoencapsulation, nanogels, nanosensors, nanomedicine, micelles, SLNs, nanostructured lipid carriers, liquid crystals, liposomes, molecular inclusion complexes and microemulsions. Overall, countries are finding alluring options using nanovectors for safe, effective, controlled and long-term release of nutraceutical management of epigenetic factors for the regulation and potential treatment of cancer. However, the United States, China and India are focusing on the epigenetic mode of treatment for cancers with the help of nano-nutraceuticals. This falls into the domain of nutri-epigenomics, which is a currently booming field dealing with various mechanisms through which highly nutritive foods affect gene regulation and expression. Some of the nanocarrier formulations for transport and delivery systems are frequently used for research into solid lipid nanoparticles, gold nanoparticles, nanoencapsulation and nanoemulsion formulations of flavonoids and polyphenols.

Besides this, multi-drug therapy, the utilization of more than one drug, is a budding area of research in this field. The combination of drugs and nutraceuticals has been termed "Phantraceuticals" and can possibly improve drug bioavailability and lessen toxicity in the body. Some of the common nutraceuticals used are curcumin, chrysin, quercetin, thymoquinone and caffeic acid.

Nanocarriers modify the pharmacokinetics and pharmacodynamics of the nutraceuticals. Overall, these approaches require detailed toxicology tests. Combinational therapy/multi-drug therapy should be developed in a responsible manner, keeping in view the potential synergistic or antagonistic interactions amongst a combination of ingredients leading to impacts on stability, bioavailability, safety and efficacy in the local environment; their scientific rationale based on the scientific literature; and finally, approval from the relevant authorized agency, which will help to gain public acceptance, a crucial factor for mass production.

REFERENCES

Agame-Lagunes, B., Alegria-Rivadeneyra, M., Quintana-Castro, R., Torres-Palacios, C., Grube-Pagola, P., Cano-Sarmiento, C., Garcia-Varela, R., Alexander-Aguilera, A., & García, H. S. (2020). Curcumin nanoemulsions stabilized with modified phosphatidylcholine on skin carcinogenesis protocol. *Current Drug Metabolism*, *21*(3), 226–234. https://doi.org/10.2174/1389200221666200429111928.

Al-Attar, T., & Madihally, S. V. (2020). Recent advances in the combination delivery of drug for leukemia and other cancers. *Expert Opinion on Drug Delivery*, *17*(2), 213–223. https://doi.org/10.1080/17425247.2020.1715938

Arora, D., & Jaglan, S. (2016). Nanocarriers based delivery of nutraceuticals for cancer prevention and treatment: A review of recent research developments. *Trends in Food Science & Technology*, *54*, 114–126. https://doi.org/10.1016/j.tifs.2016.06.003

Bunkar, N., Shandilya, R., Bhargava, A., Samarth, R. M., Tiwari, R., Mishra, D. K., & Mishra, P. K. (2019). Nano-engineered flavonoids for cancer protection. *Frontiers in Bioscience*, *24*, 1097–1157. https://doi.org/10.2741/4771

Bansode, P. A., Patil, P. V., Birajdar, A. R., Somasundaram, I., Bachute, M. T., & Rashinkar, G. S. (2019). Anticancer, antioxidant and antiangiogenic activities of nanoparticles of bioactive dietary nutraceuticals. *Chemistry Select*, *4*(47), 13792–13796. https://doi.org/10.1002/slct.201903946

Chakraborty, A., & Dhar, P. (2017). A review on potential of proteins as an excipient for developing a nanocarrier delivery system. *Critical Reviews in Therapeutic Drug Carrier Systems*, *34*(5), 453–488. https://doi.org/10.1615/CritRevTherDrugCarrierSyst.2017018612

Ernest, U., Chen, H. Y., Xu, M. J., Taghipour, Y. D., Asad, M., Rahimi, R., & Murtaza, G. (2018). Anticancerous potential of polyphenol-loaded polymeric nanotherapeutics. *Molecules*, *23*(11), 2787. https://doi.org/10.3390/molecules23112787

Ferlay, J., Colombet, M., Soerjomataram, I., Parkin, D. M., Piñeros, M., Znaor, A., & Bray, F. (2021). Cancer statistics for the year 2020: An overview. *International Journal of Cancer*, *194*(4), 778–789, https://doi.org/10.1002/ijc.33588,

Golonko, A., Lewandowska, H., Świsłocka, R., Jasińska, U. T., Priebe, W., & Lewandowski, W. (2019). Curcumin as tyrosine kinase inhibitor in cancer treatment. *European Journal of Medicinal Chemistry*, *181*, 111512. https://doi.org/10.1016/j.ejmech.2019.07.015

Huang, Q., Yu, H., & Ru, Q. (2010). Bioavailability and delivery of nutraceuticals using nanotechnology. *Journal of Food Science*, *75*(1), R50–R57. https://doi.org/10.1111/j.1750-3841.2009.01457.x

Kharat, M., & McClements, D. J. (2019). Recent advances in colloidal delivery systems for nutraceuticals: A case study–delivery by design of curcumin. *Journal of Colloid and Interface Science*, *557*, 506–518.

Kunwar, A., Barik, A., Pandey, R., & Priyadarsini, K. I. (2006). Transport of liposomal and albumin loaded curcumin to living cells: An absorption and fluorescence spectroscopic study. *Biochimica et Biophysica Acta*, *1760*(10), 1513–1520. https://doi.org/10.1016/j.bbagen.2006.06.012

Leena, M. M., Silvia, M. G., Vinitha, K., Moses, J. A., & Anandharamakrishnan, C. (2020). Synergistic potential of nutraceuticals: Mechanisms and prospects for futuristic medicine. *Food & Function*, *11*(11), 9317–9337. https://doi.org/10.1039/D0FO02041A

Li, Y., Xiao, H., & McClements, D. J. (2012). Encapsulation and delivery of crystalline hydrophobic nutraceuticals using nanoemulsions: Factors affecting polymethoxyflavone solubility. *Food Biophysics*, *7*(4), 341–353. https://doi.org/10.1007/s11483-012-9272-1

Lu, M., Ho, C. T., & Huang, Q. (2017). Improving quercetin dissolution and bioaccessibility with reduced crystallite sizes through media milling technique. *Journal of Functional Foods*, *37*, 138–146. https://doi.org/10.1016/j.jff.2017.07.047

Li, R., Jia, K., Chen, X. G., & Xiao, H. T. (2017). A novel curcuminoid-tea polyphenol formulation: Preparation, characterization, and in vitro anti-cancer activity. *Journal of Food Biochemistry*, *41*(2), e12332. https://doi.org/10.1111/jfbc.12332

Lydia, D. E., Khusro, A., Immanuel, P., Esmail, G. A., Al-Dhabi, N. A., & Arasu, M. V. (2020). Photoactivated synthesis and characterization of gold nanoparticles from *Punica granatum* L. seed oil: An assessment on antioxidant and anticancer properties for functional yoghurt nutraceuticals. *Journal of Photochemistry and Photobiology B: Biology*, *206*, 111868. https://doi.org/10.1016/j.jphotobiol.2020.111868

Manzoor, M., Singh, J., Bandral, J. D., Gani, A., & Shams, R. (2020). Food hydrocolloids: Functional, nutraceutical and novel applications for delivery of bioactive compounds. *International Journal of Biological Macromolecules*, *165*(Pt A), 554–567. https://doi.org/10.1016/j.ijbiomac.2020.09.182\

McClements, D. J., & Xiao, H. (2017). Designing food structure and composition to enhance nutraceutical bioactivity to support cancer inhibition. In *Seminars in cancer biology* (Vol. 46, pp. 215–226). Academic Press. https://doi.org/10.1016/j.semcancer.2017.06.003

McClements, D. J. (2020). Nano-enabled personalized nutrition: Developing multicomponent-bioactive colloidal delivery systems. *Advances in Colloid and Interface Science*, 102211. https://doi.org/10.1016/j.cis.2020.102211

Mock, C. D., Jordan, B. C., & Selvam, C. (2015). Recent advances of curcumin and its analogues in breast cancer prevention and treatment. *RSC Advances*, *5*(92), 75575–75588. https://doi.org/10.1039/C5RA14925H

Nair, H. B., Sung, B., Yadav, V. R., Kannappan, R., Chaturvedi, M. M., & Aggarwal, B. B. (2010). Delivery of anti-inflammatory nutraceuticals by nanoparticles for the prevention and treatment of cancer. *Biochemical Pharmacology*, *80*(12), 1833–1843. https://doi.org/10.1016/j.bcp.2010.07.021

Nafee, N., Gaber, D. M., Elzoghby, A. O., Helmy, M. W., & Abdallah, O. Y. (2020). Promoted antitumor activity of myricetin against lung carcinoma via nano encapsulated phospholipid complex in respirable microparticles. *Pharmaceutical Research*, *37*(4), 1–24. https://doi.org/10.1007/s11095-020-02794-z

Narayanan, S., Binulal, N. S., Mony, U., Manzoor, K., Nair, S., & Menon, D. (2010). Folate targeted polymeric 'green'nanotherapy for cancer. *Nanotechnology*, *21*(28), 285107. https://doi.org/10.1088/0957-4484/21/28/285107

Ning, F., Wang, X., Zheng, H., Zhang, K., Bai, C., Peng, H., & Xiong, H. (2019). Improving the bioaccessibility and in vitro absorption of 5-demethylnobiletin from chenpi by se-enriched peanut protein nanoparticles-stabilized pickering emulsion. *Journal of Functional Foods*, *55*, 76–85. https://doi.org/10.1016/j.jff.2019.02.019

Ouyang, Y., Qiu, Y., Liu, Y., Zhu, R., Chen, Y., El-Seedi, H. R., ... & Zhao, C. (2021). Cancer-fighting potentials of algal polysaccharides as nutraceuticals. *Food Research International*, *147*, 110522. https://doi.org/10.1016/j.foodres.2021.110522

Rahim, M., Iram, S., Syed, A., Ameen, F., Hodhod, M. S., & Khan, M. S. (2018). Nutratherapeutics approach against cancer: Tomato-mediated synthesised gold nanoparticles. *IET Nanobiotechnology*, *12*(1), 1–5. https://doi.org/10.1049/iet-nbt.2017.0068

Rana, A., Rana, S., & Kumar, S. (2021). Phytotherapy with active tea constituents: A review. *Environmental Chemistry Letters*, *19*, 2031–2041. https://doi.org/10.1007/s10311-020-01154-y

Sadeghi, R., Kalbasi, A., Emam-jomeh, Z., Razavi, S. H., Kokini, J., & Moosavi-Movahedi, A. A. (2013). Biocompatible nanotubes as potential carrier for curcumin as a model bioactive compound. *Journal of Nanoparticle Research*, *15*(11), 1–11. https://doi.org/10.1007/s11051-013-1931-8

Salama, L., Pastor, E. R., Stone, T., & Mousa, S. A. (2020). Emerging nano pharmaceuticals and nano nutraceuticals in cancer management. *Biomedicines*, *8*(9), 347. https://doi.org/10.3390/biomedicines8090347

Senapati, S., Mahanta, A. K., Kumar, S. et al. (2018) Controlled drug delivery vehicles for cancer treatment and their performance. *Signal Transduction and Targeted Therapy*, *3*, 7.

Sheokand, S., Navik, U., & Bansal, A. K. (2019). Nanocrystalline solid dispersions (NSD) of hesperetin (HRN) for prevention of 7, 12-dimethylbenz[a]anthracene (DMBA)-induced breast cancer in Sprague-Dawley (SD) rats. *European Journal of Pharmaceutical Sciences: Official Journal of the European Federation for Pharmaceutical Sciences*, *128*, 240–249. https://doi.org/10.1016/j.ejps.2018.12.006

Shende, P., & Mallick, C. (2020). Nanonutraceuticals: A way towards modern therapeutics in healthcare. *Journal of Drug Delivery Science and Technology*, *58*, 101838.

Squillaro, T., Schettino, C., Sampaolo, S., Galderisi, U., Di Iorio, G., Giordano, A., & Melone, M.A. (2018). Adult-onset brain tumors and neurodegeneration: Are polyphenols protective? *Journal of Cellular Physiology*, *233*(5), 3955–3967. https://doi.org/10.1016/j.bcp.2018.05.016

Tang, C. (2020). Nanocomplexation of proteins with curcumin: From interaction to nanoencapsulation (A review). *Food Hydrocolloids*, *109*, 106106. https://doi.org/10.1016/j.foodhyd.2020.106106

Taylor, T. M., Weiss, J., Davidson, P. M., & Bruce, B. D. (2005). Liposomal nanocapsules in food science and agriculture. *Critical Reviews in Food Science and Nutrition*, *45*, 587–605. https://doi.org/10.1080/10408390591001135

Thipe, V. C., Amiri, K. P., Bloebaum, P., Karikachery, A. R., Khoobchandani, M., Katti, K. K., & Katti, K. V. (2019). Development of resveratrol-conjugated gold nanoparticles: Interrelationship of increased resveratrol corona on anti-tumor efficacy against breast, pancreatic and prostate cancers. *International Journal of Nanomedicine*, *14*, 4413.

Thipparaboina, R., Chavan, R. B., Kumar, D., Modugula, S., & Shastri, N. R. (2015). Micellar carriers for the delivery of multiple therapeutic agents. *Colloids and Surfaces B: Biointerfaces*, *135*, 291–308. https://doi.org/10.1016/j.colsurfb.2015.07.046

Van Eck N. J., Waltman L. (2010). Software survey: VOS viewer, a computer program for bibliometric mapping. *Scientometrics*, *84*(2), 523–38.

Wang, C., Chen, X., Nakamura, Y., Yu, C., & Qi, H. (2020). Fucoxanthin activities motivate its nano/microencapsulation for food or nutraceutical application: A review. *Food & Function*, *11*(11), 9338–9358. https://doi.org/10.1039/D0FO02176H

Wang, G., & Sukumar, S. (2020). Characteristics and antitumor activity of polysorbate 80 curcumin micelles preparation by cloud point cooling. *Journal of Drug Delivery Science and Technology*, *59*, 101871. https://doi.org/10.1016/j.jddst.2020.101871

Wang, Y. H., Yuan, Y., Yang, X. Q., Wang, J. M., Guo, J., & Lin, Y. (2016). Comparison of the colloidal stability, bioaccessibility and antioxidant activity of corn protein hydrolysate and sodium caseinate stabilized curcumin nanoparticles. *Journal of Food Science and Technology*, *53*(7), 2923–2932. https://doi.org/10.1007/s13197-016-2257-1

Wang, Z., Zhang, R. X., Zhang, C., Dai, C., Ju, X., & He, R. (2019). Fabrication of stable and self-assembling rapeseed protein nanogel for hydrophobic curcumin delivery. *Journal of Agricultural and Food Chemistry*, *67*(3), 887–894. https://doi.org/10.1021/acs.jafc.8b05572

World Health Organization (WHO). (2020). *Global health estimates 2020: Deaths by cause, age, sex, by Country and by region, 2000–2019*. WHO. Accessed December 11, 2020. who.int/data/gho/data/themes/mortality-and-global-health-estimates/ghe-leading-causes-of-death

Yadav, R., Kumar, D., Kumari, A., & Yadav, S. K. (2014). Encapsulation of catechin and epicatechin on BSA NPs improved their stability and antioxidant potential. *EXCLI Journal*, *13*, 331. PMID: 26417264 PMCID: PMC4462830

Yang, C. S., Wang, H., & Hu, B. (2013). Combination of chemo preventive agents in nanoparticles for cancer prevention. *Cancer Prevention Research*, *6*(10), 1011–1014. https://doi.org/10.1158/1940-6207.CAPR-13-0312

Zhang, L., Qi, Z., Huang, Q., Zeng, K., Sun, X., Li, J., & Liu, Y. N. (2014). Imprinted-like biopolymeric micelles as efficient nanovehicles for curcumin delivery. *Colloids and Surfaces B: Biointerfaces*, *123*, 15–22. https://doi.org/10.1016/j.colsurfb.2014.08.033

Zheng, J., Li, Y., Song, M., Fang, X., Cao, Y., McClements, D. J., & Xiao, H. (2014). Improving intracellular uptake of 5-demethyltangeretin by food grade nanoemulsions. *Food Research International*, *62*, 98–103. https://doi.org/10.1016/j.foodres.2014.02.013

4 Strategic Design of a Nanocarrier System for Nutraceuticals

Chenmala Karthika, Mehrukh Zehravi, Rokeya Akter, and Md. Habibur Rahman[*]

CONTENTS

4.1 Introduction ... 79
4.1.1 Nutraceuticals and Their Concept ... 80
4.2 Antioxidants ... 80
4.3 Dietary Supplements ... 82
4.4 Probiotics ... 83
4.5 Design of Nutraceutical Delivery Vehicles ... 84
4.5.1 Polymers ... 84
4.5.2 Biodegradable Polymers ... 84
4.5.3 Proteins ... 84
4.5.4 Polysaccharides ... 85
4.5.5 Polymers Containing Ester, Anhydride, and Amide Connections ... 85
4.5.6 Temperature-Sensitive Polymers ... 86
4.5.7 pH-Sensitive Polymers ... 86
4.5.8 Micelles ... 87
4.5.9 Liposomes ... 87
4.6 Probiotics and Nutraceutical Delivery ... 88
4.7 Conclusions ... 89
References ... 89

4.1 INTRODUCTION

Food scientists and engineers formerly evaluated meals based on their flavor, which included sensory (taste and nutritional value) and texture (which comprised carbohydrate, water, fat, protein, vitamins, and minerals) (Onabanjo et al. 2009; Sharoba 2014; Dewettinck et al. 2008). However, there is a growing suggestion that chemical components of food can play an important part in human health. Probiotics, vitamins, antioxidants, and phytochemicals are chemical components taken from plant, nutritional, and microbial sources that deliver long-term health benefits. The concept of nutraceuticals was developed in 1989 by Stephen L. DeFelice, MD of the Foundation for Innovation in Medicine (New York, United States) as a mix of the words *nutrition* and *medicine*. Nutraceuticals, often known as activity foods, are foods or dietary components that provide medicinal or therapeutic advantages, such as chronic illness prevention and therapy, in addition to regular nutrition. Nutraceuticals include separate nutrients, refined foods, multivitamins, herbal items, and specialized

* Corresponding author: pharmacisthabib@gmail.com

DOI: 10.1201/9781003244721-4

diets, as well as genetically altered foods such micronutrients, minerals, grains, herbal blends, milk, stews, and drinks. For a long time, there has been a link between diet and health (Carlsen et al. 2010; Singh and Sinha 2012). "Let food be thy medicine, and medicine be thy nourishment," Hippocrates, the father of modern medicine, declared roughly 2500 years ago. Old Chinese medicine, Korean traditional medicine, Ayurvedic, Siddha medicines, homeopathy, Islamic healthcare, ancient medicine, Muti, Ifá, and African traditional medicine have all been used as folk medicine or folk medicines for a long time in Asia and Africa (Polito et al. 2016; Abdullahi 2011).

4.1.1 Nutraceuticals and Their Concept

The significant increase in human life expectancy during the twentieth century resulted in a greater number of individuals affected by aging-related chronic conditions such as diabetes, cardiovascular disease (CVDs), forms of cancer, and neurodegenerative disorders, bringing naturopathic remedies, or functional foods, to the forefront as an alternative therapy. Inflammation is induced by the aging and disease mechanisms; it reduces physical function and leads to excessive aging-related illnesses (Cai et al. 2019; Russo et al. 2020). The initial response of the defense system to disease or irritation caused by infectious pathogens, significant injury, hypoxia, or clinical manifestations is inflammatory. Redness, heat, discomfort, and inflammation are all symptoms of irritation. All inflammatory diseases exhibit features of inflammation at the cellular and molecular levels, such as oxidative stress, extracellular matrix remodeling, angiogenesis, and fibrosis. Anti-inflammatory nutraceuticals have potential for reducing the aging cycle and disease recurrence. Phytoconstituent nutraceuticals, in particular, are employed as an effective force in sustaining wellness and fighting acute and chronic illnesses, fostering optimal health, lifespan, and wellbeing. Natural foods with antioxidants or vitamins, prebiotics, vital nutrients, phytonutrients, functional food ingredients, and probiotics are examples of nutritional supplements with medical benefits (Panghal et al. 2018; Fernandez and Marette 2017; Raman et al. 2019).

4.2 ANTIOXIDANTS

Reactive oxygen species (ROS) are formed during normal cellular metabolism, particularly in the mitochondria. ROS are chemical forms of oxygen that are extremely reactive and capable of destroying DNA, proteins, and lipids. Superoxide, hydrogen peroxide, and the hydroxyl radical are a few examples. ROS are extremely efficient at killing biological pathogens because they damage proteins, lipids, and DNA. ROS, on the other hand, are not pathogen-specific, and excessive ROS formation can be harmful to host cells if there aren't enough antioxidants available. Antioxidants are substances that protect cells from the oxidative damage produced by ROS. Catalase, superoxide dismutase (SOD), and peroxidase are antioxidant enzymes that may scavenge ROS *in vivo*. The conversion of oxygen to superoxide is carried out by metabolic enzymes such as nicotinamide adenine dinucleotide phosphate (NADPH) oxidase and xanthine oxidase (Koju et al. 2019; Fragoso-Morales et al. 2021; Laddha and Kulkarni 2020).

Oxidative stress refers to the quantity of oxidative damage induced by ROS in a cell, tissue, or organ at any one time, and it increases with age or exposure to external stimuli such as cigarette smoke and radiation. Numerous diseases, such as cancer, Alzheimer's disease (AD), arthritis, atherosclerosis, and diabetes, have been linked to oxidative stress. Antioxidant enzymes, nonenzymatic antioxidants such as catechin, quercetin, and anthocyanidin, vitamins A, C, and E, carotenoids, polyphenols, and flavonoids such as catechin, quercetin, and anthocyanidin can all help to reduce oxidative stress (Lyu et al. 2019; Ezealigo 2017).

Antioxidant enzymes such as catalase, SOD, and glutathione peroxidase (GPx) are spontaneously produced by young individuals to defend against free radicals created during regular biological activity. As humans age, their levels of catalase, SOD, and other antioxidant enzymes decrease, resulting in an epidemic of age-related disorders. Catalase 14 394.9 22.8 IU/g Hb and SOD 14 5981

375 IU/g Hb values in young people (22 healthy boys, age 14 28 12 years) are catalase 14 394.9 22.8 IU/g Hb and SOD 14 5981 375 IU/g Hb values in young people (22 healthy boys, age 14 28 12 years) are catalase 14 394.9 22.8 IU/g Hb values in young people (22 healthy boys. Catalase 14 52.9 17.1 IU/g Hb and SOD 14 995 145 IU/g Hb levels diminish in the elderly (1780 healthy males aged 14 55e69 years). Most vegetables and fruits are high in antioxidants. GPx may be found in a range of fruits and vegetables, including spinach, asparagus, watermelon, tomato, broccoli, and avocado (Turcov et al., 2020; Hegazy et al. 2019).

Vitamin E (α-tocopherol) can convert free radicals to α-tocoquinone by neutralizing their unpaired electrons. Vitamin C may scavenge free radicals by contributing a single electron due to resonance delocalization, resulting in ascorbyl radicals, which are more stable than the original radical species. Citrus fruits, broccoli, green leafy vegetables, green peppers, raw cabbage, strawberries, and potatoes are high in vitamin C. Vitamin E is abundant in nuts, seeds, wheat germ, vegetable oil, whole grains, green leafy vegetables, and fish liver oil. Cranberries, beets, berries, kale, red and black grapes, lemons, oranges, grapefruits, and green tea are high in flavonoids (Premi and Khan 2019).

"Any meal or food item that delivers a healthiness benefit in adding to the standard nutrients it includes," is how functional foods are defined. A functional food is a natural or processed food that comprises physiologically active ingredients that can aid in the prevention, management, and mitigation of chronic illnesses. Soybeans' two primary components are lipids and proteins. Soybean oil is rich in vitamin E and includes a variety of important fatty acids. It is also low in saturated fat. Soybean has been revealed to lessen the risk of heart disease, cancer, and osteoporosis, as well as to improve menopausal symptoms. Oats offer various physiological advantages, including the decrease of hyperglycemia, hyperinsulinemia, and hypercholesterolemia. β-Glucan and saponin, two phytochemicals present in oats, are both helpful. The soluble fiber β-glucan decreases total cholesterol and low-density lipoprotein levels (LDL). As a result, there is less danger of coronary heart disease (Kent et al. 2017; Vallejo-Vaz et al. 2017; Peloso et al. 2019).

Flaxseed has a lot of omega-3 fatty acids, a lot of α-linolenic acid, and a lot of lignans, which are phenolic compounds. Because it decreases total and LDL cholesterol levels, flaxseed possesses cardioprotective as well as hypocholesterolemic effects. It is also valuable to the kidneys. In multiple studies, flaxseed eating has been associated with a number of hormonal changes that are connected to a decreased risk of breast cancer (Parikh et al. 2019).

Tomatoes and tomato-related goods include the following. Potassium and vitamins A, C, and E as well as folate are abundant in many tomato products. Tomatoes also include phytochemicals that are useful to the body, including carotenoids and polyphenols, such as β-carotene, a provitamin A molecule, and red-pigmented lycopene. Tomato phytochemicals assist in reducing the menace of cardiovascular disease and prostate cancer. Lycopene diminishes prostate, colon, breast, and lung cancers, as well as cardiovascular problems. Lycopene is an extremely prevalent antioxidant and the most efficient singlet oxygen quencher identified in biological systems. Spinach, collard greens, kale, broccoli, broccoli sprouts, broccoli rabe, and arugula are examples of leafy greens. Carotenoids, apigenin, sulforaphanes, and lutein/zeaxanthin are among the phytochemicals found in them (Ashok, Ravivarman, and Kayalvizhi 2020; Ademowo et al. 2019).

Carotenoids are antioxidants that prevent carcinogens from entering cells. Sulforaphanes and apigenin, which prevents cancer, are antioxidants and anti-inflammatory substances that protect the heart. Lutein lowers the risk of blindness in the elderly, and zeaxanthin boosts immunity (Bruins et al. 2019).

Garlic has two physiologically dynamic organosulfur components: allicin and allylic sulfides. Garlic has been demonstrated in clinical research to have a minor effect on reducing blood pressure. Garlic intake lowers the chance of developing some stomach and colon cancers (Shang et al. 2019).

Eicosatetraenoic acid (EPA) and docosahexaenoic acid (DHA) are polyunsaturated fatty acids obtained mostly from fish oil and are key components of fish. Omega-3 fatty acids may be helpful in the prevention of cardiovascular disease (Manson et al. 2020).

Red wine and grape juice, especially red wine, are high in polyphenols. Phenolic and flavonoid acids, polyphenols, alkaloids, tannins, and stilbenes are all flavonoids. Polyphenols appear to be playing a function in stroke prevention. Wine phytochemicals inhibit both LDL oxidation and platelet activation. Resveratrol, which may have been present in high quantities in grape skins as well as red wines and also in lesser percentages in red wine, mulberries, and peanuts, is becoming incredibly common. Resveratrol is an antibiotic that plants produce to aid in disease protection, and it has been utilized as an antioxidant to heal a range of illnesses, including inflammation and cardiovascular disease. Cranberry is frequently used to treat and protect urinary tract infections. Cranberry juice prevents *Escherichia coli* from adhering to uroepithelial cells. Cranberries are high in salicylic acid, which is found in aspirin. Salicylic acid has anticancer activities as well as the ability to stop blood clots and reduce edema (Mouly et al. 2017; Tohge and R Fernie 2017).

Green tea has a lot of antioxidants as well as anti-inflammatory, antibacterial, and anti-mutagenic qualities. Green tea polyphenols constitute 30% of the dry weight of green tea leaves. Catechins are by far the most frequent polyphenols found in tea. Green tea contains four catechins: epigallocatechin-3-gallate, epicatechin, epigallocatechin, and epicatechin-3-gallate. It was formerly used to treat common chronic disorders such as cancer. Tea has been revealed to lower the risk of heart disease (Durazzo et al. 2019).

4.3 DIETARY SUPPLEMENTS

Dietary supplements are concentrated versions of nutrients or other substances that are added to a normal diet to provide nutritional or physiological advantages. Amino acids are chemical compounds that include the functional groups amine (-NH_2) and carboxylic acid (-COOH) as well as an amino acid–specific side chain. Essential amino acids, nonessential amino acids, and qualified amino acids are the three types of amino acids. The body does not produce essential amino acids; thus, they must be obtained through the diet. The body creates nonessential amino acids from essential amino acids or from the regular breakdown of proteins. Unless you're unwell, stressed, or have a long-term medical problem, additional amino acids are typically unnecessary. Humans do not manufacture the amino acids phenylalanine, threonine, lysine, valine, tryptophan, leucine, methionine, isoleucine, and histidine. Amino acids are abundant in eggs, meats, soybeans, and quinoa. Inadequate amounts of essential amino acids influence the immune system, brain and stomach mucosal function, permeability, and renal function, among other organs and systems (Yang et al. 2018; González-González et al. 2019; Martin et al. 2018).

Humans and other animals must ingest essential fatty acids, since their bodies are unable to produce them. Linoleic acid and α-linolenic acid are two examples (both omega-3 fatty acids). Almonds, walnuts, soybeans, cold water fatty fish (tuna, salmon, cod, and blue fish), and flaxseed oil are high in essential fatty acids. Essential fatty acids help in blood coagulation and cholesterol breakdown, as well as blood pressure, immunological reactions, and liver function. They also help you look better, since a diet lacking in these fatty acids has been related to skin diseases such as dermatitis, dandruff, damaged nails, and brittle hair. A lack of these fatty acids causes a variety of symptoms and illnesses, such as liver and kidney failure, delayed development, depression, decreased immune function, and skin dryness (Chen et al. 2017; Mezzetti et al. 2019; Puzanov et al. 2017).

Taurine, also known as 2-aminoethanesulfonic acid, is a bile component that helps in bile acid conjugation, antioxidation, osmoregulation, membrane integrity, and calcium signaling modulation, among other things. The cysteine sulfinic acid pathway is utilized in animals to produce taurine in the pancreas. Taurine is a common ingredient in energy drinks, and the best sources of taurine in the diet are fish and beef. Taurine is required for cardiovascular function, skeletal muscle development and function, retinal function, and central nervous system (CNS) function. Chondroitin sulfate is a sulfated glycosaminoglycan with an alternating sugar chain (N-acetyl galactosamine and glucuronic acid). Chondroitin sulfate is a structural component of cartilage that aids in its resistance to compression. Chondroitin sulfate is found in bovine sources (cow), chicken, porcine sources (pig), and shark. Chondroitin sulfate has been shown to aid in the prevention of osteoarthritis, brittle bones (osteoporosis), heart disease, and excessive cholesterol levels (Salehi et al. 2020; Zhang et al. 2019; Sukhikh et al. 2020; Schlenk and Shi 2019). Although excessive use of dietary supplements may be harmful, some supplements, such as carnitine, creatine, and glucosamine, can be extremely useful.

4.4 PROBIOTICS

Probiotics, a term that derives from the Greek word for "life," are living microorganisms, such as bacteria and yeast, that aid in the health of your digestive system. Probiotic bacteria present in fermented foods can help boost the immune system. Several probiotic strains increase phagocytic and natural killer (NK) cell activity. Probiotics may aid the gut flora in a range of functions. The gastrointestinal (GI) microflora (microbiota) is a complex ecology that coexists peacefully with the host. The human body has around 10^{14} bacterial cells, which is 10 times the number of human cells. Due to the low intragastric pH, only a few bacterial species (0–103 colony-forming units (CFU/mL) may grow in the stomach (Hagner et al. 2017). Streptococci, Enterobacteriaceae, lactobacilli, staphylococci, and yeasts are the most common bacteria in the stomach. The gut microbiota is home to around 500 distinct bacterial species. In the ileum, bacteria multiply at a pace of 0 to 105–108 CFU/mL, whereas in the colon, bacteria multiply at a rate of 1010 to 1012 CFU/mL. Lactobacilli and bifidobacteria are important components of the microbiome. Bifidobacterium and Lactobacillus, both carbohydrate-fermenting bacteria, have no known hazardous species. Breast-fed newborns have a microbiome rich in bifidobacteria, which is considered to help with immune system priming and defense against disease (Thomas et al. 2017). Breast-fed children's microbiomes comprise 99% bifidobacteria, whereas formula-fed neonates' gut fauna is far more diverse. Bifidus factors are oligosaccharides and other components found in breast milk that increase the viability and activity of bifidobacteria. Lactic acid–producing bacteria, such as bifidobacteria and lactobacilli, may play a role in maintenance of colonization resistance through a number of processes. Short-chain fatty acids (propionate, acetate, and butyrate) produced during carbohydrate fermentation benefit the host's health as well (LeBlanc et al. 2017). Butyrate is a short-chain fatty acid produced by bacteria during dietary fiber fermentation. Butyrate can be utilized as a source of nutrition by intestinal epithelial cells (colonocytes). It promotes colonic cell proliferation and differentiation while also providing other health advantages. Butyrate inhibits the IFN-γ/STAT1 signaling pathway, which decreases inflammation and oxidative stress while also protecting against colorectal cancer (Liu et al. 2020). Probiotics are ubiquitous in yoghurt, but they may also be found in fermented cheese and fermented vegetables. This microbiota is fed by the first prebiotics, oligosaccharides found in breast milk. Prebiotics are nondigestible oligosaccharides used as a food source by probiotic microorganisms (Zartl et al. 2018). In general, oligosaccharides are sugar combinations with varying degrees of polymerization. Prebiotics include fructooligosaccharides, lactulose, inulin, galactooligosaccharides, and maltooligosaccharides. Asparagus, bananas, maple syrup, Jerusalem artichokes, oats, honey, red wine, and legumes are examples of foods and plants that contain prebiotics. Microbiological production, enzymatic synthesis, and polysaccharide enzymatic breakdown are three methods for producing prebiotic oligosaccharides (Panesar et al. 2018). Human breast milk oligosaccharides are the model prebiotics, promoting the growth of bifidobacteria and lactobacilli in the colons of exclusively breast-fed neonates. Synbiotics, which include both probiotics and prebiotics, are commonly used in food products to take advantage of their synergistic characteristics. Prebiotics like RaftilosesP95 improve viability at 4°C for 4 weeks when combined with *Bifidobacterium* spp., *Lactobacillus acidophilus*, *Lactobacillus casei*, and *Lactobacillus rhamnosus*. Probiotics, prebiotics, and synbiotics may be useful medicines for treating and preventing immunological disorders such as allergies due to their immunomodulatory effects (Dargahi et al. 2019).

4.5 DESIGN OF NUTRACEUTICAL DELIVERY VEHICLES

Hydrogels, nanoparticles, and microparticles are examples of nanotechnology-based delivery systems with significant promise in the food and pharmaceutical sciences because they provide tools and nanostructured materials that increase the performance of nutraceuticals and pharmaceutical medications (Afreen et al. 2018). Up to 60% of probiotic bacteria, for example, cannot survive in the stomach environment, and the stability of flavonoids and anthocyanidins is pH dependent. As a result, establishing a nutraceutical delivery strategy that protects probiotic bacteria or nutraceuticals from the hostile environment of the stomach is crucial (M. Singh et al. 2020). Figure 4.1 depicts the many types of nano-based delivery systems.

4.5.1 Polymers

Polymers can shield nutraceuticals from the severe environment of the gastrointestinal system while also allowing them to be integrated into polymeric particles. Nutraceuticals can be delivered via natural or synthetic polymers containing nano- or microparticles (Zia et al. 2018; Tabasum et al. 2018; Rostamabadi, Falsafi, and Jafari 2019).

4.5.2 Biodegradable Polymers

Biodegradable polymers, both natural and synthetic, have been intensively explored for medical applications such as drug or nutraceutical administration and tissue engineering due to their biocompatibility. Biodegradable polymers dissolve in the body by chemical or biological processes, with the vast majority of synthetic biodegradable polymers intended to break down through hydrolysis. Nutraceuticals are released when biodegradable polymers break down. Physiologically appropriate breakdown products must be used (Gao et al. 2017; Jamaledin et al. 2020; Perinelli et al. 2019; Zare et al. 2021).

4.5.3 Proteins

Proteins are biological macromolecules required by the human body to operate. They're constructed of one or more lengthy amino acid chains. Some proteins, in addition to supplying basic nutrition, are used as nutraceutical delivery vehicles by coating or encapsulating nutraceuticals. Plant proteins like soy glycinin and wheat gliadin, as well as animal proteins like gelatin, casein, albumin, collagen, and whey protein, have all been examined as potential delivery vehicles. Because of

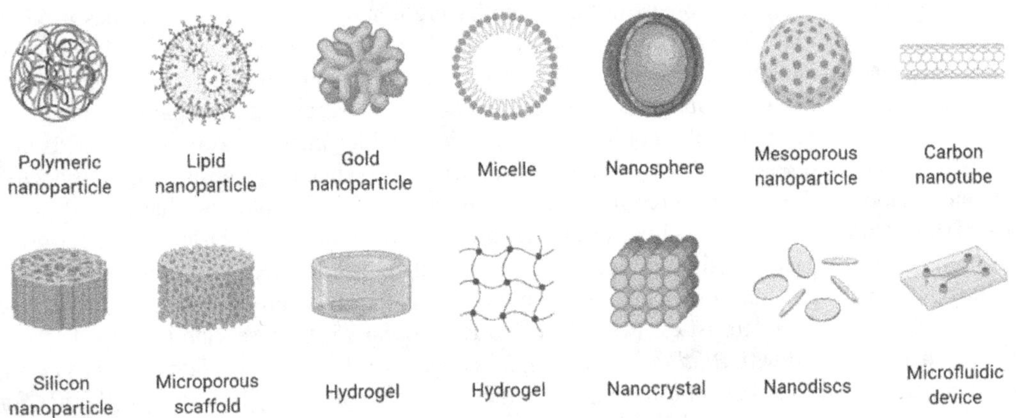

FIGURE 4.1 Types of nano-based drug delivery systems.

their biodegradability and non-immunogenicity, collagen and gelatin have been widely employed in pharmaceutical and nutraceutical delivery as well as tissue engineering. Collagen may be formed into particles, gels, sponges, and films to aid in the dispersion of antibiotics, nutraceuticals, and anti-inflammatory drugs. Because albumin is nontoxic, biocompatible, nonimmunogenic, and biodegradable, it looks to have therapeutic or nutraceutical potential. Previously, albumin nanoparticles were utilized to carry anticancer pharmaceuticals like noscapine and doxorubicin for breast cancer therapy, as well as anti-inflammatory medications like SB202190, which inhibited p38 mitogen-activated protein kinase and inflammatory cytokine production (Manthey et al. 1998; S.-W. Wang et al. 1999; Lappas et al. 2007; Yoshino et al. 2003).

Gliadin and zein, two prolamin proteins employed by plants for storage, were used to create oral delivery vehicles. Hordein from barley, gliadin from wheat, and zein from maize are all prolamin proteins. Catalase and superoxide dismutase (SOD) are antioxidant proteins with several medicinal uses. Antioxidant proteins, on the other hand, are rapidly damaged in the gastrointestinal (GI) tract due to the harsh environment, which includes a low pH and protein-degrading enzymes such as pepsin and trypsin. Antioxidant proteins were encapsulated in gliadin or zein, and prolamin proteins successfully protected antioxidant proteins from the hostile GI tract environment (Shishir et al. 2018; Martínez-López et al. 2020; Lim et al. 2019).

4.5.4 Polysaccharides

Polysaccharides are polymeric carbohydrate molecules composed of long chains of monosaccharide units, and their primary function in living organisms is to build or store energy. Polysaccharides may be found in both plants (e.g., inulin, fiber, pectin, and starch) and mammals (e.g., glycogen). Polysaccharides (such as glycogen, chitosan, and chondroitin sulfate) are broken down into smaller components by gut bacteria. Nutraceuticals are protected from the hostile environment of the GI tract by polysaccharide-based delivery methods. In the colon, they are hydrolyzed, and the delivery mechanism distributes nutraceuticals. Probiotics such as bifidobacteria and lactobacilli are commonly delivered using polysaccharide-based delivery methods such as synthesized biodegradable polymers (Kwiecień and Kwiecień 2018; Asgari et al. 2020; Liu et al. 2020; Tan et al. 2020).

The use of synthetic biodegradable polymers as nutrition carriers may boost delivery efficiency by delaying the release of encapsulated nutraceuticals or bioactive chemicals for days to weeks. Because of the use of organic solvents and harsher formulation conditions, natural polymers have a shorter drug release period. Synthetic polymers have an advantage over natural polymers in that they can regulate or sustain the release of bioactive chemicals over a period of days to weeks. On the other hand, synthetic polymers have the potential to cause toxicity and chronic inflammation. As a result, the toxicity and immunogenicity of synthetic polymers used as delivery vehicles must be assessed (Shiraishi and Yokoyama 2019; Hoang Thi et al. 2020; Naahidi et al. 2017; Laubach et al. 2021).

4.5.5 Polymers Containing Ester, Anhydride, and Amide Connections

Polyglycolic acid (PGA), poly (lactic-co-glycolic) acid (PLGA), polyanhydrides, polyorthoesters, polylactic acid (PLA), and polyamide are examples of synthetic biodegradable polymers. PLA, PGA, and PLGA are examples of commonly used ester-based polymers. These polymers are used in a range of biological applications, including as drug delivery carriers, resorbable sutures, and artificial-tissue materials. The polymer characteristics of PLGA, a copolymer of lactic acid and glycolic acid, are determined by the composition of the two monomers. PGA is a glycolic acid polymer that is crystalline, hard, and long-lasting (Emami et al. 2019; Arun et al. 2021; Tian and Bilal 2020; Vert 2017). PGA has good fiber-forming characteristics, but it is insoluble in most polymer solvents, limiting its value as a drug carrier because it cannot be produced into films, rods, or capsules. PLA is a lactic acid–based biodegradable thermoplastic polymer that may be hydrolyzed and destroyed.

To treat chronic inflammatory disorders, anti-inflammatory medicines such as dexamethasone are administered over time using PLGA (Gonzalez-Pizarro et al. 2018).

Biocompatible polyanhydrides can be produced into pharmaceutical or nutraceutical delivery discs, microspheres, coatings, and tubes. Polyanhydrides are used to carry non-steroidal anti-inflammatory drugs (NSAIDs), such as salicylic acid, and antibiotics, such as ampicillin, that are released when polymers degrade. At room temperature, polyorthoesters are anhydrous and stable. Polyorthoesters release NSAIDs as a result of surface erosion. One advantage of these polymers is that the time it takes for medicine to be released from polyorthoester nano-/micro-spheres might range from a few days to months. Despite its biodegradability, the use of polyamide has been restricted due to its immunogenicity and poor mechanical properties. Polymers that respond to stimuli are known as smart polymers (Moradali and Rehm 2020; Jacob et al. 2018; Oryan et al. 2018).

4.5.6 Temperature-Sensitive Polymers

Temperature-sensitive hydrogels are intelligent gels that undergo sol–gel transition at specific temperatures. Polymeric hydrogels are three-dimensional, loosely crosslinked polymeric networks that absorb a large amount of water by hydration with a large number of hydrophilic groups. Sol is a liquid that contains a stable suspension of colloidal solid particles (0.1–1 mm) or polymers. Poly(N-isopropylacrylamide) (PNIPAAm) is a temperature-sensitive hydrogel that is now in use. Because the temperature surpasses the lower critical solution temperature (LCST), when the phase of PNIPAAm shifts from a bloated hydrated form to a shrunken dehydrated form, it becomes soluble in water below the LCST. At 32 °C, the hydrogel loses around 90% of its capacity (LCST for PNIPAAm). In the context of inflammation, PNIPAAm is beneficial for temperature-controlled drug or nutraceutical delivery. Poly[2-(dimethylamino) ethylmethacrylate] (PDMAEMA), poly(N-vinylcaprolactam) (PVCL), and poly(N,N-diethylacrylamide) are examples of thermoresponsive polymers containing LCST (PDEAAm) (Brugnoni 2020; Merati et al. 2019). The polar opposite phase transition of thermoresponsive polymers with an upper critical solution temperature (UCST) hydrogels. UCST hydrogels dissolve and swell when exposed to water. Below the UCST, the hydrogel solution will be hazy. Polyacrylamide (PAAm), poly (acrylic acid) (PAA), and polyurethane (PU) hydrogels (acrylamide-co-butyl methacrylate) are examples of UCST hydrogels (Ghaeini-Hesaroeiye et al. 2020). Because of the nature of hydrogels, smart hydrogels respond not only to temperature but also to pH, enzymes, and electrical potential (Qu et al. 2018; Hoque et al. 2019).

4.5.7 pH-Sensitive Polymers

pH-sensitive polymers have properties that are sensitive to their environment's pH. To respond to variations in ambient pH, the most pH-sensitive polymers include either acidic (carboxylic/sulfonic acid) or basic (ammonium salts) functional groups. PAA and poly (sulfonic acid) polymers are examples of polyanions utilized in medication delivery. At neutral, alkaline, or high pH, acidic functional groups ionize, causing polymers to dissolve or swell more. Polycations, on the other hand, such as poly(N,N9-diethylaminoethyl methacrylate) (PDEAEM), dissolve or swell more at low pH. Because polycationic hydrogels grow faster in acidic environments, they have been utilized to transport drugs to the stomach. This type of hydrogel might be used to transport drugs to the stomach, such as amoxicillin and metronidazole, to treat *Helicobacter pylori* infection. Polyketals are acid-sensitive polymers with backbone ketal connections. Polyketals have been created to hydrolyze in the acidic environment of the phagosome following macrophage phagocytosis, hence enhancing intracellular delivery of phagocytosed medicinal medicines. Microparticles of poly(cyclohexane-1,4-diyl acetone dimethylene ketal) (PCADK) significantly improved the ability of SOD, a superoxide-scavenging enzyme, to eliminate ROS generated by macrophages. Tumor necrosis factor-alpha small interfering RNA (TNF-α siRNA) was successfully delivered to Kupffer cells *in vivo* by another polyketal, PK3, which reduced gene expression in the liver (Lenders et al. 2020).

4.5.8 Micelles

For a long time, polymer micelles have been employed to transport hydrophobic medicines and nutraceuticals. Polymeric micelles may increase the efficacy of therapeutic drugs used to treat CNS illnesses such as Alzheimer's, which are hampered by the blood–brain barrier (BBB). Because of the BBB's restrictive nature, just 5% of 7000 medicines tried for CNS diseases such as depression, schizophrenia, and sleeplessness were successful (Krogmann et al. 2019). Because of their inability to cross the BBB, almost all hydrophilic and large-molecular medications are rendered useless. Despite the fact that only minuscule molecules with exceptional lipid solubility and low molecular mass (Mr 400–500 Da) may pass across the BBB, medicines that are excessively hydrophobic are insoluble in physiological fluids and poorly absorbed across cell membranes. For hydrophobic medicines, micellar drug delivery systems clearly outperform free hydrophobic drugs. The hydrophilic head region's water-bound barrier prevents opsonin-like plasma proteins from detecting hydrophobic medications. Opsonins, on the other hand, bind hydrophobic medications and remove them from circulation in seconds to minutes through the reticuloendothelial system (RES) (Mozar and Chowdhury 2018). RES and nanosized micelles obstruct the elimination of biological waste during renal filtration (10–200 nm in diameter). The surface of micellar drug delivery devices is modified by specific ligands, which improves targeting while minimizing cell toxicity. Polyethylene glycol (PEG), for example, must be transformed to a micelle to increase the duration of a medicine's circulation. Folate ligands are also widely used to target cancer cells, which have double the amount of folate receptor overexpression as normal tissue. Incorporating folate ligands into micelles increases medication transport to cancer cells and may improve treatment outcomes (Senapati et al. 2018; Avramović et al. 2020; Wang et al. 2018; Dong et al. 2019).

Micelles have the potential to be effective drug delivery vehicles for hydrophobic treatments, but they must overcome two major obstacles: drug loading efficiency and serum stability. Because polymer micelles are dissolved by blood following systemic distribution, they have poor serum stability. Micelles become unstable when their concentration falls below a critical micelle concentration (Shao et al. 2019), defined as the smallest quantity of polymer required to generate micelles.

4.5.9 Liposomes

A liposome is a lipid bilayer made up of phosphatidylcholine-rich phospholipids. Lipids or lipid chains, such as cholesterol, long-chain fatty acids, sphingolipids, and phosphatidylethanolamine, make up liposomes. Liposomes are bilayers of watery spherical vesicles separated in the middle by a hydrophobic membrane. The mononuclear phagocyte system, which is naturally targeted by mononuclear phagocyte system (MPS) cells, particularly macrophages, and distributes medications to MPS cells effectively, cleans liposomes *in vivo*. It has been demonstrated that liposome pegylation improves blood circulation and RES clearance. As a result, pegylation improves liposomes' capacity to transport drugs. Liposomes can also be coupled with a targeting molecule, like an antibody or a ligand, and directed to certain cells or organs. Liposomes are used to deliver therapeutic medications to patients suffering from CNS illnesses such as brain tumors, ischemia, and infection. Liposome-encapsulated therapeutic medicines can be administered intravenously, intracerebrally, or intraventricularly. Liposomes are one of the most studied delivery technologies because of their low immunogenicity and biocompatibility; nonetheless, they have limitations, such as poor stability, quick RES clearance after intravenous injection, limited shelf life, and high production costs (Attia et al. 2019; Zhang et al. 2017; Muralidhara and Wong 2020; Xu et al. 2018). Probiotics are one example of nutraceutical delivery. A description of a nano-based delivery system is depicted in Figure 4.2.

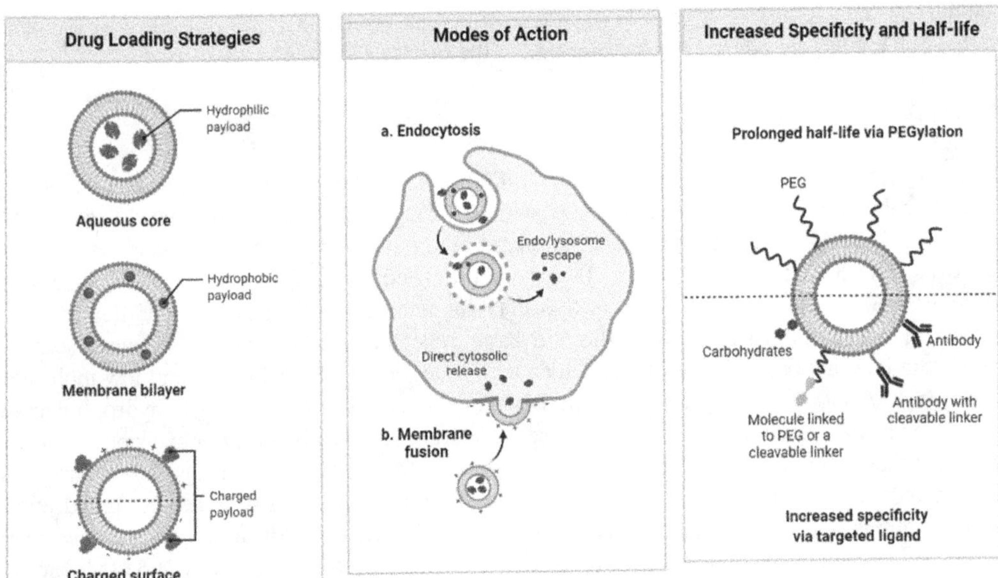

FIGURE 4.2 Liposome-based drug delivery.

4.6 PROBIOTICS AND NUTRACEUTICAL DELIVERY

By rebalancing the microbiota, probiotics enhance gut health. In contrast, probiotic bacteria are unable to survive in the stomach environment in up to 60% of instances. Due to the low intragastric pH, only a few bacterial species (0–103 CFU/mL) may survive. The most prevalent bacteria in the stomach include streptococci, lactobacilli, Enterobacteriaceae, staphylococci, and yeasts. There are around 500 different bacterial species in the gut microbiota. More probiotic bacteria may survive in the small intestine, and their numbers increase from 0–105 CFU/g in the duodenum to 108 CFU/g in the ileum and 1010–1012 CFU/g in the colon. As a result, it is vital to establish a nutraceutical delivery strategy that protects probiotic bacteria or nutraceuticals from the stomach's hostile environment. Microencapsulation of probiotic bacteria protects biological cells from potentially harmful circumstances. To entrap probiotic bacteria in the gel matrix, a variety of gel manufacturing procedures are employed. To encapsulate probiotic bacteria, extrusion, emulsion, or spray drying can all be utilized. Extrusion is the most common and oldest method of producing probiotics. Extrusion is used to make a cell suspension for alginate capsules by adding probiotic cells to a hydrocolloid solution. After passing through the syringe needle, the cell suspension is dripped into the hardening solution, which contains cations such as calcium. The alginate polymers in the cell suspension are crosslinked by cations in the hardening solution, resulting in the creation of an alginate capsule. The alginate capsule is extracted and dried using the proper process (Atencio et al. 2020; Martins et al. 2017; Patole and Pandit 2018; Németh et al. 2018; Belhouchat et al. 2017; Xu et al. 2018; Fu et al. 2019). Lactic acid bacteria may be encapsulated using emulsion. This process emulsifies a dispersed phase made up of a small quantity of cell and polymer slurry into a continuous phase made up of a large amount of vegetable oil, such as soy oil, maize oil, or sunflower oil, or light paraffin oil. Crosslinking processes such as ionic, enzymatic, and interfacial polymerization are used to make emulsion gels (Gadeyne et al. 2017).

Gum Arabic and starches tend to form spherical microparticles during the drying process. Probiotic cells are produced by spray drying a dissolved polymer matrix containing gum Arabic or starches. Although spray drying can produce microparticles, probiotic cells may be eliminated as a result of heat generation and physical damage to microparticles during the drying process. To limit probiotic cell loss, spray drying input and exit temperatures should be managed, and an appropriate cryoprotectant should be used during freeze drying. Probiotic bacteria are important because they are thought to aid in the functioning of the immune system. To effectively propagate probiotic bacteria, appropriate encapsulation procedures and encapsulating materials should be employed. The probiotic bacteria should be subjected to variations in pH, mechanical stress, temperature, enzymatic activity, duration, and osmotic force in the environment. To enhance the lifespan of probiotic bacteria, several factors, such as heat generation, should be addressed throughout the manufacturing process (Fenster et al. 2019; Terpou et al. 2019; Siwek et al. 2018).

4.7 CONCLUSIONS

Atherosclerosis, cancer, Alzheimer's disease, rheumatoid arthritis, asthma, and diabetes are examples of chronic inflammatory diseases induced by oxidative stress, aging, and high blood sugar levels. Although not all causes or treatments for inflammatory diseases are fully understood, reducing inflammation, which is frequently related to disease processes, is helpful. Both anti-inflammatory medications and nutraceuticals can help reduce inflammation. Nutraceuticals are foods that, in addition to basic nutrition, provide pharmacological or health advantages such as illness prevention and therapy. Natural foods that include antioxidants or vitamins, functional foods, vital minerals, prebiotics, phytochemicals, and probiotics are known as nutraceuticals. To guarantee effective distribution, nutraceuticals would be encapsulated in a safe, biocompatible, target-specific delivery device. Polymers offer unique properties as a delivery medium that no other material can match. Biodegradable and stimuli-responsive polymers, in particular, have demonstrated efficacy in managing medicine release rates. Despite the fact that amphiphilic carriers like micelles and liposomes have low serum stability, they have been widely used to carry antisense nucleotides, small molecules, proteins, and siRNA. Nanotechnology, including the development of biomaterials and the design of delivery mechanisms, has enormous promise for treating inflammatory disorders associated with aging and a wide spectrum of diseases.

REFERENCES

Abdullahi, Ali Arazeem. 2011. "Trends and Challenges of Traditional Medicine in Africa." *African Journal of Traditional, Complementary and Alternative Medicines* 8 (Supplement, 5S), 115–123.

Ademowo, O Stella, H K Irundika Dias, Chathyan Pararasa, and Helen R Griffiths. 2019. "Nutritional Hormesis in a Modern Environment." In *The Science of Hormesis in Health and Longevity*, 75–86. Elsevier.

Afreen, Shagufta, Rishabh Anand Omar, Neetu Talreja, Divya Chauhan, and Mohammad Ashfaq. 2018. "Carbon-Based Nanostructured Materials for Energy and Environmental Remediation Applications." In *Approaches in Bioremediation*, 369–92. Springer.

Arun, Yuvaraj, Radhakanta Ghosh, and Abraham J Domb. 2021. "Biodegradable Hydrophobic Injectable Polymers for Drug Delivery and Regenerative Medicine." *Advanced Functional Materials* 31 (44): 2010284. https://doi.org/10.1002/adfm.202010284.

Asgari, Shadi, Ali Pourjavadi, Tine Rask Licht, Anja Boisen, and Fatemeh Ajalloueian. 2020. "Polymeric Carriers for Enhanced Delivery of Probiotics." *Advanced Drug Delivery Reviews* 161–162: 1–21. https://doi.org/10.1016/j.addr.2020.07.014.

Ashok, A, J Ravivarman, and K Kayalvizhi. 2020. "Nutraceutical Value of Salad Vegetables to Combat COVID 19." *Journal of Pharmacognosy and Phytochemistry* 9 (3): 2144–48.

Atencio, Sharmaine, Alicia Maestro, Esther Santamaria, José María Gutiérrez, and Carmen Gonzalez. 2020. "Encapsulation of Ginger Oil in Alginate-Based Shell Materials." *Food Bioscience* 37: 100714.

Attia, Mohamed F, Nicolas Anton, Justine Wallyn, Ziad Omran, and Thierry F Vandamme. 2019. "An Overview of Active and Passive Targeting Strategies to Improve the Nanocarriers Efficiency to Tumour Sites." *Journal of Pharmacy and Pharmacology* 71 (8): 1185–98.

Avramović, Nataša, Boris Mandić, Ana Savić-Radojević, and Tatjana Simić. 2020. "Polymeric Nanocarriers of Drug Delivery Systems in Cancer Therapy." *Pharmaceutics* 12 (4): 298.

Belhouchat, N, H Zaghouane-Boudiaf, and César Viseras. 2017. "Removal of Anionic and Cationic Dyes from Aqueous Solution with Activated Organo-Bentonite/Sodium Alginate Encapsulated Beads." *Applied Clay Science* 135: 9–15.

Brugnoni, Monia. 2020. "Synthesis and Structure of Deuterated Ultra-Low Cross-Linked poly(N-isopropylacrylamide) Microgels." 207.

Bruins, Maaike J, Peter Van Dael, and Manfred Eggersdorfer. 2019. "The Role of Nutrients in Reducing the Risk for Noncommunicable Diseases during Aging." *Nutrients* 11 (1): 85.

Cai, Zhonglin, Jianzhong Zhang, and Hongjun Li. 2019. "Selenium, Aging and Aging-Related Diseases." *Aging Clinical and Experimental Research* 31 (8): 1035–47.

Carlsen, Monica H, Bente L Halvorsen, Kari Holte, Siv K Bøhn, Steinar Dragland, Laura Sampson, Carol Willey, Haruki Senoo, Yuko Umezono, and Chiho Sanada. 2010. "The Total Antioxidant Content of More than 3100 Foods, Beverages, Spices, Herbs and Supplements Used Worldwide." *Nutrition Journal* 9 (1): 1–11.

Chen, Kang, Wei-Dan Jiang, Pei Wu, Yang Liu, Sheng-Yao Kuang, Ling Tang, Wu-Neng Tang, Yong-An Zhang, Xiao-Qiu Zhou, and Lin Feng. 2017. "Effect of Dietary Phosphorus Deficiency on the Growth, Immune Function and Structural Integrity of Head Kidney, Spleen and Skin in Young Grass Carp (Ctenopharyngodon Idella)." *Fish & Shellfish Immunology* 63: 103–26.

Dargahi, Narges, Joshua Johnson, Osaana Donkor, Todor Vasiljevic, and Vasso Apostolopoulos. 2019. "Immunomodulatory Effects of Probiotics: Can They Be Used to Treat Allergies and Autoimmune Diseases?" *Maturitas* 119: 25–38.

Dewettinck, Koen, Filip Van Bockstaele, Bianka Kühne, Davy Van de Walle, T M Courtens, and Xavier Gellynck. 2008. "Nutritional Value of Bread: Influence of Processing, Food Interaction and Consumer Perception." *Journal of Cereal Science* 48 (2): 243–57.

Dong, Peng, K P Rakesh, H M Manukumar, Yasser Hussein Eissa Mohammed, C S Karthik, S Sumathi, P Mallu, and Hua-Li Qin. 2019. "Innovative Nano-Carriers in Anticancer Drug Delivery: A Comprehensive Review." *Bioorganic Chemistry* 85: 325–36.

Durazzo, Alessandra, Massimo Lucarini, Eliana B Souto, Carla Cicala, Elisabetta Caiazzo, Angelo A Izzo, Ettore Novellino, and Antonello Santini. 2019. "Polyphenols: A Concise Overview on the Chemistry, Occurrence, and Human Health." *Phytotherapy Research* 33 (9): 2221–43.

Emami, Fakhrossadat, Seyed Jamaleddin Mostafavi Yazdi, and Dong Hee Na. 2019. "Poly(Lactic Acid)/Poly(Lactic-Co-Glycolic Acid) Particulate Carriers for Pulmonary Drug Delivery." *Journal of Pharmaceutical Investigation* 49 (4): 427–42. https://doi.org/10.1007/s40005-019-00443-1.

Ezealigo, Uchechukwu Stella. 2017. "Antioxidant Activity, Total Phenolic and Flavonoid Contents of Whole Plant Methanol Extract and Solvent Fractions of *Desmodium ramosissimum* G. Don."

Fenster, Kurt, Barbara Freeburg, Chris Hollard, Connie Wong, Rune Rønhave Laursen, and Arthur C Ouwehand. 2019. "The Production and Delivery of Probiotics: A Review of a Practical Approach." *Microorganisms* 7 (3): 83.

Fernandez, Melissa Anne, and André Marette. 2017. "Potential Health Benefits of Combining Yogurt and Fruits Based on Their Probiotic and Prebiotic Properties." *Advances in Nutrition* 8 (1): 155S–164S.

Fragoso-Morales, Leticia Guadalupe, José Correa-Basurto, and Martha Cecilia Rosales-Hernández. 2021. "Implication of Nicotinamide Adenine Dinucleotide Phosphate (Nadph) Oxidase and Its Inhibitors in Alzheimer's Disease Murine Models." *Antioxidants* 10 (2): 218.

Fu, Yinghao, Congming Xiao, and Juan Liu. 2019. "Facile Fabrication of Quaternary Water Soluble Chitosan-Sodium Alginate Gel and Its Affinity Characteristic toward Multivalent Metal Ion." *Environmental Technology & Innovation* 13: 340–45.

Gadeyne, Frederik, Nympha De Neve, Bruno Vlaeminck, and Veerle Fievez. 2017. "State of the Art in Rumen Lipid Protection Technologies and Emerging Interfacial Protein Cross-linking Methods." *European Journal of Lipid Science and Technology* 119 (5): 1600345.

Gao, Chengde, Shuping Peng, Pei Feng, and Cijun Shuai. 2017. "Bone Biomaterials and Interactions with Stem Cells." *Bone Research* 5 (1): 17059. https://doi.org/10.1038/boneres.2017.59.

Ghaeini-Hesaroeiye, Sobhan, Hossein Razmi Bagtash, Soheil Boddohi, Ebrahim Vasheghani-Farahani, and Esmaiel Jabbari. 2020. "Thermoresponsive Nanogels Based on Different Polymeric Moieties for Biomedical Applications." *Gels* 6 (3): 20.

González-González, Marianela, Camilo Díaz-Zepeda, Johana Eyzaguirre-Velásquez, Camila González-Arancibia, Javier A Bravo, and Marcela Julio-Pieper. 2019. "Investigating Gut Permeability in Animal Models of Disease." *Frontiers in Physiology* 9: 1–10.

Gonzalez-Pizarro, Roberto, Marcelle Silva-Abreu, Ana Cristina Calpena, María Antonia Egea, Marta Espina, and María Luisa García. 2018. "Development of Fluorometholone-Loaded PLGA Nanoparticles for Treatment of Inflammatory Disorders of Anterior and Posterior Segments of the Eye." *International Journal of Pharmaceutics* 547 (1–2): 338–46.

Hagner, A, M Underhill, F Rossi, and J Biernaskie. 2017. "Normal and Cancer Stem Cells: Discovery, Diagnosis and Therapy International Scientific Conference." *Experimental Oncology* 39 (3): 234–56.

Hegazy, Amany M, Eman M El-Sayed, Khadiga S Ibrahim, and Amal S Abdel-Azeem. 2019. "Dietary Antioxidant for Disease Prevention Corroborated by the Nrf2 Pathway." *Journal of Complementary and Integrative Medicine* 16 (3): 20180161.

Hoang Thi, Thai Thanh, Emily H Pilkington, Dai Hai Nguyen, Jung Seok Lee, Ki Dong Park, and Nghia P Truong. 2020. "The Importance of Poly(Ethylene Glycol) Alternatives for Overcoming PEG Immunogenicity in Drug Delivery and Bioconjugation." *Polymers* 12 (2). https://doi.org/10.3390/polym12020298.

Hoque, Jiaul, Nivedita Sangaj, and Shyni Varghese. 2019. "Stimuli-Responsive Supramolecular Hydrogels and Their Applications in Regenerative Medicine." *Macromolecular Bioscience* 19 (1): 1800259.

Jacob, Jaicy, Namdev More, Kiran Kalia, and Govinda Kapusetti. 2018. "Piezoelectric Smart Biomaterials for Bone and Cartilage Tissue Engineering." *Inflammation and Regeneration* 38 (1): 2. https://doi.org/10.1186/s41232-018-0059-8.

Jamaledin, Rezvan, Pooyan Makvandi, Cynthia K Y Yiu, Tarun Agarwal, Raffaele Vecchione, Wujin Sun, Tapas Kumar Maiti, Franklin R Tay, and Paolo Antonio Netti. 2020. "Engineered Microneedle Patches for Controlled Release of Active Compounds: Recent Advances in Release Profile Tuning." *Advanced Therapeutics* 3 (12): 2000171. https://doi.org/10.1002/adtp.202000171.

Kent, Shia T, Robert S Rosenson, Christy L Avery, Yii-Der I Chen, Adolfo Correa, Steven R Cummings, L Adrienne Cupples, Mary Cushman, Daniel S Evans, and Vilmundur Gudnason. 2017. "PCSK9 Loss-of-Function Variants, Low-Density Lipoprotein Cholesterol, and Risk of Coronary Heart Disease and Stroke: Data from 9 Studies of Blacks and Whites." *Circulation: Cardiovascular Genetics* 10 (4): e001632.

Koju, Nirmala, Abdoh Taleb, Jifang Zhou, Ge Lv, Jie Yang, Xian Cao, Hui Lei, and Qilong Ding. 2019. "Pharmacological Strategies to Lower Crosstalk between Nicotinamide Adenine Dinucleotide Phosphate (NADPH) Oxidase and Mitochondria." *Biomedicine & Pharmacotherapy* 111: 1478–98.

Krogmann, Amanda, Luisa Peters, Laura Von Hardenberg, Katja Bödeker, Viktor B Nöhles, and Christoph U Correll. 2019. "Keeping up with the Therapeutic Advances in Schizophrenia: A Review of Novel and Emerging Pharmacological Entities." *CNS Spectrums* 24 (S1): 38–69.

Kwiecień, Iwona, and Michał Kwiecień. 2018. "Application of Polysaccharide-Based Hydrogels as Probiotic Delivery Systems." *Gels* 4 (2). https://doi.org/10.3390/gels4020047.

Laddha, Ankit P, and Yogesh A Kulkarni. 2020. "NADPH Oxidase: A Membrane-Bound Enzyme and Its Inhibitors in Diabetic Complications." *European Journal of Pharmacology* 881: 173206.

Lappas, M, M Permezel, and G E Rice. 2007. "Mitogen-Activated Protein Kinase Proteins Regulate LPS-Stimulated Release of Pro-Inflammatory Cytokines and Prostaglandins from Human Gestational Tissues." *Placenta* 28 (8): 936–45. https://doi.org/10.1016/j.placenta.2007.02.009.

Laubach, Joseph, Meerab Joseph, Timothy Brenza, Venkata Gadhamshetty, and Rajesh K Sani. 2021. "Exopolysaccharide and Biopolymer-Derived Films as Tools for Transdermal Drug Delivery." *Journal of Controlled Release* 329: 971–87. https://doi.org/10.1016/j.jconrel.2020.10.027.

LeBlanc, Jean Guy, Florian Chain, Rebeca Martín, Luis G Bermúdez-Humarán, Stéphanie Courau, and Philippe Langella. 2017. "Beneficial Effects on Host Energy Metabolism of Short-Chain Fatty Acids and Vitamins Produced by Commensal and Probiotic Bacteria." *Microbial Cell Factories* 16 (1): 1–10.

Lenders, Vincent, Xanthippi Koutsoumpou, Ara Sargsian, and Bella B Manshian. 2020. "Biomedical Nanomaterials for Immunological Applications: Ongoing Research and Clinical Trials." *Nanoscale Advances* 2 (11): 5046–89.

Lim, Loong-Tak, Ana C Mendes, and Ioannis S Chronakis. 2019. "Electrospinning and Electrospraying Technologies for Food Applications." In *Food Applications of Nanotechnology*, edited by Loong-Tak Lim, and Michael B T, Advances in Food and Nutrition Research Rogers, 88, 167–234. Academic Press. https://doi.org/10.1016/bs.afnr.2019.02.005.

Liu, Huan, Mingyong Xie, and Shaoping Nie. 2020. "Recent Trends and Applications of Polysaccharides for Microencapsulation of Probiotics." *Food Frontiers* 1 (1): 45–59. https://doi.org/10.1002/fft2.11.

Liu, Yunfeng, Pinghua Wei, Zhaozhi Qiu, Xingang Shen, Lin Gao, and Lihua Chen. 2020. "Study on Mechanism of Shufeng Jiedu Granules in Treating Novel Coronavirus Pneumonia Based on Network Pharmacology." In AIP Conference Proceedings, 2252, 20013. AIP Publishing LLC.

Lyu, Xiaomei, Jaslyn Lee, and Wei Ning Chen. 2019. "Potential Natural Food Preservatives and Their Sustainable Production in Yeast: Terpenoids and Polyphenols." *Journal of Agricultural and Food Chemistry* 67 (16): 4397–4417.

Manson, JoAnn E, Shari S Bassuk, Nancy R Cook, I-Min Lee, Samia Mora, Christine M Albert, Julie E Buring, and VITAL Research Group. 2020. "Vitamin D, Marine n-3 Fatty Acids, and Primary Prevention of Cardiovascular Disease Current Evidence." *Circulation Research* 126 (1): 112–28.

Manthey, Carl L, Shen-Wu Wang, Stephen D Kinney, and Zhengbin Yao. 1998. "SB202190, a Selective Inhibitor of P38 Mitogen-Activated Protein Kinase, Is a Powerful Regulator of LPS-Induced MRNAs in Monocytes." *Journal of Leukocyte Biology* 64 (3): 409–17. https://doi.org/10.1002/jlb.64.3.409.

Martin, C R, V Osadchiy, A Kalani, and E A Mayer. 2018. "The Brain-Gut-Microbiome Axis." In *Cellular and Molecular Gastroenterology and Hepatology* 6 (2): 133–148.

Martínez-López, Ana L, Cristina Pangua, Cristian Reboredo, Raquel Campión, Jorge Morales-Gracia, and Juan M Irache. 2020. "Protein-Based Nanoparticles for Drug Delivery Purposes." *International Journal of Pharmaceutics* 581: 119289. https://doi.org/10.1016/j.ijpharm.2020.119289.

Martins, Evandro, Denis Renard, Zenia Adiwijaya, Emre Karaoglan, and Denis Poncelet. 2017. "Oil Encapsulation in Core–Shell Alginate Capsules by Inverse Gelation. I: Dripping Methodology." *Journal of Microencapsulation* 34 (1): 82–90.

Merati, Ali Akbar, Nahid Hemmatinejad, Mina Shakeri, and Azadeh Bashari. 2019. "Preparation, Classification, and Applications of Smart Hydrogels." *Advanced Functional Textiles and Polymers: Fabrication, Processing and Applications* 1: 337–64, editor Shahid ul-Islam and B.S. Butola.

Mezzetti, M, A Minuti, F Piccioli-Cappelli, M Amadori, M Bionaz, and E Trevisi. 2019. "The Role of Altered Immune Function during the Dry Period in Promoting the Development of Subclinical Ketosis in Early Lactation." *Journal of Dairy Science* 102 (10): 9241–58.

Moradali, M Fata, and Bernd H A Rehm. 2020. "Bacterial Biopolymers: From Pathogenesis to Advanced Materials." *Nature Reviews Microbiology* 18 (4): 195–210. https://doi.org/10.1038/s41579-019-0313-3.

Mouly, Stéphane, Célia Lloret-Linares, Pierre-Olivier Sellier, Damien Sene, and J-F Bergmann. 2017. "Is the Clinical Relevance of Drug-Food and Drug-Herb Interactions Limited to Grapefruit Juice and Saint-John's Wort?" *Pharmacological Research* 118: 82–92.

Mozar, Fitya Syarifa, and Ezharul Hoque Chowdhury. 2018. "Pegylation of Carbonate Apatite Nanoparticles Prevents Opsonin Binding and Enhances Tumor Accumulation of Gemcitabine." *Journal of Pharmaceutical Sciences* 107 (9): 2497–2508.

Muralidhara, Bilikallahalli K, and Marcus Wong. 2020. "Critical Considerations in the Formulation Development of Parenteral Biologic Drugs." *Drug Discovery Today* 25 (3): 574–81.

Naahidi, Sheva, Mousa Jafari, Megan Logan, Yujie Wang, Yongfang Yuan, Hojae Bae, Brian Dixon, and P Chen. 2017. "Biocompatibility of Hydrogel-Based Scaffolds for Tissue Engineering Applications." *Biotechnology Advances* 35 (5): 530–44. https://doi.org/10.1016/j.biotechadv.2017.05.006.

Németh, Bence, Ágnes S Németh, Aurél Ujhidy, Judit Tóth, László Trif, János Gyenis, and Tivadar Feczkó. 2018. "Fully Bio-Originated Latent Heat Storing Calcium Alginate Microcapsules with High Coconut Oil Loading." *Solar Energy* 170: 314–22.

Onabanjo, O.O, C. O Akinyemi, and A.C Agbon. 2009. "Characteristics of Complementary Foods Produced from Sorghum, Sesame, Carrot and Crayfish." *Journal of Natural Sciences Engineering and Technology* 8 (1): 71–83. https://doi.org/10.51406/jnset.v8i1.977.

Oryan, Ahmad, Amir Kamali, Ali Moshiri, Hossien Baharvand, and Hamed Daemi. 2018. "Chemical Crosslinking of Biopolymeric Scaffolds: Current Knowledge and Future Directions of Crosslinked Engineered Bone Scaffolds." *International Journal of Biological Macromolecules* 107: 678–88. https://doi.org/10.1016/j.ijbiomac.2017.08.184.

Panesar, Parmjit S, Rupinder Kaur, Ram S Singh, and John F Kennedy. 2018. "Biocatalytic Strategies in the Production of Galacto-Oligosaccharides and Its Global Status." *International Journal of Biological Macromolecules* 111: 667–79.

Panghal, Anil, Sandeep Janghu, Kiran Virkar, Yogesh Gat, Vikas Kumar, and Navnidhi Chhikara. 2018. "Potential Non-Dairy Probiotic Products–A Healthy Approach." *Food Bioscience* 21: 80–89.

Parikh, Mihir, Thane G Maddaford, J Alejandro Austria, Michel Aliani, Thomas Netticadan, and Grant N Pierce. 2019. "Dietary Flaxseed as a Strategy for Improving Human Health." *Nutrients* 11 (5): 1171.

Patole, Vinita C, and Ashlesha P Pandit. 2018. "Mesalamine-Loaded Alginate Microspheres Filled in Enteric Coated HPMC Capsules for Local Treatment of Ulcerative Colitis: In Vitro and in Vivo Characterization." *Journal of Pharmaceutical Investigation* 48 (3): 257–67.

Peloso, Gina M, Akihiro Nomura, Amit V Khera, Mark Chaffin, Hong-Hee Won, Diego Ardissino, John Danesh, Heribert Schunkert, James G Wilson, and Nilesh Samani. 2019. "Rare Protein-Truncating Variants in APOB, Lower Low-Density Lipoprotein Cholesterol, and Protection against Coronary Heart Disease." *Circulation: Genomic and Precision Medicine* 12 (5): e002376.

Perinelli, Diego Romano, Marco Cespi, and Giulia Bonacucina. 2019. "Nanostructures of Chemical Biodegradable Polymers and Their Derivatives for Encapsulation of Food Ingredients." In *Biopolymer Nanostructures for Food Encapsulation Purposes*, edited by Seid Mahdi Jafari, 581–606. Nanoencapsulation in the Food Industry. Academic Press. https://doi.org/10.1016/B978-0-12-815663-6.00019-7

Polito, Letizia, Massimo Bortolotti, Stefania Maiello, Maria Giulia Battelli, and Andrea Bolognesi. 2016. "Plants Producing Ribosome-Inactivating Proteins in Traditional Medicine." *Molecules* 21 (11): 1560.

Premi, Monica, and Khursheed Alam Khan. 2019. "Antioxidants in Fruits and Vegetables: Role in the Prevention of Degenerative Diseases." In *Processing of Fruits and Vegetables: From Farm to Fork*, 3–22. CRC Press.

Puzanov, I, A Diab, K Abdallah, C O 3rd Bingham, C Brogdon, R Dadu, L Hamad, S Kim, M E Lacouture, and N R LeBoeuf. 2017. "Managing Toxicities Associated with Immune Checkpoint Inhibitors: Consensus Recommendations from the Society for Immunotherapy of Cancer (SITC) Toxicity Management Working Group." *Journal for Immunotherapy of Cancer* 5 (1): 1–28.

Qu, Jin, Xin Zhao, Peter X Ma, and Baolin Guo. 2018. "Injectable Antibacterial Conductive Hydrogels with Dual Response to an Electric Field and pH for Localized "Smart" Drug Release." *Acta Biomaterialia* 72: 55–69.

Raman, Maya, Padma Ambalam, and Mukesh Doble. 2019. "Probiotics, Prebiotics, and Fibers in Nutritive and Functional Beverages." In *Nutrients in Beverages*, 315–67. Elsevier.

Rostamabadi, Hadis, Seid Reza Falsafi, and Seid Mahdi Jafari. 2019. "Starch-Based Nanocarriers as Cutting-Edge Natural Cargos for Nutraceutical Delivery." *Trends in Food Science & Technology* 88: 397–415. https://doi.org/10.1016/j.tifs.2019.04.004.

Russo, Gian Luigi, Carmela Spagnuolo, Maria Russo, Idolo Tedesco, Stefania Moccia, and Carmen Cervellera. 2020. "Mechanisms of Aging and Potential Role of Selected Polyphenols in Extending Healthspan." *Biochemical Pharmacology* 173: 113719.

Salehi, Bahare, Antonio Rescigno, Tinuccia Dettori, Daniela Calina, Anca Oana Docea, Laxman Singh, Fatma Cebeci, Beraat Özçelik, Mohammed Bhia, and Amirreza Dowlati Beirami. 2020. "Avocado–Soybean Unsaponifiables: A Panoply of Potentialities to Be Exploited." *Biomolecules* 10 (1): 130.

Schlenk, Elizabeth A, and Xiaojun Shi. 2019. "Evidence-Based Practices for Osteoarthritis Management: Self-Management Education Is the Foundation of Osteoarthritis Care." *American Nurse Today* 14 (5): 22–29.

Senapati, Sudipta, Arun Kumar Mahanta, Sunil Kumar, and Pralay Maiti. 2018. "Controlled Drug Delivery Vehicles for Cancer Treatment and Their Performance." *Signal Transduction and Targeted Therapy* 3 (1): 1–19.

Shang, Ao, Shi-Yu Cao, Xiao-Yu Xu, Ren-You Gan, Guo-Yi Tang, Harold Corke, Vuyo Mavumengwana, and Hua-Bin Li. 2019. "Bioactive Compounds and Biological Functions of Garlic (Allium Sativum L.)." *Foods* 8 (7): 246.

Shao, Ping, Yong Liu, Christos Ritzoulis, and Ben Niu. 2019. "Preparation of Zein Nanofibers with Cinnamaldehyde Encapsulated in Surfactants at Critical Micelle Concentration for Active Food Packaging." *Food Packaging and Shelf Life* 22: 100385.

Sharoba, Ashraf M. 2014. "Nutritional Value of Spirulina and Its Use in the Preparation of Some Complementary Baby Food Formulas." *Journal of Food and Dairy Sciences* 5 (8): 517–38.

Shiraishi, Kouichi, and Masayuki Yokoyama. 2019. "Toxicity and Immunogenicity Concerns Related to PEGylated-Micelle Carrier Systems: A Review." *Science and Technology of Advanced Materials* 20 (1): 324–36. https://doi.org/10.1080/14686996.2019.1590126.

Shishir, Mohammad Rezaul Islam, Lianghua Xie, Chongde Sun, Xiaodong Zheng, and Wei Chen. 2018. "Advances in Micro and Nano-Encapsulation of Bioactive Compounds Using Biopolymer and Lipid-Based Transporters." *Trends in Food Science & Technology* 78: 34–60. https://doi.org/10.1016/j.tifs.2018.05.018.

Singh, Jagtar, and Shweta Sinha. 2012. "Classification, Regulatory Acts and Applications of Nutraceuticals for Health." *International Journal of Pharma and Bio Sciences* 2 (1): 177–87.

Singh, Mahendra, Navneeta Singh, Balakumar Chandrasekaran, and Pran Kishore Deb. 2020. "Nanomaterials in Nutraceuticals Applications." In *Integrative Nanomedicine for New Therapies*, edited by Anand Krishnan and Anil Chuturgoon, 405–35. Springer.

Siwek, M, A Slawinska, K Stadnicka, J Bogucka, A Dunislawska, and M Bednarczyk. 2018. "Prebiotics and Synbiotics–in Ovo Delivery for Improved Lifespan Condition in Chicken." *BMC Veterinary Research* 14 (1): 1–17.

Sukhikh, Stanislav, Olga Babich, Alexander Prosekov, Nikolai Patyukov, and Svetlana Ivanova. 2020. "Future of Chondroprotectors in the Treatment of Degenerative Processes of Connective Tissue." *Pharmaceuticals* 13 (9): 220.

Tabasum, Shazia, Aqdas Noreen, Muhammad Farzam Maqsood, Hijab Umar, Nadia Akram, Shahzad Ali Shahid Chatha, and Khalid Mahmood Zia. 2018. "A Review on Versatile Applications of Blends and Composites of Pullulan with Natural and Synthetic Polymers." *International Journal of Biological Macromolecules* 120: 603–32.

Tan, Kei-Xian, Vidya N Chamundeswari, and Say Chye Joachim Loo. 2020. "Prospects of Kefiran as a Food-Derived Biopolymer for Agri-Food and Biomedical Applications." *RSC Advances* 10 (42): 25339–51. https://doi.org/10.1039/D0RA02810J.

Terpou, Antonia, Aikaterini Papadaki, Iliada K Lappa, Vasiliki Kachrimanidou, Loulouda A Bosnea, and Nikolaos Kopsahelis. 2019. "Probiotics in Food Systems: Significance and Emerging Strategies towards Improved Viability and Delivery of Enhanced Beneficial Value." *Nutrients* 11 (7): 1591.

Thomas, Sunil, Jacques Izard, Emily Walsh, Kristen Batich, Pakawat Chongsathidkiet, Gerard Clarke, David A Sela, Alexander J Muller, James M Mullin, and Korin Albert. 2017. "The Host Microbiome Regulates and Maintains Human Health: A Primer and Perspective for Non-Microbiologists." *Cancer Research* 77 (8): 1783–1812.

Tian, Kangming, and Muhammad Bilal. 2020. "Research Progress of Biodegradable Materials in Reducing Environmental Pollution." In *Abatement of Environmental Pollutants Borthakur*, edited by Pardeep Singh, Ajay Kumar, and Anwesha B T, 313–30. Elsevier. https://doi.org/10.1016/B978-0-12-818095-2.00015-1.

Tohge, Takayuki, and Alisdair R Fernie. 2017. "An Overview of Compounds Derived from the Shikimate and Phenylpropanoid Pathways and Their Medicinal Importance." *Mini Reviews in Medicinal Chemistry* 17 (12): 1013–27.

Turcov, Delia, Lăcrămioara Rusu, Anca Zbranca, and D Suteu. 2020 "New Dermatocosmetic Formulations Using Bioactive Compounds from Indigenous Natural Sources." *Buletinul Institutului Politehnic Din Iaşi* 66 (70): 2.

Vallejo-Vaz, Antonio J, Michele Robertson, Alberico L Catapano, Gerald F Watts, John J Kastelein, Chris J Packard, Ian Ford, and Kausik K Ray. 2017. "Low-Density Lipoprotein Cholesterol Lowering for the Primary Prevention of Cardiovascular Disease among Men with Primary Elevations of Low-Density Lipoprotein Cholesterol Levels of 190 Mg/DL or above: Analyses from the WOSCOPS (West of Scotland Coronary Prevention Study) 5-Year Randomized Trial and 20-Year Observational Follow-Up." *Circulation* 136 (20): 1878–91.

Vert, Michel. 2017. "Degradable, Biodegradable, and Bioresorbable Polymers for Time-Limited Therapy." In *Bioresorbable Scaffolds: From Basic Concept to Clinical Applications*, 6. CRC Press.

Wang, S.-W., J Pawlowski, S T Wathen, S D Kinney, H S Lichenstein, and C L Manthey. 1999. "Cytokine MRNA Decay Is Accelerated by an Inhibitor of P38-Mitogen-Activated Protein Kinase." *Inflammation Research* 48 (10): 533–38. https://doi.org/10.1007/s000110050499.

Wang, Zhe, Xiangping Deng, Jinsong Ding, Wenhu Zhou, Xing Zheng, and Guotao Tang. 2018. "Mechanisms of Drug Release in pH-Sensitive Micelles for Tumour Targeted Drug Delivery System: A Review." *International Journal of Pharmaceutics* 535 (1–2): 253–60.

Xu, Congcong, Farzin Haque, Daniel L Jasinski, Daniel W Binzel, Dan Shu, and Peixuan Guo. 2018. "Favorable Biodistribution, Specific Targeting and Conditional Endosomal Escape of RNA Nanoparticles in Cancer Therapy." *Cancer Letters* 414: 57–70.

Xu, S, A Tabaković, X Liu, and E Schlangen. 2018. "Calcium Alginate Capsules Encapsulating Rejuvenator as Healing System for Asphalt Mastic." *Construction and Building Materials* 169: 379–87.

Yang, Tao, Elaine M Richards, Carl J Pepine, and Mohan K Raizada. 2018. "The Gut Microbiota and the Brain–Gut–Kidney Axis in Hypertension and Chronic Kidney Disease." *Nature Reviews Nephrology* 14 (7): 442–56.

Yoshino, Osamu, Yutaka Osuga, Yasushi Hirota, Kaori Koga, Tetsuya Hirata, Tetsu Yano, Takuya Ayabe, Osamu Tsutsumi, and Yuji Taketani. 2003. "Endometrial Stromal Cells Undergoing Decidualization Down-Regulate Their Properties to Produce Proinflammatory Cytokines in Response to Interleukin-1β via Reduced P38 Mitogen-Activated Protein Kinase Phosphorylation." *Journal of Clinical Endocrinology & Metabolism* 88 (5): 2236–41. https://doi.org/10.1210/jc.2002-021788.

Zare, Mina, Karolina Dziemidowicz, Gareth R Williams, and Seeram Ramakrishna. 2021. "Encapsulation of Pharmaceutical and Nutraceutical Active Ingredients Using Electrospinning Processes." *Nanomaterials* 11 (8). https://doi.org/10.3390/nano11081968.

Zartl, Barbara, Karina Silberbauer, Renate Loeppert, Helmut Viernstein, Werner Praznik, and Monika Mueller. 2018. "Fermentation of Non-Digestible Raffinose Family Oligosaccharides and Galactomannans by Probiotics." *Food & Function* 9 (3): 1638–46.

Zhang, Fan, J Trent Magruder, Yi-An Lin, Todd C Crawford, Joshua C Grimm, Christopher M Sciortino, Mary Ann Wilson, Mary E Blue, Sujatha Kannan, and Michael V Johnston. 2017. "Generation-6 Hydroxyl PAMAM Dendrimers Improve CNS Penetration from Intravenous Administration in a Large Animal Brain Injury Model." *Journal of Controlled Release* 249: 173–82.

Zhang, Wei, William Brett Robertson, Jinmin Zhao, Weiwei Chen, and Jiake Xu. 2019. "Emerging Trend in the Pharmacotherapy of Osteoarthritis." *Frontiers in Endocrinology* 10: 431.

Zia, Khalid Mahmood, Shazia Tabasum, Muhammad Faris Khan, Nadia Akram, Naheed Akhter, Aqdas Noreen, and Mohammad Zuber. 2018. "Recent Trends on Gellan Gum Blends with Natural and Synthetic Polymers: A Review." *International Journal of Biological Macromolecules* 109: 1068–87.

5 Nanostructured Lipids as a Bioactive Compound Carrier

Peeyush Kaushik, Deepali Tomar, Anjoo Kamboj, Hitesh Malhotra, and Rupesh K. Gautam[*]

CONTENTS

5.1 Introduction 98
5.2 Structural Model of NLCs 100
5.3 Types of NLCs 100
5.3.1 Type I NLC (Imperfect Crystal Model) 100
5.3.2 NLC Type II (Multiple Types) 100
5.3.3 NLC Type III (Amorphous Model) 101
5.4 Advantages 101
5.5 Disadvantages 101
5.6 Components 103
5.6.1 Lipids 103
5.6.1.1 Solid Lipids 104
5.6.1.2 Liquid Lipids 104
5.6.2 Emulsifying Agents – Surfactants 104
5.7 Methods of Preparation 105
5.7.1 High-Pressure Homogenization 106
5.7.1.1 Advantages of High-Pressure Homogenization Method 106
5.7.2 Solvent Injection Method 106
5.7.2.1 Advantages 106
5.7.2.2 Disadvantages 106
5.7.3 Phase Inversion 106
5.7.3.1 Advantages 107
5.7.3.2 Disadvantages 107
5.7.4 Solvent Emulsification-Evaporation Technique 108
5.7.4.1 Advantages 108
5.7.4.2 Disadvantages 109
5.7.5 Solvent Emulsification-Diffusion Technique 109
5.7.6 Melting Dispersion Method 109
5.7.7 High-Shear Homogenization or Ultrasonication Technique 109
5.7.8 Double Emulsion Technique 110
5.8 Stability 111
5.8.1 Strategies Engaged for Overcoming the Problems Associated with the Stability of NLCs 111
5.8.1.1 Spray Drying 111
5.8.1.2 Lyophilization 111
5.8.2 Stabilizing Agents 111

[*] Corresponding author: drrupeshgautam@gmail.com

DOI: 10.1201/9781003244721-5

5.8.2.1 Poloxamers ... 111
5.8.2.2 Polyethylene Glycol ... 112
5.9 Characterization ... 113
5.9.1 Particle Size Analysis/PCS (Photon Correction Spectroscopy) ... 113
5.9.2 Zeta Potential Measurement/Laser Doppler Electrophoresis Technique ... 113
5.9.3 Transmission Electron Microscopy (TEM) ... 113
5.9.4 Scanning Electron Microscopy (SEM) ... 113
5.9.5 Atomic Force Microscopy (AFM) ... 113
5.9.6 Confocal Laser Scanning Microscopy (CLSM) ... 114
5.9.7 Differential Scanning Calorimetry (DSC) ... 114
5.9.8 Wide-Angle X-ray Diffraction (XRD) ... 114
5.9.9 Rheological Study ... 114
5.9.10 Drug Entrapment Efficiency ... 114
5.9.11 Ultrafiltration ... 115
5.9.12 High-Performance Liquid Chromatographic (HPLC) Analysis ... 115
5.9.13 pH Analysis ... 115
5.9.14 Nuclear Magnetic Resonance (NMR) ... 115
5.10 Drug Encapsulation ... 116
5.11 Drug Release ... 116
5.11.1 Factors Affecting Drug Release ... 117
5.11.1.1 Particle Size ... 117
5.11.1.2 Lipid Matrix ... 117
5.11.1.3 Surfactant ... 117
5.11.1.4 Drug Loading ... 117
5.11.1.5 Drug Type ... 117
5.12 Applications ... 117
5.13 Conclusion ... 117
References ... 118

5.1 INTRODUCTION

According to the National Nanotechnology Initiative, nanotechnology is typically defined as "the overview, use and manage of compounds on nanoscale (1 and a hundred nm) where one-of-a-kind phenomena empower novel programs. Likewise, Nanoparticle (NPs) is a discrete substance that has at least one quantity of the order of 100 nm or a great deal much low". The physicochemical properties such as appearance, solubility, material electricity, consistency, diffusivity and toxicity as well as natural occurrence of structural frameworks at the nanoscale differ notably from the microscale and are enhanced due to the interactions of individual particles and atoms, which provides new and realistic applications in the drug delivery field and improvements in medicine (Tamjidi et al. 2013). The development of different types of nanoparticles has improved the delivery of many drugs and given alternative inventive answers to overcome many of the difficulties related to their safety and production.

A delivery system is a method to control the release rate of a bioactive compound by encapsulating it in a carrier. Nanoparticulate carriers have demonstrated exceptional potential in medication delivery in recent years due to their nanoscale measurements and distinct characteristics. They have thermodynamic and kinetic stability and are quasi-monodispersing. These systems have a high surface area and may be able to boost solubility, resulting in improved bioavailability of bioactive compounds. Nanoparticulate delivery frameworks have led to controlled and site-specific release as well as different benefits, such as avoiding change to the active ingredient due to moisture, physiological pH and catalysts, improved bioavailability, increased potency, sustained release of drug, enhanced intracellular penetration and site-specific delivery by modifications on the carrier's surface (see Table 5.5).

Likewise, they play vital roles as carriers for an assortment of particles together with peptides and proteins, RNA, markers, antibodies, and so on (Saha et al. 2010; Tagami and Ozeki 2017). A range of nanocarriers, for example, nanotubes, nanocrystals (Müller et al. 2006; Junghanns and Müller 2008), nanowires (Bianco et al. 2005; Karimi et al. 2015; Peng et al. 2014), polymeric nanoparticles (Cheng et al. 2015; Masood 2016), liposomes (Narayan et al. 2016; Allen and Cullis 2013), dendrimers (Majoros et al. 2008), hydrogels (Hamidi et al. 2008; Hoffman 2012), and lipid nanoparticles (Battaglia and Gallarate 2012; Schwarz 1999), have been designed for drug delivery system and as analytical tools.

Lipoidal drug delivery systems have rapidly increased in popularity in comparison to other polymeric and inorganic nanoparticulate drug delivery structures, as because of their lipophilic nature, they can permeate confounding physiological obstacles, specifically the blood–brain barrier (BBB), resulting in higher biocompatibility. Furthermore, they are simple to prepare. These delivery methods are becoming increasingly appealing because of their cost-effectiveness and ease of large-scale manufacturing (Gaba et al. 2015; Tapeinos et al. 2017). Lipid carriers are classified into numerous kinds depending on their manufacturing technique and physicochemical properties. Liposomes, nanosomes, solid lipid nanoparticles (SLNs), and nanostructured lipid carriers (NLCs) are a few examples of nanosized lipid carriers.

NLCs are novel medicinal preparations that combine physiologically compatible lipids with surfactants/co-surfactants. NLCs are second-generation lipid nanoparticles that have been created to solve the limitations of the previous generation (SLNs).

Incorporating liquid lipids (oil) into solid lipids causes structural defects, resulting in a distorted arrangement of atoms in a crystal lattice that ensures greater entrapment of drug (Naseri et al. 2015; López-García and Ganem-Rondero 2015). In recent years, researchers have been interested in NLCs as a possible replacement for polymeric nanoparticles, micro-emulsions, SLNs, liposomes, and other nanoparticles (Figure 5.1) (Jain et al. 2017). These nanocarriers are capable of transporting both hydrophilic and lipophilic medications. NLCs have emerged as a possible delivery system for medications administered by oral, parenteral, ophthalmic, pulmonary, topical, and transdermal routes. Neurological therapy, chemotherapy, gene targeting, the food industry, and cosmeceutical and nutraceutical shipping have all recently employed NLCs (Jaiswal et al. 2016).

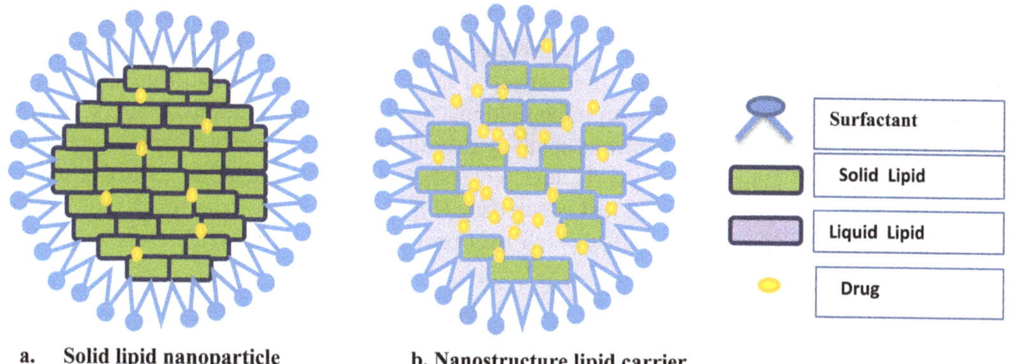

FIGURE 5.1 Structural model of lipid nanoparticles. (a) Solid lipid nanoparticles, depicts compact arrangement of solid lipids and less drug entrapment. (b) Nanostructure lipid carrier, depicts disorganized arrangement of solid and liquid lipids that results in high drug loading.

5.2 STRUCTURAL MODEL OF NLCs

NLCs are a type of lipidic nanoparticle, developed to cope with the drawbacks of the first era, i.e., SLNs. NLCs were created as an unstructured stable lipid matrix using numerous combinations of stable and liquid lipids (70:30–99.90 w/v) in aqueous solution with a surfactant/s (0.5–5% w/v). NLCs are made with biodegradable and appropriate lipids (solid and liquid), together with emulsifiers.

In order to produce an amorphous lattice with structural flaws in solid lipids, liquid lipids (oil) must be included. This results in a less ordered crystalline association, which will limit the leakage of drug from the crystal lattice, supply a high drug load, and keep the medication stable within the body.

Solid lipids include tri-stearin, waxes (e.g., carnauba wax), fatty acids (stearic acid), and steroids (e.g., low-density lipoprotein [LDL], cholesterol), whereas liquid lipids consist of oils derived from natural sources (e.g., olive oil, oleic acids and Miglyol 812). The solubility of such materials improves while liquid lipid tiers upward thrust, taking into account extra encapsulation (Gomaa et al. 2022).

5.3 TYPES OF NLCs

NLCs are categorized into three classes depending on the form of lipid content used:

- Imperfect
- Amorphous
- Multiple

5.3.1 Type I NLC (Imperfect Crystal Model)

Imperfect crystals, also referred to as type 1 NLCs, are formed from quite a disordered matrix with several gaps and voids, permitting extra drug molecules to fit into amorphous clusters (Haider et al. 2020). The solid lipids must be mixed in a sufficient quantity of liquefied lipids to create these crystal order defects (oils). The NLC matrix is not able to develop a regular form due to the chain lengths of the fatty acids and additionally, the mixture of mono-, di-, and triacylglycerols. The capacity of the drug payload can be multiplied by blending various lipids geomatrically, although this approach has low entrapment effectiveness. A disarranged lipid structure formed between the crystal and the liquid lipid permits medicines to penetrate more deeply into the lipid layer (Figure 5.2).

5.3.2 NLC Type II (Multiple Types)

For lipophilic medications, liquid lipids are preferred solvents than solid lipids. Various types of NLCs with a higher content of liquid lipid material have been created due to this concept. Oil moieties are efficaciously dispersed within the lipid matrix at low concentrations. Phase separation happens when oil is present in excess of its solubility, resulting in small micro-compartments of oil surrounded by a solid matrix. The type II device has several benefits, including effective entrapment, controlled release, and minimal leakage of drug (Figure 5.3). The form is much like W/O/W micro-emulsions (Van Tran et al. 2019), which may provide delayed drug release and high drug loading while avoiding decomposition.

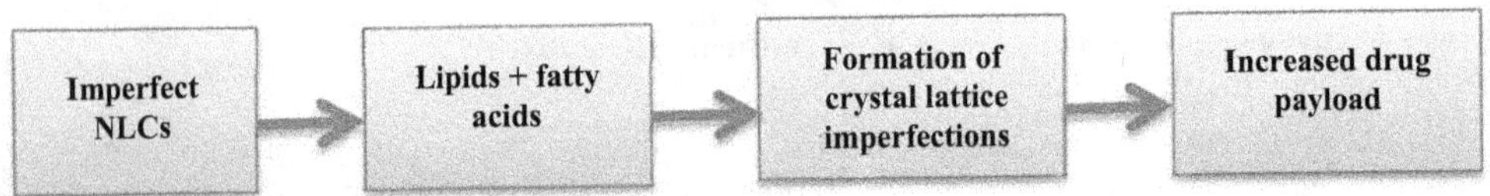

FIGURE 5.2 Imperfect type of nanostructured lipid carrier.

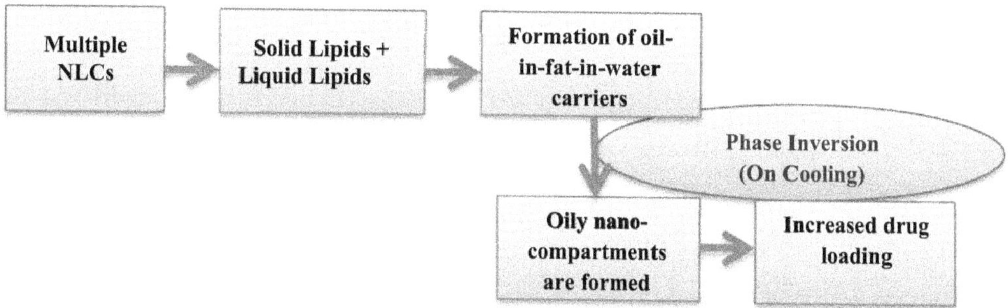

FIGURE 5.3 Multiple type of nanostructured lipid carrier.

5.3.3 NLC Type III (Amorphous Model)

Amorphous (non-crystalline state) NLCs are designed by deliberately combining lipids in such a way that drug leakage due to crystallization is minimized. Because crystals develop during the cooling process, the inclusion of a lipid mixture inhibits crystal formation (Figure 5.4a). Solid but non-crystalline particles are produced with the aid of lipids together with hydroxyloctacosanyl hydroxystearate, isopropyl myristate, hydroxyl stearate, and dibutyl adipate. The lipid matrix is both homogeneous and amorphous in character (Mohammadi et al. 2019).

The nature and properties of all types of NLC are summarized in Figure 5.4b (Chauhan et al. 2020).

5.4 ADVANTAGES

- Improved bioactive compound solubility.
- Controlled particle size.
- Better physical stability (Patidar et al. 2010).
- Easy to prepare for large-scale production (Singh et al. 2017).
- For some medicines, more loading capacity is required.
- Sustained release of bioactive compounds (Purohit 2016).
- Water-based technology; reduced solvent use (Lingayat et al. 2017).
- Possibilities of loading medicines that are both lipophilic and hydrophilic.
- More economically effective than other delivery systems (Pardeike et al. 2009).
- Biodegradable and biocompatible lipids are used (Nguyen et al. 2012).
- Reduce the amount of water in the dispersion.
- Controlled and precise drug release.
- Easier to qualify, validate, and get regulatory approval.
- Boost the benefit-to-risk ratio.
- In an aqueous medium, increased dispersibility.
- A sophisticated and effective carrier method for chemicals in particular.
- Increased occlusion of the skin.

5.5 DISADVANTAGES

- Cytotoxic effects due to lipid matrix type and concentration.
- Surfactant, irritant, and sensitizer activity.
- Application and efficacy in protein and peptide therapies, as well as gene delivery techniques, need to be investigated further.
- Lipid stability.

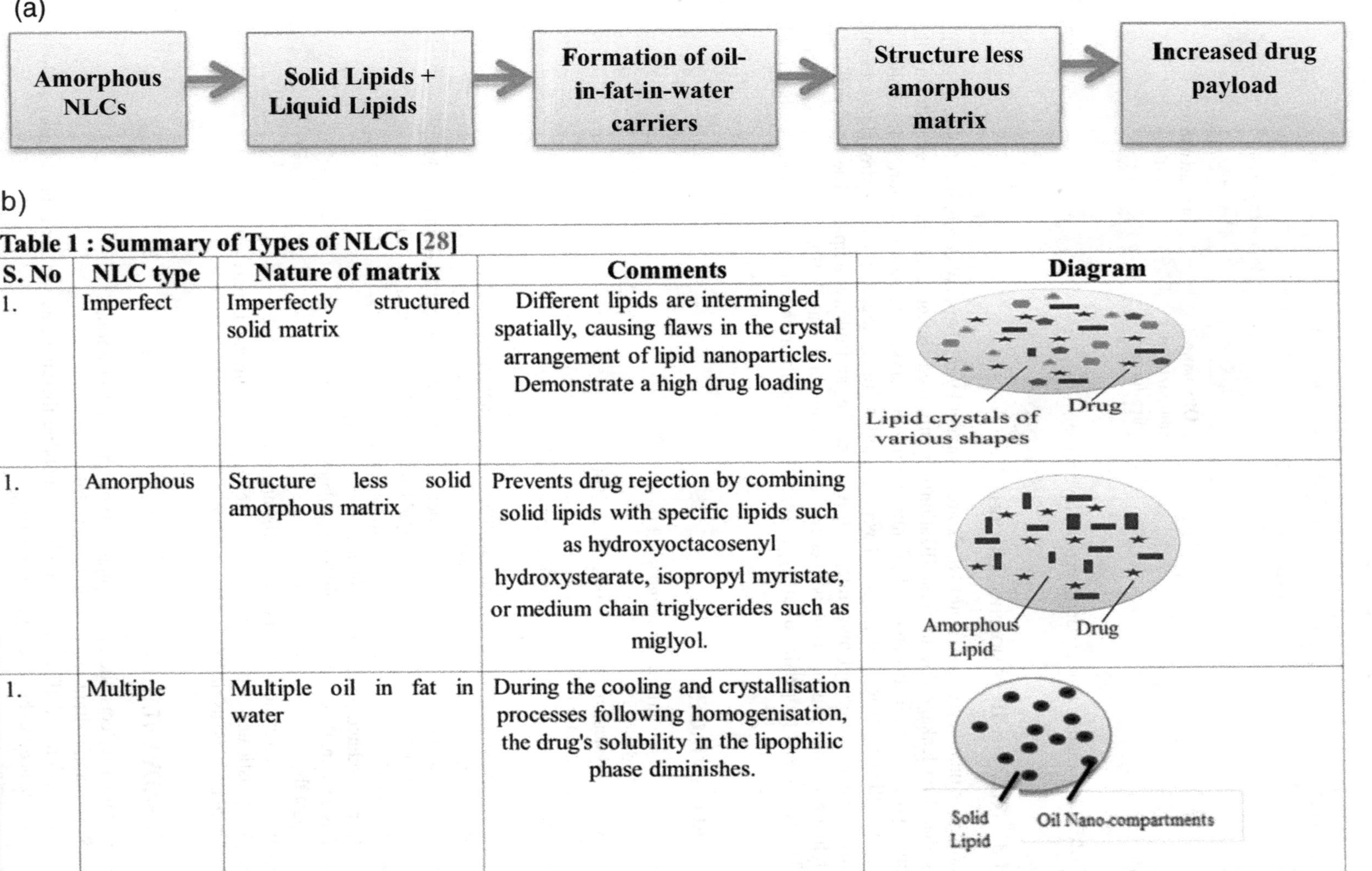

Table 1 : Summary of Types of NLCs [28]

S. No	NLC type	Nature of matrix	Comments	Diagram
1.	Imperfect	Imperfectly structured solid matrix	Different lipids are intermingled spatially, causing flaws in the crystal arrangement of lipid nanoparticles. Demonstrate a high drug loading	
1.	Amorphous	Structure less solid amorphous matrix	Prevents drug rejection by combining solid lipids with specific lipids such as hydroxyoctacosenyl hydroxystearate, isopropyl myristate, or medium chain triglycerides such as miglyol.	
1.	Multiple	Multiple oil in fat in water	During the cooling and crystallisation processes following homogenisation, the drug's solubility in the lipophilic phase diminishes.	

FIGURE 5.4 (a) Amorphous type of nanostructured lipid carrier. (b) Nature and properties of all types of nanostructured lipid carriers.

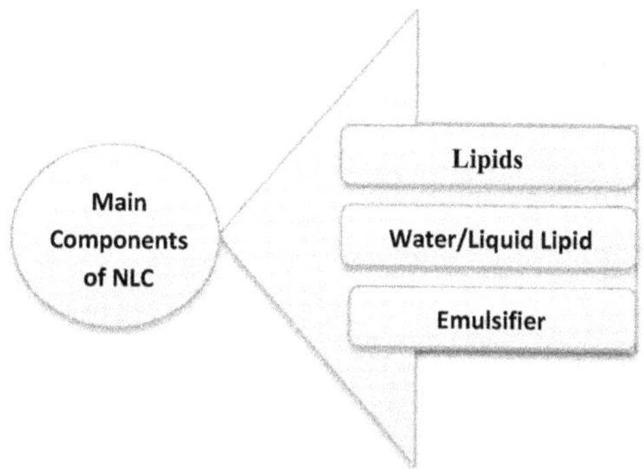

FIGURE 5.5 Main components of nanostructured lipid carriers.

5.6 COMPONENTS

The main components of NLCs are (Figure 5.5):

- Lipid
- Water
- Emulsifiers

5.6.1 Lipids

Lipids are the most vital component of NLCs, as it influences drug loading ability, stability, and continual drug loading behavior. Diverse lipid components, like fatty acids, glycerides, and waxes, are used to create lipid nanoparticle dispersions. Apart from cetyl palmitate, the maximum of these lipids have been deemed "generally regarded as safe" (GRAS) and are physiologically well characterized. It is vital to select the proper lipids before utilizing them to make dispersions of lipoid nanoparticles. Even though no precise guideline exists, variables like drug solubility in lipids have been presented as useful criteria for lipid crystallization selection (Chauhan et al. 2020). Elongated fatty acid chains crystallize more slowly than shorter chains.

Although wax-based NLCs are physically more stable, their more crystalline structure causes considerable drug ejection. NLCs have been developed utilizing a twofold combination of two spatially distinctive solid lipid matrices to overcome lipid crystallinity and polymorphism problems. All the types of lipids employed in the development of NLCs are listed in Table 5.1.

TABLE 5.1
List of Lipids Employed in Nanostructured Lipid Carrier (NLC) Preparation

Lipids	Examples
Fatty acids	Dodecanoic acid, Myristic acid, Palmitic acid, Linoleic acid and Stearic acid
Monoglycerides	Glyceryl monostearate, Glyceryl palmitostearate and Glyceryl behenate
Diglycerides	Glyceryl palmitostearate and Glyceryl dibehenate
Triglycerides	Caprylate triglyceride, Caprate triglyceride, Glyceryl and tribehenate/Tribehenin
Waxes	Cetyl Palmitate, Carnauba and Beeswax
Lipids	Soya bean oil, Oleic acid, Medium-chain triglyceride (MCT)/caprylic and capric triglycerides, Tocopherol/Vitamin E, Squalene, Hydroxyoctacosanyl hydroxy stearate and isopropyl myristate
Cationic lipids	Cetyl pyridinum chloride (hexadecyl pyridinum chloride, CPC), Cetrimide (tetradecyl trimethyl ammonium bromide, CTAB)
Steroids	Cholesterol

5.6.1.1 Solid Lipids

Solid lipids are an assortment of various organic complexes with a melting point above 40°C. At room temperature, these solid lipids are stable, and they are well tolerated (Mukherjee et al. 2009).

- Also biodegradable in *in vivo* conditions
- Accepted for human use

5.6.1.2 Liquid Lipids

These are oils extracted from plant sources. These oils are easily digested and well tolerated, making them suitable for human intake. A few medium-chain triglycerides, along with Miglyol® 812, which have structural similarities to Compritol®, can be used as liquid lipid components. Paraffin oil, propylene glycol dicaprylocaprate (Labrafac®), isopropyl myristate, 2-octyl dodecanol, and squalene are among the other oily ingredients (Poonia et al. 2016).

5.6.2 Emulsifying Agents – Surfactants

Emulsifiers are employed to decrease the surface tension of the lipid dispersions, which helps to stabilize them. The utilization of a mixture of emulsifiers prevents particle aggregation more effectively (Figure 5.6). Polyethylene glycol (PEG), which is sometimes added to NLCs, exists on the outside of the nanoparticulate shell to limit absorption by the reticuloendothelial system and increase drug circulation time (Xu et al. 2013; Bunjes et al. 2003). A list of emulsifiers that can be employed for manufacture is shown in Table 5.2.

Surfactants for NLCs are chosen depending on a variety of criteria, comprising the surfactant's hydrophilic-lipophilic balance (HLB) value and the method of administration of NLCs. Table 5.3 lists several kinds of surfactants and co-surfactants.

The concentration and type of surface-active agents affect NLC characteristics such as particle length and particle size distribution. Due to the static and steric repulsion among the particles, the NLC is solid.

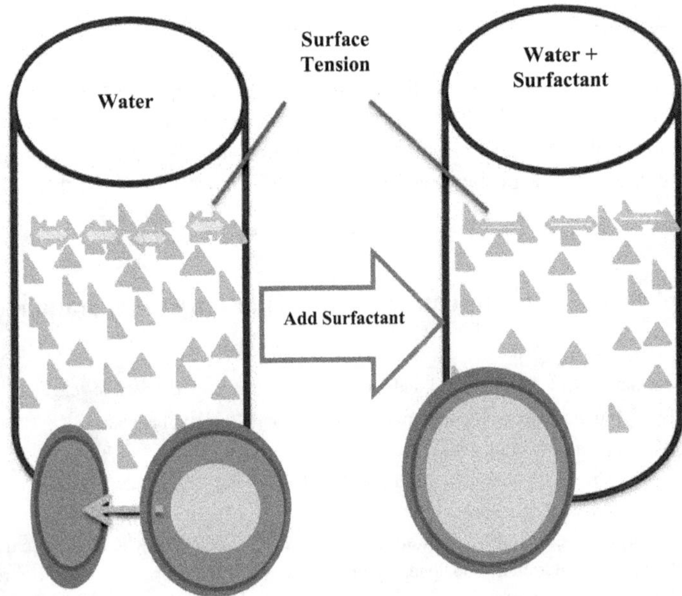

FIGURE 5.6 Mechanisms of surfactants: Presence of surfactant results in decreased surface tension and increased stability.

TABLE 5.2
Emulsifiers Employed for Fabrication

Types of emulsifiers	Examples
Hydrophobic emulsifiers	Pluronic F68 (poloxamer 188), polysorbates (20), polyvinyl alcohol, and sodium deoxycholate
Lipophilic emulsifiers	Span 80, Myverol®18-04K, Span 60, lecithin
Amphiphilic emulsifiers	Egg lecithin, soya lecithin, phosphatidylcholines, phosphatidylethanolamines, Gelucire®50/13

TABLE 5.3
Types of Surfactants and Co-Surfactants Used in the Preparation of NLCs

Types of surfactants	Examples
Ionic surfactants	Sodium taurodeoxycholate, Sodium oleate, Sodium dodecyl sulfate
Non-ionic surfactants	Span 20, 80, 85, 20, Tyloxapol, Poloxamer 188, Poloxamer 407, Solutol HS15
Amphoteric surfactants	Egg phospholipids (Lipoid E80, Lipid E80 S) Soy Hydrogenated soy phosphatidylcholine (Lipoid S PC-3) Hydrogenated egg phosphatidylcholine (Lipoid E PC-3) Phospholipon 80H, Phospholipon 90H
Co-surfactants	Butanol, Butyric acid

TABLE 5.4
Agents for Preparation of NLCs

Surface modifiers

- Dipalmitoyl-phosphatidyl ethanolamine conjugated with polyethylene glycol 2000 (DPPE-PEG2000)
- Distearoyl-phosphatidyl ethanolamine-N-poly (ethylene glycol) 2000 (DSPE-PEG2000)
- Stearic acid-PEG 2000 (SA-PEG2000)
- α-methoxy-PEG 2000-carboxylic acid-α-lipoamino acids (mPEG2000-C-LAA18)
- Ionic polymers: Dextran sulphate sodium salt

The density of the surfactant molecules is determined by the inner fluid droplets and the outer aqueous segment, the thickness of the layers of absorbent surfactant, and the dimension of the core water droplets and oil globules, all determined by steric interplay (J. Cheng et al. 2005). Stronger steric contact may be established by impenetrable layers, efficiently preventing conjugation among the interior liquid droplets and likewise, the outer aqueous phase. Small oil globules and larger inner aqueous droplets may form a more stable double emulsion, as steric repulsion decreases with larger oil globules. Poly-hydroxyl surfactants stabilize systems by creating spatial exclusion and have zero or very low zeta potential due to their non-ionic nature. The ionic energy of the continual phase is due to the rate density on the surface of water and fat, affect the stability of the nanolipid carriers (Triplett 2004).

Surfactants and surface modifiers that can be employed for NLC preparation are listed in Table 5.3 and Table 5.4, respectively.

5.7 METHODS OF PREPARATION

In the literature, many ways of generating NLCs have been documented. High-pressure homogenization (HPH), micro-emulsion, probe sonication, solvent diffusion, solvent emulsion evaporation, solvent injection/solvent displacement, and phase inversion are only a few of these techniques. The

HPH technique, which is an efficient, simple, and scalable process, is described in depth, as well as a review of many of these approaches.

5.7.1 High-Pressure Homogenization

Lipids are driven through a tiny space, a few micrometers in diameter, at high pressure (100–200 bar) in the HPH method. Shear stress and cavitation are the factors responsible for particle breakup in the submicron range. The majority of the time, the lipid content is between 5% and 10%. In other words, HPH is the process of splitting particles into nanosize fragments and producing a stable emulsion (Mu€ller et al. 2006).

There are two distinct types of homogenizers on the market (Duong et al. 2020):

- Jet-stream homogenizers
- Piston-gap homogenizers

The three most common HPH strategies for generating NLCs are illustrated in Figure 5.7. The medication is dissolved within the lipid become liquefied at 510 °C higher than the temperature in each hot and cold homogenization method (Teng et al. 2018).

- Hot homogenization
- Cold homogenization
- Micro-emulsion

5.7.1.1 Advantages of High-Pressure Homogenization Method

- Used in the food and dairy sectors in addition to the cosmetics business.
- Increase the shelf life, stability, and digestibility of the product.
- Complements the taste of the formulation.
- Substantially decreases the quantity of excipients.
- In the beauty industry, it's crucial for product quality and consistency.
- Improved bioavailability of formulated products.
- Less microbial contamination.
- Cost-effective.
- Dependable and powerful technique for generating NLCs on a mass scale.

5.7.2 Solvent Injection Method (Gomaa et al. 2022)

A solvent that spreads rapidly in water (dimethylsulfoxide [DMSO], ethanol) is employed in this method. The lipid in solvent aggregate is injected immediately into an aqueous surfactant solution using an injection needle (Figure 5.8). Rapid displacement of solvent from water leads to fast precipitation of lipid particles in aqueous medium.

5.7.2.1 Advantages

- Simple and fast production technique
- No need for high heat, shear pressure, or modern instruments

5.7.2.2 Disadvantages

- Use of organic solvents
- Low particle concentration

5.7.3 Phase Inversion (Mahant et al. 2018)

In this technique, the drug, lipid, water, and surfactant are gently combined and heated above the surfactant's phase inversion temperature. During the warming stage (above the inversion

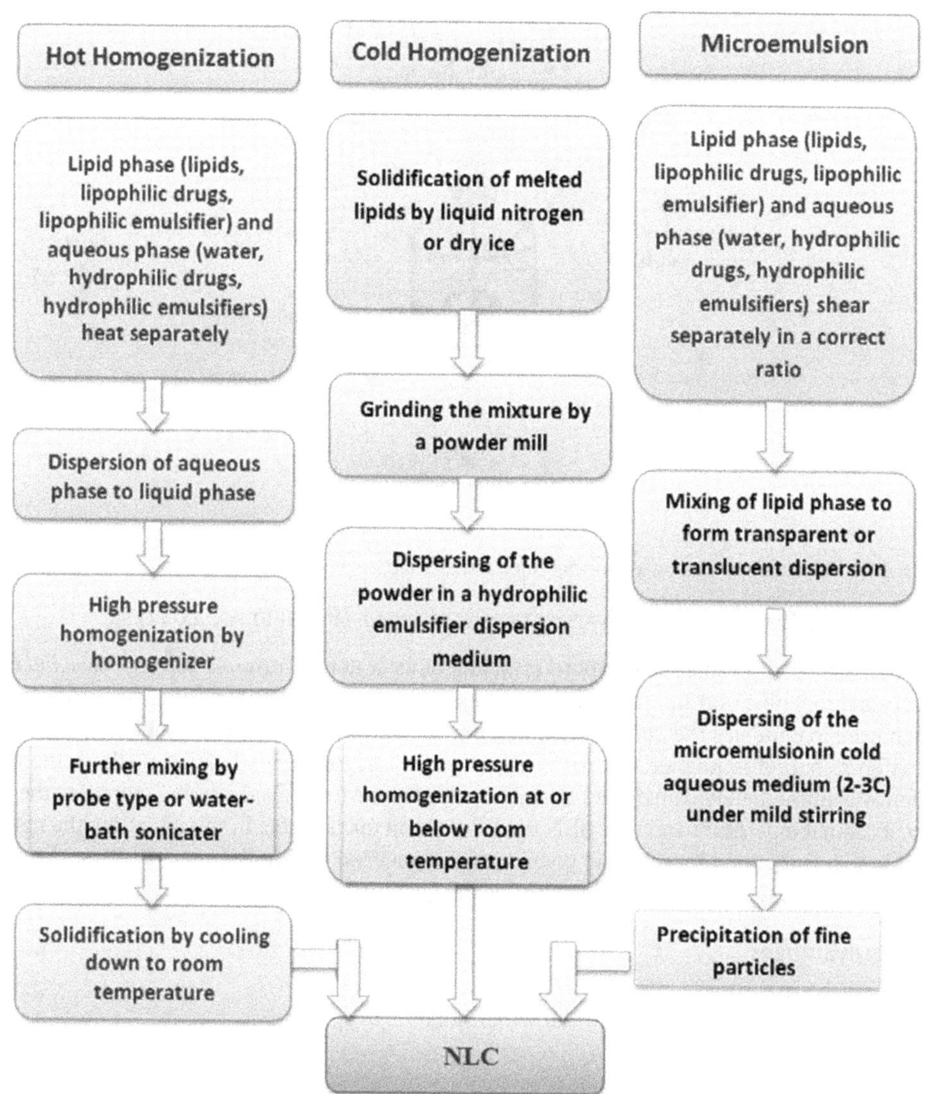

FIGURE 5.7 Preparation procedures of NLCs by various homogenization techniques. (From Gomaa, E., et al., *Methods*, 199, 3–8, 2022.)

temperature), the surfactant is dried, converting its hydrophilic-lipophilic equilibrium and, thus, its partiality for every stage, completing with inside the emulsion reversal. The surfactant returns to a hydrophilic state after brief cooling (e.g., in an ice bath), resulting in the precipitation of NLCs.

5.7.3.1 Advantages

- Low-energy input
- Avoid solvent use avoided

5.7.3.2 Disadvantages

- Poor stability
- Numerous temperature stages required

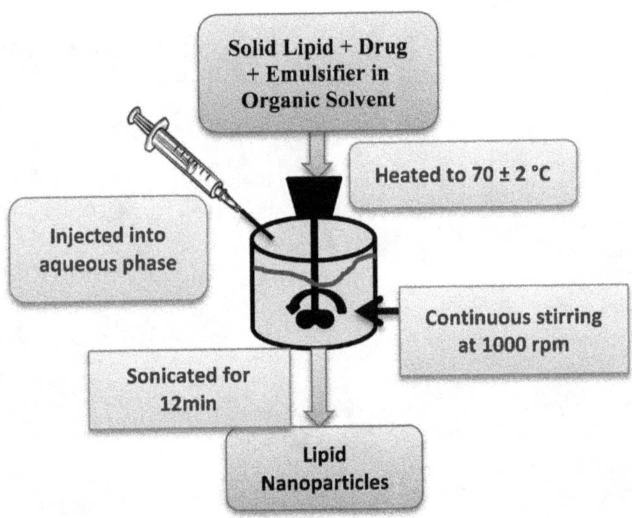

FIGURE 5.8 Schematic overview of solvent injection method.

5.7.4 Solvent Emulsification-Evaporation Technique (Parhi et al. 2012)

Hexane, toluene, dichloromethane, chloroform, and other water-immiscible solvents have been used to liquefy hydrophobic and lipophilic medications. The solvent emulsification-evaporation method was then used to emulsify the combination in a liquid solution with the use of a high-speed homogenizer (Figure 5.9). To enhance the efficacy and quality of the emulsification method, the coarse emulsion was immediately transferred to a micro-fluidizer. Mechanical stirring increases the evaporation at room temperature and thus SLN lipid formation takes place. In this situation, the resulting molecular size is dictated by the lipid content of the material within the organic phase. For drugs with a low lipid content (5%), it is considered as the best method.

5.7.4.1 Advantages

- Avoids thermal stress
- Suitable for the assimilation of highly thermolabile drugs

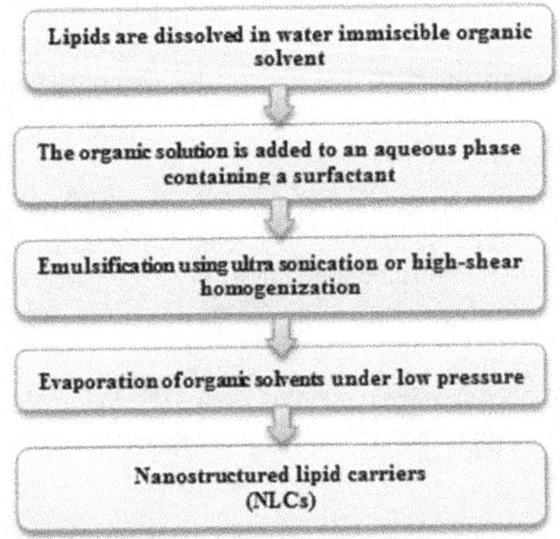

FIGURE 5.9 Schematic overview of solvent emulsification-evaporation technique.

5.7.4.2 Disadvantages

- Organic solvents required
- Solubility issues

5.7.5 Solvent Emulsification-Diffusion Technique (Li et al. 2017)

The solvent used in the solvent emulsification-diffusion method has to be partly miscible with water (e.g., benzyl alcohol, butyl lactate, ethyl acetate, isopropyl acetate, methyl acetate), and the operation may be performed in both an aqueous and an oil phase. To make certain that the solvent and the water are both in thermodynamic equilibrium, each solution is first saturated. The saturation stage ended at the temperature at which heating became critical to solubilize the lipid. The internal organic phase was emulsified using a solvent-saturated liquid solution containing emulsifier via a mechanical stirrer, after which the lipoid drug was dissolved in a water-saturated solvent (Figure 5.10).

5.7.6 Melting Dispersion Method (Hielscher Ultrasound Technology 2019)

After the formation of an oil in water emulsion, water (dilution medium) was introduced into the container at ratio from 1:5 to 1:10, to permit solvent penetration into the continuous phase, resulting in lipid aggregation within the nanoparticles. Diffusion became both at room temperature and on the temperature at which the lipid became dissolved. Throughout the procedure, there was constant stirring. Finally, vacuum distillation or lyophilization was used to remove the dispersed solvent.

5.7.7 High-Shear Homogenization or Ultrasonication Technique (Fan et al. 2014; Pamudji et al. 2016; Awadeen et al. 2020)

Ultrasonication employs a cavitation process. The drug was initially added to a previously melted solid lipid. The heated aqueous phase was injected into the melted lipid and emulsified using an ultrasonic probe or high-speed homogenization, or the aqueous solution was added using a dropper

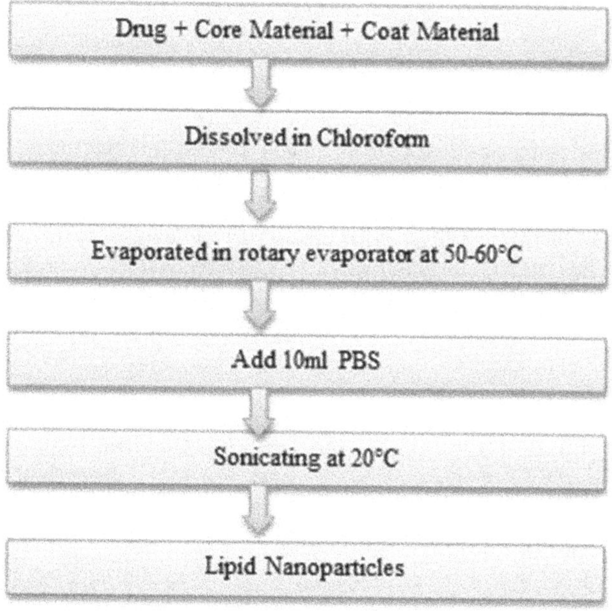

FIGURE 5.10 Schematic procedure of solvent emulsification-evaporation.

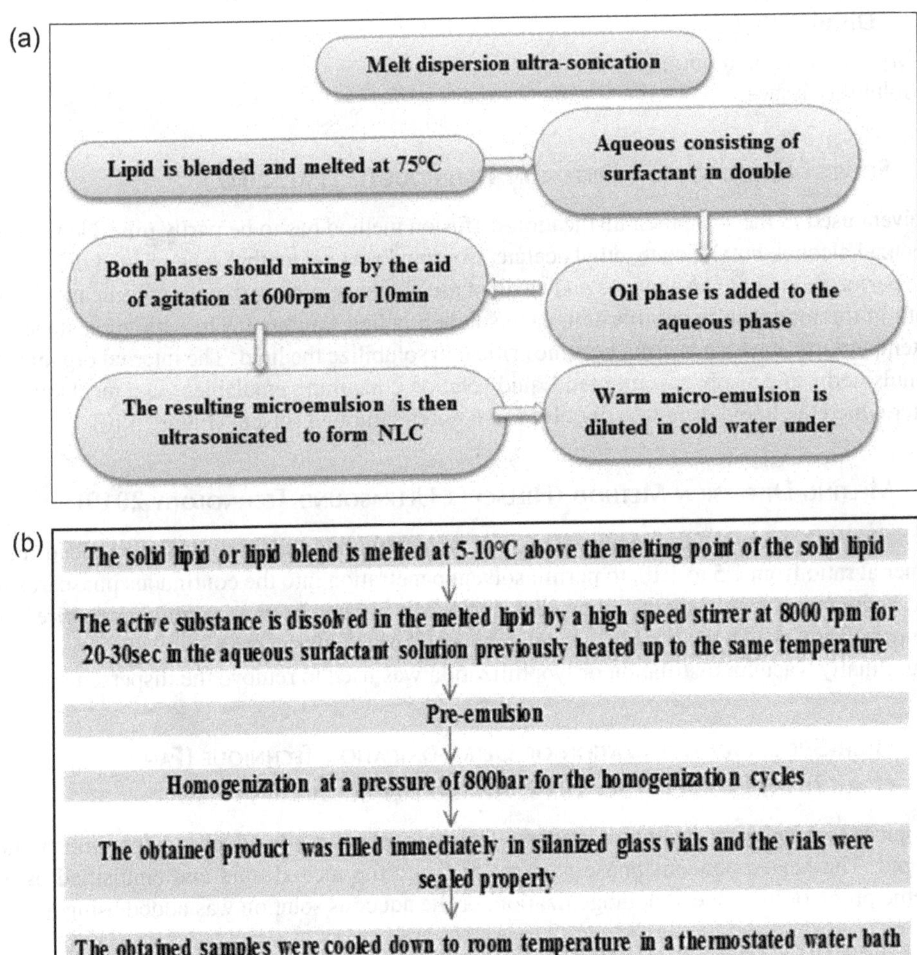
(a)
Melt dispersion ultra-sonication
Lipid is blended and melted at 75°C
Aqueous consisting of surfactant in double
Both phases should mixing by the aid of agitation at 600rpm for 10min
Oil phase is added to the aqueous phase
The resulting microemulsion is then ultrasonicated to form NLC
Warm micro-emulsion is diluted in cold water under
(b)
The solid lipid or lipid blend is melted at 5-10°C above the melting point of the solid lipid
The active substance is dissolved in the melted lipid by a high speed stirrer at 8000 rpm for 20-30sec in the aqueous surfactant solution previously heated up to the same temperature
Pre-emulsion
Homogenization at a pressure of 800bar for the homogenization cycles
The obtained product was filled immediately in silanized glass vials and the vials were sealed properly
The obtained samples were cooled down to room temperature in a thermostated water bath to 15°C
NLC will obtained

FIGURE 5.11 (a) Preparation procedures of NLCs: Melt dispersion ultra-sonication method. (b) Preparation of aqueous NLC dispersion.

into the lipid solution in the second stage, accompanied by magnetic stirring. A probe sonicator was used to ultrasonicate the obtained pre-emulsion in a water bath (at 0 °C). To avoid recrystallization, the temperature was maintained at a minimum of 5 °C above the lipid melting point during the operation. A filtration device was used to remove impurities from the product.

Types of ultrasonic techniques:

- Melt dispersion ultrasonic method (Figure 5.11a)
- Aqueous dispersion ultrasonic method (Figure 5.11b)

5.7.8 Double Emulsion Technique (Gaba et al. 2015)

The drug (often a hydrophilic pharmaceutical) is liquefied in the aqueous phase and eventually blended into melting fats within the double emulsion system. By dispensing a stabilizer in an aqueous phase encompassing a hydrophilic emulsifier, agitating it, and then filtering it, the primary emulsion is stabilized. The use of a double emulsion method removes the need to soften the lipid to shape peptide-loaded lipid nanoparticles (PLLN), and the nanoparticle surface may be modified to make them sterically strong by including lipid-PEG derivatives.

5.8 STABILITY (Shegokar et al. 2011)

Micelles, liposomes, and nanoemulsions are colloidal systems that are useful resources for stabilizing NLCs. Particle size increase, dispersion gelation, and drug ejection from the lipid matrix are all key concerns during storage. The development of a network and lipid bridges among the particles causes gelation. Zeta potential (ZP), particle size (photon correlation spectroscopy, photon correction spectroscopy [PCS]; laser diffraction [LD]), and temperature examinations are frequently used to study the physical stability of these dispersions (differential scanning calorimetry [DSC]).

5.8.1 Strategies Engaged for Overcoming the Problems Associated with the Stability of NLCs (Haider et al. 2020)

5.8.1.1 Spray Drying

Spray drying SLNs/NLC dispersions can improve their stability in addition to providing appropriate storage conditions (Figure 5.12). For spray drying, the melting point of the lipid matrix must be greater than 70 °C (Reithmeier et al. 2001).

5.8.1.2 Lyophilization

Another method for increasing stability is lyophilization. The final product of lyophilizing SLNs without a cryoprotectant usually results in particle aggregation. Trehalose, sorbitol, glucose, sucrose, mannose, and maltose are some of the most often utilized cryoprotectants. It was discovered that trehalose is the most efficient cryoprotectant in terms of preventing particle formation (Molpeceres et al. 1996).

5.8.2 Stabilizing Agents

5.8.2.1 Poloxamers

Poloxamer 188 was utilized to create a formula that was clinically tested in both human plasma and blood. The fibrin fiber configuration was found to be connected to greater whole blood permeability

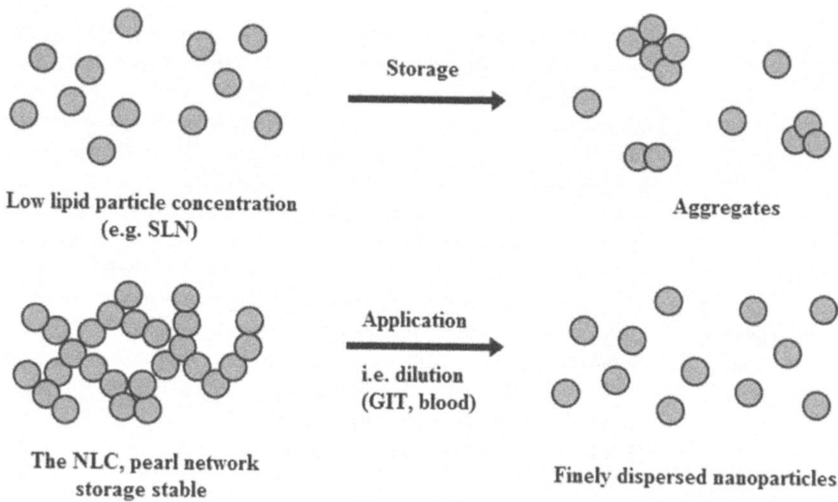

FIGURE 5.12 Stabilization effect in highly concentrated lipid particle dispersions.

in networks. Increased mechanical stability is primarily caused by fibrin alterations, which adds to the antithrombotic and rheological advantages (Yong et al. 2004; Nogueiras-Nieto et al. 2012). The stability of gels was enhanced by mixing.

Poloxamer 407 self-assembles into two thermodynamically stable liquid crystal forms in the presence of those natural solvents: micellar cubic and hexagonal systems. When used in tandem with a liposome, Poloxamer 407 increased liposome formulation stability by prolonging half-life and inhibiting the aggregation and fusion of phosphatidylcholine multilamellar vesicles (Kaul et al. 2011).

A mixture of poloxamer 407 with 17.5 W/W% acrylate and thiol groups was created at the formulation temperature due to the poor water stability of poloxamer hydrogel. The property of poloxamer 407 was altered by an instantaneous crosslinking between acrylate and thiol, resulting in a fourfold increase in drug stability and its prospective application in controlled drug release (Jain et al. 2006).

5.8.2.2 Polyethylene Glycol (Date et al. 2011)

In general, it has been demonstrated that coating colloidal particles with a hydrophilic substance such as PEG offers the following advantages:

- Enhancing colloidal stability and bioavailability
- Increasing the presence of colloids in the bloodstream for systemic application
- Promoting colloid stability in body fluids such as gastrointestinal fluid
- Accelerating colloid transport across the epithelium
- Increasing biocompatibility and decreasing thermogenicity of drug carriers by modifying colloids' interaction with mucosa (Tables 5.5 and 5.6)

TABLE 5.5
Excipients and Their Specifications for Nanostructured Lipid Carriers

Solidification test (oil in the solid lipid matrix) smearing	This can be done by smearing a piece of the solid mixture on a filter paper and observing if there are any oil spots on the filter paper
Calorimetric analysis	Can be done on the solid solutions obtained using a differential scanning calorimeter, these analyses will detect any presence of crystalline active and also can show if there is an unincorporated part of an active ingredient in the lipid matrix (i.e., oil)
Temperature	The lipid (oil) phase and the aqueous surfactant solution were heated upto about 80 °C. Temperature also affects the zeta potential. NLC can be stored at a temperature between 5–25 °C. The particle size of NLCs varies depending on the storage temperature
Particle size	Very important for the stability of NLC. Techniques were direct measurement (microscopy) and indirect measurement (lases diffractometry and photon correlation spectroscopy)
Homogenization Speed: 8000–16000 rpm Pressure: 800 bar	The homogenization speed ranges were selected based on instrument limitation and trial batches. The homogenization speed less than 8000 rpm leads to large particle size and polydisperse colloidal system. However, the upper range should be selected e.g., 16,000 rpm. Homogenization pressure of 800 bar and two homogenization cycles
Homogenization temperature	Homogenization should be performed at higher temperature (80–90 °C)
Sonication time: 5–15 min	The time duration for sonication was selected based on the literature and trial batches. Moreover, longer duration of sonication was avoided due to leaching of the drug from the matrix and possible metal contamination
Physical stability	Physical stability can be evaluated by measuring the zeta potential • Zeta potential—30 mV: Colloidal system is in stable state • Zeta potential—60 mV: Super high stability

TABLE 5.6
Process Variables and Their Role in the Preparation of NLCs

Process variables	Step involved	Process responses
Speed and Time	Mixing	Particles shape, Particle size
Temperature	Melting	Phase transition, Solubility
Speed and Time	Stirring	Particles shape, Particle size
Speed, Temperature, Pressure	Homogenization	Particles shape, Particle size

5.9 CHARACTERIZATION

5.9.1 Particle Size Analysis/PCS (Photon Correction Spectroscopy) (International Standard ISO13321 1996; ISO22412 2008)

The measurements were done in triplicate (n = 3) at a constant angle of 173° and at a temperature of 25 °C, and standard deviations were calculated. To avoid backscattering, the aqueous NLCs were diluted with distilled water before examination. PCS based on laser light diffraction is a suitable approach for study and may be used on particles as small as 200 nm and as large as 1000 micrometer. Rayleigh's hypothesis states that scattering power is proportional to the sixth efficiency of particle diameter for particles smaller than 200 nm. Both Fraunhofer's and Rayleigh's theories are modification of Mie's theory, which states that scattering depth is decided by scattering perspective, particle absorption, and length, in addition to the refractive indices of the particles and the dispersion medium.

5.9.2 Zeta Potential Measurement/Laser Doppler Electrophoresis Technique

The electrostatic charge of particles was determined using this approach. The check is frequently done in triplicate at a pH of 7.26, which is intended to imitate physiological pH (Doktorovova and Souto 2009).

5.9.3 Transmission Electron Microscopy (TEM)

A drop of diluted NLC dispersion was deposited on the surface of a copper grid covered with carbon,. The grid, with a mesh size of 300, had been dyed with 2% phosphotungstic acid (w/v) for 120 seconds and dried at room temperature. TEM was used to analyze the NLC samples, which had been positioned on sample holders (Hitachi H-7100, Japan) (Jenning et al. 2000).

5.9.4 Scanning Electron Microscopy (SEM)

A scanning electron microscope (JEOL-JSM-6360, Japan) was used to analyze the morphological characteristics of NLCs. A single drop of sample was positioned on a slide, and any extra water was allowed to evaporate at room temperature. The slide was taped to the specimen holder, coated with gold under vacuum for 10 min with a sputter coater (Model JFC-1100, JEOL, Japan), and then inspected at 20 kV (Nasr et al. 2008).

5.9.5 Atomic Force Microscopy (AFM)

AFM micrographs of NLCs were obtained before and after lyophilization to observe changes in morphology and particle length. A Nanosurf cellular S AFM was used to perform the assessments (nanosurfe AG, Liestal, Switzerland). Measurements of the forces between the tip and the sample surface were used to provide the graph. The experiment was performed in non-contact mode in ambient air at a temperature (250 °C). Drops of suspension were deposited on a tiny mica disk. The measurements were taken at some special locations. The amplitude AFM pictures were collected before and after freeze-drying NLCs at a freezing temperature of -80°C for 24 hours and a 48-hour sublimation time under the most efficient freeze-drying conditions. The Easyscan 2 software was used to take a look at the picture data (Varshosaz et al. 2012).

5.9.6 Confocal Laser Scanning Microscopy (CLSM)

A drop of glycerol introduced to samples and determined with a confocal FV-a thousand stations established on an inverted microscope IX-eighty-one to research the shape of NLCs (Olympus, Tokyo, Japan). A spectral detector channel was used to detect the emitted fluorescence (Vitorino et al. 2013).

5.9.7 Differential Scanning Calorimetry (DSC)

By monitoring glass and melting element temperatures at their particular enthalpies, DSC is used to study the crystal lattice and polymorphism of excipients and drug nanoparticles. DSC is a kind of measurement calorimeter (822e, Mettler Toledo, Greifensee, Switzerland) (Mukherjee et al. 2009). Around 10 mg of bulk lipid, drug, and lyophilized NLCs were heated in aluminum pans with lids. An empty aluminum pan was used as a control. In natural ultra-dry nitrogen, DSC curves were acquired throughout a temperature range of 20–80 °C at a consistent linear heating rate of 5 °C per minute. The evaluation was performed in triplicate, and the mean of the three was taken as the result. Finally, the enthalpies were determined using Mettler Star software (Wa Kasongo et al. 2011; Montenegro et al. 2011).

5.9.8 Wide-Angle X-ray Diffraction (XRD)

Wide-angle XRD may be employed to evaluate the degree of crystallinity by using geometric smattering of moderate from crystal planes interiors of nanoparticle dispersion. The crystallinity of the lyophilized NLC was assessed using an X-ray diffractometer (Philips, Hamburg, Germany) with a copper anode for radiation. Powdered lipid, drug, and lyophilized NLC samples with a duration of 10 mm were positioned on top of X-ray plates at room temperature and exposed to a voltage of 45 kV and a current of 40 mA, with a scanning rate of 5° per minute and a scanning range of 2°. The XRD styles were visible at temperatures from 20 to 80 °C (Eid et al. 2011; Patel et al. 2012).

5.9.9 Rheological Study

A Brookfield viscometer (Brookfield LV-DV II+, United States) was used to assess the rheological parameters of the produced lipid nanoparticles (Lee et al. 2009). The sample (20 g) was placed in a beaker and given 5 min to equilibrate. The checks were accomplished at room temperature with the correct spindle. Before measuring the size, the spindle speed was varied in ascending order from 1 to 100 rpm, then in descending order from 100 to 1 rpm, with each placing held consistent for 10 seconds (Moreno 2001; Marcotte et al. 2001).

5.9.10 Drug Entrapment Efficiency

A 2 mL volume of every drug-loaded sample was centrifuged at 12500 rpm for 45 mins to separate the lipid and aqueous stages (Microfuge, Remi motors, Mumbai). The supernatant was diluted with methanol and filtered using 40 micron filter paper, and the drug concentration was determined using a UV-visible spectrophotometer (UV1800, Shimadzu, Japan) at 273 nm (Hi-media, Mumbai). The following formulae were used to calculate the NLCs' entrapment efficiency (Doktorovová et al. 2010; Hu et al. 2005).

$$\mathrm{EE} = \mathrm{Wa} - \mathrm{Ws}/\mathrm{Wa} \times 100$$

$$\mathrm{DL} = \mathrm{Wa} - \mathrm{Ws}/\mathrm{Wa} - \mathrm{Ws} + \mathrm{W1} \times 100$$

where EE = entrapment efficiency
DL = drug loading
Wa = mass of aceclofenac added to the formulation
Ws = analyzed weight of the drug in supernatant
W = weight of lipid added

5.9.11 Ultrafiltration

As the pore size of conventional filter paper is large, colloidal dispersions can flow through it. The pore size of this filter paper shrinks as it is impregnated with colloid ion. Ultrafilters are filter sheets that have been modified in this way. To eliminate all electrolytes, the colloidal dispersion is filtered via an ultrafilter. Colloidal particles cling to the filter paper like slime. They are composed and distributed in a clean dispersion to produce a solution. Because ultrafiltration is a slow process, pressure or suction is used to speed up the process (Sawant et al. 2010).

5.9.12 High-Performance Liquid Chromatographic (HPLC) Analysis

A Hitachi L-2130 pump, a Hitachi L-2200 pattern processor, and a Hitachi L-2400 UV detector were used in the HPLC device. The column was a stainless-steel C18 column with a 25-cm length and a 4-mm internal diameter (Merck, Darmstadt, Germany). For calcipotriol, the mobile phase was methanol:water (80:20), and for methotrexate, the mobile phase was acetonitrile:water (15:85) at pH 2.7, adjusted with phosphoric acid. The flow rate was 1 mL/min. For calcipotriol and methotrexate, the UV detector was adjusted to wavelengths of 265 and 303 nm, respectively (Lin et al. 2010).

5.9.13 pH Analysis

Because the pH of a formula designed for cutaneous applications has to be well matched with the pH of the application site, it is crucial to get the pH right. The natural pH of the pores and skin is partly determined by perspiration and sebaceous gland secretions, in addition to lactic acid generation, which results in the production of a barrier film, referred to as the hydrolipidic film, that completely covers the pores and the skin surface. The normal pH of the pores and skin is 5.5, but this varies widely depending on the area of the body. The pH of all created HGs was evaluated on days 7 and 30 following storage at various temperatures. A glass pH electrode was used for this (Basic 20; Crison Instruments, Barcelona, Spain). The pH electrode was directly immersed in semisolid formula in a beaker (Basic 20; Crison Instruments, Barcelona, Spain). All the analyses were performed in triplicate, and the results were expressed as mean and standard deviation (Schmid-Wendtner and Korting 2006; Stefaniak et al. 2013).

5.9.14 Nuclear Magnetic Resonance (NMR)

The size and qualitative nature of nanoparticles might also be determined using NMR. Chemical shift selectivity interacts with molecular mobility sensitivity to provide information on the physicochemical range of additives within the nanoparticle. The width of the signal at 1/2 of amplitude is associated with the mobility of solid and liquid lipids (Wissing et al. 2004). Molecules with restricted mobility and robust interactions indicated by slight amplitudes. In assessment of the physical mixture of substances contained in NLCs, the expanded line width of NLCs exhibits the interplay of liquid oil with stable lipid. NLCs show greater immobilization of nanoparticles than SLNs with absolutely crystalline cores.

5.10 DRUG ENCAPSULATION

Encapsulating or integrating drugs into lipid nanoparticles or NLCs may be achieved by three approaches: a stable responsive matrix, a drug-enriched shell, and a drug-enriched center.

(a) *Homogeneous, stable, responsive matrix*: The drug is randomly dispersed in the lipid matrix of the debris in this encapsulation technique, and the drug is delivered through diffusion.
(b) *Drug-enriched shell*: The drug is concentrated on the lipid nanoparticles' outermost layer or shell on this technique. These nanoparticles administer the medication in bursts because of the precipitation and solubilization tactics.
(c) *Drug-enriched center*: This technique produces extended release because of the drug's saturation solubility inside the lipid.

In the case of NLCs, the rate of breakdown and diffusion determines drug release from a matrix. It is extensively documented in the literature that particular and controlled release, further to dispersion and degradation, is required. When a particle is delivered, an impulse ought to be used to provoke the discharge.

5.11 DRUG RELEASE

Controlled or sustained drug release from NLCs can result in extended half-life and delayed enzymatic attack within the systemic circulation. The temperature at which NLCs are made, the emulsifier composition, and the quantity of oil within the lipid matrix all impact drug release behavior (Figure 5.13). The drug on the outer shells of nanoparticles and at the particulate surface is released in a burst; however, the drug contained in the particulate center is released over time. The delayed release of the drug may be defined by considering how each drug partitions between the lipid matrix and water, in addition to the interfacial membrane's barrier function. Two strategies for assessing drug release from nanoparticles *in vitro* are the Franz diffusion cell and the dialysis system. Keep in mind the specific circumstances of the *in vivo* setting while analyzing *in vitro* medication release patterns. The composition of lipid nanoparticles has a substantial influence on enzyme degradation

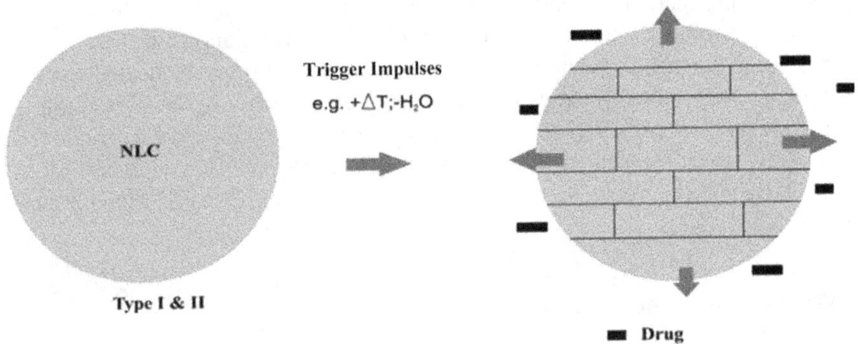

FIGURE 5.13 Altering the disordered lipoid structure to a more ordered structure enhances the drug release from NLCs.

(Salvi and Pawar 2019; Jenning and Gohla 2000; Purohit 2016; Üner 2006; Pezeshki 2017; McEvoy et al. 2016; Anuradha and Kumar 2014).

5.11.1 Factors Affecting Drug Release

The reason for studying release pattern is to determine the ability of numerous substances to keep the drug for an extended time period while at the same time gradually releasing it from the nanoparticles' lipid matrix. A number of factors should affect the drug's release profile from the NLC device. It is reasonable to analyze the effects of particle size, lipid matrix, surfactant, drug concentration withinside the lipid matrix, and drug kind.

5.11.1.1 Particle Size

The particle size of a colloidal device (e.g. NLC) is a vital factor for the release of the materials(s) included within the matrix.

5.11.1.2 Lipid Matrix

Different lipid matrices have different release profiles. For instance, the lipids have particular crystallization orders and modifications, melting temperatures, and HLB values: HLB Apifil = 94, HLB Compritol 888 = 2 and HLB Compritol 888 = 2. As a result, the drug's proclivity to end up captured within the lipoid matrix differs from lipid to lipid.

5.11.1.3 Surfactants

Surfactants, which can be used to stabilize solid materials in dispersion environments, might also influence lipid nanoparticles to regulate their shape (or emulsify oil in water). The interplay of the emulsifying agent molecules with the lipid molecules is the reason for this. The HLB of the surfactant and the molecular weight of the surfactant molecules dictate the surfactant's affinity for the lipid. Surfactant molecules within the lipid matrix might also interfere with lipid crystallization, making holes in the lipid lattice. Increased drug loading ability and drug integration in particle matrix flaws may be viable way to overcome those gaps. The ability of the surfactant to stabilize oil droplets (from the lipid melted state all through homogenization) and to generate NLCs permits it to govern the dimensions of the lipid debris produced. The form of surfactant used has a massive impact on the physicochemical properties of NLCs.

5.11.1.4 Drug Loading

By increasing the drug loading, its release profile can be improved. This is determined using the drug's propensity for blending with the lipid and being encased within the matrix.

5.11.1.5 Drug Type

Because various drug compositions have varying affinities for the lipid matrix, the drug type has an impact on the release profile. NLCs offer unique properties that can help a variety of integrated medication forms operate better.

5.12 APPLICATIONS

There are countless applications of NLCs in the pharmaceutical field. Some of them are shown in Figure 5.14.

5.13 CONCLUSION

NLCs have a large influence on the massive barriers that are encountered in developing a stable drug delivery system. The use of nanocarriers in transdermal drug delivery has established a new

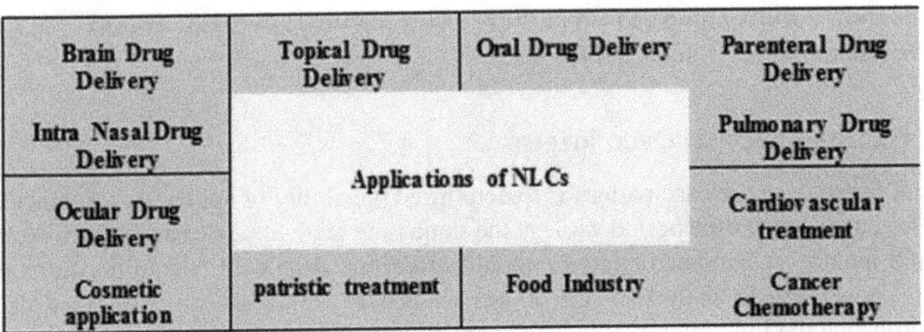

FIGURE 5.14 Applications of NLCs.

domain in drug delivery. NLCs are chemically and physically stable systems that have better drug incorporation and bioavailability. The growing interest in lipid carrier systems in industry has resulted in significant advancements in recent years. Currently, more than 30 commercial NLC formulations containing drug and cosmetic ingredients are available on the market. NLCs, as a smarter generation of lipid nanoparticles, are promising candidates for skin targeting, occlusive effect, and prolonged release.

Lipid nanocarriers are gaining industrial attention due to qualified and validated scale-up technology, GRAS status of excipients, and ease of large-scale production. Because of their numerous advantages over first-generation systems, future NLC formulations have the potential to increase the success of the lipid carrier system. Future concerns include determining the toxicity and health risks associated with nanostructures. More pre-clinical and clinical research will pave the way for nanolipid structures to succeed. Achievement in this field is possible if the pharmaceutical industry takes up the academic research to design this carrier system for various therapeutic and cosmetic agents.

REFERENCES

Allen, Theresa M, and Pieter R Cullis. 2013. "Liposomal Drug Delivery Systems: From Concept to Clinical Applications." *Advanced Drug Delivery Reviews* 65 (1): 36–48.

Anuradha, Kush, and Senthil Kumar. 2014. "Development of Lacidipine Loaded Nanostructured Lipid Carriers (NLCs) for Bioavailability Enhancement." *International Journal of Pharmaceutical and Medicinal Research* 2 (2): 50–57.

Awadeen, Randa Hanie, Mariza Fouad Boughdady, and Mahasen Mohamed Meshali. 2020. "Quality by Design Approach for Preparation of Zolmitriptan/Chitosan Nanostructured Lipid Carrier Particles: Formulation and Pharmacodynamic Assessment." *International Journal of Nanomedicine* 15 (November): 8553–68. https://doi.org/10.2147/IJN.S274352.

Battaglia, Luigi, and Marina Gallarate. 2012. "Lipid Nanoparticles: State of the Art, New Preparation Methods and Challenges in Drug Delivery." *Expert Opinion on Drug Delivery* 9 (5): 497–508.

Bianco, Alberto, Kostas Kostarelos, and Maurizio Prato. 2005. "Applications of Carbon Nanotubes in Drug Delivery." *Current Opinion in Chemical Biology* 9 (6): 674–79.

Bunjes, Heike, Michel H J Koch, and Kirsten Westesen. 2003. "Influence of Emulsifiers on the Crystallization of Solid Lipid Nanoparticles." *Journal of Pharmaceutical Sciences* 92 (7): 1509–20. https://doi.org/10.1002/jps.10413.

Chauhan, Iti, Mohd Yasir, Madhu Verma, and Alok Pratap Singh. 2020. "Nanostructured Lipid Carriers: A Groundbreaking Approach for Transdermal Drug Delivery." *Advanced Pharmaceutical Bulletin* 10 (2): 150.

Cheng, Christopher J, Gregory T Tietjen, Jennifer K Saucier-Sawyer, and W Mark Saltzman. 2015. "A Holistic Approach to Targeting Disease with Polymeric Nanoparticles." *Nature Reviews Drug Discovery* 14 (4): 239–47.

Cheng, Jing, Shiai Xu, Lixiong Wen, and Jianfeng Chen. 2005. "Steric Repulsion between Internal Aqueous Droplets and the External Aqueous Phase in Double Emulsions." *Langmuir* 21 (25): 12047–52. https://doi.org/10.1021/la051906r.

Date, Abhijit A, Nimish Vador, Aarti Jagtap, and Mangal S Nagarsenker. 2011. "Lipid Nanocarriers (GeluPearl) Containing Amphiphilic Lipid Gelucire 50/13 as a Novel Stabilizer: Fabrication, Characterization and Evaluation for Oral Drug Delivery." *Nanotechnology* 22 (27): 275102. https://doi.org/10.1088/0957-4484/22/27/275102.

Doktorovova, Slavomira, and Eliana B Souto. 2009. "Nanostructured Lipid Carrier-Based Hydrogel Formulations for Drug Delivery: A Comprehensive Review." *Expert Opinion on Drug Delivery* 6 (2): 165–76. https://doi.org/10.1517/17425240802712590.

Doktorovová, Slavomíra, Joana Araújo, Marisa L Garcia, Erik Rakovský, and Eliana B Souto. 2010. "Formulating Fluticasone Propionate in Novel PEG-Containing Nanostructured Lipid Carriers (PEG-NLC)." *Colloids and Surfaces B: Biointerfaces* 75 (2): 538–42. https://doi.org/10.1016/j.colsurfb.2009.09.033.

Duong, Van-An, Thi-Thao-Linh Nguyen, and Han-Joo Maeng. 2020. "Preparation of Solid Lipid Nanoparticles and Nanostructured Lipid Carriers for Drug Delivery and the Effects of Preparation Parameters of Solvent Injection Method." *Molecules*. https://doi.org/10.3390/molecules25204781.

Eid, Eltayeb E M, Ahmad Bustamam Abdul, Fakhr Eldin O Suliman, Mohd A Sukari, A Rasedee, and Safa S Fatah. 2011. "Characterization of the Inclusion Complex of Zerumbone with Hydroxypropyl-β-Cyclodextrin." *Carbohydrate Polymers* 83 (4): 1707–14. https://doi.org/10.1016/j.carbpol.2010.10.033.

Fan, Hengfeng, Guoqing Liu, Yiqing Huang, Yan Li, and Qiang Xia. 2014. "Development of a Nanostructured Lipid Carrier Formulation for Increasing Photo-Stability and Water Solubility of Phenylethyl Resorcinol." *Applied Surface Science* 288: 193–200. https://doi.org/10.1016/j.apsusc.2013.10.006.

Gaba, Bharti, Mohammad Fazil, Asgar Ali, Sanjula Baboota, Jasjeet K Sahni, and Javed Ali. 2015. "Nanostructured Lipid (NLCs) Carriers as a Bioavailability Enhancement Tool for Oral Administration." *Drug Delivery* 22 (6): 691–700.

Gomaa, Eman, Heba A Fathi, Noura G Eissa, and Mahmoud Elsabahy. 2022. "Methods for Preparation of Nanostructured Lipid Carriers." *Methods* 199: 3–8.

Haider, Mohamed, Shifaa M Abdin, Leena Kamal, and Gorka Orive. 2020. "Nanostructured Lipid Carriers for Delivery of Chemotherapeutics: A Review." *Pharmaceutics* 12 (3): 288.

Hamidi, Mehrdad, Amir Azadi, and Pedram Rafiei. 2008. "Hydrogel Nanoparticles in Drug Delivery." *Advanced Drug Delivery Reviews* 60 (15): 1638–49.

Hielscher Ultrasond Technology. 2019. *Ultrasonic Formulation of Nanostructured Lipid Drug Carriers*. Hielscher Ultrasonics GmbH. 2019. https://www.hielscher.com/ultrasonic-formulation-of-nanostructured-lipid-drug-carriers.htm.

Hoffman, Allan S. 2012. "Hydrogels for Biomedical Applications." *Advanced Drug Delivery Reviews* 64: 18–23.

Hu, Fu-Qiang, Sai-Ping Jiang, Yong-Zhong Du, Hong Yuan, Yi-Qing Ye, and Su Zeng. 2005. "Preparation and Characterization of Stearic Acid Nanostructured Lipid Carriers by Solvent Diffusion Method in an Aqueous System." *Colloids and Surfaces B: Biointerfaces* 45 (3): 167–73. https://doi.org/10.1016/j.colsurfb.2005.08.005.

International Standard ISO13321. 1996. *Methods for Determination of Particle Size Distribution Part 8: Photon Correlation Spectroscopy*. International Organisation for Standardisation (ISO).

ISO22412, International Standard. 2008. *Particle Size Analysis: Dynamic Light Scattering*. International Organisation for Standardisation.

Jain, Priyanka, Prerna Rahi, Vikas Pandey, Saket Asati, and Vandana Soni. 2017. "Nanostructure Lipid Carriers: A Modish Contrivance to Overcome the Ultraviolet Effects." *Egyptian Journal of Basic and Applied Sciences* 4 (2): 89–100.

Jain, Sunil K, Govind P Agrawal, and Narendra K Jain. 2006. "A Novel Calcium Silicate Based Microspheres of Repaglinide: In Vivo Investigations." *Journal of Controlled Release: Official Journal of the Controlled Release Society* 113 (2): 111–16. https://doi.org/10.1016/j.jconrel.2006.04.005.

Jaiswal, Piyush, Bina Gidwani, and Amber Vyas. 2016. "Nanostructured Lipid Carriers and Their Current Application in Targeted Drug Delivery." *Artificial Cells, Nanomedicine, and Biotechnology* 44 (1): 27–40.

Jenning, Volkhard, and Sven Gohla. 2000. "Comparison of Wax and Glyceride Solid Lipid Nanoparticles (SLN®)." *International Journal of Pharmaceutics* 196 (2): 219–22. https://doi.org/10.1016/S0378-5173(99)00426-3.

Jenning, Volkhard, Andreas F Thünemann, and Sven H Gohla. 2000. "Characterisation of a Novel Solid Lipid Nanoparticle Carrier System Based on Binary Mixtures of Liquid and Solid Lipids." *International Journal of Pharmaceutics* 199 (2): 167–77. https://doi.org/10.1016/S0378-5173(00)00378-1.

Junghanns, Jens-Uwe A H, and Rainer H Müller. 2008. "Nanocrystal Technology, Drug Delivery and Clinical Applications." *International Journal of Nanomedicine* 3 (3): 295.

Karimi, Mahdi, Navid Solati, Amir Ghasemi, Mehrdad Asghari Estiar, Mahshid Hashemkhani, Parnian Kiani, Elmira Mohamed, Ahad Saeidi, Mahdiar Taheri, and Pinar Avci. 2015. "Carbon Nanotubes Part II: A Remarkable Carrier for Drug and Gene Delivery." *Expert Opinion on Drug Delivery* 12 (7): 1089–105.

Kaul, Goldi, Jun Huang, Ramarao Chatlapalli, Krishnendu Ghosh, and Arwinder Nagi. 2011. "Quality-by-Design Case Study: Investigation of the Role of Poloxamer in Immediate-Release Tablets by Experimental Design and Multivariate Data Analysis." *AAPS PharmSciTech* 12 (4): 1064–76. https://doi.org/10.1208/s12249-011-9676-0.

Lee, Chi H, Venkat Moturi, and Yugyung Lee. 2009. "Thixotropic Property in Pharmaceutical Formulations." *Journal of Controlled Release* 136 (2): 88–98. https://doi.org/10.1016/j.jconrel.2009.02.013.

Li, Qianwen, Tiange Cai, Yinghong Huang, Xi Xia, Susan P C Cole, and Yu Cai. 2017. "A Review of the Structure, Preparation, and Application of NLCs, PNPs, and PLNs." *Nanomaterials*. https://doi.org/10.3390/nano7060122.

Lin, Yin-Ku, Zih-Rou Huang, Rou-Zi Zhuo, and Jia-You Fang. 2010. "Combination of Calcipotriol and Methotrexate in Nanostructured Lipid Carriers for Topical Delivery." *International Journal of Nanomedicine* 5 (March): 117–28. https://doi.org/10.2147/ijn.s9155.

Lingayat, Vishal J, Nilesh S Zarekar, and Rajan S Shendge. 2017. "Solid Lipid Nanoparticles: A Review." *Nanoscience and Nanotechnology Research* 2: 67–72.

López-García, R, and A Ganem-Rondero. 2015. "Solid Lipid Nanoparticles (SLN) and Nanostructured Lipid Carriers (NLC): Occlusive Effect and Penetration Enhancement Ability." *Journal of Cosmetics, Dermatological Sciences and Applications* 5 (02): 62.

Mahant, Sheefali, Rekha Rao, and Sanju Nanda. 2018. "Chapter 3 - Nanostructured Lipid Carriers: Revolutionizing Skin Care and Topical Therapeutics." In *Design of Nanostructures for Versatile Therapeutic Applications*, edited by Alexandru Mihai Grumezescu, 97–136. William Andrew Publishing. https://doi.org/10.1016/B978-0-12-813667-6.00003-6.

Majoros, Istvan J, Christopher R Williams, and James R Baker Jr. 2008. "Current Dendrimer Applications in Cancer Diagnosis and Therapy." *Current Topics in Medicinal Chemistry* 8 (14): 1165–79.

Marcotte, Michèle, Ali R Taherian Hoshahili, and H S Ramaswamy. 2001. "Rheological Properties of Selected Hydrocolloids as a Function of Concentration and Temperature." *Food Research International* 34 (8): 695–703. https://doi.org/10.1016/S0963-9969(01)00091-6.

Masood, Farha. 2016. "Polymeric Nanoparticles for Targeted Drug Delivery System for Cancer Therapy." *Materials Science and Engineering: C* 60: 569–78.

McEvoy, Dustin S, Dean F Sittig, Thu-Trang Hickman, Skye Aaron, Angela Ai, Mary Amato, David W Bauer, et al. 2016. "Variation in High-Priority Drug-Drug Interaction Alerts across Institutions and Electronic Health Records." *Journal of the American Medical Informatics Association* 24 (2): 331–38. https://doi.org/10.1093/jamia/ocw114.

Mohammadi, Maryam, Elham Assadpour, and Seid Mahdi Jafari. 2019. "Encapsulation of Food Ingredients by Nanostructured Lipid Carriers (NLCs)." In *Lipid-Based Nanostructures for Food Encapsulation Purposes*, edited by M. Mohammadi, 217–70. Elsevier.

Molpeceres, Jesús, Manuel Guzmán, Pilar Bustamante, and M del Rosario Aberturas. 1996. "Exothermic-Endothermic Heat of Solution Shift of Cyclosporin a Related to Poloxamer 188 Behavior in Aqueous Solutions." *International Journal of Pharmaceutics* 130 (1): 75–81. https://doi.org/10.1016/0378-5173(95)04294-6.

Montenegro, L, M G Sarpietro, S Ottimo, G Puglisi, and F Castelli. 2011. "Differential Scanning Calorimetry Studies on Sunscreen Loaded Solid Lipid Nanoparticles Prepared by the Phase Inversion Temperature Method." *International Journal of Pharmaceutics* 415 (1): 301–306. https://doi.org/10.1016/j.ijpharm.2011.05.076.

Moreno, R. 2001. "Rheology." In *Encyclopedia of Materials: Science and Technology*, edited by K H Jürgen Buschow, Robert W Cahn, Merton C Flemings, Bernhard Ilschner, Edward J Kramer, Subhash Mahajan, and Patrick Veyssière, 8192–96. Elsevier. https://doi.org/10.1016/B0-08-043152-6/01468-6.

Mukherjee, S, S Ray, and R S Thakur. 2009. "Solid Lipid Nanoparticles: A Modern Formulation Approach in Drug Delivery System." *Indian Journal of Pharmaceutical Sciences* 71 (4): 349.

Müller, R H, S Runge, V Ravelli, W Mehnert, Andreas F Thünemann, and E B Souto. 2006. "Oral Bioavailability of Cyclosporine: Solid Lipid Nanoparticles (SLN®) versus Drug Nanocrystals." *International Journal of Pharmaceutics* 317 (1): 82–89.

Narayan, Reema, Mohan Singh, OmPrakash Ranjan, Yogendra Nayak, Sanjay Garg, Gopal V Shavi, and Usha Y Nayak. 2016. "Development of Risperidone Liposomes for Brain Targeting through Intranasal Route." *Life Sciences* 163: 38–45.

Naseri, Neda, Hadi Valizadeh, and Parvin Zakeri-Milani. 2015. "Solid Lipid Nanoparticles and Nanostructured Lipid Carriers: Structure, Preparation and Application." *Advanced Pharmaceutical Bulletin* 5 (3): 305.

Nasr, Maha, Samar Mansour, Nahed D Mortada, and A A El Shamy. 2008. "Lipospheres as Carriers for Topical Delivery of Aceclofenac: Preparation, Characterization and In Vivo Evaluation." *AAPS PharmSciTech* 9 (1): 154–62. https://doi.org/10.1208/s12249-007-9028-2.

Nguyen, H M, I C Hwang, Jae W Park, and Hyun Jin Park. 2012. "Enhanced Payload and Photo-Protection for Pesticides Using Nanostructured Lipid Carriers with Corn Oil as Liquid Lipid." *Journal of Microencapsulation* 29 (6): 596–604.

Nogueiras-Nieto, Luis, Eduardo Sobarzo-Sánchez, José Luis Gómez-Amoza, and Francisco J Otero-Espinar. 2012. "Competitive Displacement of Drugs from Cyclodextrin Inclusion Complex by Polypseudorotaxane Formation with Poloxamer: Implications in Drug Solubilization and Delivery." *European Journal of Pharmaceutics and Biopharmaceutics : Official Journal of Arbeitsgemeinschaft Fur Pharmazeutische Verfahrenstechnik e.V* 80 (3): 585–95. https://doi.org/10.1016/j.ejpb.2011.12.001.

Pamudji, Jessie Sofia, Rachmat Mauludin, and Nasya Indriani. 2016. "Development of Nanostructured Lipid Carrier Formulation Containing of Retinyl Palmitate." *International Journal of Pharmacy and Pharmaceutical Sciences* 8 (2): 256–60. https://innovareacademics.in/journals/index.php/ijpps/article/view/9813.

Pardeike Jana, Aiman Hommoss, and Rainer H Müller. 2009. "Lipid Nanoparticles (SLN, NLC) in Cosmetic and Pharmaceutical Dermal Products." *International Journal of Pharmaceutics* 366 (1–2): 170–84.

Parhi, Rabinarayan, Suresh, Padilama. 2012. "Preparation and Characterization of Solid Lipid Nanoparticles: A Review." *Current Drug Discovery Technologies* 9 (1): 2–16. https://doi.org/10.2174/157016312799304552.

Patel, Dilip, Sandipan Dasgupta, Sanjay Dey, Y Roja Ramani, Subhabrata Ray, and Bhaskar Mazumder. 2012. "Nanostructured Lipid Carriers (NLC)-Based Gel for the Topical Delivery of Aceclofenac: Preparation, Characterization, and In Vivo Evaluation." *Scientia Pharmaceutica* 80 (3): 749–64. https://doi.org/10.3797/scipharm.1202-12.

Patidar, Ajay, Devendra Singh Thakur, Peeyush Kumar, and Jhageshwar Verma. 2010. "A Review on Novel Lipid Based Nanocarriers." *International Journal of Pharmacy and Pharmaceutical Sciences* 2 (4): 30–35.

Peng, Fei, Yuanyuan Su, Xiaoyuan Ji, Yiling Zhong, Xinpan Wei, and Yao He. 2014. "Doxorubicin-Loaded Silicon Nanowires for the Treatment of Drug-Resistant Cancer Cells." *Biomaterials* 35 (19): 5188–95.

Pezeshki, Akram. 2017. "Effect of Surfactant Concentration on the Particle Size, Stability and Potential Zeta of Beta Carotene Nano Lipid Carrier Original Research Article Effect of Surfactant Concentration on the Particle Size, Stability and Potentia." *International Journal of Current Microbiology and Applied Sciences* 4 (September): 924–932.

Poonia, Neelam, Rajeev Kharb, Viney Lather & Deepti Pandita. 2016. "Nanostructured Lipid Carriers: Versatile Oral Delivery Vehicle." *Future Science OA* 2 (3): 135.

Purohit, Dhruv K. 2016. "Nano-Lipid Carriers for Topical Application: Current Scenario." *Asian Journal of Pharmaceutics (AJP)* 10 (1): S1–S9.

Rainer, H Mu€ller, Jan Mo€schwitzer, and Faris Nadiem Bushrab. 2006. "Manufacturing of Nanoparticles by Milling and Homogenization Techniques." In *Nanoparticle Technology for Drug Delivery*, edited by Uday B Kompella Ram, and B Gupta, 1st ed., 32. CRC Press.

Reithmeier, H, J Herrmann, and A Göpferich. 2001. "Lipid Microparticles as a Parenteral Controlled Release Device for Peptides." *Journal of Controlled Release: Official Journal of the Controlled Release Society* 73 (2–3): 339–50. https://doi.org/10.1016/s0168-3659(01)00354-6.

Saha, Ranendra N, Sekar Vasanthakumar, Girish Bende, and Movva Snehalatha. 2010. "Nanoparticulate Drug Delivery Systems for Cancer Chemotherapy." *Molecular Membrane Biology* 27 (7): 215–31.

Salvi, Vedanti R, and Pravin Pawar. 2019. "Nanostructured Lipid Carriers (NLC) System: A Novel Drug Targeting Carrier." *Journal of Drug Delivery Science and Technology* 51: 255–67. https://doi.org/10.1016/j.jddst.2019.02.017.

Sawant, Rupa R, Onkar S Vaze, Karen Rockwell, and Vladimir P Torchilin. 2010. "Palmitoyl Ascorbate-Modified Liposomes as Nanoparticle Platform for Ascorbate-Mediated Cytotoxicity and Paclitaxel Co-Delivery." *European Journal of Pharmaceutics and Biopharmaceutics* 75 (3): 321–26. https://doi.org/10.1016/j.ejpb.2010.04.010.

Schmid-Wendtner, M-H, and H C Korting. 2006. "The pH of the Skin Surface and Its Impact on the Barrier Function." *Skin Pharmacology and Physiology* 19 (6): 296–302. https://doi.org/10.1159/000094670.

Schwarz, C. 1999. "Solid Lipid Nanoparticles (SLN) for Controlled Drug Delivery II. Drug Incorporation and Physicochemical Characterization." *Journal of Microencapsulation* 16 (2): 205–13.

Shegokar, R, K K Singh, and R H Müller. 2011. "Production & Stability of Stavudine Solid Lipid Nanoparticles—From Lab to Industrial Scale." *International Journal of Pharmaceutics* 416 (2): 461–70. https://doi.org/10.1016/j.ijpharm.2010.08.014.

Singh, Parveen, Rajiv Kumar Gupta, Rayaz Jan, and Sunil Kumar Raina. 2017. "Adherence for Medication among Self-Reporting Rural Elderly with Diabetes and Hypertension." *Journal of Medical Society* 31 (2): 86.

Stefaniak, Aleksandr B, Johan du Plessis, Swen M John, Fritz Eloff, Tove Agner, Tzu-Chieh Chou, Rosemary Nixon, Markus F C Steiner, Irena Kudla, and D Linn Holness. 2013. "International Guidelines for the In Vivo Assessment of Skin Properties in Non-Clinical Settings: Part 1. pH." *Skin Research and Technology* 19 (2): 59–68. https://doi.org/10.1111/srt.12016.

Tagami, Tatsuaki, and Tetsuya Ozeki. 2017. "Recent Trends in Clinical Trials Related to Carrier-Based Drugs." *Journal of Pharmaceutical Sciences* 106 (9): 2219–26.

Tamjidi, Fardin, Mohammad Shahedi, Jaleh Varshosaz, and Ali Nasirpour. 2013. "Nanostructured Lipid Carriers (NLC): A Potential Delivery System for Bioactive Food Molecules." *Innovative Food Science & Emerging Technologies* 19: 29–43.

Tapeinos, Christos, Matteo Battaglini, and Gianni Ciofani. 2017. "Advances in the Design of Solid Lipid Nanoparticles and Nanostructured Lipid Carriers for Targeting Brain Diseases." *Journal of Controlled Release* 264: 306–32.

Teng, Zaijin, Miao Yu, Yang Ding, Huaqing Zhang, Yan Shen, Menglao Jiang, Peixin Liu, Yaw Opoku-Damoah, Thomas J Webster, and Jianping Zhou. 2018. "Preparation and Characterization of Nimodipine-Loaded Nanostructured Lipid Systems for Enhanced Solubility and Bioavailability." *International Journal of Nanomedicine* 14 (December): 119–33. https://doi.org/10.2147/IJN.S186899.

Tran, Vinh Van, Tuan Loi Nguyen, Ju-Young Moon, and Young-Chul Lee. 2019. "Core-Shell Materials, Lipid Particles and Nanoemulsions, for Delivery of Active Anti-Oxidants in Cosmetics Applications: Challenges and Development Strategies." *Chemical Engineering Journal* 368: 88–114.

Triplett MD, Michael D. 2004. "Enabling Solid Lipid Nanoparticle Drug Delivery Technology by Investigating Improved Production Techniques."

Üner, Melike. 2006. "Preparation, Characterization and Physico-Chemical Properties of Solid Lipid Nanoparticles (SLN) and Nanostructured Lipid Carriers (NLC): Their Benefits as Colloidal Drug Carrier Systems." *Pharmazie* 61 (5): 375–86.

Varshosaz, Jaleh, Sharareh Eskandari, and Majid Tabbakhian. 2012. "Freeze-Drying of Nanostructure Lipid Carriers by Different Carbohydrate Polymers Used as Cryoprotectants." *Carbohydrate Polymers* 88 (4): 1157–63. https://doi.org/10.1016/j.carbpol.2012.01.051.

Vitorino, C, J Almeida, L M Gonçalves, A J Almeida, J J Sousa, and A A C C Pais. 2013. "Co-Encapsulating Nanostructured Lipid Carriers for Transdermal Application: From Experimental Design to the Molecular Detail." *Journal of Controlled Release* 167 (3): 301–14. https://doi.org/10.1016/j.jconrel.2013.02.011.

Wa Kasongo, Kasongo, Ranjita Shegokar, Rainer H Müller, and Roderick B Walker. 2011. "Formulation Development and in Vitro Evaluation of Didanosine-Loaded Nanostructured Lipid Carriers for the Potential Treatment of AIDS Dementia Complex." *Drug Development and Industrial Pharmacy* 37 (4): 396–407. https://doi.org/10.3109/03639045.2010.516264.

Wissing, Sylvia A, Rainer H Müller, Lars Manthei, and Christian Mayer. 2004. "Structural Characterization of Q10-Loaded Solid Lipid Nanoparticles by NMR Spectroscopy." *Pharmaceutical Research* 21 (3): 400–405. https://doi.org/10.1023/B:PHAM.0000019291.36636.c1.

Xu, Qingguo, Nicholas J Boylan, Shutian Cai, Bolong Miao, Himatkumar Patel, and Justin Hanes. 2013. "Scalable Method to Produce Biodegradable Nanoparticles That Rapidly Penetrate Human Mucus." *Journal of Controlled Release* 170 (2): 279–86. https://doi.org/10.1016/j.jconrel.2013.05.035.

Yong, Chul Soon, Yu-Kyoung Oh, Se Hyun Jung, Jong-Dal Rhee, Ho-Dong Kim, Chong-Kook Kim, and Han-Gon Choi. 2004. "Preparation of Ibuprofen-Loaded Liquid Suppository Using Eutectic Mixture System with Menthol." *European Journal of Pharmaceutical Sciences* 23 (4): 347–53. https://doi.org/10.1016/j.ejps.2004.08.008.

6 Role of β-Glucans in Dyslipidemia and Obesity

*Hitesh Malhotra, Peeyush Kaushik, Anjoo Kamboj, and Rupesh K. Gautam**

CONTENTS

6.1 Introduction 123
6.2 Fibers in the Prevention and Treatment of Metabolic Disorders 125
6.3 β-Glucans in Metabolic Syndrome 125
  6.3.1 β-Glucans and Insulin Resistance 126
  6.3.2 Role of β-Glucans in Dyslipidemia 127
  6.3.3 Role of β-Glucans in Regulating Blood Pressure 129
  6.3.4 Role of β-Glucans in Obesity 129
6.4 Mechanisms of Action 133
  6.4.1 Clinical Studies on Hypocholesterolemic Capability of β-Glucans 133
  6.4.2 Role of Dietary Fibers in Bile Acid Cycle and Cholesterol Homeostasis 135
  6.4.3 Role of Short-Chain Fatty Acids 138
  6.4.4 Role of Microbial Polysaccharides in Lipid Metabolism/Homeostasis 139
  6.4.5 Microbiome-Mediated Lipid Assimilation and Biotransformation 139
6.5 Conclusion 139
References 140

6.1 INTRODUCTION

Metabolic syndrome includes several correlated conditions, such as obesity, hyperglycemia, hypertension and dyslipidemia, as illustrated in Figure 6.1. The existence of any or a combination of these conditions is considered to be metabolic syndrome (Torpy, Lynm & Glass, 2006). Obesity is today considered a pandemic, as around 1 billion of the population are affected by this disorder. Obesity is associated with numerous comorbidities, such as ischemic heart disease, acute coronary syndrome, hyperinsulinemia, diabetes mellitus and vascular disorders (Fujioka, 2002). Clinical and pre-clinical studies reveal that diet is considered to be an important determinant of metabolic disorder as well as a first-line approach in its prevention and treatment. Numerous dietary ingredients and practices are used for regulating blood glucose, blood pressure, insulin resistance, obesity and dyslipidemia. Dietary fiber present in fruits, vegetables, legumes and cereals is one of the major ingredients (Vrolix & Mensink, 2010).

β-Glucan is a by-product obtained after milling of oat and barley bran and consumed as a food adjuvant and health promoter. β-Glucan content varies with environmental conditions due to the presence of an endosperm development enzyme, i.e. glucan hydrolase or licheninase, which facilitates cell wall degradation during germination (Charalampopoulos *et al.*, 2002). The maximum content of β-glucan has been reported in barley, i.e. around 20 g per 100 g of water-soluble fraction, while for other cereals, the contents are as follows: sorghum 1.3–2.7 g, rye 1.1–6.2 g, triticale

* Corresponding author: drrupeshgautam@gmail.com

DOI: 10.1201/9781003244721-6

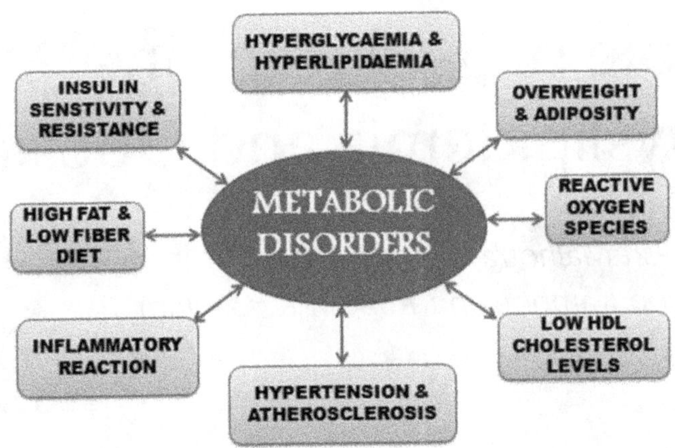

FIGURE 6.1 Types of metabolic disorders.

0.3–1.2 g, maize 0.8–1.7 g and wheat 0.5–1.0 g (Bacic, Fincher & Stone, 2009). Canada has been the major producer of oats and barley since 2005 and is thus involved in the mass production of β-glucans through dry milling (Lazaridou & Biliaderis, 2007).

Fibers in general are defined based on source, chemistry, metabolism, digestion and physiological significance. "Dietary fibers are the main component of food mainly derived from the cellular walls of plants having a poor digestive value in human beings" (Trowell, 1972). Later, due to certain limitations, this definition become non-centralized, as fibers are also extracted from animal, fungal, bacterial and synthetic sources. Furthermore, fibers are classified on the basis of chain length and the linkages involved in monomeric units. Also, chemical bonding and linkages between the monomeric units are considered as the basis of classification. Physiologically, fibers referred to as non-digestible molecules possess some metabolic outcomes. The outcomes of non-digestible fibers depend upon their physicochemical properties, such as viscosity in the alimentary canal, fermentation and prebiotic effects (Jenkins, Wolever & Leeds, 1978; Wood *et al.*, 1994). For instance, carbohydrates are divided on the basis of chain length, i.e. oligosaccharides and polysaccharides. Oligosaccharides, further, are of two types: maltodextrins (α-glucans); and raffinose and stachyose, fructo- and galactooligosaccharides or other oligosaccharides (non-α-glucan). Polysaccharides are sub-classified into starch (α-1,4 and 1,6 glucans) and non-starch compounds such as cellulose, hemicelluloses and pectin.

In certain cases, fibers are distinguished as soluble and insoluble dietary fibers based on solubility in buffer solutions. For dietary fibers, solubility means that the fibers can be easily dissolved in a buffer solution (Cho, De Vries & Prosky, 1997). The insoluble fibers are mainly composed of cellulose, hemicellulose, starch and chitin, while β-glucans, mucilages, pectins and galactomannan gums are the main components of soluble fibers. Insoluble fibers can also be termed bulk fibers, as they increase fecal bulk and reduce the transit time in the intestine. In contrast, soluble fibers increase transit time by interfering with gastric emptying time and absorption (Cummings & Stephen, 2007).

The digestion of carbohydrates depends upon the emptying rate of carbohydrates in the stomach and their availability for absorption as well as transport of released sugars (Englyst, Liu & Englyst, 2007). Besides, the rate at which carbohydrates are released from the food matrix and amylase activity also affect the absorption rate of glucose. Carbohydrates possessing cell wall polysaccharides, gums, fructans, maltodextrins and resistant starch are categorized as resistant carbohydrates due to negligible digestion in the large intestine but may be fermented by intestinal microbiota. However, fibers like lignins, cellulose, etc., are neither digested nor fermented in the large intestine (Cummings, 1981).

6.2 FIBERS IN THE PREVENTION AND TREATMENT OF METABOLIC DISORDERS

Currently, dietary fibers are recommended by many medical practitioners for the prevention and treatment of metabolic disorders. Dietary fibers play a significant role in the treatment of obesity, cardiac diseases and diabetes mellitus. Furthermore, a fiber-rich diet lowers glucose in diabetic patients and reduces low-density lipoprotein levels in dyslipidemia patients (Brennan, 2005). Surveys conducted by various international bodies reveal that subjects consuming high-fiber or cereal diets are less susceptible to cardiac and metabolic disorders. Furthermore, a whole grain–rich diet negatively impacts metabolic syndrome (Liu *et al.*, 2003). In the diverse list of dietary fibers, α-cyclodextrins are considered to be of great significance, especially in dyslipidemia, where they reduce cholesterol and triglyceride levels (Artiss *et al.*, 2006). α-Cyclodextrin in diabetics improves insulin resistance and leptin level, thus controlling the plasma glucose level. Psyllium (*Plantago ovata*) husk and methylcellulose, when taken for 2 months, ameliorate the lipid profile in dyslipidemia and modulate the secretion of inflammatory molecules such as tumor necrosis alpha (TNF-α) and adiponectin by adipocytes in obesity. By activating the AMP-protein kinase system, *Plantago ovata* suppresses the accumulation of lipids in the hepatocytes and improves hyperinsulinemia as well as hyperleptinemia in diabetic and obese patients (Galisteo *et al.*, 2010). Numerous clinical studies depict that dietary fibers in the diet reduce the chances of metabolic syndrome, such as diabetes mellitus, obesity and hyperlipidemia (Steemburgo *et al.*, 2009). Besides dietary fibers also enhance the level of adiponectin in diabetic patients and are thus responsible for improving glycemic control and insulin resistance (Mantzoros *et al.*, 2005). β-Glucan soluble fiber is also consumed by patients suffering from dyslipidemia, obesity, insulin resistance and hypertension.

6.3 β-GLUCANS IN METABOLIC SYNDROME

β-Glucan is gaining importance in developed as well as developing countries as a dietary fiber due to its profound therapeutic and bioactive significance with negligible side effects (Ripsin, Keenan & Jacobs, 1992). Glucans are glucose polymers and are classified based on chain linkages, i.e. α or β-linked. β-Glucans are heterogeneous polysaccharides containing D-glucose monomers linked by β-glycosidic bonds (Barsanti *et al.*, 2011). β-(1,3)-D-Glucan is the simplest form, present in cell walls of most prokaryotes and eukaryotes, while in cereal grains, linear β-(1,3;1,4)-D-glucans are present. Glucans are also present in the cell walls of yeasts and fungi, and certain bacteria, such as *Streptococcus pneumoniae*, possess β-(1,3;1,2)-D-glucan. In plants and microorganisms, glucans are involved in signaling processes and interaction with the environment (McIntosh, Stone & Stanisich, 2005). Besides, the biological activities of glucan depend upon frequency and length of branching, molecular mass, conformations and solubility.

β-Glucans are non-digestible molecules fermented mainly in the colon and caecum of the large intestine. β-Glucans profoundly induce the proliferation and growth of gut bacteria (Topping & Clifton, 2001; Kedia, Vazquez & Pandiella, 2008). Moreover, the solubility of β-glucans varies with linkage properties. For instance, (1→ 3)-β-glucans are insoluble in water with a degree of polymerization of more than 100. With a reduction in the degree of polymerization, the solubility of a particular glucan is enhanced. Furthermore, the substituents attached to the side chain also influence the solubility. For example, branched (1→ 6)-β-glucan shows a higher solubility profile as compared with the unbranched molecule.

On the basis of physicochemical characteristics, the biochemical properties of β-glucans vary. β-Glucans are considered to be most effective in infectious diseases and neoplasm. β-Glucans are potent immunostimulants with targeted activity (Sonck *et al.*, 2010). Pre-clinical and clinical studies reveal that (1→ 3)-β-glucans significantly improve immune responses by humoral as well as cell-mediated immune cells (Vetvicka, Dvorak & Vetvickova, 2007). Furthermore, β-glucans upregulate the activity of macrophages, mononuclear cells and neutrophils, and thus, possess anti-microbial activity (Tzianabos, 2000). β-Glucans when administered in animals significantly enhance microbial clearance and thus, reduce the mortality rate (Hetland *et al.*, 2000). Numerous clinical studies reveal that β-(1→ 3;1→ 6)-D-glucan suppresses post-surgical yeast growth and shortens length of stay in the intensive care unit as well as improving survival efficacy (Babineau *et al.*, 1994).

β-Glucan has gained noteworthy interest among practitioners in the management and prevention of metabolic syndrome. When β-glucan was administered in obese dyslipidemic subjects for 2 months, a significant improvement was observed in the lipid profile, mainly increased high-density lipoprotein (HDL) levels (Nicolosi *et al.*, 1999). Also, prolonged consumption of β-glucan in mice improves the metabolic abnormalities that occur due to a high-fat diet. The cell wall contains the prime source of glucans, i.e. branched chitin-β-1,3 glucan, which significantly reduces body weight, plasma glucose level, triglyceride accumulation in the liver and dyslipidemia in subjects on a high-fat diet, as depicted in Figure 6.2. Moreover, β-glucan also upregulates the microbiota in the large intestine, which is considered as the main mechanism behind the health-promoting action of β-glucan. Also, β-glucan suppresses post-meal plasma glucose and cholesterol levels as well as improving insulin activity. Due to its linear, unbranched and non-starch polysaccharide structure, β-glucan forms a highly viscous solution, and the viscosity also depends on the molecular weight, solubility and concentration. The highly viscous β-glucan solution possesses significant health benefits as compared with other forms of β-glucan (Wood & Beer, 1998).

6.3.1 β-Glucans and Insulin Resistance

Insulin resistance due to hyperglycemia or diabetes mellitus type II occurs as a result of metabolic abnormalities in obese subjects (Xu, Song & You, 2010). Today, various dietary fibers are used to overcome postprandial hyperglycemia, insulin resistance and insulin desensitization in obese patients (Hanai, Ikuma & Sato, 1997). For instance, a solution containing dextrin or corn fibers significantly attenuates postprandial hyperglycemia and insulin activity (Kendall, Esfahani & Hoffman, 2008). In animals, psyllium administration prevents insulin resistance as well as stroke frequency in obese subjects (Song *et al.*, 2000). β-Glucan shows significant improvement in glycemic control by interacting with food digestion and absorption mechanisms. As compared with other dietary fibers, β-glucan produces a profound hypoglycemic effect as well as improving insulin sensitivity at a low dose (Makelainen, Anttila & Sihvonen, 2007; Maki, Galant & Samuel, 2007). A clinical study in which subjects suffering from type II diabetes mellitus were administered β-glucan

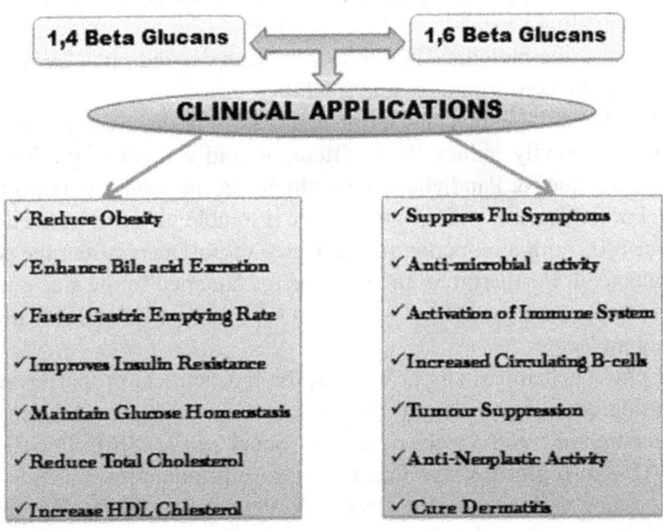

FIGURE 6.2 Therapeutic applications of β-glucans.

at varying doses revealed that β-glucan decreased plasma glucose levels when compared with the control group (Tappy, Gugolz & Wursch, 1996). A linear relation was seen between the dose of β-glucan and the regulation of glucose level. Also, oat bran contains 7.3 g β-glucan as compared with cereals, which contain only 3.5 g, and thus, oat bran is of great value in metabolic disorders (Jenkins *et al.*, 2002). Furthermore, oat bran flour contains 9.4 g of β-glucan and is thus useful in reducing postprandial hyperglycemia (Tapola *et al.*, 2005). β-Glucan at a dose of 5 g/day attenuates plasma glucose and insulin responses in hypercholesterolemia as compared with untreated subjects (Biorklund *et al.*, 2005). Indeed, in normal individuals, β-glucan when given at a low dose maintains the glucose level without affecting postprandial glycemia. Food significantly influences β-glucan-mediated glucose homeostasis. Further, it has been seen that insulin response, as well as release, consistently declines after β-glucan ingestion in healthy and obese individuals (Granfeldt, Nyberg & Bjorck, 2008). Likewise, when rye bread containing β-glucan was ingested in healthy individuals, a significant reduction in insulin release was seen after the meal without affecting the glycemic control.

Numerous reasons have been proposed to elucidate the insulin- and glucose-lowering effect of β-glucan. The prime reason behind this includes the formation of a viscous solution, which retards gastric emptying and thus reduces digestion and absorption (Braaten *et al.*, 1991). The highly viscous solution retards enzyme diffusion and forms a non-stirred water layer, thus reducing glucose transportation through intestinal epithelial cells. Thus, it has been stated that β-glucan-mediated viscosity accounts for glucose homeostasis (Schneeman & Gallaher, 1985). Pre-clinical and clinical studies reveal that consumption of β-glucan lowers the plasma glucose level by 18% as compared with control in obese subjects (Nazare *et al.*, 2009). Besides, short-chain fatty acids (SCFAs), produced by microfloral fermentation in the large intestine, are considered as another mechanism for maintaining the glucose balance in the body by β-glucan. SCFAs such as butyric and propionic acid enhance the expression of glucose transport type-4 (GLUT-4) through peroxisome proliferator-activated receptor (PPAR-γ) activation. Enhanced expression of GLUT-4 controls hyperglycemia and improves insulin sensitivity, especially in muscle cells (Song *et al.*, 2000). Thus, the fermentability and viscosity mediated by β-glucan significantly work against hyperglycemia and insulin resistance.

6.3.2 Role of β-Glucans in Dyslipidemia

People suffering from any metabolic syndrome generally appear with atherosclerosis, elevated triglycerides and reduced HDL levels. The altered lipid profile leads to a higher risk of cardiovascular disease, mainly in overweight individuals. Dietary fibers are considered as the most reliable component present in a meal for maintaining cholesterol metabolism. A clinical study has shown that fibers such as psyllium, guar gum, pectin and bran significantly reduce the total cholesterol and HDL level. In diabetic and hypercholesterolemic patients, β-glucan and other fibers lower the concentration of low-density lipoprotein (LDL) by 8–10%. A clinical study involving obese subjects administered psyllium husk for 8 weeks shows a profound reduction in total cholesterol and LDL serum levels without altering the HDL level (Sola, Bruckert & Valls, 2010; Abumweis, Jew & Ames, 2010). Another study includes diabetic individuals who ingested dietary fibers for 6 weeks, and the concentration of triglycerides decreased by around 10% (Chandalia *et al.*, 2000). Similarly, psyllium and arabinoxylan consumption significantly improves the postprandial serum concentration of triglycerides as well as glucose tolerance in obese subjects and thus, reduces the probability of cardiovascular mortality and morbidity (Sola, Godas & Ribalta, 2007). The effect of dietary fibers depends upon the degree of solubility and viscosity, molecular structure and frequency of administration of the fiber.

The effect of β-glucan on lipid parameters depends upon various factors, for instance, origin, molecular weight and source of the β-glucan, meal composition, method adopted for the preparation of food as well as β-glucan, the disease state of the individual and study design (Talati *et al.*,

2009). β-Glucan obtained from barley shows limited variation in lipid content, and one of the major determinants is dose. A randomized clinical study demonstrates that ingestion of barley β-glucan for 4–6 weeks lowers the LDL and total cholesterol level. Another clinical study, including individuals receiving 3–10 g of β-glucan once a day for around 3 months, reveals that significant reductions in cholesterol, triglycerides and LDL concentration were seen as compared with the control group (Talati *et al.*, 2009). Furthermore, a clinical trial was conducted in Japan in which hypercholesterolemic patients were treated with pearl barley containing a significant amount of β-glucan for 3 months. A significant reduction in total cholesterol and LDL serum concentration shows the hypolipidemic activity of β-glucan (Shimizu, Kihara & Aoe, 2008). Furthermore, barley β-glucan given at a dose of 6 gm/day caused marked reduction in both LDL and total cholesterol serum level in dyslipidemic patients (Behall, Scholfield & Hallfrisch, 2004). On the other hand, β-glucan ingested in the form of beverages, bread, cakes or muffins produced no significant reduction in serum cholesterol or lipoprotein level (Keogh, Cooper & Mulvey, 2003). Therefore, regulation of lipid serum concentration by β-glucan also depends upon the vehicle or delivery system. As per the U.S. Food and Drug Administration (FDA), a daily intake of 3 g of oat β-glucans is effective in maintaining lipid profile. Clinical studies clearly demonstrate that 2–10 gm/day of β-glucans causes net reduction of total cholesterol and LDL cholesterol by 12.4 mg/dl and 11.3 mg/dl, respectively. A more significant reduction was observed in total cholesterol level as compared with other lipids in serum after the ingestion of 4 g of β-glucans. Later, another study was conducted in which a beverage obtained from oats containing 5 g of β-glucans was administered to hypercholesterolemic obese patients for 2 months. A marked reduction in total cholesterol, as well as LDL cholesterol, was observed, and significant improvement was seen in HDL cholesterol level and total cholesterol/HDL cholesterol ratio as well as LDL cholesterol/HDL cholesterol ratio in overweight and hypercholesterolemic patients (Reyna-Villasmil, Bermudez-Pirela & Mengual-Moreno, 2007). Moreover, a β-glucan intake of 4 g/day when incorporated in a soup meal for 5 weeks reduces the serum LDL levels by around 4.0% as compared with a normal diet (Biorklund & Holm, 2008), but β-glucan ingestion in the form of oat bran cereal for 10–12 weeks showed no marked improvement in the lipid profile of the subjects. Thus, along with dose, the food vehicle also has a considerable effect on the therapeutic efficacy of β-glucans (Lovegrove *et al.*, 2000).

Furthermore, the efficacy of β-glucans is also determined by the mode of administration, as variation in results occurs even after using the same vehicle with different methods of administration. Also, the ingestion of β-glucans via different food vehicles, like cereals and muffins, markedly reduces LDL cholesterol, indicating the significance of molecular weight and structure (Kerckhoffs, Hornstra & Mensink, 2003). In contrast, the consumption of β-glucans in bread did not produce any significant hypocholesterolemia, as the bread formation process leads to depolymerization of β-glucan (Trogh *et al.*, 2004).

The variation in response to β-glucans could be attributed to the molecular weight, structure, viscosity and solubility of the dietary fibers, which in turn depend upon the genome of the source from which the β-glucans are obtained (Burkus & Temelli, 1998). For example, β-glucans obtained from oat are readily soluble in water as compared with those from barley because β-glucans from oat have a higher molecular weight, which is attributed to more side chains and a high degree of polymerization, resulting in high water solubility (Wood, Weisz & Mahn, 1991). Secondly, the lower the solubility and/or the molecular weight, the lower will be the viscosity, resulting in poor hypocholesterolemic activity. Therefore, β-glucans having a high molecular weight and greater water solubility lead to a marked reduction in LDL cholesterol as compared with low-weight β-glucans (Theuwissen & Mensink, 2008). This is why β-glucans obtained from oats produce a profound effect on lipid profile as compared with barley β-glucans. The mechanism behind β-glucan-mediated hypercholesterolemia could be altered bile composition and release pattern. In general, fibers enhance bile acid excretion and thus activate cholesterol hydroxylase, the enzyme involved in the elimination of cholesterol (Goel *et al.*, 1999). In addition, β-glucans inhibit the reabsorption of bile through hepatoportal circulation and consequently promote the metabolism of bile by the microflora and increase the excretion of bile as well. Also, enhanced excretion of bile via GIT stimulates the conversion of serum cholesterol to bile in the liver (Dongowski, Huth & Gebhardt, 2003). Some studies also report that β-glucans and some other soluble fibers promote the excretion of cholesterol by the level of deoxycholic acid in the intestine, which inhibits the absorption of cholesterol in the systemic circulation (Marlett *et al.*, 1994). Fermentation of β-glucans alters the concentration of bile acids as well as the production of SCFAs, which influences fat metabolism. For instance, propionic acid suppresses cholesterol synthesis and thus contributes to hypercholesterolemia. Similar results are depicted after the production of acetate as a by-product of fermentation (Lin *et al.*, 1995).

Besides hypercholesterolemia, β-glucans are also responsible for overcoming hypertriglyceridemia and hyperglycemia by reducing the rate of absorption of triglycerides and glucose in the intestine. It is well documented that glucose induces hypertriglyceridemia through de novo lipogenesis. Thus, inhibition of lipogenesis caused by soluble fibers is proposed as one of the mechanisms of β-glucan-mediated hypotriglyceridemia. For instance, oligofructose inhibits lipogenesis in the liver by modulating fatty acid synthetase activity and leads to hypotriglyceridemia. In addition to the inhibition of uptake of long-chain fatty acids and cholesterol, β-glucans also cause downregulation of genes related to lipid mobilization and synthesis (Ebihara & Schneeman, 1989; Liljeberg & Bjorck, 2000). Thus, from this discussion, it is concluded that β-glucans possess a significant hypocholesterolemic property with mild hypotriglyceridemic activity.

6.3.3 Role of β-Glucans in Regulating Blood Pressure

Increased blood pressure is considered a major outcome of metabolic disorder, resulting in life-threatening conditions like ischemic heart disease, stroke and renal diseases (Chobanian, Bakris & Black, 2003). β-Glucan is being studied by many researchers for its hypotensive activity. A clinical study conducted on hypertensive subjects reveals that by increasing dietary fiber consumption, a marked reduction in blood pressure was seen (Whelton *et al.*, 2005). Furthermore, in a study on dyslipidemic patients, by increasing the intake of dietary fiber such as β-glucan or psyllium for 4 weeks, a significant reduction was seen in total cholesterol level and blood pressure (Jenkins, Kendall & Vuksan, 2002). A comparative randomized study was conducted on hyperinsulinemic and hypertensive patients, in which one group received 5.52 g/day of β-glucan and the second group received <1 g/day. The result shows that the group receiving 5.5 g/day β-glucan experienced a marked reduction in systolic and diastolic arterial pressure as compared with the second group (Keenan *et al.*, 2002). In a similar way, another study included untreated stage I hypertensive patients receiving 8 g/day of oat bran for 12 weeks, showing a significant reduction in blood pressure (He *et al.*, 2004).

Numerous hypotensive mechanisms have been proposed for β-glucan. Modulation of insulin activity and improvement in insulin resistance are considered a major mechanism contributing to the hypotensive activity of β-glucan. In addition, β-glucan regulates insulin metabolism in the body (Ferrannini, Buzzigoli & Bonadonna, 1987). Moreover, intake of β-glucan causes a marked reduction in serum cholesterol level, which leads to significant improvement in endothelium-mediated arterio-venodilation (Crago *et al.*, 1998). Thus, β-glucan produces satisfactory hypotensive activity by multiple pathways.

6.3.4 Role of β-Glucans in Obesity

Obesity is considered the main component of metabolic disorders and responsible for serious consequences as depicted in Figure 6.3. Presently, obesity is treated by either modifying meal components or using drugs. Practitioners generally advise obese patients to take a meal rich in fibers, as they reduce energy intake by enhancing satiety. Numerous trials have evidenced that by increasing the consumption of dietary fibers like β-glucan, a significant reduction in body weight was reported. A trial including 22 obese subjects concluded that after the intake of 12 g/day fiber for around 4 months, a 2 kg reduction in body weight was reported (Howarth, Saltzman & Roberts, 2001).

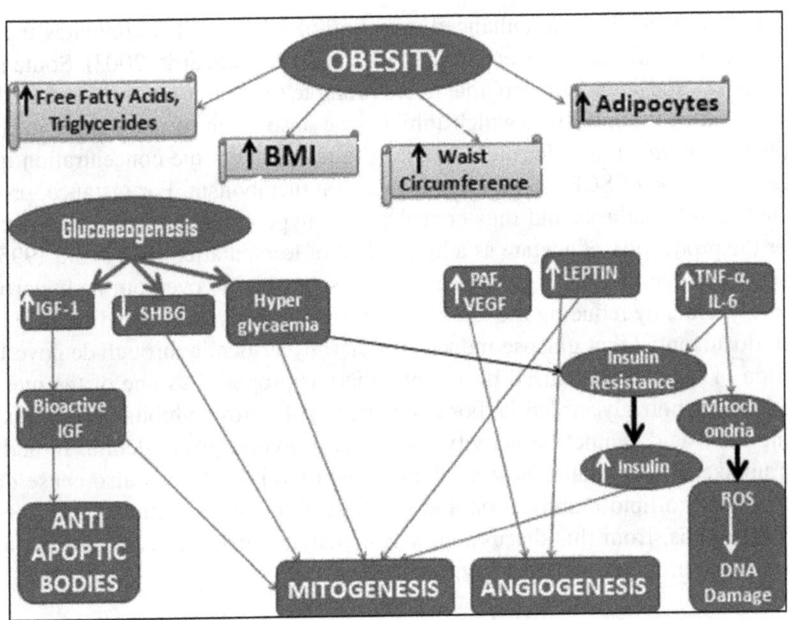

FIGURE 6.3 Complications of obesity.

Furthermore, administration of 1.25 g/day of glucomannan for 1 month causes a significant reduction in body weight due to high water-holding capacity (Birketvedt *et al.*, 2005).

For reduction in body weight, fibers such as β-glucan, psyllium and pectin are preferred, as they impart high viscosity. For instance, guar gum, when ingested along with a semi-solid meal, reduces appetite and desire for frequent food intake in obese subjects (Kovacs *et al.*, 2001). Similarly, dextrins administered along with desserts produce a dose-dependent improvement in satiety (Raben *et al.*, 1997). β-Glucan, either alone or in combination with diet, significantly enhances satiating properties and thus diminishes hunger (Schroeder *et al.*, 2009). Barley is considered as the best source of β-glucan, and barley-based meals produce a prompt effect as compared with normal, high-glucose or non-fibrous meals (Granfeldt *et al.*, 1994). Furthermore, β-glucan in beverages at a concentration of 10.5 g/400 g or 2.5–5.0 g/300 g produces fullness sensations as compared with non-fiber beverages (Lyly *et al.*, 2009). Also, biscuits containing β-glucan markedly suppress the appetite in healthy subjects without altering the afternoon meal. β-Glucan-rich bread also diminishes hunger by enhancing the satiety pathway and thus significantly reduces the energy intake (Vitaglione *et al.*, 2009). In contrast, β-glucan consumption through barley or oats by healthy subjects did not significantly modify appetite and satiety scores (Kim *et al.*, 2006). Thus, the efficacy of β-glucan in controlling hunger and satiety depends upon numerous factors such as dose, food vehicle, time of administration, etc. The dose is considered a major determining factor. For instance, 2.5 g of β-glucan in 5 g of oat barley produces a highly significant reduction in hunger as well as improvement in satiety as compared with a fiber-free diet, while by doubling the dose of β-glucan/barley, no supra-additive effect was observed (Lyly, Ohls & Lahteenmaki, 2010). In addition to dose, β-glucan effects also depend upon solubility and molecular size. β-Glucans with molecular weight between 31 and 3100 kDa are considered suitable for weight reduction in obese patients. Lastly, the carrier used for intake of β-glucan also plays an important role. For instance, administration of β-glucan via a solid or semi-solid diet does not produce any significant effect as compared with a liquid meal as a carrier. Thus, using β-glucan via solid meals masks the satiating activity of β-glucans (Kirkmeyer & Mattes, 2000).

In a clinical study, diabetic patients were supplemented with β-glucans at a dose of 9 g/day for 6 months. No significant improvement in hunger patterns was observed (Pick *et al.*, 1996).

Similarly, no reductions or alterations in hunger patterns were reported in dyslipidemic patients receiving β-glucans at 8 g/day for 4 weeks as compared with the control group. Thus, administration of β-glucans does not always result in weight reduction or alteration in hunger patterns; rather, the effect is limited to improvement in lipid and glucose profiles. Even at a dose range of 5–9 g/day, the addition of β-glucans to an energy-deficient diet does not produce any significant weight loss in obese women (Beck *et al.*, 2010). However, administration of 7 g/day of β-glucans to hypercholesterolemic patients in Japan for 4 months caused a profound reduction in the mass index, body circumference and fat depots as compared with patients consuming only normal meals. The reasons behind the variation in effect may be dose of administration, time of administration and regulatory factors of body weight. The mechanism for the improvement of satiety by dietary fibers is related to appetite regulation pathways such as taste, gastric emptying, fermentation and absorption (Burton-Freeman, 2000). Viscosity also plays a crucial role in stimulating satiety. Highly viscous β-glucan-containing beverages produce significant stimulation of satiety in overweight individuals. However, the viscosity-imparting β-glucans retard gastric emptying by slowing the digestion and absorption of nutrients as well as suppressing the enzymatic activity responsible for digestive purposes (Isaksson Lundquist & Ihse, 1982), which leads to satiety sensations. For instance, gastric emptying was slower after oat bran beverages as compared with low-viscosity drinks as shown by the paracetamol absorption test (Juvonen, Purhonen & Salmenkallio-Marttila, 2009).

Apart from viscosity, palatability also plays a significant role, as the lower the palatability, the lower will be the efficacy of fibers on satiety. Thus, an inverse association is observed between satiation and palatability (Ellis, Dawoud & Morris, 1991). Thirdly, lower insulinemic and glycemic response to fibers such as β-glucan may contribute to the better satiety property. Therefore, an inverse relationship exists between glucose and insulin response and satiation (Holt *et al.*, 1992). β-Glucans are responsible for the release of putative peptides responsible for maintaining satiety, which is considered as the fourth mechanism that demarcates the efficacy of SCFAs in appetite regulation. Furthermore, SCFAs regulate the release of hormones from the intestine that modulate the satiety signaling pathway.

The mechanism of appetite regulation by SCFAs is elucidated further. In general, fibers do not undergo any digestive process in the small intestine; rather, after they reach the colon, the microflora convert some of the dietary fibers into SCFAs through the process of fermentation. Soluble fibers such as pectin, β-glucan, gums and inulin are easily fermented as compared with non-soluble fibers. The main SCFAs generated in the gut are butyric acid, acetic acid and propionic acid (Cummings *et al.*, 1987). Butyric acid formed in the large intestine is rapidly absorbed in the colonocytes and utilized as an energy substrate (Roediger, 1980), while propionic acid and acetic acid are readily absorbed in the systemic circulation, maintain energy homeostasis and produce satiety effects (Hamer *et al.*, 2008). SCFAs control appetite by suppressing gastric motility, consequently retarding the digestion and absorption of nutrients and showing an anorexic effect. Both pre-clinical and clinical studies prove the activity of SCFAs in regulating the transit time in the gastrointestinal tract (GIT). However, the effect is more profound in humans than in animals due to altered physiology. SCFAs retard gastric motility either by activation of the vagus nerve in the gut, regulating intestinal smooth muscle tone, or enhancing the release of Peptide YY (PYY) and other related peptides responsible for gastric motility (Cherbut, 1995). Besides, SCFAs regulate gastric motility by modulating the release pattern of serotonin from enterochromaffin-like cells of enterocytes through the stimulation of the free fatty acid receptor (FFAF-2). In the nervous system, serotonin generally modulates appetite, mood and behavior (Berger, Gray & Roth, 2009). Thus, the release of serotonin from either the central or the peripheral system leads to the stimulation of vagal neurons and thus extends the transit time (Zhu *et al.*, 2001). Serotonin regulates appetite also by regulating the secretion of appetite-related hormones, for instance, cholecystokinin, PYY, ghrelin and glucagon-like peptide-1, in the entire gut (Dumoulin *et al.*, 1998). β-Glucan produces a profound effect on satiety hormones and subsequently, on food and appetite intake.

PYY, a 36–amino acid molecule, is mainly isolated from the porcine small intestine and exists in two forms, i.e. PYY1-36 and PYY3-36. The level of PYY reaches its maximum value after the ingestion of a meal and is maintained for several hours. In a pre-clinical study, the administration of PYY to rats significantly decreased food intake and body weight (Batterham, Cowley & Small, 2002), while in humans, PYY markedly suppresses hunger and thus, food intake. The addition of dietary fibers such as β-glucan, psyllium, etc., to the meal increases the PYY response in both healthy and obese volunteers (Karhunen, Juvonen & Flander, 2010). The poly fiber complex PolyGlycopleX, when administered for 1 month, increases the secretion of PYY while fasting as well as after the meal and retards food intake due to its gel-forming property (Reimer, Pelletier & Carabin, 2010). β-Glucan when administered to healthy subjects causes a 15% increase in PYY activity as compared with normal individuals. Similarly, when β-glucan is administered in a dose range of 2.16–5.45 g to obese patients, a dose-dependent increase in PYY response is observed (Beck *et al.*, 2009). The production of SCFAs after the fermentation of β-glucan is thought to be the mechanism behind increased PYY activity (Longo *et al.*, 1991). The SCFA-induced enhanced PYY activity is mainly attributed to FFA2 or GPR43 receptor present on the colonocytes (Karaki, Mitsui & Hayashi, 2006).

Glucagon-like peptide 1 (GLP-1) is secreted along with PYY from enterocytes and stimulates the release of insulin from pancreatic cells following nutrient ingestion to maintain energy homeostasis (Holst, 2007). Administration of GLP-1 to rodents causes a marked reduction in food intake and thus, leads to a reduction in body weight (Davis, Mullins & Pines, 1998). In addition, intravenous administration of GLP causes marked dose-dependent reduction in food intake in both normal and obese individuals (Verdich, Flint & Gutzwiller, 2001). The addition of dietary fibers such as β-glucan and oligofructose to a meal increases the expression of GLP-producing cells as well as the production of GLP in the colon (Cani *et al.*, 2007). Similarly, the addition of guar gum to breakfast also increases the GLP release as compared with control (Adam & Westerterp-Plantenga, 2005). In contrast, psyllium addition to breakfast did not alter the GLP release in healthy or overweight subjects (Frost *et al.*, 2003). Such variation is attributed to the difference in the food origin, the structure of the fiber, and the mode as well as the dose of administration. Release of GLP mainly occurs when fermented fibers like inulin and β-glucans are supplemented in a meal, as the fermentation by-products produced in the colon enhance the mRNA expression of proglucagon, and consequently, the plasma level of GLP is elevated. For instance, when fructans were administered to rats for 3 weeks the mRNA expression in the colon as well as the GLP level in plasma increases as compared with normal subjects (Cani, Dewever & Delzenne, 2004). In a similar way, administration of 16 g/day of fructans for 15 days significantly increases the level of GLP-1 by upregulation of mRNA expression in normal-weight adults (Cani, Lecourt & Dewulf, 2009). However, with only 6 days of supplementation, no effect was observed. Thus, the duration of supplementation plays an important role in fiber-mediated GLP release. As per the studies, the minimum time required to produce the effect, i.e. enhanced mRNA expression and increased level of plasma GLP, is 15–21 days (Gee & Johnson, 2005).

Cholecystokinin (CCK) is primarily responsible for the modulation of food intake and is secreted from intestinal cells in response to food intake. Pre-clinical and clinical data revealed that the level of CCK reached a maximum 15 minutes after food ingestion and was responsible for the reduction in food intake as well as an increase in satiety (Kissileff *et al.*, 1981). Dietary fibers regulate the production and release of CCK. β-Glucans (15.7 g) and guar gum (20 g), when administered in healthy and obese adults along with a meal, lead to significant, long-lasting production of CCK levels as compared with control subjects (Heini *et al.*, 1998). Besides, ingestion of β-glucans by obese females produces dose-dependent improvement in CCK levels during the initial 4 hours after the meal.

Ghrelin is a potent appetite stimulant that acts via secretagogue receptor (GH-SR), whose secretion rises rapidly before the meal and rapidly declines after food intake (Kojima *et al.*, 1999). Consumption of β-glucans and other soluble fibers suppresses the ghrelin levels significantly in healthy individuals (Nedvıdkova, Krykorkova & Bartak, 2003). Similarly, administration of guar gum at a dose of 21 g suppresses plasma ghrelin as compared with a fat, protein and carbohydrate diet (Erdmann, Lippl & Schusdziarra, 2003). Furthermore, when compared with an arabinoxylan-rich diet, β-glucans produce marked and prolonged suppression of ghrelin in comparison to the control group eating raw bread. The most prominent mechanism of suppression of ghrelin by fibers is the upregulation of plasma PYY levels. In addition, only fiber that undergoes fermentation significantly participates in ghrelin suppression. Secondly, fibers cause significant release of somatostatin and GLP-1, which in turn, leads to the suppression of ghrelin in healthy and overweight subjects (Shimada, Date & Mondal, 2003).

6.4 MECHANISMS OF ACTION

Numerous studies have reported that addition of oats to a meal causes a marked reduction in serum LDL cholesterol and consequently, a lower incidence of cardiac disorders (Tiwari & Cummins, 2011). Oats are considered the most reliable source of β-glucans, xyloglucan, arabinoxylan, insoluble fiber, proteins, vitamins and phenolic compounds. The constituents present in the oats impact the serum cholesterol level by either increasing the viscosity of the gut fluid or enhancing the elimination of cholesterol and bile acid via the fecal route (Wolever *et al.*, 2010). Governing bodies such as the World Health Organization (WHO), the USFDA and the European Commission claim that the intake of at least 3 g of β-glucans significantly lowers the cholesterol level. Pre-clinical studies reveal that the lipid-lowering capability of β-glucans can be attributed not only to viscosity enhancement but also to a potentiating effect on the gut microflora that maintains cholesterol homeostasis (Connolly *et al.*, 2016). The mechanism behind the cholesterol-lowering capability of β-glucans involves the metabolism of bile acids via bile acid hydroxylase (BSH) activity and subsequently, their excretion (Begley, Hill & Gahan, 2006). Bile acid metabolism is indirectly linked with the hypocholesterolemic property of β-glucans, because cholesterol is the immediate precursor of bile in the liver (Jones *et al.*, 2014). Secondly, the regulation of cholesterol metabolism through farnesoid X receptor (FXR) also influences cholesterol level by modulating cholesterol transportation in the gut (de Boer *et al.*, 2017). Furthermore, the synthesis of SCFAs such as propionic acid also interferes with cholesterol metabolism. Oats and their related products regulate the microflora in the gut and thus, cholesterol levels in the body (Connolly, Lovegrove & Tuohy, 2010). In a nutshell, β-glucans impact microbiota and consequently, regulates the bile acid pathway and the SCFA pathway involved in cholesterol homeostasis.

6.4.1 Clinical Studies on Hypocholesterolemic Capability of β-Glucans

A number of clinical studies (depicted in Table 6.1) have established the significant role of β-glucan-containing meals in lowering LDL cholesterol and thus, cardiac complications (Ripsin *et al.*, 1992). Tiwari and Cummins conducted a study including 126 subjects, in which the impact of β-glucans was investigated by measuring the level of total cholesterol and LDL cholesterol as well as plasma glucose levels. The study demonstrated that there is a significant dose-dependent reduction in serum level of total cholesterol, triglyceride and LDL cholesterol, while the HDL cholesterol level is increased. A meta-analysis by Whitehead *et al* also concluded that β-glucans or oats at 3 g/day markedly suppress both LDL and cholesterol levels without altering the HDL cholesterol. Thus, both studies support the recommendation that at a dose of 3 g/day, β-glucans significantly correct the lipid profile of dyslipidemic patients. In contrast, Abu Mweis *et al.*, studying various trials, concluded that not only dose but duration, sample size, population genome and source of β-glucans also influence the effect of β-glucans on cholesterol levels (Abu Mweis, Jew & Ames, 2010). There is no doubt that β-glucans at 3 g/day produce a hypolipidemic effect, but proportionate activity is not seen with an increase in dose. The reason behind this may be the physicochemical properties of

TABLE 6.1
Clinical Studies for Metabolic Syndrome

S. no.	Source of β-glucan	Study protocol	Outcomes	References
1	Oat	Subjects of either sex, 30–60 years old and having high levels of LDL-C	Significant reduction in LDL cholesterol and total cholesterol	Davidson *et al.* (1991)
2		Subjects with borderline hyperlipidemia and elevated BMI	Marked improvement in lipid profile observed at higher doses	Lovegrove *et al.* (2000)
3		Subjects 30–80 years old suffering high-grade hyperlipidemia	Significant reduction in LDL and total cholesterol, improvement in HDL level, and patients less prone to cardiac disorders	Jenkins *et al.* (2002)
4		Hypercholesterolemics 40–65 years of age with BMI around 30 kg/m^2	Significant reduction in LDL-C with no change in HDL-C and triglycerides	Kerckhoffs *et al.* (2003)
5		Subjects of either sex suffering from moderate to severe hypertension	Marked improvement in lipid profile as well as systolic and diastolic blood pressure	He *et al.* (2004)
6		Subjects aged more than 40 years suffering from dyslipidemia and hypertension	Marked suppression in blood pressure and insulin level is seen with non-significant effect on biomarkers of oxidative stress	Maki *et al.* (2007)
7		Subjects with average age of 50 years suffering from non-insulin-dependent diabetes mellitus (NIDDM) and BMI around 35 kg/m^2	Increased postprandial insulin secretion and sensitivity	Tappy *et al.* (1996)
8		Geriatric diabetic subjects with BMI around 30 kg/m^2	Marked reduction in postprandial glucose level with increase in insulin activity	Tapola *et al.* (2005)
9		Healthy adult volunteers	Reduction in insulin and glucose index	Makelainen *et al.* (2007), Granfeldt *et al.* (2008)
10		Obese subjects suffering from diabetes	Marked reduction in plasma glucose level with increase in insulin activity. No improvement observed in dyslipidemia	Cugnet-Anceau *et al.* (2010)
11	Barley	Subjects on average 30 years old suffering from hypercholesterolemia and hypertriglyceridemia, BMI of around 22–32 kg/m^2	Enhanced cholesterol transport mechanism, increased level of cholecystokinin and improvement in lipid profile	Bourdon *et al.* (1999)
12		Non-diabetic subjects, 25 years old	Dose-dependent improvement in plasma glucose and postprandial glucose index	Cavallero *et al.* (2002)
13		Dyslipidemic adult males with average BMI >30 kg/m^2	No significant reduction was seen in serum bad cholesterol levels	Keogh *et al.* (2003)
14		Subjects of all ages with moderate hypercholesterolemia	Marked reduction in triglycerides and LDL-C observed with reduction in HDL-C	Behall *et al.* (2004)
15		Hypercholesterolemics in the age group 30–60 with BMI around 22 kg/m^2	Significant reduction in serum LDL-C and visceral fat area	Shimizu *et al.* (2008)
16	Yeast	Hypercholesterolemic obese subjects in the age group 21–60	Reduction in total cholesterol, LDL-C and VLDL-C with improvement in HDL-C	Nicolosi *et al.* (1999)

β-glucans obtained from different sources. For instance, highly water-soluble and high–molecular weight β-glucans produce marked hypolipidemic activity as compared with low-MW and poorly water-soluble β-glucans (Theuwissen & Mensink, 2008). Furthermore, some studies suggest that variation in activity occurs due to differences in the food vehicle used, as it influences the physico-chemical properties.

6.4.2 Role of Dietary Fibers in Bile Acid Cycle and Cholesterol Homeostasis

Bile is mainly synthesized in the liver by cytochrome enzymes from cholesterol and stored in the gall bladder as chenodeoxycholic acid and cholic acid, the primary bile acids. Under normal physiological conditions, most of the bile released after a meal in the first segment of the intestine is reabsorbed from the last segment of the ileum and reused for digestion (Russell, 2003). The postprandial release of bile acid in the small intestine is modified by bile salt hydrolase released by microflora and consequently, the production of cognate amino acids, lithocholic acid and deoxycholic acid. Thus, gut microbiota, mainly Eubacteria and Clostridia, are responsible for bile metabolism, and alteration in the gut microflora may lead to certain disease states due to improper bile acid metabolism and thus, cholesterol homeostasis. Dietary fibers such as β-glucans markedly impact the bile acid profile via the regulation of microflora (Joyce & Gahan, 2016).

Moreover, FXR, a bile acid sensor/nuclear receptor, is widely distributed in the intestine and liver. FXR regulates energy intake and metabolism by controlling bile acid synthesis through fibroblast growth factor (FGR)-19, which acts on hepatocyte receptors and reduces the synthesis of bile acid. In addition, the modified bile acid in the intestine binds to FXR to produce the specific response (Sayin *et al.*, 2013). A study in mice revealed that defects in the FXR gene cause marked elevation in LDL cholesterol, while excessive activation leads to hypercholesterolemia. Furthermore, Fexaramine, an FXR agonist, induces fibroblast activity, which causes a reduction in body weight and serum cholesterol by regulating fat metabolism (Fang *et al.*, 2015).

In the GIT microenvironment, β-glucans form a viscous liquid suspension, and the degree of viscosity depends upon the solubility, concentration and molecular weight of β-glucans. Various pre-clinical and clinical studies reveal that β-glucan-mediated viscosity plays a crucial role in lowering cholesterol levels. Randomized trial studies have shown that the lipid-lowering capability of β-glucans was directly linked with molecular weight, i.e. β-glucans with higher molecular weight produce a profound effect as compared with lower–molecular weight β-glucans. In addition to lipid level, the plasma glucose level is also regulated after β-glucan administration due to the same mechanism (Wood *et al.*, 1994). Some researchers also reported that the high efficacy of high–molecular weight β-glucans is associated with their ability to form a viscous solution. Further, β-glucan-mediated increase in viscosity modulates bile acid metabolism and interferes with its reabsorption from the gut (McRorie & McKeown, 2017). As a result, enhanced excretion of bile acid takes place, which leads to greater consumption of cholesterol for bile acid synthesis in the liver via de novo synthesis. Several animal and human studies support that chronic consumption of β-glucans enhances the excretion of bile acid (Andersson *et al.*, 2010). The key marker of bile acid synthesis in humans is 7 alpha-hydroxy-4-cholesterin-3-one, whose measurement indicates the rate of de novo synthesis (Andersson, Ellegard & Andersson, 2002). Later, a study in pigs confirmed that there is a marked elevation in bile acid excretion during initial intake of β-glucans, but after prolonged administration, the bile acid excretion decreases due to adaptation. The reasons include modulation of gut microflora and reduction in bile acid uptake as well as cholesterol absorption (Gunness *et al.*, 2016). Therefore, β-glucans significantly overcome dyslipidemia by influencing bile acid excretion, cholesterol absorption and de novo synthesis of bile acid in the liver, mainly by acting on the microflora.

Many studies also indicate that treatment of mice with antibiotics alters cholesterol metabolism and consequently, hypercholesterolemia (Wainfan *et al.*, 1952). Another study demonstrates that in microbiota-free rodents, there is over-accumulation of cholesterol in the body as well as suppression of catabolism of dietary lipids, mainly cholesterol, and a marked reduction in the elimination of bile acid and cholesterol (Kellogg & Wostmann, 1969). Various studies have evidenced that upregulation of microflora through the administration of probiotics and dietary fibers such as β-glucans is responsible for hypocholesterolemia. Thus, by regulating the microflora, cholesterol homeostasis can be maintained, indicating that the microflora plays a key role in cholesterol metabolism. Studies show that there is a direct link between the diversity of microbial genes and metabolic state. An individual having a low content of microflora is more prone to develop adiposity and dyslipidemia (Le Chatelier *et al.*, 2013). β-Glucans show significant tolerance against pancreatic and gastric juice enzymes and transit directly to the colon, where their fermentation by microflora will take place. Deviation in bile acid metabolism influences the microbiota population and thus, is sometimes considered the main mechanism behind β-glucan-mediated alteration in gut microflora (Islam *et al.*, 2011).

Furthermore, *in vitro* and *in vivo* studies reveal that oat fibers containing β-glucans significantly influence the composition of gut microflora (Jayachandran *et al.*, 2018). *In vitro* fermentation studies that mimic the environment of the human colon reveal that a β-glucan-containing diet promotes the proliferation of gut microflora, mainly the *Clostridium histolyticum* subgroup and Bacteroidetes/Prevotella (Hughes *et al.*, 2008). *In vitro* studies demonstrate that oats containing β-glucans stimulate the proliferation of Bacteroidetes (Kristek *et al.*, 2019). Similarly, the administration of oat flakes in rodents and humans promotes the growth of the Bifidobacterium group and thus enhances bile acid excretion. Further studies show that β-glucans also promote the growth of Prevotella and Roseburia species and thus enhance the production of SCFAs like propionic acid and butyric acid (Fehlbaum *et al.*, 2018).

A study on mice reveals that the administration of β-glucans causes a marked increase in the population of Prevotella and Bacteroides species, while the count of Firmicutes declines (Luo *et al.*, 2017). Furthermore, Zhou *et al.* proved that whole grain administration in animals produces profound alterations in microflora diversity, mainly the Lactobacillaceae, Prevotellaceae and Alcaligenaceae families. This alteration in gut microflora significantly improves the total cholesterol and lipoprotein levels (Zhou *et al.*, 2015). A study conducted by Ryan *et al.* on apo E-deficient mice demonstrated a marked reduction in key markers of cardiac disorders after ingestion of β-glucans due to enhanced expression of Verrucomicrobiata and butyric acid in the gut. *Akkermansia muciniphila*, a member of Verrucomicrobiata, is responsible for improving gut barrier function, reducing obesity and enhancing metabolism. In contrast, clinical studies reveal that low-dose β-glucans did not affect the microflora composition, while at high doses, significant stimulation of gut microbiota appears, especially Firmicutes and Bacteroidetes. Further, a high dose of β-glucan, when administered for approximately 1 month, causes a reduction in body mass index (BMI), blood pressure, total cholesterol and LDL cholesterol (Wang *et al.*, 2016). Similar studies were conducted using the fluorescent *in situ* hybridization technique, and a marked increment in Lactobacillus and Bifidobacterium species and hypocholesterolemia was observed after the ingestion of oat-based β-glucans. The principal connection here is the overexpression of BSH activity produced by Lactobacillus and Bifidobacterium species in the host. Therefore, the consumption of β-glucans promotes the composition of gut microflora and thus, controls dyslipidemia.

In vitro and *in vivo* studies have confirmed that the consumption of β-glucan-containing fibers significantly increases the count of bacteria, such as Lactobacillus, Bacteroides and Bifidobacterium, responsible for BSH activity. Such bacterial species or genera possessing BSH activity are termed BSH positive species (Jones *et al.*, 2008). Thus, β-glucans strongly impact the BSH activity via modulation of the host microbiome. BSH-producing probiotics significantly reduce serum LDL cholesterol and regulate cholesterol levels in the host, which is considered as the prime mechanism for overcoming dyslipidemia via FXR modulation. BSH causes the deconjugation of bile acids and thus inhibits the capability of bile to form micelles with lipids. Hence, cholesterol absorption from enterocytes is reduced. This mechanism was proved by the administration of a BSH-positive *L. reuteri* strain in humans, which causes a significant reduction in total cholesterol and LDL cholesterol by interfering with the cholesterol in the gut. Thus, enhanced BSH activity is directly associated with suppression of cholesterol intake from the intestine and is considered as a precise mechanism of regulation of cholesterol level in the systemic circulation. Furthermore, the formation of unconjugated bile acids and an increase in de novo synthesis are also linked with the mechanism. Besides, Joyce *et al.* showed that enhancing the level of a BSH-active *L. salivarius* strain caused a marked improvement in dyslipidemia by decreasing the levels of serum triglycerides, total cholesterol and LDL-C. In a similar way, *L. acidophilus* administration in humans for 6 weeks significantly suppresses atherogenic cholesterol (Ooi & Liong, 2010). *L. reuteri*, a BSH-active strain, significantly treats hypercholesterolemia when ingested by humans by enhancing the elimination of bile acids.

Moreover, increased elimination and reduced reabsorption of bile suppress its interaction with FXR and thus, inhibit the stimulation of FXR. The formation of unconjugated bile acids further supports the bacterial conversion of bile to secondary bile molecules, which act as an FXR agonist. Thus, BSH-active strains promote the activation of BSH in the gut as well as the expression of FGRs by enteric cells, which leads to a marked reduction in bile acid production by hepatic cells. Thus, two possible hypotheses for the association of increased BSH activity with improvement in dyslipidemia are first, an increase in bile acid synthesis without FXR activity and second, stimulation of FXR as well as the release of fibroblast-induced bile acid synthesis in hepatocytes in the host following dietary intake of fibers containing β-glucans.

Hypothesis 1: Enhanced Activity of BSH Increases Bile Acid Synthesis and Elimination without Engaging FXR

The absorption of conjugated bile acids will take place actively through a specific transportation system in the ileum, while the unconjugated bile acids are absorbed through passive diffusion but at a very slow rate (Aldini *et al.*, 1992). BSH-active microflora markedly reduce the level of conjugated bile acids by converting them into the unconjugated form, and thus, reabsorption into enteric cells declines (Degirolamo *et al.*, 2014). The unconjugated bile acids, after conversion into secondary bile acids, are excreted via the fecal route. The decrease in bile acid absorption lowers the FXR activation in the last segment of the ileum. Various studies demonstrate that after oral ingestion of BSH-active probiotics, downregulation of FXR nuclear receptors is seen due to alteration in the bile acid profile. Further, a BSH-active strain reduced the FGR level and enhanced the Cyp7a1 and 8b1 expression in hepatocytes as well as bile acid synthesis (Degirolamo *et al.*, 2014). Similarly, oral administration of *L. plantarum* in mice elevated bile acid synthesis via the FXTR-FGF signaling pathway and thus, hypocholesterolemia (Zhai *et al.*, 2019). Besides, BSH-active *L. rhamnose* and *L. plantarum*, when administered to mice on a high-fat diet, reduced the serum cholesterol level by downregulating FXR translocation in enteric cells and enhancing the expression of Cyp7a1 in the liver. Furthermore, *L. reuteri* containing probiotics also contributes to hypocholesterolemia by enhancing the de novo synthesis of bile acids (Kim *et al.*, 2016).

The liver X nuclear receptor (LXR) is responsible for regulating FXR signaling in humans. Activation of LXR causes marked stimulation of the cholesterol efflux system in enteric cells and consequently, increases cholesterol excretion (Repa *et al.*, 2002). In support of this, *L. plantarum* administered to adult mice significantly suppresses the serum total cholesterol by enhancing the expression of Cyp7A1 in the liver and LXR in the gut and consequently, modulates the expression of ABCG5/8 and NPC 1L1 genes, responsible for cholesterol uptake. Thus, BSH-active probiotics cause the elimination of lipids through LXR activity (Heo *et al.*, 2018). Simultaneously, bile acid synthesis regulated by the FXR-FGF axis in the gut and hepatic cells is considered as another mechanistic approach for altering the circulating bile acids, as the concentration of bile acid entering the portal circulation influences the de novo bile acid synthesis as well as cholesterol metabolism via FXR signaling (Thomas *et al.*, 2008). Therefore, hypocholesterolemia is attributed to enhanced elimination of bile acid due to downregulation of FXR activity in the liver and reduction in bile acid re-circulation. The downregulation of FXR causes a reduction in small heterodimer partner (SHP), which stimulates LXR and consequently, enhances the expression of the ABCG5/8 gene in hepatic cells, responsible for the elimination of cholesterol as well as bile acids, and also lowers the serum cholesterol level (Yu *et al.*, 2002).

Hypothesis 2: FXR-FGF-Mediated Downregulation of Bile Acid via Enhanced BSH Activity

Certainly, conjugated and unconjugated bile salts activate the FXR; unconjugated bile acid produced by BSH-active strains enters the targeted cells for the regulation of bile acid and cholesterol without any specific transport system (Parks *et al.*, 1999). In a clinical study, administration of *L. reuteri* suppresses LDL cholesterol with an increase in FGF levels. Therefore, despite evidence for bile acid synthesis by probiotics, involvement of the FXR-FGF axis might be linked with regulation of bile acid synthesis and hypolipidemia. The involvement of the FXR-FGF axis and the role of BSH-active microbes in FXR signaling has been proved by many studies; for instance, administration of antimicrobials caused marked reduction in FXR activation due to the diminished expression of BSH-active microbes in the gut (Miyata *et al.*, 2009). But on administration of β-glucans to such BSH-deficit subjects, enhanced expression of FXR is seen, which maintains cholesterol homeostasis. Similarly, FXR agonists when administered orally to mice significantly lower the total cholesterol and LDL cholesterol in the systemic circulation (Kellogg & Wostmann, 1969). Besides, FXR agonists decrease the quantity of atherosclerotic plaque in the systemic circulation and thus, reduce the frequency of stroke (Wainfan *et al.*, 1952). Therefore, a possible mechanism by which FXR upregulation controls systemic cholesterol metabolism includes regulation of LDL cholesterol cellular uptake, reduction in very low-density lipoprotein (VLDL)-C, improvement in HDL-C level and regulation of cholesterol transportation as well as excretion. Further Joyce *et al.* demonstrated that the administration of β-glucans and probiotics enhances the cholesterol transportation through macrophages and excretion of cholesterol by the fecal route by modulating the bile signaling. The ABCG5/8 cholesterol transport system is tightly controlled by FXR as well as LXR, and enhanced expression of ABCG5/8 in the liver promotes bile secretion, while in the gut, it stimulates the elimination of bile and cholesterol via trans intestinal cholesterol excretion (TICE) activity (Repa *et al.*, 2002).

6.4.3 Role of Short-Chain Fatty Acids

SCFAs are produced after the fermentation of dietary fibers containing β-glucan-like molecules by the enteric microbiota. The most common SCFA produced in the human colon is propionic acid (Hughes *et al.*, 2007). Pre-clinical studies reveal that fermentation of oats enhances the production of SCFAs, i.e. propionate and butyrate, by regulating microflora activity in the colon. Similarly, chronic administration of β-glucan to pigs significantly increases the level of butyric acid (Hooda *et al.*, 2010). Hence, β-glucan intake in animals influences gut microflora, which causes marked production of SCFAs. The further effect of β-glucan on the microbial population of the human colon was examined, particularly butyric acid. In a clinical study, β-glucan administration caused a marked reduction in serum cholesterol level by enhancing the production of butyric acid (Queenan *et al.*, 2007). Similarly, administration of β-glucan-containing fibers for 2 months to humans lowers the serum cholesterol level and treats dyslipidemia by altering the microbiota and hence, enhancing the production of SCFAs (Velikonja *et al.*, 2019). Besides, increased SCFA production after ingestion of high–molecular weight β-glucan increases bile acid excretion as compared with low–molecular weight β-glucan (Thandapilly *et al.*, 2018), while certain studies demonstrate that whole-grain β-glucan significantly lowers the serum total cholesterol and LDL-C without enhancing the production of SCFAs.

Certainly, SCFAs are readily absorbed by epithelial cells of the intestine, and their level is considered as a key marker of microbial activity in the gut. For this, fecal studies were conducted, which represent the exact impact of gut microbes on SCFA production. Further, Carlson *et al* proved that β-glucan present in oats sufficiently enhances the production of propionic acid in the human gut by the microbial fermentation system (Carlson *et al.*, 2017). In support of this, *in vitro* studies were conducted by adding β-glucan to the microbial fermentation system, and a significant increase in SCFAs was observed (Sayar, Jannink & White, 2007). Moreover, SCFAs acting via specialized receptors, i.e. GPR41 and GPR43, regulate the production of local hormones, satiety and intestinal transit time (Koh *et al.*, 2016). Besides, the interaction of SCFAs with GPR43 expressed on Treg cells and GPR109A on dendritic cells exerts anti-inflammatory activity (Smith *et al.*, 2013). Propionic acid also plays a significant role in modulating lipid metabolism and is thus responsible for lipid-lowering effects. Also, when rat hepatocytes are exposed to propionate-rich culture, marked suppression of cholesterol synthesis will occur by reducing the activity of acetyl-CoA synthetase (Lin *et al.*, 1995). Thus, a number of pre-clinical and clinical studies reveal that there is a proportionate relationship between elevated SCFAs (mainly propionates and butyrates) by dietary fibers such as β-glucan and systemic hypocholesterolemia (Wang *et al.*, 2012). For instance, rodents receiving supplements containing SCFAs such as propionates, butyrates, etc., were observed to show a significant reduction in total cholesterol and LDL cholesterol as well as an increase in HDL-C. The rectification of dyslipidemia was linked with enhanced bile acid elimination and production.

6.4.4 Role of Microbial Polysaccharides in Lipid Metabolism/Homeostasis

Besides regulation of bile acid concentration and SCFA production, gut microflora also influences host cholesterol homeostasis via toll-like receptor agonists and bacterial exopolysaccharides (EPS). Exopolysaccharides are repeating units of carbohydrates linked by peptidoglycans and mainly produced by Bifidobacterium and Lactobacillus species (Kleerebezem *et al.*, 2010). It is predicted that the population of EPS-producing bacteria is significantly altered by the consumption of β-glucans, and thus, β-glucans influence lipid homeostasis via EPS production. EPS, in addition to regulation of lipid metabolism, also protect the bacterial cells from stress conditions and increase their life expectancy in the host gut. In addition, EPS production by microflora is associated with an immunoregulatory property of bacterial strains (Rodriguez *et al.*, 2009). EPS produced by *Pediococcus parvulus* show a structural resemblance to β-glucans and possess cholesterol-lowering properties in human volunteers (Martensson *et al.*, 2005). London *et al.* demonstrated a significant improvement in lipid profile after the administration of genetically modified Lactobacillus species to mice. Similarly, an *L. mucosae* EPS-producing strain profoundly lowers the cholesterol level in the mouse (London *et al.*, 2014). *In vitro* adipocyte studies demonstrate that EPS obtained from Lactobacillus inhibit the accumulation of triacylglycerol lipid and cholesterol (Zhang *et al.*, 2016). Thus, from the preceding discussion, it is demonstrated that enhanced EPS level in the gut has the potential to regulate the lipid metabolism via toll-like receptors in the host.

6.4.5 Microbiome-Mediated Lipid Assimilation and Biotransformation

Many bacterial strains present in the gut are responsible for cholesterol metabolism (Garcia, Uhia & Galan, 2012). *Eubacterium coprostanoligenes* present in the gut is potentially involved in cholesterol metabolism and transportation. *E. coprostanoligenes* converts cholesterol to coprostanol in the alimentary canal (Li *et al.*, 1996). *Lactobacillus acidophilus, L. casei* and *L. bulgaricus* regulate the activity of cholesterol reductase, responsible for cholesterol metabolism. Thus, when administered to rats, Lactobacillus significantly lowers the serum cholesterol level by regulating cholesterol metabolism along with the production of SCFAs and bile acid excretion. Furthermore, Bacteroides in the gut produce a chemical known as commendamide, responsible for the degradation of cholesterol in the gut microenvironment (Lynch *et al.*, 2017). Clinical studies reveal that the consumption of β-glucan increases the population of Bacteroidetes and thus, enhances cholesterol degradation and metabolism. Therefore, by altering the gut microflora population in the gut, the physiological function can be modified.

6.5 CONCLUSION

Thus, β-glucan is considered as an important component present in food, responsible for the modulation of metabolic deregulations involved in various disorders like diabetes mellitus, obesity, dyslipidemia, etc. The therapeutic outcomes of β-glucan mainly rely upon its structural and physicochemical properties as well as its potential to interact with the gastric microenvironment. In addition to these, the easy availability and handling of β-glucan lead to its enhanced consumption. Furthermore, the capability of β-glucan to alter the satiety and generation of SCFAs via colonic fermentation further explains the reasons behind its significant use.

REFERENCES

Abumweis, S.S., Jew, S. & Ames, N.P. (2010). Beta-glucan from barley and its lipid-lowering capacity: a meta-analysis of randomized, controlled trials. *European Journal of Clinical Nutrition*, 64(12), 1472–1480.

Adam, T.C.M. & Westerterp-Plantenga, M.S. (2005). Nutrient stimulated GLP-1 release in normal-weight men and women. *Hormone and Metabolic Research*, 37(2), 111–117.

Aldini, R., Roda, A., Lenzi, P.L., Ussia, G., Vaccari, M.C. & Mazzella, G. (1992). Bile acid active and passive ileal transport in the rabbit: effect of luminal stirring. *European Journal of Clinical Investigation*, 22, 744–750.

Andersson, K.E., Svedberg, K.A., Lindholm, M.W., Oste, R. & Hellstrand, P. (2010). Oats (*Avena sativa*) reduce atherogenesis in LDL-receptor-deficient mice. *Atherosclerosis*, 212, 93–99.

Andersson, M., Ellegard, L. & Andersson, H. (2002). Oat bran stimulates bile acid synthesis within 8 h as measured by 7alpha-hydroxy-4-cholesten-3-one. *American Journal of Clinical Nutrition*, 76, 1111–1116.

Artiss, J.D., Brogan, K., Brucal, M., Moghaddam, M. & Jen, K.L.C. (2006). The effects of a new soluble dietary fiber on weight gain and selected blood parameters in rats. *Metabolism*, 55(2), 195–202.

Babineau, T.J., Marcello, P., Swails, W., Kenler, A., Bistrian, B. & Forse, R.A. (1994). Randomized phase I/II trial of a macrophage-specific immunomodulator (PGG-glucan) in high-risk surgical patients. *Annals of Surgery*, 220(5), 601–609.

Bacic, A., Fincher, G.B. & Stone, B.A. (2009). *Chemistry, Biochemistry and Biology of (1–3)-[beta]-Glucans and Related Polysaccharides*, Academic Press, Amsterdam, The Netherlands, 1st edition.

Barsanti, L., Passarelli, V., Evangelista, V., Frassanito, A.M. & Gualtieri, P. (2011). Chemistry, physico-chemistry and applications linked to biological activities of β-glucans. *Natural Product Reports*, 28(3), 457–466.

Batterham, R.L., Cowley, M.A. & Small, C.J. (2002). Gut hormone PYY3-36 physiologically inhibits food intake. *Nature*, 418(6898), 650–654.

Beck, E.J., Tapsell, L.C., Batterham, M.J., Tosh, S.M. & Huang, X.F. (2009). Increases in peptide Y-Y levels following oat beta-glucan ingestion are dose-dependent in overweight adults. *Nutrition Research*, 29(10), 705–709.

Beck, E.J., Tapsell, L.C., Batterham, M.J., Tosh, S.M. & Huang, X.F. (2010). Oat β-glucan supplementation does not enhance the effectiveness of an energy-restricted diet in overweight women. *British Journal of Nutrition*, 103(8), 1212–1222.

Begley, M., Hill, C. & Gahan, C.G (2006). Bile salt hydrolase activity in probiotics. *Applied and Environmental Microbiology*, 72, 1729–1738.

Behall, K.M., Scholfield, D.J. & Hallfrisch, J. (2004). Lipids significantly reduced by diets containing Barley in moderately hypercholesterolemic men. *Journal of the American College of Nutrition*, 23(1), 55–62.

Berger, M., Gray, J.A. & Roth, B.L. (2009). The expanded biology of serotonin. *Annual Review of Medicine*, 60, 355–366.

Biorklund, M. & Holm, J. (2008). Serum lipids and postprandial glucose and insulin levels in hyperlipidemic subjects after consumption of an oat β-glucan-containing ready meal. *Annals of Nutrition and Metabolism*, 52(2), 83–90.

Biorklund, M., van Rees, A., Mensink, R.P. & Onning, G. (2005). Changes in serum lipids and postprandial glucose and insulin concentrations after consumption of beverages with β-glucans from oats or barley: a randomised dose-controlled trial. *European Journal of Clinical Nutrition*, 59(11), 1272–1281.

Birketvedt, G.S., Shimshi, M., Thom, E. & Florholmen, J. (2005). Experiences with three different fiber supplements in weight reduction. *Medical Science Monitor*, 11(1), PI5–PI8.

Bourdon, I., Yokoyama, W., Davis, P., Hudson, C., Backus, R., Richter, D., Knuckles, B. & Schneeman, B.O. (1999). Postprandial lipid, glucose, insulin, and cholecystokinin responses in men fed barley pasta enriched with glucan. *American Journal of Clinical Nutrition*, 69, 55–63.

Braaten, J.T., Wood, P.J., Scott, F.W., Riedel, K.D., Poste, L.M. & Collins, M.W. (1991). Oat gum lowers glucose and insulin after an oral glucose load. *American Journal of Clinical Nutrition*, 53(6), 1425–1430.

Brennan, C.S. (2005). Dietary fibre, glycaemic response, and diabetes. *Molecular Nutrition and Food Research*, 49(6), 560–570.

Burkus, Z. & Temelli, F. (1998). Effect of extraction conditions on yield, composition, and viscosity stability of barley β-glucan gum. *Cereal Chemistry*, 75(6), 805–809.

Burton-Freeman, B. (2000). Dietary fiber and energy regulation. *Journal of Nutrition*, 130(2), 272S–275S.

Carlson, J.L., Erickson, J.M., Hess, J.M., Gould, T.J. & Slavin, J.L. (2017). Prebiotic dietary fiber and gut health: comparing the in vitro fermentations of beta-glucan, inulin and xylooligosaccharide. *Nutrients*, 9, 1361.

Cani, P.D., Dewever, C. & Delzenne, N.M. (2004). Inulin-type fructans modulate gastrointestinal peptides involved in appetite regulation (glucagon-like peptide-1 and ghrelin) in rats. *British Journal of Nutrition*, 92(3), 521–526.

Cani, P.D., Hoste, S., Guiot, Y. & Delzenne, N.M. (2007). Dietary non-digestible carbohydrates promote L-cell differentiation in the proximal colon of rats. *British Journal of Nutrition*, 98(1), 32–37.

Cani, P.D., Lecourt, E. & Dewulf, E.M. (2009). Gut microbiotafermentation of prebiotics increases satietogenic and incretin gut peptide production with consequences for appetite sensationand glucose response after a meal.*American Journalof Clinical Nutrition*, 90(5), 1236–1243.

Cavallero, A., Empilli, S., Brighenti, F. & Stanca, A.M. (2002). High glucan barley fractions in bread making and their effects on human glycemic response. *Journal of Cereal Science*, 36, 59–66.

Chandalia, M., Garg, A., Lutjohann, D., Von Bergmann, K., Grundy, S.M. & Brinkley, L.J. (2000). Beneficial effects of high dietary fiber intake in patients with type 2 diabetes mellitus. *The New England Journal of Medicine*, 342(19), 1392–1398.

Charalampopoulos, D., Wang, R., Pandiella, S.S. & Webb, C (2002). Application of cereals and cereal components in functional foods: a review. *International Journal of Food Microbiology*, 79(1), 131–141.

Cherbut, C. (1995). Effects of short-chain fatty acids on gastrointestinal motility, in *Physiological and Clinical Aspects of Short-Chain Fatty Acids*, J.H. Cummings, J.L. Rombeau & T. Sakata, Eds., Cambridge University Press, Cambridge, UK.

Cho, S., De Vries, J.W. & Prosky, L. (1997). *Dietary Fiber Analysis and Applications*, AOAC International, Gaithersburg.

Chobanian, A.V., Bakris, G.L. & Black, H.R. (2003). The seventh report of the joint national committee on prevention, detection, evaluation, and treatment of high blood pressure: the JNC 7 report. *Journal of the American Medical Association*, 289(19), 2560–2572.

Connolly, M.L., Lovegrove, J.A. & Tuohy, K.M. (2010). In vitro evaluation of the microbiota modulation abilities of different sized whole oat grain flakes. *Anaerobe*, 16, 483–488.

Connolly, M.L., Tzounis, X., Tuohy, K.M. & Lovegrove, J.A. (2016). Hypocholesterolemic and prebiotic effects of a whole-grain oat-based granola breakfast cereal in a cardio-metabolic "At Risk" population. *Frontier in Microbiology*, 7, 1675.

Crago, M.S., West, S.D., Hoeprich, K.D., Michaelis, K.J. & McKenzie, J.E. (1998). Effects of hyperlipidemia on blood pressure and coronary blood flow in swine. *The FASEB Journal*, 12(4), A238.

Cugnet-Anceau, C., Nazare, J.A., Biorklund, M., le Coquil, E., Sassolas, A., Sothier, M., Holm, J., Landin-Olsson, M., Onning, G. & Laville, M. (2010). A controlled study of consumption of _-glucan-enriched soups for 2 months by type 2 diabetic free-living subjects. *British Journal of Nutrition*, 103, 422–428.

Cummings, J.H. (1981). Short chain fatty acids in the human colon. *Gut*, 22(9), 763–779.

Cummings, J.H. & Stephen, A.M. (2007). Carbohydrate terminology and classification. *European Journal of Clinical Nutrition*, 61(1) Supplement 1, S5–S18.

Cummings, J.H., Pomare, E.W., Branch, W.J., Naylor, C.P.E. & Macfarlane, G.T. (1987). Short chain fatty acids in human large intestine, portal, hepatic and venous blood. *Gut*, 28(10), 1221–1227.

Davidson, M.H., Dugan, L.D., Burns, J.H., Bova, J., Story, K. & Drennan, K.B. (1991). The hypocholesterolemic effects of glucan in oatmeal and oat bran. A dose-controlled study. *Journal of the American Medical Association*, 265, 1833–1839.

Davis, H.R., Mullins, D.E. & Pines, J.M. (1998). Effect of chronic central administration of glucagon-like peptide-1 (7–36) amide on food consumption and body weight in normal and obese rats. *Obesity Research*, 6(2), 147–156.

de Boer, J.F., Schonewille, M., Boesjes, M., Wolters, H., Bloks, V.W. & Bos, T. (2017). Intestinal farnesoid X receptor controls transintestinal cholesterol excretion in mice. *Gastroenterology*, 152, 1126–1138.

Degirolamo, C., Rainaldi, S., Bovenga, F., Murzilli, S. & Moschetta, A. (2014). Microbiota modification with probiotics induces hepatic bile acid synthesis via downregulation of the Fxr-Fgf15 axis in mice. *Cell Reports*, 7, 12–18.

Dongowski, G., Huth, M. & Gebhardt, E. (2003). Steroids inthe intestinal tract of rats are affected by dietary-fibre-rich barley-based diets. *British Journal of Nutrition*, 90(5), 895–906.

Dumoulin, V., Moro, F., Barcelo, A., Dakka, T. & Cuber, J.C. (1998). Peptide YY, glucagon-like peptide-1, and neurotensin responses to luminal factors in the isolated vascularly perfused rat ileum. *Endocrinology*, 139(9), 3780–3786.

Ebihara, K. & Schneeman, B.O. (1989). Interaction of bile acids, phospholipids, cholesterol and triglyceride with dietary fibers in the small intestine of rats. *Journal of Nutrition*, 119(8), 1100–1106.

Ellis, P.R., Dawoud, F.M. & Morris, E.R (1991). Blood glucose, plasma insulin and sensory responses to guar-containing wheat breads: effects of molecular weight and particle size of guar gum. *British Journal of Nutrition*, 66(3), 363–379.

Englyst, K.N., Liu, S. & Englyst, H.N. (2007). Nutritional characterization and measurement of dietary carbohydrates. *European Journal of Clinical Nutrition*, 61(1) Supplement 1, S19–S39.

Erdmann, J., Lippl, F. & Schusdziarra, V. (2003). Differential effect of protein and fat on plasma ghrelin levels in man. *Regulatory Peptides*, 116(1–3), 101–107.

Fang, S., Suh, J.M., Reilly, S.M., Yu, E., Osborn, O. & Lackey, D. (2015). Intestinal FXR agonism promotes adipose tissue browning and reduces obesity and insulin resistance. *Nature Medicine*, 21, 159–165.

Fehlbaum, S., Prudence, K., Kieboom, J., Heerikhuisen, M., van den Broek, T. & Schuren, F.H.J. (2018). In vitro fermentation of selected prebiotics and their effects on the composition and activity of the adult gut microbiota. *International Journal of Molecular Sciences*, 19, 3097.

Ferrannini, E., Buzzigoli, G. & Bonadonna, R. (1987). Insulin resistance in essential hypertension. *The New England Journal of Medicine*, 317(6), 350–357.

Frost, G.S., Brynes, A.E., Dhillo, W.S., Bloom, S.R. & McBurney, M.I. (2003). The effects of fiber enrichment of pasta and fat content on gastric emptying, GLP-1, glucose, and insulin responses to a meal. *European Journal of Clinical Nutrition*, 57(2), 293–298.

Fujioka, K. (2002). Management of obesity as a chronic disease:non-pharmacologic, pharmacologic, and surgical options. *Obesity Research*, 10, 2.

Galisteo, M., Moron, R., Rivera, L., Romero, R., Anguera, A. & Zarzuelo, A. (2010). *Plantago ovata* husks-supplemented diet ameliorates metabolic alterations in obese Zucker rats through activation of AMP-activated protein kinase. Comparative study with other dietary fibers. *Clinical Nutrition*, 29(2), 261–267.

Garcia, J.L., Uhia, I. & Galan, B. (2012). Catabolism and biotechnological applications of cholesterol degrading bacteria. *Microbiology and Biotechnology*, 5, 679–699.

Gee, J.M. & Johnson, I.T. (2005). Dietary lactitol fermentation increases circulating peptide YY and glucagon-like peptide-1 in rats and humans. *Nutrition*, 21(10), 1036–1043.

Goel, V., Cheema, S.K., Agellon, L.B., Ooraikul, B. & Basu, T.K. (1999). Dietary rhubarb (*Rheum rhaponticum*) stalk fibre stimulates cholesterol 7α-hydroxylase gene expression and bile acid excretion in cholesterol-fed C57BL/6J mice. *British Journal of Nutrition*, 81(1), 65–71.

Granfeldt, Y., Liljeberg, H., Drews, A., Newman, R. & Bjorck, I. (1994). Glucose and insulin responses to barley products: influence of food structure and amylose-amylopectin ratio. *American Journal of Clinical Nutrition*, 59(5), 1075–1082.

Granfeldt, Y., Nyberg, L. & Bjorck, I. (2008). Muesli with 4 goat β-glucans lowers glucose and insulin responses after a bread meal in healthy subjects. *European Journal of Clinical Nutrition*, 62(5), 600–607.

Gunness, P., Michiels, J., Vanhaecke, L., De Smet, S., Kravchuk, O. & Van de Meene, A. (2016). Reduction in circulating bile acid and restricted diffusion across the intestinal epithelium are associated with a decrease in blood cholesterol in the presence of oat beta-glucan. *FASEB Journal*, 30, 4227–4238.

Hamer, H.M., Jonkers, D., Venema, K., Vanhoutvin, S., Troost, F.J. & Brummer, R.J. (2008). Review article: the role of butyrate on colonic function. *Alimentary Pharmacology and Therapeutics*, 27(2), 104–119.

Hanai, H., Ikuma, M. & Sato, Y. (1997). Long-term effects of water-soluble corn bran hemicellulose on glucose tolerance in obese and non-obese patients: improved insulin sensitivity and glucose metabolism in obese subjects. *Bioscience, Biotechnology and Biochemistry*, 61(8), 1358–1361.

He, J., Streiffer, R.H., Muntner, P., Krousel-Wood, M.A. & Whelton, P.K. (2004). Effect of dietary fiber intake on blood pressure: a randomized, double-blind, placebo-controlled trial. *Journal of Hypertension*, 22(1), 73–80.

Heini, A.F., Lara-Castro, C., Schneider, H., Kirk, K.A., Considine, R.V. & Weinsier, R.L. (1998). Effect of hydrolyzed guar fiber on fasting and postprandial satiety and satietyhormones: a double-blind, placebo-controlled trial during controlled weight loss. *International Journal of Obesity*, 22(9), 906–909.

Heo, W., Lee, E.S., Cho, H.T., Kim, J.H., Lee, J.H. & Yoon, S.M. (2018). *Lactobacillus plantarum* LRCC 5273 isolated from Kimchi ameliorates diet-induced hypercholesterolemia in C57BL/6 mice. *Bioscience, Biotechnology, and Biochemistry*, 82, 1964–1972.

Hetland, G., Ohno, N., Aaberge, I.S. & Løvik, M. (2000). Protective effect of β-glucan against systemic *Streptococcus pneumonia* infection in mice. *FEMS Immunology and Medical Microbiology*, 27(2), 111–116.

Holst, J.J. (2007). The physiology of glucagon-like peptide 1. *Physiological Reviews*, 87(4), 1409–1439.

Holt, S., Brand, J., Soveny, C. & Hansky, J. (1992). Relationship of satiety to postprandial glycaemic, insulin and cholecystokinin responses. *Appetite*, 18(2), 129–141.

Hooda, S., Matte, J.J., Vasanthan, T. & Zijlstra, R.T. (2010). Dietary oat beta-glucan reduces peak net glucose flux and insulin production and modulates plasma incretin in portal-vein catheterized grower pigs. *Journal of Nutrition*, 140, 1564–1569.

Howarth, N.C., Saltzman, E. & Roberts, S.B. (2001). Dietary fiber and weight regulation. *Nutrition Reviews*, 59(5), 129–139.

Hughes, S.A., Shewry, P.R., Li, L., Gibson, G.R., Sanz, M.L. & Rastall R.A. (2007). In vitro fermentation by human fecal microflora of wheat arabinoxylans. *Journal of Agricultural and Food Chemistry*, 55, 4589–4595.

Hughes, S.A., Shewry, P.R., Gibson, G.R., Mc Cleary, B.V. & Rastall, R.A. (2008). In vitro fermentation of oat and barley derived beta-glucans by human faecal microbiota. *FEMS Microbiology Ecology*, 64, 482–493.

Isaksson, G., Lundquist, I. & Ihse, I. (1982). Effect of dietary fiber on pancreatic enzyme in vitro. *Gastroenterology*, 82(5), 918–924.

Islam, K.B., Fukiya, S., Hagio, M., Fujii, N., Ishizuka, S. & Ooka, T. (2011). Bile acid is a host factor that regulates the composition of the cecal microbiota in rats. *Gastroenterology*, 141, 1773–1781.

Jayachandran, M., Chen, J., Chung, S.S.M. & Xu, B. (2018). A critical review on the impacts of beta-glucans on gut microbiota and human health. *Journal of Nutritional Biochemistry*, 61, 101–110.

Jenkins, A.L., Jenkins, D.J.A., Zdravkovic, U., Wursch, P. & Vuksan, V. (2002). Depression of the glycemic index by high levels of β-glucan fiber in two functional foods tested in type 2 diabetes. *European Journal of Clinical Nutrition*, 56(7), 622–628.

Jenkins, D.J.A., Wolever, T.M.S. & Leeds, A.R. (1978). Dietary fibres, fibre analogues, and glucose tolerance: importance of viscosity. *British Medical Journal*, 1(6124), 1392–1394.

Jenkins, D.J.A., Kendall, C.W.C. & Vuksan, V. (2002). Soluble fiber intake at a dose approved by the US Food and Drug Administration for a claim of health benefits: serum lipid risk factors for cardiovascular disease assessed in a randomized controlled crossover trial. *American Journal of Clinical Nutrition*, 75(5), 834–839.

Jones, B.V., Begley, M., Hill, C., Gahan, C.G. & Marchesi, J.R. (2008). Functional and comparative metagenomic analysis of bile salt hydrolase activity in the human gut microbiome. *Proceedings of the National Academy of Sciences of the United States of America*, 105, 13580–13585.

Jones, M.L., Martoni, C.J., Ganopolsky, J.G., Labbe, A. & Prakash, S. (2014). The human microbiome and bile acid metabolism: dysbiosis, dysmetabolism, disease and intervention. *Expert Opinion on Biological Therapy*, 14, 467–482.

Joyce, S.A. & Gahan, C.G. (2016). Bile acid modifications at the microbe-host interface: potential for nutraceutical and pharmaceutical interventions in host health. *Annual Review of Food Science and Technology*, 7, 313–333.

Juvonen, K.R., Purhonen, A.K. & Salmenkallio-Marttila, M. (2009). Viscosity of oat bran-enriched beverages influences gastrointestinal hormonal responses in healthy humans. *Journal of Nutrition*, 139(3), 461–466.

Karaki, S.I., Mitsui, R. & Hayashi, H. (2006). Short-chain fatty acid receptor, GPR43, is expressed by enteroendocrine cells and mucosal mast cells in rat intestine. *Cell and Tissue Research*, 324(3), 353–360.

Karhunen, L.J., Juvonen, K.R. & Flander, S.M. (2010). A psyllium fiber-enriched meal strongly attenuates postprandial gastrointestinal peptide release in healthy young adults. *Journal of Nutrition*, 140(4), 737–744.

Kedia, G., Vazquez, J.A. & Pandiella, S.S. (2008). Evaluation of the fermentability of oat fractions obtained by debranning using lactic acid bacteria. *Journal of Applied Microbiology*, 105(4), 1227–1237.

Keenan, J.M., Pins, J.J., Frazel, C., Moran, A. & Turnquist, L. (2002). Oat ingestion reduces systolic and diastolic blood pressure in patients with mild or borderline hypertension: a pilottrial. *Journal of Family Practice*, 51(4), 369.

Kellogg, T.F. & Wostmann, B.S. (1969). Fecal neutral steroids and bile acids from germfree rats. *Journal of Lipid Research*, 10, 495–503.

Kendall, C.W.C., Esfahani, A. & Hoffman, A.J. (2008). Effect of novel maize-based dietary fibers on postprandial glycemia and insulinemia. *Journal of the American College of Nutrition*, 27(6), 711–718.

Keogh, G.F., Cooper, G.J.S. & Mulvey, T.B. (2003). Randomized controlled crossover study of the effect of a highly β-glucan-enriched barley on cardiovascular disease risk factors in mildly hypercholesterolemic men. *American Journal of Clinical Nutrition*, 78(4), 711–718.

Kerckhoffs, D.A.J.M., Hornstra, G. & Mensink, R.P. (2003). Cholesterol-lowering effect of β-glucan from oat bran in mildly hypercholesterolemic subjects may decrease when β-glucan is incorporated into bread and cookies. *American Journal of Clinical Nutrition*, 78(2), 221–227.

Kim, B., Park, K.Y., Ji, Y., Park, S., Holzapfel, W. & Hyun, C.K. (2016). Protective effects of Lactobacillus rhamnosus GG against dyslipidemia in high-fat diet induced obese mice. *Biochemical and Biophysical Research Communications*, 473, 530–536.

Kim, H., Behall, K.M., Vinyard, B. & Conway, J.M. (2006). Short term satiety and glycemic response after consumption of whole grains with various amounts of β-glucan. *Cereal Foods World*, 51(1), 29–33.

Kirkmeyer, S.V. & Mattes, R.D. (2000). Effects of food attributes on hunger and food intake. *International Journal of Obesity*, 24(9), 1167–1175.

Kissileff, H.R., Pi-Sunyer, F.X., Thornton, J. & Smith, G.P. (1981). C-terminal octapeptide of cholecystokinin decreases food intake in man. *American Journal of Clinical Nutrition*, 34(2), 154–160.

Kleerebezem, M., Hols, P., Bernard, E., Rolain, T., Zhou, M. & Siezen, R.J. (2010). The extracellular biology of the lactobacilli. *FEMS Microbiology Reviews*, 34, 199–230.

Koh, A., De Vadder, F., Kovatcheva-Datchary, P. & Backhed F. (2016). From dietary fiber to host physiology: short-chain fatty acids as key bacterial metabolites. *Cell*, 165, 1332–1345.

Kojima, M., Hosoda, H., Date, Y., Nakazato, M., Matsuo, H. & Kangawa, K. (1999). Ghrelin is a growth-hormone-releasing acylated peptide from stomach. *Nature*, 402(6762), 656–660.

Kovacs, E.M.R., Westerterp-Plantenga, M.S., Saris, W.H.M., Goossens, I., Geurten, P. & Brouns, F. (2001). The effect of addition of modified guar gum to a low-energy semi-solid meal on appetite and body weight loss. *International Journal of Obesity*, 25(3), 307–315.

Kristek, A., Wiese, M., Heuer, P., Kosik, O., Schar, M.Y. & Soycan, G (2019). Oat bran, but not its isolated bioactive beta-glucans or polyphenols, have a bifidogenic effect in an in vitro fermentation model of the gut microbiota. *British Journal of Nutrition*, 121, 549–559.

Lazaridou, A. & Biliaderis, C.G. (2007). Molecular aspects of cereal β-glucan functionality: physical properties, technological applications and physiological effects. *Journal of Cereal Science*, 46(2), 101–118.

Le Chatelier, E., Nielsen, T., Qin, J., Prifti, E., Hildebrand, F. & Falony, G. (2013). Richness of human gut microbiome correlates with metabolic markers. *Nature*, 500, 541–546.

Li, L., Baumann, C.A., Meling, D.D., Sell, J.L. & Beitz D.C. (1996). Effect of orally administered Eubacterium coprostanoligenes ATCC 51222 on plasma cholesterol concentration in laying hens. *Poultry Science*, 75, 743–745.

Liljeberg, H. & Bjorck, I (2000). Effects of a low-glycaemic index spaghetti meal on glucose tolerance and lipaemia at a subsequent meal in healthy subjects. *European Journal of Clinical Nutrition*, 54(1), 24–28.

Lin, Y., Vonk, R.J., Slooff, M.J., Kuipers, F. & Smit, M.J. (1995). Differences in propionate induced inhibition of cholesterol and triacylglycerol synthesis between human and rat hepatocytes in primary culture. *British Journal of Nutrition*, 74, 197–207.

Liu, S., Sesso, H.D., Manson, J.E., Willett, W.C. & Buring, J.E. (2003). Is intake of breakfast cereals related to total and cause-specific mortality in men? *American Journal of Clinical Nutrition*, 77(3), 594–599.

London, L.E., Kumar, A.H., Wall, R., Casey, P.G., O'Sullivan, O. & Shanahan F. (2014). Exopolysaccharide-producing probiotic Lactobacilli reduce serum cholesterol and modify enteric microbiota in ApoE-deficient mice. *Journal of Nutrition*, 144, 1956–1462.

Longo, W.E., Ballantyne, G.H., Savoca, P.E., Adrian, T.E., Bilchik, A.J. & Modlin, I.M. (1991). Short-chain fatty acid release of peptide YY in the isolated rabbit distal colon. *Scandinavian Journal of Gastroenterology*, 26(4), 442–448.

Lovegrove, J.A., Clohessy, A., Milon, H. & Williams, C.M. (2000). Modest doses of β-glucan do not reduce concentrations of potentially atherogenic lipoproteins. *American Journal of Clinical Nutrition*, 72(1), 49–55.

Luo, Y., Zhang, L., Li, H., Smidt, H., Wright, A.G. & Zhang, K. (2017). Different types of dietary fibers trigger specific alterations in composition and predicted functions of colonic bacterial communities in BALB/c mice. *Frontier in Microbiology*, 8, 966.

Lyly, M., Liukkonen, K.H., Salmenkallio-Marttila, M., Karhunen, L., Poutanen, K. & Lahteenmaki, L. (2009). Fibre in beverages can enhance perceived satiety. *European Journal of Nutrition*, 48(4), 251–258.

Lyly, M., Ohls, N. & Lahteenmaki, L. (2010). The effect of fibre amount, energy level and viscosity of beverages containing oat fibre supplement on perceived satiety. *Food and Nutrition Research*, 54(1), 1–8.

Lynch, A., Crowley, E., Casey, E., Cano, R., Shanahan, R. & McGlacken, G. (2017). The bacteroidales produce an N-acylated derivative of glycine with both cholesterol-solubilising and hemolytic activity. *Scientific Reports*, 7, 13270.

Makelainen, H., Anttila, H. & Sihvonen, J. (2007). The effect of β-glucan on the glycemic and insulin index. *European Journal of Clinical Nutrition*, 61(6), 779–785.

Maki, K.C., Galant, R. & Samuel, P. (2007). Effects of consuming foods containing oat β-glucan on blood pressure, carbohydrate metabolism and biomarkers of oxidative stress in men and women with elevated blood pressure. *European Journal of Clinical Nutrition*, 61(6), 786–795.

Mantzoros, C.S., Li, T., Manson, J.E., Meigs, J.B. & Hu, F.B. (2005). Circulating adiponectin levels are associated with better glycemic control, more favorable lipid profile, and reduced inflammation in women with type 2 diabetes. *Journal of Clinical Endocrinology and Metabolism*, 90(8), 4542–4548.

Marlett, J.A., Hosig, K.B., Vollendorf, N.M., Shinnick, F.L., Haack, V.S. & Story, J.A. (1994). Mechanism of serum cholesterol reduction by oat bran. *Hepatology*, 20(6), 1450–1457.

Martensson, O., Biorklund, M., Lambo, A.M., Dueñas-Chasco, M., Irastorza, A. & Holst, O. (2005). Fermented, ropy, oat-based products reduce cholesterol levels and stimulate the bifidobacteria flora in humans. *Nutrition Research*, 25, 429–442.

McIntosh, M., Stone, B.A. & Stanisich, V.A. (2005). Curdlan and other bacterial (1→3)-β-D-glucans. *Applied Microbiology and Biotechnology*, 68(2), 163–173.

McRorie, J.W. Jr. & McKeown, N.M. (2017). Understanding the physics of functional fibers in the gastrointestinal tract: an evidence-based approach to resolving enduring misconceptions about insoluble and soluble fiber. *Journal of Academy of Nutrition and Dietetics*, 117, 251–264.

Miyata, M., Takamatsu, Y., Kuribayashi, H. & Yamazoe, Y. (2009). Administration of ampicillin elevates hepatic primary bile acid synthesis through suppression of ileal fibroblast growth factor 15 expression. *Journal of Pharmacology and Experimental Therapeutics*, 331, 1079–1085.

Nazare, J.A., Normand, S., Triantafyllou, A.O., DeLa Perrière, A.B., Desage, M. & Laville, M. (2009). Modulation of the postprandial phase by β-glucan in overweight subjects: effects on glucose and insulin kinetics. *Molecular Nutrition and Food Research*, 53(3), 361–369.

Nedvıdkova, J., Krykorkova, I. & Bartak, V. (2003). Loss of meal-induced decrease in plasma ghrelin levels in patients with anorexia nervosa. *Journal of Clinical Endocrinology andMetabolism*, 88(4), 1678–1682.

Nicolosi, R., Bell, S.J., Bistrian, B.R., Greenberg, I., Forse, R.A. & Blackburn, G.L. (1999). Plasma lipid changes after supplementation with β-glucan fiber from yeast. *American Journal of Clinical Nutrition*, 70(2), 208–212.

Ooi, L-G. & Liong, M-T (2010). Cholesterol-lowering effects of probiotics and prebiotics: a review of in vivo and in vitro findings. *International Journal of Molecular Sciences*, 11, 2499–522.

Parks, D.J., Blanchard, S.G., Bledsoe, R.K., Chandra, G., Consler, T.G. & Kliewer, S.A. (1999). Bile acids: natural ligands for an orphan nuclear receptor. *Science*, 284, 1365–1368.

Pick, M.E., Hawrysh, J.Z., Gee, M.I., Toth, E., Garg, M.L. & Hardin, R.T. (1996). Oat bran concentrate bread products improve long-term control of diabetes: a pilot study. *Journal of the American Dietetic Association*, 96(12), 1254–1261.

Queenan, K.M., Stewart, M.L., Smith, K.N., Thomas, W., Fulcher, R.G. & Slavin, J.L. (2007). Concentrated oat beta-glucan, a fermentable fiber, lowers serum cholesterol in hypercholesterolemic adults in a randomized controlled trial. *Nutrition Journal*, 6, 6.

Raben, A., Andersen, K., Karberg, M.A., Holst, J.J. & Astrup, A. (1997). Acetylation of or β-cyclodextrin addition to potato starch: beneficial effect on glucose metabolism and appetite sensations. *American Journal of Clinical Nutrition*, 66(2), 304–314.

Reimer, R.A., Pelletier, X. & Carabin, I.G. (2010). Increased plasma PYY levels following supplementation with the functional fiber PolyGlycopleX in healthy adults. *European Journal of Clinical Nutrition*, 64(10), 1186–1191.

Repa, J.J., Berge, K.E., Pomajzl, C., Richardson, J.A., Hobbs, H. & Mangelsdorf, D.J. (2002). Regulation of ATP-binding cassette sterol transporters ABCG5 and ABCG8 by the liver X receptors alpha and beta. *Journal of Biological Chemistry*, 277, 18793–18800.

Reyna-Villasmil, N., Bermudez-Pirela, V. & Mengual-Moreno, E. (2007). Oat-derived β-glucan significantly improves HDL-C and diminishes LDLC and non-HDL cholesterol in overweight individuals with mild hypercholesterolemia. *American Journal of Therapeutics*, 14(2), 203–212.

Ripsin, C.M., Keenan, J.M. & Jacobs, D.R. (1992). Oat products and lipid lowering: a meta-analysis. *Journal of the American Medical Association*, 267(24), 3317–3325.

Rodriguez, C., Medici, M., Rodriguez, A.V., Mozzi, F. & Font de Valdez, G. (2009). Prevention of chronic gastritis by fermented milks made with exopolysaccharide-producing *Streptococcus thermophilus* strains. *Journal of Dairy Sciences*, 92, 2423–2434.

Roediger, W.E.W. (1980). Role of anaerobic bacteria in the metabolic welfare of the colonic mucosa in man. *Gut*, 21(9), 793–798.

Russell, D.W. (2003). The enzymes, regulation, and genetics of bile acid synthesis. *Annual Review of Biochemistry*, 72, 137–174.

Sayar, S., Jannink, J.L. & White, P.J. (2007). Digestion residues of typical and high-beta glucan oat flours provide substrates for in vitro fermentation. *Journal of Agricultural and Food Chemistry*, 55, 5306–5311.

Sayin, S.I., Wahlström, A., Felin, J., Jäntti, S., Marschall, H.-U. & Bamberg, K. (2013). Gut microbiota regulates bile acid metabolism by reducing the levels of taurobeta-muricholic acid, a naturally occurring FXR antagonist. *Cell Metabolism*, 17, 5–35.

Schneeman, B.O. & Gallaher, D. (1985). Effects of dietary fiberon digestive enzyme activity and bile acids in the small intestine. *Proceedings of the Society for Experimental Biology and Medicine*, 180(3), 409–414.

Schroeder, N., Gallaher, D.D., Arndt, E.A. & Marquart, L. (2009). Influence of whole grain barley, whole grain wheat, and refined rice-based foods on short-term satiety and energy intake. *Appetite*, 53(3), 363–369.

Shimada, M., Date, Y. & Mondal, M.S. (2003). Somatostatin suppresses ghrelin secretion from the rat stomach. *Biochemical and Biophysical Research Communications*, 302(3), 520–525.

Shimizu, C., Kihara, M. 7 Aoe, S. (2008). Effect of high β-glucan barley on serum cholesterol concentrations and visceral fat area in Japanese men: A randomized, double-blinded, placebo-controlled trial. *Plant Foods for Human Nutrition*, 63(1), 21–25.

Smith, P.M., Howitt, M.R., Panikov, N., Michaud, M., Gallini, C.A. & Bohlooly, Y.M. (2013). The microbial metabolites, short-chain fatty acids, regulate colonic Treg cell homeostasis. *Science*, 341, 569–573.

Sola, R., Godas, G. & Ribalta, J. (2007). Effects of soluble fiber (*Plantago ovata* husk) on plasma lipids, lipoproteins and apolipoproteins in men with ischemic heart disease. *American Journal of Clinical Nutrition*, 85(4), 1157–1163.

Sola, R., Bruckert, E. & Valls, R.M. (2010). Soluble fibre (*Plantago ovata* husk) reduces plasma low-density lipoprotein (LDL) cholesterol, triglycerides, insulin, oxidised LDL and systolic blood pressure in hypercholesterolaemic patients: a randomised trial. *Atherosclerosis*, 211(2), 630–637.

Sonck, E., Stuyven, E., Goddeeris, B. & Cox, E. (2010). The effect of β-glucans on porcine leukocytes. *Veterinary Immunology and Immunopathology*, 135(3–4), 199–207.

Song, Y.J., Sawamura, M., Ikeda, K., Igawa, S. & Yamori, Y. (2000). Soluble dietary fibre improves insulin sensitivity by increasing muscle GLUT-4 content in stroke-prone spontaneously hypertensive rats. *Clinical and Experimental Pharmacology and Physiology*, 27(1–2), 41–45.

Steemburgo, T., Dall'Alba, V., Almeida, J.C., Zelmanovitz, T., Gross, J.L. & de Azevedo, M.L. (2009). Intake of soluble fibers has a protective role for the presence of metabolic syndrome in patients with type 2 diabetes. *European Journal of Clinical Nutrition*, 63(1), 127–133.

Talati, R., Baker, W.L., Pabilonia, M.S., White, C.M. & Coleman, C.I. (2009). The effects of barley-derived soluble fiber on serum lipids. *Annals of Family Medicine*, 7(2), 157–163.

Tapola, N., Karvonen, H., Niskanen, L., Mikola, M. & Sarkkinen, E. (2005). Glycemic responses of oat bran products in type 2 diabetic patients. *Nutrition, Metabolism and Cardiovascular Diseases*, 15(4), 255–261.

Tappy, L., Gugolz, E. & Wursch, P. (1996). Effects of breakfastcereals containing various amounts of β-glucan fibers onplasma glucose and insulin responses in NIDDM subjects. *Diabetes Care*, 19(8), 831–834.

Thandapilly, S.J., Ndou, S.P., Wang, Y., Nyachoti, C.M. & Ames, N.P. (2018). Barley beta glucan increases fecal bile acid excretion and short chain fatty acid levels in mildly hypercholesterolemic individuals. *Food and Function*, 9, 3092–3096.

Theuwissen, E. & Mensink, R.P. (2008). Water-soluble dietary fibers and cardiovascular disease. *Physiology & Behavior*, 94, 285–292.

Thomas, C., Pellicciari, R., Pruzanski, M., Auwerx, J. & Schoonjans, K. (2008). Targeting bile-acid signaling for metabolic diseases. *Nature Reviews Drug Discovery*, 7, 678–793.

Tiwari, U. & Cummins, E. (2011). Meta-analysis of the effect of beta-glucan intake on blood cholesterol and glucose levels. *Nutrition*, 27, 1008–1016.

Topping, D.L. & Clifton, P.M. (2001). Short-chain fatty acids and human colonic function: roles of resistant starch and non starch polysaccharides. *Physiological Reviews*, 81(3), 1031–1064.

Torpy, J.M., Lynm, C. & Glass, R.M. (2006). JAMA patient page the metabolic syndrome. *Journal of the American Medical Association*, 295(7), 850.

Trogh, I., Courtin, C.M., Andersson, A.A.M., Aman, P., Sørensen, J.F. & Delcour, J.A. (2004). The combined use of hulllessbarley flour and xylanase as a strategy for wheat/hullless barley flour breads with increased arabinoxylan and (1→3,1→4)-β-D-glucan levels. *Journal of Cereal Science*, 40(3), 257–267.

Trowell, H. (1972). Ischemic heart disease and dietary fiber. *American Journal of Clinical Nutrition*, 25(9), 926–932.

Tzianabos, A.O. (2000). Polysaccharide immunomodulators as therapeutic agents: structural aspects and biologic function. *Clinical Microbiology Reviews*, 13(4), 523–533.

Velikonja, A., Lipoglavsek, L., Zorec, M., Orel, R. & Avgustin, G. (2019). Alterations in gut microbiota composition and metabolic parameters after dietary intervention with barley beta glucans in patients with high risk for metabolic syndrome development. *Anaerobe*, 55, 67–77.

Verdich, C., Flint, A. & Gutzwiller, J.P. (2001). A meta-analysis of the effect of glucagon-like peptide-1 (7–36) amide on ad libitum energy intake in humans. *Journal of Clinical Endocrinology and Metabolism*, 86(9), 4382–4389.

Vetvicka, V., Dvorak, B. & Vetvickova, J. (2007). Orally administered marine (1→3)-β-d-glucan Phycarine stimulates both humoral and cellular immunity. *International Journal of Biological Macromolecules*, 40(4), 291–298.

Vitaglione, P., Lumaga, R.B., Stanzione, A., Scalfi, L. & Fogliano, V. (2009). β-Glucan enriched bread reduces energy intakeand modifies plasma ghrelin and peptide YY concentrations in the short term. *Appetite*, 53(3), 338–344.

Vrolix, R. & Mensink, R.P. (2010). Effects of glycemic loadon metabolic risk markers in subjects at increased risk of developing metabolic syndrome. *American Journal of Clinical Nutrition*, 92(2), 366–374.

Wainfan, E., Henkin, G., Rice, L.I. & Marx, W. (1952). Effects of antibacterial drugs on the total cholesterol balance of cholesterol-fed mice. *Archives of Biochemistry and Biophysics*, 38, 187–193.

Wang, J., Zhang, H., Chen, X., Chen, Y., Menghebilige & Bao, Q. (2012). Selection of potential probiotic lactobacilli for cholesterol-lowering properties and their effect on cholesterol metabolism in rats fed ahigh-lipid diet. *Journal of Dairy Science*, 95, 1645–1654.

Wang, Y., Ames, N.P., Tun, H.M., Tosh, S.M., Jones, P.J. & Khafipour, E. (2016). High molecular weight barley b-glucan alters gut microbiota toward reduced cardiovascular disease risk. *Frontier in Microbiology*, 7, 129.

Whelton, S.P., Hyre, A.D., Pedersen, B., Yi, Y., Whelton, P.K. & He, J. (2005). Effect of dietary fiber intake on blood pressure: a meta-analysis of randomized, controlled clinical trials. *Journal of Hypertension*, 23(3), 475–481.

Wolever, T.M., Tosh, S.M., Gibbs, A.L., Brand-Miller, J., Duncan, A.M. & Hart, V. (2010). Physicochemical properties of oat beta-glucan influence its ability to reduce serum LDL cholesterol in humans: a randomized clinical trial. *American Journal of Clinical Nutrition*, 92, 723–732.

Wood, P.J. & Beer, M.U. (1998). Functional oat products, in *Functional Foods, Biochemical and Processing Aspects*, J. Mazza, Ed., Technomic Publishing Company, Lancester, UK.

Wood, P.J., Weisz, J. & Mahn, W. (1991). Molecular characterization of cereal β-glucans. II. Size-exclusion chromatography for comparison of molecular weight. *Cereal Chemistry*, 68, 530–536.

Wood, P.J., Braaten, J.T., Scott, F.W., Riedel, K.D., Wolynetz, M.S. & Collins, M.W. (1994). Effect of dose and modificationof viscous properties of oat gum on plasma glucose and insulin following an oral glucose load. *British Journal of Nutrition*, 72(5), 731–743.

Xu, H., Song, Y. & You, N.C. (2010). Prevalence and clustering of metabolic risk factors for type 2 diabetes among Chinese adults in Shanghai, China. *BMC Public Health*, 10, 683.

Yu, L., Li-Hawkins, J., Hammer, R.E., Berge, K.E., Horton, J.D. & Cohen, J.C. (2002). Overexpression of ABCG5 and ABCG8 promotes biliary cholesterol secretion and reduces fractional absorption of dietary cholesterol. *Journal of Clinical Investigation*, 110, 671–680.

Zhai, Q., Liu, Y., Wang, C., Qu, D., Zhao, J. & Zhang, H. (2019). *Lactobacillus plantarum* CCFM8661 modulates bile acid enterohepatic circulation and increases lead excretion in mice. *Food & Function*, 10, 1455–1464.

Zhang, Z., Zhou, Z., Li, Y., Zhou, L., Ding, Q. & Xu, L. (2016). Isolated exopolysaccharides from *Lactobacillus rhamnosus* GG alleviated adipogenesis mediated by TLR2 in mice. *Scientific Reports*, 6, 36083.

Zhou, A.L., Hergert, N., Rompato, G. & Lefevre, M. (2015). Whole grain oats improve insulin sensitivity and plasma cholesterol profile and modify gut microbiota composition in C57BL/6J mice. *Journal of Nutrition*, 145, 222–230.

Zhu, J.X., Wu, X.Y., Owyang, C. & Li, Y. (2001). Intestinal serotonin acts as a paracrine substance to mediate vagal signal transmission evoked by luminal factors in the rat. *Journal of Physiology*, 530(3), 431–442.

7 Nanotechnology-Based Delivery of Non-Opioid Therapy for Opioid Addiction

*Mohd Adzim Khalili Rohin**, *Atif Amin Baig, Norhaslinda Ridzwan, and Norhayati Abd Hadi*

CONTENTS

7.1 Introduction 149
7.2 Pomegranate-Derived Anthocyanin: Non-Opioid Therapy 150
7.2.1 Composition of Pomegranate (*Punica granatum*) 151
7.2.2 Opioids 152
7.2.3 Opioid System and Brain Pathway 152
7.2.4 Molecular Changes in Opioid Addiction and Dependence 153
7.2.5 Regulation of MOR Proteins in Opioid Addiction and Dependence 154
7.2.6 Non-Opioid Agents from Natural Sources 155
7.3 Approaches to Nanotechnology-Based Non-Opioid Delivery 157
7.3.1 Brain Barriers: CNS Protection 158
7.3.2 Nanocarriers across the BBB 159
7.3.2.1 Poly (Lactic-co-Glycolic Acid) (PLGA) 159
7.3.2.2 Liposome-Based 163
7.3.2.3 Chitosan 165
7.3.2.4 Zinc Oxide (ZNO_2) 165
7.3.2.5 Titanium Dioxide (TiO_2) 167
7.3.2.6 Silica 168
7.4 Conclusion and Recommendations 169
References 169

7.1 INTRODUCTION

Opioids are strong and effective analgesics (Dang & Christie, 2012), which humans have been using for thousands of years (Chartoff & Connery, 2014). Opioids may be endogenous, occurring naturally in the brain, or synthetic compounds made to attach to and activate μ-opioid receptors (MORs). Opiates are a sub-class of opioids that come naturally from derivatives of the opium plant, *Papaver somniferum*, and include codeine and morphine, two major metabolites of heroin (Caputi *et al.*, 2013). Morphine is widely used in many clinical settings (Caputi *et al.*, 2013) and is the most potent analgesic (Almeida *et al.*, 2014). It has been extensively used for dealing with moderate to severe pain in diverse pathologies and clinical situations (Almeida *et al.*, 2014). Morphine primarily activates MORs, and its chronic administration induces tolerance, so that dosage needs to be increased to maintain the required anti-nociceptive effects (Caputi *et al.*, 2013). It also induces opioid

* Corresponding author: mohdadzim@unisza.edu.my

DOI: 10.1201/9781003244721-7

dependence, in which the subject feels compelled to repeatedly administer opioids (Al-Hasani & Bruchas, 2011).

Current opioid agonists such as methadone, naloxone, and buprenorphine are expensive, but they also cause many side effects of opioid dependence, such as severe withdrawal effects, either physical or/and mental (George, 2015; Borg *et al.*, 2014). In response to this problem, researchers have proposed to employ non-opioid therapy from natural sources for opioid dependence. The pomegranate (*Punica granatum*) fruit is proposed as a non-opioid therapy to reduce withdrawal symptoms by inducing pleasurable sensations that modulate central reward pathways in brain areas (Zubeldia *et al.*, 2013). Pomegranate extract and juices have been proven effective to deal with chronic diseases (Zhao *et al.*, 2017; Alvarez-Suarez *et al.*, 2014), to improve cognitive performance (Ilyas *et al.*, 2017), and to provide anti-nociceptive effects (Chen *et al.*, 2017), and have not shown toxicity in any human intervention studies (Wallace & Giusti, 2015; Patel *et al.*, 2008).

Due to the rapid growth of nanoscience and the superior performance of nanomaterials, nanotechnology has arisen as a unique solution to the non-opioid therapeutic bottleneck. Nano-drug delivery systems are a form of nanomaterial that can improve the stability and water solubility of drugs, increase the absorption rate of target cells or tissues, and reduce enzyme inhibition, all of which contribute to better non-opioid pharmaceutical safety and efficacy (Gupta *et al.*, 2019; Quan *et al.*, 2015). The implications of nanotechnology for non-opioid delivery methods might lead to the creation of innovative and better formulations to improve therapeutic agent administration across the blood–brain barrier (BBB) (Abou *et al.*, 2017; Davoodi & Saghavaz, 2017; Pignatello *et al.*, 2017). Furthermore, pomegranate extract, as a potential contender for non-opioid replacement treatment, may cross the BBB due to protein modulation (Neshatdoust *et al.*, 2016; Abbott *et al.*, 2006).

It might be recommended to carry out non-opioid treatment complemented by nanotechnology to achieve superior drug administration outcomes across the BBB due to the associated pharmacological activities in the brain. The BBB is distinguished by its distinct structure and tightly regulated interactions between its cellular and acellular components (Teleanu *et al.*, 2018). The BBB's primary function is to provide an ideal atmosphere for the suitable functioning of the neuronal net by sustaining brain homeostasis, adaptable fluid entry and outflow, and defending the brain from infective agents (Khanna & Farag, 2017) with a responsive, dynamic interplay of molecular, vascular, cellular, and ionic components. Numerous studies have previously concentrated on the development of techniques for treatments and drug transport systems to penetrate the brain capillary endothelium safely and effectively via paracellular or transcellular routes (Barar *et al.*, 2016).

Nanotechnology is a new field that combines information from a variety of fields. In at least one dimension, chemistry, quantum mechanics, industrial processes, and ecology all include the production and manipulation of resources at the 1–100-nm size scale (Ayodele *et al.*, 2017; Faisal & Kumar, 2017; Husain, 2017). Using nanoparticles, nano-based transport methods have become a new dawn for medication delivery across the BBB. Numerous studies have previously concentrated on the investigation of drug delivery methods based on nanotechnology for pharmaceutical agents, plus organic nanomaterials such as polymers, liposomes, dendrimers, and micelles, as well as mineral nanomaterials such as gold, silica, and carbon nanotubes (Teleanu *et al.*, 2018; Cetin *et al.*, 2017). As a result, the suggested nanotechnology-based delivery of non-opioid treatment for opioid addiction will be the major focus.

7.2 POMEGRANATE-DERIVED ANTHOCYANIN: NON-OPIOID THERAPY

Since the usage of opioid agonists and antagonists reveals high relapse rates and causes severe withdrawal effects to the subject (Borg *et al.*, 2014), non-opioids based on natural sources are being introduced, as they can reduce withdrawal symptoms by inducing a pleasurable sensation that modulates central reward pathways (Zubeldia *et al.*, 2013). Previous studies have shown pomegranate to be a fruit with numerous health benefits, involving analgesic (Chen *et al.*, 2017; Chan *et al.*, 2015) and anti-nociceptive effects (Boutemak *et al.*, 2016; Saad *et al.*, 2014) and potential use in neurodegenerative diseases (Ilyas *et al.*, 2017; Subash *et al.*, 2014).

7.2.1 Composition of Pomegranate (*Punica granatum*)

Seeds (3% of the fruit weight), arils (30% of the fruit weight), and peels (including the internal network) are the three components of the pomegranate fruit (Fernandes *et al.*, 2014), as shown in Figure 7.1. Polyphenol compounds, which are defined by numerous phenol rings with many hydroxyl groups, make up the greatest composition of phytochemicals in pomegranates. A recent study found that polyphenols contained in commercial pomegranate juice extract are above 2 g/L, the highest among all fruit extracts (Oziyci *et al.*, 2012).

Hydrolyzable tannins and anthocyanin are two primary kinds of polyphenol chemicals present in pomegranates, and they account for most of the fruit's antioxidant activity (Mirsaeedghazi *et al.*, 2014). Comparing fruit juices, green tea, and red wine, it was previously discovered that commercial pomegranate juice extract possessed the most antioxidant activity (Farag *et al.*, 2014; El-falleh *et al.*, 2012). Pomegranate has greater cellular antioxidant activity and overall phenolic content than widely eaten fruits including blueberry, apple, red grape, cherry, and lemon (Anahita *et al.*, 2015).

Flavonoids are among the polyphenols encountered as a major component in pomegranate juice, and common types of flavonoids are anthocyanin, catechins, and other complex flavonoids (Glazer *et al.*, 2012). Meanwhile, pomegranate peel constitutes approximately 50% of the whole fruit and is considered an inedible by-product obtained through juice extract processing. It is characterized by the significant presence of ellagic acid and its derivatives, such as the ellagitannins, punicalagin, punicalin, gallic acid, and flavonoids (Fernandes *et al.*, 2014).

In pomegranate, arils comprise 40% juice extract sacs and 10% seeds (Mphahlele *et al.*, 2016; Oziyci *et al.*, 2012), with 10% total sugars and 85% water and organic acids such as ascorbic acid and malic acid. Pomegranate juice is a possible source of anthocyanin, ellagic acid, tannins, flavonoids, and organic acids, of which several are sources of antioxidants (Mphahlele *et al.*, 2016). Meanwhile, 100 g of pomegranate arils can provide 72 kcal of energy, 16.6 g carbohydrate, 1.0 g protein, 379 mg potassium, 1 mg sodium, 12 mg magnesium, 0.7 mg iron, 13 mg calcium, 7 mg vitamin C, and 0.3 mg niacin (Viuda-Martos *et al.*, 2013).

FIGURE 7.1 Pomegranate fruit showing the peel (outer skin).

7.2.2 Opioids

All natural and synthetic forms linked functionally to the poppy plant and endogenous neuropeptides are referred to as opioids (Caputi *et al.*, 2013). Opioids include prescription pain medicines like morphine, codeine, oxycodone, and fentanyl, as well as illicit narcotics like heroin (National Institute on Drug Abuse [NIDA], 2015). Opium, on the other hand, is a natural combination made from the juice of the opium poppy (*Papaver somniferum*) (Caputi *et al.*, 2013). There are three kinds of opioids, natural, semi-synthetic, and synthetic opioids, as depicted in Figure 7.2. Natural opioids, such as codeine and morphine, come from opium poppy plants (Borg *et al.*, 2014). Meanwhile, semi-synthetic opioids, such as oxycodone, hydrocodone, and heroin, are naturally occurring substances that are processed in a laboratory setting to create opium derivatives (Borg *et al.*, 2014).

Synthetic opioids such as methadone and fentanyl are man-made. They do not come from the opium poppy, and there is no added opium or opioid derivatives. However, they still has a similar effect to opioids (American Society of Addiction Medicine, 2011). Clinically, opioids are used as pain relievers and as opioid therapy for addiction. Hydrocodone and morphine have been used extensively for both minor and severe pain (Borg *et al.*, 2014). Besides their medicinal use, morphine-containing drugs and pure morphine serve as recreational agents, which contribute to illegal usage (Byrne *et al.*, 2012). Subjects may deliberately abuse morphine for euphoria and combine it with several opioids or chemicals to produce more potent effects and more advanced euphoria (Byrne *et al.*, 2012).

As a natural product from the poppy plant, morphine is an alkaloid compound that belongs to the phenanthrenes class and is described mainly as a high-potency analgesic (Byrne *et al.*, 2012). Opioids have been described by NIDA (2015) as being chemically linked and affecting the nervous system by operating on opioid receptors on nerve cells in the brain to reduce pain and provide pleasurable sensations. However, compulsively seeking opioids and taking them to obtain reward relief is recognized as opioid addiction, a primary, chronic, and relapsing brain disease (Spagnolo & Goldman, 2017).

7.2.3 Opioid System and Brain Pathway

Opioid dependence has been characterized by enormously unpleasant emotional and physical behavioral states that require continued use of opioids for their rewarding effects or to avoid withdrawal symptoms (Pascoli *et al.*, 2014). The mesolimbic dopamine system, which comprises the ventral tegmental area (VTA) and nucleus accumbens (NAc, a section of ventral striatum), is responsible for opioid dependency and withdrawal indicators for dopamine neurons and medium spiny neurons, respectively (Benyhe *et al.*, 2015).

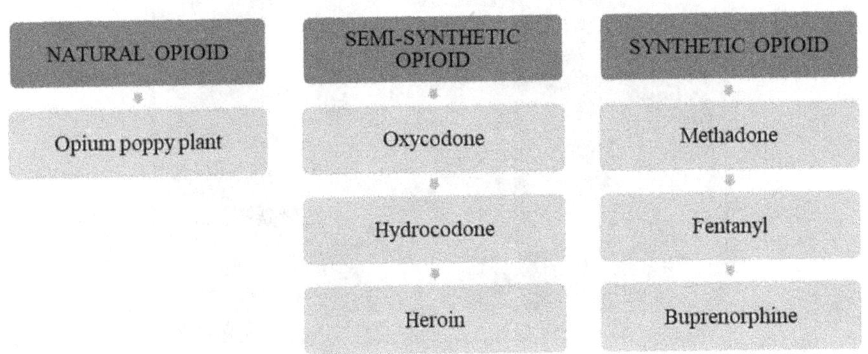

FIGURE 7.2 Schematic diagram of types of opioids.

Studies by Parsadaniantz *et al.* (2015) and Robison *et al.* (2013) have demonstrated that the VTA innervates many other regions of the forebrain, comprising the ventral hippocampus, basolateral amygdala, and prefrontal cortex (PFC). These regions were known to respond to the NAc by circulating reward effects after chronic opioid exposure. However, these pathways did not specifically mediate the NAc signals underlying withdrawal (Robison *et al.*, 2013). Interestingly, Zhu *et al.* (2016) identified the paraventricular (PVT) nucleus of the thalamus as a noticeable response to the NAc for mediating the manifestations of opioid withdrawal, initiating the aversive memory and physical signs.

Opioid withdrawal has shown high expression of proteins in PVT-NAc pathway projection neurons (Zhu *et al.*, 2016). The NAc consists of two major subtypes of medium spiny neurons (MSNs), distinguished by expression of dopamine 1 (D1) or dopamine 2 (D2) receptors, which are suggested to have contrasting roles in behavior and reward-mediated learning (Sillivan *et al.*, 2013; Kravitz *et al.*, 2012). Meanwhile, the synaptic plasticity of D1 and D2 MSNs is involved in behavioral adaptations to the state of chronic pain and opioid addiction, respectively (Pascoli *et al.*, 2014; Schwartz *et al.*, 2014; Pascoli *et al.*, 2012).

Chronic opioid exposure has been revealed to selectively potentiate excitatory transmission between the PVT and D2 MSNs through synaptic attachment in GluA2-lacking AMPA receptors (Zhu *et al.*, 2016). This result showed an underlying relation between the agility of synapses at PVT-D2 and the severe behavioral manifestation states accompanying opioid withdrawal. Thus, this synaptic pathway is compulsory for states of unpleasant expression linked to opioid withdrawal (Schwartz *et al.*, 2014). Opioid withdrawal in subjects results from reduction or cessation of opioid usage, and this process can also be shown in animal studies (Vargas-Perez *et al.*, 2007).

The locus coeruleus (LC) is another area in the brain that participates in the progression of opioid tolerance and dependence (Yamasaki *et al.*, 2017). In the brain, the LC is located on the floor of the fourth ventricle in the anterior pons, and it is the main site of norepinephrine, also known as noradrenaline, a hormone and a neurotransmitter (Berridge & Waterhouse, 2003). The LC is the area targeted by several endogenous and stress-related peptides (Reyes *et al.*, 2009), and its plasticity has been observed to change during chronic stress and chronic exposure to exogenous opioids (Barker *et al.*, 2015).

Once endogenous or exogenous opioids target the brain areas of the NAc and LC, several intracellular neuro-adaptative changes occur. These include activating MORs and inhibiting adenosine 3, 5 cyclic monophosphates (cAMP) protein (Chartoff & Connery, 2014). Consequently, regulation of the cAMP pathway will phosphorylate and activate various substrates, including the cAMP response element-binding (CREB) (Bian *et al.*, 2012) and brain-derived neurotropic factor (BDNF) proteins, which are also involved in the opioid addiction and dependence mechanism (Wang *et al.*, 2013).

7.2.4 Molecular Changes in Opioid Addiction and Dependence

Nowadays, many studies have focused on the nervous processes that lead to opioid addiction, tolerance, dependence, and withdrawal (Yan *et al.*, 2018; Taylor *et al.*, 2016; Trang *et al.*, 2015; Ghelardini *et al.*, 2015), as depicted in summary in Figure 7.3. The opioid receptors can activate both G protein-coupled receptor (GPCR) and G protein activatory/G protein inhibitory (G_i/G_o) with similar potencies and subsequently regulate adenylyl cyclase (AC) and mitogen-activated protein kinases (MAPK) (Chartoff & Connery, 2014). Meanwhile, the induction of the opioid properties of analgesia, tolerance, and dependence also leads to the involvement of each component: AC, cAMP, and protein kinase A (PKA) (Zhang *et al.*, 2016).

During stimulation of opioid receptors, inhibition of AC by the Gα subunit occurs, which reduces the production of intracellular cAMP, leading to decreased activity of PKA (Ghelardini *et al.*, 2015). However, the AC activity is differentially altered with acute and chronic opioid exposure. In acute opioid exposure, AC activity is reduced, and cAMP production and eventually, PKA activity

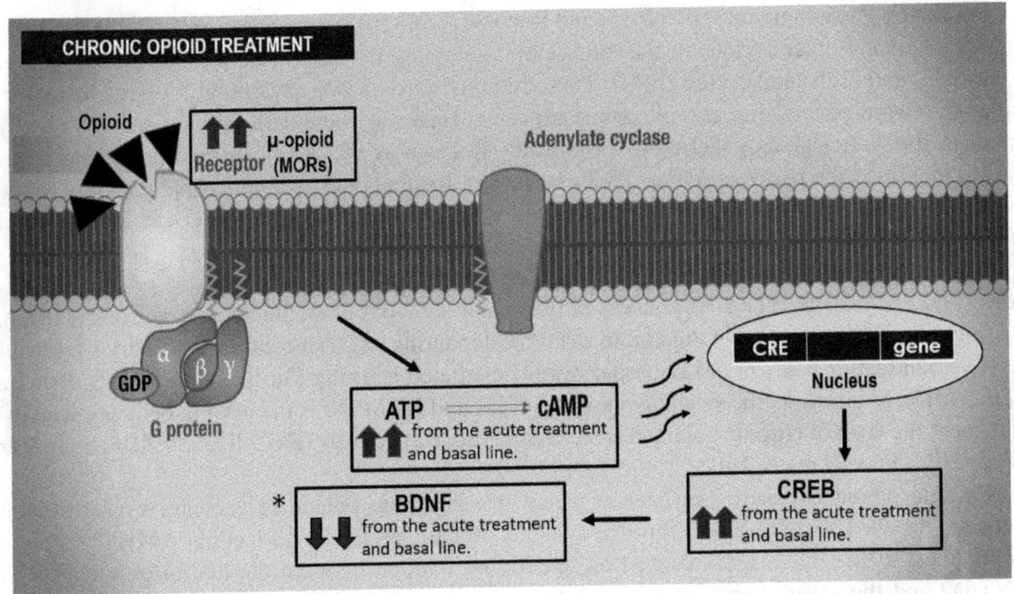

FIGURE 7.3 Summary of opioid receptor signaling. Activation of MORs receptor induces cAMP, CREB, and BDNF protein signaling.

decrease. In chronic opioid exposure, it was found that AC activity returned toward basal levels, causing increased cAMP level and PKA activity (Yan *et al.*, 2018).

The activation of PKA phosphorylation leads to enzyme activation, which has various roles in the cell, such as increased sodium ion-dependent inward current, upregulation of tyrosine kinase, and expression of the gene for CREB (Zhao *et al.*, 2013; Ren *et al.*, 2013). Nevertheless, the level of CREB phosphorylation is also differentially altered with acute and chronic exposure to opioids and in specific regions of the brain. Following extinction in opioid-induction, the level of CREB returns toward basal amounts in the hippocampus, NAc, and PFC (Song *et al.*, 2017).

However, CREB activity is reported to increase signal-regulated protein kinase (ERK) extracellularly for extinction in opioid withdrawal (Song *et al.*, 2017). Therefore, ERK can be a powerful tool for drug improvement in the therapy of opioid dependence and opioid craving. In addition, ERK is essential for neuronal plasticity and may depend on comparable molecular mechanisms operated by BDNF protein in the learning and memory aspect (Nettiksimmons *et al.*, 2014). Evans & Cahill (2016) and Williams *et al.* (2013) reported that mice BDNF deficient in LC neurons had a markedly abnormal reaction to opioid treatment, as evidenced by a significant dysregulation of cAMP-mediated cellular signaling and a failure to upregulate the expression of tyrosine hydroxylase (TH). Generally, these abnormalities happened in combination with an intense reaction to attenuated opioid withdrawal and suggest that BDNF is crucial for opioid-related plasticity in the noradrenergic system (Williams *et al.*, 2013).

7.2.5 Regulation of MOR Proteins in Opioid Addiction and Dependence

The Oprm1 gene encodes MORs, which are the molecular targets of opioid-mediated analgesia and belong to the GPCR family, qualified by a seven-transmembrane-span construction (Trang *et al.*, 2015; Gaveriaux-Ruff, 2013). The opioid receptors consist of μ (Mu), δ (Delta), and κ (Kappa) receptors (Chartoff & Connery, 2014). Gene knock-out research showed that physiological effects of opioids are absent in MOR-deficient mice, providing vital support that MORs are essential for the clinical effects of opioids (Caputi *et al.*, 2013). MORs are generally pre-synaptic, but this receptor can be found post-synaptically on dendrites to regulate neuronal excitability and pre-synaptically on axon terminals to inhibit neurotransmitters (Chartoff & Connery, 2014). Song *et al.* (2013) detailed that activation of MORs and their subsequent regulation and signaling are strongly dependent on the agonist. Opioid agonists at the MOR binding site may eventually modulate the effectors in intracellular control over the inhibitory Go/Gi proteins (Song *et al.*, 2013). Receptor signaling is readily terminated by diverse cellular regulation processes comprising desensitization, phosphorylation, and endocytosis (Song *et al.*, 2012).

Nevertheless, Yan *et al.* (2014) examined a causal link between both; the efficacy of agonists and regulatory processes induced by agonists; involving DAMGO ([D-Ala^2, N-Me-Phe^4, Gly^5-ol]-enkephalin) and enkephalins as peptidic opioids and alkaloid-type compounds such as morphine, fentanyl, and methadone. The result shows that morphine induces the receptor for internalization and phosphorylation less than other peptidic opioids, but morphine can primarily activate protein kinase C (PKC) and other molecules, leading to the development of opioid tolerance (Yan *et al.*, 2014; Zheng *et al.*, 2012). Formerly, quantitative models were used to measure the rapid desensitization of MOR–effector coupling and short-range and long-range tolerance to morphine (Williams, 2014). It was reported that a loss of 80% to 95% in MORs function was needed to justify the perceived shift in morphine concentration–response curves (Bailey *et al.*, 2009a). In addition, comparable estimations of loss of MOR function suggest that tolerance represents desensitization at the level of MORs (Williams *et al.*, 2013). However, qualitatively, recovery from rapid desensitization occurs in about 1 hour (Williams *et al.*, 2013).

After long-term treatment with opioid agonists, MOR function eventually recovers in two phases. In the first phase, recovery of MOR function and desensitization is believed to happen due to the removal of opioid agonist or inhibition of PKC within 2 hours after the agonist has been removed from the receptor (Bailey *et al.*, 2009a). Meanwhile, the next phase continues for several hours and characterizes opioid tolerance (Zhang *et al.*, 2016). In addition, other related studies have reported the inhibition of GABA-ergic synaptic transmission in the nerve terminals of the periaqueductal grey for animals chronically treated with morphine (Fyfe *et al.*, 2010).

Importantly, Bailey *et al.* (2009a) observed that different opioid agonists produce different biological effects in terms of regulating receptor phosphorylation and internalization. In cells transfected with MORs, Halls *et al.* (2016) revealed that DAMGO stimulated MORs within the plasma membrane and increased ERK activity in both cytosol and nucleus. Meanwhile, morphine stimulated a PKC-dependent pathway that restricted MORs and increased cytosolic ERK activity. This suggests that differences in MOR regulation may underlie the differential effects of its opioid agonist *in vivo*.

7.2.6 Non-Opioid Agents from Natural Sources

Ibrahim *et al.* (2018), Adnan *et al.* (2016), and Jamil *et al.* (2013) investigated the capability of dates (*Phoenix dactylifera*), thymoquinone from black seeds (*Nigella sativa*), and mitragynine from kratom leaves (*Mitragynina speciose*) to be used as a non-opioid treatment in chronic opioid tolerance and dependence (see Figure 7.4). Intriguingly, in animal studies, date extract had been explained as a calcium channel antagonist (Ibrahim *et al.*, 2018), thymoquinone had been shown to have an effect as an opioid receptor stimulating compound with 45% ligand displacement at MORs (Nutten *et al.*, 2012), and mitragynine had been recognized as a substance with opioid effects (Khor *et al.*, 2011). As for pomegranate, the extract significantly reduced the MOR level at a higher concentration and showed antagonistic regulation of MORs compared with morphine (Ridzwan *et al.*, 2020). A previous study by Benyhe *et al.* (2015) detailed the potential of bioactive compounds from natural sources as safer agonists or antagonists, with the interaction between ligands and MORs being evaluated in radio ligand binding tests or radio receptor binding assays.

Chan *et al.* (2017) observed that the ligands might be agonists or antagonists to the binding site from a pharmacological standpoint. Furthermore, the general method whereby antagonists respond in the central nervous system (CNS) is competitive binding, which suppresses the agonist.

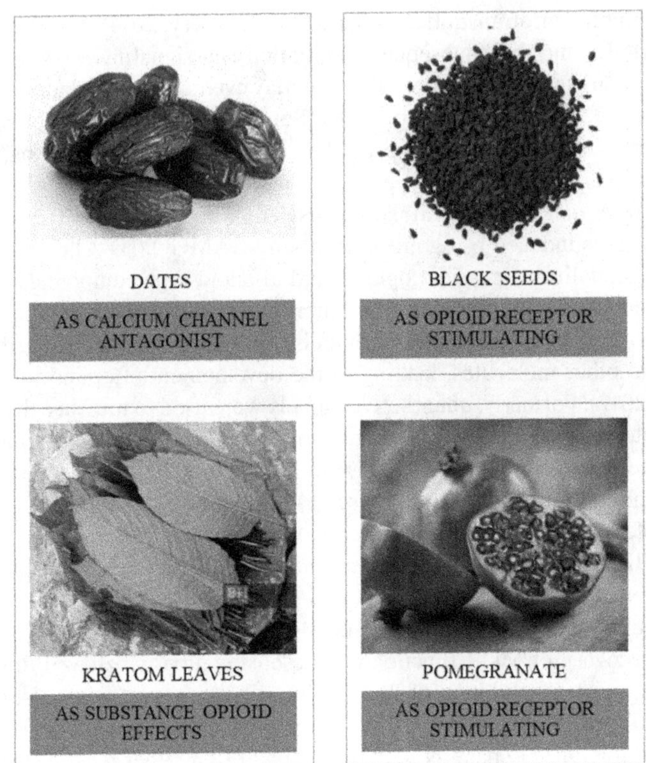

FIGURE 7.4 Examples of non-opioids from natural sources.

In co-treatment research conducted by Ridzwan *et al.* (2020), pomegranate extract therapy significantly reduced MORs and cAMP levels when matched with morphine treatment only. As a result, the study showed that pomegranate extract, through increasing MORs and cAMP levels under prolonged morphine administration, might be a safer non-opioid treatment.

Anthocyanin derived from pomegranate extracts, on the other hand, has never been studied in relation to receptor–ligand interactions at MORs. The C ring in the anthocyanin structure has a positive charge and an electron deficit, allowing it to act as an antioxidant to remove free radicals in the body by donating a hydrogen atom and an electron (Biedermann *et al.*, 2013). Meanwhile, morphine has agonist properties or is chemically sensitive to its receptor because it possesses functional groupings at the C3 phenolic and C6 secondary alcoholic sites (Borg *et al.*, 2014). Both the anthocyanin compound and morphine have a carbon ring in their chemical structure, which provides them with reactivity that influences stability structure towards antioxidant ability of free radical scavenging. Adaptive alterations in MORs and cAMP imply that anthocyanin generated from pomegranate extracts might counteract morphine. The findings provide light on the impacts of pomegranate extract as a substitute treatment for opiate addiction on the cellular level.

The efficacy of pomegranate extract *in vivo* has been examined, since the anthocyanin has been shown to activate the ERK-CREB-BDNF protein pathway, which is involved in neurodegenerative research (Ilyas *et al.*, 2017; Subash *et al.*, 2014). Through direct pharmaceutical effects on the neurotransmitter system, opioid addiction and dependence can take over normal learning and memory systems, significantly reaffirming the capacity of opioids to impact learning and memory pathways (Taylor *et al.*, 2016). According to Ridzwan *et al.* (2020), rats in the pomegranate extract treatment group showed improvement in cognitive impairment in the Morris Water Maze (MWM) technique, which could be explained by the establishment of reference memory in this group as compared with the morphine group. The findings also indicated that pomegranate extract therapy might improve learning and memory behavior in mice, confirming that decreased distance through the MWM technique was not related to motor activity (Ridzwan *et al.,* 2020; Ilyas *et al.*, 2017).

In the processing of learning and memory in the brain, long-term potentiation (LTP) is known to be one of the primary mechanisms (Nettiksimmons *et al.*, 2014). LTP happens when there is an increased connection between two neurons, and a continuous increase of chemical strength in the synapse leads to synaptic plasticity, a process understood to cause memory (Bissonette & Roesch, 2016). Various signaling pathways are related to the fusion of brand-new proteins in the LTP conditions, synaptic plasticity, and memory, including cAMP-dependent PKA, protein kinase B, PKC, and ERK (Park & Poo, 2013; Numakawa *et al.*, 2010).

All these pathways eventually lead to the CREB protein, a transcription factor that binds to the promoter areas of numerous genes involved in synaptic re-modeling, flexibility, and memory (Song *et al.*, 2017). According to Vauzour *et al.* (2008), CREB activation is important for the start of long-term alterations in synaptic plasticity and memory; when CREB activation was disrupted, long-term memory formation was blocked. Previously, the CREB protein was known as a key transcription factor for neurotrophins, including BDNF, which controls the survival, differentiation, and purpose of neurons (Wang *et al.*, 2012).

Further, Ridzwan *et al.* (2020) observed a reduction of CREB level and an elevated BDNF level in the serum of chronic morphine-induced rats treated with pomegranate extract compared with rats on prolonged morphine alone. A study by Spencer (2008) detailed that interaction of flavonoids in the body through the ERK pathway could affect memory, as ERK is often linked with pro-survival and pro-neurotrophin signaling in response to CREB stimulation (Vauzour *et al.*, 2008). This was supported by Maher *et al.* (2006) proving that flavonoids could recover object recognition and LTP by a dependent mechanism of ERK-CREB activation.

Pre-clinical evidence showed that flavonoids could have decisive effects on mammalian cognitive tasks, such as improving cognitive functions, learning, and memory (Socci *et al.*, 2017; Angeloni *et al.*, 2012). The most relevant properties of these compounds to protect the brain from memory impairment are their ability to pass through the BBB (Neshatdoust *et al.*, 2016), scavenging pathological concentrations of reactive oxygen species (ROS), and activating the brain's major antioxidant enzymes (Faria *et al.*, 2012). Several studies have indicated that some dietary anthocyanidins can cross the BBB (Abbott *et al.*, 2006; Mandel *et al.*, 2006). Spencer (2008) strengthened this indication by detecting and identifying anthocyanin in rat brain areas after feeding rats an anthocyanin-rich diet. Faria *et al.* (2011) detected a flavonoid and its metabolites in the brains of rats following oral administration of green tea, suggesting that it is brain permeable.

Nonetheless, one of the most important aspects of anthocyanin's bioavailability to the brain is the route by which anthocyanin is transported across the BBB (Faria *et al.*, 2012). It is currently uncertain whether or not the anthocyanin or other bioactive compounds may affect the contact of anthocyanin in the CNS. Along with this, Youdim *et al.* (2003) tested an *in vitro* model BBB of ECV304, which represents the peripheral cells of the BBB, co-cultured with glioma cells of C6, bEND5, and RBE4 cells, which represent the CNS side. The results revealed that flavonols, flavan-3-ols, and anthocyanin can cross the RBE4 cells in a time-dependent manner (Faria *et al.*, 2010; Youdim *et al.*, 2003). In addition, the primary metabolites of flavonoids that had been in the blood circulation also could be found integrated in the CNS cells (Ishisaka *et al.*, 2011). These studies proved the potency of anthocyanin derived from pomegranate extract to stimulate the ERK-CREB-BDNF protein pathway in opioid addiction.

7.3 APPROACHES TO NANOTECHNOLOGY-BASED NON-OPIOID DELIVERY

According to epidemiologic and pharmacologic studies, plants contain physiologically active substances that may give health advantages against degenerative illnesses, with a focus on antioxidants found in fruits (Shiban *et al.*, 2012). Anti-nociceptive properties, enhanced cognitive function, and activated protein signaling in the opioid system via the BBB are all proven health benefits of anthocyanin derived from pomegranate (Hasnaoui *et al.*, 2014). Despite its numerous health advantages, anthocyanin has been demonstrated to have low plasma bioavailability (less than 6% of the original dosage) following consumption of anthocyanin-rich food (Xie *et al.*, 2016; He *et al.*, 2009; Pascual-Teresa & Sanchez-Ballesta, 2007; Talavera *et al.*, 2004). Due to its high reactivity, anthocyanin is sensitive to destruction under a selection of external circumstances, such as enzymes, pH, oxygen, light, temperature, and water activity (Tsuda, 2012; Patras *et al.*, 2010).

Anthocyanin appears to be quickly absorbed from the bloodstream into the tissues, where this compound accumulated from the onset though the availability to target tissues appears to be limited (Vanzo *et al.*, 2011; He & Giusti, 2010; Talavera *et al.*, 2005). In a 15-day *in vivo* investigation, rats given an anthocyanin-rich diet were found to have 68.6 nmoL/g anthocyanins in stomach tissue and 605 nmoL/g in the jejunum, 0.38 nmoL/g in the liver, 3.27 nmoL/g in the kidney, and 0.25 nmoL/g in the brain (Talavera *et al.*, 2005). Following this, Passamonti *et al.* (2005) discovered anthocyanin in rat brain 10 minutes after consumption of 8 mg/kg anthocyanin. This demonstrates that anthocyanin may travel quickly and selectively from the stomach to the brain across the BBB. Furthermore, Andres-Lacueva *et al.* (2005) observed that anthocyanins were discovered in the cerebellum, striatum, and hippocampus of rat brain after feeding an anthocyanin diet for 70 days.

Nanotechnology has been promoted as a feasible tool for the impediment, treatment, and delivery of therapies for a wide range of pathophysiological disorders (Martinez-Ballesta *et al.*, 2018). Nanotechnological presentations in biology and medicine could display bulk mesoscale and macroscale chemical or physical characteristics that are unique to the designed material or device and are not necessarily possessed by molecules alone (Feng *et al.*, 2004). This encourages the development of nanotechnologies that can perform several specialized tasks at the same time or in a predetermined order, which is a requirement for clinically effective opioid and non-opioid chemical delivery to the CNS and brain through the BBB (Silva, 2008).

Nanotechnology in opioid delivery systems, on the other hand, has the potential to develop more soluble and stable structures, nanocarriers, with the capability to cross cell barriers, therefore improving delivery efficiency and decreasing side effects (Craparo *et al.*, 2011; Ulbrich *et al.*, 2009; Bhaskar *et al.*, 2010). Nonetheless, because of its capability to cross the BBB, a nanodelivery–opioid complex would be perfect for systemic administration, generating low systemic effects in the CNS and then delivering an opioid to a particular area of the brain (Silva, 2008). Drug payload conveyance, BBB-impermeant medications, good chemical and biological strength, the possibility of integrating both hydrophilic and hydrophobic medications, and the ability to be administered by oral, inhalation, or parenteral routes are all predicted (Masserini, 2013; Petkar *et al.*, 2011) (see Figure 7.5.) Nanoparticles' high surface-area-to-volume ratio permits copies of ligands like antibodies, proteins, or aptamers to affect tissues, and covalent conjugation may be employed to alter them (Montet *et al.*, 2006).

However, nanoparticles' ability to cross the BBB does not always imply drug delivery to the brain. By boosting drug strength within or on the luminal surface of BBB cells, nanoparticles may have a role in producing a significant concentration gradient between blood and brain (Haque *et al.*, 2012). Nanoparticles that target brain capillary endothelial cells and then cross the BBB transcellularly also make it feasible for their therapeutic payload to migrate into the CNS (Martin-Banderas *et al.*, 2011).

7.3.1 Brain Barriers: CNS Protection

The BBB is a kind of microvasculature made up of highly specialized endothelial cells linked by zonula occludens, claudins, and adherens joints (Anna *et al.*, 2020). The BBB's tight connections keep paracellular penetrability by creating a high electrical transmembrane resistance of >1500 cm^2, which is the highest among endothelial barriers (Masserini, 2013; Trapani *et al.*, 2011). The tightness of the BBB is controlled by a range of cell types, including microglial cells, pericytes, astrocytes, and neurons, which form the neurovascular section with endothelial cells (Winger *et al.*,

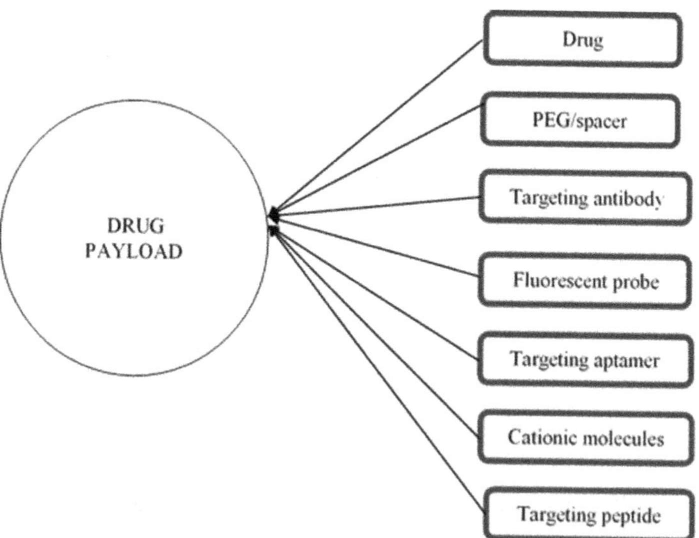

FIGURE 7.5 Multifunctionalized nanoparticles and drug payload.

2014). Finally, efflux pumps such as P-glycoprotein are added to the BBB; these expel harmful chemicals and a variety of therapeutic actions to the outside of the brain, and function as the primary barrier to medication delivery to the brain (Moura *et al.*, 2019). Furthermore, virtually all macromolecules, such as antibiotics, are unable to penetrate the BBB; only 2% of small molecules, such as opiates, are able to do so (Masserini, 2013; Pardridge, 2007).

7.3.2 Nanocarriers across the BBB

Nanocarriers have been utilized in recent years to increase anthocyanin stability and bioavailability by generating a stable habitat for anthocyanin that shields it from environmental factors including temperature, light, oxygen, and pH reactivity (Betz *et al.*, 2012). Nanocarriers are tiny particles that are enclosed with chosen carriers and adsorbed or conjugated onto their surface for the regulated administration of medicinal opioids (Parboosing *et al.*, 2012). As shown in Table 7.1, materials that have been used as nanocarriers for anthocyanin derivatives include poly (lactic-co-glycolic acid) (PLGA) (Zhang *et al.*, 2021; Beconcini *et al.*, 2019; Amin *et al.*, 2017), liposomes (Homayoonfal *et al.*, 2021; Zhao & Temelli, 2016), chitosan (Raman *et al.*, 2020; Sun *et al.*, 2020), titanium dioxide (TiO_2) (Hashemi *et al.*, 2018; Septiani *et al.*, 2017), zinc oxide (ZnO_2) (Wang *et al.*, 2020; Zyoud *et al.*, 2018), and silica (Chu *et al.*, 2017; Yesil-Celiktas *et al.*, 2017). Small particle diameter, non-toxicity, biodegradability and biocompatibility, stability in blood circulation, targeting the BBB, non-inflammatory nature, and extended circulation time are all typical criteria for nanocarriers (Parboosing *et al.*, 2012; Bhaskar *et al.*, 2010).

From 2017 until the present, the keywords "nanoparticles" and "anthocyanin" generated 117 hits in the PubMed database. Only two research publications from 2017 to 2021 that incorporated "anthocyanin" with "chitosan, TiO2, PLGA, zinc oxide, silica, and liposome loaded nanoparticles" are presented in Table 7.1.

7.3.2.1 Poly (Lactic-co-Glycolic Acid) (PLGA)

PLGA is a linear co-polymer formed by the arbitrary ring-opening co-polymerization of dual discrete monomers, glycolic acid, and lactic acid cyclic dimers (1, 4-dioxane-2, 5-diones) (Mirakabad *et al.*, 2014) (see Figure 7.6). The monomer ratios used in PLGA may be used to differentiate between different forms of PLGA, such as PLGA 75:25, which is a co-polymer created of 75% lactic acid

TABLE 7.1
Characteristics, Functions, and Application of Anthocyanin Nanoparticles

Type	Characteristics	Experiment model/dose/route	Functions	Application	References
PLGA	SZ: 206.1 ± 1.8 nm ZP: −8.36 ± 1.07 mV EE: 88.6 ± 6.2%	*In vitro* on HUVEC cell lines • 100 mg PLGA 50:50 • Inflammatory cytokines (IL-6, IL-10, PGE2, TNF-α) • NLRP3 inflammasome ELISA kit • Scanning microscopy and flow-cytometry	Anti-inflammatory activities Intestinal absorption Bioavailability	For anti-inflammatory effect on HUVEC cells	Beconcini *et al.* (2019)
	SZ: 120–165 nm ZP: −12 mV EE: 60%	*In vitro* on SH-SY5Y cancer cell lines • MTT assay with 50 to 200 μg/mL anthocyanin • Oxidative stress detection with 50, 100, 200 μg/mL anthocyanin • ApoTox-Glo triplex, Western blot, TUNEL assays, and immunofluorescence staining	Bioavailability Stability Enhanced cellular uptake Bioactivity	Enhance the bioavailability of hydrophilic drugs	Amin *et al.* (2017)
Liposome	SZ: 141–196 nm ZP: −42.85 ± 0.77 mV EE: 47.19% PY: 0.27 ± 0.04	*In vitro* on MSC and fibroblast cell lines • Synthesized using hydration and ultrasound combined method • Anthocyanin concentration (4.5, 6.75, 9% w/v) • Maximum compatibility MSC is 0.5 mL/mg • Maximum compatibility fibroblast is 10.3 mL/μg • TEM images show spherical shape	Biocompatibility Bioavailability Sphericity	To compare bioavailability of anthocyanin-loaded nanoliposomes	Homayoonfal *et al.* (2021)
	EE: 52.2 ± 2.1% ZP: −44.3 ± 2.9 mV PY: 0.26 ± 0.01	Experimental study • Soy lecithin for liposome • Anthocyanin (76%) from bilberry • Using the supercritical carbon dioxide method	Intactness Sphericity Uniformity Stability	To produce liposomes to deliver bioactives for food applications	Zhao & Temelli (2016)
Chitosan	N/A	An *in vivo* study using male Wistar rats • Encapsulated using conventional ionotropic gelation method • Manifestation of anti-inflammatory cytokines, IL-4 • Inhibition of pro-inflammatory cytokines, IFN-y	Bioavailability Gastro-protective	Therapeutic potential of anthocyanin nanoparticles in induced gastric ulcer	Raman *et al.* (2020)
	SZ: 180 nm ZP: −19 mV EE: 53.88%	Experimental study • Established by three types of chitosan capsules • Cytotoxicity study using 5, 10, 15, 20, 50 μmol/mL nanoparticles • Morphology study using 1 mL of 0.02 mg/mL nanoparticles	Stability Sphericity Biosafety Enhancement value	Adding bioactive enhancements to the usage in food processing	Sun *et al.* (2020)

(*Continued*)

TABLE 7.1 (CONTINUED)
Characteristics, Functions, and Application of Anthocyanin Nanoparticles

Type	Characteristics	Experiment model/dose/route	Functions	Application	References
ZnO_2	SZ: 41 nm	Experimental study • Concentration of nanoparticles: 0, 2457, 4915 μM • At 2457 and 4915 μM, levels of chlorophyll a, b, and carotenoid decreased • Level of anthocyanin and PAL activity increased • Level of proline decreased • MDA, CAT, and H_2O_2 increased	Level of anthocyanin increased Level of proline decreased	Effect of different concentrations of nanoparticles synthesized by olive extract	Hashemi *et al.* (2018)
	SZ: ~30 nm	Experimental study • 100 g of black rice was smashed • The crystallinity of zinc oxide was studied using X-ray diffraction and SEM microscopy	As a capping agent To change the shape Crystallinity	Utilizing anthocyanin as a capping agent	Septiani *et al.* (2017)
TiO_2	SZ: 420 ± 5.75 nm ZP: −22.83 ± 0.25 mV EE: 88.6 ± 6.2% PD: 0.27 ± 0.01	*In vitro* on CaCo-2 cancer cell lines • Using six different types of anthocyanin • 50 μM anthocyanin used in characteristics study • 4 mg/mL used in quenching fluorescence • 50 μM nanoparticles with 50–200 μM anthocyanin used for staining	Binding affinity Stability	Interactions of nanoparticles and anthocyanin with human intestinal cells	Wang *et al.* (2020)
	SZ: ~100 nm ZP: 3 eV	Experimental study • 30 g commercial roselle dried flowers • Photocatalyst with 50 mL AC/TiO_2/anthocyanin (0.1 g) • Catalyst stability with 50 mL, 2×10^{-5} M anthocyanin	Efficiency of photocatalyst Stability	The enhanced catalytic efficiency of the recovered TiO_2	Zyoud *et al.* (2018)
Silica	SZ: 100–500 nm	Experimental Stober strategy • 30 μL of black carrot solution • APTES/ATH volume ratio 1/30 • Morphology of ATH-encapsulated is confirmed by FTIR assay	Stability Sensitivity	As a borate-sensing application	Chu *et al.* (2017)
	SZ: 134.8 nm ZP: +19.78 mV PY: 0.338	*In vitro* on neuroblastoma cell lines • Anthocyanin from black carrots • 6.25 μg/mL nanoparticles for cytotoxicity • Electron microscopy for characteristics study	Provide routes in the industry	Create a design for making silica dendrimer hybrids	Yesil-Celiktas *et al.* (2017)

Abbreviations: PGE: prostaglandin E; TNF: tumor necrosis factor; NLRP: nucleotide-binding oligomerization domain; ELISA: enzyme-linked immunosorbent assay; MTT: (3-[4, 5-dimethylthiazol-2-yl]-2,5 diphenyl tetrazolium bromide); TUNEL: terminal deoxynucleotidyl transferase dUTP nick end labelling; TEM: transmission electron microscopy; IFN: interferond; PAL: physical activity level; MDA: malonaldehyde; CAT: catalase: EE: encapsulation efficiency; AC: anthocyanins; APTES: (3-aminopropyl)triethoxysilane; ATH: alumina trihydrate; FTIR: fourier-transform infrared spectroscopy; PY/PD: polydispersity index.

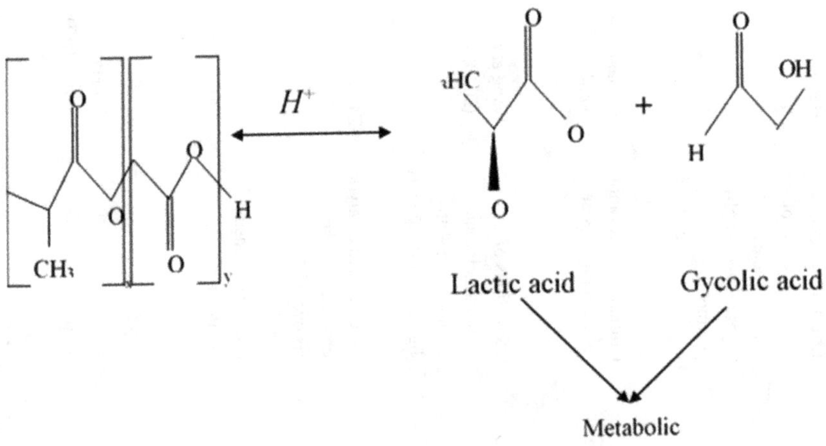

FIGURE 7.6 Chemical structure of PLGA and its monomers.

and 25% glycolic acid (Gentile *et al.*, 2014). The European Medicines Agency (EMA) and the U.S. Food and Drug Administration (FDA) have approved PLGA for use in several biomedical devices, including sutures, implants, bone tissue engineering, drug delivery systems, and biomaterials such as human screws (Gentile *et al.*, 2014; Hines & Kaplan, 2013). It's also popular because of its biocompatibility, biodegradability, tolerability, extensive track record in biological activities, well-documented value for longer drug release compared with traditional devices that last for several days, weeks, or months, and the convenience of parenteral administration by injection (Zhi *et al.*, 2021; Locatelli *et al.*, 2012; Acharya & Sahoo, 2011).

Due to the hydrolytic de-esterification process, followed by the clearance of their monomeric anions, glycolate and lactate, and final metabolism into carbon dioxide to be excreted through the lungs and water to be excreted through the renal system, PLGA co-polymers are non-toxic and biodegradable (Danhier *et al.*, 2012; Makadia & Siegel, 2011; Silva *et al.*, 2015). The different molecular weights and co-polymer compositions among lactate/glycolate acids have the greatest impact on PLGA breakdown rates (Dinarvand *et al.*, 2011; Schliecker *et al.*, 2003). A greater lactate/glycolate acid ratio results in higher hydrophobicity, which leads to slower removal and drug delivery kinetics (Engineer *et al.*, 2011), whereas a higher glycolate content contributes to a more hydrophilic PLGA, which leads to considerably quicker removal and drug delivery kinetics (Xu *et al.*, 2017).

PLGAs have been widely employed to cross the BBB due to their drug administration features, for example high loading ability, high strength, self-regulated drug delivery, and targeting efficacy (Gong *et al.*, 2017; Kumari *et al.*, 2010). However, to penetrate the BBB and deliver opioids to the brain, the produced PLGA nanoparticles require multiple surface modification methods (Del Amo *et al.*, 2021). The half-life of particles in the blood circulation and the number of particles entrapped by the target cell region are both adversely affected by these PLGA surface approaches (Sah *et al.*, 2013). Tween 80 (Gelperina *et al.*, 2002), polyethylene glycol (Tang *et al.*, 2007), polyethylene oxide (Soppimath *et al.*, 2001), poloxamers (Mozafari *et al.*, 2009), polysorbate 80, and polysaccharides like dextran (Mahapatro & Singh, 2011) were used to coat the PLGA nanoparticles with specific proteins, peptides, or monoclonal antibodies. Surface modification of PLGA might also be done via the terminal carboxylic acid groups, resulting in di-block co-polymers like polyethylene glycol (PEG)-b-PLGA or tri-block co-polymers like PLGA-b-PEG-b-PLGA (Zhang *et al.*, 2014).

According to the processes, PLGA nanoparticles can function in two ways, depending on the methods used: modified surface PLGA and unmodified surface PLGA. As depicted in Figure 7.7, there are three active endocytosis pathways that the modified PLGA nanoparticles can use to cross the BBB (Pulgar, 2018; Ding & Zhu, 2018; Herve *et al.*, 2008). Unmodified PLGA nanoparticles, on the other hand, penetrate the BBB largely by passive absorption based on size, resulting in limited

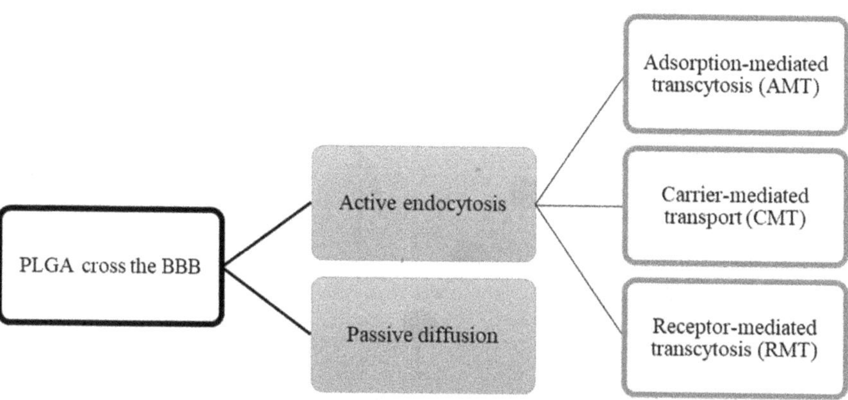

FIGURE 7.7 Transport mechanisms for PLGA nanoparticles to cross the BBB.

brain uptake (Pulgar, 2018; Herve *et al.*, 2008). Positive charges on PLGA nanoparticle surfaces are attracted to negatively charged portions of luminal surfaces, allowing PLGA to pass across the BBB. PLGA is modified or covalently coupled with ligands that affect cell surface receptors, which are identified to be BBB-carrying pathways for carrier-mediated transport.

In clinical investigations, Mi *et al.* (2020) effectively targeted a prostate-specific membrane antigen in prostate cancer cells using PLGA nanoparticles loaded with docetaxel. Monge-Fuentes *et al.* (2021) demonstrated that dopamine encapsulated in albumin/PLGA nanosystems at the maximum dose enabled effective crossing of the BBB for medication administration and significantly improved symptoms in mice with untreated Parkinson's disease. An *in vivo* study of Angiopep-2 as a main targeting ligand for the release of nanomedicines into the brain indicated potential BBB crossing and highlighted neuron accumulation in the cortex and hippocampus following conjugation with PLGA nanoparticles. Modified PLGA nanoparticles have therefore been shown to improve the cellular absorption and therapeutic effectiveness of medicines delivered across the BBB.

7.3.2.2 Liposome-Based

Liposomes are spherical synthetic cells made up of single amphiphilic lipid bilayers that can encapsulate medicinal compounds; examples are medicines, vaccines, nucleic acids, and proteins (Abbina & Parambath, 2018). This form of liposome is composed of uni- or oligolamellar phospholipid vesicles that are sustained by cholesterol and have stealth properties due to PEG chains connected to their sides (Mahringer *et al.*, 2021). The potential for these vesicles to be used as a carrier system for the treatment and/or diagnostics of illness was realized immediately after their discovery in the early 1960s (Micheli *et al.*, 2012; Schnyder & Huwyler, 2005). Liposomal formulations have previously been used to provide anticancer medicines; examples are methotrexate (Hu *et al.*, 2017), 5-fluorouracil (Lakkadwala & Singh, 2018), paclitaxel (Peng *et al.*, 2018), doxorubicin (Lakkadwala & Singh, 2019; Zhan & Wang, 2018), and erlotinib (Lakkadwala & Singh, 2019).

Liposomes can include hydrophilic, lipophilic, and hydrophobic therapeutic compounds due to their distinct physicochemical features (Vieira & Gamarra, 2016). Hydrophilic compounds can be trapped in liposomes' watery cores or detected at the interface between the lipid bilayer and the external phase. In liposomes, lipophilic or hydrophobic medications are virtually fully entrapped in the hydrophobic core of lipid bilayers. Liposomes have several advantages, including high biodegradability and biocompatibility, minimal toxicity, drug-targeted transport, and regulated drug transport (Noble, 2014; Torchilin, 2005). However, one of the earliest discoveries made by Postmes *et al.* (1980) revealed that thyrotropin-releasing hormone contained in liposomes did not have a surface decoration and had only a small effect once it crossed the BBB. As a result, different surface changes have been developed to facilitate liposomal transfer across the BBB.

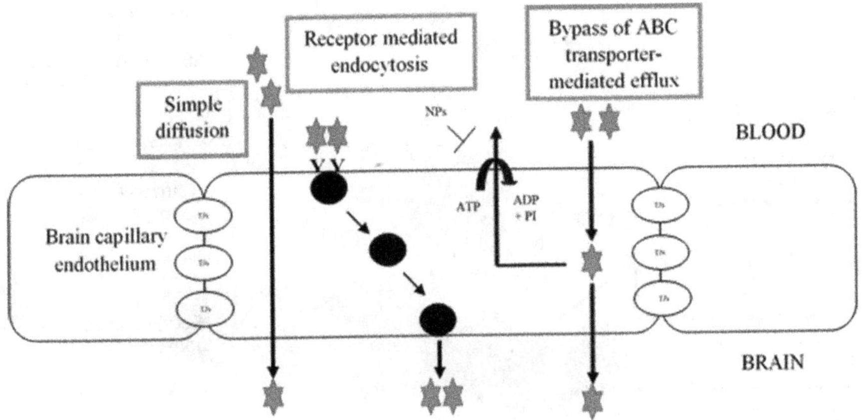

FIGURE 7.8 Interface of drug-loaded liposome nanoparticles across the BBB.

Pinzon-Daza *et al.* (2013) previously demonstrated that nanoparticles penetrate the BBB due to passive (non-specific endocytosis) or active (binding to receptors on the luminal surface of BBB cells) targeting (Pinzon-Daza *et al.*, 2013) (see Figure 7.8). It has long been recognized that the surface of the BBB contains several receptors, including those for various proteins, peptides, antibodies, and other molecules (Spuch & Navarro, 2011). These compounds serve as surface-active ligands that aid translocation through receptor-mediated transcytosis. Furthermore, cationic liposomes might traverse the BBB via absorption-mediated transcytosis as well as carrier-mediated transcytosis, which uses nutrients like glucose and glutathione to attach to the liposome's side and promote translocation (Vieira & Gamarra, 2016; Noble, 2014). Liposomes can be covered with a range of molecules, such as PEG, which extend the blood circulation time of the formulations (Hu *et al.*, 2017), transferrin, which improved carrier translocation across the BBB (Lakkadwala & Singh, 2018, 2019), and the glucose–vitamin C complex, which enhanced liposome growth at the affected site (Peng *et al.*, 2018).

Previously, Lewicky *et al.* (2020) employed novel mannose-linked liposomes to deliver dynantin, a powerful synthetic polypeptide opioid receptor antagonist with antidepressant and anxiolytic-like properties. To transport opioids to the brain, researchers used antibodies, liposomes, nanoparticles, and tiny nucleic acid molecules to decorate or co-decorate liposomes with cell-penetrating peptides (Mahringer *et al.*, 2021). The cell-penetrating peptides are short cationic peptides of fewer than 30 amino acids that have the potential to permeate BBB cells, albeit the exact mechanism is unknown. Furthermore, leptin-conjugated liposomes have been employed as opioid transporters for beneficial substances like resveratrol (RES) and epigallocatechin gallate (EGCG) to save dopaminergic neurons that have deteriorated (Kuo *et al.*, 2021). The leptin receptor, which governs appetite and fullness and is a key regulator of body weight, is one of the main concerns in the BBB (Tamaru *et al.*, 2010). According to the findings, Lep/RES-EGCG liposomes have a higher capacity to cross an *in vitro* BBB model than non-modified liposomes.

Anthocyanin-loaded liposomes are also necessary as nanotechnology-based techniques to bypass the BBB, preserving the effect of anthocyanin from different unfavorable circumstances and maintaining its stability and efficacy (Zhao *et al.*, 2017; Zhao & Temelli, 2016; Tsuda, 2012). A small quantity of cholesterol is added into liposome membranes to enhance the filling of phospholipids, minimize membrane absorbency, and limit anthocyanin outflow from the vesicles, which improves encapsulation efficiency and bioactive loading (Bozzuto & Molinari, 2015). When compared with free anthocyanin concentration, high anthocyanin intensity caused a large rise in SZ (particle size) and PY (polydispersity index), but cholesterol addition resulted in a relatively minor increase (Zhao *et al.*, 2017; Zhao & Temelli, 2016). The anthocyanin-loaded liposome's greater anthocyanin content (30–40%) changed the wrapping of bilayer membranes and caused enhanced irregularity in the liposome shape, making it more favorable to cross the BBB cells.

7.3.2.3 Chitosan

Chitosan is a polysaccharide discovered in many crab shells that has been shown to be a viable material for nanoparticle production due to its biodegradability, biocompatibility, non-toxicity, and low cost (Sonia & Sharma, 2011). It's made by partially N-deacetylating chitin in hot alkaline conditions, yielding a co-polymer made up of N-acetyl-glucosamine and N-glucosamine units connected by β-(1,4)-glycosidic linkages (Anna *et al.*, 2020). Chitosan's application in encapsulation was previously limited, since it can only be dissipated in a few dilute acid liquids and is insoluble at physiological pH (Alves & Mano, 2008).

Chitosan-based nanoparticles are a potential choice due to their biocompatibility and unique properties such as positive charge, which allows ionotropic gelation and the formation of ionic connections with anions on the cell surface (Sahin *et al.*, 2017; Bugnicourt *et al.*, 2014). Previous findings have demonstrated that chitosan-based nanoparticles may transport drugs to the brain (Karatas *et al.*, 2009), and *in vitro* tests employing brain endothelial cells have been performed to assess their effectiveness (Aparicio-Blanco *et al.*, 2016; Obermeier *et al.*, 2016). Electrostatic connections between the positive charges on chitosan and the negative charges on a non-toxic cross-linker, such as sodium tripolyphosphate (TPP), are used to create nanoparticles by ionic gelation (Bugnicourt & Ladaviere, 2016; Bugnicourt *et al.*, 2014). TPP is the chemical most often utilized for cross-linking chitosan with a wide range of medications and macromolecules that are either hydrophobic or hydrophilic, making it a potential drug delivery vehicle (Bugnicourt & Ladaviere, 2016). When using this approach to optimize nanoparticle physicochemical properties like SZ and ZP, the chitosan/TPP ratio, the chitosan concentration, and the ionic strength of the solution form are the most essential parameters to consider (Hejjaji *et al.*, 2018; Jonassen *et al.*, 2012). Because nanoparticles must be reduced below 200 nm and have a positive ZP value to cross the BBB, the SZ and ZP play a crucial part in their endocytosis by brain endothelial cells (Wang *et al.*, 2010; Lu *et al.*, 2005).

Sahin *et al.* (2017) tested brain-targeted chitosan nanoparticles via BBB endothelial cells. TfRmAB-conjugated nanoparticles had positive SZ, ZP, and PY values of 284.158.6 nm, 34.42.1 mV, and 0.480.13, respectively, compared with control. Jahromi *et al.* (2019), on the other hand, investigated the mechanistic transport of methotrexate-loaded chitosan-created hydrogel nanoparticles across the BBB. Methotrexate is a chemotherapeutic drug that cannot cross the BBB because it is a substrate of numerous BBB effluence transporters, including the P-glycoprotein transporter (Fraser *et al.*, 2015). It acquires a positive value of SZ and ZP after being loaded with chitosan nanoparticles, which is necessary to traverse the BBB and to prevent clearance by the reticuloendothelial cells (Jahromi *et al.*, 2019; Jo *et al.*, 2016).

Based on non-opioid administration, the use of anthocyanin chitosan nanoparticles was investigated for gastrointestinal fluid degradation and therapeutic potential in induced gastric ulcers (Raman *et al.*, 2020). According to He *et al.* (2016), anthocyanin-loaded chitosan nanoparticles were made utilizing two distinct water-soluble chitosan derivatives: negatively charged carboxymethyl chitosan (CMC) and positively charged chitosan hydrochloride (CHC). Minimum SZ (211–751 nm) and maximum EE (38.2–64.2%) were achieved with the optimum CMC and CHC concentrations and anthocyanin quantity, while encapsulated anthocyanin nanoparticles loaded with chitosan polymers on alcohol-HCl reduced the induced gastric ulcer and reliably restored the mucosal layer (Raman *et al.*, 2020). The encapsulated anthocyanin produced substantially more of the anti-inflammatory cytokine interleukin (IL)-4 and had significantly better absorption than the free anthocyanin. As a result, the functionalization of chitosan nanoparticles with opioid and anthocyanin targeting, in addition to chitosan's muco-adhesive characteristics, can improve nanocarrier transit across the BBB.

7.3.2.4 Zinc Oxide (ZNO_2)

ZnO_2 nanoparticles are frequently used in toothpaste, cosmetic products, sunblock, fabrics, wall paints, and construction tools due to their unique physical and chemical features of tiny size and high surface area (Tian *et al.*, 2015). Zinc is well recognized as an essential trace element that may be found in all human tissues, including brain, muscular tissue, bone, and skin. Zinc is an essential component of many enzyme systems in the body, and it is involved in protein and nucleic acid production, hematopoiesis, and neurogenesis (Ruszkiewicz *et al.*, 2017; Kolodziejczak-Radzimska & Jesionowski, 2014). An *in vivo* study discovered that ZnO_2 could penetrate the BBB after oral and inhalation administration (Lee *et al.*, 2012; Kao *et al.*, 2012b), induce changes in spatial understanding and memory capability by altering synaptic flexibility (Han *et al.*, 2011), and interact with different influenced of the zinc toxicity in the plasma and brain (Shim *et al.*, 2014).

ZnO_2 nanoparticles with a thickness of <100 nm are deemed biocompatible, which supports their biomedical uses in drug delivery by possibly lowering the total quantity of medicines needed and avoiding unwanted side effects (Shim *et al.*, 2014; Erathodiyil & Ying, 2011). Various medicines, including doxorubicin, paclitaxel, curcumin, baicalin, and DNA fragments, were packed onto ZnO_2 nanoparticles, resulting in improved solubility, increased toxicity, and efficient transport into cancer cells when compared with individual compounds (Wang *et al.*, 2017; Ghaffari *et al.*, 2017; Li *et al.*, 2017). ZnO_2 nanoparticles, on the other hand, have been effectively produced using anthocyanin, a flavonoid molecule, as a capping agent via a method involving thermal breakdown of a precursor (Septiani *et al.*, 2017). The addition of anthocyanin reduced the size of nanoparticles from 30 to 16 nm while having no effect on the crystalline form of ZnO_2.

Hariharan *et al.* (2012) employed the co-precipitation method to create PEG 600 solution–modified ZnO_2 nanoparticles, which were then loaded with doxorubicin to produce DOX-ZnO_2/PEG nanocomposites for drug administration. DOX-ZnO_2/PEG demonstrated an effective drug delivery system and substantially increased cellular absorption of the medication after culture with hepatocarcinoma cells. Kao *et al.* (2012b) revealed that the introduction of ZnO_2 nanoparticles into the brain can be a strong way of zinc delivery to elevate zinc ion and overall control zinc homeostasis in the CNS based on the reference via the old factory bulb in rats and in the cultured cells tested. Following that, ZnO_2 is taken up by cells through endocytosis and dissolution in acid partitions, which converts zinc oxide nanoparticles to zinc ions (Kao *et al.*, 2012a). Despite the benefits of using ZnO_2 nanoparticles for drug distribution across the BBB, high doses of this nanocarrier can cause apoptosis in lung (Kao *et al.*, 2012a) and neural stem cells (Deng *et al.*, 2009) and hinder ion channel activities in cultured rat hippocampal neurons (Zhao *et al.*, 2009).

Toxicity is not observed at low doses of ZnO_2 nanoparticles, such as 70000 mM (Deng *et al.*, 2009) or 10000 mM (Kao *et al.*, 2012b). ZnO_2 nanoparticles do not cause cell damage at concentrations lower than 70 mM, but at concentrations greater than 148 mM, they significantly damage and trigger apoptosis in rats' neural stem cells after 24 hours of exposure (Deng *et al.*, 2009). The dissolved zinc ion in the culture medium or inside cells, according to the researchers, is the source of ZnO_2 nanoparticles' toxicity to neural stem cells. When rat hippocampus neurons were subjected to commercially available ZnO_2 nanoparticles, Zhao *et al.* (2009) found that they disrupted the ion channels gated in the brain and caused death. The disturbances were triggered by the failure of cytoplasmic potassium ion due to improved potassium ion efflux via postponed rectifier potassium channels, resulting in the failure to maintain appropriate transmembrane ionic gradients, according to the researchers.

ZnO_2 nanoparticles injected directly into the brain have been demonstrated to produce irritation and neurotoxicity (Park & Poo, 2013; Jang *et al.*, 2012; Musa *et al.*, 2012), while inhaled ZnO_2 nanoparticles have been linked to oxidative-inflammatory responses at deposition spots (Oberdorster *et al.*, 2009). Zinc produced by combustion has been linked to an increase in oxidative DNA injury, which has been connected to the onset of neurodegenerative illness and failure of cognitive tasks, when inhaled (Calderon-Garciduenas *et al.*, 2004). Nonetheless, Shim *et al.* (2014) found that oral treatment with 10 mL/kg ZnO_2 nanoparticles for 28 days and dermal administration of 20 and 100 nm ZnO_2 nanoparticles for 90 days did not cause substantial neurotoxic effects in rats' brains. As a result of their physical and chemical benefits, ZnO_2 nanoparticles have been observed to have promise in biomedical applications for medication delivery to the brain, as they include functional drug carriers to improve treatment efficiency. Nonetheless, it is necessary to comprehend and anticipate their possible negative consequences (see Figure 7.9).

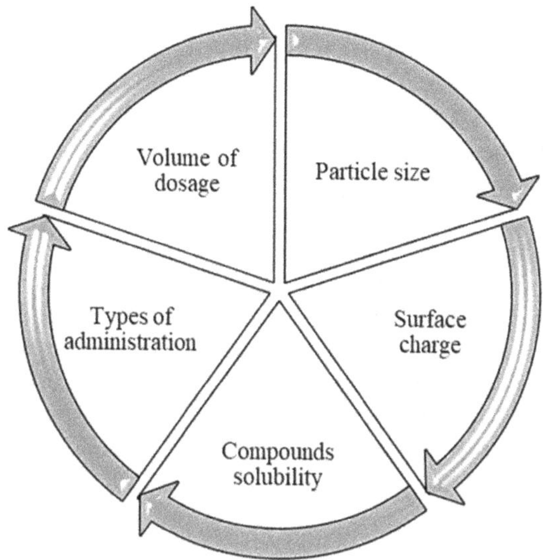

FIGURE 7.9 The factors influencing the cytotoxicity of ZnO_2 nanoparticles.

7.3.2.5 Titanium Dioxide (TiO_2)

TiO_2 is one of the very widely manufactured nanoparticles with outstanding photocatalytic activity and surface super-hydrophobic capability, allowing it to be used in paints, food additives (Lomer *et al.*, 2002), foundations (Kaida *et al.*, 2004), and purification systems in the environment (Chen & Mao, 2007). TiO_2 nanoparticles have been shown in mouse experiments to overcome epithelial barriers such as nasal, bronchial, alveolar, and gastrointestinal epithelial barriers and enter the bloodstream from the respiratory and gastrointestinal tracts (Shinohara *et al.*, 2015, 2014; Husain *et al.*, 2015). The translocation of nanoparticles into the brain for inhalation exposure, any one of precisely via the olfactory route or not directly cross the BBB, is a major source of worry (Wang *et al.*, 2008; Lockman *et al.*, 2004). TiO_2 has been found in the cerebral cortex, hippocampus, and cerebellum, and it can remain there for a long time, influencing memory and learning behavior (Skalska *et al.*, 2015; Mohammadipour *et al.*, 2014).

When the TiO_2 nanoparticles are delivered into the brain areas, disturbances in CNS processes, such as improved oxidative stress, stimulated inflammatory paths, modification of neurotransmitter release, and damage to spatial identification, might occur (Song *et al.*, 2015). Furthermore, when neurons and glial cells were treated for a long period with TiO_2, cell viability, cell cycle, cell morphology, antioxidant capabilities, and cellular elements were all altered, according to Ze *et al.* (2014, 2013). Prior to this, Wang *et al.* (2008) verified that after female mice were exposed to TiO_2 nanoparticles with dual crystal types of 80 nm for rutile and 155 nm for anatase for 30 days, the Ti intensity in the brain was considerably greater than in the control. Auffan *et al.* (2010) found Ti in comparable brain areas when TiO_2 nanoparticles were given via the blood of male mice for 4 weeks, including the hippocampus, which is responsible for memory and learning (Howland & Wang, 2008).

In addition, the consequences of primary, acute, and severe TiO_2 nanoparticle exposure on the BBB were also investigated. During primary, acute, and severe exposure, the *in vivo* models revealed a decrease in expression levels of P-glycoprotein, claudin 5, caveolin-1, and caveolin-2, which control BBB reliability (Brun *et al.*, 2012). Also, the researchers observed that levels of inflammatory indicators such as ICAM, VCAM, ADAM17, and chemokines were increased, which corresponded to reduced BBB permeability (Baello *et al.*, 2014; Louboutin & Strayer, 2013; Fang *et al.*, 2013). These data revealed that in addition to direct damage, TiO_2 nanoparticles' inflammatory impacts on BBB reliability might have unintended negative consequences. Aside from that, numerous studies have suggested that the brain damage induced by TiO_2 nanoparticles may be impacted by a variety of physical aspects such as inhalation, particle size, time of start, and others.

After 72 hours, Liu *et al.* (2013) found that TiO_2 nanoparticles with thicknesses of 10 and 20 nm caused cerebral damage in a dose-dependent manner, while 200 nm caused no major alterations in the brain through the BBB. Although TiO_2 nanoparticles can enter the brain across the BBB's tight junction and cause cerebral damage, the capacity to excrete them gives the afflicted brain areas a chance. However, due to the poor excretion of TiO_2 nanoparticles from the brain, a small dose of TiO_2 nanoparticles might accumulate in the brain with low clearance, posing a risk of damage (Song *et al.*, 2015). Zyoud *et al.* (2018) previously created anthocyanin-sensitized TiO_2 nanoparticles for photodegradation with a carbon-supported catalyst. The findings suggest that anthocyanin might be a viable alternative to other potentially dangerous sensitizers centered on artificial quantum particles or metal-based compounds. As a result, even if they cross the BBB, anthocyanin-loaded TiO_2 nanoparticles as a food application may cause brain damage, according to prior research; the anthocyanin did have potential as a photocatalyst in the photodegradation sector.

7.3.2.6 Silica

Silica-based nanoparticles are frequently utilized in biomedical nanotechnology because they are simple to create, cheap to produce, and are biocompatible, non-toxic carriers that can carry loaded medicines in live organisms (Vaculikova *et al.*, 2015; Liao *et al.*, 2014; Nehoff *et al.*, 2014). The transport of silica nanocarriers through the BBB is dependent on material properties, nanoparticle size, and surface modifications and multifunctionalized anchoring features that allow drug molecules to be conjugated to brain-specific transporters using different moieties such as PEG, glucose, oligonucleotides, amino acids, carboxyl, and/or vinyl (Tamba *et al.*, 2018). These are necessary to improve their infiltration through the BBB, plasma residence period, and clearance by the reticuloendothelial system (Georgieva *et al.*, 2014; Hanada *et al.*, 2014) (see Figure 7.10).

Song *et al.* (2017) improved receptor-mediated transcytosis through the BBB by attaching lactoferrin to the surface of PEG-silica nanoparticles with a maximum transport efficiency of 25 nm, as previously stated. Ku *et al.* (2010) investigated the capacity of manufactured PEGylated silica nanoparticles to cross the BBB and disperse within the cytoplasm of vascular endothelial cells. In terms of particle size, Liu *et al.* (2014) tested PEGylated silica nanoparticles with thicknesses of

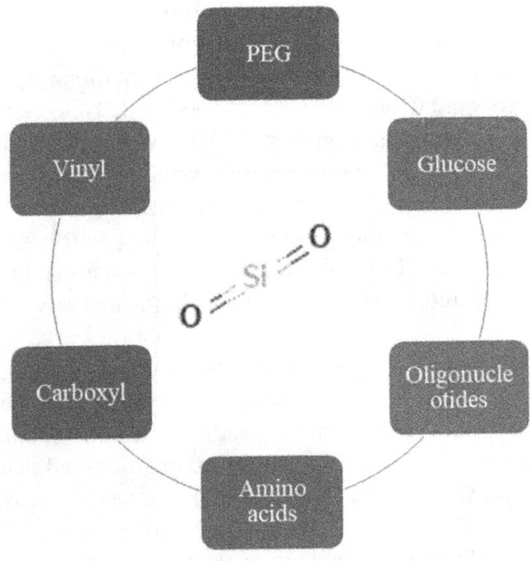

FIGURE 7.10 The multifunctionalized anchoring features of silica nanocarriers.

100, 50, and 25 nm *in vitro* and *in vivo* and found that as particle size reduces, the transport productivity and brain application improve. Even though covering the nanoparticles increased cellular absorption in *in vitro* models, Baghirov *et al.* (2016) found that the large size of silica nanoparticles ranging from 50 to 240 nm resulted in no BBB penetration in mice.

Comparison research using silica-loaded nootropics, a heterogeneous family of medicines such as piracetam, pentoxifylline, and pyridoxine referred to as memory enhancers and neuro-enhancers, was conducted to enhance the penetrability of the BBB (Jampilek *et al.*, 2015). Several nootropics, including natural and cerebral-active substances, are manmade analogs of functional molecules; examples include acetylcholine, pyridoxine, gamma-amino butyric acid (GABA), and coenzyme Q10 (Abraham & Rotella, 2010; Roth & Fenner, 2000). For each loaded medication to cross the BBB and enter the CNS, it must go through the paracellular route, which involves tight interchanges, and the transcellular pathway, which involves endothelial units (Passeleu-Le Bourdonnec *et al.*, 2013; Abbott *et al.*, 2006). The median particle size of each nootropic produced was around 120 nm, with a spherical shape that showed extremely strong sorption in silica nanocarriers as measured by liquid chromatography-high-resolution mass spectrometry.

In addition, each nootropic-loaded silica was observed to permeate rat brain tissue at 200 times greater concentrations than pentoxifylline and pyridoxine, indicating efficient transport of medicines across the BBB. The anthocyanin was additionally loaded with silica nanocarriers to function as a detector for pH and borate in the atmosphere (Chu *et al.*, 2017). To manufacture anthocyanin-loaded silica, the Stober *et al.* (1968) technique was used, since it is very easy but has limitations in particle homogeneity. The synthesized anthocyanin-doped silica nanoparticles were thoroughly tested and shown to have good storage stability and sensitivity in detecting borate via color or absorbance changes. Because of the tight connections of BBB endothelial cells, molecule targeting and transport to the brain across the BBB are difficult. Previous study has shown that silica nanoparticles may pass through the BBB and transport opioid, non-opioid, and other natural substances to the brain, assisting in the treatment of illnesses.

7.4 CONCLUSION AND RECOMMENDATIONS

Opioid drug misuse and dependence/addiction are multifaceted diseases governed by a complex web of interconnected genes and pathways that influence a variety of phenotypes. In recent years, a number of treatments for opioid-related diseases have evolved, utilizing a mix of pharmacological, psychological, and social strategies, albeit with different degrees of success. As a result, several areas have already investigated the market for non-opioid medicines derived from natural sources. Based on the regulators and adaptations of MORs and cAMP proteins, previous studies indicated the function of anthocyanin derived from pomegranate as a potential option for non-opioid replacement treatment due to its potency, which may operate antagonistically compared with morphine. Nanotechnology has emerged as a novel solution to the non-opioid therapy bottleneck thanks to the fast advancement of nanoscience and the excellent implementation of nanomaterials. These nanoparticles could administer opioids in a more targeted manner, offering fresh hope and a non-opioid therapy alternative for pain relief. As a result, combining nanotechnology with a viable non-opioid therapy is a practical step forward. It will provide medical practitioners with additional options for developing medical therapies that are effective for all patients while also decreasing the danger of opioid addiction.

REFERENCES

Abbina, S., & Parambath, A. (2018). PEGylation and its alternatives: a summary. In *Engineering of Biomaterials for Drug Delivery Systems*; Parambath A., Ed; Sawston, UK: Woodhead Publishing, 363–376.

Abbott, N. J., Ronnback, L., & Hansson, E. (2006). Astrocyte-endothelial interactions at the blood-brain barrier. *Nature Reviews Neuroscience*, *7* (1), 41–53.

Abou el Ela, A. E. S. F., El Khatib, M. M., & Salem-Bekhit, M. M. (2017). Design, characterization, and microbiological evaluation of microemulsion based gel of griseofulvin for topical delivery system. *Biointerface Research in Applied Chemistry, 7*, 2277–2285.

Abraham, D. J., & Rotella D. P. (2010). *Burger's Medicinal Chemistry, Drug Discovery, and Development.* 7th ed. New York: Wiley.

Acharya, S., & Sahoo, S. K. (2011). PLGA nanoparticles containing various anticancer agents and tumour delivery by EPR effect. *Advanced Drug Delivery Reviews, 63* (3), 170–183.

Adnan, L. H. M., Mohamad, N., Mat, K. C., Bakar, N. H. A., Mohd, K. S., & Mansor, M. I. (2016). The effect of thymoquinone on concentration of human μ-opioid receptors mediated by chronic morphine treatment in opioid receptor expressing cell (U-87 MG). *Acta Bioethica, 22* (2), 1086–1095.

Al-Hasani, R., & Bruchas, M. R. (2011). Molecular mechanisms of opioid receptor-dependent signalling and behaviour. *Anesthesiology, 115* (6), 1363–1381.

Almeida, V., Peres, F. F., Levin, R., Suiama, M. A., Calzavara, M. B., Zuardi, A. W., Hallak, J. E., Crippa, J. A., & Abilio, V. C. (2014). Effects of cannabinoid and vanilloid drugs on positive and negative-like symptoms on an animal model of schizophrenia: the SHR strain. *Schizophrenia Research, 153*, 150–159.

Alvarez-Suarez, J. M., Giampieri, F., Tulipani, S., Casoli, T., Di Stefano, G., Gonzalez-Paramas, A. M., Santos-Buelga, C., Busco, F., Quiles, J. L., Cordero, M. D., Bompadre, S., Mezzetti, B., & Battino, M. (2014). One-month strawberry-rich anthocyanin supplementation ameliorates cardiovascular risk, oxidative stress markers and platelet activation in humans. *Journal of Nutritional Biochemistry, 25*, 289–294.

Alves, N. M., & Mano, J. F. (2008). Chitosan derivatives obtained by chemical modifications for biomedical and environment applications. *International Journal of Biological Macromolecules, 43* (5), 401–414.

American Society of Addiction Medicine. (2011). Public policy statement: definition of addiction. Chevy Chase. Retrieved on 16 September 2018 from http://www.asam.org/docs/publicy-policystatements/1definition_of_addiction _long_4-11.pdf?sfvrsn=2

Amin, F. U., Shah, S. A., Badshah, H., Khan, M., & Kim, M. O. (2017). Anthocyanins encapsulated by PLGA@PEG nanoparticles potentially improved its free radical scavenging capabilities via p38/JNK pathway against $A\beta_{1-42}$ – induced oxidative stress. *Journal of Nanobiotechnology, 15*, 12.

Anahita, A., Asmah, R., & Fauziah, O. (2015). Evaluation of total phenolic content, total antioxidant activity, and antioxidant vitamin composition of pomegranate seed and juice. *International Food Research Journal, 22* (3), 1212–1217.

Andres-Lacueva, C., Shukitt-Hale, B., Galli, R. L., Jauregui, O., Lamuela-Raventos, R. M., & Joseph, J. A. (2005). Anthocyanins in aged blueberry-fed rats are found centrally and may enhance memory. *Nutritional Neuroscience, 8*, 111–120.

Angeloni, C., Pirola, L., Vauzour, D., & Muraldi, T. (2012). Dietary polyphenols and their effects on cell biochemistry and pathophysiology. *Oxidative Medicine and Cellular Longevity*, 2012, 1–3.

Anna, E. C., Peter, J. S. F., Elena, P., & Gianpiero, C. (2020). Overcoming the blood-brain barrier: functionalised chitosan nanocarriers. *Pharmaceutics, 12*, 1013.

Aparicio-Blanco, J., Martin-Sabroso, C., & Torres-Suarez, A. I. (2016). *In-vitro* screening of nanomedicines through the blood brain barrier: a critical review. *Biomaterials, 103*, 229–255.

Auffan, M., Bottero, J. Y., Chaneac, C., & Rose, J. (2010). Inorganic manufactured nanoparticles: how their physicochemical properties influence their biological effects in aqueous environments. *Nanomedicine, 5* (6), 999–1007.

Ayodele, A. T., Valizadeh, A., Adabi, M., Esnaashari, S. S., Madani, F., Khosravani, M., & Adabi, M. (2017). Ultrasound nanobubbles and their applications as theranostic agents in cancer therapy: a review. *Biointerface Research in Applied Chemistry, 7*, 2253–2262.

Baello, S., Iqbal, M., Bloise, E., Javam, M., Gibb, W., & Matthews, S. G. (2014). TGF-beta 1 regulation of multidrug resistance P-glycoprotein in the developing male blood–brain barrier. *Endocrinology, 155* (2), 475–484.

Baghirov, H., Karaman, D., Viitala, T., Duchanoy, A., Lou, Y-R., Mamaeva, V., Pryazhnikov, E., Khiroug, L., de Lange Davies, C., Sahlgren, C., & Rosenholm, J. M. (2016). Feasibility study of the permeability and uptake of mesoporous silica nanoparticles across the blood-brain barrier. *PLoS ONE, 11* (8), e0160705.

Bailey, C. P., Llorente, J., Gabra, B. H., Smith, F. L., Dewey, W. L., Kelly, E., & Henderson, G. (2009a). Role of protein kinase C and μ-opioid receptor (MORs) desensitization in tolerance to morphine in rat locus coeruleus neurons. *European Journal of Neuroscience, 29*, 307–318.

Barar, J., Rafi, M. A., Pourseif, M. M., & Omidi, Y. (2016). Blood-brain barrier transport machineries and targeted therapy of brain diseases. *BioImpacts BI, 6*, 225–248.

Barker, J. M., Taylor, J. R., De Vries, T. J., & Peters, J. (2015). Brain derived neurotropic factor and addiction: pathological versus therapeutic effects on drug seeking. *Brain Research, 1628*, 68–81.

Beconcini, D., Felice, F., Zambito, Y., Fabiano, A., Piraas, A. M., Macedo, M. H., Sarmento, B., & Di Stefano. R. (2019). Anti-inflammatory effect of cherry extract loaded in polymeric nanoparticles: relevance of particle internalization in endothelial cells. *Pharmaceutics*, *11*, 500.

Benyhe, S., Zador, F., & Otvos, F. (2015). Biochemistry of opioid (morphine) receptors: binding, structure, and molecular modelling (Review). *Acta Biologica Szagediensis*, *59* (1), 17–37.

Berridge, C. W., & Waterhouse, B. D. (2003). The locus coeruleus-noradrenergic system: modulation of behavioural state and state-dependent cognitive processes. *Brain Research Reviews*, *42*, 33–84.

Betz, M., Steiner, B., Schantz, M., Oidtmann, J., Mäder, K., Richling, E., & Kulozik, U. (2012). Antioxidant capacity of bilberry extract microencapsulated in whey protein hydrogels. *Food Research International*, *47* (1), 51–57.

Bhaskar, S., Tian, F., Stoeger, T., Kreyling, W., de la Fuente, JM., Grazu, V., Borm, P., Estrada, G., Ntziachrristosis, V., & Razansky, D. (2010). Multifunctional nanocarriers for diagnostics, drug delivery and targeted treatment across blood-brain barrier: perspectives on tracking and neuroimaging. *Particle and Fibre Toxicology*, *7*, 3.

Bian, J. M., Wu, N., Su, R., & Li, J. (2012). Opioid receptor trafficking and signalling: what happens after opioid receptor activation? *Cell Molecular Neurobiology*, *32*, 167–184.

Biedermann, L., Mwinyi, J., Scharl, M., Frei, P., Zeitz, J., Kullak-Ublick, G. A., Vricka, S. R., Fried, M., Weber, A., Humpf, H. U., Peschke, S., Jetter, A., Krammer, G., & Rogler, G. (2013). Bilberry ingestion improves disease activity in mild to moderate ulcerative colitis: an open pilot study. *Journal of Crohn's and Colitis*, *7*, 271–279.

Bissonette, G. B., & Roesch, M. R. (2016). Development and function of the midbrain dopamine system: what we know and what we need to. *Genes, Brain and Behaviour*, *15*, 62–73.

Borg, L., Buonora, M., Butelma, E. R., Ducat, E., Ray, B. M., & Kreek, M. J. (2014). The pharmacology of opioids. Section 2: pharmacology. Retrieved on 12 September 2018 from https://pdfs.semanticscholar .org/565d/51353d db36726ab8370cc72137b 56f2af7.pdf.

Boutemak, K., Safta, B., & Ayachi, N. (2016). Analgesic and antioxidant activities of the flavonic extract of *Globularia alypum L. International Journal of Agriculture and Forestry Science*, *1* (2), 7–10.

Bozzuto, G., & Molinari, A. (2015). Liposomes as nanomedical devices. *International Journal of Nanomedicine*, *2* (10), 975–999.

Brun, E., Carriere, M., & Mabondzo, A. (2012). *In-vitro* evidence of dysregulation of blood–brain barrier function after acute and repeated/long-term exposure to TiO2 nanoparticles. *Biomaterials*, *33* (3), 886–896.

Bugnicourt, E., Cinelli, P., Lazzeri, A., & Alvarez, V. (2014). Polyhydroxyalkanoate (PHA): review of synthesis, characteristics, processing and potential applications in packaging. *Polymers Letters*, *8* (11), 791–808.

Bugnicourt, L., & Ladaviere, C. (2016). Interests of chitosan nanoparticles ionically cross-linked with tripolyphosphate for biomedical applications. *Progress in Polymer Science*, *60*, 1–17.

Byrne, L. S., Peng, J., Sarkar, S., & Chang, S. L. (2012). Interleukin-1β induced up-regulation on opioid receptors in the untreated and morphine-desensitized U-87 MG human astrocytoma cells. *Journal of Neuroinflammation*, *9*, 1–12.

Calderon-Garciduenas, L., Reed, W., Maronpot, R. R., Henriquez-Roldan, C., Delgado-Chavez, R., Calderon-Garciduenas, A., Dragustinovis, I., Franco-Lira, M., Aragon-Flores, M., Solt, A. C., Altenburg, M., Torres-Jardon, R., & Swenberg, J. A. (2004). Brain inflammation and Alzheimer's-like pathology in individuals exposed to severe air pollution. *Toxicologic Pathology*, *32* (6), 1533–1601.

Caputi, F. F., Lattanzio, F., Carretta, D., Mercatelli, D., Candeletti, S., & Rmualdi, P. (2013). Morphine and fentanyl differently affect MOP and NOP gene expression in human neuroblastoma SH-SY5Y cells. *Journal of Molecular Neuroscience*, *51*, 532–538.

Cetin, M., Aytekin, E., Yavuz, B., & Bozdag-Pehlivan, S. (2017). Chapter 7—Nanoscience in targeted brain drug delivery. In *Nanotechnology Methods for Neurological Diseases and Brain Tumors*; Gursoy-Ozdemir, Y., Bozdag-Pehlivan, S., Sekerdag, E., Eds.; Cambridge, MA: Academic Press, 117–147.

Chan, H. C. S., McCarthy, D., Li, J., Palczewski, K., & Yuan, S. (2017). Designing safer analgesics via μ-opioid receptor pathways. *Trends in Pharmacology Science*, *38* (11), 1016–1037.

Chan, K. C., Huang, H. P., Ho, H. H., Huang, C. N., Lin, M. C., & Wang, C. J. (2015). Mulberry polyphenols induce cell cycle arrest of vascular smooth muscle cells by inducing NO production and activating AMPK and p53. *Journal of Functional Foods*, *15*, 604–613.

Chartoff, E. H., & Connery, H. S. (2014). It's MOR-e exciting than μ: crosstalk between μ-opioid receptors and glutamatergic transmission in the mesolimbic dopamine system (Review article). *Frontiers in Pharmacology*, *5* (116), 1–21.

Chen, H., Yu, W., Chen, G., Meng, S., Xiang, Z., & He, N. (2017). Anti-nociceptive and anti-bacterial properties of anthocyanin and flavonols from fruits of black and non-black mulberries. *Molecules*, *23* (4), 1–13.

Chen, X., & Mao, Y. (2007). Titanium dioxide nanomaterials: synthesis, properties, modifications, and applications. *Chemical Reviews, 107* (7), 2891–2959.

Chu, T. H., Nghiem, T. H. L., Nguyen, D. A., & Nguyen, L. L. (2017). Development of natural anthocyanin dye-doped silica nanoparticles for pH and borate-sensing applications. *Journal of Electronic Materials*, 6843–6847.

Craparo, E. F., Teresi, G., Bondi, M. L., Licciardi, M., & Cavallaro, G. (2011). Phospholipid-polyaspartamide micelles for pulmonary delivery of corticosteroids. *International Journal of Pharmaceutics, 406*, 135–144.

Dang, V. C., & Christie, M. J. (2012). Mechanisms of rapid opioid receptor desensitization, re-sensitization, and tolerance in brain neurons. *British Journal of Pharmacology, 165*, 704–1716.

Danhier, F., Ansorena, E., Silva, J. M., Coco, R., Le Breton, A., & Preat, V. (2012). PLGA-based nanoparticles: an overview of biomedical applications. *Journal of Controlled Release, 161*, 505–522.

Davoodi, S. D., & Saghavaz, B. H. (2017). Optimal synthesis and characterization of magnetic CuMnFe2O4 nanoparticles coated by PEG for drug delivery. *Biointerface Research in Applied Chemistry, 7*, 2249–2252.

Del Amo, L., Cano, A., Ettcheto, M., Souto, E. B., Espina, M., Camins, A., Garcia, M. L., & Sanchez-Lopez, E. (2021). Surface functionalization of PLGA nanoparticles to increase transport across the BBB for Alzheimer's disease. *Applied Science, 11*, 4305.

Deng, X., Luan, Q., Chen, W., Wang, Y., Wu, M., Zhang, H., & Jiao, Z. (2009). Nanosized zinc oxide particles induce neural stem cell apoptosis. *Nanotechnology, 20*, 1–8.

Dinarvand, R., Sepehri, N., Manoochehri, S., Rouhani, H., & Atyabi, F. (2011). Polylactide-co-glycolide nanoparticles for controlled delivery of anticancer agents. *International Journal of Nanomedicine, 6*, 877.

Ding, D., & Zhu, Q. (2018). Recent advances of PLGA micro/nanoparticles for the delivery of biomacromolecular therapeutics. *Materials Science and Engineering C-Materials for Biological Applications, 92*, 1041–1060.

El-falleh, W., Hannachi, H., Tlili, N., Yahia, Y., Nasri, N., & Ferchichi, A. (2012). Total phenolic contents and antioxidant activities of pomegranate peel, seed, leaf and flower. *Journal of Medicine Plants Resource, 6*, 4724–4730.

Engineer, C., Parikh, J., & Raval, A. (2011). Review on hydrolytic degradation behavior of biodegradable polymers from controlled drug delivery system. *Trends in Biomaterials & Artificial Organs, 25*, 79–85.

Erathodiyil, N., & Ying, J. Y. (2011). Functionalization of inorganic nanoparticles for bioimaging applications. *Accounts of Chemical Research, 44* (10), 925–935.

Evans, C. J., & Cahill, C. M. (2016). Neurobiology of opioid dependence in creating addiction vulnerability. *F1000 Research, 5 (F1000 Faculty Revision)* (1748), 1–11.

Faisal, N., & Kumar, K. (2017). Polymer and metal nanocomposites in biomedical applications. *Biointerface Research in Applied Chemistry, 7*, 2286–2294.

Fang, C., Yu R., Liu S., & Li, Y. (2013). Nanomaterials applied in asphalt modification, a review. *Journal of Materials Science and Technology, 29* (7), 589–594.

Farag, R. S., Abdel-Latif, M. S., Emam, S. S., & Tawfeek, L. S. (2014). Phytochemical screening and polyphenol constituents of pomegranate peels and leave juices. *Landmark Research Journals of Agriculture and Soil Sciences (LRJASS), 1* (6), 86–93.

Faria, A., Mateus, N., & Calhau, C. (2012). Flavonoid transport across blood-brain barrier: implication for their direct neuro-protective actions. *Nutrition and Aging*, 1 (2), 89–97.

Faria, A., Pestana, D., Teixeira, D., Azevedo, J., de Freitas, V., Mateus, N., & Calhau, C. (2010). Flavonoid transport across RBE4 cells: a blood brain barrier model. *Cellular and Molecular Biology Letters, 15* (2), 234–241.

Faria, A., Pestana, D., Teixeira, D., Courad, P. O., Romero, I., Weksler, B., de Freitas, V., Mateus, N., & Calhau, C. (2011). Insight into the putative catechin and epicatechin transport across blood-brain barrier. *Food & Function, 2*, 39–44.

Feng, Q., Liang, C. D., & Liu, G. M. (2004). Experimental study on cement-based composites with nano-SiO_2. *Cailiao Kexue Yu Gongcheng Xuebao, 22* (2), 224–227.

Feng, Y., Sun, C., Yuan, Y., Zhu, Y., Wan, J., Firempong, C. K., & Xu, X. (2016). Enhanced oral bioavailability and *in-vivo* antioxidant activity of chlorogenic acid via liposomal formulation. *International Journal of Pharmaceutics, 501* (1), 342–349.

Fernandes, I., Faria, A., & Calhau, C. (2014). Bioavailability of anthocyanin and derivatives. *Journal of Functional Foods, 7*, 54–66.

Fraser, E., Gruenberg, K., & Rubenstein, J. L. (2015). New approaches in primary central nervous system lymphoma. *Chinese Clinical Oncology, 4*, 11.

Fyfe, L. W., Cleary, D. R., Macey, T. A., Morgan, M. M, & Ingram, S. L. (2010). Tolerance to the anti-nociceptive effect of morphine in the absence of short-term presynaptic desensitization in rat periaqueductal gray neurons. *Journal of Pharmacology and Experimental Therapeutics, 335*, 674–680.

Gaveriaux-Ruff, C. (2013). Opiate-induced analgesia: contributions from μ, delta, and kappa opioid receptors mouse mutants. *Current Pharmaceutical Design, 19* (42), 7373–7381.

Gelperina, S. E., Khalansky, A. S., Skidan, I. N., Smirnova, Z. S., Bobruskin, A. I., Severin, S. E., Turowski, B., Zanella, F. E., & Kreuter, J. (2002). Toxicological studies of doxorubicin bound to polysorbate 80-coated poly(butyl cyanoacrylate) nanoparticles in healthy rats and rats with intracranial glioblastoma. *Toxicology Letters, 126* (2), 131–141.

Gentil, P., Chiono, V., Carmagnola, I., & Hatton, P. V. (2014). An overview of poly(lactic-co-glycolic) acid (PLGA)-based biomaterials for bone tissue engineering. *International Journal of Molecular Science, 15*, 3640–3659.

George, P. (2015). Outcomes from the Malaysian arm of the international survey informing greater insights in opioid treatment (INSIGHT) Project. *Medical Journal of Malaysia, 70* (2), 117–125.

Georgieva, J., Hoekstra, D., & Zuhorn, I. (2014). Smuggling drugs into the brain: an overview of ligands targeting transcytosis for drug delivery across the blood–brain barrier. *Pharmaceutics, 6*, 557–583.

Ghaffari, S. B., Sarrafzadeh, M. H., Fakhroueian, Z., Shahriari, S., & Khorramizadeh, M. R. (2017). Functionalization of ZnO nanoparticles by 3-mercaptopropionic acid for aqueous curcumin delivery: synthesis, characterization, and anticancer assessment. *Materials Science and Engineering C-Materials for Biological Applications, 79*, 465–472.

Ghelardini, C., Mannelli, L. D. C., & Bianchi, E. (2015). The pharmacological basis of opioids. *Clinical Cases in Mineral and Bone Metabolism, 12* (3), 219–221.

Glazer, I., Masaphy, S., Marciano, P., Bar-Ilan, I., Holland, D., Kerem, Z., & Amir, R. (2012). Partial identification of antifungal compounds from *Punica granatum* peel extracts. *Journal of Agriculture and Food Chemistry, 60*, 4841–4848.

Gong, Y., Chowdhury, P., Midde, N. M., Rahman, M. A., Yallapu, M. M., & Kumar, S. (2017). Novel elvitegravir nanoformulation spproach to suppress the viral load in HIV-infected macrophages. *Biochemistry and Biophysics Reports, 12*, 214–219.

Gupta, S., Dhanda, S., & Sandhir, R. (2019). 2—Anatomy and physiology of blood-brain barrier. In Brain Targeted Drug Delivery System; Gao, H., Gao, X., Eds.; Cambridge, MA: Academic Press, 7–31.

Halls, M. L., Yeatman, H. R., Nowell, C. J., Thompson, G. L., Gondin, A. B., Civciristov, S., Bunnett, N. W., Lambert, N. A., Poole, D. P., & Canals, M. (2016). Plasma membrane localization of the u-opioid receptor controls spatiotemporal signalling. *Science Signalling, 9* (414), 1–9.

Han, D., Tian, Y., Zhang, T., Ren, G., & Yang, Z. (2011). Nano-zinc oxide damages spatial cognition capability via over-enhanced long-term potentiation in hippocampus of Wistar rats. *International of Journal Nanomedicine, 6*, 1453–1461.

Hanada, S., Fujioka, K., Inoue, Y., Kanaya, F., Manome, Y., & Yamamoto, K. (2014). Cell-based *in-vitro* blood-brain barrier model can rapidly evaluate nanoparticles' brain permeability in association with particle size and surface modification. *International Journal of Molecular Sciences, 15*, 1812–1825.

Haque, S., Md, S., Alam, M. I., Sahni, J. K., Ali, J., & Baboota, S. (2012). Nanostructure-based drug delivery systems for brain targeting. *Drug Development and Industrial Pharmacy, 38* (4), 387–411.

Hariharan, R., Senthilkumar, S., Suganthi, A., & Rajarajan, M. (2012). Synthesis and characterization of doxorubicin modified ZnO/PEG nanomaterials and its photodynamic action. *Journal of Photochemistry and Photobiology B: Biology, 116*, 56–65.

Hashemi, S., Asrar, Z., Pourseyedi, S., & Nadernejad, N. (2018). Investigation of ZnO nanoparticles on proline, anthocyanin contents and photosynthetic pigments and lipid peroxidation in soybean. *IFT Nanobiotechnology, 13* (1), 66–70.

Hasnaoui, N., Jbir, R., Mars, M., Trifi, M., Kamal-Eldin, A., Melgarejo, P., & Hernandez, F. (2014). Organic acids, sugars, and anthocyanin contents in juices of Tunisian pomegranate fruits. *International Journal of Food Properties, 14* (4), 741–757.

He, B., Ge, J., Yue, P., Yue, X., Fu, R., Liang, J., & Gao, X. (2016). Loading of anthocyanins on chitosan nanoparticles influences anthocyanin degradation in gastrointestinal fluids and stability in a beverage. *Food Chemistry, 221*, 1671–1677.

He, J., & Giusti, M. M. (2010). Anthocyanins: natural colorants with health-promoting properties. *Annual Review of Food Science and Technology, 1*, 163–187.

He, J., Wallace, T. C., Keatley, K. E., Faila, M. L., & Giusti, M. M. (2009). Stability of black raspberry anthocyanins in the digestive tract lumen and transport efficiency into gastric and small intestinal tissues in the rat. *Journal of Agricultural and Food Chemistry*, *57*, 3141–3148.

Hejjaji, E. M. A., Smith, A. M., & Morris, G. A. (2018). Evaluation of the mucoadhesive properties of chitosan nanoparticles prepared using different chitosan to tripolyphosphate (CS:TPP) ratios. *International Journal of Biological Macromolecules*, *120* (Pt B), 1610–1617.

Herve, F., Ghinea, N., & Scherrmann, J. M. (2008). CNS delivery via adsorptive transcytosis. *American Association of Pharmaceutical Scientists Journal*, *10* (3), 455–472.

Hines, D. J., & Kaplan, D. L. (2013). Poly(lactic-co-glycolic) acid-controlled-release systems: experimental and modeling insights. Critical Reviews. *Therapeutic Drug Carrier Systems*, *30* (3), 257–276.

Homayoonfal, M., Mousavi, S. M., Kiani, H., Askari, G., Desobry, S., & Arab-Tehrany, E. (2021). Encapsulation of *Berberis vulgaris* anthocyanins into nanoliposome composed of rapeseed lecithin: a comprehensive study on physicochemical characteristics and biocompatibility. *Foods*, *10*, 492.

Howland, J. G., & Wang, Y. T. (2008). Synaptic plasticity in learning and memory: stress effects in the hippocampus. *Progress in Brain Research*, *169*, 145–158.

Hu, Y., Chi, C., Wang, S., Wang, L., Liang, P., Liu, F., Shang, W., Wang, W., Zhang, F., Li, S., Shen, H., Yu, X., Lio, H., & Tian, J. (2017). A comparative study of clinical intervention and interventional photothermal therapy for pancreatic cancer. *Advanced Materials*, *29* (33), 1700448.

Husain, M., Wu, D., Saber, A. T., Decan, N., Jacobsen, N. R., Williams, A., Yauk, C. L., Wallin, H., Vogel, U., & Halappanavar, S. (2015). Intratracheally instilled titanium dioxide nanoparticles translocate to heart and liver and activate complement cascade in the heart of C57BL/6 mice. *Nanotoxicology*, *9* (8), 1013–1022.

Husain, Q. (2017). Nanosupport bound lipases their stability and applications. *Biointerface Research in Applied Chemistry*, *7*, 2194–2216.

Ibrahim, H. S., Ibrahim, S., Iliyasu, Z., Umar, S. A., & Mohd, K. R. A. (2018). Antioxidant potential of *Phoenix dactylifera* Linn extract and its effects on calcium channel antagonist in the treatment of withdrawal syndrome in morphine dependent rats. *Tropical Journal of Natural Product Research*, *2* (7), 309–313.

Ilyas, E. I. I., Irawati, D., Elfiza, D., Nurlaela, C., Jusman, S. W. A., & Kartinah, N. T. (2017). The potency of Hibiscus Sabdariffa Linn. on decreased memory function related to the level of BDNF and CREB in hippocampus of over-trained rats. *International Journal of Recent Scientific Research*, *8* (5), 17097–17103.

Ishisaka, A., Ichikawa, S., Sakakibara, H., Piskula, M. K., Nakamura, T., Kato, Y., & Terao, J. (2011). Accumulation of orally administered quercetin in brain tissue and its antioxidative effects in rats. *Free Radical Biology & Medicine*, *51* (7), 1329–1336.

Jahromi, L. P., Moghaddam Panah, F., Azadi, A., & Ashrafi, H. (2019). A mechanistic investigation on methotrexate-loaded chitosan-based hydrogel nanoparticles intended for CNS drug delivery: Trojan horse effect or not? *International Journal of Biological Macromolecules*, *125*, 785–790.

Jamil, M. F. A., Subki, M. F. M., Lan, T. M., Majid, M. I. A., & Adenan, M. I. (2013). The effect of mitragynine on cAMP formation and mRNA expression of μ-opioid receptors mediated by chronic morphine treatment in SK-N-SH neuroblastoma cell. *Journal of Ethnopharmacology*, *148* (1), 135–143.

Jampilek, J., Zaruba, K., Oravec, M., Kunes, M., Babula, P., Ulbrich, P., Brezaniova, I., Opatrilova, R., Triska, J., & Suchy, P. (2015). Preparation of silica nanoparticles loaded with nootropics and they're *in-vivo* permeation through blood-brain barrier. *BioMed Research International*, 2015, 1–9.

Jang, Y. S., Lee, E. Y., Park, Y. H., Jeong, S. H., Lee, S. G., Kim, Y. R., Kim, M. K., & Son, S. W. (2012). The potential for skin irritation, phototoxicity and sensitization of ZnO nanoparticles. *Molecular and Cellular Toxicology*, *8* (2), 171–177.

Jo, H., Park, Y., Kim, S. E., & Lee, H. (2016). Exploring the intellectual structure of nanoscience and nanotechnology: journal citation network analysis. *Journal of Nanoparticle Research*, *18*, 167.

Jonassen, H., Kjoniksen, A. L., & Hiorth, M. (2012). Stability of chitosan nanoparticles cross-linked with tripolyphosphate. *Biomacromolecules*, *13*, 3747–3756.

Kaida, T., Kobayashi, K., Adachi, M., & Suzuki, F. (2004). Optical characteristics of titanium oxide interference film and the film laminated with oxides and their applications for cosmetics. *Journal of Cosmetics Science*, *55*, 219–220.

Kao, Y. Y., Chen, Y. C., Cheng, T. J., Chiung, Y. M., & Liu, P. S. (2012a). Zinc oxide nanoparticles interfere with zinc ion homeostasis to cause cytotoxicity. *Toxicological Sciences*, *125* (2), 462–472.

Kao, Y.-Y., Cheng, T.-J., Yang, D.-M., Wang, C.-T., Chiung, Y.-M., & Liu, P.-S. (2012b). Demonstration of an olfactory bulb-brain translocation pathway for ZnO nanoparticles in rodent cells *in-vitro* and *in-vivo*. *Journal of Molecular Neuroscience*, *48*, 464–471.

Karatas, H., Aktas, Y., Gursoy-Ozdemir, Y., Bodur, E., Yemisci, M., Caban, S., Vural, A., Pinarbasli, O., Capan, Y., Fernandez-Megia, E., Novoa-Carballal, R., Riguera, R., Andrieux, K., Couvreur, P., & Dalkara, T. (2009). A nanomedicine transports a peptide caspase-3 inhibitor across the blood-brain barrier and provides neuroprotection. *Journal of Neuroscience, 29*, 13761–13769.

Khanna, A. K., & Farag, E. (2017). Chapter 3—Blood-brain barrier. In *Essentials of Neuroanesthesia*; Prabhakar, H., Ed.; Cambridge, MA: Academic Press, 51–58.

Khor, B. S., Jamil, M. F., Adenan, M. I., & Shu-Chien, A. C. (2011). Mitragynine attenuates withdrawal syndrome in morphine-withdrawn zebra fish. *PLoS ONE, 6* (12), 1–8.

Kołodziejczak-Radzimska, A., & Jesionowski, T. (2014). Zinc oxide—from synthesis to application: a review. *Materials, 7* (4), 2833–2881.

Kravitz, A. V., Tye, L. D., & Kreitzer, A. C. (2012). Distinct roles for direct and indirect pathway striatal neurons in reinforcement. *Nature Neuroscience, 15*, 816–818.

Ku, S., Yan, F., Wang, Y., Sun, Y., Yang, N., & Ye, L. (2010). The blood-brain barrier penetration and distribution of PEGylated fluorescein-doped magnetic silica nanoparticles in rat brain. *Biochemical and Biophysical Research Communications, 394* (4), 871–876.

Kumari, P. M., Goswami, R., & Nimesh, S. (2010). Application of nanotechnology in diagnosis and therapeutics. *Nanotechnology for Energy and Environmental Engineering, 2010*, 413–440.

Kuo, Y. C., Wang, I. H., & Rajesh, R. (2021). Use of leptin-conjugated phosphatidic acid liposomes with resveratrol and epigallacatechin gallate to protect dopaminergic neurons against apoptosis for Parkinson's disease therapy. *Acta Biomaterialia, 119*, 360–374.

Lakkadwala, S., & Singh, J. (2018). Dual functionalized 5-fluorouracil liposomes as highly efficient nanomedicine for glioblastoma treatment as assessed in an *in vitro* brain tumor model. *Journal of Pharmaceutical Sciences, 107*, 2902–2913.

Lakkadwala, S., & Singh, J., (2019). Co-delivery of doxorubicin and erlotinib through liposomal nanoparticles for glioblastoma tumor regression using an *in vitro* brain tumor model. *Colloids Surfaces B: Biointerfaces, 173*, 27–35.

Lee, N., Cho, H. R., Oh, M. H., Lee, S. H., Kim, K., Kim, B. H., Shin, K., Ahn, T. Y., Choi, J. W., Kim, Y. W., Cjoi, S. H., & Hyeon, T. (2012). Multifunctional Fe_3O_4/TaOx core/shell nanoparticles for simultaneous magnetic resonance imaging and X-ray computed tomography. *Journal of the American Chemical Society, 134*, 10309–10312.

Lewicky, J. D., Fraleigh, N. L., Boraman, A., Martel, A. L., Nguyen, T. M., Schiller, P. W., Shiao, T. C., Roy, R., Montaut, S., & Le, H. T. (2020). Mannosylated glycoliposomes for the delivery of a peptide kappa opioid receptor antagonist to the brain. *European Journal of Pharmaceutics and Biopharmaceutics, 154*, 290–296.

Li, Y., Zhang, C., Liu, L., Gong, Y., Xie, Y., & Cao, Y. (2017). The effects of baicalein or baicalin on the colloidal stability of ZnO nanoparticles (NPs) and toxicity of NPs to Caco-2 cells. *Toxicology Mechanisms and Methods, 28* (3), 167–176.

Liao, Y. T., Liu, C. H., Yu, J., & Wu, K. C. W. (2014). Liver cancer cells: targeting and prolonged-release drug carriers consisting of mesoporous silica nanoparticles and alginate microspheres. *International Journal of Nanomedicine, 9*, 2767–2778.

Liu, D., Lin, B., Shao, W., Zhu, Z., Ji, T., & Yang, C. (2014). *In vitro* and *in vivo* studies on the transport of PEGylated silica nanoparticles across the blood-brain barrier. *ACS Applied Materials & Interfaces, 6* (3), 2131–2136.

Liu, Z., Jiang, M., Kang, T., Miao, D., Gu, G., Song, Q., Yao, L., Hu, Q., Tu, Y., Pang, Z., Chen, H., Jiang, X., Gao, X., & Chen, J. (2013). Lactoferin-modified PEG-co-PCL nanoparticles for enhanced brain delivery of NAP peptide following intranasal administration. *Biomaterials, 34*, 3870–3881.

Locatelli, E. L., Gil, L., Israel, L. L., Passoni, L., Naddaka, M., Pucci, A., Reese, T., Gomez-Vallejo, V., Milani, P., Matteoli, M., Llop, J., Lellouche, J. P, & Franchini, M. C. (2012). Biocompatible nanocomposite for PET/MRI hybrid imaging. *International Journal of Nanomedicine, 7*, 6021–6033.

Lockman, P. R., Kozaria, J. M., Mumper, R. J., Allen, D. D. (2004). Nanoparticle surface charges alter blood-brain barrier integrity and permeability. *Journal of Drug Targeting, 12* (9–10), 635–641.

Lomer, M. C., Thompson, R. P., & Powell, J. J. (2002). Fine and ultrafine particles of the diet: influence on the mucosal immune response association with Crohn's disease. *Proceedings of the Nutrition Society, 611*, 123–130.

Louboutin, J. P., & Strayer, D. S. (2013). Relationship between the chemokine receptor CCR5 and microglia in neurological disorders: consequences of targeting CCR5 on neuroinflammation, neuronal death and regeneration in a model of epilepsy. *CNS Neurological Disorders-Drug Targets, 12* (6), 815–829.

Lu, W., Zhang, Y., Tan, Y. Z., Hu, K. L., Jiang, X. G., & Fu, S. K. (2005). Cationic albumin-conjugated PEGylated nanoparticles as novel drug carrier for brain delivery. *Journal of Controlled Release, 107*, 428–448.

Mahapatro, A., & Singh, D. K. (2011). Biodegradable nanoparticles are excellent vehicle for site directed *in-vivo* delivery of drugs and vaccines. *Journal of Nanobiotechnology, 9*, 55.

Maher, P., Akaishi, T., & Abe, K. (2006). Flavonoid fisetin promotes ERK-dependent long-term potentiation and enhances memory. *Proceedings of the National Academy of Sciences of the United States of America, 103*, 16568–16573.

Mahringer, A., Puris, E., & Fricker, G. (2021). Crossing the blood-brain barrier: a review on drug delivery strategies using colloidal carrier systems. *Nanochemistry International, 147*, 105017.

Makadia, H. K., & Siegel, S. J. (2011). Poly lactic-co-glycolic acid (PLGA) as biodegradable controlled drug delivery carrier. *Polymers, 3* (3), 1377–1397.

Mandel, S., Amit, T., & Reznichenko, L. (2006). Green tea catechins as brain-permeable, natural iron chelators-antioxidants for the treatment of neurodegenerative diseases. *Molecular Nutrition & Food Research, 50*, 229–234.

Martin-Banderas, L., Holgado, M. A., Venero, J. L., Alvarez-Fuentes, J., & Fernandez-Arevalo, M. (2011). Nanostructures for drug delivery to the brain. *Current Medicinal Chemistry, 18* (34), 5303–5321.

Martinez-Ballesta, M., Gil-Izquierdo, A., Garcia-Viguera, C., & Dominguez-Perles, R. (2018). Nanoparticles and controlled delivery for bioactive compounds: outlining challenges for new "Smart-Foods" for health. *Foods, 7*, 72.

Masserini, M. (2013). Nanoparticles for brain drug delivery. *ISRN Biochemistry*, 2013, 1–18.

Mi, P., Miyata, K., Kataoka, K., & Cabral, H. (2020). Clinical translation of self-assembled cancer nanomedicines. *Advance Therapeutics, 4*, 2000159.

Micheli, M. R., Bova, R., Magini, A., Polidoro, M., & Emiliani, C. (2012). Lipid-based nanocarriers for CNS-targeted drug delivery. *Recent Patents on CNS Drug Discovery, 7*, 71–86.

Mirakabad, F. S. T., Nejati-Koshki, K., Akbarzadeh, A., Yamchi, M. R., Milani, M., Zarghami, N., Zeighamian, V., Rahimzadeh, A., Alimohammadi, S., Hanifehpour, Y., & Woo Joo, S. (2014). PLGA-based nanoparticles as cancer drug delivery systems. *Asian Pacific Journal of Cancer Prevention, 15* (2), 517–535.

Mirsaeedghazi, H., Emam-Djomeh, Z., & Ahmadkhaniha, R. (2014). Effect of frozen storage on the anthocyanin and phenolic components of pomegranate juice. *Journal Food of Science Technology, 51* (2), 382–386.

Mohammadipour, A., Fazel, A., Haghir, H., Motejaded, F., Rafatpanah, H., Zabihi, H., Hosseini, M., & Bideskan, AE. (2014). Maternal exposure to titanium dioxide nanoparticles during pregnancy; impaired memory and decreased hippocampal cell proliferation in rat offspring. *Environmental Toxicology and Pharmacology, 37*, 617–625.

Monge-Fuentes, V., Mayer, A. B., Lima, M. R., Geraldes, L. R., Zanotto, L. N., Moreira, K. G., Martins, O. P., Piva, H. L., Felipe, M. S. S., Amaral, A. C., Bocca, A. L., Tedesco, A. C., & Mortari, M. R. (2021). Dopamine-loaded nanoparticle systems circumvent the blood-brain barrier restoring motor function in mouse model for Parkinson's disease. *Nature Scientific Reports, 11*, 15185.

Montet, X., Funovics, M., Montet-Abou, K., Weissleder, R., & Josephson, L. (2006). Multivalent effects of RGD peptides obtained by nanoparticle display. *Journal of Medicinal Chemistry, 49*, 6087–6093.

Moura, R. P., Martins, C., Pinto, S., Sousa, F., & Sarmento, B. (2019). Blood-brain barrier receptors and transporters: an insight on their function and how to exploit them through nanotechnology. *Expert Opinion on Drug Delivery, 16*, 271–285.

Mozafari, M. R., Pardakhty, A., Azarmi, S., Jazayeri, J. A., Nokhodchi, A., & Omri, A. (2009). Role of nanocarrier systems in cancer nanotherapy. *Journal of Liposome Research, 19*, 310–321.

Mphahlele, R. R., Fawole, O. A., Mokwena, L. M., & Opara, U. L. (2016). Effect of extraction method on chemical, volatile composition, and antioxidant properties of pomegranate juice. *South African Journal of Botany, 103*, 135–144.

Musa, A., Ba-Abbad, M., Kadhum, A., & Mohamad, A. (2012). Photodegradation of chlorophenolic compounds using zinc oxide as photocatalyst: experimental and theoretical studies. *Research on Chemical Intermediates, 38* (3–5), 995–1005.

National Institute on Drug Abuse. (2015). *Drugs of Abuse: Opioids*. Bethesda, MD: National Institute on Drug Abuse 2015. Retrieved on 5 December 2017 from https://www.drugabuse.gov/drugs-abuse/opioids

Nehoff, H., Parayath, N. N., Domanovitch, L., Taurin, S., & Greish, K. (2014). Nanomedicine for drug targeting: strategies beyond the enhanced permeability and retention effect. *International Journal of Nanomedicine, 9*, 2539–2555.

Neshatdoust, S., Saunders, C., Castle, S. M., Vauzour, D., Williams, C., Butler, L., Lovegrove, J. A., & Spencer, J. P. E. (2016). High flavonoid intake induces cognitive improvements linked to changes in serum brain-derived neurotropic factor: two randomised, controlled trials. *Nutrition and Healthy Aging*, *4*, 81–93.

Nettiksimmons, J., Decarli, C., Landau, S., & Beckett, L. (2014). Biological heterogeneity in Alzheimer's Disease Neuroimaging Initiative (ADNI) amnestic mild cognitive impairment. *Alzheimer's & Dementia*, *10* (5), 511–521.

Noble, J. E. (2014). Quantification of protein concentration using UV absorbance and coomassie dyes. *Methods in Enzymology*, *536*, 17–27.

Numakawa, T., Suzuki, S., Kumamaru, E., Adachi, N., Richards, M., & Kunugi, H. (2010). BDNF function and intracellular signalling in neurons. *Histology and Histopathology*, *25* (2), 237–258.

Nutten, S., Philippe, D., Mercenier, A., & Duncker, S. (2012). Opioid receptors stimulating compounds (thymoquinone, *Nigella sativa*) and food allergy. *Google Patents, EP2263664A1.*

Oberdorster, G., Elder, A., & Rinderknecht, A. (2009). Nanoparticles and the brain: cause for concern? *Journal of Nanoscience and Nanotechnology*, *9*, 4996–5007.

Obermeier, B., Verma, A., & Ransohoff, R. M. (2016). Chapter 3—The blood-brain barrier. In *Handbook of Clinical Neurology*; Pittock, S. J., Vincent, A., Eds.; Amsterdam, The Netherlands: Elsevier, 133, 39–59.

Oziyci, H. R., Karhan, M., Tetik, N., & Turhan, I. (2012). Effects of processing method and storage temperature on clear pomegranate juice turbidity and color. *Journal of Food Processing and Preservation*, *37* (5), 1–8.

Parboosing, R., Maguire, G. E., Govender, P., & Kruger, H. G. (2012). Nanotechnology and the treatment of HIV infection. *Viruses*, *4*, 488–520.

Pardridge, W. M. (2007). Blood-brain barrier delivery. *Drug Discovery Today*, *12* (1–2), 54–61.

Park, H., & Poo, M. M. (2013). Neurotrophin regulation of neural circuit development and function. *Reviews in Neurosciences*, *14* (1), 7–23.

Parsadaniantz, S. M., Rivat, C., Rostene, W., & Goazigo, A. R. (2015). Opioid and chemokine receptor crosstalk: a promising target for pain therapy? *Nature Reviews Neuroscience*, *16*, 69–79.

Pascoli, V., Terrier, J., Espallerques, J., Valjent, E., O'Connor, E. C., & Luscher, C. (2014). Contrasting forms of cocaine-evoked plasticity control components of relapse. *Nature*, *509*, 459–464.

Pascoli, V., Turiault, M., & Luscher, C. (2012). Reversal of cocaine-evoked synaptic potentiation resets drug-induced adaptive behaviour. *Nature*, *481*, 71–75.

Pascual-Teresa, S., & Sanchez-Ballesta, M. T. (2007). Anthocyanins: from plant to health. *Phytochemistry Reviews*, *7* (2), 281–299.

Passamonti, S., Vrhovsek, U., Vanzo, A., & Mattivi, F. (2005). Fast access of some grape pigments to the brain. *Journal of Agricultural and Food Chemistry*, *53* (18), 7029–7034.

Passeleu-Le Bourdonnec, C., Carrupt, P. A., Scherrmann, J. M., & Martel, S. (2013). Methodologies to assess drug permeation through the blood-brain barrier for pharmaceutical research. *Pharmaceutical Research*, *30* (11), 2729–2756.

Patel, C., Dadhaniya, P., Hingorani, L., & Soni, M. G. (2008). Safety assessment of pomegranate fruit extract: acute and sub-chronic toxicity studies. *Food and Chemical Toxicology*, *46*, 2728–2735.

Patras, A., Brunton, N. P., O'Donnell, C. P., & Kumar, T. B. (2010). Effect of thermal processing on anthocyanin stability in foods: mechanisms and kinetics of degradation. *Trends in Food Science & Technology*, *21* (1), 3–11.

Peng, Y., Zhao, Y., Chen, Y., Yang, Z., Zhang, L., Xiao, W., Yang, J., Guo, L., & Wu, Y. (2018). Dual-targeting for brain-specific liposomes drug delivery system: synthesis and preliminary evaluation. *Bioorganic & Medicinal Chemistry*, *26*, 4677–4686.

Petkar, K. C., Chavhan, S. S., Agatonovik-Kustrin, S., & Sawant, K. K. (2011). Nanostructured materials in drug and gene delivery: a review of the state of the art. *Critical Reviews in Therapeutic Drug Carrier Systems*, *28*, 101–164.

Pignatello, R., Fuochi, V., Petronio, G. P., Greco, A. S., & Furneri, P. M. (2017). Formulation and characterization of erythromycin–loaded solid lipid nanoparticles. *Biointerface Research in Applied Chemistry*, *7*, 2145–2150.

Pinzon-Daza, M. L., Campia, I., Kopecka, J., Garzon, R., Ghigo, D., & Riganti, C. (2013). Nanoparticle- and liposome-carried drugs: new strategies for active targeting and drug delivery across blood-brain barrier. *Current Drug Metabolism*, *14*, 625–640.

Postmes, T. J., Hukkelhoven, M., van der Bogaard, A. E., Halders, S. G., Coenegracht, J. (1980). Passage through the blood-brain barrier of thyrotropin-releasing hormone encapsulated in liposomes. *Journal of Pharmacy and Pharmacology*, *32*, 722–724.

Pulgar, V. M. (2018). Transcytosis to cross the blood brain barrier, new advancements and challenges. *Frontiers in Neuroscience, 12*, 1019.

Quan, X., Rang, L., Yin, X., Jin, Z., & Gao, Z. (2015). Synthesis of PEGylated hyaluronic acid for loading dichloro(1,2-diaminocyclohexane)platinum(II) (DACHPt) in nanoparticles for cancer treatment. *Chinese Chemical Letters, 26*, 695–699.

Raman, S. P., Dara, P. V., Vijayan, D. K., Chatterjee, N. S., Raghavankutty, M., Mathew, S., Ravishankar, C. N., & Anandan, R. (2020). Anti-ulcerogenic potential of anthocyanin-loaded chitosan nanoparticles against alcohol-HCL induced gastric ulcer in rats. *Natural Product Research, 36* (5), 1306–1310.

Ren, X., Lutfy, K., Mangubat, M., Ferrini, M. G., Lee, M. L., Liu, Y., & Friedman, T. C. (2013). Alterations in phosphorylated CREB expression in different brain regions following short- and long-term morphine exposure: relationship to food intake. *Journal of Obesity, 2013*, 1–11.

Reyes, B. A. S., Chavkin, C., & Van Bockstaele, E. J. (2009). Subcellular targeting of kappa-opioid receptors in the rat nucleus locus coeruleus. *Journal of Comparative Neurology, 512*, 419–431.

Ridzwan, N., Jumli, M. N., Baig, A. A., & Rohin, M. A. K. (2020). Pomegranate derived anthocyanin regulates MORs-cAMP/CREB-BDNF in opioid dependence models, improved cognitive impairments? *Journal of Ayurveda and Integrative Medicine, S0975–9476* (18), 30681–30683.

Robison, A. J., Vialou, V., Mazei-Robison, M., Feng, J., Kourrich, S., Collins, M., Wee, S., Koob, S., Turecki, G., Neve, R., Thomas, M., & Nestler, E. J. (2013). Behavioural and structural responses to chronic cocaine require a feed-forward loop involving FosB and CaMKII in the nucleus accumbens shell. *Journal of Neuroscience, 33*, 4295–4307.

Roth, H. J., & Fenner, H. (2000). *Arzneistoffe*. 3rd edition. Stuttgart, Germany: Deutscher Apotheker.

Ruszkiewicz, J. A., Pinkas, A., Ferrer, B., Peres, T. V., Tsatsakis, A., & Aschner, M. (2017). Neurotoxic effect of active ingredients in sunscreen products, a contemporary review. *Toxicology Reports, 4*, 245–259.

Saad, L. B., Hwi, K. K., & Quah, T. (2014). Evaluation of the anti-nociceptive effect of the ethanolic extract of *Punica granatum. African Journal of Traditional and Complementary Medicine, 11* (3), 228–233.

Sah, R. P., Nagpal, S. J., Mukhopadhyay, D., & Chari, S. T. (2013). New insights into pancreatic cancer-induced paraneoplastic diabetes. *Nature Reviews Gastroenterology & Hepatology, 10*, 423–433.

Sahin, A., Yoyen-Ermis, D., Caban-Toktas, S., Horzum, U., Aktas, Y., Couvreur, P., Esendagli, G., & Capan, Y. (2017). Evaluation of brain-targeted chitosan nanoparticles through blood-brain barrier cerebral microvessel endothelial cells. *Journal of Microencapsulation, 34* (7), 659–666.

Schliecker, G., Schmidt, C., Fuchs, S., Wombacher, R., & Kissel, T. (2003). Hydrolytic degradation of poly(lactide-co-glycolide) films: effect of oligomers on degradation rate and crystallinity. *International Journal of Pharmaceutics, 266*, 39–49.

Schnyder, A., & Huwyler, J. (2005). Drug transport to brain with targeted liposomes. *Neurotherapeutics, 2*, 99–107.

Schwartz, N., Temkin, P., Jurado, S., Limb, K., Heifets, B. D., Polepalli, J. S., & Malenka, R. C. (2014). Decreased motivation during chronic pain requires long-term depression in the nucleus accumbens. *Science, 345*, 535–542.

Septiani, N. L., Yuliarto, W., Iqbal, B., & Nugraha, M. (2017). Synthesis of zinc oxide nanoparticles using anthocyanin as a capping agent. *IOP Conference Series: Materials Science and Engineering, 202*, 1–5.

Shiban, M. S., Al-Otaibi, M. M., & Al-Zoreky, N. S. (2012). Antioxidant activity of pomegranate (*Punica granatum* L.) fruit peels. *Food and Nutrition Sciences, 3*, 991–996.

Shim, K. H., Jeong, K. H., Bae, S. O., Kang, M. O., Maeng, E. H., Choi, C. S., Kim, Y. R., Hulme, J., Lee, E. K., Kim, M. K., & An, S. S. (2014). Assessment of ZnO and SiO_2 nanoparticle permeability through and toxicity to the blood-brain barrier using Evans blue and TEM. *International Journal of Nanomedicine, 9* (Suppl 2), 225–233.

Shinohara, N., Danno, N., Ichinose, T., Sasaki, T., Fukui, H., Honda, K., Gamo, M. (2014). Tissue distribution and clearance of intravenously administered titanium dioxide (TiO_2) nanoparticles. *Nanotoxicology, 8* (2), 132–141.

Shinohara, N., Oshima, Y., Kobayashi, T., Imatanaka, N., Nakai, M., Ichinose, T., Sasaki, T., Kawaguchi, K., Zhang, G., Gamo, M. (2015). Pulmonary clearance kinetics and extrapulmonary translocation of seven titanium dioxide nano- and submicron materials following intratracheal administration in rats. *Nanotoxicology, 9*, 1050–1058.

Sillivan, S. E., Whittard, J. D., Jacobs, M. M., Ren, Y., Mazloom, A. R., Caputi, F. F., Horvath, M., Keller, E., Ma'ayan, A., Pan, Y. X., Chiang, L. W., & Hurd, Y. L. (2013). ELK1 transcription factor linked to dys-regulated striatal μ-opioid receptor signalling network and OPRM1 polymorphism in human heroin abusers. *Biology of Psychiatry, 74*, 511–519.

Silva, A. T. C. R., Cardoso, B. C. O., Silva, M. E. S. R., Freitas, R. F. S., & Sousa, R. G. (2015). Synthesis, characterization, and study of PLGA copolymer *in-vitro* degradation. *Journal of Biomaterials and Nanobiotechnology, 6*, 8–19.

Silva, G. A. (2008). Nanotechnology approaches to crossing the blood-brain barrier and drug-delivery to the CNS. *BMC Neuroscience, 9* (3), S4.

Skalska, J., Frontczak-Baniewicz, M., & Strużyńska, L. (2015). Synaptic degeneration in rat brain after prolonged oral exposure to silver nanoparticles. *Neurotoxicology, 46*, 145–154.

Socci, V., Tempesta, D., Desideri, G., Gennaro, L. G., & Ferrara, M. (2017). Enhancing human cognition with cocoa flavonoids. *Frontiers in Nutrition, 4* (19), 1–7.

Song, B., Liu, J., Feng, X., Wei, L., & Shao, L. (2015). A review on potential neurotoxicity of titanium dioxide nanoparticles. *Nanoscale Research Letters, 10*, 342.

Song, K. Y., Choi, H. S., Law, P. Y., Wei, L. N., & Loh, H. H. (2013). Vimentin interacts with the 5′-untranslated region of mouse μ-opioid receptor (MORs) and is required for post-transcriptional regulation. *RNA Biology, 10*, 256–266.

Song, R., Zhang, H. Y., Li, X., Bi, G. H., Gardner, E. L., & Xi, Z. X. (2012). Increased vulnerability to cocaine in mice lacking dopamine D3 receptors. *Proceedings of the National Academy of Sciences, 109*, 17675–17680.

Song, X., Li, W., Shi, Y., Zhang, J., & Li, Y. (2017). Expression of protein kinase A and the kappa opioid receptor in selected brain regions and conditioned place aversion in morphine-dependent rats. *Oncotarget, 8* (47), 82632–82642.

Sonia, T. A., & Sharma, C. P. (2011). Chitosan and its derivatives for drug delivery perspective. *Advance of Polymers Science, 243*, 23–54.

Soppimath, K. S., Aminabhavi, T. M., Kulkarni, A. R., & Rudzinski, W. E. (2001). Biodegradable polymeric nanoparticles as drug delivery devices. *Journal of Controlled Release, 70* (1–2), 1–20.

Spagnolo, P. A., & Goldman, D. (2017). Neuromodulation interventions for addictive disorders: challenges, promise and roadmap for future research. *Brain, 140*, 1183–1203.

Spencer, J. P. E. (2008). Food for thought: the role of dietary flavonoids in enhancing human memory, learning and neurocognitive performance. *Proceedings of the Nutrition Society, 67*, 238–252.

Spuch, C., & Navarro, C. (2011). Liposomes for targeted delivery of active agents against neurodegenerative diseases (Alzhermeir's disease and Parkinson's disease). *Journal of Drug Delivery, 2011*, 1–12.

Stober, W., Fink, A., & Bohn, E. (1968). Controlled growth of monodisperse silica spheres in the micron size range. *Journal of Colloid and Interface Science, 26*, 62–69.

Subash, S., Essa, M. M., Al-Adawi, S., Memon, M. A., Manivasagam, T., & Akbar, M. (2014). Neuroprotective effects of berry fruits on neurodegenerative diseases. *Neural Regeneration Research, 9* (16), 1557–1566.

Sun, J., Chen, J., Mei, Z., Luo, Z., Ding, L., & Jiang, X. (2020). Synthesis, structural characterization, and evaluation of cyanidin-3-O-glucoside-loaded chitosan nanoparticles. *Food Chemistry, 330*, 127239.

Talavera, S., Felgines, C., Texier, O., Besson, C., Gil-Izquierdo, A., Lamaison, J. L., & Remesy, C. (2005). Anthocyanin metabolism in rats and their distribution to digestive area, kidney, and brain. *Journal of Agricultural and Food Chemistry, 53*, 3902–3908.

Talavera, S., Felgines, C., Texier, O., Besson, C., Manach, C., Lamaison, J. L., & Remesy, C. (2004). Anthocyanins are efficiently absorbed from the small intestine in rats. *Journal of Nutrition, 134*, 2275–2279.

Tamaru, M., Akita, H., Fujiwara, T., Kajimoto, K., & Harashima, H. (2010). Leptin-derived peptide, a targeting ligand for mouse brain-derived endothelial cells via micropinocytosis. *Biochemical and Biophysical Research Communications, 394*, 587–592.

Tamba, B. I., Streinu, V., Foltea, G., Neagu, A. N., Dodi, G., Zlei, M., Tijani, A., & Stefanescu, C. (2018). Tailored surface silica nanoparticles for blood-brain barrier penetration: preparation and *in-vivo* investigation. *Arabian Journal of Chemistry, 11* (6), 981–990.

Tang, N., Du, G., Wang, N., Liu, C., Hang, H., & Liang, W. (2007). Improving penetration in tumors with nanoassemblies of phospholipids and doxorubicin. *Journal of the National Cancer Institute, 99*, 1004–1015.

Taylor, A. M., Castonguay, A., Ghogha, A., Vayssiere, P., Pradhan, A. A., Xue, L., Mehrabani, S., Wu, J., Levitt, P., Olmstead, M. C., de Koninck, Y., Evans, C. J., & Cahill, C. M. (2016). Neuroimmune regulation of GABAergic neurons within the ventral tegmental area during withdrawal from chronic morphine. *Neuropsychopharmacology, 41* (4), 949–959.

Teleanu, D. M., Chircov, C., Grumezescu, A. M., Volceanov, A., & Teleanu, R. I. (2018). Blood-brain delivery methods using nanotechnology. *Pharmaceutics, 10*, 269.

Tian, X., Nyberg, S., Sharp, P. S., Madsen, J., Daneshpour, N., Armes, S. P., Berwick, J., Azzouz, M., Shaw, P., Abbott, N. J., & Battaglia, G. (2015). LRP-1-mediated intracellular antibody delivery to the central nervous system. *Scientific Reports*, *5*, 11990.

Torchilin, V. P. (2005). Recent advances with liposomes as pharmaceutical carriers. *Nature Reviews Drug Discovery*, *4*, 145–160.

Trang, T., Al-Hasani, R., Salvemini, D., Salter, M. W., Gutstein, H., & Cahill, C. M. (2015). Pain and poppies: the good, the bad, and the ugly of opioid analgesics. *Journal of Neuroscience*, *35*, 13879–13888.

Trapani, A., De Giglio, E., Cafagna, D., Denora, N., Agrimi, G., Cassano, T., Gaetani, S., Cuomo, V., & Trapani, G. (2011). Characterization and evaluation of chitosan nanoparticles for dopamine brain delivery. *International Journal of Pharmaceuticals*, *419*, 296–307.

Tsuda, T. (2012). Anthocyanins as functional food factors-chemistry, nutrition and health-promotion. *Food Science and Technology Research*, *18* (3), 415–324.

Ulbrich, K., Hekmatara, T., Herbert, E., & Kreuter, J. (2009). Transferrin- and transferrin-receptor-antibody-modified nanoparticles enable drug delivery across the blood-brain barrier (BBB). *European Journal of Pharmaceutics and Biopharmaceutics*, *71*, 251–256.

Vaculikova, E., Placha, D., & Jampilek, J. (2015). Toxicology of drug nanocarriers. *Chemicke Listy*, *109* (5).

Vanzo, A., Vrhovsek, U., Tramer, F., Mattivi, F., & Passamonti, S. (2011). Exceptionally fast uptake and metabolism of cyanidin 3-glucoside by rat kidneys and liver. *Journal of Natural Products*, *74* (5), 1049–1054.

Vargas-Perez, H., Ting, A., Kee, R. A., Heinmiller, A., Sturgess, J. E., & van der Kooy, D. D. (2007). A test of the opponent-process theory of motivation using lesions that selectively block morphine reward. *European Journal of Neuroscience*, *25* (12), 3713–3718.

Vauzour, D., Vafeiadou, K., Rodriguez-Mateos, A., Rendeiro, C., & Spencer, J. P. E. (2008). The neuroprotective potential of flavonoids: a multiplicity of effects. *Genes & Nutrition*, *3* (3–4), 115–126.

Vieira, D. B., & Gamarra, L. F. (2016). Getting into the brain: liposome-based strategies for effective drug delivery across the blood-brain barrier. *International Journal of Nanomedicine*, *11*, 5381–5414.

Viuda-Martos, M., Perez-Alvarez, J. A., Sendra, E., & Fernandez-Lopez, J. (2013). *In-vitro* antioxidant properties of pomegranate (*Punica granatum* L.) peel powder extract obtained as co-product in the juice extraction process. *Journal of Food Processing Preservation*, *37* (5), 772–776.

Wallace, T. C., & Giusti, M. M. (2015). Anthocyanins. *Advance Nutrition*, *6* (5), 620–622.

Wang, B., Feng, W., Wang, M., Wang, T. C., Gu, Y., Zhu, M., Ouyang, H., Shi, J., Zhang, F., Zhao, Y., Chai, Z., Wang, H., & Wang, J. (2008). Acute toxicological impact of nano- and submicro-scaled zinc oxide powder on healthy adult mice. *Journal of Nanoparticles Research*, *10*, 263–276.

Wang, H., Zhang, Y., & Qiao, M. (2013). Mechanisms of extracellular signal-regulated kinase/cAMP response element-binding protein/brain-derived neurotrophic factor signal transduction pathway in depressive disorder. *Neural Regeneration Research*, *8* (9), 843–853.

Wang, J., Lee, J. S., Kim, D., & Zhu, I. (2017). Exploration of zinc oxide nanoparticles as a multitarget and multifunctional anticancer nanomedicine. *ACS Applied Materials and Interfaces*, *9* (46), 39971–39984.

Wang, J., Zhang, J., Li, S., Huang, C., Xie, Y., & Cao, Y. (2020). Anthocyanins decrease the internalization of TiO_2 nanoparticles into 3D Caco-2 spheroids. *Food Chemistry*, *331*, 127360.

Wang, L. S., Wu, L. C., Lu, S. Y., Chang, L. L., Teng, I. T., & Yang, C. M. (2010). Bio functionalized phospholipid-capped mesoporous silica nanoshuttles for targeted drug delivery: improved water suspensibility and decreased nonspecific protein binding. *ACS Nano*, *4*, 4371–4379.

Wang, W. S., Kang, S., Liu, W. T., Li, M., Liu, Y., Yu, C., Chen, J., Chi, Z. Q., He, L., & Liu, J. G. (2012). Extinction of aversive memories associated with morphine withdrawal requires ERK-mediated epigenetic regulation of brain-derived neurotrophic factor transcription in the rat ventro-medial prefrontal cortex. *Journal of Neuroscience*, *32* (40), 13763–13775.

Williams, C., Wilson, P., Morrison, J., McMahon, A., Walker, A., Allan, L., McConnachie, A., McNeil, Y., & Tansy, L. (2013). Guided self-help cognitive behavioural therapy for depression in primary care: a randomised controlled trial. *PLoS One*, *8* (1), 1–7.

Williams, J. T. (2014). Desensitization of functional μ-opioid receptors increases agonist off-rate. *Molecular Pharmacology*, *86*, 52–61.

Winger, R. C., Koblinski, J. E., Kanda, T., Ransohoff, R. M., & Muller, W. A. (2014). Rapid remodeling of tight junctions during paracellular diapedesis in a human model of the blood–brain barrier. *Journal of Immunology*, *193*, 2427–2437.

Xie, Y., Tan, H., Ma, Z., & Huang, J. (2016). DELLA proteins promote anthocyanin biosynthesis via sequestering MYBL2 and JAZ suppressors of the MYB/bHLH/WD40 complex in *Arabidopsis thaliana*. *Molecular Plant*, *9*, 711–721.

Xu, D., Lian, D., Wu, J., Liu, Y., Zhu, M., Sun, J., He, D., & Li, L. (2017). Brain-derived neurotropic factor reduced inflammation and hippocampal apoptosis in experimental *Streptococcus pneumonia* meningitis. *Journal of Neuroinflammation, 14* (156), 1–13.

Yamasaki, T., Hoyos-Ramirez, E., Martenson, J. S., Morimoto-Tomita, M., & Tomita, S. (2017). GARLH family proteins stabilize GABAA receptors at synapses. *Neuron, 93*, 1138–1152.

Yan, K., Chen, Y. B., Wu, J. R., Li, K. D., & Cui, Y. L. (2018). Current rapid onset antidepressants and related animal models. *Current Pharmaceutical Design, 24*, 1–9.

Yan, W. S., Li, Y. H., & Sui, N. (2014). The relationship between recent stressful life events, personality traits, perceived family functioning and internet addiction among college students. *Stress Health, 30*, 3–11.

Yesil-Celiktas, O., Pala, C., Cetin-Uyanikgil, O., & Sevimli-Gur, C. (2017). Synthesis of silica-PAMAM dendrimer nanoparticles as promising carriers in neuroblastoma cells. *Analytical Biochemistry, 519*, 1–7.

Youdim, K. A., Dobbie, M. S., Kuhnle, G., Proteggente, A. R., Abbott, N. J., & Rice-Evans, C. (2003). Interaction between flavonoids and the blood-brain barrier: *in vitro* studies. *Journal of Neurochemistry, 85* (1), 180–192.

Ze, Y., Hu, R., Wang, X., Sang, X., Ze, X., Li, B., Su, J., Wang, Y., Guan, N., Zhao, X., Gui, S., Zhu, L., Cheng, Z., Cheng, J., Sheng, L., Sun, Q., Wang, L., & Hong, F. (2014). Neurotoxicity and gene-expressed profile in brain-injured mice caused by exposure to titanium dioxide nanoparticles. *Journal of Biomedical Materials Research Part A, 102* (2), 470–478.

Ze, Y., Zheng, L., Zhao, X., Gui, S., Sang, X., Su, J., Guan, N., Zhu, L., Sheng, L., Hu, R., Cheng, J., Cheng, Z., Sun, Q., Wang, I., & Hong, F. (2013). Molecular mechanism of titanium dioxide nanoparticles-induced oxidative injury in the brain of mice. *Chemosphere, 92* (9), 1183–1189.

Zhan, W., & Wang, CH. (2018). Convection enhanced delivery of liposome encapsulated doxorubicin for brain tumour therapy. *Journal of Controlled Release, 285*, 212–229.

Zhang, F., Zhang, L., Qi, Y., & Xu, H. (2016). Mitochondrial cAMP signalling. *Cellular and Molecular Life Sciences, 73* (24), 4577–4590.

Zhang, J., Zhang, X. Y., Su, H., Tao, J. Y., Xie, Y., Han, B., Lu, Y., Wei, Y., Sun, H., Wang, Y., Wu, W., Zou, S., Liang, H., Zoghbi, A. W., Tang, W., & He, J. (2014). Increased serum brain-derived neurotropic factor levels during opiate withdrawal. *Neuroscience Letters, 571*, 61–65.

Zhang, W., Mehta, A., Tong, Z., Esser, L., & Voelcker, N. H. (2021). Development of polymeric nanoparticles for blood-brain barrier transfer- strategies and challenges. *Advanced Science, 8*, 2003937–2003969.

Zhao, C. N., Meng, X., Li, S., Liu, Q., Tang, G. Y., & Li, H. B. (2017). Fruits for prevention and treatment of cardiovascular disease. *Nutrients, 9* (598), 1–29.

Zhao, J., Xu, L., Zhang, T., Ren, G., & Yang, Z. (2009). Influences of nanoparticle zinc oxide on acutely isolated rat hippocampal CA3 pyramidal neurons. *Neurotoxicology, 30*, 220–230.

Zhao, L., & Temelli, F. (2016). Preparation of anthocyanin-loaded liposomes using an improved supercritical carbon dioxide method. *Innovative Food Science & Emerging Technologies, 39*, 119–128.

Zhao, X., Yuan, Z., Fang, Y., Yin, Y., & Feng, L. (2013). Characterization and evaluation of major anthocyanin in pomegranate (*Punica granatum* L.) peel of different cultivars and their development phases. *Europe Food Resources of Technology, 236*, 109–117.

Zheng, H., Law, P. Y., & Loh, H. H. (2012). Non-coding RNAs regulating morphine function: with emphasis on the *in vivo* and *in vitro* functions of miR-190. *Frontiers in Genetics, 15*, 113.

Zhi, K., Raji, B., Nookala, A. R., Khan, M. M., Nguyen, X. H., Sakshi, S., Pourmotabbed, T., Yallapu, M. M., Kochat, H., Tadrous, E., Permell, S., & Kumar, S. (2021). PLGA nanoparticle-based formulations to cross the blood-brain barrier for drug delivery: from R&D to cGMP. *Pharmaceutics, 13*, 500.

Zhu, Y., Wienecke, C. F. R., Nachtrab, G., & Chen, X. (2016). A thalamic input to the nucleus accumbens mediates opiate dependence. *Nature, 530*, 219–239.

Zubeldia, J. M., Jimenez-del-rio, M., Perez-Lopez, V., & Hernandez-Santana, A. (2013). Fruit polyphenols can up-regulate the expression of opioid receptors (OPRD 1) in brain cells, a molecular *in vitro* and in silico study. *International Journal of Biology and Medical Research, 3* (1), 3308–3312.

Zyoud, A. H., Saleh, F., Helal, M. H., Shawahna, R., & Hilal, H. S. (2018). Anthocyanin-sensitized TiO_2 nanoparticles for phenazopyridine photodegradation under solar simulated light. *Journal of Nanomaterials, 2*, 1–14.

8 Phospholipid-Based Nanoplatforms

Evolving as Promising Carriers for Therapeutic Intervention

Amrita Chakraborty and Pubali Dhar[*]

CONTENTS

Highlights 183
8.1 Introduction 183
8.2 Phospholipid-Based Nanoemulsion 185
8.3 Liposomes: A New Frontier of the Colloidal Drug Carrier System 188
8.4 Modified Liposomes: Versatile and Flexible Nanovesicular Vehicles for Transdermal Delivery 192
8.4.1 Transferosomes 192
8.4.2 Ethosomes 194
8.5 Phospholipid-Based Solid Lipid Nanoparticles: A Novel Formulation for Various Routes of Administration 196
8.6 Multifunctional Micelles for Targeted Drug Delivery 200
8.7 Lipospheres as a Biocompatible Delivery System 204
8.8 Phospholipid–Drug Complexes for Enhancing Stability and Oral Bioavailability 205
8.8.1 Phytosomes 207
8.8.2 Pharmacosomes 209
8.9 Conclusion 211
Acknowledgement 212
References 212

HIGHLIGHTS

- Successful utilization of physicochemical properties of phospholipids for fabrication of various delivery vehicles such as nanoemulsions, liposomes, nanomicelles, solid lipid nanoparticles, lipospheres, etc.
- Improvement of dermal delivery with the modification of liposomes and invention of transferosomes and ethosomes as novel carrier systems
- Formulation of phospholipid–drug complexes such as phytosomes and pharmacosomes for enhancing stability, bioavailability, bioactivity and targeted delivery of therapeutics

8.1 INTRODUCTION

The therapeutic delivery domain, a subject of keen interest, has been addressed by new age scientific research and used to treat fatal maladies. Experimental trials have revealed that the prophylactic and

[*] Corresponding author: pubalighoshdhar23@gmail.com

DOI: 10.1201/9781003244721-8

therapeutic effectiveness of key bioactives cannot be fully attained through conventional delivery approaches due to instantaneous renal filtration, instant removal via the reticulo-endothelial system (RES), their circuitous movement from the circulation to the specific target area and the harsh acidic conditions of endo-lysosomes within the cell (Rizvi and Saleh, 2018). To overcome the unsatisfactory outcomes connected with the conventional mode of delivery, today, technological expansion has directed its focus towards the development of nanostructures as prospective delivery devices for the enhancement of therapeutic facilities. Nanodelivery strategies have emerged as promising encapsulation techniques due to their nanoscale profile, inflated surface area:volume ratio, considerable load and optimal release properties (Aklakur et al., 2016). The fabrication of stable and safe nanomaterials for efficacious therapeutic delivery remains a prime challenge to scientists, as there are very few excipients with generally regarded as safe (GRAS) status. In accordance with the growing perception of designing bio-compatible, non-toxic, natural and edible delivery vehicles for the sake of a safe and healthy lifestyle, biopolymers are increasingly gaining importance (Shit and Shah, 2014). The preference for edible polymers over synthetic surfactants is because natural biopolymers are produced entirely from an eco-friendly replenishable reservoir and hence, are anticipated to degrade more rapidly than other commercial compounds. Despite the fact that polymers are selected as vehicles because of their less detrimental effects, polyethylene oxide (PEO) polymers present hazards; for instance, they yield aldehydes after exposure to illumination and oxygen. Aldehydes causing toxicity may disrupt the physiological system with harmful effects (Chakraborty and Dhar, 2017). In current research, phospholipids have generated enormous interest as nanoformulation excipients due to their unique physio-chemical characteristics and their multi-purpose utilization in both biological and material spheres. Phospholipids, the main components of the cellular membrane, are amphipathic in nature. In the continuously expanding domain of delivery for remedial agents, the choice of phospholipids as a vehicle results from their innate biodegradability, inherent biocompatibility, nutritional merits and cost-efficient maneuverability (Li et al., 2015b). Chemically, phospholipid molecules consist of a phosphate-bearing polar head group and a non-polar hydrocarbon tail attached jointly by a glycerol or an alcohol molecule. The "head" is a water-loving moiety due to the presence of phosphate anions and glycerol, whereas the "tail" is water-repellent, as the long fatty acid tail is repelled by water and is compelled to cluster in an aqueous milieu (Singh et al., 2017b). The heterogeneous trait of the head group is due to the existence of miscellaneous functional groups linked to the phosphate groups, for example phosphatidic acid (PA), phosphatidylcholine (PC), phosphatidylethanolamine (PE) or phosphatidylserine (PS). There are two types of phospholipids: natural and synthetic (Van Hoogevest and Wendel, 2014). Natural phospholipid excipients can be isolated from various plant origins, such as flax seed, rapeseed, sunflower and soybean seed, as well as several animal sources, like egg yolk, milk and krill. For the food and pharmaceutical industry, natural phospholipids may be further reconstructed by means of hydrogenation or enzymatic modification. On the contrary, synthetic phospholipid excipients like polyethylene glycol (PEG)ylated phospholipids and the positively charged phospholipid 1,2-diacyl-P-O-ethylphosphatidylcholine are synthetic analogs of natural phospholipids. They can be designed with the help of customized chemical and/or enzymatic synthesis techniques whereby specific polar head groups or non-polar fatty acids are incorporated into the phospholipids. Lecithin is one of the major amphiphiles abundantly found in cell membranes of plants and animals. It is a blend of most of the naturally derived phospholipids, of which PC is the major component. Phospholipids can form a three-dimensional lamellar crystalline structure as well as two-dimensional lamellar crystals depending on temperature and/or hydration extent (Karamanidou et al., 2016). The phase transition is mainly instigated by thermal alteration; by raising the temperature beyond a certain extent, the hydrocarbon tail becomes liquid, which therefore promotes a transition from the solid to the liquid state. Phospholipids in the liquid state can form bilayers, generate various supramolecular structures and adopt different self-organizing molecular assemblies, such as direct and inverted micelles, emulsions, organogels or liposomes, when dispersed in aqueous media (Suriyakala et al., 2014). Hydrated phospholipids are thermodynamically stable and visco-elastic. Phospholipids have exhibited good interfacial properties, which can significantly influence emulsion stabilization (Li et al., 2014). Surface-active phospholipids can also cover the surface of crystals to augment the hydrophilic nature of lipophilic drugs and nutraceuticals (Silva et al., 2012). Along with the physicochemical functionalities, phospholipids have also shown excellent physiological benefits. Non-allergic natural phospholipids are known to be harmless permeation enhancers and hence, can be employed to regulate the absorption of active therapeutics through different mechanisms, for example altering the release of the bioactive core, increasing their bioavailability, influencing the intestinal conditions, boosting the lymphatic transport of bioactive agents, dealing with enterocyte-based metabolism, etc. (Van Hoogevest and Wendel, 2014). Phospholipids can reform the skin lipid barrier, thereby aiding the transcellular transfer of active agents and performing as penetration enhancers (Vanić, 2015). Besides their immense potential to successfully protect therapeutic molecules from instant degradation in the gastrointestinal environment and reduce undesirable aftereffects, phospholipids are generally recognized for remarkable hepatoprotection from drugs, alcohol and other toxic substances (Gundermann et al., 2011). Essential phospholipids have also been found to exhibit antilipemic and antiatherogenic activity by hindering the elevation of total lipids in dietetic hypercholesterolemia at both curative and preventive doses (Leuschner et al., 1976). Marine phospholipids have been reported to have a broad spectrum of health benefits through a synergistic mode that combines the favorable quality of long-chain n-3 polyunsaturated fatty acids (PUFAs) with the ideal stability and notable permeating abilities of amphiphilic phospholipids within biological tissues (Paul et al., 2018). High-dosage phospholipids can effectively treat neuro-degenerative diseases (Küllenberg et al., 2012). In this present scenario, this chapter intends to explore the recent evolution and expansion in the fabrication technologies as well as the versatile ameliorative application spectrum of the phospholipid-based nanovehicles. Phospholipids provide several possibilities to engineer a number of distinct nanocarriers, such as nanoemulsions, nanomicelles, nanoliposomes, solid lipid nanoparticles, phospholipid–drug complexes, lipospheres, and so on.

8.2 PHOSPHOLIPID-BASED NANOEMULSION

Nanoemulsions are submicron-sized colloidal dispersions containing tiny droplets, i.e., mean radius ranging between 10 and 100 nm (McClements, 2012). Due to the fairly small-scale droplet size in comparison with the wavelength of light ($r << \lambda$), nanoemulsions seem to be optically transparent, whereas a conventional emulsions generally appear optically turbid or opaque. Moreover, due to the very small particle radii, nanoemulsions tend to be kinetically stable against gravitational separation and aggregation (McClements and Rao, 2011). Generally, the production of nanoemulsions from natural excipients relies on high-energy emulsification approaches, whereas low-energy techniques utilize the fundamental physicochemical virtues of the excipients for spontaneous emulsification (Klang and Valenta, 2011). Figure 8.1a illustrates the high-energy approaches for fabricating a nanoemulsion.

Emulsions are shown to acquire better stability when manufactured with PC as emulsifiers compared with PS and PE, since PC is apt to organize a closely packed monolayer interface. The monolayer made from PC is in a liquid crystalline phase, while those made from PS and PE are in a liquid expanded state (Pichot et al., 2013). Zwitterionic phospholipid emulsifiers are reported to impart stability to formulations by acting simultaneously as an electrostatic and a mechanical barrier to coalescence. An interfacial barrier constructed from, at the minimum, two phospholipid layers is indispensable to ensure emulsion stability, and the emulsion stability is boosted by multiple layers of phospholipids around the droplets. Charged phospholipids intensify droplet repulsion, which results in improved emulsion stability. The selection of the lipid component seems to be a significant determining criterion for the pharmacokinetic release of active core from the nanoemulsion. Squalene has often been documented to be an exceptionally beneficial lipophilic biomolecule for the fabrication of lecithin-stabilized nanoemulsions. Squalene nanoemulsions with varied amounts of lecithin depicted satisfactory stability and the prolonged release of the hydrophobic drug rifampicin in comparison with innumerable oils *in vitro* (Chung et al., 2001). Multifaceted advantages can be obtained by utilizing lecithin-stabilized nanoemulsions for therapeutic intervention through parenteral

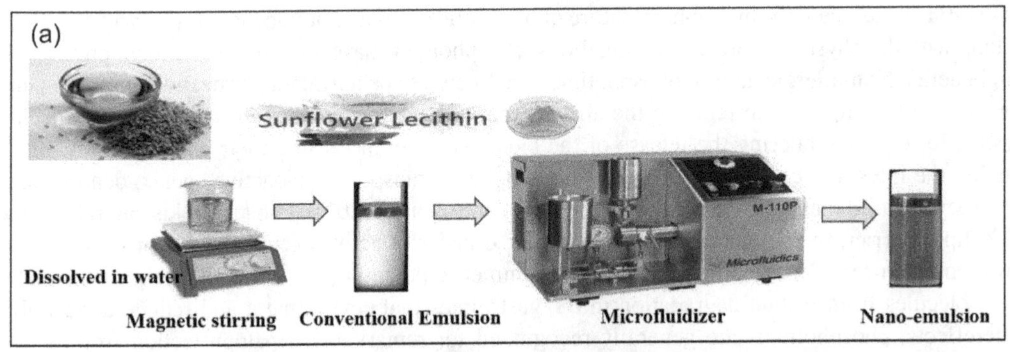

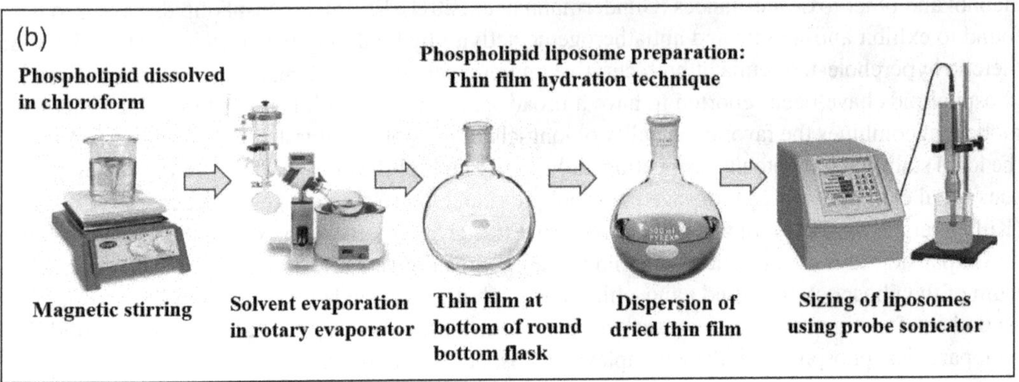

FIGURE 8.1 (a) High-energy approaches as the most commonly used technology for fabrication of nanoemulsion. (b) Thin film hydration as the simplest technique for production of phospholipid-based liposomes.

delivery. Hydrophobic therapeutics can be easily entrapped in the lipid core, and their poor water solubility is mitigated, which consequently amplifies bioavailability. Egg phospholipid–based nanoemulsion is found to enhance the analgesic potency and duration of release of nalbuphine and its prodrugs (Wang et al., 2006). For parenteral delivery, the nanoemulsion encapsulating chlorambucil was fabricated with an average size below 150 nm and encapsulation efficiency >97%. The cytotoxicity as well as the pro-apoptotic activity of chlorambucil were markedly elevated when delivered in the nanoemulsion formulation instead of aqueous solution (Ganta et al., 2008). Carbamazepine, an anticonvulsant drug, was also encapsulated in soy lecithin–based nanoemulsion for intravenous administration (Kelmann et al., 2007). The nanoemulsion formulation was found to be stable, and its characteristics were maintained in a reasonably acceptable range over a 3-month time period; for example, droplet diameter about 150 nm, zeta potential about −40 mV and drug encapsulation about 95%. *In vitro* release kinetics evaluated by dialysis bags showed that 95% of drug had been released in less than 11 hours. Nevertheless, a critical criterion limiting the global application of lecithin-stabilized nanoemulsions via the parenteral route is the fear of hemolysis, which might occur as a result of the bio-molecular interactions between phospholipids and erythrocytes. Egg phosphatidylcholine was reported to bring on moderate toxic effects in mature red blood cells *in vitro*, which was increased upon the addition of stearylamine and reduced in the presence of Brij-type surfactants. To avoid the formation of lyso-derivatives and to control the hemolytic potential of the blood, Masoumi et al. suggested using Tween 80 as a co-emulsifier. A nanoemulsion developed using lecithin and Tween 80 was found to be stable with small droplet size (Masoumi et al., 2015). On the other hand, a stable curcumin nanoemulsion can be procured by utilizing hydroxylated lecithin as emulsifying agent and using the ultrasonication technology simultaneously. The report demonstrated that the amplitude and ultrasonication time had a profound impact on the droplet diameter and the polydispersity index of the fabricated nanoemulsion (Espinosa-Andrews and Páez-Hernández, 2020). A lecithin-stabilized nanoemulsion aiming at oral administration of primaquine was shown to possess markedly better antimalarial activity and *in vivo* biodistribution compared with a readily available drug solution. Compared with the commercially available drug, the formulation was proved to penetrate into the hepatic cells with a drug concentration around 45% or above (Singh and Vingkar, 2008). The use of cefpodoxime proxetil, an oral cephalosporin antibiotic, was limited due to its inadequate water solubility along with insufficient bioavailability. A lecithin-based nanoemulsion offered amplified absorption of encapsulated cefpodoxime proxetil, which resulted in superior bioavailability (97.4%) compared with a suspension, an alcoholic solution and an emulsion encapsulating active molecule (Nicolaos et al., 2003). Additionally, this formulation was shown to shield the active component from unintended enzymatic degradation in the intestine. A recent *in vivo* pharmacokinetic investigation in a rat model demonstrated that in a lecithin-based nanoemulsion, the bioavailability and bioaccessibility of carnosic acid were about 2.2-fold and 12.6-fold higher, respectively, than those of carnosic acid suspension. Additionally, it was found that the carnosic acid–loaded nanoemulsion improved the solubility, digestion rate, retention time and sustained release of carnosic acid during digestion (Zheng et al., 2020). Paclitaxel, a poorly water-soluble chemotherapeutic agent, is a substrate of the P-glycoprotein efflux pump. Tiwari and Amiji loaded the substance into an oil-in-water nanoemulsion using egg lecithin as emulsifier (Tiwari and Amiji, 2006). The fabricated nanoemulsions had an average particle size about 90–120 nm and zeta potential ranging between −56 and +34 mV. In comparison with an aqueous solution of paclitaxel, the fabricated nanoformulation yielded a significantly elevated blood level of paclitaxel in an experimental mouse model. Nanoemulsions are being explored not only for achieving a sustained release mechanism but also for specific organ targeting. Surface alteration of the emulsion droplets, for instance with PEG derivatives, can be utilized to achieve long-term residence in circulation as well as passive targeting to inflamed areas and tumors. Lecithin-stabilized nanoemulsions with and without PEG modification were recruited for successful co-loading of paclitaxel and curcumin to conquer the multidrug resistance issues in tumor cells by stimulating apoptosis (Ganta and Amiji, 2009). *In vitro* studies have shown that ultrasound-mediated delivery of a lecithin-based nanoemulsion entrapping curcumin demonstrated significant *cellular toxicity* against OSCC-4 and OSCC-25 cell lines (Lin et al., 2012). Vecchione et al. developed a curcumin-loaded nanoemulsion utilizing lecithin as a surfactant and functionalized chitosan as a coating agent to regulate the deposition and bio-molecular interaction of the formulation with the intestinal barrier. The highest level of bioavailability was achieved in the case of the smaller nanoemulsion (average particle size 110 nm) coated with the highest degree of chitosan modification and co-delivered with piperine (Vecchione et al., 2016). Lecithin-based nanoemulsions manifested higher permeability and sustained release of resveratrol through a Caco-2 cell monolayer than those prepared by the combination of Tween 20 and glycerol monooleate (Sessa et al., 2014). This is attributed to the fact that the shell layer made of a mixture of phospholipids of soy lecithin is analogous to the phospholipid bilayer residing in the cell membrane. Therefore, absorption and entrapment of the lecithin-based nanoemulsion in the microvilli is enhanced, resulting in improved transport through the cell membrane. Heo et al. assessed the impact of nanoemulsification with lecithin on the thermal stability as well as bioavailability of conjugated linoleic acid in various free fatty acid and triacylglycerol (TG) forms. Conjugated linoleic acid (CLA) nanoemulsion in triacylglycerol (TG) form presented a smaller droplet size (70–120 nm) than CLA nanoemulsion in free fatty acid (FFA) form (230–260 nm). The *in vitro* bioavailability test revealed that in both FFA and TG forms, cellular penetration of CLA via the monolayers of Caco-2 human intestinal cells was increased after nanoemulsification. Moreover, a rat feeding study reported a remarkable increase in CLA level in plasma or small intestinal tissues after administration of a CLA nanoemulsion, demonstrating improved oral bioavailability of CLA with the assistance of nanoemulsification (Heo et al., 2016). A lecithin-based nanoemulsion was also fabricated by a spontaneous emulsification process for entrapping ferulic acid, one of the most common pharmacologically active polyphenolic compounds with poor water solubility (Ebrahimi et al., 2013). To facilitate corneal drug delivery, lecithin-based nanoemulsions can be used as penetration enhancers by detaching the mucus layer and disrupting the arrangements of tight junction proteins. Ammar et al. utilized nanoemulsions as a delivery system for dorzolamide hydrochloride in glaucoma therapy

(Ammar et al., 2009). Differential scanning calorimetry has demonstrated the alteration in fluidity of skin lipids with various lecithin-based nanoemulsion systems. Due to this improved skin lipid fluidity, lecithin might help the encapsulated active compounds to readily penetrate into the dermis. The formulation can be evenly dispersed on the skin as a result of the tiny particle size along with the uniform droplet dispersion in a nanoemulsion. The nano range of droplet sizes provides a huge interfacial area that primarily increases the skin permeation rate of active elements and simultaneously extends their residence time in the outermost epidermal layers (Sonneville-Aubrun et al., 2004). A lecithin-based nanoemulsion loaded with Nile red revealed that a nanoemulsion not only significantly improved the skin hydration of a formulation but also strengthened the skin penetration of Nile red for topical application (Zhou et al., 2010). In another study, a lecithin-based nanoemulsion entrapping progesterone was developed, with an excellent skin permeation effect (Klang et al., 2010). Silva and his team proved that a lecithin-based nanoemulsion was suitable for geneistin delivery, with mean droplet radii ranging between 230 and 280 nm along with a sustained release profile (Silva et al., 2009). Enhancement of the percutaneous absorption of flurbiprofen was observed after incorporation of the drug into an egg lecithin–based nanoemulsion (Fang et al., 2004). Komaiko et al. recently demonstrated sunflower phospholipid as a natural emulsifier for omega-3 delivery applicable in the food industry. The study revealed that increased phospholipid concentration in emulsions reduced the droplet dimensions. The fabricated emulsions were predominantly stabilized by electrostatic repulsion and consequently, were prone to aggregation in specific environments where the aqueous milieu had a high ionic strength or the droplets possessed relatively lower net charges (Komaiko et al., 2016). The lecithin utilized in the food industry is usually extracted from egg yolk, soybeans, milk, rapeseeds and sunflower (Cui and Decker, 2016). The composition of lecithin changes according to the method of extraction and refinement. Though physicochemically stabilized oil-in-water emulsions were recently shown to be fabricated from sunflower lecithins utilizing both high- and low-energy approaches, researchers also found that the behavior of the emulsion varied according to the phospholipid composition, especially the PC content in the total emulsifier (Liang et al., 2017). The unsatisfactory stability of emulsion obtained with the pure PC fraction may be due to the dearth of emulsifiers like PE (Van Hoogevest and Fahr, 2019). On the other hand, hydrogenated soybean phospholipids containing 75% or 90% PC alone were found adequate for emulsifying lipids with hydrophilic–lipophilic balance (HLB) values ranging between 4 and 11. Interestingly, the hydrogenated soybean phospholipids with 75% PC was reported to be a slightly finer emulsifier as this intermediate-grade soy phospholipid provided a perfect blend of phospholipid co-emulsifiers varying in their hydrophilic head group. Moreover, compared with saturated and monoacyl-phospholipids, unsaturated diacyl-lecithins are not adequate when used as a single stabilizer for oil-in-water emulsion. Monoacyl-phospholipids are reported to assemble in a cone-shaped configuration, which can ideally fit the curvature of the interface between oil and water in the emulsions due to the presence of substantial surface area from the hydrophilic head group and the tiny cross-sectional domain from fatty acids. The usage of biocompatible lecithin-stabilized nanoemulsions was proved to be suitable for various paths of administration, including sensitive paths such as oral, intravenous, dermal and ocular administration. Though the complex behavior of phospholipids in nanoemulsions still remains unclear, a number of innovative strategies are being continuously explored for further optimization, modification and upgradation of lecithin-based nanoemulsions. Fabrication techniques, physicochemical characteristics and therapeutic effects of phospholipid-based nanoemulsions entrapping various active compounds are briefly outlined in Table 8.1.

8.3 LIPOSOMES: A NEW FRONTIER OF THE COLLOIDAL DRUG CARRIER SYSTEM

Liposomes are familiar and extensively explored particulate nanovesicles that have been effectively recruited for the optimum release and targeted delivery of therapeutics (Akbarzadeh et al., 2013). Liposomes, spherical sacs surrounded by at least one lipid bilayer, are mainly composed of

TABLE 8.1
Method of Fabrication, Physicochemical Characteristics and Therapeutic Effects of Phospholipid-Based Nanoemulsion Entrapping Various Active Compounds

Type of phospholipid excipient	Active compound	Preparation method	Particle size (nm)	Zeta potential (mV)	Entrapment efficiency (%)	Therapeutic effect	References
Hydrogenated L-α phosphatidylcholine from egg yolk	Curcumin	Modified thin film hydration followed by sonication	47–56	–	90	Preserved cytotoxic effect to cancer cell line	Anuchapreeda et al., 2012
Soybean lecithin	Carbamazepine	Spontaneous emulsification	150	–40	95	Improved *in vitro* release profile	Kelmann et al., 2007
Soybean lecithin	Conjugated linoleic acid	High-pressure homogenization	70–120	–40	–	Improved cellular uptake and oral bioavailability	Heo et al, 2016
Lecithin, DSPE-PEG2000	Doxorubicin	Solvent diffusion	44.5	–25.6	71.2	Efficiently accumulated in targeted tumor tissue	Jiang et al., 2013
Lecithin	Aripiprazole	High-pressure homogenization	62.23	–31.6	–	Expected to treat schizophrenia efficiently	Masoumi et al., 2015
Egg phosphatidyl choline, DSPE-PEG2000	Curcumin	Coarse homogenization followed by ultrasonication	144	–44.53	97.4	Combination therapy blocked P-gp, inhibited NFκB pathway, enhanced cytotoxic efficacy and the apoptotic response	Ganta et al., 2009
Egg phosphatidyl choline, DSPE-PEG2000	Paclitaxel	Coarse homogenization followed by ultrasonication	138	–39.74	100	Egg phosphatidyl choline, DSPE-PEG2000	Paclitaxel
Egg lecithin, soyabean lecithin	Primaquine	High-speed homogenization	96.5	–	95	Improved accumulation in liver and enhanced oral bio-availability	Singh et al., 2008

naturally derived phospholipids. The physicochemical properties of the liposomes, for instance charge density, steric hindrance, membrane fluidity and rate of permeation, have been reported to decide the molecular interactions of liposomes with blood and specific tissues following administration. Liposomes are prepared to enclose both lipophilic and hydrophilic materials within the watery portion of the vesicle or into the phospholipid bilayer itself. Liposomes were first invented by Bangham in 1965, and the film hydration technique, also named the Bangham method, still presents the easiest and earliest technique in liposome formulation. Figure 8.1b portrays the thin film hydration technique for formulating phospholipid liposomes. At present, approximately 12 liposome-based therapeutics are approved for ameliorative use, and the rest remain in different stages of clinical trials (Zylberberg and Matosevic, 2016). The *stratum corneum*, a finely organized lipid matrix consisting of corneocytes, is one of the crucial obstacles to deeper penetration of therapeutics into the dermis. Pertinent literature reports that liposomal phospholipids can intensify skin penetration or dermal deposition of therapeutics due to the blending of phospholipids in the liposomes with lipid residing in the bilayers of epidermal cells (Daraee et al., 2016). Tacrolimus was successfully loaded in a saturated soy lecithin-based liposomal system for topical application. The *in vitro* investigation displayed noticeably diminished permeation of tacrolimus in the liposomal formulation in contrast to free tacrolimus delivered within propylene glycol. An *in vivo* investigation in an animal model corresponding to allergic contact dermatitis (ACD) revealed that liposomal gel loaded with 0.03% tacrolimus displayed indistinguishable efficacy in comparison to commercially available 0.03% tacrolimus ointment (Patel et al., 2010). *Orthosiphon stamineus*, a medicinal herb, possesses several active compounds with notable pharmacological attributes. Nonetheless, the low solubility of these compounds restricts their therapeutic utilization. Soybean phospholipid–based nanoliposomes containing *O. stamineus* ethanolic extract have been proved to show improved solubility, permeability, oral bioavailability and antioxidative efficacy. Transmission electron microscopy (TEM) along with dynamic light scattering (DLS) suggested the existence of spherically shaped anionic nanoliposomes possessing a particle size of about 152.5 ± 1.1 nm and zeta potential of -49.8 ± 1.0 mV, approximately (Aisha et al., 2014). To compare the characteristics and stability profile of curcumin liposomes, milk fat globule membrane (MFGM) phospholipids and soybean lecithins were both utilized individually as encapsulants in a thin film ultrasonic dispersion method. The curcumin liposomes fabricated with MFGM phospholipids exhibited greater entrapment efficiency of around 74%, a smaller particle dimension of about 212.3 nm, a higher zeta potential value of about −48.60 mV, a higher retention rate as well as a sustained *in vitro* release profile compared with soy lecithin–stabilized liposomes. Thus, MFGM phospholipids have proved to be a potent food-grade encapsulant for the fabrication of curcumin liposomes (Jin et al., 2016). Despite advantageous antitumor potential, application of L-asparaginase is restricted due to its excessive tendency to cause clinical hypersensitivity. To handle this situation, L-asparaginase-entrapped liposomes were manufactured using soy lecithin by the thin film hydration method. The mean particle dimensions of the fabricated positive, negative and neutral liposomes were shown to be 35.6, 65.8 and 43.2 μm, respectively. The percentage of drug entrapped in neutral, positive and negative liposomes was found to be 1.95%, 2.39% and 2.35%, respectively. An *in vitro* release study of L-asparaginase was executed utilizing normal saline as dissolution medium, and the release was reported to be around 78.29%, 82.04% and 86.88% for positive, negative and neutral liposomes, respectively. Furthermore, a short-term cytotoxicity study showed that the cytotoxicity concentration (CTC50) for the liposomal formulation (50 μg) was significantly lower than that of pure drug (64 μg) (De and Venkatesh, 2012). The fluorescent methyl 6-methoxy-3-(4-methoxyphenyl)-1H-indole-2-carboxylate possesses noteworthy antitumor properties against several established cell lines, such as human lung cancer cell line (e.g. NCI-H460), human malignant melanoma cell line (e.g. A375-C5) and human breast cancer cell line (e.g. MCF-7). Various nanoliposome structures comprising the fluorescent therapeutics were manufactured with egg lecithin (egg PC), cholesterol (Ch), dipalmitoyl phosphatidylglycerol (DPPG) and distearoyl phosphatidylethanolamine (DSPE) by an injection/extrusion combined process. The formulations consisting of egg PC/Ch/DPPG (6.25:3:0.75) as well as egg PC/DPPG/DSPE-PEG (5:5:1) were found to be suitable for administration due to their small hydrodynamic diameter and highly negative zeta potential along with excellent encapsulation efficiency (Abreu et al., 2011). In another study, a slightly modified method of rapid expansion of supercritical solutions (RESS) was carried out to enclose essential oil into liposomes following the oil extraction from *Atractylodes macrocephala* Koidz (Wen et al., 2010). The mean particle dimension, drug loading and entrapment efficiency of liposomes were found to be 173 nm, 5.18% and 82.18%, respectively. A freshly introduced group of sterol-modified lipids (SML) was employed to design m-PEGDSPE-containing liposomes, and it was compared with the well-established liposome phospholipid constituents mPEG-DSPE/Hydrogenated Soy PC (HSPC)/cholesterol and mPEG-DSPE/POPC/cholesterol to find the impact of the amount of entrapped cis-platinum released on C26 tumor proliferation in the murine colon carcinoma model. The three liposome formulations demonstrated better antitumor efficacy against colon cancer in comparison with the naked drug following equivalent dose administration. However, the SML liposome platinum formulation did not prove to be more effective than the HSPC formulation (Kieler-Fergusonet al., 2017). Egg phosphatidylcholine (E-PC) and E-PC/egg phosphatidylglycerol (E-PC/EPG) were also employed to fabricate Carbopol-coated mucoadhesive liposomes for delivery of the antiviral drug acyclovir (ACV). The incorporation of ACV into liposomes was found to result in markedly enhanced *in vitro* permeability compared with its aqueous solution (Naderkhani et al., 2014). As a platelet substitute, phospholipid vesicles possessing hemostatic activity were constructed by joining a dodecapeptide H12 and exploited for encapsulating adenosine 5′-diphosphate (ADP). The results of the *in vivo* investigation clearly demonstrated that the H12-(ADP)-vesicles were capable of releasing ADP, augmenting platelet aggregation and exerting substantial hemostatic action in respect of treating prolonged bleeding (Okamura et al., 2010). Zhao et al. introduced a novel thermosensitive cationic liposomal delivery system with dipalmitoylphosphatidylcholine and cholesterol for the co-administration of both therapeutics and genes to the same cell. The initial demonstration of this co-administration strategy resulted in the intended site-specific delivery potency, the temperature-sensitive release of doxorubicin (DOX) as well as SATB1 gene silencing. Furthermore, the co-administration of DOX and SATB1 shRNA (short hairpin RNA) revealed improved inhibitory activity against carcinoma propagation in a gastric cell line both *in vitro* and *in vivo* compared with a single infusion (Peng et al., 2014). A recent investigation reported that an unconventional liposome was constructed from a dual drug-tailed phospholipid as a result of coupling a prodrug and a drug carrier. The dual chlorambucil-tailed phospholipid (DCTP), consisting of a hydrophobic tail from two molecules of chlorambucil and a hydrophilic head group from one glycerol-phosphatidylcholine molecule, was organized to construct liposomes by the thin lipid film technique without any additives. The DCTP liposome, a potent vehicle of chlorambucil, demonstrated outstanding loading capacity, remarkable stability and efficacious *in vivo* antitumor activity (Fang et al., 2015). A recent study has shown that a phospholipid-rich portion was successfully utilized to formulate rutin-liposomes as a potent neuroprotective agent. The formulation attenuated glutamate-induced cytotoxicity and reduced the formation of intracellular reactive species in SH-SY5Y, a human neuroblastoma cell line (Bernardo et al., 2019). Phospholipid liposomes functionalized with protein targeted to a specific receptor of the blood–brain barrier can be exploited for administration of neurotrophic drugs in the newborn brain (Glukhova et al., 2015). By using a supercritical CO_2–based system, liposomes loaded with multivitamins were successfully prepared from MFGM phospholipid concentrate. These liposomes were found to restore the nutritional and functional attributes of active biomolecules throughout a period of prolonged heat treatment (Jash et al., 2020). Despite all the advantages, liposomes face some obstacles as a versatile delivery device. The amount of drug that can be entrapped within a liposome is often very low. In addition, oxidation and hydrolysis hamper the stability and shelf life of liposomes. Moreover, a serious concern is related to the dermal delivery of liposomes encapsulating therapeutic compounds (Van Hoogevest and Fahr, 2019). Saturated phospholipids having a phase transition temperature of about 40–60°C exhibit rigidity at the cutaneous temperature of 32 °C So, liposomes comprising saturated phospholipids are reported to shown a lower propensity to penetrate the dermis. Unsaturated phospholipids with a phase transition temperature lower than 0 °C can stay in the mobile state at skin temperature. Soybean PCs comprising unsaturated fatty acids were found effective in forming flexible liposomes that can permeate to the deeper site of the dermis in comparison to rigid liposomes rich in saturated soybean PC. Even flexible liposomes comprising unsaturated phospholipids cannot permeate to a great extent through the skin and are reported to be degraded in the stratum corneum of the epidermis. Therefore, to achieve the desirable skin penetration for topical application, the composition and flexibility of the phospholipid liposomes can be refined with the assistance of lysolecithin. Not only is permeability an issue, but also, liposomes, like all other foreign substances, encounter multiple defense systems. Following the systemic administration of the delivery vehicle, the reticulo-endothelial system (RES) is found to be the primary site of liposome accumulation (Sercombe et al., 2015). Plasma proteins have been documented to have a vital role in vesicular destabilization as well as in clearance of liposomes by the RES via opsonization. Liposomes evading opsonization cannot escape the enhanced permeation and retention (EPR) effect. Fusion of PEG polymers and liposomal membrane is a prime idea for enhancing residence time in circulation and impeding RES clearance via steric stabilization. Cholesterol internalization within the liposomal membrane imparts the required stability to the liposome by diminishing the possibility of lipid interchange with other circulating bodies such as red blood cells and lipoproteins. A thorough understanding of the advancements in liposomal technology to date will pave the path for further upgrading of the therapeutic delivery avenues by minimizing the clinical and pre-clinical challenges.

8.4 MODIFIED LIPOSOMES: VERSATILE AND FLEXIBLE NANOVESICULAR VEHICLES FOR TRANSDERMAL DELIVERY

As conventional liposomes cannot thoroughly permeate the cutaneous layers and remain restricted to the outermost layers of the stratum corneum, they are of minor importance as vehicles for transdermal administration. Today, different generations of liposomes, such as niosomes (first generation), transferosomes (second generation) and ethosomes (third generation), have been developed with different mechanisms for enhancement of drug delivery to skin. Liposomes, transferosome and ethosomes are extensively utilized as gel formulations for topical application. The mechanism behind the dermal entry of liposomes, transferosomes and ethosomes is represented in Figure 8.2 for clear understanding. Table 8.2 shows the comparative analysis of liposomes, transferosomes and ethosomes.

Liposomes facilitate the topical penetration of therapeutic molecules by distorting the finely arranged cellular lipids of the stratum corneum. Due to the presence of an edge activator, transferosomes conquer the dermal obstacle by increasing the deformability of the lipid bilayer and augmenting extracellular transport to fit themselves into it. Ethanol fluidizes both the lipid bilayers residing in ethosomal vesicles and the stratum corneum, modifying the organization and diminishing the density of skin surface lipids, i.e. lipid perturbation.

8.4.1 TRANSFEROSOMES

Transferosomes are the second generation of liposomes, ultra-deformable or ultra-flexible in nature. They can readily penetrate the stratum corneum of the epidermis in the presence of a trans-epidermal water activity gradient (Chauhan et al., 2017). These are formed by an ideal mixing of phospholipids and edge activator. An edge activator is generally a single-chain surfactant responsible for the destabilization of the lipid bilayer of the vesicle and enhancement of the vesicle elasticity or fluidity. When transferosomes come into close proximity with the skin, evaporation of the aqueous part from the formulation begins. Transferosome vesicles possess a general inclination to avoid a dry environment. Hence, the formulations are tempted by the considerable water content of the skin, leading to the spontaneous transport of the encapsulated vesicles through the cutaneous

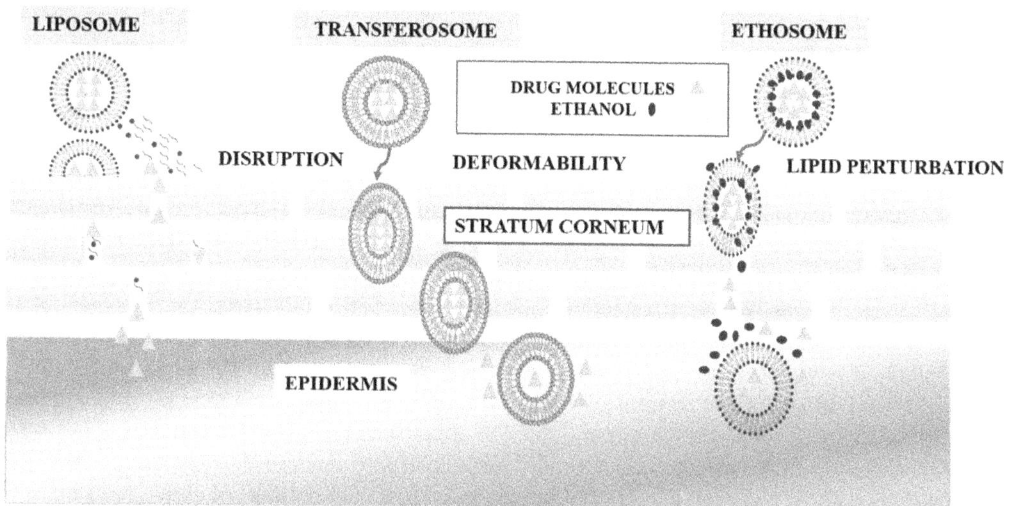

FIGURE 8.2 Possible mechanism behind the entry of bioactive agents across the skin via liposomes, transferosomes and ethosomes.

TABLE 8.2
Comparative Analysis of Liposomes, Transferosomes and Ethosomes

Characteristics	Liposome	Transferosome	Ethosome
Nature of vesicle	Bilayer lipid vesicle	Second-generation elastic lipid vesicle	Third-generation elastic lipid vesicle
Composition	Phospholipids and cholesterol	Phospholipids and edge activator	Phospholipids and ethanol
Flexibility	Microscopic vesicle, rigid in nature	Ultra-flexible liposome; high deformability	Elastic liposome; high deformability and elasticity
Extent of dermal penetration	Penetration rate is very low	Can easily penetrate through paracellular space	Can promptly invade via paracellular transport
Permeation mechanism	Diffusion, fusion, lipolysis	Deformation of vesicle	Lipid perturbation
Route of administration	Oral, parenteral, transdermal, topical	Transdermal and topical	Transdermal and topical

layer (Solanki et al., 2016). Insulin, a large peptide, is not able to penetrate easily into the dermis. Insulin entrapped within a transferosomal gel showed acceptable permeation results (*in vitro* permeation flux of 13.50 ± 0.22 μg/cm^2/h) through porcine ear skin (Malakar et al., 2012). The *in vivo* investigation of an insulin transferosome formulation reported a long-term hypoglycemic effect in alloxan-treated rats for more than 24 h following transdermal administration. For the treatment of hypertension, diltiazem HCl entrapped within transferosomes was further enclosed in a gel matrix. The diltiazem HCl–entrapping transferosomes permeated in and across rat skin successfully and showed a three- to four-times amplified sustained outcome in comparison with the naked drug (Jain et al., 2003). The transferosomal gel formulated using soy PC can be used as a potent topical delivery tool for piroxicam due to the perfect release and permeation of the drug. The transferosomal gel formulation revealed improved anti-inflammatory potential against carrageenan-induced paw edema (Shaji and Lal, 2014). The transferosome formulation encapsulating Sildenafil was created with vesicle size of about 610 nm along with remarkable entrapment efficiency of around 97.21%.

In vitro permeation of the developed transferosomes showed at least a fivefold superior permeation rate compared with the common drug suspension (Ahmed, 2015). Lornoxicam, a potent non-steroidal anti-inflammatory drug, was loaded in transferosomes for transdermal delivery (Ranade et al., 2016). The average vesicle diameter and the average drug encapsulation efficiency of the fabricated transferosomes were about 678 nm and 65.3%, respectively. The *in vitro* flux obtained for the optimized formulation was 79.1 μg/cm²/h, while that for a formulation without edge activator was found to be 70.2 μg/cm²/h. In a similar study, transferosomes entrapping pentoxifylline showed an average vesicle diameter of about 0.69 ± 0.049 μm, polydispersity index of about 0.11 ± 0.037, zeta potential of about -34.9 ± 2.2, vesicles elasticity of about 145 ± 0.6 (mg/s/cm²), drug encapsulation efficiency of about $74.9 \pm 1.6\%$ as well as permeation flux of about 56.28 ± 0.19 μg/cm²/h (Shuwaili et al., 2016). *In vivo* pharmacokinetics examination revealed that the developed transferosome formulation accelerated pentoxifylline absorption as well as lengthening its half-life compared with the commercial oral pills. Optimized transferosome formulations loading felodipine showed a particle size of about 75.71 ± 5.4 nm, polydispersity index of about 0.228 and zeta potential of about −49.8 mV. Besides, the formulation provided maximal loading and entrapment efficiency along with a high deformability index of about 119.68. In comparison to control transdermal solution, 256% inflation in permeation across rat skin (flux $= 23.72 \pm 0.64$) with the fabricated formulation was reported in an *in vitro* permeation study. The relative bioavailability of felodipine (358.42%) in a transferosome formulation compared to oral administration indicates that transdermal administration of transferosomes yielded a relatively persistent blood concentration with minimum plasma fluctuation and prompt and sustained peak time, in contrast to oral delivery (Yusuf et al., 2014).Transferosomes encapsulating asenapine maleate were optimized with an average dimension of about 126.0 nm, polydispersity index of about 0.232, zeta potential of about−43.7 mV and entrapment efficiency around 54.96%. *In vivo* pharmacokinetic investigation demonstrated a statistical inflation in bioavailability upon transdermal administration in comparison with oral delivery (Shreya et al., 2016). Exploring the deformability features of transferosomes, transferosomes entrapping Timolol maleate were found to refine the corneal transmittance and bioavailability (González-Rodríguez et al., 2016). Capsaicin-loaded transferosomes demonstrated a superior therapeutic effect against arthritis-associated inflammation compared with the marketed thermogel formulation (Sarwa et al., 2015). In another study, Clonazepam, a benzodiazepine derivative, was encapsulated in transferosomes for intra-nasal administration. The optimized formulation was responsible for minor changes in the nasal mucous layer, revealed in *ex vivo* cytotoxicity investigation, and a notable lag in the onset of epileptic seizures upon pentylenetetrazol treatment (Nour et al., 2017). Highly flexible lecithin soybean PL nanotransferosomes were recently engineered as the latest delivery device for transdermal transport of large peptide molecules, i.e., human growth hormone (Shamshiri et al., 2019). Besides these versatile trans-dermal applications, major drawbacks of transferosomes relate to the hassle of entrapping lipophilic substances into the vesicular nanostructures while keeping their flexibility and deformability intact (Priyanka and Singh, 2014). Transferosomes are lacking chemical stability due to their predisposition to oxidative stress. However, the recent successes of transferosomes show promise to handle upcoming obstacles and welcome the imminent possibilities for the establishment of novel delivery techniques with refined stability and penetration. The fabrication techniques, physicochemical characteristics and therapeutic effects of phospholipid-based transferosomes entrapping various active compounds are briefly outlined in Table 8.3.

8.4.2 Ethosomes

Ethosomes represent the third generation of elastic lipid carriers, which are actually ethanol-modified liposomes comprising hydroalcoholic or hydro/alcoholic/glycolic phospholipid (Garg et al., 2017). The high concentration of ethanol (20–50%) in ethosomal vesicles may disrupt the skin lipid's bilayer structure. Thus, these malleable vesicles can easily penetrate through the paracellular space and offer continual delivery of therapeutics to targeted sites residing in different layers of

TABLE 8.3
Method of Fabrication, Physicochemical Characteristics and Therapeutic Effects of Phospholipid-Based Transferosome Entrapping Various Active Compounds

Active compound	Preparation method	Particle size (nm)	Zeta potential (mv)	Entrapment efficiency (%)	Permeation flux (μg/cm²/h)	Therapeutic impact	References
Pentoxifylline	Modified vortexing-sonication	690	−34.9	74.9	56.28	Enhanced bioavailability	Shuwaili et al., 2016
Lornoxicam	Thin film hydration	678	−52.5	65.3	79.1	Increased skin permeability	Ranade et al., 2016
Felodipine	Rotary evaporation-sonication	75.71	−49.8	85.14	23.72	Escalated transdermal permeation	Yusuf et al., 2014
Insulin	Reverse phase evaporation	625–815	−14.30	56.55–60.23	13.50 ± 0.22	Prolonged hypoglycemic effect, increased patient compliance	Malakar et al., 2012
Asenapine maleate	Thin film hydration	126	−43.7	53.96	7.98	Elevated skin permeation	Shreya et al., 2016
Capsaicin	Thin film hydration	94	−14.5	60.34	6.28	Exhibited superior antiarthritic activity	Sarwa et al., 2014
Clonazepam	Thin film hydration	122.5	−22.95	84.22	–	Enhanced brain delivery to treat status epilepticus	Nour et al., 2017
Sildenafil	Lipid film hydration	610	1.52–3.63	97.21	–	Increased skin permeation	Ahmed, 2015

skin. A soy lecithin–based ethosomal gel formulation of Clotrimazole was developed to attain zero-order release kinetics of Clotrimazole and drug entrapment efficiency of more than 50% (Parmar et al., 2016). A soy PC formulation of Diclofenac-loaded ethosomes was optimized with a vesicular size of about 144 ± 5 nm, zeta potential of about −23.0 ± 3.76 mV, an elasticity of about 2.48 ± 0.75 and entrapment efficiency of about 71 ± 4%. The permeation flux of this formulation was found to be significantly superior to that of the drug-entrapped conventional liposomal structure, aqueous or ethanolic solution. The *in vivo* investigation revealed that the ethosomal hydrogel showed markedly superior anti-inflammatory potential in comparison to liposomal hydrogel (Jain et al., 2016). Ethosomal vesicles prepared from lecithin were also explored for enhanced topical delivery of isotretinoin (David et al., 2013). The entrapment of Etodolac in soy lecithin–based ethosomal vesicles showed efficacious delivery of active molecules at the desired area, reduced the dose and therefore, played a vital role in patient compliance (Chintala and Padmapreetha, 2014). Soy lecithin was also used for an Econazole nitrate–loaded ethosome formulation to attain prolonged release kinetics, better permeation capacity and improved antifungal activity compared with the pure drug (Verma and Pathak, 2012). Phospholipid-based piroxicam ethosomes provided several benefits in terms of quick onset and maximal release of the therapeutic molecule while minimizing undesirable complications. An ethosomal gel entrapping Gliclazide as a topical formulation generated no irritation, provided higher stability and offered a better hypoglycemic effect in comparison to an oral drug formulation (Vijayakumara et al., 2014). Osthole-loaded ethosomes formulated as an effective delivery system with a greater encapsulation efficiency of 83.3 ± 4.8% and transdermal flux of about 6.98 ± 1.6 μg/cm²/h intensified drug deposition on the cutaneous layer compared with transferosomes (1.5-fold) (Meng et al., 2016). An ethosomal formulation encapsulating Lornoxicam exhibited a small particle size (100 ± 3.9 nm), the highest percentage of drug entrapment (93.96%) and the highest percentage of drug permeation (74.18%) at the end of 24 hours. The results based on stability and anti-inflammatory activity suggested ethosomal gel as an efficient carrier of Lornoxicam (Acharya et al., 2016). Epigallocatechin gallate (EGCG), a potent antioxidant obtained from green tea, was entrapped in ethosomes with a particle size of 129.0 nm, polydispersity index of about 0.05 ± 0.00, zeta potential around −62.6 ± 5.05 mV and maximum entrapment efficiency of about 54.39 ± 0.03 which finally aided to increase EGCG penetration (Ramadon et al., 2017). Flurbiprofen-loaded ethosomes with almost 95% entrapment efficiency were found to show improved analgesic activity as well to inhibit paw edema (Paliwal et al., 2019). *In vitro* skin permeation methodology showed a release of about 82.56 ± 2.11 g/cm² in 24 hours and transdermal flux of about 226.1 μg/cm²/h. Composite phospholipid ethosomes entrapping curcumin were recently optimized with several phospholipid compositions (Li et al., 2021). The outcome of the *in vitro* skin permeation studies with the composite phospholipid ethosomal formulations indicated significantly superior penetration of curcumin across the cutaneous layer than that of the conventional ethosomes fabricated with PC or hydrogenated PC. In addition, composite phospholipid ethosomes could disrupt the lipid organization of the stratum corneum and permit curcumin to permeate effortlessly into the dermis. Interestingly, all the ethosomal formulations displayed a zero-order release profile for *in vitro* studies. Furthermore, ethosomes were found to be more stable against aggregation than liposomes due to their net negative charge. Comprehensive research on the depth of skin penetration is needed to encourage the commercial pharmaceutical domain further (Patrekar et al., 2015). Fabrication techniques, physicochemical characteristics and therapeutic effects of phospholipid-based ethosomes entrapping various active compounds are briefly outlined in Table 8.4.

8.5 PHOSPHOLIPID-BASED SOLID LIPID NANOPARTICLES: A NOVEL FORMULATION FOR VARIOUS ROUTES OF ADMINISTRATION

Solid lipid nanoparticles (SLNs) (colloidal size ranging between 50 and 1,000 nm) are colloidal carriers comprised of solid lipids that are stabilized by surface-active agents within an aqueous milieu (Mishra et al., 2018). The lipid matrix of SLNs is comprised of biological lipids, which may reduce

TABLE 8.4
Method of Fabrication, Physicochemical Characteristics and Therapeutic Effects of Phospholipid-Based Ethosomes Entrapping Various Active Compounds

Active compound	Preparation method	Particle size (nm)	Zeta potential (mv)	Entrapment efficiency (%)	Permeation flux (μg/cm²/h)	Therapeutic impact	References
Diclofenac	Rotary evaporation-sonication	144	−23.0	71	12.9	Enhanced anti-inflammatory activity	Jain et al., 2015
Methoxsalen	Thin film hydration	281.3	−2.13	67.12	5.8	Enhanced topical delivery to treat vitiligo	Garg et al., 2016a
Epigallocatechin gallate	Mechanical dispersion	129.0	−62.6	54.39	56.97	Enhanced penetration	Ramadon et al., 2017
Osthole	Modified injection	153.36	−9.51	83.3	6.98	Enhanced transdermal delivery	Meng et al., 2016
Econazole nitrate	Cold method	202.85	−75.1	81.05	0.46	Improved anti-fungal activity	Verma and Pathak 2012
Cromolyn sodium	Thin film hydration	133.8	−69.82	49.88	18.49	Enhanced skin accumulation	Rakesh and Anoop, 2012
Cetirizine dihydrochloride	Rotary evaporation-sonication	180.1	−15.7	72.85	16.3	Increased anti-allergic effect	Goindi et al., 2014
Glimepiride	Rotary evaporation-sonication	61	0.6–9.63	97.12	5.583	Sustained release of drug	Ahmed et al., 2016

the threat of chronic and acute toxic effects. SLNs can shield the core molecule from undesirable enzymatic degradation in the digestive tract and thus increase the stability of the incorporated bioactive (Ekambaram et al., 2012). Moreover, epithelial cells and the lymphoid tissues in Peyer's patches can engulf the SLNs to a certain extent (Rostami et al., 2014). Earlier, Heiati et al. prepared neutral and negatively charged SLNs for azidothymidine palmitate (AZT-P) delivery using trilaurin (TL) as the solid core material as well as a blend of dipalmitoylphosphatidylcholine (DPPC) and dimyristoylphosphatidylglycerol (DMPG) as a coating agent. They found that the entrapment and subsequent retention of AZT-P in SLNs depend primarily upon the phospholipid composition and the net surface charge of the fabricated structure. The organization of phospholipid bilayer structure and the capacity of amphiphilic drugs, for example AZT-P, to be incorporated within these two layers seems to be the factor behind the improved entrapment efficiency (EE%) (Heiati et al., 1998). Egg PC and PEGylated phospholipid were successfully utilized as stabilizers to prepare SLNs encompassing trimyristin (TM) as the solid lipid core. Irrespective of paclitaxel entrapment, the satisfactory particle sizes (around 200 nm) and zeta potentials (–38 mV) marked the SLNs as suitable for parenteral formulation. An *in vitro* release study with SLNs demonstrated prolonged release kinetics of paclitaxel from the formulation. Moreover, paclitaxel-loaded SLNs used to treat MCF-7, a well-known breast cancer cell line, and OVCAR-3, a commonly used ovarian cancer cell line, exhibited significantly higher cytotoxicity compared with those cell lines treated with the widely accepted CremophorEL-based paclitaxel formulation (Lee et al., 2007). SLNs were successfully modified with lectin to increase the oral bioavailability of insulin. These delivery systems yielded remarkable drug entrapment efficiency (>60%) and noteworthy protection from gastro-intestinal degradation (Zhang et al., 2006). In another study, utilizing soybean phosphatidylcholine (SPC) as an encapsulant, biodegradable nanoparticles entrapping an insulin phospholipid complex were fabricated by a novel reverse micelle–double emulsion technique (Cui et al., 2006). The particle size, zeta potential, drug loading capacity and entrapment efficiency were reported to be about 114.7 ± 4.68 nm, –51.36 ± 2.04 mV, 18.92 ± 0.07% and 97.78 ± 0.37%, respectively. In streptozotocin-induced diabetic rats, the administration of 20\IU/kg nanoparticles lowered fasting plasma glucose concentration by 42.6% in less than 8 h of administration with a long-term effect for 12 h. Fluorescent-labeled insulin was successfully introduced into nebulizer-compatible SLNs aiming at efficacious pulmonary drug delivery (Liu et al., 2008). The SLNs were found to be uniformly dispersed in the lung alveoli. These results unearthed the potential of SLNs as a pulmonary delivery tool for insulin by overcoming *in vitro* and *in vivo* stability concerns, enhancing bioavailability as well as augmenting hypoglycemic effect. Tween 80 and soybean lecithin were simultaneously utilized as emulsifiers to construct isotretinoin-loaded SLNs (IT-SLNs) for topical delivery (Liu et al., 2007). The *in vitro* permeation data showed a significantly enhanced skin targeting effect with the increased accumulative uptake of isotretinoin. Diacerein-loaded SLNs (DC-SLN)s were fabricated by modified high-shear homogenization as well as ultrasonication using soy lecithin as a surfactant (Jain et al., 2013). An *in vitro* drug release study exhibited controlled release kinetics, and an *in vivo* pharmacokinetic study indicated 2.7-fold improvement of oral bioavailability with a reduction in the side effects of the drug. SLNs encapsulating cisplatin were generated by a micro-emulsification technique with the help of soy lecithin and stearic acid (Doijad et al., 2008). The entrapment efficiency ranged from 47.59% to 74.53%. An *in vivo* study of cisplatin-loaded SLNs reported that cisplatin is favorably targeted to the liver, brain and lungs consecutively. Egg PC–based SLNs entrapping penciclovir with an average diameter of about 254.9 nm have been successfully formulated by a double emulsion method (W/O/W) (Lv et al., 2009). SLNs enhanced penciclovir permeation into the deeper site of the skin due to the SLNs' impact on diminishing or inhibiting the skin dehydration and hence, disturbing the lamellar arrangement of the lipids and loosening the structure of the stratum corneum. Similarly, soy lecithin–stabilized SLNs incorporating adapalene were established to be able to target epidermis (Jain et al., 2014). A rat skin model confirmed that the fabricated system can evade the systemic uptake of adapalene in cutaneous layers and can accumulate the drug in epidermis. Catalase, a potent hydroxyl radical scavenger, is effectively loaded in PC-stabilized SLNs with the sequential application of the double emulsion method and solvent evaporation techniques (Qi et al., 2011). Catalase loaded into SLNs is found to be protected from proteolysis. The capacity of catalase to degrade hydrogen peroxide diffusing through the polymer shell was fully retained even after encapsulation. SLNs were successfully formulated with egg PC and Tween 80 for delivering all-trans-retinoic acid (ATRA) (Lim and Kim, 2002). The determined mean particle dimension and the zeta potential of ATRA-loaded SLNs were about 154.9 nm and around –38.18 mV, respectively. Phospholipid complexes technology was successfully utilized to fabricate diclofenac sodium (DS)-loaded SLNs, aiming to increase the lipophilicity of the water-soluble drug DS (Liu et al., 2011). The assistance of phospholipid complexes changed the zeta potential and restricted the rapid core release from the SLNs, which suggested a multi-layer formation of phospholipid enclosing the solid lipid core of the SLNs. SLNs are also suitable for formulating low-molecular weight compounds other than peptide drugs with high molecular weight and poor aqueous solubility, for example pentoxifylline (Varshosaz et al., 2010). The bioavailability in rats was found to be notably improved in comparison to that of pentoxifylline solution. Praziquantel, an anthelmintic drug potent against flatworms, was formulated with an average diameter of about 110 nm, a zeta potential of around −66.3 mV and an encapsulation efficiency of approximately 80% (Yang et al., 2009). The average residence period of the drug was also markedly augmented after delivery of SLNs, leading to a twofold amplification in comparison to the administration of tablets. SLNs entrapping the cholesterol-lowering drug Lovastatin boosted the bioavailability by up to 173% compared with a Lovastatin suspension in animals (Suresh et al., 2007). Total flavonoid extract from *Dracocephalum moldavica* L. (TFDM) has been successfully encapsulated in soy lecithin–stabilized SLNs, which are round in shape with mean particle dimension of about 104.83 nm (Tan et al., 2017). The findings of pharmacodynamics revealed a considerably higher protective effect from TFDM-SLNs than from TFDM alone, possibly due to the infarct area, histopathological study, cardiac enzyme levels and inflammatory marker analysis in serum. For versatile topical applications, soy PC has been utilized to encapsulate trans-resveratrol into SLNs (Rigon et al., 2016). The resveratrol-loaded SLNs with a mean diameter less than 200 nm were found to be more useful than kojic acid at inhibiting tyrosinase without any significant cellular toxicity towards HaCat keratinocytes. SLNs co-loaded with ferulic acid (FA) and tocopherol (Toc) were fabricated using soy lecithin with the aid of a hot homogenization technique (Oehlke et al., 2017). The various formulations entrapping up to 2.8 mg/g of FA or Toc exhibited unique stability for a minimum of 15 weeks of storage at ambient temperature with retention and/or enhancement of their antioxidative activity. PC-stabilized PEGylated Docetaxel-loaded SLNs (DCX-SLNs) exhibited notably superior cellular toxicity towards several human and murine carcinoma cell lines as well as a stronger antitumor activity in a mouse model, because the DCX-SLNs markedly augmented the deposition of the DCX at the tumor site (Naguib et al., 2014). Furthermore, the reduced deposition of DCX in essential organs following intravenous administration of DCX-SLNs compared with that of DCX in Tween 80/ethanol mixed solution indicates the utility of DCX-SLNs as a favorable safe formulation. In another study, hydrogenated soy PC acted as a stabilizer in a curcumin-loaded SLN system (Mulik et al., 2010). Flow cytometric data analysis affirmed superior anticancer activity of curcumin entrapped in transferrin-mediated curcumin-loaded SLNs towards the MCF-7 breast cancer cell line in comparison to curcumin solubilized surfactant solution (CSSS). A novel SLN of 4-(*N*)-docosahexaenoyl 2′, 2′-difluorodeoxycytidine (DHA-dFdC) was recently reported with increased aqueous solubility, chemical stability and *in vivo* efficacy. The SLNs were fabricated from lecithin/glycerol monostearate-in-water emulsions using (TPGS) and Tween 20 as emulsifiers (Valdes et al., 2019). The SLNs can be delivered by several routes, for example topical, oral, parenteral, ocular, pulmonary, etc. The greatest benefits of lipid vehicles are their excellent stability, biocompatibility, biodegradability, maneuverability and scalability as well as controlled and refined release kinetics. However, the pitfalls of SLN production are their inadequate drug loading capacity, drug exclusion after polymeric transition throughout storage, comparatively higher aqueous content of the formulation, initial "burst effect", agglomeration and RES clearance (Geszke-Moritz and Moritz, 2016). The emergence of several promising approaches regarding SLNs would promote the development of nanostructured lipid carriers as well as the second generation of lipid-based nanocarriers. Fabrication techniques, physicochemical characteristics and therapeutic effects of phospholipid-based SLNs entrapping various active compounds are briefly outlined in Table 8.5.

8.6 MULTIFUNCTIONAL MICELLES FOR TARGETED DRUG DELIVERY

Polymeric micelles having a particle size within the 5–50 nm range represent the simplest self-assembly structures in spherical shapes spontaneously formed by amphiphiles in an aqueous solution. Phospholipids become more water soluble following conjugation with PEG and form micelles above the critical micellar concentration (CMC) when dispersed in water (Lim et al., 2012). The hydrophobic domains of the molecule are arranged in a cluster and directed away from the aqueous medium, while the hydrophilic parts are directed toward the water. The hydrophilic PEG coating on micelles imparts stability against aggregation, reduces susceptibility to reticulo-endothelial opsonization, enhances residence time in circulation and ensures pH-responsive drug release at the active site (Kalepu and Nekkanti, 2015). PEGylated lipids such as DSPE-PEG (Distearoyl-sn-glycero-3-phosp hoethanolamine conjugated to PEG) are also utilized to fabricate sterically stabilized micelles (SSMs) (Gülçür et al., 2013). Conjugation of PEG to the phospholipid builds a dense brush of highly hydrated chains, forming a conical structure for each monomer, which consequently promotes self-assembly of a micellar structure that is not absolutely spherical in shape. Though the peptides display enormous potential benefits and therapeutic capabilities for several clinical disorders, they suffer from numerous physicochemical and biological concerns, for instance aggregation, chemical degradation *in vitro*, and immediate proteolysis as well as renal clearance *in vivo*. Phospholipid micelles have been used to address these delivery issues by improving their stability against rapid proteolysis and increasing the circulation time. Human pancreatic peptide loaded into SSMs was found to be stable against protease activity, and its bioactivity was retained *in vitro* (Banerjee and Önyüksel, 2012). A novel, self-associated, stable PEGylated phospholipid nanomicelle of vasoactive intestinal peptide (VIP) having a mean size of ~18 nm was developed to amplify vasodilation in the *in situ* peripheral microcirculation model (Önyüksel et al., 1999). Sterically stabilized micelles are found to be attractive nanocarriers for encapsulating camptothecin (CPT), a topoisomerase I inhibitor. The average dimension of CPT-SSM was about ~14 nm with a narrow size distribution. At a concentration of 15 mmol/L PEGylated phospholipids, camptothecin solubilization in SSM was found to be approximately 25-fold superior to a buffer solution of camptothecin (Koo et al., 2005). These micelles are sufficiently small for ample extravasation through the leaky microvessels of tumors, resulting in substantial drug deposition in tumors as well as minimum drug toxicity to the healthy tissues (Koo et al., 2006). To achieve active targeting, the surface of CPT-SSM was altered with VIP. For the first time, CPT was reported to be effective against collagen-induced arthritis (CIA) at a considerably lower concentration compared with the standard anticancer dose. Moreover, only one subcutaneous administration of CPT-SSM-VIP (0.1 mg/kg) to CIA mice alleviated rheumatoid arthritis–associated joint swelling and inflammation for a minimum of 32 days without general systemic toxic effects, whereas CPT alone required a minimum 10 times higher dose to attain the same result (Koo et al., 2011). GLP1-SSMs constructed from human glucagon-like peptide 1 (GLP-1) and PEGylated phospholipid micelles were found to exhibit a hydrodynamic size range of ~15 nm and to exert protection against acute pulmonary inflammation in lipopolysaccharide-treated mice (Lim et al., 2011). Spontaneous self-association of human neuropeptide Y (NPY) with biocompatible and biodegradable SSMs was found to stabilize and protect the peptide in active monomeric form and thus, to promote inhibition of cAMP production in SK-N-MC neuroepithelioma cells *in vitro* (Kuzmis et al., 2011). 17-Allylamino-17-demethoxy geldanamycin (17-AAG), the heat shock protein 90 (Hsp90) inhibitor, is already established as an antitumor drug for breast adenocarcinoma treatment, though insufficient water solubility and hepatotoxicity limit its use worldwide. For active targeting, VIP surface-grafted SSMs entrapping 17-AAG were developed with particle size range about 16 ± 1 nm

TABLE 8.5
Method of Fabrication, Physicochemical Characteristics and Therapeutic Effects of Phospholipid-Based Solid Lipid Nanoparticles Entrapping Various Active Compounds

Active compound	Preparation method	Particle size (nm)	Zeta potential (mV)	Entrapment efficiency (%)	Route of administration	Therapeutic impact	References
Paclitaxel	Hot homogenization	200	−38	80	Parenteral	Boosted cytotoxicity to MCF-7 breast carcinoma cell line	Lee et al., 2007
Praziquantel	Ultrasound technique	110	−66.3	80	Oral	Increased bioavailability	Yang et al., 2009
Lovastatin	Hot homogenization	60–119	−16 to −21	99%	Intraduodenal	Increased bioavailability	Suresh et al., 2006
Diacerein	High-shear homogenization and ultrasonication	370–510	−14 to −20	88.1	Oral	Improved bioavailability minimizing side effects	Jain et al., 2013
Adapalene	Solvent injection	148.3	−12	89.9	Topical	Improved therapeutic efficacy with minimal penetration	Jain et al., 2014
Methotrexate	Emulsification-solvent diffusion	174.51	10.21	84.3	Intravenous	Increased antitumor efficacy against breast carcinoma cell line	Garg et al., 2016b
Resveratrol	Solvent diffusion-solvent evaporation	134	−34.3	88.9	Oral	Enhanced bioavailability with sustained release	Pandita et al., 2014
Candesartan cilexetil	Hot homogenization followed by ultrasonication	180–220	−28 to −29	91–96	Oral	Increased anti-hypertensive effect	Dudhipala and Veerabrahma, 2016

and drug content approximately 97 ± 2%. The conservation of cellular toxicity of these fabricated nanomicelles to MCF-7 breast cancer cells suggests overexpression of high-affinity receptors for VIP in these cells (Önyüksel et al., 2009b). Using CdSe/ZnS quantum dots (QD), Rubinstein et al. suggested that a substantial amount of QD-loaded VIP-sterically stabilized mixed micelles (SSMMs) are deposited at high speed in MCF-7 breast carcinoma cells compared with QD-loaded SSMMs without grafting ($p < 0.05$) (Rubinstein et al., 2008). For oral insulin delivery, Wang et al. designed phospholipid-based reverse micelles that are sufficiently stable at 4 °C for about 6 months (Wang et al., 2010). Mixed micelles fabricated from phospholipid and Tween 80 were explored as an injectable drug-loaded nanovehicle for the naturally derived chemotherapeutic agent plumbagin. The micelles possessed a particle size range of about 46 nm, zeta potential of about 5.04 mV, entrapment efficiency of approximately 98.38% and 2.1-fold increase in antitumor efficacy against the MCF-7 cell line *in vitro* (Bothiraja et al., 2013). Self-association of Polymyxin B with SSMs is found to alleviate the antibacterial activity *in vitro* owing to the presence of PEG layer surrounding the micelle. Hindrance of electrostatic interactions between positively charged polymyxin B and the negatively charged lipopolysaccharide on microbial cell wall is thought to be the reason behind this observation (Brandenburg et al., 2012). Aerosolized, rehydrated sterically stabilized phospholipid nanomicelles encapsulating Beclomethasone Dipropionate (BDP) suitable for pulmonary administration were formulated with adequate reproducibility as well as uniform lung deposition. The droplet dimension and zeta potential of this formulation were found to be around 19.89 ± 0.67 nm and about 28.03 ± 2.05 mV, respectively. The aerodynamic values of the BDP-SSMs were found to be within the acceptable range, and the fabricated nanomicelles exhibited prolonged release of BDP (Sahib et al., 2012). Curcumin-entrapped phospholipid-sodium deoxycholate-mixed micelles (CUR-PC-SDC-MMs) were prepared to improve the solubility and anti-proliferative and apoptotic effects of curcumin. Notably high encapsulation efficiency, lower IC_{50} values and sustained release kinetics proved curcumin-loaded micelles to be an effective nanoformulation (Duan et al., 2015). Sirolimus-loaded micelles in the particle size range of 14 nm were found to reduce vascular stenosis by 42% and to provide better antirestenosis effects than PEGylated liposomes (Haeri et al., 2013). Inhaled corticosteroids are anticipated to be suitable for local intervention in chronic obstructive pulmonary disease or asthma. Still, the administration of poorly aqueous soluble drugs via a nebulizer remains insufficient, and sufferers have to depend on colossal doses to achieve optimal relief. Administration of PEG5000-DSPE polymeric micelles entrapping budesonide (BUD-SSMs) was found to considerably reduce inflammatory cell numbers in broncho-alveolar lavage fluid at an optimum dose compared with Pulmicort Respules® (Sahib et al., 2011). For oral 20(S)-protopanaxadiol (PPD) delivery, Xia et al developed novel mixed micelles comprising PPD–phospholipid complexes and Labrasol® and assessed their efficacy in antitumor therapies. These formulations were found to markedly increase solubility, enhance absorption and improve bioavailability (Xia et al., 2013). Baohuoside I is a powerful anticancer drug, but its worldwide use is restricted due to its inadequate aqueous solubility and removal mechanism from the body. To enhance the solubility, increase permeability and restrict efflux of baohuoside I, mixed micelles were successfully fabricated from phospholipid complex and D-α-tocopherol polyethylene glycol 1000 succinate (TPGS) (Jin et al., 2013). One of the problems with SSMs is limited solubilization potential, which is controlled by the total number of micelles in the system. The solubilization potential of SSMs was improved by increasing the hydrophobic core of each SSM, and thus, a novel delivery system, i.e., SSMMs, was formed through the incorporation of a water-insoluble phospholipid such as PC. SSMMs containing paclitaxel demonstrated superior solubilization potential of the drug and greater cellular toxicity towards MCF-7, a breast carcinoma cell line, in comparison with SSMs (Krishnadas et al., 2003). By contrast, SSMs can solubilize indisulam to a substantially greater extent (at least 40 times) compared with SSMMs (Cesur et al., 2009). PEG 2000-grafted 1,2-distearoyl phosphoethanolamine (DSPE-PEG2000) alone was utilized to develop SSMs, and egg PC in addition to DSPE-PEG2000 was used to fabricate SSMMs. The considerable variation in the dissolution potential of SSMs and SSMMs can be explained by assuming that indisulam settles at the interface between the lipophilic interior and the comparatively hydrophilic exterior. In SSMMs, the majority of this area may be occupied by egg PC molecules, permitting only a few indisulam molecules to colonize, resulting in reduced drug solubilization in SSMMs. Using the MCF-7 breast cancer cell line, an *in vitro* study indicated that indisulam-SSM was more efficacious than indisulam in dimethyl sulfoxide (DMSO). The multi-drug resistance property of breast carcinoma remains a hindrance to successful chemotherapy even now. Nanosized polymeric micelles promise better accumulation in tumor via EPR. In drug-sensitive MCF-7cells, SSMMs (P-SSMMs) as well as VIP-grafted SSMMs (P-SSMM-VIP) encapsulating paclitaxel was found to markedly inhibit cell proliferation in a dose-dependent manner ($p<0.05$). Both nanomicelles were found to be seven times more efficacious than paclitaxel in DMSO solution(P-DMSO) (Önyüksel et al., 2009a). On the other hand, in drug-resistant BC19/3 cells, P-SSMM-VIP was found to be markedly more efficacious than both P-SSMM and P-DMSO (two and five times, respectively; $p<0.05$). These observations have proved paclitaxel-loaded SSMM-VIP as a suitable nanocarrier for active targeting and treating multiple drug-resistant breast cancer. Figure 8.3 shows the possible mechanism of SSMs inhibiting P-glycoprotein efflux and restricting multi-drug resistance in neoplastic cells. In another study, for loading paclitaxel, novel polymeric micelles were engineered by utilizing PEs and hyaluronic acid (HA). *In vivo* real-time imaging indicated accumulation of the micellar system mostly in heart, liver and spleen (Saadat et al., 2014). Following administration, functional paclitaxel nanomicelles with particle size in the range of 15 nm considerably increased the intracellular uptake of paclitaxel, accumulated selectively in mitochondria and endoplasmic reticulum, and showed intense inhibitory activity against MCF-7 and MCF-7/Adr cells. They also intensively invaded the interior of the MCF-7 and MCF-7/Adr spheroids and markedly reduced the size of the spheroids. Freeze-dried nanomicelles (QUE-FD-NMs) entrapping quercetin were recently developed with a particle size ranging from 20 to 80 nm, a time-dependent slow-release profile, enhanced intracellular assimilation inside human intestinal Caco-2 cells with minimum cellular toxicity, sublime penetration across the blood–brain barrier, and last but not least, substantial cytotoxicity to C6 glioma cells.

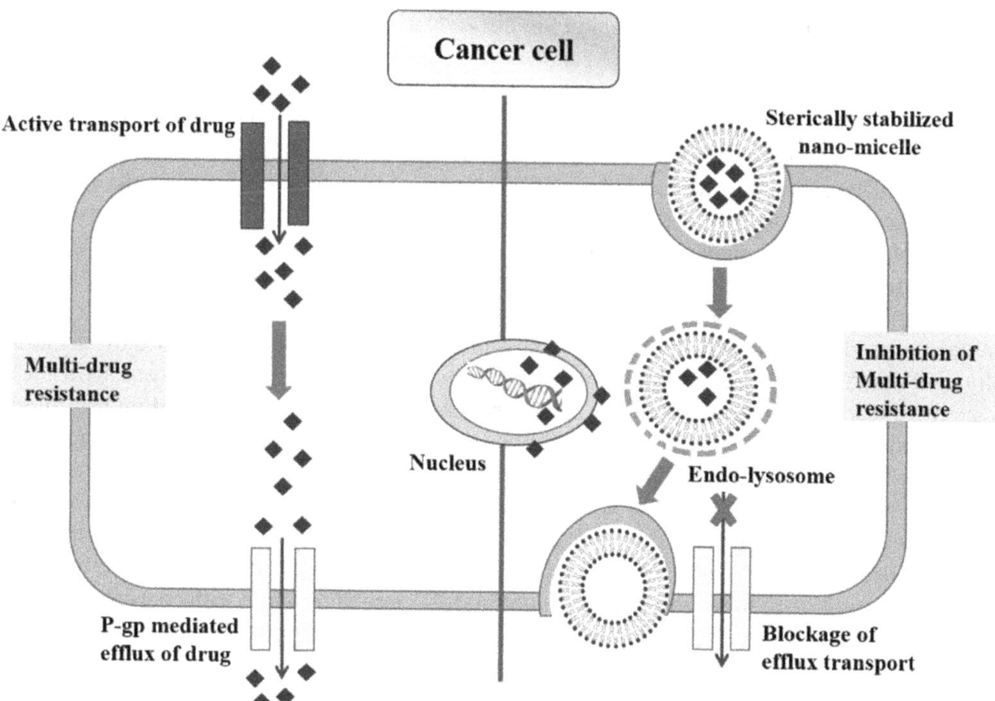

FIGURE 8.3 Inhibition of P-glycoprotein efflux by sterically stabilized phospholipid nanomicelle restricting the multi-drug resistance property in neoplastic cells.

QUE-FD-NMs were shown to accumulate in brain tumor tissues and demonstrated profound antitumor effects in glioma-bearing mice (Wang et al., 2016). Labrasol/Pluronic F68 modified phospholipid-based nanomicelles entrapping flavonoids were recently designed for targeting brain tumors. The findings indicated a uniform distribution of nanomicelles entrapping myricetin in specific tissues, thus improving *in vivo* antitumor efficacy without exceptional cellular toxicity to Caco-2 cells (Wang et al., 2020). Monoacyl-phospholipids (lyso-phospholipids), robust cone-shaped surfactants, can successfully produce micelles in a water dispersion. However, substances comprising monoacyl- as well as diacyl-phospholipids with about 60% lyso-PC can be regarded as intermediate emulsifiers to produce mixed micelles (Van Hoogevest and Fahr, 2019). Micelles have been extensively explored as a delivery device due to their pH-responsive release property of therapeutics at targeted tissues and active targeting strategy utilizing specific ligands. Nowadays, biocompatible nanomicelles as oral and parenteral formulations are attracting pharmaceutical companies to provide better clinical support.

8.7 LIPOSPHERES AS A BIOCOMPATIBLE DELIVERY SYSTEM

The greatest drawback of polymeric drug delivery devices, for example nanoemulsions, nanoparticles and nanoliposomes, relates to polymer degradation. Hence, the organic solvent present in the delivery vehicle may circulate in the physiological system, resulting in various toxicological hazards. To combat these problems, lipid microspheres, also known as lipospheres, have been preferred as a novel class of lipid-based system for encapsulating therapeutic molecules (Swain et al., 2016). Lipospheres comprise a solid lipid matrix surrounded by a layer of phospholipids as an external coat (Jadhav et al., 2014). The bioactive core is dissolved (lipophilic) or dispersed (hydrophobic) in lipospheres. Due to the particle dimensions, with a range from 0.01 to 100 μm in diameter, lipospheres can successfully deliver both hydrophilic (proteins and peptides) and lipophilic therapeutics into deep and targeted tissue areas, for instance the central nervous system and cerebrospinal fluid, with enhanced efficacy and stability (Dudala et al., 2014). Nasr et al. formulated lipospheres with soy lecithin and egg lecithin for topical application of angiotensin-converting enzyme (ACE) and documented improved drug loading capacity, elevated stability and sustained anti-inflammatory potency. Egg lecithin with a lower amount of unsaturated fatty acid manifested higher entrapment efficiency for aceclofenac than soybean PC consisting predominantly of unsaturated fatty acids, e.g. linoleic acid (67%). The greater membrane fatty acid unsaturation was found to amplify the membrane fluidity, one of the major criteria determining drug entrapping efficiency. Unsaturated fatty acids of phospholipid chains are unable to pack tightly in comparison to saturated fatty acids, which causes more leakage of the entrapped core. The percentages of drug released after 8 h from egg lecithin–based lipospheres were observed to be much lower compared with a system prepared with soybean lecithin under identical circumstances, since the efflux transport of drugs is increased by membrane fatty acid unsaturation. The anti-inflammatory potency of fabricated lipospheres was shown to be higher than that of commercial products in an experiment on carrageenan-induced rat paw edema. The lipospheres were also found to be capable of conserving their physical properties during storage at 2–8 °C (Nasr et al., 2008). Using lecithin as surfactant, Benzocaine lipospheres were prepared to enhance permeation and drug retention in skin and thus, to improve local anesthetic performance. The onset of anesthetic effect in rabbit cornea was found to be 6 min with lipospheres in comparison to 17 min observed with the conventional emulsion system. Furthermore, the anesthetic effect lasted much longer, for 53 min, with lipospheres in contrast to 11 min and 18 min with conventional emulsion (Bhatia et al., 2007). Phospholipon 90H, phospholipin 80H and soy lecithin were employed for the development of Fenofibrate lipospheres to enhance the bioavailability of the drug (Saroja and Lakshmi, 2013). Phospholipid lipospheres loaded with rifampicin displayed improved efficacy *towards* $H_{37}Rv$ strain *in vitro*. Thus, the findings have shown the possibilities of lipospheres for pulmonary administration of rifampicin (Singh et al., 2015). Upendra et al. fabricated Glimepiride lipospheres with particle size varying about 25.68 ± 0.18 μm, entrapment efficiency of 85.37 ±

2.50%, drug content of 85.13 ± 2.35% and drug release of 81.19 ± 3.91% within 8 hours (Upendra et al., 2015). Similarly, for the treatment of diabetes, pioglitazone hydrochloride lipospheres were optimized with spherical-shaped particles in the size range of 23.74 ±0.35 μm in diameter, entrapment efficiency of 79.69 ± 1.35%, drug content of 94.63 ± 2.10% and drug release of 96.06 ± 0.54% within 8 hours (Bhosale et al., 2016). Formulated naproxen-entrapped lipospheres were found to be free-flowing and spherical in shape. Drug entrapment efficiency and *in vitro* drug release content were found to vary from 80% to 90% and from 80% to 85%, respectively (Satheesh et al., 2014). Ofloxacin was also successfully loaded in lipospheres to enhance oral bioavailability, minimize toxicity and attain better patient compliance. Pharmacokinetics studies performed in rabbits stated that the lipospheres improved the bioavailability of ofloxacin 2.45-fold following oral administration (Satheesh et al., 2016). Avramoff et al. formulated pro-dispersion lipospheres for delivery of cyclosporine to improve bioavailability and thus, potential clinical use (Avramoff et al., 2012). Lecithin-based lipospheres entrapping lamivudine with particle size ranging from 68.27 to 173. m, encapsulation efficiency of 84.87–88.14% and drug release of 49.72–76.83% were explored for the treatment of AIDS (Saritha and Ravikumar, 2015). Lipospheres have been effectively utilized for oral, parenteral and topical delivery. Phospholipon 90H–based ibuprofen lipospheres fabricated by a homogenization technique exhibited superior anti-inflammatory and analgesic activity in comparison to a conventional system (Momoh et al., 2015). Novel co-spray-dried rifampicin (RMP) phospholipid lipospheres were considered as a promising carrier system, showing higher peak plasma concentration and thus, improved antimycobacterial activity compared with pure rifampicin (Singh et al., 2017a). The fusion of the solid interior with phospholipid at the external surface of lipospheres presents various satisfactory outcomes compared with conventional microspheres, for instance considerable dispensability in an aqueous environment and distinct release kinetics of the core, controlled by the phospholipid covering (Ganesan and Allimalarkodi, 2015). To overcome the delivery issues targeting breast carcinoma, lipospheres loading cabazitaxel and thymoquinone together were fabricated from egg PC aided by a melt dispersion method. Enhanced efficacy of the formulated lipospheres was reported, probably due to diffusion-controlled release of drugs from the liposphere matrix for prolonged period followed by faster cellular internalization. Compared with treatment by a solution of the combination, major morphological alterations in cancer cells, for instance nuclear fragmentation, enhanced Sub G1 phase arrest and apoptotic cell death, were perceived upon the administration of drug-loaded lipospheres (Kommineni et al., 2019). Lipospheres showed improved shelf life and thus, stability as compared with emulsions and liposomes. Lipospheres possessing a solid matrix can minimize the risk of interaction of core molecules with the encapsulant compared with a conventional system. In the liposphere, the solubilization of the active substance is not mandatory, since it may be dispersed within the solid matrix. Despite all these advantages, lipospheres suffer from several drawbacks, such as limited drug entrapment efficiency for proteins, high-pressure–mediated drug degradation, etc. To enhance the efficiency of lipospheres as a promising delivery tool, pharmaceutical scientists should pay attention to the biochemical interaction and mechanism of action of the encapsulated therapeutics in a physiological system. Fabrication techniques, physicochemical characteristics and therapeutic effects of phospholipid-based lipospheres entrapping various active compounds are briefly outlined in Table 8.6.

8.8 PHOSPHOLIPID–DRUG COMPLEXES FOR ENHANCING STABILITY AND ORAL BIOAVAILABILITY

Several active constituents demonstrating robust *in vitro* therapeutic effects suffer from inadequate oral bioavailability and poor permeability through the physiological barriers. In the absence of a sufficient amount of drug in the gastro-intestinal (GI) tract, the drug cannot efficiently penetrate through the epithelia of the GI system, leading to insufficient systemic absorption. To overcome this problem, pharmacosomes and phytosomes have begun to attract attention for the delivery of conventional commercial drugs and natural phyto-compounds, respectively (Abhinav et al., 2016).

TABLE 8.6
Method of Fabrication, Physicochemical Characteristics and Therapeutic Effects of Phospholipid-Based Lipospheres Entrapping Various Active Compounds

Type of phospholipid	Active compound	Preparation method	Particle size (µm)	Entrapment efficiency (%)	Net release content (%)	Therapeutic impact	Reference
Soybean lecithin, egg lecithin	Aceclofenac	Melt method, solvent evaporation method	0.69	94.60	85.52	Augmented anti-inflammatory activity	Nasr et al., 2008
Soy lecithin, phospholipon 90H, phospholipon 80H	Fenofibrate	Melt dispersion technique	30–45	85.98	87%	Elevated activity against hyperlipidemia	Saroja and Lakshmi, 2013
Phospholipon 90G	Glimepiride	Melt dispersion technique	25.68	85.37	81.19	Expected to increase anti-diabetic activity	Upendra et al., 2015
Phospholipon 90G	Pioglitazone hydrochloride	Melt dispersion technique	23.74	79.69	96.06	Expected to increase anti-diabetic activity	Bhosale et al., 2016
Phospholipon 80H	Naproxen	Melt dispersion technique	41–48	80–90	80–85	Minimized local side effects	Satheesh et al., 2014
Phospholipon 90H	Ibuprofen	Hot homogenization technique	101–178	89.4–97.9	75–96.9	Increased anti-inflammatory and analgesic activity	Momoh et al., 2015
Lecithin	Lamivudine	Melt dispersion	68.27–173.47	84.87–88.14	49.72–76.83	Expected to effectively treat HIV	Saritha and Ravikumar, 2015
Phospholipon 90H	Gentamicin	Solvent-melting	271.0–290.1	86.32	83	Enhanced antibacterial efficacy	Momoh and Esimone, 2012

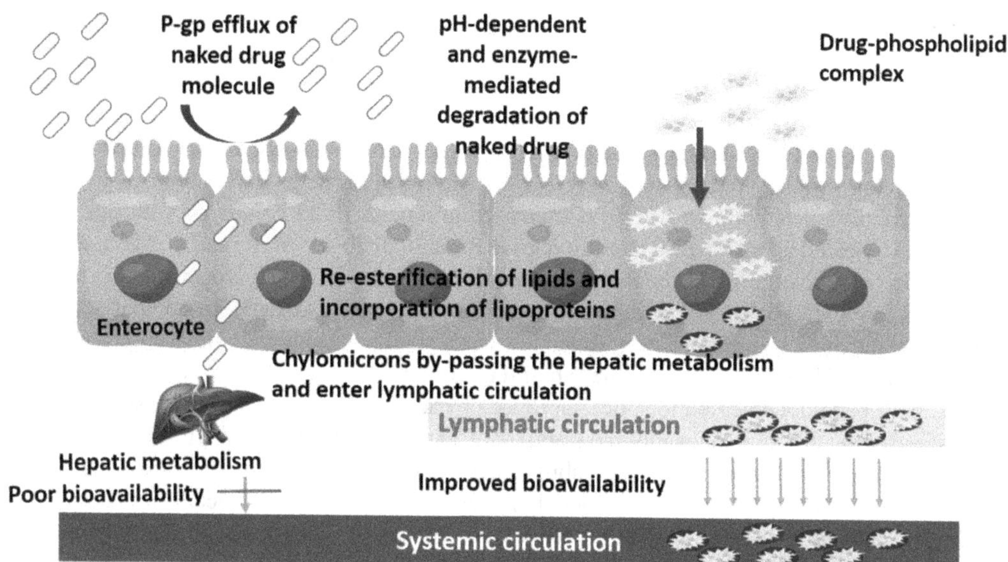

FIGURE 8.4 Entry of phospholipid–drug complex into the systemic circulation via enterocytes to overcome the burden of poor bioavailability. Drug–phospholipid complex evades the P-glycoprotein efflux and first-pass hepatic metabolism that must be faced by naked drug molecules.

Figure 8.4 depicts the mechanism of action of the phospholipid–drug complex entering the enterocytes and reaching the systemic circulation. PC is the major constituent of the pharmacosome and phytosome structure formation. Phytosomes and pharmacosomes can self-assemble similarly to liposomes when dispersed in an aqueous medium, but there are some differences in their chemical structure. In liposomes, hydrophilic bioactive agents are entrapped in the aqueous inner core, whereas in phospholipid-based phytosomes and pharmacosomes, weak hydrogen bonds and strong covalent bonds, respectively, form between the phospholipid and the active compound (Grimaldi et al., 2016). The comparative representation of phytosomes and pharmacosomes is shown in detail in Table 8.7.

8.8.1 Phytosomes

Active phytocompounds such as polyphenols, flavonoids, terpenoids and alkaloids suffer from poor lipophilicity and bioavailability due to their high molecular weight and multiple ring structure. These features limit their ability to cross cellular membranes or to be absorbed by simple diffusion.

TABLE 8.7
Comparative Representation of Phytosomes and Pharmacosomes

Characteristics	Phytosomes	Pharmacosomes
Bond	Weak bonding; hydrogen bond	Strong bonding; covalent bond
Drug leakage	Less	No
Entrapment efficacy	Low	High
Drug release	By bilayer diffusion, surface deposition or degradation	By hydrolysis
Stability	Not very stable, shorter shelf life	Highly stable, longer shelf life
Membrane fluidity	Takes place and influences the rate of release	Does not take place and does not restrict the rate of release

Bifunctional PC possesses a hydrophilic choline moiety and a lipophilic phosphatidyl constituent. In the phyto-phospholipid complex, a weak hydrogen bond is formed between the choline molecule and the active hydrogen atom of a polar phytoactive constituent. Then, the lipophilic phosphatidyl constituent joins the hydrophilic choline-bound complex, leading to the formation of a phytosome (Gnananath et al., 2017). Luteolin–phospholipid complex (LPC) was successfully prepared with a drug-encapsulating ability of approximately 72.64% and mean particle size of about 152.6 nm. The aqueous solubility of luteolin in LPC was increased around 2.5-fold compared with that of pure luteolin. LPC yielded about 95.12% luteolin release through diffusion at the end of 2 h, resulting in increased therapeutic efficacy in the *in vivo* system (Khan et al., 2014). Hydrogenated PC has been explored to fabricate phytosome complexes from *Terminalia arjuna* bark. In comparison to the pure extract, the phytosomes have been found to exert a significantly higher antiproliferative effect against the MCF-7 breast cancer cell line (Shalini et al., 2015). Rutin-loaded phytosomes prepared from PC have been documented to show improved physical stability over a 3-week storage period (Hooresfand et al., 2015). A phospholipid stabilized Centella extract (SCE) phytosome was formulated with a markedly higher aqueous solubility and improved sustained release kinetics. *Ex vivo* permeation studies using the everted sac model in intestine revealed the enhancement in permeation of SCE phytosomes (82.8 ± 3.7 %w/w), and *in vivo* efficacy assessments with the Morris Water Maze test depicted a pronounced augmentation in spatial learning and memory in elderly mice after phytosome treatment (Saoji et al., 2016). Rosmarinic acid–phospholipid complex was found to demonstrate elevated oral bioavailability as well as improved preventive effects against oxidative stress–mediated cellular damage compared with unformulated rosmarinic acid (Yang et al., 2015). Optimized pomegranate extract–phospholipid complex (SPEPC) showed greater permeability than the pure extract in an everted sac model. Dissolution studies revealed that the phospholipid coating might restrict the core to be free in intestinal environment and arrest their degradation by gut microbiome (Vora et al., 2015). For dermal delivery, *Phyllanthus emblica* extract–phospholipid complex has shown better skin retention and prolonged antioxidant effect compared with a conventional dosage form (Pereira and Mallya, 2015). The lipophilicity, intestinal absorption and thus, oral bioavailability of Echinacoside were statistically proved to be functional in an Echinacoside–phospholipid phytosome formulation (Li et al., 2015a). Solidified powder of oleanolic acid–phospholipid complex was found to be suitable to increase the solubility of oleanolic acid (Yang et al., 2016). *In vitro* studies indicated notably superior antioxidant properties of (–)-epigallocatechin gallate (EGCG)–phospholipid complex compared with butylated hydroxytoluene after a 15-day storage period (Chen et al., 2015). Silymarin–phospholipid complex was documented to have the best physical attributes, with average vesicle diameter of about 133.534 ± 8.76 nm, polydispersity index of around 0.339 ± 0.078, encapsulation efficiency of approximately 97.169 ± 2.412%, loading capacity of about 12.18 ± 0.30% and satisfactory stability (Maryana et al., 2016). A phospholipid complex was found to be a favorable encapsulant for increasing the oral bioavailability of dihydromyricetin due to the enhancement of the water solubility of the complex, escalation of absorption, and an anticipated reduction in hepatic and intestinal metabolism (Zhao et al., 2019). In a recent study, a piperine–phospholipid complex (PPC) was developed and thoroughly characterized as a novel nanocarrier to overcome the hindrances associated with the conventional delivery strategies. *In vitro* as well as *in vivo* investigations exhibited notably higher hepatoprotection by the fabricated PPC than the free piperine due to the improved bioavailability and pharmacokinetic profile (Biswas et al., 2020). A novel self-nanoemulsifying drug delivery system (SNEDDS) enclosing phospholipid complex was successfully formulated to enhance the oral bioavailability of norisoboldine through increasing intestinal lymphatic transport and inhibiting its metabolism in liver (Zhang et al., 2019). Based on previous reports, phytosomes were considered to protect herbal compounds from digestive degradation and gut bacteria, provide better absorption and improve bioavailability. On the contrary, a recent study utilizing a TNO intestinal model-1 (TIM-1) revealed the impact of phospholipid on the bioaccessibility of polar compounds (Huang et al., 2019). It showed that complexation with phospholipids diminished the bioaccessibility of rosmarinic acid in the jejunum whilst conserving it in the ileum of the small intestine. The all-inclusive bioaccessibility of the rosmarinic acid–phospholipid complex was found to be lower than for rosmarinic acid alone. The enhanced oral absorption found in the preceding *in vivo* research may be attributed to the overwhelming power of improved intestinal permeability over reduced bioaccessibility. Surprisingly, thousands of PC molecules were found to encompass the water-soluble component in liposomes. In contrast, the PC and the active phytoconstituent form phytosomes by complexing in a 1:1 or a 2:1 ratio. Thus, phytosomes offer better stability and absorption over liposomes (Karimi et al., 2015). The dose requirement is decreased and the therapeutic effect intensified by the use of a phytosomal drug delivery system. Despite various advantages of phytosomes, some fatal complications have been observed in phytosomal delivery systems (Singh et al., 2018). Phospholipid (lecithin) has been documented to promote proliferation in MCF-7, a human breast adenocarcinoma cell line (Chivte et al., 2017). A crucial drawback of phytosomes is leaching of the phytoactive compounds, which was reported to reduce the effective concentration. Phytosome formulations can be administered via both oral and topical routes to achieve maximum bioavailability and thus, beneficial efficacy. As the phytosome formulation methodology has proved to be a simple one, it can be commercially scaled up for the pharmaceutical, nutraceutical or cosmetic industry in the near future. Fabrication techniques, physicochemical characteristics and therapeutic effects of phospholipid-based phytosomes entrapping various active compounds are briefly outlined in Table 8.8.

8.8.2 Pharmacosomes

Pharmacosomal therapeutic carriers are presently emerging for the refined delivery of several therapeutics, such as cancer drugs, non-steroidal anti-inflammatory drugs (NSAIDs), etc. A drug with an active hydrogen atom moiety or a free carboxyl group can be esterified to the hydroxyl group of a phospholipid molecule, resulting in an amphiphilic prodrug (Kapoor et al., 2018). There are several advantages of pharmacosomes over liposomes. Unlike liposomes, the volume captured does not alter the encapsulation efficiency of pharmacosomes. In pharmacosomes, there is no requirement for the monotonous as well as laborious step of eliminating the unbound, unencapsulated drug from the system, which is necessary for liposomes. As the active molecule is covalently conjugated with phospholipid in pharmacosomes, the leakage of the active core, which is common in liposomes, does not occur. Pharmacosomes can be administered through the oral, topical and intra-vascular route (Pandita and Sharma, 2013). In two different studies, to recover the aqueous solubility and bioavailability and to minimize the toxic effects of the drugs on the GI system, pharmacosomes were fabricated by conjugating PC with aceclofenac and ketoprofen in various ratios (Semalty et al., 2010; Kamalesh et al., 2014). Furosemide pharmacosome formulations were found to exhibit a five-fold elevation in dissolution and a pronounced enhancement in permeability in comparison to the pure drug (Semalty et al., 2014). *In vitro* release kinetics also indicated a prolonged drug release profile. The solubility of pranlukast–phospholipid complex was reported to be increased about 150-fold, resulting in a 20-fold improvement of its bioavailability in comparison to its native crystalline state (Hao et al., 2015). Unsatisfactory bioavailability of rifampicin due to its inadequate water solubility was recently reported to be overcome by constructing a complex with PC (Singh et al., 2014). This was observed because of the increased solubility accompanied by the enhanced absorption of rifampicin. The complex was found to increase the oral bioavailability of rifampicin by enhancing absorption, thereby decreasing its metabolism. In another recent report, it was found that the low solubility of atorvastatin could be resolved by forming a phospholipid complex (Qin et al., 2018). The oral adsorption of atorvastatin–phospholipid complex showed a 2.58-fold increase in plasma concentration with superior pharmacokinetics compared with Lipitor (a commercial product). A parallel investigation performed by another batch of researchers was the fabrication of Erlotinib–phospholipid complex to challenge the drug solubility issues as well as to promote bioavailability and reduce cytotoxicity (Dora et al., 2017). The developed Erlotinib–phospholipid complex in the form of nanostructures achieved 1.7-fold higher bioavailability and superior efficacy of Erlotinib

TABLE 8.8
Method of Fabrication, Physicochemical Characteristics and Therapeutic Effects of Phytosome Complexing with Various Active Compounds

Type of phospholipid	Active compound	Preparation method	Particle size (nm)	Zeta-potential (mV)	Entrapment efficiency (%)	Therapeutic impact	References
Phospholipon 90H	Luteolin	Solvent evaporation	152.6	–27.6	72.64	Enhanced anti-inflammatory activity	Khan et al., 2014
Phosphatidylcholine	Rutin	Thin layer hydration	99–123	–45.2	72–80	Expected to increase antioxidative property and bioavailability	Hooresfand et al., 2015
Soy phosphatidylcholine	Silymarin	Solvent evaporation	133.534	–	97.169	Improved oral bioavailability	Maryana et al., 2016
Soy phosphatidylcholine	Ashwagandha extract	Ethanol method and reflux method	98.4	–28.7	90.1	Augmented antioxidative property	Keerthi et al., 2014
Phospholipid	Evodiamine	Solvent evaporation	246.1	–26.94	–	Elevated oral bioavailability	Tan et al., 2012
Phosphatidylcholine	Quercetin	Thin layer hydration method	266.6	-29.43 ± 0.75	96.57	Expected to enhance bioavailability	Lestari et al., 2017
Soy phosphatidylcholine	Mitomycin	Solvent evaporation	210.87	–33.38	–	Exhibited superior antitumor activity	Hou et al., 2012
Soy phosphatidylcholine	Curcumin	Co-solvent method	185.3	–15.7	92.5	Enhanced tumor accumulation	Xie et al., 2017

due to amorphization and enhanced solubility in comparison to free Erlotinib. A geniposide pharmacosome formulation was optimized to increase lipophilicity to about 20 times higher than that of the pure material (Yue et al., 2012). Complex formation of raloxifene with phospholipids demonstrated significant promise in augmenting therapeutic efficacy because of its refined biopharmaceutical properties along with improved oral bioavailability. Van der Waals interactions and other electrostatic forces were found to be responsible for the complex formation between raloxifene and phospholipid (Jain et al., 2019). To increase solubility and bioavailability as well as to diminish GI irritation, an aspirin–phospholipid complex was fabricated. The developed formulation showed 90.93% core release at the end of 10 h in an acidic medium (Semalty et al., 2010). Cholesteryl-phosphonyl gemcitabine (CPNG), an amphiphilic prodrug of gemcitabine, was fabricated by conjugating the drug with cholesterol with the aid of diphenyl phosphonate. The nanoassemblies yielded three to six times greater cellular toxicity compared with the naked drug, probably due to the phosphonyl substitution as well as the amphiphilic nature of CPNG. After bolus intravenous administration, CPNG nanoassemblies were mostly dispersed throughout the mononuclear phagocyte system prevailing in liver and spleen. The nanoassemblies entrapping a considerable dose of CPNG showed markedly superior *in vivo* chemotherapeutic activity compared with gemcitabine (Li et al., 2015c). Despite the appreciable stability, greater shelf life and sustained release kinetics *in vivo*, pharmacosomes may also suffer from fusion, aggregation and chemical hydrolysis during storage (Sailaja, 2016). To fabricate pharmacosomes as a productive delivery device, bulk interaction of the drug with phospholipid is most important. Besides this, research groups across the world have tried to explore various nanoplatforms manufactured with phospholipid in an effort to conquer all the above-mentioned obstacles (Kuche et al., 2019). Micro/nanoemulsion, self-emulsifying delivery devices, micelles and several other nanodelivery tools were employed synergistically with drug–phospholipid complex to mitigate the bioavailability issues and successfully improve the physicochemical attributes of the therapeutic agents.

8.9 CONCLUSION

In the pharmaceutical industry, phospholipids are globally used as surface active agents, coating agents, wetting agents, carriers, solubilizers and permeation enhancers. But, the physicochemical properties of heterogeneous phospholipid mixtures obtained from various natural resources may differ from each other depending on the nature of the water-loving head group as well as the long fatty acid tail. So, the choice of the right phospholipid excipient to fabricate a novel formulation and to accommodate the intended benefits of the therapeutic agents requires a deep understanding of the composition of the phospholipid mixture. Therapeutic delivery avenues are currently experiencing a substantial expansion due to the unified interests of researchers, industrialists and governing bodies. Though natural phospholipids should be chosen as excipients for fabricating nanostructures due to their countless inherent beneficial properties, as mentioned throughout this chapter, current research is focusing on synthetic phospholipids, possibly due to the fear of undiscovered impurities, allergens, etc., existing in naturally available phospholipids. Along with the active targeting ability and the enhanced permeation–retention effect, phospholipid-based nanovehicles are anticipated to flourish soon as one of the safest and most favorable delivery methods. Because of their crucial contribution in constructing and maintaining biological interfaces, the interaction of phospholipids with the physiological system has earned a lot of consideration from contemporary researchers. Phospholipids, especially PC, can be employed to integrate into the cell membrane, substituting for cellular phospholipids as well as modifying the fluidity of the membrane. The complex biomolecular interactivity between the phospholipid vehicle and the membrane phospholipid, as well as its influence on cellular signaling, still remains stochastic. Furthermore, detailed comprehension would also be required to correlate the unique stability of the phospholipid nanostructure and its sustained release profile. To explore the exciting possibilities in the frontier area of therapeutic delivery, more holistic study of phospholipid-based nanovehicles will be required.

ACKNOWLEDGEMENT

The authors would like to convey their wholehearted indebtedness to colleagues and co-researchers, especially Dr. Tanmoy Kumar Dey, for their constant effort and support.

REFERENCES

Abhinav, M., Neha, J., Anne, G. and Bharti, V., 2016. Role of novel drug delivery systems in bioavailability enhancement: At a glance. *International Journal of Drug Delivery Technology*, *6*(1), pp.7–26.

Abreu, A.S., Castanheira, E.M., Queiroz, M.J.R., Ferreira, P.M., Vale-Silva, L.A. and Pinto, E., 2011. Nanoliposomes for encapsulation and delivery of the potential antitumoral methyl 6-methoxy-3-(4-methoxyphenyl)-1 H-indole-2-carboxylate. *Nanoscale Research Letters*, *6*(1), p.482.

Acharya, A., Ahmed, M.G. and Rao, B.D., 2016. Development and evaluation of ethosomal gel of lornoxicam for transdermal delivery: in-vitro and in-vivo evaluation. *Manipal Journal of Pharmaceutical Sciences*, *2*(1), p.3.

Ahmed, T.A., 2015. Preparation of transfersomes encapsulating sildenafil aimed for transdermal drug delivery: Plackett–Burman design and characterization. *Journal of Liposome Research*, *25*(1), pp.1–10.

Ahmed, T.A., Khalid, M., Aljaeid, B.M., Fahmy, U.A. and Abd-Allah, F.I., 2016. Transdermal glimepiride delivery system based on optimized ethosomalnano-vesicles: Preparation, characterization, in vitro, ex vivo and clinical evaluation. *International Journal of Pharmaceutics*, *500*(1–2), pp.245–254.

Aisha, A.F., Majid, A.M.S.A. and Ismail, Z., 2014. Preparation and characterization of nano liposomes of Orthosiphon stamineus ethanolic extract in soybean phospholipids. *BMC Biotechnology*, *14*(1), p.23.

Akbarzadeh, A., Rezaei-Sadabady, R., Davaran, S., Joo, S.W., Zarghami, N., Hanifehpour, Y., Samiei, M., Kouhi, M. and Nejati-Koshki, K., 2013. Liposome: Classification, preparation, and applications. *Nanoscale Research Letters*, *8*(1), p.102.

Aklakur, M., Asharf Rather, M. and Kumar, N., 2016. Nanodelivery: An emerging avenue for nutraceuticals and drug delivery. *Critical Reviews in Food Science and Nutrition*, *56*(14), pp.2352–2361.

Ammar, H.O., Salama, H.A., Ghorab, M. and Mahmoud, A.A., 2009. Nanoemulsion as a potential ophthalmic delivery system for dorzolamide hydrochloride. *AAPS Pharmscitech*, *10*(3), p.808.

Anuchapreeda, S., Fukumori, Y., Okonogi, S. and Ichikawa, H., 2012. Preparation of lipid nanoemulsions incorporating curcumin for cancer therapy. *Journal of Nanotechnology*, *2012*, p.11.

Avramoff, A., Khan, W., Ezra, A., Elgart, A., Hoffman, A. and Domb, A.J., 2012. Cyclosporin pro-dispersion liposphere formulation. *Journal of Controlled Release*, *160*(2), pp.401–406.

Banerjee, A. and Onyuksel, H., 2012. Peptide delivery using phospholipid micelles. *Wiley Interdisciplinary Reviews: Nanomedicine and Nanobiotechnology*, *4*(5), pp.562–574.

Bernardo, J., Videira, R.A., Valentão, P., Veiga, F. and Andrade, P.B., 2019. Extraction of phospholipid-rich fractions from egg yolk and development of liposomes entrapping a dietary polyphenol with neuroactive potential. *Food and Chemical Toxicology*, *133*, p.110749.

Bhatia, A., Singh, B., Rani, V. and Katare, O.P., 2007. Formulation, characterization, and evaluation of benzocaine phospholipid-tagged lipospheres for topical application. *Journal of Biomedical Nanotechnology*, *3*(1), pp.81–89.

Bhosale, U.M., Galgatte, U.C. and Chaudhari, P.D., 2016. Development of pioglitazone hydrochloride lipospheres by melt dispersion technique: Optimization and evaluation. *Journal of Applied Pharmaceutical Science*, *6*(01), pp.107–117.

Biswas, S., Mukherjee, P.K., Kar, A., Bannerjee, S., Charoensub, R. and Duangyod, T., 2020. Optimized piperine-phospholipid complex with enhanced bioavailability and hepatoprotective activity. *Pharmaceutical Development and Technology*, *26*(1), pp.69–80.

Bothiraja, C., Kapare, H.S., Pawar, A.P. and Shaikh, K.S., 2013. Development of plumbagin-loaded phospholipid–Tween® 80 mixed micelles: Formulation, optimization, effect on breast cancer cells and human blood/serum compatibility testing. *Therapeutic Delivery*, *4*(10), pp.1247–1259.

Brandenburg, K.S., Rubinstein, I., Sadikot, R.T. and Önyüksel, H., 2012. Polymyxin B self-associated with phospholipid nanomicelles. *Pharmaceutical Development and Technology*, *17*(6), pp.654–660.

Cesur, H., Rubinstein, I., Pai, A. and Önyüksel, H., 2009. Self-associated indisulam in phospholipid-based nanomicelles: A potential nanomedicine for cancer. *Nanomedicine: Nanotechnology, Biology and Medicine*, *5*(2), pp.178–183.

Chakraborty, A. and Dhar, P., 2017. A review on potential of proteins as an excipient for developing a nanocarrier delivery system. *Critical Reviews™ in Therapeutic Drug Carrier Systems*, *34*(5), pp.453–488.

Chauhan, N., Kumar, K. and Pant, N.C.C., 2017. An updated review on transfersomes: A novel vesicular system for transdermal drug delivery. *Universal Journal of Pharmaceutical Research*, *2*(4), pp.42–45.

Chen, J.Y., Zhang, X., Wu, Z.F. and Weng, P.F., 2015. Antioxidant activity of (–)-epigallocatechin gallate–phospholipid complex. *Modern Food Science and Technology*, *31*, pp.137–143.

Chintala, P.K. and Padmapreetha, J., 2014 Feb 1. Formulation and in-vitro evaluation of gel containing ethosomes entrapped with etodolac. *International Journal of Pharmaceutical Science and Research*, *5*(2), 630–635.

Chivte, P.S., Pardhi, V.S., Joshi, V.A. and Rani, A., 2017. A review on therapeutic applications of phytosomes. *Journal of Drug Delivery and Therapeutics*, *7*(5), pp.17–21.

Chung, H., Kim, T.W., Kwon, M., Kwon, I.C. and Jeong, S.Y., 2001. Oil components modulate physical characteristics and function of the natural oil emulsions as drug or gene delivery system. *Journal of Controlled Release*, *71*(3), pp.339–350.

Cui, F., Shi, K., Zhang, L., Tao, A. and Kawashima, Y., 2006. Biodegradable nanoparticles loaded with insulin–phospholipid complex for oral delivery: Preparation, in vitro characterization and in vivo evaluation. *Journal of Controlled Release*, *114*(2), pp.242–250.

Cui, L. and Decker, E.A., 2016. Phospholipids in foods: Prooxidants or antioxidants? *Journal of the Science of Food and Agriculture*, *96*(1), pp.18–31.

Daraee, H., Etemadi, A., Kouhi, M., Alimirzalu, S. and Akbarzadeh, A., 2016. Application of liposomes in medicine and drug delivery. *Artificial Cells, Nanomedicine, and Biotechnology*, *44*(1), pp.381–391.

David, S.R.N., Hui, M.S., Pin, C.F., Ci, F.Y. and Rajabalaya, R., 2013. Formulation and in vitro evaluation of ethosomes as vesicular carrier for enhanced topical delivery of isotretinoin. *International Journal of Drug Delivery*, *5*(1), pp.28–34.

De, A. and Venkatesh, D.N., 2012. Design and evaluation of liposomal delivery system for L-Asparaginese. *Journal of Applied Pharmaceutical Science*, *2*(8), p.112.

Doijad, R.C., Manvi, F.V., Godhwani, D.M., Joseph, R. and Deshmukh, N.V., 2008. Formulation and targeting efficiency of cisplatin engineered solid lipid nanoparticles. *Indian Journal of Pharmaceutical Sciences*, *70*(2), p.203.

Dora, C.P., Kushwah, V., Katiyar, S.S., Kumar, P., Pillay, V., Suresh, S. and Jain, S., 2017. Improved oral bioavailability and therapeutic efficacy of erlotinib through molecular complexation with phospholipid. *International Journal of Pharmaceutics*, *534*(1–2), pp.1–13.

Duan, Y., Wang, J., Yang, X., Du, H., Xi, Y. and Zhai, G., 2015. Curcumin-loaded mixed micelles: Preparation, optimization, physicochemical properties and cytotoxicity in vitro. *Drug Delivery*, *22*(1), pp.50–57.

Dudala, T.B., Yalavarthi, P.R., Vadlamudi, H.C., Thanniru, J., Yaga, G., Mudumala, N.L. and Pasupati, V.K., 2014. A perspective overview on lipospheres as lipid carrier systems. *International Journal of Pharmaceutical Investigation*, *4*(4), p.149.

Dudhipala, N. and Veerabrahma, K., 2016. Candesartan cilexetil loaded solid lipid nanoparticles for oral delivery: Characterization, pharmacokinetic and pharmacodynamic evaluation. *Drug Delivery*, *23*(2), pp.395–404.

Ebrahimi, P., Ebrahim-Magham, B., Pourmorad, F. and Honary, S., 2013. Ferulic acid lecithin-based nanoemulsions prepared by using spontaneous emulsification process. *Iranian Journal of Chemistry and Chemical Engineering (IJCCE)*, *32*(3), pp.17–25.

Ekambaram, P., Sathali, A.A.H. and Priyanka, K., 2012. Solid lipid nanoparticles: A review. *Scientific Reviews and Chemical Communications*, *2*(1), pp.80–102.

Espinosa-Andrews, H. and Páez-Hernández, G., 2020. Optimization of ultrasonication curcumin-hydroxylated lecithin nanoemulsions using response surface methodology. *Journal of Food Science and Technology*, *57*(2), pp.549–556.

Fang, J.Y., Leu, Y.L., Chang, C.C., Lin, C.H. and Tsai, Y.H., 2004. Lipid nano/submicron emulsions as vehicles for topical flurbiprofen delivery. *Drug Delivery*, *11*(2), pp.97–105.

Fang, S., Niu, Y., Zhu, W., Zhang, Y., Yu, L. and Li, X., 2015. Liposomes assembled from a dual drug-tailed phospholipid for cancer therapy. *Chemistry: An Asian Journal*, *10*(5), pp.1232–1238.

Ganesan, V. and Allimalarkodi, S., 2015. Liposphere: A versatile drug delivery system: An over view. *Research Journal of Pharmaceutical Biological and Chemical Sciences*, *6*(4), pp.1753–1762.

Ganta, S. and Amiji, M., 2009. Coadministration of paclitaxel and curcumin in nanoemulsion formulations to overcome multidrug resistance in tumor cells. *Molecular Pharmaceutics*, *6*(3), pp.928–939.

Ganta, S., Devalapally, H., Baguley, B.C., Garg, S. and Amiji, M., 2008. Microfluidic preparation of chlorambucil nanoemulsion formulations and evaluation of cytotoxicity and pro-apoptotic activity in tumor cells. *Journal of Biomedical Nanotechnology*, *4*(2), pp.165–173.

Garg, B.J., Garg, N.K., Beg, S., Singh, B. and Katare, O.P., 2016a. Nanosized ethosomes-based hydrogel formulations of methoxsalen for enhanced topical delivery against vitiligo: Formulation optimization, in vitro evaluation and preclinical assessment. *Journal of Drug Targeting*, *24*(3), pp.233–246.

Garg, N.K., Singh, B., Jain, A., Nirbhavane, P., Sharma, R., Tyagi, R.K., Kushwah, V., Jain, S. and Katare, O.P., 2016b. Fucose decorated solid-lipid nanocarriers mediate efficient delivery of methotrexate in breast cancer therapeutics. *Colloids and Surfaces B: Biointerfaces*, *146*, pp.114–126.

Garg, V., Singh, H., Bimbrawh, S., Kumar Singh, S., Gulati, M., Vaidya, Y. and Kaur, P., 2017. Ethosomes and transfersomes: Principles, perspectives and practices. *Current Drug Delivery*, *14*(5), pp.613–633.

Geszke-Moritz, M. and Moritz, M., 2016. Solid lipid nanoparticles as attractive drug vehicles: Composition, properties and therapeutic strategies. *Materials Science and Engineering: C*, *68*, pp.982–994.

Glukhova, O.E., Savostyanov, G.V. and Grishina, O.A., 2015, March. Phospholipid liposomes functionalized by protein. In *Reporters, Markers, Dyes, Nanoparticles, and Molecular Probes for Biomedical Applications VII* (Vol. 9339, p.93390S). International Society for Optics and Photonics.

Gnananath, K., Nataraj, K.S. and Rao, B.G., 2017. Phospholipid complex technique for superior bioavailability of phytoconstituents. *Advanced Pharmaceutical Bulletin*, *7*(1), p.35.

Goindi, S., Dhatt, B. and Kaur, A., 2014. Ethosomes-based topical delivery system of antihistaminic drug for treatment of skin allergies. *Journal of Microencapsulation*, *31*(7), pp.716–724.

González-Rodríguez, M.L., Arroyo, C.M., Cózar-Bernal, M.J., González-R, P.L., León, J.M., Calle, M., Canca, D. and Rabasco, A.M., 2016. Deformability properties of timolol-loaded transfersomes based on the extrusion mechanism. Statistical optimization of the process. *Drug Development and Industrial Pharmacy*, *42*(10), pp.1683–1694.

Grimaldi, N., Andrade, F., Segovia, N., Ferrer-Tasies, L., Sala, S., Veciana, J. and Ventosa, N., 2016. Lipid-based nanovesicles for nanomedicine. *Chemical Society Reviews*, *45*(23), pp.6520–6545.

Gülçür, E., Thaqi, M., Khaja, F., Kuzmis, A. and Önyüksel, H., 2013. Curcumin in VIP-targeted sterically stabilized phospholipid nanomicelles: A novel therapeutic approach for breast cancer and breast cancer stem cells. *Drug Delivery and Translational Research*, *3*(6), pp.562–574.

Gundermann, K.J., Kuenker, A., Kuntz, E. and Droździk, M., 2011. Activity of essential phospholipids (EPL) from soybean in liver diseases. *Pharmacological Reports*, *63*(3), pp.643–659.

Haeri, A., Sadeghian, S., Rabbani, S., Anvari, M.S., Lavasanifar, A., Amini, M. and Dadashzadeh, S., 2013. Sirolimus-loaded stealth colloidal systems attenuate neointimal hyperplasia after balloon injury: A comparison of phospholipid micelles and liposomes. *International Journal of Pharmaceutics*, *455*(1–2), pp.320–330.

Hao, Y., Wang, L., Li, J., Liu, N., Feng, J., Zhao, M. and Zhang, X., 2015. Enhancement of solubility, transport across Madin-Darby canine kidney monolayers and oral absorption of pranlukast through preparation of a pranlukast-phospholipid complex. *Journal of Biomedical Nanotechnology*, *11*(3), pp.469–477.

Heiati, H., Tawashi, R. and Phillips, N.C., 1998. Drug retention and stability of solid lipid nanoparticles containing azidothymidine palmitate after autoclaving, storage and lyophilization. *Journal of Microencapsulation*, *15*(2), pp.173–184.

Heo, W., Kim, J.H., Pan, J.H. and Kim, Y.J., 2016. Lecithin-based nano-emulsification improves the bioavailability of conjugated linoleic acid. *Journal of Agricultural and Food Chemistry*, *64*(6), pp.1355–1360.

Hooresfand, Z., Ghanbarzadeh, S. and Hamishehkar, H., 2015. Preparation and characterization of rutin-loaded nanophytosomes. *Pharmaceutical Sciences*, *21*(3), pp.145–51.

Hou, Z., Li, Y., Huang, Y., Zhou, C., Lin, J., Wang, Y., Cui, F., Zhou, S., Jia, M., Ye, S. and Zhang, Q., 2012. Phytosomes loaded with mitomycin C–soybean phosphatidylcholine complex developed for drug delivery. *Molecular Pharmaceutics*, *10*(1), pp.90–101.

Huang, J., Chen, P.X., Rogers, M.A. and Wettig, S.D., 2019. Investigating the phospholipid effect on the bioaccessibility of rosmarinic acid-phospholipid complex through a dynamic gastrointestinal in vitro model. *Pharmaceutics*, *11*(4), p.156.

Jadhav, S.V., Sadgir, D.P., Patil, M.P. and Jagtap, R.M., 2014. Liposphere review: Methods and its applications in bio-compatible drug delivery system. *World Journal of Pharmacy and Pharmaceutical Sciences*, *3*(9), pp.1023–1043.

Jain, A., Saini, S., Kumar, R., Sharma, T., Swami, R., Katare, O.P. and Singh, B., 2019. Phospholipid-based complex of raloxifene with enhanced biopharmaceutical potential: Synthesis, characterization and preclinical assessment. *International Journal of Pharmaceutics*, *571*, p.118698.

Jain, A., Singh, S.K., Singh, Y. and Singh, S., 2013. Development of lipid nanoparticles of diacerein, an antiosteoarthritic drug for enhancement in bioavailability and reduction in its side effects. *Journal of Biomedical Nanotechnology*, *9*(5), pp.891–900.

Jain, A.K., Jain, A., Garg, N.K., Agarwal, A., Jain, A., Jain, S.A., Tyagi, R.K., Jain, R.K., Agrawal, H. and Agrawal, G.P., 2014. Adapalene loaded solid lipid nanoparticles gel: An effective approach for acne treatment. *Colloids and Surfaces B: Biointerfaces, 121*, pp.222–229.

Jain, S.K., Chourasia, M.K., Sabitha, M., Jain, R., Jain, A.K., Ashawat, M. and Jha, A.K., 2003. Development and characterization of transdermal drug delivery systems for diltiazem hydrochloride. *Drug Delivery, 10*(3), pp.169–177.

Jain, S., Patel, N., Madan, P. and Lin, S., 2015. Quality by design approach for formulation, evaluation and statistical optimization of diclofenac-loaded ethosomes via transdermal route. *Pharmaceutical Development and Technology, 20*(4), pp.473–489.

Jash, A., Ubeyitogullari, A. and Rizvi, S.S., 2020. Synthesis of multivitamin-loaded heat stable liposomes from milk fat globule membrane phospholipids by using a supercritical-CO 2 based system. *Green Chemistry, 22*(16), pp.5345–5356.

Jiang, S.P., He, S.N., Li, Y.L., Feng, D.L., Lu, X.Y., Du, Y.Z., Yu, H.Y., Hu, F.Q. and Yuan, H., 2013. Preparation and characteristics of lipid nanoemulsion formulations loaded with doxorubicin. *International Journal of Nanomedicine, 8*, p.3141.

Jin, H.H., Lu, Q. and Jiang, J.G., 2016. Curcumin liposomes prepared with milk fat globule membrane phospholipids and soybean lecithin. *Journal of Dairy Science, 99*(3), pp.1780–1790.

Jin, X., Zhang, Z.H., Sun, E., Tan, X.B., Zhu, F.X. and Jia, X.B., 2013. A novel drug–phospholipid complex loaded micelle for baohuoside I enhanced oral absorption: In vivo and in vivo evaluations. *Drug Development and Industrial Pharmacy, 39*(9), pp.1421–1430.

Kalepu, S. and Nekkanti, V., 2015. Insoluble drug delivery strategies: Review of recent advances and business prospects. *Acta Pharmaceutica Sinica B, 5*(5), pp.442–453.

Kamalesh, M., Diraj, D.B., Kiran, B. and Wagh, K., 2014. Formulation and evaluation of pharmacosomes of ketoprofen. *IAJPR, 4*, pp.1363–1368.

Kapoor, B., Gupta, R., Singh, S.K., Gulati, M. and Singh, S., 2018. Prodrugs, phospholipids and vesicular delivery-An effective triumvirate of pharmacosomes. *Advances in Colloid and Interface Science, 253*, pp.35–65.

Karamanidou, T., Bourganis, V., Kammona, O. and Kiparissides, C., 2016. Lipid-based nanocarriers for the oral administration of biopharmaceutics. *Nanomedicine, 11*(22), pp.3009–3032.

Karimi, N., Ghanbarzadeh, B., Hamishehkar, H., Keivani, F., Pezeshki, A. and Gholian, M.M., 2015. Phytosome and liposome: The beneficial encapsulation systems in drug delivery and food application. *Applied Food Biotechnology, 2*(3), pp.17–27.

Kateh Shamshiri, M., Momtazi-Borojeni, A.A., Khodabandeh Shahraky, M. and Rahimi, F., 2019. Lecithin soybean phospholipid nano-transfersomes as potential carriers for transdermal delivery of the human growth hormone. *Journal of Cellular Biochemistry, 120*(6), pp.9023–9033.

Keerthi, B., Pingali, P.S. and Srinivas, P., 2014. Formulation and evaluation of capsules of ashwagandha phytosomes. *International Journal of Pharmaceutical Sciences Review and Research, 29*(2), p.140.

Kelmann, R.G., Kuminek, G., Teixeira, H.F. and Koester, L.S., 2007. Carbamazepine parenteral nanoemulsions prepared by spontaneous emulsification process. *International Journal of Pharmaceutics, 342*(1–2), pp.231–239.

Khan, J., Alexander, A., Saraf, S. and Saraf, S., 2014. Luteolin–phospholipid complex: Preparation, characterization and biological evaluation. *Journal of Pharmacy and Pharmacology, 66*(10), pp.1451–1462.

Kieler-Ferguson, H.M., Chan, D., Sockolosky, J., Finney, L., Maxey, E., Vogt, S. and Szoka Jr, F.C., 2017. Encapsulation, controlled release, and antitumor efficacy of cisplatin delivered in liposomes composed of sterol-modified phospholipids. *European Journal of Pharmaceutical Sciences, 103*, pp.85–93.

Klang, V. and Valenta, C., 2011. Lecithin-based nanoemulsions. *Journal of Drug Delivery Science and Technology, 21*(1), pp.55–76.

Klang, V., Matsko, N., Zimmermann, A.M., Vojnikovic, E. and Valenta, C., 2010. Enhancement of stability and skin permeation by sucrose stearate and cyclodextrins in progesterone nanoemulsions. *International Journal of Pharmaceutics, 393*(1–2), pp.153–161.

Komaiko, J., Sastrosubroto, A. and McClements, D.J., 2016. Encapsulation of ω-3 fatty acids in nanoemulsion-based delivery systems fabricated from natural emulsifiers: Sunflower phospholipids. *Food Chemistry, 203*, pp.331–339.

Kommineni, N., Saka, R., Bulbake, U. and Khan, W., 2019. Cabazitaxel and thymoquinone co-loaded lipospheres as a synergistic combination for breast cancer. *Chemistry and Physics of Lipids, 224*, p.104707.

Koo, O.M.Y., Rubinstein, I. and Onyuksel, H., 2005. Camptothecin in sterically stabilized phospholipid micelles: A novel nanomedicine. *Nanomedicine: Nanotechnology, Biology and Medicine, 1*(1), pp.77–84.

Koo, O.M.Y., Rubinstein, I. and Onyuksel, H., 2006. Camptothecin in sterically stabilized phospholipid nano-micelles: A novel solvent pH change solubilization method. *Journal of Nanoscience and Nanotechnology*, *6*(9–10), pp.2996–3000.

Koo, O.M.Y., Rubinstein, I. and Önyüksel, H., 2011. Actively targeted low-dose camptothecin as a safe, long-acting, disease-modifying nanomedicine for rheumatoid arthritis. *Pharmaceutical Research*, *28*(4), pp.776–787.

Krishnadas, A., Rubinstein, I. and Önyüksel, H., 2003. Sterically stabilized phospholipid mixed micelles: In vitro evaluation as a novel carrier for water-insoluble drugs. *Pharmaceutical Research*, *20*(2), pp.297–302.

Kuche, K., Bhargavi, N., Dora, C.P. and Jain, S., 2019. Drug-phospholipid complex: A go through strategy for enhanced oral bioavailability. *AAPS PharmSciTech*, *20*(2), p.43.

Küllenberg, D., Taylor, L.A., Schneider, M. and Massing, U., 2012. Health effects of dietary phospholipids. *Lipids in Health and Disease*, *11*(1), p.3.

Kuzmis, A., Lim, S.B., Desai, E., Jeon, E., Lee, B.S., Rubinstein, I. and Önyüksel, H., 2011. Micellar nanomedicine of human neuropeptide Y. *Nanomedicine: Nanotechnology, Biology and Medicine*, *7*(4), pp.464–471.

Lee, M.K., Lim, S.J. and Kim, C.K., 2007. Preparation, characterization and in vitro cytotoxicity of paclitaxel-loaded sterically stabilized solid lipid nanoparticles. *Biomaterials*, *28*(12), pp.2137–2146.

Lestari, A., Anwar, E. and Harahap, Y., Design and formulation quercetin formula in the phytosomes system as novel. *Drug Delivery*, *10*(6); pp.148–151.

Leuschner, F., Wagener, H.H. and Neumann, B., 1976. The anti-hyperlipemic and anti-atherogenic effect of "essential" phospholipids: A pharmacologic trial. *Arzneimittel-Forschung*, *26*(9a), pp.1743–1772.

Li, F., Yang, X., Yang, Y., Li, P., Yang, Z. and Zhang, C., 2015. Phospholipid complex as an approach for bioavailability enhancement of echinacoside. *Drug Development and Industrial Pharmacy*, *41*(11), pp.1777–1784.

Li, J., Wang, X., Zhang, T., Wang, C., Huang, Z., Luo, X. and Deng, Y., 2015. A review on phospholipids and their main applications in drug delivery systems. *Asian Journal of Pharmaceutical Sciences*, *10*(2), pp.81–98.

Li, M., Qi, S., Jin, Y. and Dong, J., 2015. Self-assembled drug delivery systems. Part 8: In vitro/in vivo studies of the nanoassemblies of cholesteryl-phosphonyl gemcitabine. *International Journal of Pharmaceutics*, *478*(1), pp.124–130.

Li, Y., Xu, F., Li, X., Chen, S.Y., Huang, L.Y., Bian, Y.Y., Wang, J., Shu, Y.T., Yan, G.J., Dong, J. and Yin, S.P., 2021. Development of curcumin-loaded composite phospholipid ethosomes for enhanced skin permeability and vesicle stability. *International Journal of Pharmaceutics*, *592*, p.119936.

Liang, L., Chen, F., Wang, X., Jin, Q., Decker, E.A. and McClements, D.J., 2017. Physical and oxidative stability of flaxseed oil-in-water emulsions fabricated from sunflower lecithins: Impact of blending lecithins with different phospholipid Profiles. *Journal of Agricultural and Food Chemistry*, *65*(23), pp.4755–4765.

Lim, S.B., Banerjee, A. and Önyüksel, H., 2012. Improvement of drug safety by the use of lipid-based nanocarriers. *Journal of Controlled Release*, *163*(1), pp.34–45.

Lim, S.B., Rubinstein, I., Sadikot, R.T., Artwohl, J.E. and Önyüksel, H., 2011. A novel peptide nanomedicine against acute lung injury: GLP-1 in phospholipid micelles. *Pharmaceutical Research*, *28*(3), pp.662–672.

Lim, S.J. and Kim, C.K., 2002. Formulation parameters determining the physicochemical characteristics of solid lipid nanoparticles loaded with all-trans retinoic acid. *International Journal of Pharmaceutics*, *243*(1–2), pp.135–146.

Lin, H.Y., Thomas, J.L., Chen, H.W., Shen, C.M., Yang, W.J. and Lee, M.H., 2012. In vitro suppression of oral squamous cell carcinoma growth by ultrasound-mediated delivery of curcumin microemulsions. *International Journal of Nanomedicine*, *7*, p.941.

Liu, D., Jiang, S., Shen, H., Qin, S., Liu, J., Zhang, Q., Li, R. and Xu, Q., 2011. Diclofenac sodium-loaded solid lipid nanoparticles prepared by emulsion/solvent evaporation method. *Journal of Nanoparticle Research*, *13*(6), pp.2375–2386.

Liu, J., Gong, T., Fu, H., Wang, C., Wang, X., Chen, Q., Zhang, Q., He, Q. and Zhang, Z., 2008. Solid lipid nanoparticles for pulmonary delivery of insulin. *International Journal of Pharmaceutics*, *356*(1–2), pp.333–344.

Liu, J., Hu, W., Chen, H., Ni, Q., Xu, H. and Yang, X., 2007. Isotretinoin-loaded solid lipid nanoparticles with skin targeting for topical delivery. *International Journal of Pharmaceutics*, *328*(2), pp.191–195.

Lv, Q., Yu, A., Xi, Y., Li, H., Song, Z., Cui, J., Cao, F. and Zhai, G., 2009. Development and evaluation of penciclovir-loaded solid lipid nanoparticles for topical delivery. *International Journal of Pharmaceutics*, *372*(1–2), pp.191–198.

Malakar, J., Sen, S.O., Nayak, A.K. and Sen, K.K., 2012. Formulation, optimization and evaluation of transferosomal gel for transdermal insulin delivery. *Saudi Pharmaceutical Journal*, *20*(4), pp.355–363.

Maryana, W., Rachmawati, H. and Mudhakir, D., 2016. Formation of phytosome containing silymarin using thin layer-hydration technique aimed for oral delivery. *Materials Today: Proceedings*, *3*(3), pp.855–866.

Masoumi, H.R.F., Basri, M., Samiun, W.S., Izadiyan, Z. and Lim, C.J., 2015. Enhancement of encapsulation efficiency of nanoemulsion-containing aripiprazole for the treatment of schizophrenia using mixture experimental design. *International Journal of Nanomedicine*, *10*, p.6469.

McClements, D.J., 2012. Nanoemulsions versus microemulsions: Terminology, differences, and similarities. *Soft Matter*, *8*(6), pp.1719–1729.

McClements, D.J. and Rao, J., 2011. Food-grade nanoemulsions: Formulation, fabrication, properties, performance, biological fate, and potential toxicity. *Critical Reviews in Food Science and Nutrition*, *51*(4), pp.285–330.

Meng, S., Zhang, C., Shi, W., Zhang, X.W., Liu, D.H., Wang, P., Li, J.X. and Jin, Y., 2016. Preparation of osthole-loaded nano-vesicles for skin delivery: Characterization, in vitro skin permeation and preliminary in vivo pharmacokinetic studies. *European Journal of Pharmaceutical Sciences*, *92*, pp.49–54.

Mishra, V., Bansal, K., Verma, A., Yadav, N., Thakur, S., Sudhakar, K. and Rosenholm, J., 2018. Solid lipid nanoparticles: Emerging colloidal nano drug delivery systems. *Pharmaceutics*, *10*(4), p.191.

Momoh, M.A. and Esimone, C.O., 2012. Phospholipon 90H (P90H)-based PEGylated microscopic lipospheres delivery system for gentamicin: An antibiotic evaluation. *Asian Pacific Journal of Tropical Biomedicine*, *2*(11), p.889.

Momoh, M.A., Kenechukwu, F.C., Gwarzo, M.S. and Builders, P.F., 2015. Formulation and evaluation of ibuprofen loaded lipospheres for effective oral drug delivery. *Dhaka University Journal of Pharmaceutical Sciences*, *14*(1), pp.17–27.

Mulik, R.S., Mönkkönen, J., Juvonen, R.O., Mahadik, K.R. and Paradkar, A.R., 2010. Transferrin mediated solid lipid nanoparticles containing curcumin: Enhanced in vitro anticancer activity by induction of apoptosis. *International Journal of Pharmaceutics*, *398*(1–2), pp.190–203.

Naderkhani, E., Erber, A., Škalko-Basnet, N. and Flaten, G.E., 2014. Improved permeability of acyclovir: Optimization of mucoadhesive liposomes using the phospholipid vesicle-based permeation assay. *Journal of Pharmaceutical Sciences*, *103*(2), pp.661–668.

Naguib, Y.W., Rodriguez, B.L., Li, X., Hursting, S.D., Williams III, R.O. and Cui, Z., 2014. Solid lipid nanoparticle formulations of docetaxel prepared with high melting point triglycerides: In vitro and in vivo evaluation. *Molecular Pharmaceutics*, *11*(4), pp.1239–1249.

Nasr, M., Mansour, S., Mortada, N.D. and El Shamy, A.A., 2008. Lipospheres as carriers for topical delivery of aceclofenac: Preparation, characterization and in vivo evaluation. *AAPS PharmSciTech*, *9*(1), pp.154–162.

Nicolaos, G., Crauste-Manciet, S., Farinotti, R. and Brossard, D., 2003. Improvement of cefpodoxime proxetil oral absorption in rats by an oil-in-water submicron emulsion. *International Journal of Pharmaceutics*, *263*(1–2), pp.165–171.

Nour, S.A., Abdelmalak, N.S. and Naguib, M.J., 2017. Transferosomes for trans-nasal brain delivery of clonazepam: Preparation, optimization, ex-vivo cytotoxicity and pharmacodynamic study. *Journal of Pharmaceutical Research*, *1*(2), pp.1–15.

Oehlke, K., Behsnilian, D., Mayer-Miebach, E., Weidler, P.G. and Greiner, R., 2017. Edible solid lipid nanoparticles (SLN) as carrier system for antioxidants of different lipophilicity. *PloS One*, *12*(2), p.e0171662.

Okamura Y, Katsuno S, Suzuki H, Maruyama H, Handa M, Ikeda Y, Takeoka, S., 2010 Dec. Release abilities of adenosine diphosphate from phospholipid vesicles with different membrane properties and their hemostatic effects as a platelet substitute. *Journal of Controlled Release*, *148*(3):373–379.

Önyüksel, H., Ikezaki, H., Patel, M., Gao, X.P. and Rubinstein, I., 1999. A novel formulation of VIP in sterically stabilized micelles amplifies vasodilation in vivo. *Pharmaceutical Research*, *16*(1), pp.155–160.

Önyüksel, H., Jeon, E. and Rubinstein, I., 2009a. Nanomicellar paclitaxel increases cytotoxicity of multidrug resistant breast cancer cells. *Cancer Letters*, *274*(2), pp.327–330.

Önyüksel, H., Mohanty, P.S. and Rubinstein, I., 2009b. VIP-grafted sterically stabilized phospholipid nanomicellar 17-allylamino-17-demethoxy geldanamycin: A novel targeted nanomedicine for breast cancer. *International Journal of Pharmaceutics*, *365*(1–2), pp.157–161.

Paliwal, S., Tilak, A., Sharma, J., Dave, V., Sharma, S., Yadav, R., Patel, S., Verma, K. and Tak, K., 2019. Flurbiprofen loaded ethosomes-transdermal delivery of anti-inflammatory effect in rat model. *Lipids in Health and Disease*, *18*(1), p.133.

Pandita, A. and Sharma, P., 2013. Pharmacosomes: An emerging novel vesicular drug delivery system for poorly soluble synthetic and herbal drugs. *ISRN Pharmaceutics*, *2013*, p.10.

Pandita, D., Kumar, S., Poonia, N. and Lather, V., 2014. Solid lipid nanoparticles enhance oral bioavailability of resveratrol, a natural polyphenol. *Food Research International*, *62*, pp.1165–1174.

Parmar, P., Mishra, A. and Pathak, A., 2016. Preparation and evaluation of ethosomal gel of clotrimazole for fungal infection by mechanical dispersion method. *Current Research in Pharmaceutical Sciences*, *6*(2), pp.5–49.

Patel, S.S., Patel, M.S., Salampure, S., Vishwanath, B. and Patel, N.M., 2010. Development and evaluation of liposomes for topical delivery of tacrolimus (Fk-506). *Journal of Scientific Research*, *2*(3), p.585.

Patrekar, P.V., Inamdar, S.J., Mali, S.S., Mujib, M.T., Ahir, A.A. and Hosmani, A.H., 2015. Ethosomes as novel drug delivery system: A review. *The Pharma Innovation*, *4*(9, Part A), p.10.

Paul, D., Dey, T.K., Chakraborty, A. and Dhar, P., 2018. Promising functional lipids for therapeutic applications. In *Role of Materials Science in Food Bioengineering* (pp.413–449). Academic Press.

Peng, Z., Wang, C., Fang, E., Lu, X., Wang, G. and Tong, Q., 2014. Co-delivery of doxorubicin and SATB1 shRNA by thermosensitive magnetic cationic liposomes for gastric cancer therapy. *PLoS One*, *9*(3), p.e92924.

Pereira, A. and Mallya, R., 2015. Formulation and evaluation of a photoprotectant cream containing Phyllanthus emblica extract-phospholipid complex. *Journal of Pharmacognosy and Phytochemistry*, *4*(2), pp.232–240.

Pichot, R., Watson, R. and Norton, I., 2013. Phospholipids at the interface: Current trends and challenges. *International Journal of Molecular Sciences*, *14*(6), pp.11767–11794.

Priyanka, K. and Singh, S., 2014. A review on skin targeted delivery of bioactives as ultradeformable vesicles: Overcoming the penetration problem. *Current Drug Targets*, *15*(2), pp.184–198.

Qi, C., Chen, Y., Jing, Q.Z. and Wang, X.G., 2011. Preparation and characterization of catalase-loaded solid lipid nanoparticles protecting enzyme against proteolysis. *International Journal of Molecular Sciences*, *12*(7), pp.4282–4293.

Qin, L., Niu, Y., Wang, Y. and Chen, X., 2018. Combination of phospholipid complex and submicron emulsion techniques for improving oral bioavailability and therapeutic efficacy of water-insoluble drug. *Molecular Pharmaceutics*, *15*(3), pp.1238–1247.

Rakesh, R. and Anoop, K.R., 2012. Formulation and optimization of nano-sized ethosomes for enhanced transdermal delivery of cromolyn sodium. *Journal of Pharmacy & Bioallied Sciences*, *4*(4), p.333.

Ramadon, D., Goldie, A.W. and Anwar, E., 2017. Novel transdermal ethosomal gel containing green tea (*Camellia sinensis* L. Kuntze) leaves extract: Formulation and in vitro penetration study. *Journal of Young Pharmacists*, *9*(3), p.336.

Ranade, S.Y. and Gaud, R.S., 2016. Lornoxicam loaded transfersome: Formulation and evaluation. *American Journal of Pharmtech Research*, *6*(3), pp.223–238.

Rigon, R.B., Fachinetti, N., Severino, P., Santana, M.H. and Chorilli, M., 2016. Skin delivery and in vitro biological evaluation of trans-resveratrol-loaded solid lipid nanoparticles for skin disorder therapies. *Molecules*, *21*(1), p.116.

Rizvi, S.A. and Saleh, A.M., 2018. Applications of nanoparticle systems in drug delivery technology. *Saudi Pharmaceutical Journal*, *26*(1), pp.64–70.

Rostami, E., Kashanian, S., Azandaryani, A.H., Faramarzi, H., Dolatabadi, J.E.N. and Omidfar, K., 2014. Drug targeting using solid lipid nanoparticles. *Chemistry and Physics of Lipids*, *181*, pp.56–61.

Rubinstein, I., Soos, I. and Onyuksel, H., 2008. Intracellular delivery of VIP-grafted sterically stabilized phospholipid mixed nanomicelles in human breast cancer cells. *Chemico-Biological Interactions*, *171*(2), pp.190–194.

Saadat, E., Amini, M., Khoshayand, M.R., Dinarvand, R. and Dorkoosh, F.A., 2014. Synthesis and optimization of a novel polymeric micelle based on hyaluronic acid and phospholipids for delivery of paclitaxel, in vitro and in-vivo evaluation. *International Journal of Pharmaceutics*, *475*(1–2), pp.163–173.

Sahib, M.N., Darwis, Y., Peh, K.K., Abdulameer, S.A. and Fung Tan, Y.T., 2012. Incorporation of beclomethasone dipropionate into polyethylene glycol-diacyl lipid micelles as a pulmonary delivery system. *Drug Development Research*, *73*(2), pp.90–105.

Sahib, M.N., Darwis, Y., Peh, K.K., Abdulameer, S.A. and Tan, Y.T.F., 2011. Rehydrated sterically stabilized phospholipid nanomicelles of budesonide for nebulization: Physicochemical characterization and in vitro, in vivo evaluations. *International Journal of Nanomedicine*, *6*, p.2351.

Sailaja, A.K., 2016. Pharmacosomes: A novel carrier for drug delivery. *Innoriginal: International Journal of Sciences*, *3*(6), pp.8–10.

Saoji, S.D., Raut, N.A., Dhore, P.W., Borkar, C.D., Popielarczyk, M. and Dave, V.S., 2016. Preparation and evaluation of phospholipid-based complex of standardized Centella extract (SCE) for the enhanced delivery of phytoconstituents. *AAPS Journal*, *18*(1), pp.102–114.

Saritha, A. and Ravi Kumar, P., 2015. Formulation and evaluation of pegylated lipospheric delivery system of lamivudine. *Journal of Scientific Research in Pharma*, *4*(1), pp.44–47.

Saroja, C. and Lakshmi, P.K., 2013. Formulation and optimization of fenofibrate lipospheres using Taguchi's experimental design. *Acta Pharmaceutica*, *63*(1), pp.71–83.

Sarwa, K.K., Mazumder, B., Rudrapal, M. and Verma, V.K., 2015. Potential of capsaicin-loaded transfersomes in arthritic rats. *Drug Delivery*, *22*(5), pp.638–646.

Satheesh, B.N., Senthil, R.D., Prabakaran, L., Venkata, S.M. and Suriyakala, P.C., 2014. Formulation optimization, scale up technique and stability analysis of naproxen loaded lipospheres. *Asian Journal of Pharmaceutical and Clinical Research*, *7*(2), pp.121–126.

Semalty, A., Semalty, M., Rawat, B.S., Singh, D. and Rawat, M.S.M., 2010. Development and evaluation of pharmacosomes of aceclofenac. *Indian Journal of Pharmaceutical Sciences*, *72*(5), p.576.

Semalty, A., Semalty, M., Singh, D. and Rawat, M.S.M., 2010. Development and characterization of aspirin-phospholipid complex for improved drug delivery. *International Journal of Pharmaceutical Sciences and Nanotechnology*, *3*(2), pp.940–947.

Semalty, M., Badoni, P., Singh, D. and Semalty, A., 2014. Modulation of solubility and dissolution of furosemide by preparation of phospholipid complex. *Drug Development and Therapeutics*, *5*(2), p.172.

Sercombe, L., Veerati, T., Moheimani, F., Wu, S.Y., Sood, A.K. and Hua, S., 2015. Advances and challenges of liposome assisted drug delivery. *Frontiers in Pharmacology*, *6*, p.286.

Sessa, M., Balestrieri, M.L., Ferrari, G., Servillo, L., Castaldo, D., D'Onofrio, N., Donsì, F. and Tsao, R., 2014. Bioavailability of encapsulated resveratrol into nanoemulsion-based delivery systems. *Food Chemistry*, *147*, pp.42–50.

Shaji, J.E.S.S.Y. and Lal, M.A.R.I.A., 2014. Preparation, optimization and evaluation of transferosomal formulation for enhanced transdermal delivery of a COX-2 inhibitor. *International Journal of Pharmacy and Pharmaceutical Sciences*, *6*(1), pp.467–477.

Shalini, S., Kumar, R.R. and Birendra, S., 2015. Antiproliferative effect of Phytosome complex of Methanolic extact of *Terminalia arjuna* bark on Human Breast Cancer Cell Lines (MCF-7). *International Journal of Drug Delivery Research*, *7*(1), pp.173–182.

Shit, S.C. and Shah, P.M., 2014. Edible polymers: Challenges and opportunities. *Journal of Polymers*, *2014*, p.13.

Shreya, A.B., Managuli, R.S., Menon, J., Kondapalli, L., Hegde, A.R., Avadhani, K., Shetty, P.K., Amirthalingam, M., Kalthur, G. and Mutalik, S., 2016. Nano-transfersomal formulations for transdermal delivery of asenapine maleate: In vitro and in vivo performance evaluations. *Journal of Liposome Research*, *26*(3), pp.221–232.

Shuwaili, A.H.A., Rasool, B.K.A. and Abdulrasool, A.A., 2016. Optimization of elastic transfersomes formulations for transdermal delivery of pentoxifylline. *European Journal of Pharmaceutics and Biopharmaceutics*, *102*, pp.101–114.

Silva, A.C., Santos, D., Ferreira, D. and M Lopes, C., 2012. Lipid-based nanocarriers as an alternative for oral delivery of poorly water-soluble drugs: Peroral and mucosal routes. *Current Medicinal Chemistry*, *19*(26), pp.4495–4510.

Silva, A.P., Nunes, B.R., De Oliveira, M.C., Koester, L.S., Mayorga, P., Bassani, V.L. and Teixeira, H.F., 2009. Development of topical nanoemulsions containing the isoflavone genistein. *Die Pharmazie: An International Journal of Pharmaceutical Sciences*, *64*(1), pp.32–35.

Singh, B., Awasthi, R., Ahmad, A. and Saifi, A., 2018. Phytosome: Most significant tool for herbal drug delivery to enhance the therapeutic benefits of phytoconstituents. *Journal of Drug Delivery and Therapeutics*, *8*(1), pp.98–102.

Singh, C., Bhatt, T.D., Gill, M.S. and Suresh, S., 2014. Novel rifampicin–phospholipid complex for tubercular therapy: Synthesis, physicochemical characterization and in vivo evaluation. *International Journal of Pharmaceutics*, *460*(1–2), pp.220–227.

Singh, C., Koduri, L.S.K., Bhatt, T.D., Jhamb, S.S., Mishra, V., Gill, M.S. and Suresh, S., 2017a. In vitro-in vivo evaluation of novel co-spray dried rifampicin phospholipid lipospheres for oral delivery. *AAPS PharmSciTech*, *18*(1), pp.138–146.

Singh, C., Koduri, L.S.K., Dhawale, V., Bhatt, T.D., Kumar, R., Grover, V., Tikoo, K. and Suresh, S., 2015. Potential of aerosolized rifampicin lipospheres for modulation of pulmonary pharmacokinetics and biodistribution. *International Journal of Pharmaceutics*, *495*(2), pp.627–632.

Singh, K.K. and Vingkar, S.K., 2008. Formulation, antimalarial activity and biodistribution of oral lipid nanoemulsion of primaquine. *International Journal of Pharmaceutics*, *347*(1–2), pp.136–143.

Singh, R.P., Gangadharappa, H.V. and Mruthunjaya, K., 2017b. Phospholipids: Unique carriers for drug delivery systems. *Journal of Drug Delivery Science and Technology*, *39*, pp.166–179.

Solanki, D., Kushwah, L., Motiwale, M. and Chouhan, V., 2016. Transferosomes: A Review. *World Journal of Pharmacy and Pharmaceutical Sciences*, *5*, pp.435–449.

Sonneville-Aubrun, O., Simonnet, J.T. and L'alloret, F., 2004. Nanoemulsions: A new vehicle for skincare products. *Advances in Colloid and Interface Science*, *108*, pp.145–149.

Suresh, G., Manjunath, K., Venkateswarlu, V. and Satyanarayana, V., 2007. Preparation, characterization, and in vitro and in vivo evaluation of lovastatin solid lipid nanoparticles. *AAPS Pharmscitech*, *8*(1), pp.E162–E170.

Suriyakala, P.C., Babu, N.S., Rajan, D.S. and Prabakaran, L., 2014. Phospholipids as versatile polymer in drug delivery systems. *International Journal of Pharmacy and Pharmaceutical Sciences*, *6*, pp.8–11.

Swain, S., Beg, S. and Babu, S.M., 2016. Liposheres as a novel carrier for lipid-based drug delivery: Current and future directions. *Recent Patents on Drug Delivery & Formulation*, *10*(1), pp.59–71.

Tan, M.E., He, C.H., Jiang, W., Zeng, C., Yu, N., Huang, W., Gao, Z.G. and Xing, J.G., 2017. Development of solid lipid nanoparticles containing total flavonoid extract from *Dracocephalum moldavica* L. and their therapeutic effect against myocardial ischemia–reperfusion injury in rats. *International Journal of Nanomedicine*, *12*, p.3253.

Tan, Q., Liu, S., Chen, X., Wu, M., Wang, H., Yin, H., He, D., Xiong, H. and Zhang, J., 2012. Design and evaluation of a novel evodiamine-phospholipid complex for improved oral bioavailability. *AAPS Pharmscitech*, *13*(2), pp.534–547.

Tiwari, S.B. and Amiji, M.M., 2006. Improved oral delivery of paclitaxel following administration in nanoemulsion formulations. *Journal of Nanoscience and Nanotechnology*, *6*(9–10), pp.3215–3221.

Upendra, C.G., Umesh, M.B. and Pravin, D.C., 2015. Formulation development, optimization and in-vitro evaluation of glimepiride lipospheres. *International Journal of Pharmaceutical Sciences Review and Research*, *34*(2), pp.157–162.

Valdes, S.A., Alzhrani, R.F., Rodriguez, A., Lansakara-P, D.S., Thakkar, S.G. and Cui, Z., 2019. A solid lipid nanoparticle formulation of 4-(N)-docosahexaenoyl 2′, 2′-difluorodeoxycytidine with increased solubility, stability, and antitumor activity. *International Journal of Pharmaceutics*, *570*, p.118609.

van Hoogevest, P. and Fahr, A., 2019. Phospholipids in cosmetic carriers. In *Nanocosmetics* (pp.95–140). Springer.

Van Hoogevest, P. and Wendel, A., 2014. The use of natural and synthetic phospholipids as pharmaceutical excipients. *European Journal of Lipid Science and Technology*, *116*(9), pp.1088–1107.

Vanić, Ž., 2015. Phospholipid vesicles for enhanced drug delivery in dermatology. *Journal of Drug Discovery, Development and Delivery*, *2*(1), p.1.

Varshosaz, J., Minayian, M. and Moazen, E., 2010. Enhancement of oral bioavailability of pentoxifylline by solid lipid nanoparticles. *Journal of Liposome Research*, *20*(2), pp.115–123.

Vecchione, R., Quagliariello, V., Calabria, D., Calcagno, V., De Luca, E., Iaffaioli, R.V. and Netti, P.A., 2016. Curcumin bioavailability from oil in water nano-emulsions: In vitro and in vivo study on the dimensional, compositional and interactional dependence. *Journal of Controlled Release*, *233*, pp.88–100.

Verma, P. and Pathak, K., 2012. Nanosized ethanolic vesicles loaded with econazole nitrate for the treatment of deep fungal infections through topical gel formulation. *Nanomedicine: Nanotechnology, Biology and Medicine*, *8*(4), pp.489–496.

Vijayakumar, K.S., Parthiban, S., Senthilkumar, G.P. and Tamizmani, T., 2014. Formulation and evaluation of gliclazide loaded ethosomes as transdermal drug delivery carriers. *Asian Journal of Research in Biological and Pharmaceutical Sciences*, *2*, pp.89–98.

Vora, A.K., Londhe, V.Y. and Pandita, N.S., 2015. Preparation and characterization of standardized pomegranate extract-phospholipid complex as an effective drug delivery tool. *Journal of Advanced Pharmaceutical Technology & Research*, *6*(2), p.75.

Wang, G., Wang, J. and Guan, R., 2020. Novel phospholipid-based labrasol nanomicelles loaded flavonoids for oral delivery with enhanced penetration and anti-brain tumor efficiency. *Current Drug Delivery*, *17*(3), pp.229–245.

Wang, G., Wang, J.J., Chen, X.L., Du, L. and Li, F., 2016. Quercetin-loaded freeze-dried nanomicelles: Improving absorption and anti-glioma efficiency in vitro and in vivo. *Journal of Controlled Release*, *235*, pp.276–290.

Wang, J.J., Sung, K.C., Hu, O.Y.P., Yeh, C.H. and Fang, J.Y., 2006. Submicron lipid emulsion as a drug delivery system for nalbuphine and its prodrugs. *Journal of Controlled Release*, *115*(2), pp.140–149.

Wang, T., Wang, N., Hao, A., He, X., Li, T. and Deng, Y., 2010. Lyophilization of water-in-oil emulsions to prepare phospholipid-based anhydrous reverse micelles for oral peptide delivery. *European Journal of Pharmaceutical Sciences*, *39*(5), pp.373–379.

Wen, Z., Liu, B., Zheng, Z., You, X., Pu, Y. and Li, Q., 2010. Preparation of liposomes entrapping essential oil from *Atractylodes macrocephala* Koidz by modified RESS technique. *Chemical Engineering Research and Design*, *88*(8), pp.1102–1107.

Xia, H.J., Zhang, Z.H., Jin, X., Hu, Q., Chen, X.Y. and Jia, X.B., 2013. A novel drug–phospholipid complex enriched with micelles: Preparation and evaluation in vitro and in vivo. *International Journal of Nanomedicine*, *8*, p.545.

Xie, J., Li, Y., Song, L., Pan, Z., Ye, S. and Hou, Z., 2017. Design of a novel curcumin-soybean phosphatidylcholine complex-based targeted drug delivery systems. *Drug Delivery*, *24*(1), pp.707–719.

Yang, J.H., Zhang, L., Li, J.S., Chen, L.H., Zheng, Q., Chen, T., Chen, Z.P., Fu, T.M. and Di, L.Q., 2015. Enhanced oral bioavailability and prophylactic effects on oxidative stress and hepatic damage of an oil solution containing a rosmarinic acid–phospholipid complex. *Journal of Functional Foods*, *19*, pp.63–73.

Yang, L., Geng, Y., Li, H., Zhang, Y., You, J. and Chang, Y., 2009. Enhancement the oral bioavailability of praziquantel by incorporation into solid lipid nanoparticles. *Die Pharmazie-An International Journal of Pharmaceutical Sciences*, *64*(2), pp.86–89.

Yue, P.F., Zheng, Q., Wu, B., Yang, M., Wang, M.S., Zhang, H.Y., Hu, P.Y. and Wu, Z.F., 2012. Process optimization by response surface design and characterization study on geniposide pharmacosomes. *Pharmaceutical Development and Technology*, *17*(1), pp.94–102.

Yusuf, M., Sharma, V. and Pathak, K., 2014. Nanovesicles for transdermal delivery of felodipine: Development, characterization, and pharmacokinetics. *International Journal of Pharmaceutical Investigation*, *4*(3), p.119.

Zhang, J., Wen, X., Dai, Y. and Xia, Y., 2019. Mechanistic studies on the absorption enhancement of a self-nanoemulsifying drug delivery system loaded with norisoboldine-phospholipid complex. *International Journal of Nanomedicine*, *14*, p.7095.

Zhang, N., Ping, Q., Huang, G., Xu, W., Cheng, Y. and Han, X., 2006. Lectin-modified solid lipid nanoparticles as carriers for oral administration of insulin. *International Journal of Pharmaceutics*, *327*(1–2), pp.153–159.

Zhao, X., Shi, C., Zhou, X., Lin, T., Gong, Y., Yin, M., Fan, L., Wang, W. and Fang, J., 2019. Preparation of a nanoscale dihydromyricetin-phospholipid complex to improve the bioavailability: In vitro and in vivo evaluations. *European Journal of Pharmaceutical Sciences*, *138*, p.104994.

Zheng, H., Wijaya, W., Zhang, H., Feng, K., Liu, Q., Zheng, T., Yin, Z., Cao, Y. and Huang, Q., 2020. Improving the bioaccessibility and bioavailability of carnosic acid using a lecithin-based nanoemulsion: Complementary in vitro and in vivo studies. *Food & Function*, *11*(9), pp.8141–8149.

Zhou, H., Yue, Y., Liu, G., Li, Y., Zhang, J., Gong, Q., Yan, Z. and Duan, M., 2010. Preparation and characterization of a lecithin nanoemulsion as a topical delivery system. *Nanoscale Research Letters*, *5*(1), p.224.

Zylberberg, C. and Matosevic, S., 2016. Pharmaceutical liposomal drug delivery: A review of new delivery systems and a look at the regulatory landscape. *Drug Delivery*, *23*(9), pp.3319–3329.

9 Investigating the Potential of Multifunctional Nanoparticle-Based Nutraceuticals in Targeted Therapeutics

*Giselle Amanda Borges e Soares**,
Ali H. Hamzah, and Tanima Bhattacharya

CONTENTS

Abbreviations 224
9.1 Introduction 225
9.2 Nutraceuticals: Present-Day Practices and Challenges 227
9.3 Nanocarriers for Drug Delivery 229
9.3.1 Liposomes 231
9.3.2 Solid Lipid Nps (SLNs) 231
9.3.3 Polymeric Micelles (PMs) 232
9.3.4 Dendrimers 232
9.3.5 Virus-Based Nps (VNps) 232
9.3.6 Carbon Nanotubes (CNTs) 233
9.3.7 Metallic Nps 233
9.3.8 Quantum dots (QDs) 233
9.3.9 Hybrid Nanocarriers 233
9.4 Novel Multifunctional Therapeutics Development: Np-Nutraceutical (Np-Nu) Conjugation Techniques 234
9.5 Role of Np-Nus in Neurodegenerative Disorders 235
9.5.1 Np–Curcumin Conjugates 238
9.5.2 Np–Resveratrol Conjugates 239
9.5.3 Np–Quercetin Conjugates 239
9.5.4 Np-Scylloinositol Conjugates 239
9.5.5 Np–Lycopene Conjugates 240
9.5.6 Np–Bryostatin Conjugates 240
9.5.7 Np–Asiatic Acid Conjugates 240
9.5.8 Np–Huperzine A Conjugates 240
9.6 Role of Np-Nus in Cancer Management 241
9.6.1 Np–Thymoquinone Conjugates 242
9.6.2 Np–Dihydroartemisin Conjugates 242
9.6.3 Np–Eugenol Conjugates 243
9.6.4 Np–Resveratrol Conjugates 243
9.6.5 Np–Naringenin Conjugates 243

* Corresponding author: giselleamanda.borgesesoares@rockets.utoledo.edu

DOI: 10.1201/9781003244721-9

9.7 Role of Np-Nus in Oxidative Stress 243
9.7.1 Np–Curcumin Conjugates 246
9.7.2 Np–Epigallocatechin Gallate Conjugates 247
9.7.3 Np–Lutein Conjugates 247
9.8 Safety Evaluation/Toxicity Concerns 247
9.9 Research Gaps and Future Perspective of Np-Nus 249
9.10 Conclusion 250
Acknowledgements 250
References 250

ABBREVIATIONS

5-FU 5-fluorouracil
AChE acetylcholine esterase
AD Alzheimer's disease
ADME absorption, distribution, metabolism, and elimination
BBB blood–brain barrier
CAT catalase
CCMV cowpea chlorotic mottle virus
CDC Centers for Disease Control
CMC critical micelle concentration
CNS central nervous system
CNTs carbon nanotubes
COMT catechol o-methyltransferase
COX-2 cyclooxygenase-2
CPMV cowpea mosaic virus
CT computer tomography
CVDs cardiovascular diseases
DLS dynamic light scattering
DNA deoxyribonucleic acid
DDS drug delivery system
DR diabetic retinopathy
EGCG epigallocatechin gallate
EFSA European Food Safety Authority
EU European Union
FDA Food and Drug Administration
GI gastrointestinal tract
GRAS generally recognized as safe
GSH-P glutathione peroxidase
GSH-R glutathione reductase
H_2O_2 hydrogen peroxide
HO-1 heme oxygenase-1
HRECs human retinal endothelial cells
Hup A Huperzine A
iNOS nitric oxide synthase
LDL low-density lipoprotein
MNps magnetic nanoparticles
MSNs microporous silica nanoparticles
MWCNTs multi-walled carbon nanotubes
NADPH nicotinamide adenine dinucleotide phosphate
NMDA N-methyl-d-aspartate

NNI	National Nanotechnology Initiative
NO	nitric oxide
NOX	nicotinamide adenine dinucleotide phosphate oxidase
Np	nanoparticle
Np-Nu	nanoparticle nutraceutical conjugate
Nrf2	erythroid-2-related factor-2
Nu	nutraceutical
O_2^-	superoxide anion
-OH	hydroxyl radical
PAMAM	poly amidoamine
PD	Parkinson's disease
PEG	polyethylene glycol
PEG-PA	PEGylated polylactic acid
PEG-PAA	PEGylated poly aspartic acid
PEG-PCL	PEGylated polycaprolactone
PEG-PLGA	PEGylated poly (lactic-*co*-glycolic acid)
PKC	protein kinase C
PMs	polymeric micelles
PRX	peroxiredoxin
QD	quantum dot
RBC	red blood cell
RCNMV	red clover necrotic mosaic virus
RES	reticuloendothelial system
RNps	radical-containing nanoparticles
RO	alkoxyl radical
ROO	peroxyl radical
ROS	reactive oxygen species
siRNA	small interfering ribonucleic acid
SLNs	solid lipid nanoparticles
SOD	superoxide dismutase
SPION	superparamagnetic iron oxide nanoparticle
SSRIs	selective serotonin reuptake inhibitors
SWCNTs	single-walled carbon nanotubes
TCA	tricyclic antidepressant
TEG	thromboelastographic
TMV	tobacco mosaic virus
tPA	tissue plasminogen activator
TPP	triphenyl phosphonium
TQ	thymoquinone
Trx	thioredoxin
UNFAO	United Nations Food and Agricultural Organization
USFDA	United States Food and Drug Administration
VA	vanillyl alcohol
VNps	virus-based nanoparticles
WHO	World Health Organization

9.1 INTRODUCTION

Since prehistoric times, humans have utilized plants and animal products derived from natural sources for the treatment of various ailments. Advances in research and technology have caused the design and development of many therapeutics derived from plants. Currently, natural resources and their derivatives constitute about 25% of the compounds marketed by pharmaceutical companies (Mohanty et al., 2017; Swamy & Sinniah, 2016).

Natural compounds that possess varying structures and functional groups have served and continue to serve as starting points for the development and discovery of novel therapeutics. Over the years, several natural bioactive compounds have been thoroughly investigated and screened for their ability to act as lead molecules that can be synthetically modified to produce efficient and safe therapeutics. Their possession of bioactivity coupled with targeted delivery and lower toxicity allows them to act as favorable leads in drug discovery and development (Bhattacharya et al., 2021a; Rodrigues et al., 2016; Siddiqui et al., 2014). The use of computational studies and docking software has also aided in understanding the molecular interactions of these bioactive molecules with target proteins within the body (enzymes, receptors, channels). To improve the efficacy and binding of these bioactive molecules to their targets, scientists have tried altering structural parameters of these molecules, such as rational functional group modification through understanding the structure–activity relationships (SAR), changes in stereochemistry, heterocyclic ring variations, etc. Derivatives of lead molecules known to possess biological activity have also been subjected to chemical modification to develop novel therapeutics (Gaude et al., 2017).

Natural products offer many advantages, such as low cost, good therapeutic potential, lower side effects, and lower toxicity, and are being screened for the treatment of diseases such as diabetes, inflammatory and cardiovascular diseases (CVDs), cancer, etc. (Bhattacharya et al., 2021a). Although natural products are safe and offer several advantages, pharmaceuticals are more invested in exploring and modifying already available compounds as therapeutics, since natural product screening and discovery is time-consuming, expensive and does not always result in the discovery of a novel lead molecule. However, they possess certain drawbacks and have difficulty moving past clinical trials due to their low solubility, large dimensions, poor absorption, and lack of target specificity, to name a few. Therefore, over the years, scientists have tried to overcome these challenges through the development of novel nanotechnology-based DDS (Bonifacio et al., 2014; Jahangirian et al., 2017; Martinho et al., 2011; Thilakarathna & Rupasinghe, 2013; Watkins et al., 2015). Successful coupling of nanotechnology with drug delivery has resulted in the generation of newer nanocarriers or nanoformulations that offer site-specific targeted delivery and controlled drug release.

Nanotechnology has bridged the challenges faced by scientists in the biological and physical sciences through the application and development of nanostructures and nanoparticles (Nps) in drug delivery, biosensors, microfluidics, and tissue engineering (Liu et al., 2009). Nanomaterials and nanostructures possess sizes and dimensions ranging from 1 to 100 nm (Arayne et al., 2007; Joseph & Venkatraman, 2017; Jayanta Kumar et al., 2014). Due to their small dimensions, they can penetrate the body better and can remain in circulation for extended periods in comparison to larger-sized drugs. Besides their structural features, they also possess unique chemical, magnetic, mechanical, electric, and biological properties. The ability to encapsulate large-sized molecules as well as lipophilic and hydrophilic drug molecules, which generally have low bioavailability, has led to nanomedicines being well appreciated as drug delivery agents (Jahangirian et al., 2017; Lam et al., 2017). Nanomedicine is a rapidly emerging field that encompasses the use of nanoscience-based techniques and principles in medical biology and disease management/prevention. For instance, Haba et al. developed gold Nps that had photothermal as well as imaging capability relevant to cancer diagnosis and treatment (Haba et al., 2007). These nanosystems show significantly higher oral bioavailability in comparison to the free drug forms, as they exhibit an absorptive endocytosis-based uptake mechanism. In addition, nanoformulations prevent early degradation of the drug by gastrointestinal (GI) tract enzymes and thus, aid in hydrophobic drug delivery to the required site. Figure 9.1 illustrates the current applications of nanotechnology in healthcare settings.

The use of nano-based systems and the choice of nanocarriers are dependent on the physicochemical aspects of the drug. Natural bioactive compounds have been widely investigated over the years due to their ability to cure and manage inflammation and other disease states as well as the ability to induce tumor-suppressant and antimicrobial, antibacterial, and antiviral action

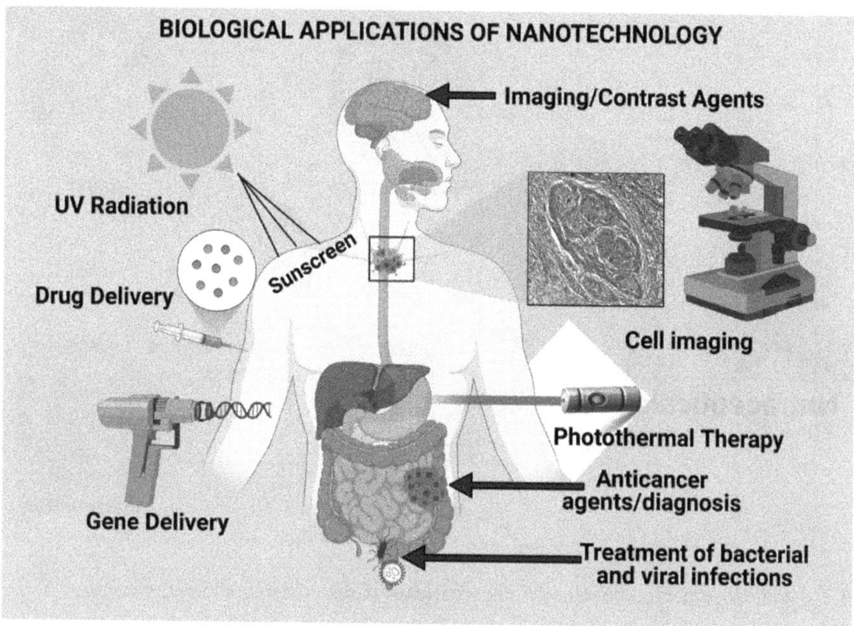

FIGURE 9.1 Current biological applications of nanotechnology. (Created using Biorender.com.)

(T. Bhattacharya et al., 2021b). Cinnamaldehyde, curcumin, and eugenol possess antimicrobial activity (Ouattara et al., 1997; Sharma et al., 2014), while curcumin and caffeine facilitate autophagy (Wang & Feng, 2015). The encapsulation of these bioactive constituents into nanocarriers has led to the enhancement of their biological activities as well as properties such as controlled release, site-specific delivery, and bioavailability. Thymoquinone (TQ) isolated from *Nigella sativa*, a flowering plant that is well known for its benefits in asthma, inflammation, and bronchitis, was encapsulated in a lipid nanocarrier by Abdelwahab et al. The researchers observed a sixfold enhancement of bioavailability in comparison to free TQ, and this was found extremely beneficial in gastric ulcer treatment. It also enhanced the pharmacokinetic profile, rendering better therapeutic efficiency (Abdelwahab et al., 2013).

Inadequate research and lack of knowledge about nanodrug toxicity is a major concern of nanosystems and therefore, needs further exploration for the improvement of safety and practical administration of these medicines. Expertise, as well as cautious designing of nanoformulations, could assist in challenges pertaining to their utilization.

This chapter is intended to shed light on the various nanocarriers used in drug delivery, techniques employed for the conjugation of nutraceuticals and Nps (Np-Nus), as well as the current practices and challenges faced in the nutraceutical industry. An extensive review of the various Np-Nus so far investigated for the treatment of neurodegenerative disease, cancer, and oxidative stress is presented. All figures for illustrative purposes were created using Biorender.com. Finally, concerns pertaining to safety evaluation, toxicity, and a future perspective to bridge research gaps in the field are addressed.

9.2 NUTRACEUTICALS: PRESENT-DAY PRACTICES AND CHALLENGES

Nutraceuticals have gained significant importance in the last decade due to their nutritional and medicinal value (Nasri et al., 2014). By definition, they possess physiological benefits or provide protection against chronic diseases (Kalra, 2003). The natural constituents of nutraceuticals have also been shown to have curative effects. For instance, polyunsaturated omega-3 fatty acid, a constituent found in salmon and flax seeds, has been found to regulate brain function, immune

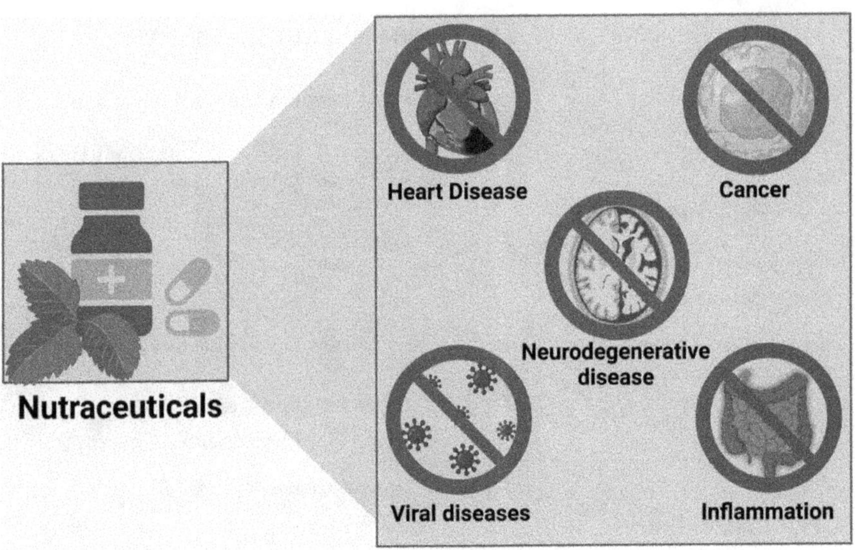

FIGURE 9.2 Well-documented benefits of nutraceuticals. (Created using Biorender.com.)

responses, and deposition of cholesterol (Avallone et al., 2019). Tannin oil obtained from lavender (*Lavandula angustifolia*) has been found to be useful in hysteria and paralysis-based central nervous system (CNS) disorders (B. Ali et al., 2015; Donelli et al., 2019). Lavender oil and rosemary–based aromatherapy has been found to be advantageous in dementia management and also enhanced cognition in Alzheimer's patients due to the free radical scavenging activity of the tannin oils (Atsumi & Tonosaki, 2007; Soares et al., 2021). Moreover, the beneficial effects of curcumin, present in turmeric, in neurodegenerative disease are well known (R. B. Mythri & Bharath, 2012; Mythri & Bharath, 2019). Although well documented, the study of nutraceutical compositions, their chemistry, and their mechanism of action as therapeutics poses multiple challenges in the field of medicinal chemistry, thereby making their clinical effectiveness during trials difficult to assess (Sut et al., 2016).

Due to their possession of anti-inflammatory, anticancer, anti-hypertensive, antioxidant effects, nutraceuticals have been widely accepted as therapeutics (Makkar et al., 2020). Figure 9.2 illustrates the beneficial biological activities of nutraceuticals harnessed currently. Nutraceuticals such as gallic acid, curcumin, caffeine have shown anti-aging and antioxidant activity (Williams et al., 2015). Nutraceuticals have also been found to enhance cholinergic production and acetylcholinesterase inhibition, thereby behaving as anti-inflammatory and anti-hypercholesterolemic agents (Ghabaee et al., 2010). In the last decade, several studies of the benefits of fish oil, rich in polyunsaturated fatty acid methyl esters, have been pursued by researchers worldwide due to its ability to improve cognitive function and reduce heart disease (Sokola-Wysoczanska et al., 2018).

Despite all their advantages, major drawbacks associated with nutraceuticals include diminished levels of absorption, bioavailability, and permeability through the blood–brain barrier (BBB) (C. Braithwaite et al., 2014; Pandareesh et al., 2015). Decreased absorption has been attributed to the type of active ingredients contained within the nutraceutical tested (Keservani et al., 2017). Moreover, a major concern with respect to the composition of nutraceuticals coupled with other ingredients/chemicals or drugs is their ability to interact in an antagonistic manner, resulting in therapeutic ineffectiveness or undesired/untoward physiological effects (Bushra et al., 2011). With respect to bioavailability, characteristics such as diminished stability, permeability, and solubility of the bioactive component in the GI tract need to be considered. For instance, curcumin has been shown to possess low peroral bioavailability, which has been attributed to partial tissue distribution, poor serum levels, short half-life, and rapid first-pass metabolism (Zaki, 2014). While optimization to cope with low hydrophilicity and poor dissolution in GI fluids would be beneficial for administration of coenzyme Q10, a lipophilic antioxidant, considerations of permeability into tissues, efflux of P-glycoprotein and active transport would be necessary for larger moieties (Balakrishnan et al., 2009). Therefore, based on the active ingredient and the accessory chemicals/ingredients contained in the nutraceutical, various parameters and several approaches need to be navigated to ensure targeted delivery and efficient bioavailability.

The oral delivery of drug moieties that possess low hydrophilicity and stability is another barrier (Pathak & Raghuvanshi, 2015). However, Np delivery of these molecules can improve bioavailability and therapeutic effectiveness. An example of the utilization of nanotechnology for drug delivery is that of nitrendipine and nimodipine, calcium channel blockers that have an oral bioavailability of approximately 10% and 20% due to the first-pass metabolism. Coupling of these agents with solid lipid Nps (SLNs) containing triglyceride, monoglyceride, and wax, using a 23-factorial design, was found to improve the bioavailability of both nimodipine and nitrendipine (Chalikwar et al., 2012; Kumar et al., 2007).

Park *et al.* in 2013 investigated a novel method pertaining to drug delivery. Generally speaking, nanocarriers coupled with drugs were able to enter the blood circulation and therefore reach the target site; however, only a small fraction of these drugs penetrated the brain due to the BBB. This could be bypassed through an increase in drug loading capacity (Park, 2013). Gao et al., suggested that the formulation of a successful nanoformulation should be based on the consideration that only a small fraction (approximately 5%) of the total administered Nps reach the target site, as the general drug loading rate in a nanocarrier was approximately 10%. Therefore, to ensure a delivery rate of 25%, an enhancement of drug loading capacity by a factor of 5 could provide a greater amount of drug at the target site (L. Gao et al., 2012). To provide enhanced physiochemical stability and cell affinity of the conjugated moiety, the surface properties can be altered using surfactants or polymer functionalization (Guerrini et al., 2018; Mozetic, 2019). Further, drug delivery into the brain can be enhanced by the optimization and development of Nps with an enhanced drug loading capacity and controlled drug release (Masserini, 2013).

Although many approaches, such as chemical modification along with liposomal-based delivery, have been studied over the years, Nps coupled with nutraceuticals (Np-Nus) have been shown to be promising in enhancing targeted drug delivery as well as delivering the drug across the BBB. Being a novel solution to bypass lower bioavailability and biodistribution of currently existing marketed preparations, the *in vivo* therapeutic efficiency in clinical trials, efficacy, quality, and safety of these Np-Nu conjugates need consideration (Moradi et al., 2020). The benefits of coupling nanotechnology with nutraceuticals for efficient and targeted drug delivery, along with its limitations, will be discussed in subsequent sections.

9.3 NANOCARRIERS FOR DRUG DELIVERY

According to the National Nanotechnology Initiative (NNI), Nps are defined as chemical structures possessing sizes ranging from 1 to 100 nm (Wilczewska et al., 2012). As compared with bulky large chemical moieties, nanocarrier delivery of therapeutics offers better targeted delivery as well as a more efficient take-up by cells (Suri et al., 2007). The past few decades have witnessed extensive studies on nanocarriers as agents to provide efficient delivery of drugs due to their enhanced surface area to volume ratio, modification of basic properties, and drug bioactivity (How et al., 2013; Mishra et al., 2010). Other favorable features provided by nanocarriers include enhanced biodistribution and pharmacokinetics, reduced toxicity, solubility and stability enhancement, and controlled and targeted delivery (Mishra et al., 2010). Moreover, through composition alteration (organic, inorganic, or hybrid), alterations in size (small or large), shape (sphere, rod, or cube), and surface property (surface charge, functional groups, surface coating, or PEGylation), the physicochemical properties can be varied, thus making them highly versatile (T. Sun et al., 2014). Figure 9.3 and Figure 9.4 showcase the structural attributes of the organic and inorganic nanocarriers described

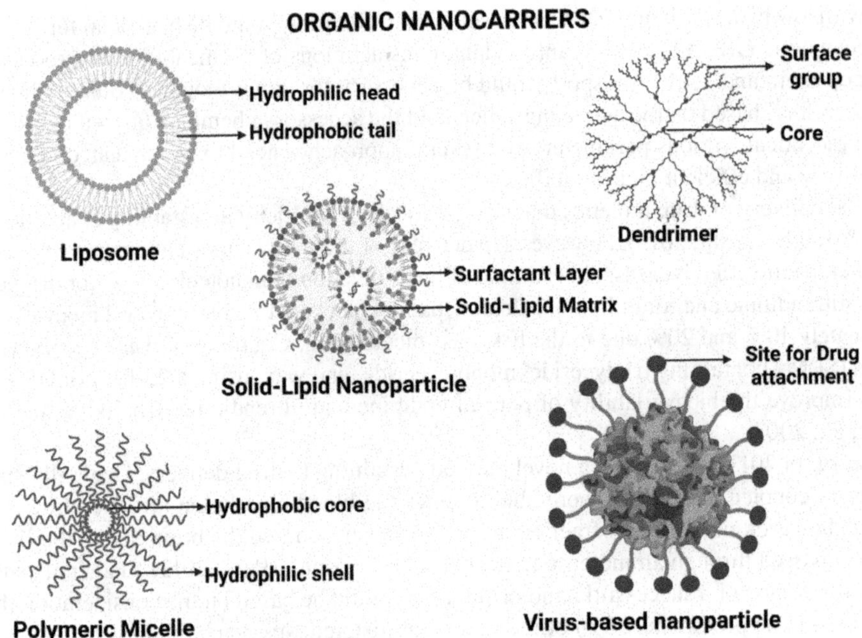

FIGURE 9.3 Structures of organic nanocarriers used for drug delivery. (Created using Biorender.com.)

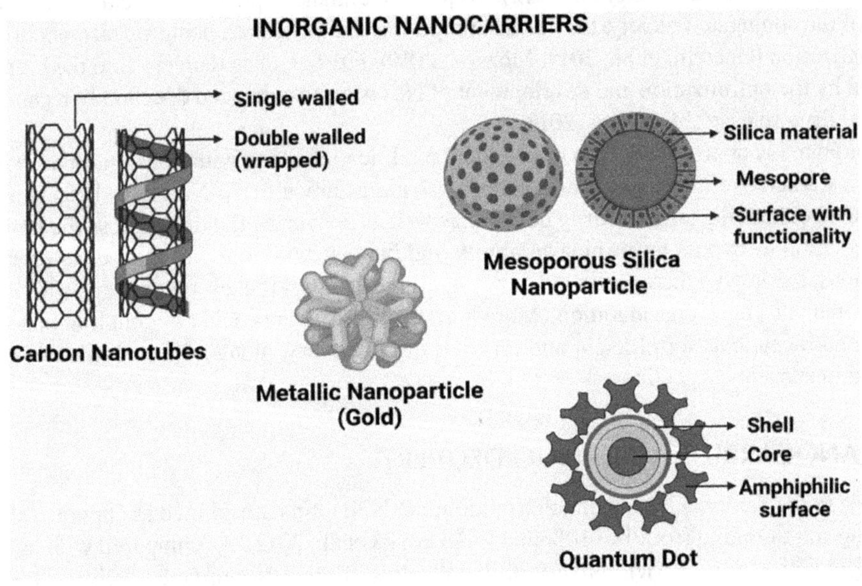

FIGURE 9.4 Structures of inorganic nanocarriers used for drug delivery. (Created using Biorender.com.)

in this chapter, respectively, while Table 9.1 provides a summary of the various characteristics and types of nanocarriers used for delivery of drugs.

Nanocarrier classification comprises (T. Sun et al., 2014):

1) Organic Nps: Liposomes, SLNs, polymeric micelles, dendrimers, and virus-based Nps
2) Inorganic Nps: Carbon nanotubes, metallic Nps, mesoporous solid Nps, and quantum dots
3) Organic/inorganic hybrid nanocarriers

TABLE 9.1
Summary of the Nanocarriers Used and the Benefits Offered

Nanocarrier type	Classification	Beneficial characteristics	References
Liposomes	Organic	Enhanced drug solubility, drug metabolism, and pharmacokinetics	(Suri et al., 2007)
SLNs	Organic	Controlled drug release and improved physical stability	(Mudshinge et al., 2011)
PMs	Organic	Better delivery of hydrophilic drug molecules with improvement in stability and bioavailability	(Amin, Butt et al., 2017; Liu et al., 2020)
Dendrimers	Organic	Better delivery of hydrophobic molecules along with addition of other agents (APIs, excipients)	(Lombardo et al., 2019)
VNps	Organic	Hydrophilicity, better entrapment of drug, and biocompatibility	(Ma et al., 2012)
CNTs	Inorganic	High drug payload, structural flexibility, surface functionalization, and stability	(Din et al., 2017)
Metallic Nps	Inorganic	Contrast agents for MRI, targeted and gene delivery, and imaging	(Mody et al., 2010)
QDs	Inorganic	Real-time visualization of tumor along with monitoring release of drug at target site	(Zayed et al., 2019)
Organic/Inorganic Hybrid	Hybrid	Increased cellular uptake, efficient delivery, controlled delivery of siRNA, high payload, and targeted delivery	(Han et al., 2015; Prabhakar et al., 2016; Xia et al., 2009)

9.3.1 Liposomes

Liposomes have dimensions ranging between 80 and 300 nm (Suri et al., 2007) and are promising carriers that can enable the enhancement of pharmacokinetic properties, drug solubility, and effectiveness of chemotherapeutic agents, and are rapidly metabolized with a lower propensity for side effects *in vivo* (Deng et al., 2019). They can be used for loading both hydrophobic and hydrophilic drug moieties. *In vivo* studies indicate that liposomes are rapidly captured by macrophages circulating in the CNS, which allows cell-specific adsorption as well as targeted release (Fontes et al., 2018). Liposomal modifications could also achieve higher efficiencies in drug delivery. For example, surface modification of liposomes with polyethylene glycol (PEG) allows prolonged periods of circulation in the blood, along with a reduction in uptake by liver and splenic mononuclear phagocytes (Feng, 2006).

9.3.2 Solid Lipid Nps (SLNs)

These consist of a matrix that is based on solid lipid and were developed in the early 1990s. They possess dimensions ranging from 50 to 1,000 nm (Muller et al., 2000). The preparation of SLNs involves the dispersion in water of melted solid lipid(s), such as mixtures of glyceride, mono-, di- and triglycerides, free fatty acids, free fatty alcohols, waxes, and steroids that are coupled with surfactant(s) or emulsifier(s) for ensuring stabilization of the dispersion. The techniques used for the preparation are high-pressure homogenization and micro-emulsification. SLNs offer plenty of advantages, such as controlled drug delivery, minimal to zero biotoxicity, enhanced bioavailability of poorly hydrophilic drugs, protection of the drug from degradation, and stability(Mehnert & Mader, 2001; Mudshinge et al., 2011; Zeb et al., 2017), along with economically facile large-scale production (Tekade et al., 2017). Several proposals have been made for the incorporation of drug moieties into SLNs. However, based on the composition (lipid, drug, and surfactant) of SLNs as well as the conditions of production (homogenization using heat/cold), the drug can be (1) dispersed homogeneously in a matrix of lipid, also known as the solid solution or homogeneous matrix model, (2) introduced inside a shell that surrounds a core comprising lipid, also known as the drug-enriched shell model, or (3) introduced into a core that is further encapsulated by a lipid shell, also known as the drug-enriched core model (Muller et al., 2002). Due to their enormous benefits, SLNs have been widely utilized for anticancer therapeutic development. Challenges that require to be overcome include rapid reticuloendothelial system (RES) elimination from the bloodstream, encapsulation of drugs with high water solubility/ionic strength, and control over drug release at the target site (Wong et al., 2007). Examples where the use of these systems was successful include docetaxel incorporation (Qureshi et al., 2017), coupling to doxorubicin (Battaglia et al., 2014) and paclitaxel (Kakkar et al., 2015), and the incorporation of 5-fluorouracil (5-FU) and methotrexate (Patel et al., 2014).

9.3.3 Polymeric Micelles (PMs)

These were developed for the delivery of hydrophobic drugs that possess low solubility and bioavailability (Deng et al., 2019). They comprise two functional regions: an amphiphilic block polymer region that organizes itself into a shell-like structure in the presence of an aqueous medium, thereby increasing the overall stability of the system, and an internal center region that allows the binding of lipophilic drug molecules due to its hydrophobicity (J. K. Patra et al., 2018). The absorption, distribution, metabolism and elimination (ADME) characteristics of Docetaxel, an antineoplastic drug, were found to be enhanced in a rodent model using (ethylene glycol)-b-poly(epsilon-caprolactone) micelles (Mikhail & Allen, 2010). Targeted delivery using ultrasound was achieved using a nanocarrier based on poly butyl cyanoacrylate–based polymeric micro-bubbles (M. Liu et al., 2020). In Canada, Europe, and the United States, n-butyl cyanoacrylate is commonly used to produce PMs. Examples include Indermil® and liquiband®, both of which are tissue adhesive agents (S. Gao et al., 2015).

9.3.4 Dendrimers

The presence of unique surface functional groups coupled with a hydrophobic nature makes dendrimers excellent and highly promising drug delivery vesicles. Their nature allows them to improve hydrophobic drug solubility and bioavailability (Matea et al., 2017). Dendrimers such as PAMAM (poly amidoamine) and poly propylene amine have commonly been used for drug delivery and imaging (Madaan et al., 2014). The ability to add and modify surface functional groups allows targeted delivery as well as the incorporation or conjugation of drugs/antibodies. However, their clinical use is limited due to the toxicity observed resulting from the presence of an amine group (Zayed et al., 2019).

9.3.5 Virus-Based NPs (VNPs)

VNps or virus-like particles have dimensions of less than 100 nm and consist of self-assembled, robust protein cages with a well-defined and uniform geometry (Manchester & Singh, 2006; Singh et al., 2007). VNps have been shown to be promising nanosystems for imaging, vaccination, and gene therapy. (Pattenden et al., 2005). The protein cage that makes up the core of this nanostructure is obtained from various viruses. Plant viruses such as the tobacco mosaic virus (TMV), cowpea mosaic virus (CPMV), animal viruses including the adenovirus and polyomavirus, and bacteriophages such as MS2, M13, etc., have been investigated for the development of VNps (Ma et al., 2012; Manchester & Singh, 2006; Singh et al., 2007). Advantages offered by this nanocarrier include morphological uniformity, biocompatibility, facile functionalization of the surface, and availability to modify the size and shape (Ma et al., 2012). The ability to perform chemical as well as genetic modification of surface functional groups allows features such as hydrophilicity coupled with better entrapment of drug and biocompatibility. Moreover, the addition of PEG to the surface enhances the circulation and distribution time in the host (Jabir et al., 2012; Ma et al., 2012).

9.3.6 Carbon Nanotubes (CNTs)

In 1991, CNTs, which are hollow, nanosized, tube-like assemblies, were discovered by Iijima (Iijima, 1991). CNTs are developed through rolling graphene sheet(s) into a column/tube-like structure (Bianco, 2004). CNTs can either consist of a roll-up of one graphene sheet, termed single-walled (SWCNTs), or comprise a roll-up of multiple graphene sheets, termed multi-walled (MWCNTs). A benefit that CNTs offer is that they have small dimensions while also being capable of extension by over 10^3 times their diameter. The diameter of both kinds of CNTs usually ranges from 0.4 to 2 nm and from 2 to 100 nm, respectively (Madani et al., 2011). Techniques used for the development of CNTs include arc discharge, chemical vapor deposition enhanced by heat or plasma, and laser ablation (Din et al., 2017; Yan et al., 2007).

9.3.7 Metallic Nps

Extensive research has been conducted on these Nps, including gold, silver, and magnetic Nps, for the development of diagnostic and therapeutic agents (Mody et al., 2010). Moreover, these Nps have been widely utilized as contrast agents in diagnostics such as computed tomography (CT) and X-ray as well as Raman spectroscopy. Photodynamic therapy using gold Nps as cytotoxic agents has been investigated (Elahi et al., 2018; Jeong et al., 2019). Other metals have also been investigated and developed as metallic Nps. Superparamagnetic iron oxide Nps (SPIONs) with a modified surface have become a very promising candidate for biomedicinal applications such as drug delivery, magnetic resonance imaging (MRI), gene therapy, detection of inflammation, imaging, and hyperthermia (Dulinska-Litewka et al., 2019; Idris et al., 2018; Yoffe et al., 2013).

9.3.8 Quantum dots (QDs)

These inorganic semiconducting fluorescent-based nanosystems have been widely investigated for their use in imaging and efficient drug delivery (Patra et al., 2018). QDs vary in size and usually possess dimensions ranging from 2 to 10 nm (Matea et al., 2017). Coupled with quantum effects, their small size allows the visualization of tumors as well as real-time monitoring of targeted drug release (Barry et al., 2015). QDs can also provide sustained and controlled release of medication using external stimulants including heat, light, magnetic fields, or radiofrequency (Karimi et al., 2016).

9.3.9 Hybrid Nanocarriers

To take advantage of the benefits offered by both organic and inorganic Nps, hybrid nanocarriers were developed. The approach of coupling organic functional groups of organic compounds on inorganic Np surfaces has been investigated for many years to provide better efficiency and selectivity of antitumor agents, thereby reducing the side effects usually observed with these classes of molecules. An example is surface coating with polyethyleneimine (PEI), which was found to enhance uptake of microporous silica Nps (MSNs) into cells while also helping to generate cationic regions on the surface for efficient nucleic acid delivery (Xia et al., 2009). The coupling of hyperbranched PEI with MSNs was also found to be advantageous, resulting in sustained and controlled delivery of short interfering RNA (siRNA) as well as a high payload (Prabhakar et al., 2016). These hybrid nanosystems reached the site of the tumor successfully and migrated from the endosomes to the cytoplasm. Desai et al. further investigated a hybrid system comprising MSNs and lipids in breast cancer and observed efficient delivery and retention of zoledronic acid intracellularly (D. Desai et al., 2017). Han et al. formulated a doxorubicin-based hybrid coupled with lipid-capped MSNs that had the ability to release drug molecules based upon pH and redox state. These hybrids showed targeted delivery to the tumor site and better uptake efficiency of doxorubicin, as well as possessing better accumulation intracellularly as compared with the free drug solution (Han et al., 2015).

9.4 NOVEL MULTIFUNCTIONAL THERAPEUTICS DEVELOPMENT: Np-NUTRACEUTICAL (Np-Nu) CONJUGATION TECHNIQUES

The effectiveness of targeted delivery of drug moieties is highly dependent on the process used for nanosystem production as well as the technique employed for coupling (Seid, 2017). Addressed in this section are the processes currently used for the encapsulation and embedding of bioactive materials into nanosystems that can further be used for Np-Nu development.

For the encapsulation of heat-sensitive compounds, spray-drying is a very efficient technique, the principle of which is illustrated in Figure 9.5. The technique employs a hot gaseous stream that causes the atomization of a liquid product, leading to the formation of powdered particles of high quality (Lidia et al., 2019). Illustrated in Figure 9.6 is another technique known as spray-chilling, which is used to encapsulate bioactive agents with lipids (Khorasani et al., 2018). It employs an extrusion method that utilizes a small nozzle to expel small drops of an encapsulating agent. As a result, the dimensions of the nanodroplet formed are based on the nozzle diameter and size (Khinast et al., 2013). An alternative approach is that of nanoprecipitation, illustrated in Figure 9.7, wherein an organic solution-based polymer is maintained with a surfactant in an aqueous medium by diffusion (Miladi et al., 2017). This technique is often utilized for hydrophobic drugs due to solvent miscibility. An example of the successful utilization of this approach is for the synthesis and development of TQ Nps. These Nps showed enhanced bioavailability coupled with antitumor activity (El-Far et al., 2018). Another approach is that of nanoemulsion. It involves creating a mixture of oil droplets in water, which offers high bioavailability and stability (McClements, 2013; Aboalnaja et al., 2016;

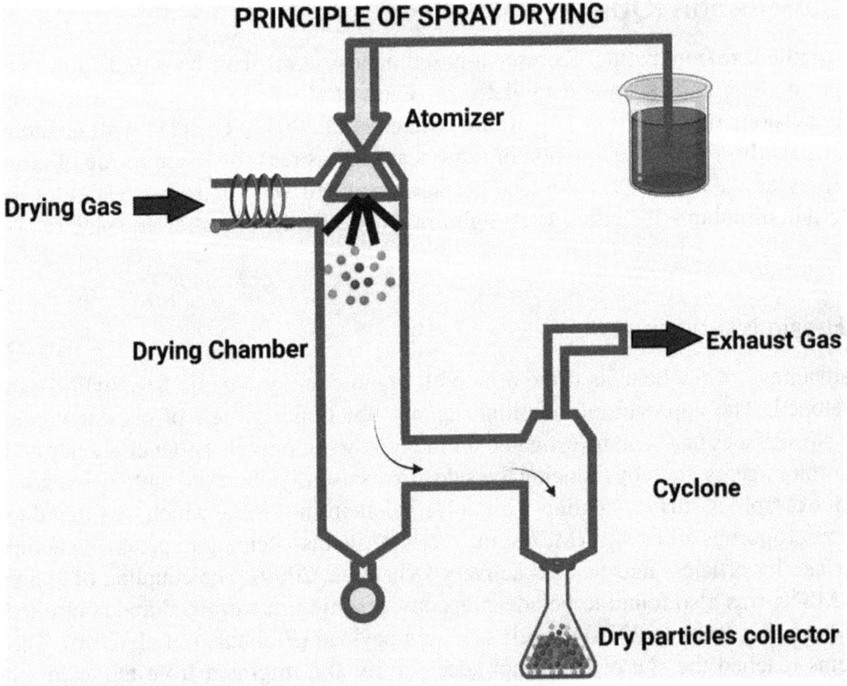

FIGURE 9.5 Principle of spray-drying. (Created using Biorender.com.)

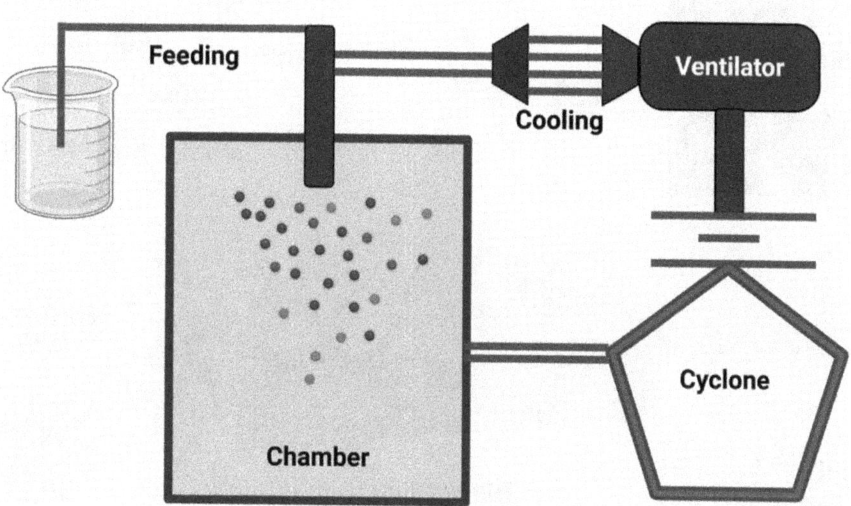

FIGURE 9.6 Principle of spray-chilling technique. (Created using Biorender.com.)

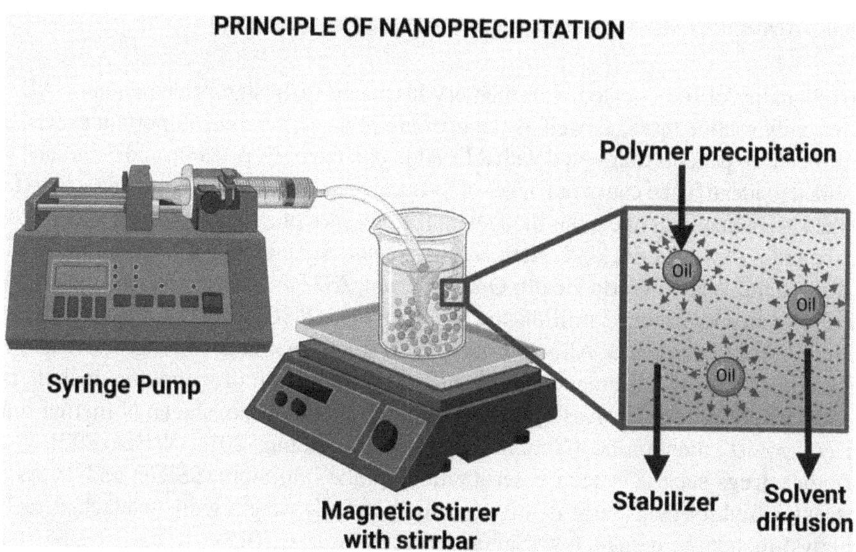

FIGURE 9.7 Principle involved in the nanoprecipitation technique. (Created using Biorender.com.)

Seid, 2017). As suggested by various studies, high stability and pH levels have been reported with polyphenols that are encapsulated, such as epigallocatechin gallate (EGCG) and curcumin (Wang et al., 2009). It is possible to control the size of a nanosystem, drug release, and drug yield by tuning synthesis parameters. However, other factors that can influence delivery systems, such as stability, processing technique, pH, temperature, light, etc., are beyond the scope of this chapter.

9.5 ROLE OF NP-NUs IN NEURODEGENERATIVE DISORDERS

Neurodegenerative diseases are disorders wherein the central and peripheral nervous system undergo progressive structural and functional degeneration. The most common include dementia, Alzheimer's disease (AD), other memory disorders, and Parkinson's disease (PD).

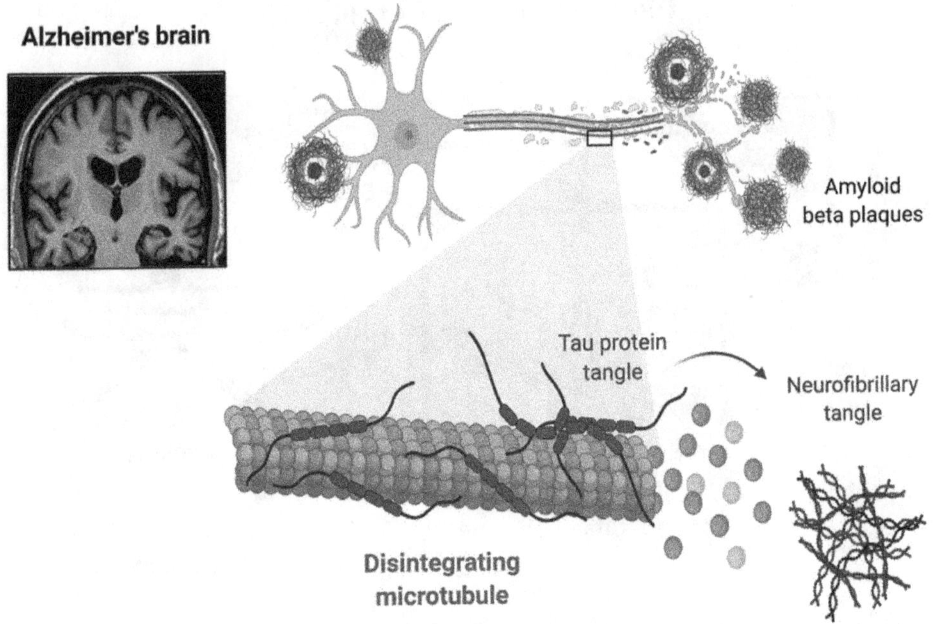

FIGURE 9.8 Alzheimer's disease pathology. (Created using Biorender.com.)

Disrupted quality of life coupled with memory loss is an early sign of dementia or AD. The performance of daily routine tasks, as well as the tracking of dates, seasons, important events, and time, becomes difficult for patients diagnosed with AD. Although ongoing, research and continued investigation have failed to identify the cause or trigger of its occurrence in patients. The disease is so far associated with the loss of neural connections along with the presence of amyloid plaques and neurofibrillary tangles in the brain (Association, 2021; National Institute on Ageing, 2021), as illustrated in Figure 9.8.

According to the WHO (World Health Organization), AD has affected 50 million people worldwide and is projected to affect 82 million and 152 million in 2030 and 2050, respectively, as it is the most common form of dementia. Although therapy for total treatment of progressive dementia and AD is yet to be established, therapeutics that aid in the reduction of symptoms include inhibitors of acetylcholinesterase, such as rivastigmine, donepezil, galantamine, and an N-methyl-d-aspartate (NMDA) antagonist, memantine (National Institute of Ageing, 2018; WHO, 2021b). The side effects of other drugs such as selective serotonin reuptake inhibitors (SSRIs) and TCAs (tricyclic antidepressants) used for neurodegenerative disorders include weight gain, headaches, tachycardia, and sexual dysfunction, to name a few (Santarsieri & Schwartz, 2015).

PD is a disorder of the CNS that is progressive and primarily affects movement. The disease pathology is illustrated in Figure 9.9. Symptoms of PD include tremor, rigidity, cognitive impairment, bradykinesia, mood disorders, gait and balance problems, and sleep disturbances (Hermanowicz et al., 2019; Postuma et al., 2015).

As with AD, the cause or trigger of this disease, as well as a cure, remains unknown. Current therapy involves symptomatic treatment through administration of medication such as levodopa, carbidopa, dopamine agonists, and catechol o-methyltransferase (COMT) inhibitors such as entacapone, which are associated with undesirable side effects affecting the quality of life (Stoker et al., 2018). In the United States, the Centers for Disease Control and Prevention (CDC) ranked PD-related complications as the 14th most common cause of death (Xu et al., 2018).

Therefore, due to the projected increase in the number of cases as well as the lack of therapeutics capable of treating these disorders, there is a growing need to develop novel therapeutics with an acceptable and suitable benefit–risk ratio.

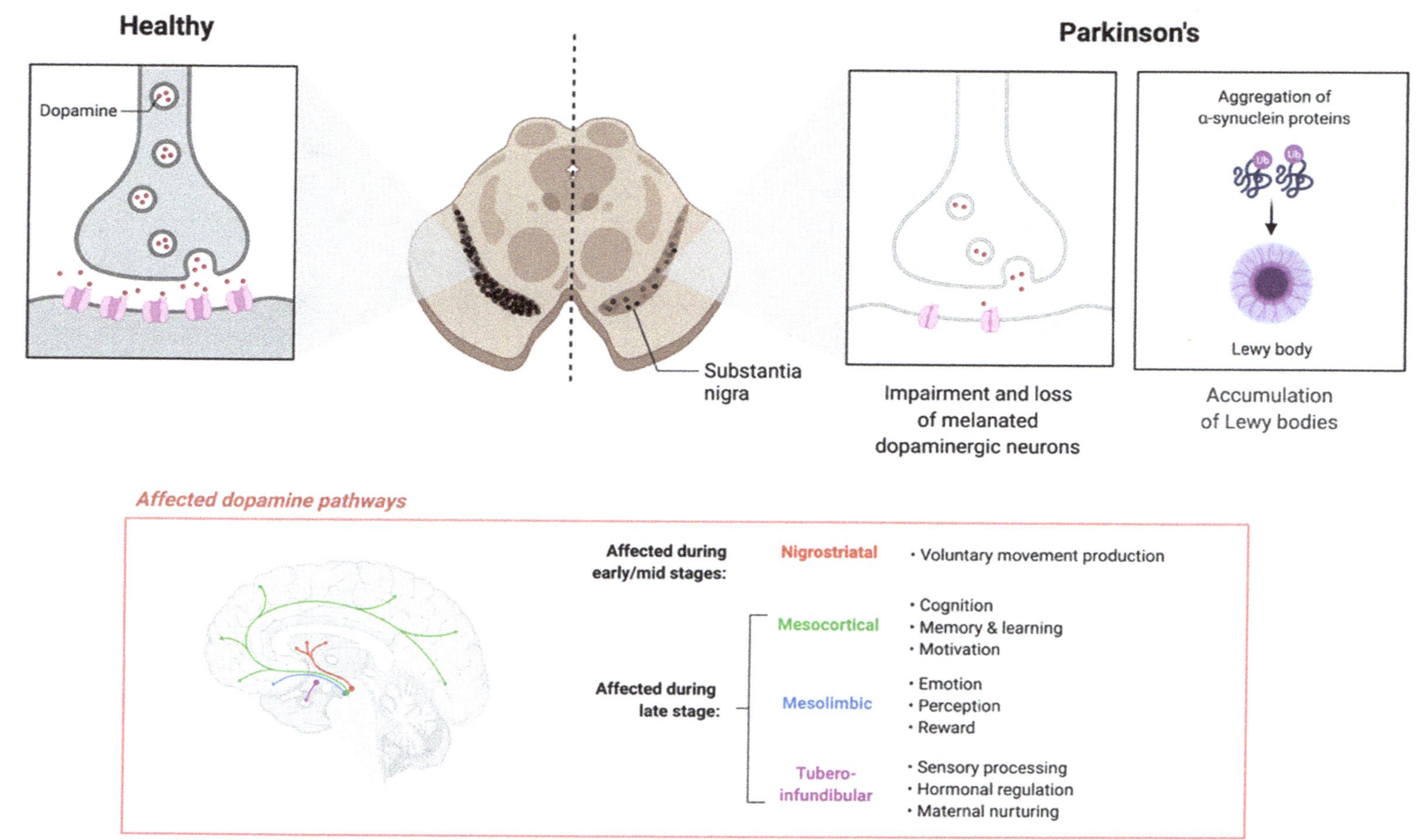

FIGURE 9.9 Pathology of Parkinson's disease. (Created using Biorender.com.)

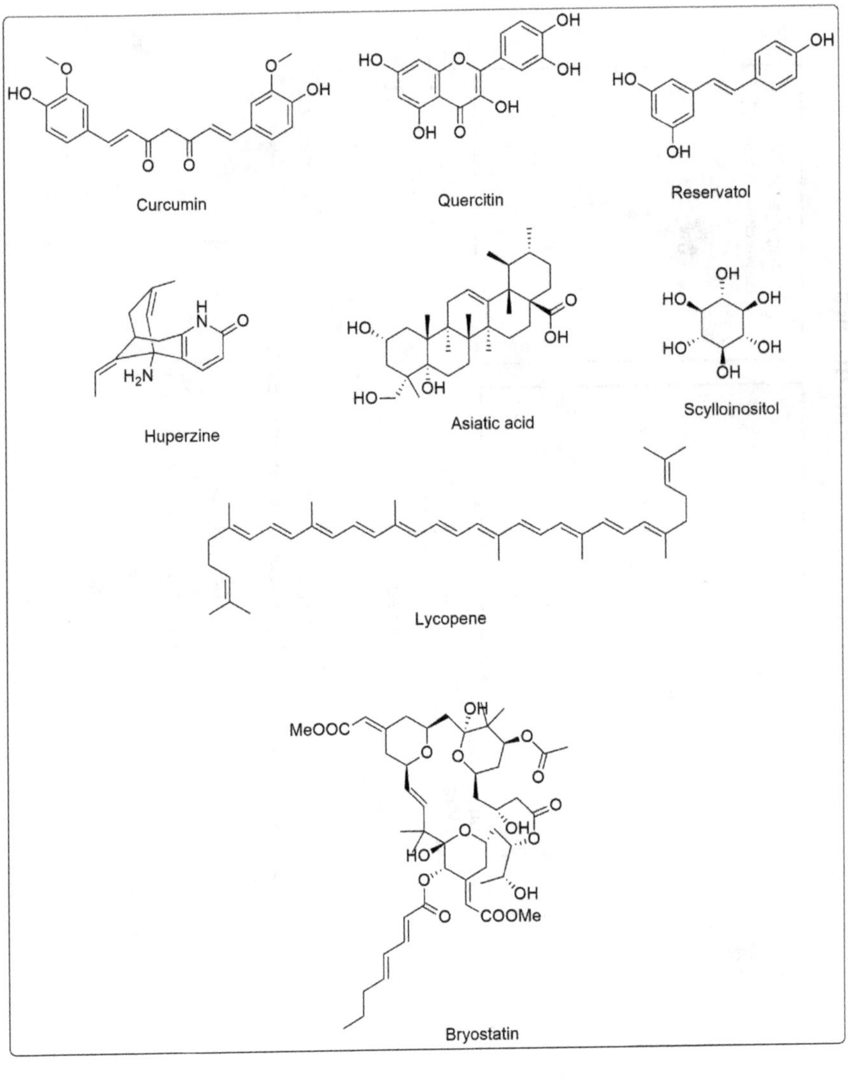

FIGURE 9.10 Chemical structures of nutraceuticals used for neurodegenerative diseases. (Created using Biorender.com.)

The ability of nanotechnology to improve penetrability into tissues as well as targeted drug delivery led scientists to investigate and develop Np-Nus that could penetrate the BBB as potential therapeutics for neurodegenerative diseases (Bhattacharya et al., 2022). This section will focus on recent studies of these attempts and the conclusions generated. Figure 9.10 illustrates the chemical structures of the nutraceuticals discussed in the following sections.

9.5.1 Np–Curcumin Conjugates

Due to its traditional, well-documented antioxidant and anti-inflammatory properties, curcumin has been thoroughly investigated as a promising bioactive agent for neurodegenerative disorders. However, its hydrophobicity and poor solubility make curcumin-based therapeutics very challenging to develop. To attain high bioavailability and to harness its beneficial effects, curcumin is often encapsulated with nanoformulations (Bollimpelli et al., 2016; Hitesh Chopra et al., 2021). Curcumin's effects on AD were studied using a formulation consisting of curcumin with poly lactide-co-glycolide-polyethylene glycol (PLGA-PEG) and coupled to B6 peptide by Fan et al. (2018). Following administration into HT22 cells and APP/PS1 AI transgenic mice, good blood compatibility as well as a decrease in particle size of curcumin was observed, thereby enhancing cellular uptake. Mice also showed an improvement in spatial learning and memory capability in comparison to those administered native curcumin. Ex vivo assays performed by the group indicated a reduction in β-amyloid formation and deposits in the hippocampus along with protein tau hyperphosphorylation following administration of curcumin-PLGA-PEG-B6, suggesting its promise as a candidate for use in AD (Fan et al., 2018). Another group of scientists performed studies on selenium Nps that were encapsulated with PLGA nanospheres coupled with curcumin for AD treatment. A decline in inflammation, Aβ aggregation, and systemic toxicity was observed following administration of this nanosystem (Huo et al., 2019).

A study pertaining to the conjugation of Nps and curcumin was conducted by Dey and Sreenivasan in 2015. Hyaluronic acid–coated gold Nps conjugated with curcumin showed enhanced aqueous solubility, bioavailability, and pH-responsive drug delivery, thus being able to successfully circumvent the physical limitations of curcumin (Dey & Sreenivasan, 2015). Yavarpour-Bali *et al.* in 2019 performed a detailed review on the benefits of curcumin in PD due to its ability to cause a downregulation of apoptosis and oxidative stress. The review also attempted to explore curcumin's mechanism of action in preventing AD progression (Yavarpour-Bali et al., 2019). Using sol-oil chemistry, curcumin-loaded lactoferrin showed neuroprotective effects in the PD rotenone model (Bollimpelli et al., 2016). In 2014–2017, research on curcumin-based nanoformulations indicated promising ability to inhibit β-amyloid (H. Chopra et al., 2021; Tang & Taghibiglou, 2017; Yao & Xue, 2014).

9.5.2 Np–Resveratrol Conjugates

Resveratrol, a natural polyphenolic phytoalexin, has shown beneficial effects in decreasing β-amyloid peptide formation, commonly observed in the pathology of AD (Rahman et al., 2020; Richard et al., 2011). In an attempt to improve bioavailability, Neves et al. in 2016 developed apolipoprotein E–modified SLNs encapsulated with resveratrol. These particles showed better BBB permeability through low-density lipoprotein (LDL) receptors in comparison to free resveratrol Nps (Neves et al., 2016). Studies by Loureiro et al. performed through the administration of resveratrol intravenously with SLNs functionalized with OX26, a monoclonal antibody, also showed significant BBB permeation (Loureiro et al., 2017).

9.5.3 Np–Quercetin Conjugates

A bioflavonoid that is well known for its antioxidant action, quercetin has been investigated thoroughly for its therapeutic application in neurodegenerative disorders. Using the micro-emulsification technique, SLNs conjugated with quercetin were shown to possess significant antioxidant action after intravenous administration along with an enhancement in therapeutic efficiency in AD (Dhawan et al., 2011). In another study, quercetin-based Nps were found to show neuroprotection against the behavioral, biochemical, and histological changes produced by scopolamine (Palle & Neerati, 2017). Encapsulated nanocarrier-based quercetin following oral administration was found to be highly beneficial in ischemia and reperfusion–induced neuronal damage in young and aged rat models (Ghosh et al., 2013). SPIONs coupled with quercetin also showed beneficial effects. Lower concentrations of this nanosystem were found to be superior in improving learning and memory in comparison to free quercetin in healthy rats, suggesting a higher level of *in vivo* circulation and bioavailability (Amanzadeh et al., 2019). Research by Bagad et al. in 2015 also suggested that quercetin coupled with poly n-butyl cyanoacrylate and coated with polysorbate-80 (P-80) seemed promising for the delivery of water-soluble drugs to the brain, thereby enhancing the peroral bioavailability of quercetin (Bagad & Khan, 2015).

9.5.4 Np-Scylloinositol Conjugates

Being an endogenous stereoisomer of inositol found naturally and abundantly in coconut palms, studies on scylloinositol indicate that it can aid in inhibition of Aβ aggregation and formation of Aβ fibrils *in vitro*. However, the mechanism is unclear (A. Y. Lai & McLaurin, 2012; McLaurin et al., 2000). It has also been found to lower Aβ levels and deposition of plaques, and to cause significant improvement in cognitive deficits (Hawkes et al., 2010). Lee et al. developed and investigated a novel derivative consisting of guanidine coupled with scylloinositol named AAD-66. Studies in the 5xFAD mice disease model concluded that the derivative caused a downregulation of gliosis, Aβ and plaque deposits, as well as memory deficits (D. Lee et al., 2017).

9.5.5 Np–Lycopene Conjugates

Lycopene comprises a complex of carotenoids and is widely found in tomatoes, papayas, pink guavas, and watermelons (P. R. Desai et al.). Due to the presence of polyunsaturated hydrocarbons in its structure, it possesses several activities. Noteworthy applications include cytotoxicity, apoptosis, and effects as an anti-inflammatory and anti-ischemic agent (Carvalho et al., 2020). The ability of lycopene to penetrate the BBB has made it a very promising agent in the field of neurodegenerative drug discovery (Nazemiyeh et al., 2016). Guo et al. administered microemulsions loaded with lycopene and observed higher bioavailability and biodistribution in the brain (Guo et al., 2019). Zhao et al., following the administration of liposomes that encapsulated lycopene, indicated its neuroprotective mechanism. Scientists performing this study concluded that lycopene possessed beneficial effects in an ischemic brain due to its ability to regulate iron metabolism (Zhao et al., 2018).

9.5.6 Np–Bryostatin Conjugates

Being derivatives of marine natural products that possess a macrocyclic polyketide structure, bryostatins have been found to modulate the activity of protein kinase C (PKC) (Sun & Alkon, 2006). Studies conducted by Kovochich et al. in 2011 on lipid-based Nps coupled with bryostatin-2 target showed *in vitro* CD4+T-cell activation along with latent infected cell ex vivo activation in SCID-hu (a humanized mouse model) (Kovochich et al., 2011). Moreover, a company named Aphios Corporation was granted a patent (US10485766B2) by the United States Patent Office for Bryostatin-1 Nps administration perorally in CNS disorders due to its ability for rapid cognitive performance restoration in AD, studied using transgenic mouse models (Schrott et al., 2015).

9.5.7 Np–Asiatic Acid Conjugates

Asiatic acid is a pentacyclic triterpene that has been isolated from plants because of its ability to produce neuroprotective effects (Ahmad Rather et al., 2018). Five hours post-administration, Nps with a coating of conjugates comprising bovine serum albumin (BSA) and glutathione coupled to asiatic acid were found to significantly enhance targeted delivery to the brain, thereby showing enhanced bioavailability as compared with the administration of a solution of asiatic acid (Raval et al., 2015). Moreover, the ability of asiatic acid to act on glioma cells through the administration of SLNs coupled with asiatic acid makes this nanoformulation promising for use in brain cancer therapy. The formulation also showed a dose-dependent apoptosis of glioma cells when studied by Garanti et al. in 2016 (Garanti et al., 2016).

9.5.8 Np–Huperzine A Conjugates

Huperzine A (HupA) is a promising candidate for neurodegenerative diseases, being a reversible acetylcholinesterase (AChE) inhibitor. As a result, in behavioral animal models, the administration of Hup A has been found to enhance memory and cognitive function (Farooqui, 2019; Sathya & Devi, 2018). A lactoferrin-based nanoformulation with conjugates of modified polylactide-coglycolide (PLGA) and N-trimethylated chitosan encapsulating Hup A as a mucoadhesive was investigated by Meng et al. in 2018 for intranasal delivery of Hup A to the brain of Alzheimer's patients

(Meng et al., 2018). In another attempt to formulate a Hup A–based nanosystem, PLGA Nps encapsulating Hup A (HupA-PLGA-Nps) were synthesized through the solvent evaporation of the O/W emulsion formed. The results obtained show beneficial use against toxicity in soman-induced mice as well as against poisoning due to chemical warfare–based nerve agents (Zhang et al., 2015).

9.6 ROLE OF NP-NUs IN CANCER MANAGEMENT

Cancer is defined as a disease that causes the uncontrolled growth of abnormal cells, which further spreads to other regions or organs of the body. The spread and growth of tumors in other locations that are different from that of the primary site of cancer is termed *metastasis* (Sun et al., 2014). During the progression of cancer, tumors become heterogeneous, thus leading to a mixed population of cells that differ in their molecular features, which further affects their responses to anticancer therapeutics (Pucci et al., 2019). As a result, complete elimination of the resistant phenotype of cells is a challenge in cancer treatment. The WHO estimates that, worldwide, cancer accounted for approximately 10 million deaths in 2020 (WHO, 2021a).

Current cancer therapies include radiation therapy, immunotherapy, hormone therapy, stem cell transplant, targeted therapy, and surgery. Chemotherapeutics such as doxorubicin, fluorouracil (5-FU), carboplatin, etc., have been found to be efficacious but also produce a lot of side effects (Institute, 2021). Therefore, efforts to find lead molecules or active compounds that are safe while being efficacious as antitumor agents are of utmost significance. Recently, several phytochemicals and natural antioxidants were introduced as adjuvant anticancer therapies due to their anti-proliferative and pro-apoptotic activities (Pucci et al., 2019).

Nanocarrier-based systems have allowed the targeted delivery of anticancer therapeutics through exploitation of the pathophysiology of the tumor microenvironment, thereby improving therapeutic outcomes and aiding in cancer management. Tumor cells bear overexpressed receptors on their surface, which have been targeted by nanosystems encapsulating or bearing specific ligands on their surface capable of binding to the receptor very efficiently. Many nanocarrier-based anticancer therapeutics have obtained approval for utilization by regulatory bodies, some of which are summarized in Table 9.2, while some are still in different phases of clinical trials (Din et al., 2017; Martins

TABLE 9.2
Approved Nanocarrier-Based Anticancer Therapeutics

Product name	Company	Anticancer drug	Nanoformulation	Applications	Country approval	Year
Myocet	Teva	Doxorubicin	Non-PEGylated liposome	Metastatic breast cancer	European Union (EU)	2000
Doxil	Janssen	Doxorubicin HCl	PEGylated liposome	Multiple myeloma, ovarian cancer	United States	1995
Marqibo	Spectrum	Vincristine sulfate	Non-PEGylated liposome	Acute lymphoid leukemia	United States	2012
DaunoXome	Galen	Daunorubicin citrate	Non-PEGylated liposome	Kaposi sarcoma	United States	1996
Mepact	Takeda	Mifamurtide	Liposome	High-grade non-metastatic osteosarcoma	EU	2009
NanoTherm	MagForce	Iron oxide and amino silane	Metal-based system	Pancreatic cancer, glioblastoma, prostate cancer	EU	2013

TABLE 9.3
Other Np-Nus Investigated as Antitumor Therapeutics

Nutraceutical	Nanocarrier	Applications in cancer therapy	References
Ellagic acid	PLGA-PEG Nps, MSN's	Breast cancer, prostate cancer, melanoma, non-small cell lung cancer, ovarian cancer	(Ali et al., 2020; Ceci et al., 2018; Neamatallah et al., 2020)
β-Lapachone	PEG-PLA PMs, Gold Nps	Prostate cancer, breast cancer	(Blanco et al., 2007; Jeong et al., 2019; Ma et al., 2015; Yang et al., 2017)
Daidzein	SLNs	Breast cancer, CVDs	(Gao et al., 2008; Jin et al., 2010)
Gambogic acid	Chitosan-based Nps, CTs, magnetic Nps	Pancreatic cancer, breast cancer	(Saeed et al., 2014; Salama et al., 2020; Wang et al., 2020)
Ferulic acid	Nps, SLN coated with chitosan	Colon cancer, pancreatic cancer	(Thakkar et al., 2015; Zheng et al., 2019)
Honokiol	QDs and Nps	Breast cancer, melanoma, liver cancer, glioblastoma, colorectal cancer	(AbdElhamid et al., 2018; Gao et al., 2016; Ponnurangam et al., 2012; Tang et al., 2018)
Curcumin	Alginate Nps, soy protein Nps, magnetic Nps, PVP–conjugated micelles, thermosensitive polymer Nps, silver Nps	Brain cancer, prostrate and breast cancer, neuroblastoma, leukemia, colorectal cancer	(Bisht et al., 2007; Das et al., 2010; Marjaneh et al., 2018; Mukerjee & Vishwanatha, 2009; Bomb et al., 2020; Soto-Quintero et al., 2019; Tsai et al., 2011)
Ursolic acid	PEG-modified liposomes, chitosan Nps, MSNs, PLGA-Nps coupled with gold	Cervical cancer, lung cancer, breast cancer, prostate cancer, bladder carcinoma	(Caldeira de Araujo Lopes et al., 2013; Jin et al., 2016; Wang et al., 2017; Zhao et al., 2015)

et al., 2015; Salama et al., 2020; Sun et al., 2014). Novel Np-Nu conjugates investigated for their anticancer effect and efficacy are described in the following sections and in Table 9.3.

9.6.1 Np–Thymoquinone Conjugates

As discussed earlier, TQ, obtained from *Nigella sativa*, is an anti-inflammatory agent, effective in gastric ulcers. Shahein et al. developed TQ-encapsulated MSNs whose shell consisted of gum Arabic and whey protein. Its efficacy as a cytotoxic agent was evaluated on SW1088 and A172 (brain cancer cells), and the mechanism of action was studied through cell cycle arrest, caspase 3, and apoptosis. At acidic pH, which is observed in tumor environments, TQ release was enhanced by the nanosystem as compared with free TQ. Moreover, the nanoformulation also causes the activation of caspase 3, G2/M phase cell cycle arrest, triggering of cytochrome c, and induced apoptosis in comparison to free TQ (Shahein et al., 2019). Alam et al. developed chitosan Nps of TQ using the ionic gelation technique. Physical attributes were analyzed through differential light scattering (DLS) and X-ray diffractometry (XRD). They observed enhanced mucoadhesive properties of the nanosystem, which showed enhanced brain targeting in comparison to TQ solution administered intravenously or intranasally (Alam et al., 2012).

9.6.2 Np–Dihydroartemisin Conjugates

Dihydroartemisin is an active metabolite in artemisinin compounds and is widely used for the treatment of malaria. Its effect as an anticancer agent was investigated by Guo et al. in 2020. Magnetic

Nps loaded with dihydroartemisin showed superior cytotoxicity against MCF-7 breast cancer cells and multi-drug resistant (MDR) breast cancer cells. Its effect was attributed to the production of ferrous ions in the acidic tumor microenvironment, thereby producing a high concentration of reactive oxygen species (ROS), leading to apoptosis (S. Guo et al., 2020).

9.6.3 Np–Eugenol Conjugates

Eugenol is an aromatic oil isolated from clove and is well known for the treatment of toothache. It also possesses antiviral, antioxidant, anti-inflammatory, and antibacterial effects (Xu et al., 2013). Eugenol-loaded nanoemulsions were evaluated for their cytotoxic activity against liver (HB8065) and colon (HTB37) cancer cells. The nanoformulation showed higher apoptotic cell percentages observed through microscopy and flow cytometry. The mechanism for its cytotoxic activity was believed to be due to ROS generation. Majeed et al. were the first to identify the anti-proliferative effect of eugenol coupled to a nanoemulsion (Majeed et al., 2014).

9.6.4 Np–Resveratrol Conjugates

Thipe et al. developed resveratrol-conjugated gold Nps and tested the efficacy of the nanoformulation on breast, pancreatic, and prostate cancers. The synthesis of this formulation was performed at room temperature using gum Arabic, thus providing better encapsulation of resveratrol and increasing the stability of the system. They observed increased uptake after 24 hours of incubation, indicating promising use as an anticancer formulation (Thipe et al., 2019). Resveratrol-coupled SLNs studied by Teskac et al. were found to have better cellular uptake and enhanced distribution. Moreover, the cytostatic effect of the nanosystem was higher than that seen following administration of free resveratrol solution, indicating its effectiveness in decreasing cell proliferation, thereby preventing the progression of skin cancer (Teskac & Kristl, 2010).

9.6.5 Np–Naringenin Conjugates

Naringenin is a flavonoid that is predominantly found in grapefruit, tomatoes, and other fruits. It has been found useful in obesity, hypertension, diabetes, and cardiovascular disease treatment (Salehi et al., 2019). Silk fibroin Nps loaded with naringenin were tested for their controlled release and anticancer effects on HeLa and EA.hy926 cancer cell lines. In both cell lines, this nanosystem showed superior anticancer activity in comparison to free naringenin (Fuster et al., 2020). MWCNTs functionalized with naringenin were tested for their antitumor effects on A569 human lung cancer cell line. This system showed a pH-responsive and sustained release in the tumor environment. It also showed lower levels of cytotoxicity on non-malignant cells as compared with free naringenin (Morais et al., 2020). PLGA-Nps encapsulating naringenin were developed and tested for their cytotoxic effect in pancreatic cancer. The results of this study indicated sustained release behavior and higher cytotoxicity as compared with free naringenin (Akhter et al., 2020).

9.7 ROLE OF Np-Nus IN OXIDATIVE STRESS

Oxidative stress occurs through the overproduction of ROS coupled with a reduction in biological antioxidizing system activity. Generally, ROS include radical species such as superoxide anion (O_2^-), peroxyl radicals (ROO), hydroxyl radical (OH), alkoxyl (RO), nitric oxide (NO), singlet oxygen, hydrogen peroxide H_2O_2, ozone, etc. Under normal metabolic conditions, moderate concentrations of ROS act as signaling messengers responsible for facilitating normal physiological functions. In contrast, a high concentration of ROS is destructive to biological processes, including cell proliferation, lipid peroxidation, cellular acidosis, differentiation, migration, and signal transduction; increases the expression of enzymes that produce ROS, like nicotinamide adenine dinucleotide

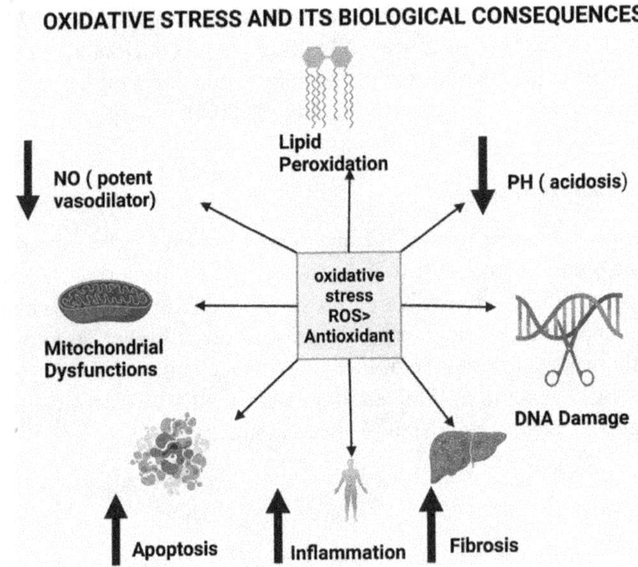

FIGURE 9.11 Oxidative stress and its biological consequences. (Created using Biorender.com.)

phosphate oxidase (NOX); and creates harmful effects such as programmed cell death, illustrated in Figure 9.11 (Kowluru & Chan, 2007).

ROS are very reactive due to the presence of multiple unpaired electrons, which are unstable. Their accumulation attacks cellular components and causes severe damage to molecules such as lipids, proteins, and nucleic acids (Calderon et al., 2017). Redox hemostasis is achieved by the action of endogenous antioxidant defense enzymes such as superoxide dismutase (SOD), glutathione peroxidase (GSH-P), peroxiredoxin (PRX), glutathione reductase (GSH-R), thioredoxin (Trx), and catalase (CAT). These enzymes play an essential role in ROS decomposition (Tokarz et al., 2013). Administration of exogenous antioxidants such as vitamin C, vitamin E, and β-carotene can counterbalance oxidative stress through direct free radical scavenger action, cut off free radical chain reactions, or both. Besides that, exogenous antioxidants can increase the expression of endogenous antioxidant enzymes and suppress ROS production (Ahmadinejad et al., 2017; Nimse & Pal, 2015).

Mounting evidence reveals that oxidative stress plays a significant and irreversible role in the pathology of several diseases. A lack of target specificity characterizes the use of exogenous antioxidants. For these reasons, it is necessary to develop non-enzymatic, natural antioxidant-based therapies such as nutraceuticals for the preservation of intracellular redox hemostasis (Kim et al., 2019; Rossino & Casini, 2019; Wang & Kang, 2020). The potential benefits of nanomedicines attributed to ROS scavengers and catalytic activity represent innovative strategies to treat oxidative stress diseases. Nps of metal oxide are an excellent example. Nps can be used as carriers to overcome common problems such as solubility, cell membrane permeability, and serum degradation accompanied by oxidative stress treatment with clinical drugs or recombinant enzyme therapy (Mauricio et al., 2018). pH-sensitive Nps, such as radical-containing Np (RNp) micelles, are another promising strategy for treating oxidative stress disease states (Chonpathompikunlert et al., 2015). Mitochondrial dysfunction is responsible for colossal ROS production and oxidative stress. Mitochondrial-targeted Nps constitute a novel approach for chemotherapeutic drug delivery directly into the mitochondria. Conjugated PLGA-b-PEG Nps with triphenylphosphonium (TPP), which have the presence of a lipophilic cation with high antioxidant activity, have been generated to deliver the TPP precisely into the mitochondria (Marrache & Dhar, 2012).

Oxidative stress mediates the progress of various diseases such as CVDs, as illustrated in Figure 9.12. There is a reversible relationship between CVDs and oxidative stress. The potential uses of Nps in the treatment and diagnosis of CVDs have been investigated and are described in the following list.

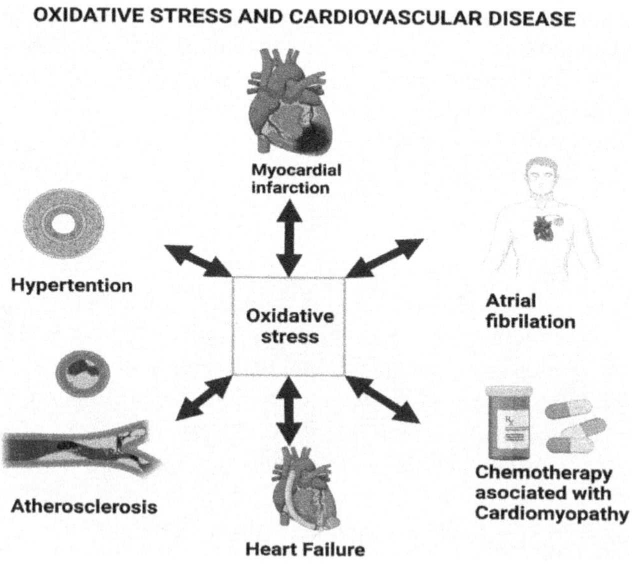

OXIDATIVE STRESS AND CARDIOVASCULAR DISEASE

Myocardial infarction

Hypertention

Oxidative stress

Atrial fibrilation

Atherosclerosis

Chemotherapy asociated with Cardiomyopathy

Heart Failure

FIGURE 9.12 Relationship between oxidative stress and cardiovascular disease. (Created using Biorender.com.)

A—Ischemic and reperfusion injuries

After a cardiac ischemic attack, the principal aim is to restore the blood supply to the injured area and reverse cellular damage such as free radical production, which is suspected to be the leading cause of reperfusion injuries and cell death. Novel H_2O_2-responsive Nps produced from co-polyoxalate and vanillyl alcohol (VA) have been investigated in animal models of acute myocardial infarction due to the ester linkage being very sensitive and degradable at an excessive level of H_2O_2. After the ester bonds are cleaved, the potent component (VA) exerts anti-apoptotic and anti-inflammatory effects through cyclooxygenase-2 (COX-2) downregulation coupled with the expression of iNOS (Bae et al., 2016; Dongwon Lee et al., 2013). Innovative targeted Nps coated with nicotinamide adenine dinucleotide phosphate oxidase (NOX2) coupled with small interfering RNA (siRNA) were generated for drug delivery and to increase ROS clearance after injection to the site of myocardium injury in mice. These Nps demonstrated improved cardiac function in a short period after the surgery (Somasuntharam et al., 2013). Overexpression of intracellular N-acetylcysteine and superoxide dismutase 1 (SOD1) has been shown to possess beneficial protective effects in mouse ischemia models. Nps designed to carry N-acetylcysteine or SOD1 have significantly depleted cardiac fibrosis and improved cardiac function, respectively (Gray et al., 2011; Seshadri et al., 2010).

B—Thromboembolic conditions

Thrombus formation is the leading pathological cause of many CVDs, like stroke, pulmonary embolism, and deep vein thrombosis. Oxidative stress is a significant factor for thrombi formation. Due to clinical drugs possessing off-target effects, the use of Nps as a carrier to deliver a thrombolytic drug to its site of action, in a targeted manner, after thrombotic events is a major achievement in drug delivery (Jain et al., 2015). In addition, Nps have been used to investigate the clot formation processes, especially the primary step, which is the conversion of fibrinogen to fibrin. EP-2104R, an MRI contrast agent, is an Np used for targeting the image of clot formation in animal and human models (Spuentrup et al., 2005). Some polypeptides bind specifically to the clotting factors using Nps as a carrier for these amino acids to detect new clot formation (Kang et al., 2017). The detection of microthrombi was challenging with regular MR agents, and so researchers developed Nps designed to carry P-selectin, a small cellular adhesion protein responsible for platelet activation and subsequent plaque formation. These have shown comparable results to localize and obtain images of small thrombi in animal models (Wang et al., 2010). In patients with acute stroke, hemorrhage is a frequent complication. The effects of tissue plasminogen activator (tPA) and urokinase in the treatment of thrombolysis in acute stroke were found to be significantly higher when they were conjugated to magnetic Nps (MNps) in comparison with the administration of free drug (Bi et al., 2009; Chen et al., 2012).

C—Atherosclerosis

Elevated levels of cholesterol in the blood result in the accumulation of fatty acids inside the arteries, followed by an immune response, connective tissue deposition, and thromboembolic ischemia, which further leads to atheromatous plaque rupture. H_2O_2 overproduction leads to endothelial cell dysfunction due to the loss of its signal messenger activity. The imaging of macrophage activation at the atherosclerosis site and enzymatic release of myeloperoxidase, the source of ROS, makes this an excellent therapeutic target to deliver ROS scavengers through Np design. Numerous studies have been performed to detect atherosclerosis plaque formation using nanomedicine techniques. PEGylated and gold coated Nps are good examples of this (Weissleder et al., 2014).

D—Diabetic retinopathy

A complication related to diabetes, diabetic retinopathy (DR) is characterized by an irregular growth of retinal vessels, the leading cause of vision loss in late disease stages. Neuronal and vascular changes are involved in disease progression; however, the exact underlying pathophysiological mechanism is still unknown. Many reasons make the retina highly susceptible to oxidative stress, such as extended exposure to light, high concentration of polyunsaturated fatty acids, glucose oxidation, lower levels of glutathione (GSH), and increased oxygen uptake (Wu et al., 2018). Different metabolic pathways accelerate oxidative stress in DR because of hyperglycemia, including the accumulation of receptor-acting glycation end-products (RAGEs), stimulation of PKC, and the activation of the hexosamine pathway. The consequences of retinal oxidative stress increase ROS, which further creates a propagating cycle that continues the disease progression of DR (Brownlee, 2005; Kowluru & Chan, 2007). Several studies both *in vitro* and *in vivo* have revealed the benefits of using natural dietary compounds as Np-Nus to correct retinal oxidative stress typical of DR through direct scavenger activity leading to the enhancement of the antioxidant effect of the endogenous system or through the upregulation of antioxidant enzymes.

Figure 9.13 illustrates the benefits that nutraceuticals offer to minimize oxidative stress. The Np-Nus investigated in recent years to decrease oxidative stress are discussed below.

9.7.1 Np–Curcumin Conjugates

Curcumin, a yellowish polyphenolic substance extracted from *Curcuma longa*, has direct and indirect antioxidant activity in addition to an anti-inflammatory effect. The natural antioxidant capability of curcumin has been studied extensively over the years; intracellular ROS production has been revealed to decrease significantly after treatment with millimicron concentrations of curcumin both for ARPE-19 (retinal pigment epithelial cell line) and in HRECs in the hyperglycemic model (Platania et al., 2018; Premanand et al., 2006). The indirect antioxidant activity of curcumin can be explained by simultaneous induction of antioxidant enzyme expression following the reduced ROS level. A protective effect against the stress following activation is provided by heme oxygenase-1 (HO-1) (Yang et al., 2009). Following curcumin treatment, induction of nuclear factor erythroid-2-related factor-2 (Nrf2) occurs, which causes the transcription of genes coding for antioxidant enzymes (Li et al., 2013). Inhibition of the accumulation of retinal stress biomarkers such as malondialdehyde along with the protein damage marker nitrotyrosine has been demonstrated following administration of curcumin in diabetic rat models. In addition, it

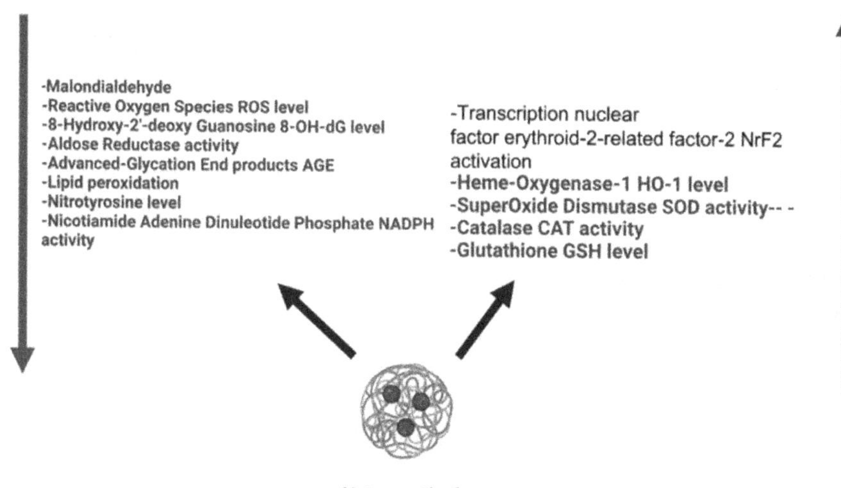

FIGURE 9.13 Effects of nutraceuticals in oxidative stress management. (Created using Biorender.com.)

prevented the increase in 8-hydroxy-20-deoxyguanosine, an oxidative DNA damage marker (Kowluru et al., 2007). Np-Nu therapy can be utilized to overcome the disadvantages of free curcumin administration, such as poor solubility, insufficient absorption, and high metabolic susceptibility, which limit its clinical application. Curcumin's low water solubility can be improved by using metal oxide Nps, nanogel, polymeric, and lipid Np encapsulation (Shome et al., 2016). Similarly, the rapid biological degradation of curcumin can be avoided through Np encapsulation with a calix or arene nanoassembly. In both *in vivo* and *in vitro* trials, these curcumin-based Np-Nus exhibited a comparable increase in the antioxidant and anti-inflammatory activity of curcumin (Granata et al., 2017). Resveratrol is another example of a polyphenol, mainly found in peanuts, grapes, and berries. The antioxidant effect of this compound can be boosted by conjugation with Nps like cyclodextrins, polymeric, and solid lipid Nps (Bonechi et al., 2012).

9.7.2 Np–Epigallocatechin Gallate Conjugates

Epigallocatechin gallate, a flavonoid polyphenol, is most particularly found in green tea and has potent antioxidant activity by causing aldose reductase inhibition, thereby affecting the polyol pathway and reducing the retinal accumulation of AGE (Sampath et al., 2016). The poor liposomal delivery of epigallocatechin gallate can be solved by using inorganic and polymeric-based Nps. (Granja et al., 2017).

9.7.3 Np–Lutein Conjugates

Lutein, a carotenoid present in oranges, yellow fruits, and green leafy vegetables, has been studied to prevent retinal oxidative damage by decreasing nitrotyrosine levels, lipid peroxidation, and retinal DNA modifications. This protective effect can be achieved by using it together with zeaxanthin, a carotenoid compound found in the same fruits. The poor ocular tissue delivery of lutein and other undesired physicochemical properties can be overcome by using lutein encapsulated in hyaluronic acid–coated PLGA Nps (Chittasupho et al., 2019).

9.8 SAFETY EVALUATION/TOXICITY CONCERNS

Recent innovations in the field of nanotherapeutic development have fostered an increase in the market as well as consumer acceptance of nanoformulations coupled with nutraceuticals. However, this surge in the production, manufacturing, development, and marketing of Np-Nus has also caused a rise in environmental and public safety, ethics, and regulatory concerns. The ability to manipulate nanocarriers in a variety of ways physiochemically and biologically is a boon for scientists aiming to bypass challenges faced in drug delivery but at the same time a bane due to the unpredictable hazards that can occur due to this feature of nanocarrier development.

Although a wide variety of advantageous therapeutics for neurodegenerative diseases, cancer therapy, and oxidative stress can be developed and utilized through the coupling of the benefits of nutraceuticals as well as the physicochemical characteristics of nanocarriers, scientists have also questioned the toxicity and safety features of these systems. Their ability for facile penetration through the BBB and targeted drug delivery in certain organs could also be detrimental and could cause untoward effects if overdosed or administered in an unethical manner.

Studies based on the ADME characteristics (absorption, distribution, metabolism, and elimination) as well as toxicity studies in biological systems are limited (Lai, 2012). Their mechanism of action also needs elucidation. This can be studied through the interaction of these formulations with biosystems *in vivo* (Gonçalves et al., 2018).

Zhang et al. in 2018 illustrated the unwanted ability of Nps to cause inhibition of the hydroxyl radical scavenging activity of vitamin C, gallic acid, and epigallocatechin gallate, leading to the generation of ROS and apoptosis (Zhang et al., 2018). Surfactant presence in the nanoformulation as part of the functionalization on the surface of the nanocarrier could also affect BBB integrity and toxicity (Chassaing et al., 2015; F.S.Gonçalves et al., 2018). Antibodies, drugs, bioactive molecules, linkers, or probes could also act as antigenic species, thereby causing toxicity and neuronal damage (Mahmoud et al., 2020). Also, as mentioned earlier, due to the complexity of conjugate formation, it is difficult to ascertain the mechanism of action by which these Np-Nus show activity. Research, however, reports an alteration in metabolic pathways as well as effects on transporters of drugs following administration of Np-Nus (Levy et al., 2017).

Guidelines for toxicity evaluation of Nps proposed by the European Food Safety Authority (EFSA) (Committee et al., 2018) led to a decline in drug withdrawal risks to 20% from the initial 50% (Meenambal & Bharath, 2020). The toxicity assessment is based on fundamental physicochemical properties of Nps, such as size, shape, hydrophobicity, dispersion state, etc. (Sukhanova et al., 2018) and the evaluation for toxicity is performed on many cellular and animal models *in vitro*. Data such as analysis of toxicity comprising of *in vitro* uptake and localization, cell viability assays inclusive of cell proliferation, apoptosis, necrosis, oxidative stress, and DNA damage, are collected through mechanistic studies for toxicity determination (Arora et al., 2012). The understanding of clinical applications pertaining to nanoformulation dosage and distribution is based on *in vivo* studies performed in rodent models (Adamcakova-Dodd et al., 2011). Assessments include studies such as variations in tissue structure, cellular apoptosis, and organ inflammation of heart, brain, spleen, kidney, etc. (Yuan Yang et al., 2016). Many *in vitro* and *in vivo*–based studies were investigated using various models of disease; however, systemic assessment, as well as the evaluation of toxicity rendered by nanoformulations, remains uncertain and indefinite. Moreover, assessment protocols and guidelines for the determination of adverse effects and toxicity following administration of these nanosystems have not been established (Kumar et al., 2017). Guidelines and standards followed to produce marketed nutraceuticals are also hazardous. In this regard, regulatory bodies such as the WHO and the United Nations Food and Agricultural Organization (UNFAO) set guidelines, standards, and restrictions to ensure the quality and safety of products released globally. However, policies governing the manufacturing process as well as the marketing strategies employed by nutraceutical companies are not set in place (Santini et al., 2018). Therefore, around the globe, regulatory bodies and authorities need to impose strict and standardized regulations along with well-defined policies pertaining to the manufacturing and sale of nutraceuticals (Jampilek et al., 2019; J. Singh & Sinha).

The oral administration of Np-Nus is also a concern even though the concept of encapsulation of drugs/active pharmaceutical ingredients (APIs) has been extensively studied (He & Hwang, 2016).

This is because the fate of these nanosystems is highly dependent on the ability of the digestive enzymes, as well as the conditions of the GI tract, to hydrolyze the system (McClements & Xiao, 2017). Usually, nanocarriers have facile penetration through intestinal and cellular barriers, thereby leading to an increased bioavailability or drug accumulation in blood cells and tissues. Therefore, the drug loading amount into the nanocarrier needs to be given consideration in such a way as to prevent drug overdose or toxicity.

The utilization of organic solvents coupled with stabilizers and emulsifiers for nanocarrier preparation is another hazard. The removal of organic solvents through evaporation could result in residual solvent presence in the final product, thus affecting the safety of the system if the concentration is unknown. The WHO, USFDA (United States Food and Drug Administration), and the EFSA have therefore classified and defined toxic solvents and emulsifiers, and have also set standards and regulations pertaining to their safe usage levels within nanosystems (McClements & Rao, 2011).

9.9 RESEARCH GAPS AND FUTURE PERSPECTIVE OF Np-Nus

The advantages as well as the massive potential and benefits of the conjugation of nutraceuticals and nanocarriers have been previously described in this chapter. However, even if the constituents used in the formulation have GRAS status (generally recognized as safe) designated by the USFDA, toxicity is a huge concern. The favorable physicochemical properties coupled with the low particulate size of nanosystems could allow undesired penetration and permeation into unwanted sites, thereby leading to acute toxicity, DNA damage, oxidative stress, and inflammation along with cellular and tissue apoptosis. Surface charges present on the surface of nanocarriers pose a huge risk, especially if they are positively charged, due to the ability for higher cellular interactions and better internalization (Podila & Brown, 2013). Therefore, there is a need to perform further investigation as well as a complete physicochemical characterization for toxicity estimation. Utilization of both *in vitro* and *in vivo* disease models along with cellular and biological systems for understanding the mechanistic pathway of nanosystems also needs to be investigated. Microscopy could also be employed to gain insights into cellular viability and apoptosis following administration of nanoformulations. It will also help determine whether any untoward effects occur as a function of the physicochemical attributes of the nanosystem (Arora et al., 2012; Ciappellano et al., 2016).

Rat and mouse models are widely used for testing nutraceuticals, as they provide insights and help in ADME prediction as well as dose–response relationships. Further studies to provide dosing schedules, as well as the frequency of dosing of Np-Nus, need consideration to avoid untoward effects (F.S. Gonçalves et al., 2018). Following the GI absorption of these nanosystems coupled to nutraceuticals, conformational changes in the structure could result in variations in the ADME profile (Jampilek & Kralova, 2020). Some Np-Nus have been developed to assist in permeation. As a result, their formulation includes the presence of permeation enhancers. However, the challenge associated with these types of systems is the formulation of nanosystems that can only cause the improvement of permeability characteristics of the GI barrier to specific molecules instead of also allowing the permeation of allergens, toxins, and bacteria (Maher et al., 2016). Damage caused due to this feature of Np-Nus also needs to be reversed easily and resolved quickly if such a situation were to arise. Therefore, further research and investigation of the mechanisms of GI barrier repair are needed. Deeper investigation pertaining to the mechanism involved in barrier repair is necessary (Cristiano et al., 2020). Another factor that needs to be investigated is the possible immunotoxicity that could arise in nanosystems coupled with antibody fragments and peptides behaving as antigenic species following administration (Renukuntla et al., 2013).

Following oral administration of Np-Nus, the interaction of its constituents with the gut microbiota, along with the consequences pertaining to biosafety and bio-efficacy, needs exploration and poses additional challenges. The phytochemical and drug interaction with gut microbiota, along with the repercussions of these interactions on bio-efficacy, remains undiscovered and unexplored. Therefore, further research on *in vitro* and *in vivo* models of disease along with *in vitro*–based cellular models would provide opportunities to gain better insights that can further be used to make interactions between phytochemicals and gut microbiota more advantageous (Karavolos & Holban, 2016; Paolino et al., 2021).

The scope for the development of Np-Nus is immense, especially in interventions pertaining to medicine and therapeutics, research, and commercialization. Although the fundamentals pertaining to experimental design followed in pre-clinical trials are well documented and evident, the success of their progress to clinical stages has been challenging. Scientifically sound evidence and experimental data can assist in validating novel nano-nutraceutical formulations. Protocols, experimental design, and clinical trial standardization in the course of time can assist with the safety and efficacy studies of Np-Nus. Future research and investigation involve the utilization of novel breakthrough innovative technology for the development of novel Np-Nus with an enhanced safety, efficacy, quality stability, and stability profile (Singh & Sinha, 2012; Wiwanitkit, 2012). As mentioned earlier, issues pertaining to nanoformulation efficacy in ensuring targeted delivery as well as nanotoxicity issues need to be investigated. A rationally designed animal model needs to be established for the assessment of drug delivery efficiency of nanoformulations. Though toxicity data and delivery efficiencies need to be evaluated, and regulations need to be imposed by regulatory agencies, it must be emphasized that Np-Nus have tremendous potential as breakthrough therapies for several disease states that still today have no cure (Prasad et al., 2018).

9.10 CONCLUSION

The design, formulation, and development of Np-Nus to meet the growing need for therapeutics of neurological diseases, cancer management, and oxidative stress has become a highly significant research area in the domain of drug delivery. As these nanosystems are revolutionary, they can target tumors and certain organs within the body with enhanced permeability and can penetrate the BBB with ease. Technological advancements in nanotechnology and drug discovery and development have given patients with these ailments hope of a better quality of life and improved life expectancy. Nutraceuticals on their own possess low water solubility and bioavailability; they are unstable, and their ADME profiles are erratic. The coupling of nanocarriers to nutraceuticals provides a solution to these physio-chemical challenges. Further research on improvements in quality, efficacy, safety, bioavailability, and the targeted, controlled, and sustained drug delivery of these nanoformulations needs to be conducted before initiating clinical trials. In the future, management of the amount of drug loaded into these nanocarriers to ensure highest efficacy with minimal to no side effects, coupled with the standardization of policies, regulations, and guidelines by regulatory agencies, will ensure the wide acceptance and use of these novel formulations and will also enhance patient compliance.

ACKNOWLEDGEMENTS

All diagrams were created using ChemDraw and Biorender.com.

REFERENCES

AbdElhamid, A. S., Zayed, D. G., Helmy, M. W., Ebrahim, S. M., Bahey-El-Din, M., Zein-El-Dein, E. A., . . . Elzoghby, A. O. (2018). Lactoferrin-tagged quantum dots-based theranostic nanocapsules for combined COX-2 inhibitor/herbal therapy of breast cancer. *Nanomedicine, 13*(20), 2637–2656. doi:10.2217/nnm-2018-0196

Abdelwahab, S. I., Sheikh, B. Y., Taha, M. M., How, C. W., Abdullah, R., Yagoub, U., . . . Eid, E. E. (2013). Thymoquinone-loaded nanostructured lipid carriers: preparation, gastroprotection, in vitro toxicity, and pharmacokinetic properties after extravascular administration. *Int J Nanomed, 8*, 2163–2172. doi:10.2147/IJN.S44108

Aboalnaja, K. O., Yaghmoor, S., Kumosani, T. A., & McClements, D. J. (2016). Utilization of nanoemulsions to enhance bioactivity of pharmaceuticals, supplements, and nutraceuticals: nanoemulsion delivery systems and nanoemulsion excipient systems. *Expert Opin Drug Deliv, 13*(9), 1327–1336. doi:10.1517/17425247.2016.1162154

Adamcakova-Dodd, A., Thorne, P. S., & Grassian, V. H. (2011). In vivo toxicity studies of metal and metal oxide nanoparticles. *Gen Appl Syst Toxicol.* doi:10.1002/9780470744307.gat244

Ahmad Rather, M., Justin Thenmozhi, A., Manivasagam, T., Dhivya Bharathi, M., Essa, M. M., & Guillemin, G. J. (2018). Neuroprotective role of Asiatic acid in aluminium chloride induced rat model of Alzheimer's disease. *Front Biosci, 10*, 262–275. doi:10.2741/s514

Ahmadinejad, F., Geir Møller, S., Hashemzadeh-Chaleshtori, M., Bidkhori, G., & Jami, M.-S. J. A. (2017). Molecular mechanisms behind free radical scavengers function against oxidative stress. *6*(3), 51.

Akhter, M. H., Kumar, S., & Nomani, S. (2020). Sonication tailored enhance cytotoxicity of naringenin nanoparticle in pancreatic cancer: design, optimization, and in vitro studies. *Drug Dev Ind Pharm, 46*(4), 659–672. doi:10.1080/03639045.2020.1747485

Alam, S., Khan, Z. I., Mustafa, G., Kumar, M., Islam, F., Bhatnagar, A., & Ahmad, F. J. (2012). Development and evaluation of thymoquinone-encapsulated chitosan nanoparticles for nose-to-brain targeting: a pharmacoscintigraphic study. *Int J Nanomed, 7*, 5705–5718. doi:10.2147/IJN.S35329

Ali, B., Al-Wabel, N. A., Shams, S., Ahamad, A., Khan, S. A., & Anwar, F. (2015). Essential oils used in aromatherapy: a systemic review. *Asian Pac J Trop Biomed, 5*(8), 601–611. doi:10.1016/j.apjtb.2015.05.007

Ali, O. M., Bekhit, A. A., Khattab, S. N., Helmy, M. W., Abdel-Ghany, Y. S., Teleb, M., & Elzoghby, A. O. (2020). Synthesis of lactoferrin mesoporous silica nanoparticles for pemetrexed/ellagic acid synergistic breast cancer therapy. *Colloids Surf B Biointerfaces, 188*, 110824. doi:10.1016/j.colsurfb.2020.110824

Amanzadeh, E., Esmaeili, A., Abadi, R. E. N., Kazemipour, N., Pahlevanneshan, Z., & Beheshti, S. (2019). Quercetin conjugated with superparamagnetic iron oxide nanoparticles improves learning and memory better than free quercetin via interacting with proteins involved in LTP. *Sci Rep, 9*(1), 6876. doi:10.1038/s41598-019-43345-w

Amin, M. C. I. M., Butt, A. M., Amjad, M. W., & Kesharwani, P. (2017). Polymeric Micelles for drug targeting and delivery. In *Nanotechnology-Based Approaches for Targeting and Delivery of Drugs and Genes* (pp. 167–202): Academic Press.

Arayne, M. S., Sultana, N., & Qureshi, F. (2007). Review: nanoparticles in delivery of cardiovascular drugs. *Pak J Pharm Sci, 20*(4), 340–348. Retrieved from https://www.ncbi.nlm.nih.gov/pubmed/17604260

Arora, S., Rajwade, J. M., & Paknikar, K. M. (2012). Nanotoxicology and in vitro studies: the need of the hour. *Toxicol Appl Pharmacol, 258*(2), 151–165. doi:10.1016/j.taap.2011.11.010

Association, A. S. (2021). What is Alzheimer's disease? Retrieved from https://www.alz.org/alzheimers-dementia/what-is-alzheimers

Atsumi, T., & Tonosaki, K. (2007). Smelling lavender and rosemary increases free radical scavenging activity and decreases cortisol level in saliva. *Psychiatry Res, 150*(1), 89–96. doi:10.1016/j.psychres.2005.12.012

Avallone, R., Vitale, G., & Bertolotti, M. (2019). Omega-3 fatty acids and neurodegenerative diseases: new evidence in clinical trials. *Int J Mol Sci, 20*(17). doi:10.3390/ijms20174256

Badar, A., Pachera, S., Ansari, A. S., & Lohiya, N. K. (2019). Nano based drug delivery systems: present and future prospects. *Nanomed Nanotechnol J, 2*, 121.

Bae, S., Park, M., Kang, C., Dilmen, S., Kang, T. H., Kang, D. G., . . . Kang, P. M. (2016). Hydrogen peroxide-responsive nanoparticle reduces myocardial ischemia/reperfusion injury. *J Am Heart Assoc, 5*(11), e003697.

Bagad, M., & Khan, Z. A. (2015). Poly(n-butylcyanoacrylate) nanoparticles for oral delivery of quercetin: preparation, characterization, and pharmacokinetics and biodistribution studies in Wistar rats. *Int J Nanomed, 10*, 3921–3935. doi:10.2147/IJN.S80706

Balakrishnan, P., Lee, B. J., Oh, D. H., Kim, J. O., Lee, Y. I., Kim, D. D., . . . Choi, H. G. (2009). Enhanced oral bioavailability of Coenzyme Q10 by self-emulsifying drug delivery systems. *Int J Pharm, 374*(1–2), 66–72. doi:10.1016/j.ijpharm.2009.03.008

Barry, H., Gettens, R., & English, A. (2015). *A quantum dot nano-carrier system for targeted drug delivery.* Paper presented at the Annual Northeast Bioengineering Conference, Troy.

Battaglia, L., Gallarate, M., Peira, E., Chirio, D., Muntoni, E., Biasibetti, E., . . . Riganti, C. (2014). Solid lipid nanoparticles for potential doxorubicin delivery in glioblastoma treatment: preliminary in vitro studies. *J Pharm Sci, 103*(7), 2157–2165. doi:10.1002/jps.24002

Bhattacharya, T., Dey, P. S., Akter, R., Kabir, M. T., Rahman, M. H., & Rauf, A. (2021a). Effect of natural leaf extracts as phytomedicine in curing geriatrics. *Exp Gerontol*, *150*, 111352. doi:10.1016/j.exger.2021.111352

Bhattacharya, T., Rather, G., Akter, R., Kabir, M. T., Rauf, A., & Rahman, M. (2021b). Nutraceuticals and bio-inspired materials from microalgae and their future perspectives. *Curr Top Med Chem*, *21*(12), 1037–1051. doi:10.2174/1568026621666210524095925

Bhattacharya, T., Soares, G. A. B. e., Chopra, H., Rahman, M. M., Hasan, Z., Swain, S. S., & Cavalu, S. (2022). Applications of phyto-nanotechnology for the treatment of neurodegenerative disorders. *Materials*, *15*(3), 804. doi:10.3390/ma15030804

Bi, F., Zhang, J., Su, Y., Tang, Y.-C., & Liu, J.-N. J. B. (2009). Chemical conjugation of urokinase to magnetic nanoparticles for targeted thrombolysis. *30*(28), 5125–5130.

Bianco, A. (2004). Carbon nanotubes for the delivery of therapeutic molecules. *Expert Opin Drug Deliv*, *1*(1), 57–65. doi:10.1517/17425247.1.1.57

Bisht, S., Feldmann, G., Soni, S., Ravi, R., Karikar, C., Maitra, A., & Maitra, A. (2007). Polymeric nanoparticle-encapsulated curcumin ("nanocurcumin"): a novel strategy for human cancer therapy. *J Nanobiotechnol*, *5*, 3. doi:10.1186/1477-3155-5-3

Blanco, E., Bey, E. A., Dong, Y., Weinberg, B. D., Sutton, D. M., Boothman, D. A., & Gao, J. (2007). Beta-lapachone-containing PEG-PLA polymer micelles as novel nanotherapeutics against NQO1-overexpressing tumor cells. *J Control Release*, *122*(3), 365–374. doi:10.1016/j.jconrel.2007.04.014

Bollimpelli, V. S., Kumar, P., Kumari, S., & Kondapi, A. K. (2016). Neuroprotective effect of curcumin-loaded lactoferrin nano particles against rotenone induced neurotoxicity. *Neurochem Int*, *95*, 37–45. doi:10.1016/j.neuint.2016.01.006

Bonechi, C., Martini, S., Ciani, L., Lamponi, S., Rebmann, H., Rossi, C., & Ristori, S. (2012). Using liposomes as carriers for polyphenolic compounds: the case of trans-resveratrol.

Bonifacio, B. V., Silva, P. B., Ramos, M. A., Negri, K. M., Bauab, T. M., & Chorilli, M. (2014). Nanotechnology-based drug delivery systems and herbal medicines: a review. *Int J Nanomed*, *9*, 1–15. doi:10.2147/IJN.S52634

Braithwaite,M. C., Tyagi, C., Tomar, L. K., Kumar, P., Choonara, Y. E., & Pillay, V. (2014). Nutraceutical-based therapeutics and formulation strategies augmenting their efficiency to complement modern medicine: an overview. *J Funct Foods*, *6*, 82–99. doi:10.1016/j.jff.2013.09.022

Brownlee, M. J. d. (2005). The pathobiology of diabetic complications: a unifying mechanism. *54*(6), 1615–1625.

Bushra, R., Aslam, N., & Khan, A. Y. (2011). Food-drug interactions. *Oman Med J*, *26*(2), 77–83. doi:10.5001/omj.2011.21

Caldeira de Araujo Lopes, S., Vinicius Melo Novais, M., Salviano Teixeira, C., Honorato-Sampaio, K., Tadeu Pereira, M., Ferreira, L. A., . . . Cristina Oliveira, M. (2013). Preparation, physicochemical characterization, and cell viability evaluation of long-circulating and pH-sensitive liposomes containing ursolic acid. *Biomed Res Int*, *2013*, 467147. doi:10.1155/2013/467147

Calderon, G., Juarez, O., Hernandez, G., Punzo, S., & De la Cruz, Z. D. (2017). Oxidative stress and diabetic retinopathy: development and treatment. *Eye*, *31*(8), 1122–1130.

Carvalho, G. C., Sabio, R. M., & Chorilli, M. (2020). An overview of properties and analytical methods for lycopene in organic nanocarriers. *Crit Rev Anal Chem*, 1–13. doi:10.1080/10408347.2020.1763774

Ceci, C., Lacal, P. M., Tentori, L., De Martino, M. G., Miano, R., & Graziani, G. (2018). Experimental evidence of the antitumor, antimetastatic and antiangiogenic activity of ellagic acid. *Nutrients*, *10*(11). doi:10.3390/nu10111756

Chalikwar, S. S., Belgamwar, V. S., Talele, V. R., Surana, S. J., & Patil, M. U. (2012). Formulation and evaluation of Nimodipine-loaded solid lipid nanoparticles delivered via lymphatic transport system. *Colloids Surf B Biointerfaces*, *97*, 109–116. doi:10.1016/j.colsurfb.2012.04.027

Chassaing, B., Koren, O., Goodrich, J. K., Poole, A. C., Srinivasan, S., Ley, R. E., & Gewirtz, A. T. (2015). Dietary emulsifiers impact the mouse gut microbiota promoting colitis and metabolic syndrome. *Nature*, *519*(7541), 92–96. doi:10.1038/nature14232

Chen, J.-P., Yang, P.-C., Ma, Y.-H., Tu, S.-J., & Lu, Y.-J. (2012). Targeted delivery of tissue plasminogen activator by binding to silica-coated magnetic nanoparticle. *Int J Nanomed*, *7*, 5137.

Chittasupho, C., Posritong, P., & Ariyawong, P. (2019). Stability, cytotoxicity, and retinal pigment epithelial cell binding of hyaluronic acid-coated PLGA nanoparticles encapsulating lutein. *AAPS PharmSciTech*, *20*(1), 1–13.

Chonpathompikunlert, P., Yoshitomi, T., Vong, L. B., Imaizumi, N., Ozaki, Y., & Nagasaki, Y. (2015). Recovery of cognitive dysfunction via orally administered redox-polymer nanotherapeutics in SAMP8 mice. *PLoS One*, *10*(5), e0126013.

Chopra, H., Dey, P. S., Das, D., Bhattacharya, T., Shah, M., Mubin, S., . . . Alamri, B. M. (2021). Curcumin nanoparticles as promising therapeutic agents for drug targets. *Molecules, 26*(16), 4998. Retrieved from https://www.mdpi.com/1420-3049/26/16/4998

Ciappellano, S. G., Tedesco, E., Venturini, M., & Benetti, F. (2016). In vitro toxicity assessment of oral nano-carriers. *Adv Drug Deliv Rev, 106*(Pt B), 381–401. doi:10.1016/j.addr.2016.08.007

Committee, E. S., Hardy, A., Benford, D., Halldorsson, T., Jeger, M. J., Knutsen, H. K., . . . Mortensen, A. (2018). Guidance on risk assessment of the application of nanoscience and nanotechnologies in the food and feed chain: part 1, human and animal health. *EFSA J, 16*(7), e05327. doi:10.2903/j.efsa.2018.5327

Cristiano, M. C., Froiio, F., Mancuso, A., Iannone, M., Fresta, M., Fiorito, S., . . . Paolino, D. (2020). In vitro and in vivo trans-epidermal water loss evaluation following topical drug delivery systems application for pharmaceutical analysis. *J Pharm Biomed Anal, 186*, 113295. doi:10.1016/j.jpba.2020.113295

Das, R. K., Kasoju, N., & Bora, U. (2010). Encapsulation of curcumin in alginate-chitosan-pluronic composite nanoparticles for delivery to cancer cells. *Nanomedicine, 6*(1), 153–160. doi:10.1016/j.nano.2009.05.009

Deng, Y., Zhang, X., Shen, H., He, Q., Wu, Z., Liao, W., & Yuan, M. (2019). Application of the nano-drug delivery system in treatment of cardiovascular diseases. *Front Bioeng Biotechnol, 7*, 489. doi:10.3389/fbioe.2019.00489

Desai, D., Zhang, J., Sandholm, J., Lehtimaki, J., Gronroos, T., Tuomela, J., & Rosenholm, J. M. (2017). Lipid bilayer-gated mesoporous silica nanocarriers for tumor-targeted delivery of zoledronic acid in vivo. *Mol Pharm, 14*(9), 3218–3227. doi:10.1021/acs.molpharmaceut.7b00519

Desai, P. R., Holihosur, P. D., Marathe, S. V., Kulkarni, B. B., Kalebar, V., Kamble, G., . . . Hireath, S. Studies on isolation and quantification of lycopene from tomato and papaya and its antioxidant and antifungal properties. *Agric Innov Res, 6*, 257–260.

Dey, S., & Sreenivasan, K. (2015). Conjugating curcumin to water soluble polymer stabilized gold nanoparticles via pH responsive succinate linker. *J Mater Chem B, 3*(5), 824–833. doi:10.1039/c4tb01731e

Dhawan, S., Kapil, R., & Singh, B. (2011). Formulation development and systematic optimization of solid lipid nanoparticles of quercetin for improved brain delivery. *J Pharm Pharmacol, 63*(3), 342–351. doi:10.1111/j.2042-7158.2010.01225.x

Din, F. U., Aman, W., Ullah, I., Qureshi, O. S., Mustapha, O., Shafique, S., & Zeb, A. (2017). Effective use of nanocarriers as drug delivery systems for the treatment of selected tumors. *Int J Nanomed, 12*, 7291–7309. doi:10.2147/IJN.S146315

Donelli, D., Antonelli, M., Bellinazzi, C., Gensini, G. F., & Firenzuoli, F. (2019). Effects of lavender on anxiety: a systematic review and meta-analysis. *Phytomedicine, 65*, 153099. doi:10.1016/j.phymed.2019.153099

Dulinska-Litewka, J., Lazarczyk, A., Halubiec, P., Szafranski, O., Karnas, K., & Karewicz, A. (2019). Superparamagnetic iron oxide nanoparticles-current and prospective medical applications. *Materials, 12*(4). doi:10.3390/ma12040617

El-Far, A. H., Al Jaouni, S. K., Li, W., & Mousa, S. A. (2018). Protective roles of thymoquinone nanoformulations: potential nanonutraceuticals in human diseases. *Nutrients, 10*(10). doi:10.3390/nu10101369

Elahi, N., Kamali, M., & Baghersad, M. H. (2018). Recent biomedical applications of gold nanoparticles: a review. *Talanta, 184*, 537–556. doi:10.1016/j.talanta.2018.02.088

Fan, S., Zheng, Y., Liu, X., Fang, W., Chen, X., Liao, W., . . . Liu, J. (2018). Curcumin-loaded PLGA-PEG nanoparticles conjugated with B6 peptide for potential use in Alzheimer's disease. *Drug Deliv, 25*(1), 1091–1102. doi:10.1080/10717544.2018.1461955

Farooqui, A. A. (2019). Potential treatment strategies for the treatment of dementia with chinese medicinal plants. In *Molecular Mechanisms of Dementia* (pp. 251–286): Academic Press.

Feng, S. S. (2006). New-concept chemotherapy by nanoparticles of biodegradable polymers: where are we now? *Nanomedicine, 1*(3), 297–309. doi:10.2217/17435889.1.3.297

Fontes, M. A. P., Vaz, G. C., Cardoso, T. Z. D., de Oliveira, M. F., Campagnole-Santos, M. J., Dos Santos, R. A. S., . . . Frezard, F. (2018). GABA-containing liposomes: neuroscience applications and translational perspectives for targeting neurological diseases. *Nanomedicine, 14*(3), 781–788. doi:10.1016/j.nano.2017.12.007

Fuster, M. G., Carissimi, G., Montalban, M. G., & Villora, G. (2020). Improving anticancer therapy with naringenin-loaded silk fibroin nanoparticles. *Nanomaterials, 10*(4). doi:10.3390/nano10040718

Gao, D. Q., Qian, S., & Ju, T. (2016). Anticancer activity of Honokiol against lymphoid malignant cells via activation of ROS-JNK and attenuation of Nrf2 and NF-kappaB. *J BUON, 21*(3), 673–679. Retrieved from https://www.ncbi.nlm.nih.gov/pubmed/27569089

Gao, L., Liu, G., Ma, J., Wang, X., Zhou, L., & Li, X. (2012). Drug nanocrystals: in vivo performances. *J Control Release, 160*(3), 418–430. doi:10.1016/j.jconrel.2012.03.013

Gao, S., Xu, Y., Asghar, S., Chen, M., Zou, L., Eltayeb, S., . . . Xiao, Y. (2015). Polybutylcyanoacrylate nanocarriers as promising targeted drug delivery systems. *J Drug Target*, *23*(6), 481–496. doi:10.3109/1061186X.2015.1020426

Gao, Y., Gu, W., Chen, L., Xu, Z., & Li, Y. (2008). The role of daidzein-loaded sterically stabilized solid lipid nanoparticles in therapy for cardio-cerebrovascular diseases. *Biomaterials*, *29*(30), 4129–4136. doi:10.1016/j.biomaterials.2008.07.008

Garanti, T., Stasik, A., Burrow, A. J., Alhnan, M. A., & Wan, K. W. (2016). Anti-glioma activity and the mechanism of cellular uptake of asiatic acid-loaded solid lipid nanoparticles. *Int J Pharm*, *500*(1–2), 305–315. doi:10.1016/j.ijpharm.2016.01.018

Gaude, T. T., Soares, G. A. B. e., Priolkar, R. N. S., Biradar, B., & Mamledesai, S. (2017). Synthesis of 4-Hydroxy-1-(phenyl/methyl)-3-[3-(substituted amino)-2 nitropropanoyl]quinolin-2(1H)-ones as Antimicrobial and Antitubercular Agents. *Indian J Heterocycl Chem*, *27*, 223–228.

Ghabaee, M., Jabedari, B., Al, E. E. N., Ghaffarpour, M., & Asadi, F. (2010). Serum and cerebrospinal fluid antioxidant activity and lipid peroxidation in Guillain-Barre syndrome and multiple sclerosis patients. *Int J Neurosci*, *120*(4), 301–304. doi:10.3109/00207451003695690

Ghosh, A., Sarkar, S., Mandal, A. K., & Das, N. (2013). Neuroprotective role of nanoencapsulated quercetin in combating ischemia-reperfusion induced neuronal damage in young and aged rats. *PLoS One*, *8*(4), e57735. doi:10.1371/journal.pone.0057735

Gonçalves, F.S.R., Martins, T. J., Duarte, M. M. C., Vicente, A. A., & Pinheiro, C. A. (2018). Advances in nutraceutical delivery systems: from formulation design for bioavailability enhancement to efficacy and safety evaluation. *Trends Food Sci Technol*, *78*, 270–291. doi:10.1016/j.tifs.2018.06.011

Granata, G., Paterniti, I., Geraci, C., Cunsolo, F., Esposito, E., Cordaro, M., . . . Consoli, G. M. J. M. p. (2017). Potential eye drop based on a calix [4] arene nanoassembly for curcumin delivery: enhanced drug solubility, stability, and anti-inflammatory effect. *14*(5), 1610–1622.

Granja, A., Frias, I., Neves, A. R., Pinheiro, M., & Reis, S. (2017). Therapeutic potential of epigallocatechin gallate nanodelivery systems. *BioMed Res Int*, *2017*.

Gray, W. D., Che, P., Brown, M., Ning, X., Murthy, N., & Davis, M. E. (2011). N-acetylglucosamine conjugated to nanoparticles enhances myocyte uptake and improves delivery of a small molecule p38 inhibitor for post-infarct healing. *J Cardiovasc Trans Res*, *4*(5), 631–643.

Guerrini, L., Alvarez-Puebla, R. A., & Pazos-Perez, N. (2018). Surface modifications of nanoparticles for stability in biological fluids. *Materials*, *11*(7). doi:10.3390/ma11071154

Guo, S., Yao, X., Jiang, Q., Wang, K., Zhang, Y., Peng, H., . . . Yang, W. (2020). Dihydroartemisinin-loaded magnetic nanoparticles for enhanced chemodynamic therapy. *Front Pharmacol*, *11*, 226. doi:10.3389/fphar.2020.00226

Guo, Y., Mao, X., Zhang, J., Sun, P., Wang, H., Zhang, Y., . . . Liu, X. (2019). Oral delivery of lycopene-loaded microemulsion for brain-targeting: preparation, characterization, pharmacokinetic evaluation and tissue distribution. *Drug Deliv*, *26*(1), 1191–1205. doi:10.1080/10717544.2019.1689312

Haba, Y., Kojima, C., Harada, A., Ura, T., Horinaka, H., & Kono, K. (2007). Preparation of poly(ethylene glycol)-modified poly(amido amine) dendrimers encapsulating gold nanoparticles and their heat-generating ability. *Langmuir*, *23*(10), 5243–5246. doi:10.1021/la0700826

Han, N., Zhao, Q., Wan, L., Wang, Y., Gao, Y., Wang, P., . . . Wang, S. (2015). Hybrid lipid-capped mesoporous silica for stimuli-responsive drug release and overcoming multidrug resistance. *ACS Appl Mater Interfaces*, *7*(5), 3342–3351. doi:10.1021/am5082793

Hawkes, C. A., Deng, L. H., Shaw, J. E., Nitz, M., & McLaurin, J. (2010). Small molecule beta-amyloid inhibitors that stabilize protofibrillar structures in vitro improve cognition and pathology in a mouse model of Alzheimer's disease. *Eur J Neurosci*, *31*(2), 203–213. doi:10.1111/j.1460-9568.2009.07052.x

He, X., & Hwang, H. M. (2016). Nanotechnology in food science: functionality, applicability, and safety assessment. *J Food Drug Anal*, *24*(4), 671–681. doi:10.1016/j.jfda.2016.06.001

Hermanowicz, N., Jones, S. A., & Hauser, R. A. (2019). Impact of non-motor symptoms in Parkinson's disease: a PMDAlliance survey. *Neuropsychiatr Dis Treat*, *15*, 2205–2212. doi:10.2147/NDT.S213917

How, C. W., Rasedee, A., Manickam, S., & Rosli, R. (2013). Tamoxifen-loaded nanostructured lipid carrier as a drug delivery system: characterization, stability assessment and cytotoxicity. *Colloids Surf B Biointerfaces*, *112*, 393–399. doi:10.1016/j.colsurfb.2013.08.009

Huo, X., Zhang, Y., Jin, X., Li, Y., & Zhang, L. (2019). A novel synthesis of selenium nanoparticles encapsulated PLGA nanospheres with curcumin molecules for the inhibition of amyloid beta aggregation in Alzheimer's disease. *J Photochem Photobiol B*, *190*, 98–102. doi:10.1016/j.jphotobiol.2018.11.008

Idris, M. I., Zaloga, J., Detsch, R., Roether, J. A., Unterweger, H., Alexiou, C., & Boccaccini, A. R. (2018). Surface modification of SPIONs in PHBV microspheres for biomedical applications. *Sci Rep, 8*(1), 7286. doi:10.1038/s41598-018-25243-9

Iijima, S. (1991). Helical microtubules of graphitic carbon. *Nature*, 56–58.

Institute, N. C. (2021). *Types of Cancer Treatment.*

Jabir, N. R., Tabrez, S., Ashraf, G. M., Shakil, S., Damanhouri, G. A., & Kamal, M. A. (2012). Nanotechnology-based approaches in anticancer research. *Int J Nanomed, 7*, 4391–4408. doi:10.2147/IJN.S33838

Jahangirian, H., Lemraski, E. G., Webster, T. J., Rafiee-Moghaddam, R., & Abdollahi, Y. (2017). A review of drug delivery systems based on nanotechnology and green chemistry: green nanomedicine. *Int J Nanomed, 12*, 2957–2978. doi:10.2147/IJN.S127683

Jain, K. A., Mehra, K. N., & Swarnakar, K. N. (2015). Role of antioxidants for the treatment of cardiovascular diseases: challenges and opportunities. *21*(30), 4441–4455.

Jampilek, J., & Kralova, K. (2020). Potential of nanonutraceuticals in increasing immunity. *Nanomaterials, 10*(11). doi:10.3390/nano10112224

Jampilek, J., Kos, J., & Kralova, K. (2019). Potential of Nanomaterial Applications in Dietary Supplements and Foods for Special Medical Purposes. *Nanomaterials, 9*(2). doi:10.3390/nano9020296

Jeong, H.-H., Choi, E., Ellis, E., & Lee, T.-C. (2019). Recent advances in gold nanoparticles for biomedical applications: from hybrid structures to multi-functionality. *J Mater Chem B, 7*(22), 3480–3496. doi:10.1039/C9TB00557A

Jin, H., Pi, J., Yang, F., Wu, C., Cheng, X., Bai, H., . . . Chen, Z. W. (2016). Ursolic acid-loaded chitosan nanoparticles induce potent anti-angiogenesis in tumor. *Appl Microbiol Biotechnol, 100*(15), 6643–6652. doi:10.1007/s00253-016-7360-8

Jin, S., Zhang, Q. Y., Kang, X. M., Wang, J. X., & Zhao, W. H. (2010). Daidzein induces MCF-7 breast cancer cell apoptosis via the mitochondrial pathway. *Ann Oncol, 21*(2), 263–268. doi:10.1093/annonc/mdp499

Joseph, R. R., & Venkatraman, S. S. (2017). Drug delivery to the eye: what benefits do nanocarriers offer? *Nanomedicine, 12*(6), 683–702. doi:10.2217/nnm-2016-0379

Kakkar, D., Dumoga, S., Kumar, R., Chuttania, K., & Mishra, A. K. (2015). PEGylated solid lipid nanoparticles: design, methotrexate loading and biological evaluation in animal models. *MedChemComm, 6*(8). doi:10.1039/C5MD00104H

Kalra, E. K. (2003). Nutraceutical: definition and introduction. *AAPS PharmSci, 5*(3), E25. doi:10.1208/ps050325

Kang, C., Gwon, S., Song, C., Kang, P. M., Park, S.-C., Jeon, J., . . . Lee, D. J. A. N. (2017). Fibrin-targeted and H2O2-responsive nanoparticles as a theranostics for thrombosed vessels. *11*(6), 6194–6203.

Karavolos, M., & Holban, A. (2016). Nanosized drug delivery systems in gastrointestinal targeting: interactions with microbiota. *Pharmaceuticals, 9*(4). doi:10.3390/ph9040062

Karimi, M., Ghasemi, A., Sahandi Zangabad, P., Rahighi, R., Moosavi Basri, S. M., Mirshekari, H., . . . Hamblin, M. R. (2016). Smart micro/nanoparticles in stimulus-responsive drug/gene delivery systems. *Chem Soc Rev, 45*(5), 1457–1501. doi:10.1039/c5cs00798d

Keservani, K. R., Kesharwani, K. R., Sharma, K. A., Gautam, P. S., & Verma, K. S. (2017). Nutraceutical Formulations and Challenges. In *Developing New Functional Food and Nutraceutical Products* (pp. 161–177): Academic Press.

Khinast, J., Baumgartner, R., & Roblegg, E. (2013). Nano-extrusion: a one-step process for manufacturing of solid nanoparticle formulations directly from the liquid phase. *AAPS PharmSciTech, 14*(2), 601–604. doi:10.1208/s12249-013-9946-0

Khorasani, S., Danaei, M., & Mozafari, M.R. (2018). Nanoliposome technology for the food and nutraceutical industries. *Trends Food Sci Technol, 79*, 106–115. doi:10.1016/j.tifs.2018.07.009

Kim, K. S., Song, C. G., Kang, P. M. (2019). Targeting oxidative stress using nanoparticles as a theranostic strategy for cardiovascular diseases. *Antioxid Redox Signal, 30*(5), 733–746.

Kovochich, M., Marsden, M. D., & Zack, J. A. (2011). Activation of latent HIV using drug-loaded nanoparticles. *PLoS One, 6*(4), e18270. doi:10.1371/journal.pone.0018270

Kowluru, R. A., & Chan, P.-S. (2007). Oxidative stress and diabetic retinopathy. *Exp Diabetes Res, 2007.*

Kowluru, R. A., Kanwar, M. (2007). Effects of curcumin on retinal oxidative stress and inflammation in diabetes. *Nutr. Metabol, 4*(1), 1–8.

Kumar, V., Sharma, N., & Maitra, S. S. (2017). In vitro and in vivo toxicity assessment of nanoparticles. *Int Nano Lett, 7*, 243–256.

Kumar, V. V., Chandrasekar, D., Ramakrishna, S., Kishan, V., Rao, Y. M., & Diwan, P. V. (2007). Development and evaluation of nitrendipine loaded solid lipid nanoparticles: influence of wax and glyceride lipids on plasma pharmacokinetics. *Int J Pharm, 335*(1–2), 167–175. doi:10.1016/j.ijpharm.2006.11.004

Lai, A. Y., & McLaurin, J. (2012). Inhibition of amyloid-beta peptide aggregation rescues the autophagic deficits in the TgCRND8 mouse model of Alzheimer disease. *Biochim Biophys Acta, 1822*(10), 1629–1637. doi:10.1016/j.bbadis.2012.07.003

Lai, D. Y. (2012). Toward toxicity testing of nanomaterials in the 21st century: a paradigm for moving forward. *Wiley Interdiscip Rev Nanomed Nanobiotechnol, 4*(1), 1–15. doi:10.1002/wnan.162

Lam, P. L., Wong, W. Y., Bian, Z., Chui, C. H., & Gambari, R. (2017). Recent advances in green nanoparticulate systems for drug delivery: efficient delivery and safety concern. *Nanomedicine, 12*(4), 357–385. doi:10.2217/nnm-2016-0305

Lee, D., Bae, S., Hong, D., Lim, H., Yoon, J. H., Hwang, O., . . . Kang, P. M. J. S. r. (2013). H 2 O 2-responsive molecularly engineered polymer nanoparticles as ischemia/reperfusion-targeted nanotherapeutic agents. *3*(1), 1–8.

Lee, D., Lee, W. S., Lim, S., Kim, Y. K., Jung, H. Y., Das, S., . . . Chung, S. K. (2017). A guanidine-appended scyllo-inositol derivative AAD-66 enhances brain delivery and ameliorates Alzheimer's phenotypes. *Sci Rep, 7*(1), 14125. doi:10.1038/s41598-017-14559-7

Levy, I., Attias, S., Ben-Arye, E., Goldstein, L., & Schiff, E. (2017). Adverse events associated with interactions with dietary and herbal supplements among inpatients. *Br J Clin Pharmacol, 83*(4), 836–845. doi:10.1111/bcp.13158

Li, Y., Zou, X., Cao, K., Xu, J., Yue, T., Dai, F., . . . Liu, J. (2013). Curcumin analog 1, 5-bis (2-trifluoromethylphenyl)-1, 4-pentadien–3-one exhibits enhanced ability on Nrf2 activation and protection against acrolein-induced ARPE-19 cell toxicity. *Toxicol Appl Pharmacol, 272*(3), 726–735.

Lidia, A.-V., Carlos, Z.-M., Amalia, R.-M. A. V., & Jose, V.-B. (2019). Nutraceuticals: definition, applied nanoengineering in their production and applications. *Int J Biosens Bioelectron, 5*(3). doi:10.15406/ijbsbe.2019.05.00154

Liu, M., Dasgupta, A., Koczera, P., Schipper, S., Rommel, D., Shi, Y., . . . Lammers, T. (2020). Drug Loading in Poly(butyl cyanoacrylate)-Based Polymeric Microbubbles. *Mol Pharm, 17*(8), 2840–2848. doi:10.1021/acs.molpharmaceut.0c00242

Liu, Z., Tabakman, S., Welsher, K., & Dai, H. (2009). Carbon nanotubes in biology and medicine: in vitro and in vivo detection, imaging and drug delivery. *Nano Res, 2*(2), 85–120. doi:10.1007/s12274-009-9009-8

Lombardo, D., Kiselev, M. A., & Caccamo, M. T. (2019). Smart nanoparticles for drug delivery application: development of versatile nanocarrier platforms in biotechnology and nanomedicine. *J Nanomater, 2019*. doi:10.1155/2019/3702518

Loureiro, J. A., Andrade, S., Duarte, A., Neves, A. R., Queiroz, J. F., Nunes, C., . . . Pereira, M. C. (2017). Resveratrol and grape extract-loaded solid lipid nanoparticles for the treatment of Alzheimer's disease. *Molecules, 22*(2). doi:10.3390/molecules22020277

Ma, X., Moore, Z. R., Huang, G., Huang, X., Boothman, D. A., & Gao, J. (2015). Nanotechnology-enabled delivery of NQO1 bioactivatable drugs. *J Drug Target, 23*(7–8), 672–680. doi:10.3109/1061186X.2015.1073296

Ma, Y., Nolte, R. J., & Cornelissen, J. J. (2012). Virus-based nanocarriers for drug delivery. *Adv Drug Deliv Rev, 64*(9), 811–825. doi:10.1016/j.addr.2012.01.005

Madaan, K., Kumar, S., Poonia, N., Lather, V., & Pandita, D. (2014). Dendrimers in drug delivery and targeting: drug-dendrimer interactions and toxicity issues. *J Pharm Bioallied Sci, 6*(3), 139–150. doi:10.4103/0975-7406.130965

Madani, S. Y., Naderi, N., Dissanayake, O., Tan, A., & Seifalian, A. M. (2011). A new era of cancer treatment: carbon nanotubes as drug delivery tools. *Int J Nanomed, 6*, 2963–2979. doi:10.2147/IJN.S16923

Maher, S., Mrsny, R. J., & Brayden, D. J. (2016). Intestinal permeation enhancers for oral peptide delivery. *Adv Drug Deliv Rev, 106*(Pt B), 277–319. doi:10.1016/j.addr.2016.06.005

Mahmoud, N. N., Albasha, A., Hikmat, S., Hamadneh, L., Zaza, R., Shraideh, Z., & Khalil, E. A. (2020). Nanoparticle size and chemical modification play a crucial role in the interaction of nano gold with the brain: extent of accumulation and toxicity. *Biomater Sci, 8*(6), 1669–1682. doi:10.1039/c9bm02072a

Majeed, H., Antoniou, J., & Fang, Z. (2014). Apoptotic effects of eugenol-loaded nanoemulsions in human colon and liver cancer cell lines. *Asian Pac J Cancer Prev, 15*(21), 9159–9164. doi:10.7314/apjcp.2014.15.21.9159

Makkar, R., Behl, T., Bungau, S., Zengin, G., Mehta, V., Kumar, A., . . . Oancea, R. (2020). Nutraceuticals in neurological disorders. *Int J Mol Sci, 21*(12). doi:10.3390/ijms21124424

Manchester, M., & Singh, P. (2006). Virus-based nanoparticles (VNPs): platform technologies for diagnostic imaging. *Adv Drug Deliv Rev, 58*(14), 1505–1522. doi:10.1016/j.addr.2006.09.014

Marjaneh, R. M., Rahmani, F., Hassanian, S. M., Rezaei, N., Hashemzehi, M., Bahrami, A., . . . Khazaei, M. (2018). Phytosomal curcumin inhibits tumor growth in colitis-associated colorectal cancer. *J Cell Physiol, 233*(10), 6785–6798. doi:10.1002/jcp.26538

Marrache, S., & Dhar, S. (2012). Engineering of blended nanoparticle platform for delivery of mitochondria-acting therapeutics. *Proc Nat Acad Sci, 109*(40), 16288–16293.

Martinho, N., Damg, C., & Reis, C. P. (2011). Recent advances in drug delivery systems. *J Biomater Nanobiotechnol, 2*, 510–526. doi:10.4236/jbnb.2011.225062

Martins, P., Jesus, J., Santos, S., Raposo, L. R., Roma-Rodrigues, C., Baptista, P. V., & Fernandes, A. R. (2015). Heterocyclic anticancer compounds: recent advances and the paradigm shift towards the use of nanomedicine's tool box. *Molecules, 20*(9), 16852–16891. doi:10.3390/molecules200916852

Masserini, M. (2013). Nanoparticles for brain drug delivery. *ISRN Biochem, 2013*, 238428. doi:10.1155/2013/238428

Matea, C. T., Mocan, T., Tabaran, F., Pop, T., Mosteanu, O., Puia, C., . . . Mocan, L. (2017). Quantum dots in imaging, drug delivery and sensor applications. *Int J Nanomed, 12*, 5421–5431. doi:10.2147/IJN.S138624

Mauricio, M., Guerra-Ojeda, S., Marchio, P., Valles, S., Aldasoro, M., Escribano-Lopez, I., . . . Victor, V. M. (2018). Nanoparticles in medicine: a focus on vascular oxidative stress. *Oxid Med Cell Longev, 2018.*

McClements, D. J. (2013). Nanoemulsion-based oral delivery systems for lipophilic bioactive components: nutraceuticals and pharmaceuticals. *Ther Deliv, 4*(7), 841–857. doi:10.4155/tde.13.46

McClements, D. J., & Rao, J. (2011). Food-grade nanoemulsions: formulation, fabrication, properties, performance, biological fate, and potential toxicity. *Crit Rev Food Sci Nutr, 51*(4), 285–330. doi:10.1080/10408398.2011.559558

McClements, D. J., & Xiao, H. (2017). Is nano safe in foods? Establishing the factors impacting the gastrointestinal fate and toxicity of organic and inorganic food-grade nanoparticles. *NPJ Sci Food, 1*, 6. doi:10.1038/s41538-017-0005-1

McLaurin, J., Golomb, R., Jurewicz, A., Antel, J. P., & Fraser, P. E. (2000). Inositol stereoisomers stabilize an oligomeric aggregate of Alzheimer amyloid beta peptide and inhibit abeta -induced toxicity. *J Biol Chem, 275*(24), 18495–18502. doi:10.1074/jbc.M906994199

Meenambal, R., & Bharath, M. M. S. (2020). Nanocarriers for effective nutraceutical delivery to the brain. *Neurochem Int, 140.* doi:10.1016/j.neuint.2020.104851

Mehnert, W., & Mader, K. (2001). Solid lipid nanoparticles: production, characterization and applications. *Adv Drug Deliv Rev, 47*(2–3), 165–196. doi:10.1016/s0169-409x(01)00105-3

Meng, Q., Wang, A., Hua, H., Jiang, Y., Wang, Y., Mu, H., . . . Sun, K. (2018). Intranasal delivery of Huperzine A to the brain using lactoferrin-conjugated N-trimethylated chitosan surface-modified PLGA nanoparticles for treatment of Alzheimer's disease. *Int J Nanomed, 13*, 705–718. doi:10.2147/IJN.S151474

Mikhail, A. S., & Allen, C. (2010). Poly(ethylene glycol)-b-poly(epsilon-caprolactone) micelles containing chemically conjugated and physically entrapped docetaxel: synthesis, characterization, and the influence of the drug on micelle morphology. *Biomacromolecules, 11*(5), 1273–1280. doi:10.1021/bm100073s

Miladi, K., Sfar, S., Fessi, H., & Elaissari, A. (2017). *Nanoprecipitation Process: From Particle Preparation to In Vivo Applications*: Springer.

Mishra, B., Patel, B. B., & Tiwari, S. (2010). Colloidal nanocarriers: a review on formulation technology, types and applications toward targeted drug delivery. *Nanomedicine, 6*(1), 9–24. doi:10.1016/j.nano.2009.04.008

Mody, V. V., Siwale, R., Singh, A., & Mody, H. R. (2010). Introduction to metallic nanoparticles. *J Pharm Bioallied Sci, 2*(4), 282–289. doi:10.4103/0975-7406.72127

Mohanty, S. K., Swamy, M. K., Sinniah, U. R., & Anuradha, M. (2017). *Leptadenia reticulata* (Retz.) Wight & Arn. (Jivanti): botanical, agronomical, phytochemical, pharmacological, and biotechnological aspects. *Molecules, 22*(6). doi:10.3390/molecules22061019

Moradi, S. Z., Momtaz, S., Bayrami, Z., Farzaei, M. H., & Abdollahi, M. (2020). Nanoformulations of herbal extracts in treatment of neurodegenerative disorders. *Front Bioeng Biotechnol, 8*, 238. doi:10.3389/fbioe.2020.00238

Morais, R. P., Novais, G. B., Sangenito, L. S., Santos, A. L. S., Priefer, R., Morsink, M., . . . Cardoso, J. C. (2020). Naringenin-functionalized multi-walled carbon nanotubes: a potential approach for site-specific remote-controlled anticancer delivery for the treatment of lung cancer cells. *Int J Mol Sci, 21*(12). doi:10.3390/ijms21124557

Mozetic, M. (2019). Surface modification to improve properties of materials. *Materials, 12*(3). doi:10.3390/ma12030441

Mudshinge, S. R., Deore, A. B., Patil, S., & Bhalgat, C. M. (2011). Nanoparticles: emerging carriers for drug delivery. *Saudi Pharm J, 19*(3), 129–141. doi:10.1016/j.jsps.2011.04.001

Mukerjee, A., & Vishwanatha, J. K. (2009). Formulation, characterization and evaluation of curcumin-loaded PLGA nanospheres for cancer therapy. *Anticancer Res, 29*(10), 3867–3875. Retrieved from https://www.ncbi.nlm.nih.gov/pubmed/19846921

Muller, R. H., Mader, K., & Gohla, S. (2000). Solid lipid nanoparticles (SLN) for controlled drug delivery: a review of the state of the art. *Eur J Pharm Biopharm*, *50*(1), 161–177. doi:10.1016/s0939-6411(00)00087-4

Muller, R. H., Radtke, M., & Wissing, S. A. (2002). Solid lipid nanoparticles (SLN) and nanostructured lipid carriers (NLC) in cosmetic and dermatological preparations. *Adv Drug Deliv Rev*, *54*(Suppl 1), S131–S155. doi:10.1016/s0169-409x(02)00118-7

Mythri, R. B., & Bharath, M. M. (2012). Curcumin: a potential neuroprotective agent in Parkinson's disease. *Curr Pharm Des*, *18*(1), 91–99. doi:10.2174/138161212798918995

Mythri, R. B., & Bharath, M. M. S. (2019). Omics and epigenetics of polyphenol-mediated neuroprotection: the curcumin perspective. In *Curcumin for Neurological and Psychiatric Disorders* (pp. 169–189): Academic Press.

Nasri, H., Baradaran, A., Shirzad, H., & Rafieian-Kopaei, M. (2014). New concepts in nutraceuticals as alternative for pharmaceuticals. *Int J Prev Med*, Dec;*5*(12), 1487–1499.

National institute of ageing, N. (2018). Treatment of Alzeimers disease. Retrieved from https://www.nia.nih.gov/health/how-alzheimers-disease-treated

National institute on ageing, N. (2021). Alzheimer's disease fact sheet. Retrieved from https://www.nia.nih.gov/health/alzheimers-disease-fact-sheet

Nazemiyeh, E., Eskandani, M., Sheikhloie, H., & Nazemiyeh, H. (2016). Formulation and physicochemical characterization of lycopene-loaded solid lipid nanoparticles. *Adv Pharm Bull*, *6*(2), 235–241. doi:10.15171/apb.2016.032

Neamatallah, T., El-Shitany, N., Abbas, A., Eid, B. G., Harakeh, S., Ali, S., & Mousa, S. (2020). Nano ellagic acid counteracts cisplatin-induced upregulation in OAT1 and OAT3: a possible nephroprotection mechanism. *Molecules*, *25*(13). doi:10.3390/molecules25133031

Neves, A. R., Queiroz, J. F., & Reis, S. (2016). Brain-targeted delivery of resveratrol using solid lipid nanoparticles functionalized with apolipoprotein E. *J Nanobiotechnol*, *14*, 27. doi:10.1186/s12951-016-0177-x

Nimse, S. B., & Pal, D. (2015). Free radicals, natural antioxidants, and their reaction mechanisms. *RSC Adv*, *5*(35), 27986–28006.

Ouattara, B., Simard, R. E., Holley, R. A., Piette, G. J., & Begin, A. (1997). Antibacterial activity of selected fatty acids and essential oils against six meat spoilage organisms. *Int J Food Microbiol*, *37*(2–3), 155–162. doi:10.1016/s0168-1605(97)00070-6

Palle, S., & Neerati, P. (2017). Quercetin nanoparticles attenuates scopolamine induced spatial memory deficits and pathological damages in rats. *Bull Fac Pharm Cairo Univ*, *55*(1), 101–106. doi:10.1016/j.bfopcu.2016.10.004

Pandareesh, M. D., Mythri, R. B., & Srinivas Bharath, M. M. (2015). Bioavailability of dietary polyphenols: factors contributing to their clinical application in CNS diseases. *Neurochem Int*, *89*, 198–208. doi:10.1016/j.neuint.2015.07.003

Paolino, D., Mancuso, A., Cristiano, M. C., Froiio, F., Lammari, N., Celia, C., & Fresta, M. (2021). Nanonutraceuticals: the new frontier of supplementary food. *Nanomaterials*, *11*(3). doi:10.3390/nano11030792

Park, K. (2013). Facing the truth about nanotechnology in drug delivery. *ACS Nano*, *7*(9), 7442–7447. doi:10.1021/nn404501g

Patel, M. N., Lakkadwala, S., Majrad, M. S., Injeti, E. R., Gollmer, S. M., Shah, Z. A., . . . Nesamony, J. (2014). Characterization and evaluation of 5-fluorouracil-loaded solid lipid nanoparticles prepared via a temperature-modulated solidification technique. *AAPS PharmSciTech*, *15*(6), 1498–1508. doi:10.1208/s12249-014-0168-x

Pathak, K., & Raghuvanshi, S. (2015). Oral bioavailability: issues and solutions via nanoformulations. *Clin Pharmacokinet*, *54*(4), 325–357. doi:10.1007/s40262-015-0242-x

Patra, J. K., & Baek, K.-H. (2014). Green nanobiotechnology: factors affecting synthesis and characterization techniques. *J Nanomater*, *2014*, 219.

Patra, J. K., Das, G., Fraceto, L. F., Campos, E. V. R., Rodriguez-Torres, M. D. P., Acosta-Torres, L. S., . . . Shin, H. S. (2018). Nano based drug delivery systems: recent developments and future prospects. *J Nanobiotechnology*, *16*(1), 71. doi:10.1186/s12951-018-0392-8

Pattenden, L. K., Middelberg, A. P., Niebert, M., & Lipin, D. I. (2005). Towards the preparative and large-scale precision manufacture of virus-like particles. *Trends Biotechnol*, *23*(10), 523–529. doi:10.1016/j.tibtech.2005.07.011

Platania, C., Fidilio, A., Lazzara, F., Piazza, C., Geraci, F., Giurdanella, G., . . . Bucolo, C. (2018). Retinal protection and distribution of curcumin in vitro and in vivo. *Front Pharmacol*, *9*, 670.

Podila, R., & Brown, J. M. (2013). Toxicity of engineered nanomaterials: a physicochemical perspective. *J Biochem Mol Toxicol*, *27*(1), 50–55. doi:10.1002/jbt.21442

Ponnurangam, S., Mammen, J. M., Ramalingam, S., He, Z., Zhang, Y., Umar, S., . . . Anant, S. (2012). Honokiol in combination with radiation targets notch signaling to inhibit colon cancer stem cells. *Mol Cancer Ther, 11*(4), 963–972. doi:10.1158/1535-7163.MCT-11-0999

Postuma, R. B., Berg, D., Stern, M., Poewe, W., Olanow, C. W., Oertel, W., . . . Deuschl, G. (2015). MDS clinical diagnostic criteria for Parkinson's disease. *Mov Disord, 30*(12), 1591–1601. doi:10.1002/mds.26424

Prabhakar, N., Zhang, J., Desai, D., Casals, E., Gulin-Sarfraz, T., Nareoja, T., . . . Rosenholm, J. M. (2016). Stimuli-responsive hybrid nanocarriers developed by controllable integration of hyperbranched PEI with mesoporous silica nanoparticles for sustained intracellular siRNA delivery. *Int J Nanomed, 11*, 6591–6608. doi:10.2147/IJN.S120611

Prasad, M., Lambe, U. P., Brar, B., Shah, I., J, M., Ranjan, K., . . . Prasad, G. (2018). Nanotherapeutics: an insight into healthcare and multi-dimensional applications in medical sector of the modern world. *Biomed Pharmacother, 97*, 1521–1537. doi:10.1016/j.biopha.2017.11.026

Premanand, C., Rema, M., Sameer, M. Z., Sujatha, M., & Balasubramanyam, M. (2006). Effect of curcumin on proliferation of human retinal endothelial cells under in vitro conditions. *Invest Ophthalmol Vis Sci, 47*(5), 2179–2184.

Pucci, C., Martinelli, C., & Ciofani, G. (2019). Innovative approaches for cancer treatment: current perspectives and new challenges. *Ecancermedicalscience, 13*, 961. doi:10.3332/ecancer.2019.961

Qureshi, O. S., Kim, H. S., Zeb, A., Choi, J. S., Kim, H. S., Kwon, J. E., . . . Kim, J. K. (2017). Sustained release docetaxel-incorporated lipid nanoparticles with improved pharmacokinetics for oral and parenteral administration. *J Microencapsul, 34*(3), 250–261. doi:10.1080/02652048.2017.1337247

Rahman, M. H., Akter, R., Bhattacharya, T., Abdel-Daim, M. M., Alkahtani, S., Arafah, M. W., . . . Mittal, V. (2020). Resveratrol and neuroprotection: impact and its therapeutic potential in Alzheimer's disease. *Front Pharmacol, 11*(2272). doi:10.3389/fphar.2020.619024

Raval, N., Mistry, T., Acharya, N., & Acharya, S. (2015). Development of glutathione-conjugated asiatic acid-loaded bovine serum albumin nanoparticles for brain-targeted drug delivery. *J Pharm Pharmacol, 67*(11), 1503–1511. doi:10.1111/jphp.12460

Renukuntla, J., Vadlapudi, A. D., Patel, A., Boddu, S. H., & Mitra, A. K. (2013). Approaches for enhancing oral bioavailability of peptides and proteins. *Int J Pharm, 447*(1–2), 75–93. doi:10.1016/j.ijpharm.2013.02.030

Richard, T., Pawlus, A. D., Iglesias, M. L., Pedrot, E., Waffo-Teguo, P., Merillon, J. M., & Monti, J. P. (2011). Neuroprotective properties of resveratrol and derivatives. *Ann N Y Acad Sci, 1215*, 103–108. doi:10.1111/j.1749-6632.2010.05865.x

Rodrigues, T., Reker, D., Schneider, P., & Schneider, G. (2016). Counting on natural products for drug design. *Nat Chem, 8*, 531–541.

Rossino, M. G., & Casini, G. (2019). Nutraceuticals for the treatment of diabetic retinopathy. *Nutrients, 11*(4), 771.

RS, P., Bomb, K., Srivastava, R., & Bandyopadhyaya, R. (2020). Dual drug delivery of curcumin and niclosamide using PLGA nanoparticles for improved therapeutic effect on breast cancer cells. *J Polym Res, 27*(5). doi:10.1007/s10965-020-02092-7

Saeed, L. M., Mahmood, M., Pyrek, S. J., Fahmi, T., Xu, Y., Mustafa, T., . . . Biris, A. S. (2014). Single-walled carbon nanotube and graphene nanodelivery of gambogic acid increases its cytotoxicity in breast and pancreatic cancer cells. *J Appl Toxicol, 34*(11), 1188–1199. doi:10.1002/jat.3018

Salama, L., Pastor, E. R., Stone, T., & Mousa, S. A. (2020). Emerging nanopharmaceuticals and nanonutraceuticals in cancer management. *Biomedicines, 8*(9). doi:10.3390/biomedicines8090347

Salehi, B., Fokou, P. V. T., Sharifi-Rad, M., Zucca, P., Pezzani, R., Martins, N., & Sharifi-Rad, J. (2019). The therapeutic potential of naringenin: a review of clinical trials. *Pharmaceuticals, 12*(1). doi:10.3390/ph12010011

Sampath, C., Sang, S., & Ahmedna, M. (2016). In vitro and in vivo inhibition of aldose reductase and advanced glycation end products by phloretin, epigallocatechin 3-gallate and [6]-gingerol. *Biomed Pharmacotherapy, 84*, 502–513.

Santarsieri, D., & Schwartz, T. L. (2015). Antidepressant efficacy and side-effect burden: a quick guide for clinicians. *Drugs Context, 4*, 212290. doi:10.7573/dic.212290

Santini, A., Cammarata, S. M., Capone, G., Ianaro, A., Tenore, G. C., Pani, L., & Novellino, E. (2018). Nutraceuticals: opening the debate for a regulatory framework. *Br J Clin Pharmacol, 84*(4), 659–672. doi:10.1111/bcp.13496

Sathya, S., & Devi, K. P. (2018). The use of polyphenols for the treatment of Alzheimer's disease. In *Role of the Mediterranean Diet in the Brain and Neurodegenerative Diseases* (pp. 239–252): Academic Press.

Schrott, L. M., Jackson, K., Yi, P., Dietz, F., Johnson, G. S., Basting, T. F., . . . Alexander, J. S. (2015). Acute oral Bryostatin-1 administration improves learning deficits in the APP/PS1 transgenic mouse model of Alzheimer's disease. *Curr Alzheimer Res, 12*(1), 22–31. doi:10.2174/1567205012666141218141904

Seid, J. M. (2017). *Nanoencapsulation Technologies for the Food and Nutraceutical Industries*: AcademicnPress.

Seshadri, G., Sy, J. C., Brown, M., Dikalov, S., Yang, S. C., Murthy, N., & Davis, M. E. (2010). The delivery of superoxide dismutase encapsulated in polyketal microparticles to rat myocardium and protection from myocardial ischemia-reperfusion injury. *Biomaterials*, *31*(6), 1372–1379.

Shahein, S. A., Aboul-Enein, A. M., Higazy, I. M., Abou-Elella, F., Lojkowski, W., Ahmed, E. R., . . . AbouAitah, K. (2019). Targeted anticancer potential against glioma cells of thymoquinone delivered by mesoporous silica core-shell nanoformulations with pH-dependent release. *Int J Nanomed*, *14*, 5503–5526. doi:10.2147/IJN.S206899

Sharma, G., Raturi, K., Dang, S., Gupta, S., & Gabrani, R. (2014). Combinatorial antimicrobial effect of curcumin with selected phytochemicals on *Staphylococcus epidermidis*. *J Asian Nat Prod Res*, *16*(5), 535–541. doi:10.1080/10286020.2014.911289

Shome, S., Talukdar, A. D., Choudhury, M. D., Bhattacharya, M. K., & Upadhyaya, H. (2016). Curcumin as potential therapeutic natural product: a nanobiotechnological perspective. *J Pharm Pharmacol*, *68*(12), 1481–1500.

Siddiqui, A. A., Iram, F., Siddiqui, S., & Sahu, K. (2014). Role of natural products in drug discovery process. *Int. J. Drug Dev. & Res.*, *6*(2).

Singh, J., & Sinha, S. (2012). Classification, regulatory acts and applications of nutraceuticals for health. *Int J Pharma Bio Sci*, *2*(1), 177–187.

Singh, P., Prasuhn, D., Yeh, R. M., Destito, G., Rae, C. S., Osborn, K., . . . Manchester, M. (2007). Bio-distribution, toxicity and pathology of cowpea mosaic virus nanoparticles in vivo. *J Control Release*, *120*(1–2), 41–50. doi:10.1016/j.jconrel.2007.04.003

Soares, G., Bhattacharya, T., Chakrabarti, T., Tagde, P., & Cavalu, S. (2021). Exploring pharmacological mechanisms of essential oils on the central nervous system. *Plants*, *11*(1). doi:10.3390/plants11010021

Sokola-Wysoczanska, E., Wysoczanski, T., Wagner, J., Czyz, K., Bodkowski, R., Lochynski, S., & Patkowska-Sokola, B. (2018). Polyunsaturated fatty acids and their potential therapeutic role in cardiovascular system disorders: a review. *Nutrients*, *10*(10). doi:10.3390/nu10101561

Somasuntharam, I., Boopathy, A. V., Khan, R. S., Martinez, M. D., Brown, M. E., Murthy, N., & Davis, M. E. (2013). Delivery of Nox2-NADPH oxidase siRNA with polyketal nanoparticles for improving cardiac function following myocardial infarction. *Biomaterials*, *34*(31), 7790–7798.

Soto-Quintero, A., Guarrotxena, N., Garcia, O., & Quijada-Garrido, I. (2019). Curcumin to promote the synthesis of silver NPs and their self-assembly with a thermoresponsive polymer in core-shell nanohybrids. *Sci Rep*, *9*(1), 18187. doi:10.1038/s41598-019-54752-4

Spuentrup, E., Fausten, B., Kinzel, S., Wiethoff, A. J., Botnar, R. M., Graham, P. B., . . . Manning, W. J. (2005). Molecular magnetic resonance imaging of atrial clots in a swine model. *Circulation*, *112*(3), 396–399.

Stoker, T. B., Torsney, K. M., & Barker, R. A. (2018). Emerging treatment approaches for Parkinson's disease. *Front Neurosci*, *12*, 693. doi:10.3389/fnins.2018.00693

Sukhanova, A., Bozrova, S., Sokolov, P., Berestovoy, M., Karaulov, A., & Nabiev, I. (2018). Dependence of nanoparticle toxicity on their physical and chemical properties. *Nanoscale Res Lett*, *13*(1), 44. doi:10.1186/s11671-018-2457-x

Sun, M. K., & Alkon, D. L. (2006). Bryostatin-1: pharmacology and therapeutic potential as a CNS drug. *CNS Drug Rev*, *12*(1), 1–8. doi:10.1111/j.1527-3458.2006.00001.x

Sun, T., Zhang, Y. S., Pang, B., Hyun, D. C., Yang, M., & Xia, Y. (2014). Engineered nanoparticles for drug delivery in cancer therapy. *Angew Chem Int Ed Engl*, *53*(46), 12320–12364. doi:10.1002/anie.201403036

Suri, S. S., Fenniri, H., & Singh, B. (2007). Nanotechnology-based drug delivery systems. *J Occup Med Toxicol*, *2*, 16. doi:10.1186/1745-6673-2-16

Sut, S., Baldan, V., Faggian, M., Peron, G., & Dall Acqua, S. (2016). Nutraceuticals: a new challenge for medicinal chemistry. *Curr Med Chem*, *23*(28), 3198–3223. doi:10.2174/0929867323666160615104837

Swamy, M. K., & Sinniah, U. R. (2016). Patchouli (*Pogostemon cablin* Benth.): botany, agrotechnology and biotechnological aspects. *Ind Crops Prod*, *87*, 161–176. doi:10.1016/j.indcrop.2016.04.032

Tang, M., & Taghibiglou, C. (2017). The mechanisms of action of curcumin in Alzheimer's disease. *J Alzheimers Dis*, *58*(4), 1003–1016. doi:10.3233/JAD-170188

Tang, P., Sun, Q., Yang, H., Tang, B., Pu, H., & Li, H. (2018). Honokiol nanoparticles based on epigallocatechin gallate functionalized chitin to enhance therapeutic effects against liver cancer. *Int J Pharm*, *545*(1–2), 74–83. doi:10.1016/j.ijpharm.2018.04.060

Tekade, K. R., Maheshwari, R., Tekade, M., & Chougule, B. M. (2017). Solid lipid nanoparticles for targeting and delivery of drugs and genes. In *Nanotechnology-Based Approaches for Targeting and Delivery of Drugs and Genes* (pp. 256–286): Academic Press.

Teskac, K., & Kristl, J. (2010). The evidence for solid lipid nanoparticles mediated cell uptake of resveratrol. *Int J Pharm, 390*(1), 61–69. doi:10.1016/j.ijpharm.2009.10.011

Thakkar, A., Chenreddy, S., Wang, J., & Prabhu, S. (2015). Ferulic acid combined with aspirin demonstrates chemopreventive potential towards pancreatic cancer when delivered using chitosan-coated solid-lipid nanoparticles. *Cell & Biosci, 5*.

Thilakarathna, S. H., & Rupasinghe, H. P. (2013). Flavonoid bioavailability and attempts for bioavailability enhancement. *Nutrients, 5*(9), 3367–3387. doi:10.3390/nu5093367

Thipe, V. C., Panjtan Amiri, K., Bloebaum, P., Raphael Karikachery, A., Khoobchandani, M., Katti, K. K., . . . Katti, K. V. (2019). Development of resveratrol-conjugated gold nanoparticles: interrelationship of increased resveratrol corona on anti-tumor efficacy against breast, pancreatic and prostate cancers. *Int J Nanomed, 14*, 4413–4428. doi:10.2147/IJN.S204443

Tokarz, P., Kaarniranta, K., & Blasiak, J. (2013). Role of antioxidant enzymes and small molecular weight antioxidants in the pathogenesis of age-related macular degeneration (AMD). *Biogerontology, 14*(5), 461–482.

Tsai, Y. M., Chien, C. F., Lin, L. C., & Tsai, T. H. (2011). Curcumin and its nano-formulation: the kinetics of tissue distribution and blood-brain barrier penetration. *Int J Pharm, 416*(1), 331–338. doi:10.1016/j.ijpharm.2011.06.030

Wang, N., & Feng, Y. (2015). Elaborating the role of natural products-induced autophagy in cancer treatment: achievements and artifacts in the state of the art. *Biomed Res Int, 2015*, 934207. doi:10.1155/2015/934207

Wang, P., Jiang, F., Chen, B., Tang, H., Zeng, X., Cai, D., . . . Liu, Y. (2020). Bioinspired red blood cell membrane-encapsulated biomimetic nanoconstructs for synergistic and efficacious chemo-photothermal therapy. *Colloids Surf B Biointerfaces, 189*, 110842. doi:10.1016/j.colsurfb.2020.110842

Wang, S., Meng, X., & Dong, Y. (2017). Ursolic acid nanoparticles inhibit cervical cancer growth in vitro and in vivo via apoptosis induction. *Int J Oncol, 50*(4), 1330–1340. doi:10.3892/ijo.2017.3890

Wang, W., & Kang, P. M. (2020). Oxidative stress and antioxidant treatments in cardiovascular diseases. *Antioxidants, 9*(12), 1292.

Wang, X., Wang, Y.-W., & Huang, Q. (2009). Enhancing stability and oral bioavailability of polyphenols using nanoemulsions. In *Micro/Nanoencapsulation of Active Food Ingredients* (pp. 198–212).

Wang, X.-F., Jin, P.-P., Zhou, T., Zhao, Y.-P., Ding, Q.-L., Wang, D.-B., . . . Ge, H. L. (2010). MR molecular imaging of thrombus: development and application of a Gd-based novel contrast agent targeting to P-selectin. *Clin Appl Thromb/Hemost, 16*(2), 177–183.

Watkins, R., Wu, L., Zhang, C., Davis, R. M., & Xu, B. (2015). Natural product-based nanomedicine: recent advances and issues. *Int J Nanomed, 10*, 6055–6074. doi:10.2147/IJN.S92162

Weissleder, R., Nahrendorf, M., & Pittet, M. J. (2014). Imaging macrophages with nanoparticles. *Nat Mater, 13*(2), 125–138.

WHO. (2021a, 21 September 2021). Cancer. Retrieved from https://www.who.int/news-room/fact-sheets/detail/cancer

WHO. (2021b). Dementia. Retrieved from https://www.who.int/news-room/fact-sheets/detail/dementia

Wilczewska, A. Z., Niemirowicz, K., Markiewicz, K. H., & Car, H. (2012). Nanoparticles as drug delivery systems. *Pharmacol Rep, 64*(5), 1020–1037. doi:10.1016/S1734-1140(12)70901-5

Williams, R. J., Mohanakumar, K. P., & Beart, P. M. (2015). Neuro-nutraceuticals: the path to brain health via nourishment is not so distant. *Neurochem Int, 89*, 1–6. doi:10.1016/j.neuint.2015.08.012

Wiwanitkit, V. (2012). Delivery of nutraceuticals using nanotechnology. *Int J Pharm Investig, 2*(4), 218. doi:10.4103/2230-973X.107008

Wong, H. L., Bendayan, R., Rauth, A. M., Li, Y., & Wu, X. Y. (2007). Chemotherapy with anticancer drugs encapsulated in solid lipid nanoparticles. *Adv Drug Deliv Rev, 59*(6), 491–504. doi:10.1016/j.addr.2007.04.008

Wu, M. Y., Yiang, G. T., Lai, T. T., & Li, C. J. (2018). The oxidative stress and mitochondrial dysfunction during the pathogenesis of diabetic retinopathy. *Oxid Med Cell Longev, 2018*.

Xia, T., Kovochich, M., Liong, M., Meng, H., Kabehie, S., George, S., . . . Nel, A. E. (2009). Polyethyleneimine coating enhances the cellular uptake of mesoporous silica nanoparticles and allows safe delivery of siRNA and DNA constructs. *ACS Nano, 3*(10), 3273–3286. doi:10.1021/nn900918w

Xu, J., Murphy, S. L., Kochanek, K. D., Bastian, B., & Arias, E. (2018). *Deaths: Final Data for 2016 from Center of Disease Control*.

Xu, J. S., Li, Y., Cao, X., & Cui, Y. (2013). The effect of eugenol on the cariogenic properties of *Streptococcus mutans* and dental caries development in rats. *Exp Ther Med, 5*(6), 1667–1670. doi:10.3892/etm.2013.1066

Yan, Y., Chan-Park, M. B., & Zhang, Q. (2007). Advances in carbon-nanotube assembly. *Small, 3*(1), 24–42. doi:10.1002/smll.200600354

Yang, C., Zhang, X., Fan, H., & Liu, Y. J. B. r. (2009). Curcumin upregulates transcription factor Nrf2, HO-1 expression and protects rat brains against focal ischemia. *1282*, 133–141.

Yang, Y., Qin, Z., Zeng, W., Yang, T., Cao, Y., Mei, C., & Kuang, Y. (2016). Toxicity assessment of nanoparticles in various systems and organs. *Nanotechnol Rev*, *6*(3). doi:10.1515/ntrev-2016-0047

Yang, Y., Zhou, X., Xu, M., Piao, J., Zhang, Y., Lin, Z., & Chen, L. (2017). beta-lapachone suppresses tumour progression by inhibiting epithelial-to-mesenchymal transition in NQO1-positive breast cancers. *Sci Rep*, *7*(1), 2681. doi:10.1038/s41598-017-02937-0

Yao, E. C., & Xue, L. (2014). Therapeutic effects of curcumin on Alzheimer's disease. *Adv Alzheimer's Dis*, *3*. doi:10.4236/aad.2014.34014

Yavarpour-Bali, H., Ghasemi-Kasman, M., & Pirzadeh, M. (2019). Curcumin-loaded nanoparticles: a novel therapeutic strategy in treatment of central nervous system disorders. *Int J Nanomed*, *14*, 4449–4460. doi:10.2147/IJN.S208332

Yoffe, S., Leshuk, T., Everett, P., & Gu, F. (2013). Superparamagnetic iron oxide nanoparticles (SPIONs): synthesis and surface modification techniques for use with MRI and other biomedical applications. *Curr Pharm Des*, *19*(3). doi:10.2174/1381612811306030493

Zaki, N. M. (2014). Progress and problems in nutraceuticals delivery. *J. Bioequivalence Bioavailab*, *6*, 75–77. doi:10.4172/jbb.1000183.

Zayed, D. G., AbdElhamid, A. S., Freag, M. S., & Elzoghby, A. O. (2019). Hybrid quantum dot-based theranostic nanomedicines for tumor-targeted drug delivery and cancer imaging. *Nanomedicine*, *14*(3), 225–228. doi:10.2217/nnm-2018-0414

Zeb, A., Qureshi, O. S., Kim, H. S., Kim, M. S., Kang, J. H., Park, J. S., & Kim, J. K. (2017). High payload itraconazole-incorporated lipid nanoparticles with modulated release property for oral and parenteral administration. *J Pharm Pharmacol*, *69*(8), 955–966. doi:10.1111/jphp.12727

Zhang, H., Jiang, X., Cao, G., Zhang, X., Croley, T. R., Wu, X., & Yin, J. J. (2018). Effects of noble metal nanoparticles on the hydroxyl radical scavenging ability of dietary antioxidants. *J Environ Sci Health C Environ Carcinog Ecotoxicol Rev*, *36*(2), 84–97. doi:10.1080/10590501.2018.1450194

Zhang, R. H., Li, L. Q., Wang, C., Lu, X. J., Shi, T., Xu, J. F., . . . Wang, H. F. (2015). Pretreatment with Huperzine A-loaded poly(lactide-co-glycolide) nanoparticles protects against lethal effects of soman-induced in mice. *Key Eng Mater*, *645–646*, 1374–1382. doi:10.4028/www.scientific.net/KEM.645-646.1374

Zhao, T., Liu, Y., Gao, Z., Gao, D., Li, N., Bian, Y., . . . Liu, Z. (2015). Self-assembly and cytotoxicity study of PEG-modified ursolic acid liposomes. *Mater Sci Eng C Mater Biol Appl*, *53*, 196–203. doi:10.1016/j.msec.2015.04.022

Zhao, Y., Xin, Z., Li, N., Chang, S., Chen, Y., Geng, L., . . . Chang, Y. Z. (2018). Nano-liposomes of lycopene reduces ischemic brain damage in rodents by regulating iron metabolism. *Free Radic Biol Med*, *124*, 1–11. doi:10.1016/j.freeradbiomed.2018.05.082

Zheng, Y., You, X., Guan, S., Huang, J., Wang, L., Zhang, J., & Wu, J. (2019). Poly(ferulic acid) with an anticancer effect as a drug nanocarrier for enhanced colon cancer therapy. *Adv Funct Mater*, *29*(25). doi:10.1002/adfm.201808646

10 Polysaccharide-Based Nanostructures as Nutraceutical Carriers

*V.B. Poornima, Richa Katiyar, Shivali Banerjee, Antonio Patti, and Amit Arora**

CONTENTS

10.1 Introduction 263
10.2 Techniques Used for the Production of Polysaccharide-Based Nanocarriers 264
10.3 Plant-Based Polysaccharide Nanocarriers 267
10.3.1 Cellulose 268
10.3.2 Pectin 269
10.3.3 Starch 270
10.3.4 Gums 271
10.3.5 Dextran 272
10.3.6 Inulin 273
10.3.7 Future Prospects of Starch-Based Nutraceutical Nanocarriers 273
10.4 Animal-Derived Polysaccharides 278
10.4.1 Chitosan 278
10.4.2 Hyaluronic Acid 279
10.4.3 Future Prospects of Animal-Derived Polysaccharide Nanocarriers 280
10.5 Seaweed-Derived Polysaccharides 280
10.5.1 Polysaccharides of Red Algae 281
10.5.1.1 Carrageenans 281
10.5.1.2 Agar 281
10.5.2 Polysaccharides from Green Algae 283
10.5.3 Polysaccharides of Brown Algae 283
10.5.3.1 Fucoidans 283
10.5.3.2 Alginate 284
10.5.3.3 Laminaran 284
10.5.4 Future Prospects of Seaweed-Derived Polysaccharide Nanocarriers 284
10.6 Safety of Nanocarriers for Nutraceuticals 285
10.7 Conclusion 285
References 285

10.1 INTRODUCTION

Nanotechnology studies materials at the nanoscale with at least one dimension <100 nm. Nanotechnology has applications in nutraceutical, pharmaceutical, biomedical, and electronic applications such as nanocomposites, nanocarriers, packaging films, and various others (Spizzirri

* Corresponding author: aarora@iitb.ac.in

DOI: 10.1201/9781003244721-10

et al., 2021). A nanosized system is defined as any particle with one dimension on the nanometer scale (Assadpour & Mahdi Jafari, 2019).

The term *nutraceutical* ("nutrition" and "pharmaceutical") indicates nutritional components with therapeutic benefits (Zaki, 2014). Using various techniques, natural compounds identified and obtained from plant, animal, and microbial sources have traditionally been valued for their medicinal properties in healthcare applications (Badar et al., 2019). Food bioactive compounds prevent and treat diabetes, metabolic, digestive, neurological, cardiovascular, and various deficiency disorders, and cancer (Borel & Sabliov, 2014). Food scientists have recently focused on developing nutraceutical nanodelivery systems similar to the drug delivery nanosystems (Bhushani et al., 2017). Besides, nanotechnology is also utilized in developing functional foods in a concept called "nanofood" (Dima et al., 2020). They can be incorporated to form functional foods and beverages such as probiotics, energy drinks, fortified juices, and omega fatty acid foods. Vitamins, for instance, are not commonly synthesized in the body and often need to be acquired from external sources (McClements, 2012b).

Widely used nutraceuticals include polyphenols, vitamins, minerals, fatty acids, antioxidants, carotenoids, and natural antimicrobial agents (Blanco-Padilla et al., 2014; Zaki, 2014). The absorption and delivery of these nutraceuticals are affected by their poor solubility, bioavailability, intestinal permeability, release, and stability in conditions of high temperature, pH, light, and oxygen (Bhushani et al., 2017). The efficacy of nutraceuticals reaching the site of action is dependent on their bioavailability (the rate and extent to which the nutraceutical reaches the site of action) (Rapaka & Coates, 2006). This is influenced by various factors, such as the size of the delivery system, solubility, absorption, and transformation. Nanomaterials, due to their small size, high surface/volume ratio, and high specific strength, provide an excellent alternative to micro and macro delivery systems for increasing the bioavailability and mucoadhesive properties, essential for nutraceutical delivery to target locations such as the small intestine (Assadpour & Mahdi Jafari, 2019; Singh et al., 2021). Various systems are used for nanodelivery of nutraceuticals, such as lipid-based, surfactant-based, and biopolymer-based nanodelivery systems (Bhushani et al., 2017). Polysaccharides like cellulose, starch, dextran, alginate, and many others are naturally occurring biopolymers obtained from plants, animals, and microbial sources (Wen & Oh, 2014). Natural polysaccharides have gained broad interest in developing nutraceutical delivery systems as they are renewable, biodegradable, biocompatible, and nontoxic (Wang et al., 2020). The functional groups, such as carboxyl, amino, and hydroxyl groups, present as a part of their structure, are utilized for their modifications to overcome limitations such as low solubility, stability, and bioavailability, making them suitable for preparing nanocarriers for nutraceuticals (Wen & Oh, 2014).

This chapter reviews the recent developments in research on polysaccharide-based nanocarriers, focusing on nutraceuticals, techniques, modifications, and limitations (Figure 10.1). Polysaccharide nanocarriers obtained from sources such as plants, animals, and seaweed are discussed (Figure 10.2). The potential and challenges for future development are also discussed.

10.2 TECHNIQUES USED FOR THE PRODUCTION OF POLYSACCHARIDE-BASED NANOCARRIERS

Various processes have been described to produce nanoparticles with desirable properties (Le Corre et al., 2010). The approaches used for the preparation of nanoparticles are classified into two categories: the "top-down" approach, where the breakdown of larger particles produces the nanoparticles, and the "bottom-up" approach, where smaller particles are assembled (Kim et al., 2015). Top-down methods include acid hydrolysis, reactive extrusion, ultrasonication, and high-pressure homogenization. In contrast, bottom-up approaches include techniques such as emulsification, nanoprecipitation, nano-spray drying, electrospinning, recrystallization, and γ-ray irradiation (Rostamabadi et al., 2019a). Chemical methods for the synthesis of nanoparticles include acid hydrolysis and enzymatic hydrolysis. Physical treatments include techniques such as high-pressure homogenization,

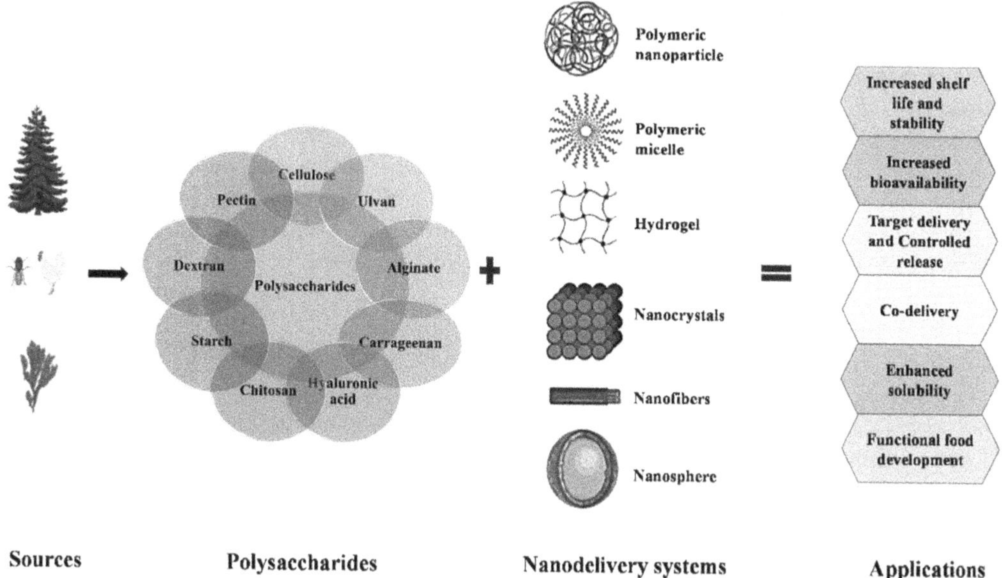

FIGURE 10.1 Schematic showing polysaccharide-based nanostructures for nutraceutical delivery.

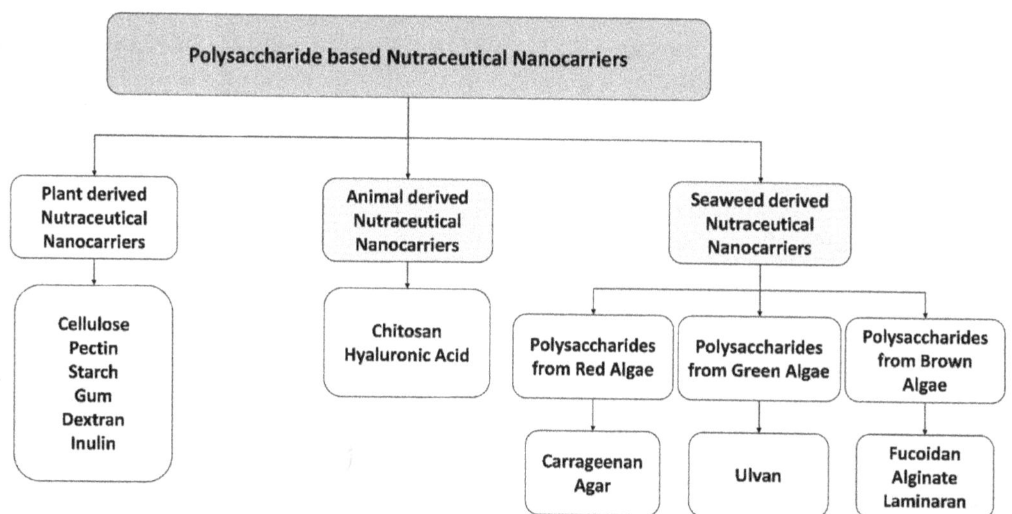

FIGURE 10.2 Classification of polysaccharides obtained from different sources used for the preparation of nanocarriers.

ultrasonication, reactive extrusion, nanoprecipitation, gamma irradiation, emulsion–cross-linking, and combination techniques (Kim et al., 2015). The mild acid hydrolysis technique that removes the amorphous regions and retains the crystalline granules is widely utilized due to its ease of use. Green techniques such as enzymatic hydrolysis, high-pressure homogenization, and microemulsion cross-linking are widely studied for the sustainable production of nanoparticles (Qiu et al., 2019). Using these methods, 10–500 nm nanoparticles can be obtained (Qiu et al., 2019).

However, nanoparticle synthesis methods such as enzymatic hydrolysis and high-pressure homogenization do not permit the particle sizes of the nanoparticles to be controlled (Chin et al., 2011). As the size of a nanoparticle is known to have implications for its final function, scientists have been studying various techniques for controlling the size of the nanoparticles (size-controlled synthesis)

(Qiu et al., 2019). Chin and colleagues synthesized starch nanoparticles within the size range of 300–400 nm using the controlled nanoprecipitation method in absolute ethanol. An approach involving varying synthesis parameters such as surfactants was utilized to control the resultant particle size (Chin et al., 2011). The bottom-up self-assembly approach has also been described as a green method for the size-controlled synthesis of nanoparticles (Qiu et al., 2019).

A widely used technique for nanocarrier synthesis is encapsulation. It is widely used to transport sensitive bioactive compounds such as folic acid, as it helps improve stability and bioavailability (Ghasemi et al., 2017). Chitosan nanoparticles with increased stability combined with sodium tripolyphosphate (TPP) and cellulose nanocrystals were used to deliver folic acid. The study concluded that such hybrid nanosystems are advantageous for the encapsulation and target delivery of sensitive bioactive compounds with high encapsulation efficiency (>90%) (Okamoto-Schalch et al., 2020). Due to higher surface area, they provide better functionality (Castro-Rosas et al., 2017). Complexation techniques involving proteins and polysaccharides are widely used to prepare nanocarriers for encapsulating flavoring agents, essential oils, vitamins, and fatty acids. This provides the combined advantages of the functional groups of both proteins and the polysaccharides involved. It is considered one of the most favorable techniques for producing food-grade delivery systems, improving the bioavailability and controlled release of nutraceuticals with poor solubility, bioavailability, and stability, such as vitamin D3 and curcumin (Jun-xia et al., 2011; Xiang et al., 2020). For proteins with a negative charge higher than the isoelectric point and a positive charge lower than the isoelectric point, when mixed with polysaccharides in an aqueous medium, there could be either attraction or repulsion, resulting in the formation of soluble/insoluble complexes or the separation of the polymers, respectively. Protein–polysaccharide complexes are formed depending on their charges, such as complexes between anionic proteins and cationic polysaccharides (Ghasemi et al., 2017). Therefore, the technique is suitable for the encapsulation of hydrophobic compounds. Coating cationic zein nanoparticles (>50% of hydrophobic amino acids and 90% solubility in ethanol) with anionic pectin for the nanoencapsulation of hydrophobic curcumin (antioxidant, anti-inflammatory, anticancer, antimicrobial, used for the treatment of neurological disorders) using an electrostatic deposition method with more than 86% loading efficiency and uniform shape overcomes the disadvantages, such as poor bioavailability and solubility (Hu et al., 2015). The compatibility and solubility of different biopolymers depend on the isoelectric charge (Ghasemi et al., 2017). Hence, encapsulation was found to be influenced by pH, method of preparation, and the ratio of pectin and protein (Raei et al., 2017).

The double emulsion technique involving emulsions, for instance, water-in-oil-in-water/oil-in-water-in-oil, containing two interfacial layers, has recently gained interest. Double emulsions are known to be less stable than single emulsions. Double emulsions (water-in-oil-in-water) prepared using pectin–whey protein nanocapsules with an outer aqueous phase of pectin and whey protein and an inner phase of Span and surfactant incorporated with spray-dried folic acid powder demonstrated an increased release rate of folic acid. However, the powders made using only whey protein capsules were smooth and uniform spherical surfaces compared with pectin–whey protein capsules, which contained dents and wrinkles (Assadpour & Mahdi Jafari, 2019). Due to large interfacial areas, they have lower thermal stability and break down easily to form single emulsions during various processing techniques. However, they have multiple advantages, such as keeping the bioactive compounds protected from environmental effects and other attacks in the inner water phase. They have been used for the entrapment and delivery of vitamins, flavors, preservatives, colors, and polyphenols (Assadpour & Mahdi Jafari, 2019). The ultrasonication technique is used to prepare bioactive compound–loaded nanoemulsions and nanodispersions. It uses a high-energy device that forms nanosized droplets and causes cavitation (Ghasemi et al., 2017). Spray drying has proven beneficial for the encapsulation technique, especially in encapsulating oils and flavoring agents. It helps prevent losses and improves the shelf life. Spray-dried emulsions were found to have enhanced morphology and higher encapsulation efficiency (Esfanjani et al., 2015). However, it has the disadvantage of the atomizer getting blocked due to the viscosity of the polymer suspension.

Polysaccharides have been proposed to improve the quality of nanoparticles obtained using this method (Akbari-Alavijeh et al., 2020). Nanoemulsions prepared using the widely used microfluidization technique were found to have better properties, such as a larger surface area (Castro-Rosas et al., 2017). High–methoxy pectin nanoemulsions used for the entrapment of essential oils from oregano, lemongrass, and thyme prepared using a microfluidization technique were found to have excellent antimicrobial activity during storage, although this was found to decrease with an increased storage time (Guerra-Rosas et al., 2017).

Nanoliposome-based delivery systems have been gaining interest for the entrapment of bioactive agents, as they can entrap hydrophilic, amphiphilic, and hydrophobic molecules in the same structure because they are made up of concentric circles of amphiphilic lipids with an aqueous core. The current limitation of using liposomes is their low stability, leading to difficulty in scale-up (Haghighi et al., 2018). A modified heating method was developed to produce and scale up liposomes. It offers the advantage of producing stable liposomes. Greener methods that did not involve using organic solvents have also been investigated to encapsulate bioactive compounds (Raei et al., 2017). A pectin-coated nanoliposomal delivery system (pectonanoliposomes) was prepared to deliver a polyphenol, phloridzin, using a green (organic solvent-free) heat stirring sonication technique. Pectin improves the stability and target delivery and is thus demonstrated as a promising technique for delivery of phloridzin, thereby enhancing its solubility, bioavailability, and stability due to its nontoxicity and biocompatibility (Haghighi et al., 2018). A nanoprecipitation technique with an organic phase consisting of dissolved nutraceutical in an organic solvent and an aqueous phase with stabilizer and surfactant is a promising green and straightforward method that avoids using high temperatures and showed high nutraceutical loading capacity in the preparation of starch nanoparticles (Qiu et al., 2019). Another widely used technique is the ionic gelation technique, involving the interaction of oppositely charged molecules to obtain nanodelivery systems with high loading capacity. It also helps overcome the limitations of sensitivity to pH and temperature (Akbari-Alavijeh et al., 2020). Phosphorylated cellulose nanocrystals (PCNs) cross-linked with glycidyltrimethylammonium chloride–chitosan were used to prepare nanocapsules loaded with vitamin C with >90% encapsulation efficiency using ionic gelation. The work demonstrated the improvement in long-term storage stability of functional foods (Baek et al., 2021)

A method that has gained broad interest because of its potential for controlled release is electrospinning. In the process, using electrostatic forces, the polymer solution is prepared using appropriate solvents extruded from the needle tip as a fine fiber through the formation of a Taylor cone, which is deposited on a grounded collector. Cellulose acetate phthalate nanofibers with high entrapment efficiency were used for loading novel curcumin using an electrospinning approach. The method was found suitable for bioactive compounds that require transdermal delivery due to degradation during oral delivery (Ravikumar et al., 2017). The fabrication of nanofibers using electrospinning has been utilized on an industrial scale for commercial applications since the 1930s. Overcoming limitations such as the development of uniform-size fibers and efficient incorporation of nutraceuticals can lead to considerable improvements in the preparation of functional foods (Ghorani & Tucker, 2015).

10.3 PLANT-BASED POLYSACCHARIDE NANOCARRIERS

Plant polysaccharides are a group of naturally occurring renewable polymers of sugar units present in leaves, stem, and bark, including cellulose, starch, pectin, inulin, and different gums. Plant-based polysaccharide nanoparticles are widely studied for drug delivery applications (Nayak & Hasnain, 2019). All plant cell walls are made up of complex polysaccharides organized in a framework made of cellulose, contributing to the structure and function of plant cell walls. These complex polysaccharides are made by the incorporation of sugars synthesized during photosynthesis. Hence, plant cells provide diverse groups of polysaccharides with properties useful for various applications.

10.3.1 Cellulose

Cellulose nanomaterials are green, nontoxic, biocompatible materials with excellent mechanical strength, making them advantageous for use as nanocarriers with high colloidal stability. They are widely used to immobilize bioactive molecules, to prepare emulsions as food additives, and for the controlled delivery of bioactive agents and nutraceuticals (Khan et al., 2018). Cellulose is an abundant structural polymer in plant sources, wood, algae, tunicates, bacteria, and agricultural waste, present along with hemicellulose and lignin (Brodeur et al., 2011; Kim et al., 2015). Linear homopolymers of anhydro glucose units (linked by β-1,4-glycosidic linkages) are joined by intermolecular hydrogen bonding side by side to form cellulose microfibrils with sequentially located crystalline and amorphous domains made up of repeating units of cellobiose (Rebouillat & Pla, 2013). The crystalline domains are made of strong hydrogen bonds, making them difficult to break. At the same time, the amorphous regions are weak spots in the fibrils with loosely packed crystals (Kim et al., 2015). Various chemical and mechanical pretreatment techniques can be applied to the cellulose microfibrils to obtain cellulose structures with one dimension in the nanoscale (usually <100 nm), called nanocelluloses (Rebouillat & Pla, 2013). Mechanical treatment of the pretreated cellulose microfibrils using techniques such as ultrasonication, high-pressure homogenization, and electrospinning leads to the formation of cellulose nanofibers (CNFs) containing both crystalline and amorphous regions with 10–100 nm diameter, a very high aspect ratio, good flexibility, and excellent mechanical properties (Brodeur et al., 2011; Dhali et al., 2021). Acid hydrolysis of the microfibrils using acids such as sulfuric acid, hydrochloric acid, and nitric acid leads to the removal of the weak amorphous regions, and the remaining crystalline rod-like structures with diameter in the range of 10–100 nm, low thermal expansion coefficient, high specific strength, and high elastic modulus are called cellulose nanocrystals (CNCs) (Brodeur et al., 2011; Dhali et al., 2021; Fortunati & Balestra, 2019).

The multiple hydroxyl groups on cellulose fibrils offer the possibility of chemical modifications such as esterification (cellulose acetate, cellulose acetate butyrate, cellulose acetate propionate, hydroxypropyl methylcellulose), etherification (methyl, ethyl, carboxymethyl, hydroxyethyl, hydroxypropyl, and hydroxypropyl methyl cellulose), and cationic and anionic modifications that increase the solubility and bioavailability of various bioactive compounds (Layek & Mandal, 2020). Of the different types of modified cellulose, cellulose acetate and methyl acetate are most commonly used for nanocarrier applications (Fox et al., 2011). Cellulose acetate (CA) nanofiber mats were used for the entrapment of curcumin (antitumor, anti-inflammatory, antioxidant) (Suwantong et al., 2007) and the loading and transdermal release of vitamin E and vitamin A using the technique of electrospinning (Taepaiboon et al., 2007). While CA nanofiber mats demonstrated the safety of curcumin–CNF transdermal patches towards human fibroblast cells, CA–phthalate nanofibers were found suitable for bioactive compounds that require transdermal delivery due to degradation during oral delivery (Ravikumar et al., 2017; Suwantong et al., 2007). Electrospun cellulose acetate nanofibers were used as nanocarriers for tannic acid and Fe^{3+} complexes (Yang et al., 2017), curcumin (Ravikumar et al., 2017), seal holly (*Acanthus ebracteatus* Vahl.) ethanolic extracts (Vongsetskul et al., 2016), and gingerol for therapeutic applications (anti-inflammatory, antihyperglycemic, antitumorigenic, antiapoptotic, analgesic) (Chantarodsakun et al., 2014). 2,3-Dialdehyde cellulose nanofibers were used to adsorb Cu(II) in a spontaneous, irreversible, exothermic reaction facilitated by the increased surface charge density and surface area (Lei et al., 2020). Higher levels of cellulose nanofibers extracted from food waste (mangosteen) impact the bioavailability and stability of vitamin D3 in oil in water emulsions. However, higher CNF levels were beneficial for preparing functional foods for satiety control, as they partially inhibit lipid digestion (Winuprasith et al., 2018).

Cellulose nanocrystal–based microemulsions have been widely used for the delivery of various nutraceuticals. They are often considered better stabilizers for oil–water-based emulsion systems than cellulose nanofibers due to their tiny, rod-like particles up to 200 nm in length (Patel, Lakshmibalasubramaniam, & Nayak, 2020a). Cetyl trimethylammonium bromide (CTAB)–modified cellulose nanocrystal microemulsions were demonstrated as a promising approach to enhance the solubility, bioavailability, and controlled target delivery of otherwise poorly soluble curcumin in a topical delivery system (Zainuddin et al., 2021). They were used for the encapsulation and controlled release of phycobiliproteins, which are sensitive to high temperature, pH, and oxidative stress (Patel et al., 2020b), and β-galactosidase in a study that provided evidence for the potential of CNCs as stabilizers in the dairy industry and for treating lactose intolerance (Deng et al., 2020). Unmodified and modified cellulose nanocrystal/chitosan hybrid systems, which provide higher surface area and improved chemical properties, were used for the encapsulation and target site delivery of curcumin (Gunathilake et al., 2017), vitamin C (Baek et al., 2021), and folic acid (Okamoto-Schalch et al., 2020) with increased encapsulation efficiency, residence time, solubility, storage stability, and bioavailability. Nanogels prepared using carboxymethyl cellulose and lysozyme for the entrapment and controlled release of 5-fluorouracil were found to offer various advantages such as increased solubility in water and biodegradability, biocompatibility, and lower immunogenicity. The study showed the potential for the preparation of CNCs with adjustable release speeds for various applications (Zhu et al., 2013).

10.3.2 Pectin

Pectin is a d-galacturonic acid–rich complex anionic hetero-polysaccharide. It is well known for pharmaceutical and food applications to prepare jams, jellies, and fruit juice as a low-cost gelling, thickening, and stabilizing agent (Chen et al., 2015). Among the various pectin sources (cell walls of plants, fruits, and vegetables), citrus peels are the most common commercial source for pectin. The source of pectin determines its degree of esterification and the ability to form gels (Srivastava & Malviya, 2011). Pectin is classified as high-methoxy pectin (highly esterified carboxyls of uronic acids) and low-methoxy pectin (partially esterified carboxyls of uronic acids) (Fathi et al., 2014; Khanvilkar et al., 2016). The tendency to form agglomerations in water affects its solubility. The structure with carboxyl and hydroxyl groups allows modifications such as alkylation, thiolation, amidation, quaternization, oxidation, sulfation, cross-linking, and grafting depolymerization, resulting in pectins with properties suitable for commercial and industrial applications (Chen et al., 2015).

Various methods exist for the extraction of pectin from different sources, such as a water-based extraction method using acidified water and temperature less than 70°C for 2–4 h followed by isopropyl alcohol or ethanol precipitation, microwave extraction, and novel green techniques such as enzymatic extraction, subcritical fluid extraction, pulsed electric field extraction, and ultrasound extraction (Adetunji et al., 2017; Srivastava & Malviya, 2011). Pectin-based nanocarriers improve the bioavailability and stability of sensitive compounds such as vitamins and polyphenols. Pectin nanoparticles combined with other polymers are used to prepare emulsions and for the nanoencapsulation of various bioactive compounds (Ghasemi et al., 2017).

Pectin-based nanoemulsions were recently found to enhance the bioavailability of lipophilic compounds such as β-carotene (Teixé-Roig et al., 2020), citrus peel flavonoids (Hu et al., 2017), and essential oils from oregano, lemongrass, and thyme (Guerra-Rosas et al., 2017). Protein–polysaccharide nanoparticles offer various advantages for improving the encapsulation efficiency, bioavailability, and sustained release of various nutraceuticals. Some of the pectin–protein systems that have been utilized for nutraceutical release include: a pectin–protein mixture for higher storage stability of orange peel oil rich in D-limonene, useful as a flavoring agent in the food industry (Ghasemi et al., 2018); ovalbumin–high-methoxy pectin for improving the storage stability of vitamin D3 (Xiang et al., 2020); quaternized curdlan–pectin complex (Wu et al., 2019), and lactoferrin–pectin complex (Yan et al., 2017) for entrapment of hydrophobic curcumin. Whey protein and zein proteins are commonly used to prepare protein–polysaccharide complexes with pectin. Cationic zein nanoparticles (>50% hydrophobic amino acids and 90% solubility in ethanol) coated with anionic pectin are widely used for the nanoencapsulation of hydrophobic nutraceuticals such as curcumin (Hu et al., 2015) and resveratrol with >86% loading efficiency, hence increasing bioavailability, solubility, shelf life, and stability (Contado et al., 2020). Pectin–whey protein systems have been used for the controlled delivery of sensitive and unstable compounds such as saffron (Esfanjani et al., 2015) and lactoferrin (Raei et al., 2017). Double emulsions (water-in-oil-in-water) prepared using pectin–whey protein nanocapsules with an outer aqueous phase of pectin and whey protein and an inner phase of Span and surfactant incorporating spray-dried folic acid powder demonstrated an increased release rate of folic acid (Assadpour & Mahdi Jafari, 2019). Highly stable double emulsions prepared with pectin–whey protein complex have been an excellent alternative to the mutagenic Tween 80 (Gharehbeglou et al., 2019).

In general, pectin-based nanodelivery systems are known to provide advantages such as improving the storage stability and sustained release of nutraceuticals (Xiang et al., 2020), enhancing bioavailability and solubility (Hu et al., 2015), improved colon-specific delivery (Contado et al., 2020), improved permeation (Hu et al., 2017), potential in the food industry (Ghasemi et al., 2018), potential as a carrier for highly sensitive compounds (Esfanjani et al., 2015), and potential for antimicrobial nutraceutical delivery (Guerra-Rosas et al., 2017).

10.3.3 Starch

Starch is plants' primary form of food reserve, manufactured by the green leaves using the surplus glucose produced during photosynthesis. The commercial form of starch is mainly obtained from starch-rich sources such as leaves, fruits, roots, and stems of higher plants such as sweet potato, potato, rice, wheat, tapioca, and corn. Fruit wastes such as mango, pineapple, banana, apple, litchi, and avocado, with a 20–60% starch content, are currently gaining popularity as sustainable sources for starch extraction (Kringel et al., 2020). Starch is made up of two polysaccharide units: amylose (a linear polymer of d-glucose units linked by (1-4) α-D-glycoside bonds) and amylopectin (a branched polymer of α-D-(1-4) glycopyranose and α-D-(1-6) glycosidic linkages) (Le Corre et al., 2010). Starch nanoparticles have various advantages, such as biodegradability, being environmentally friendly, high surface area, smaller size, reinforcement potential, abundance, high absorption capacity, digestible property, and barrier properties for their utilization as carriers for nutraceuticals.

Starch nanostructures are classified into three major forms based on crystallinity: nanocrystals, nanofibers, and nanoparticles. The terms *starch nanocrystals* and *starch nanoparticles* are often used interchangeably to describe the nanostructures that remain after the degradation of the large polymer by various methods. While some literature uses the terms to address the crystalline granules that remain after acid hydrolysis, others suggest that nanoparticles contain both crystalline and amorphous regions, while nanocrystals contain only crystalline moieties. They are usually prepared using chemical and physical degradation methods. Starch nanofibers consist of crystalline and amorphous regions and are typically prepared by electrospinning (Rostamabadi et al., 2019b). Chin and colleagues synthesized starch nanoparticles within the size range of 300–400 nm using the controlled nanoprecipitation method in absolute ethanol. An approach involving varying synthesis parameters such as surfactants was utilized to control the resultant particle size (Chin et al., 2011). A microemulsion technique (water-in-oil) was successfully performed to synthesize starch nanoparticles by controlling parameters such as surfactants, rate of stirring, and composition (Chin et al., 2014). The bottom-up self-assembly approach has also been described as a green method for the size-controlled synthesis of starch nanoparticles (Qiu et al., 2019).

Starch nanoparticles were used for the encapsulation and sustained release of tangeretin (Hu et al., 2020), essential oils (Qiu et al., 2017), quercetin (Farrag et al., 2018), *Ginkgo biloba* extracts, known for relieving Alzheimer's symptoms (Wang et al., 2018), and calcium ions (Liu et al., 2018). Modifications and cross-linking of starch nanoparticles with other compounds have improved the encapsulation efficiency compared with unmodified native starch. Modified starch nanoparticles have been shown to improve the encapsulation of polyphenols (catechin, epicatechin, epigallocatechin, epigallocatechin gallate, and proanthocyanins) (C. Liu et al., 2016; Shao et al., 2018), flavonoids

(J. Li et al., 2018), and β-carotene in comparison with unmodified starch (Santoyo-Aleman et al., 2019). Some of the modified starch nanoparticles used for nutraceutical delivery include carboxymethyl group–modified starch nanoparticles for epigallocatechin gallate controlled release (Liu et al., 2020), acetylated starch nanoparticles for improved encapsulation (>80% efficiency) of curcumin (Acevedo-Guevara et al., 2018), octenyl succinic anhydride and xanthan gum–modified starch nanoparticles for reducing losses and improved utilization of linoleic acid (Yang et al., 2020), and anionic carboxymethyl starch and cationic quaternary ammonium starch for colon-specific delivery of protein (Shao et al., 2018). Starch-based nanosystems are widely used to deliver vitamins, such as vitamin E–loaded starch nanofibers (Kheradvar et al., 2018) and Vitamin D3–loaded starch nanocarriers (Hasanvand et al., 2015). The studies have paved the way for food fortification with increased bioavailability of vitamins and other compounds using starch nanocarriers. One factor affecting the properties of starch-based nanocarriers, apart from specific modifications, is the ratio of amylose and amylopectin. Higher amylopectin-containing sources such as potato starch and pea starch showed higher releasing capacity than corn starch with higher amylose content (Farrag et al., 2018). Amylose and amylopectin ratio was also found to affect Vitamin D3 encapsulation by starch nanocarriers. A study conducted by Hasanvand et al. (2015) concluded that potato starch was shown to have higher encapsulation efficiency.

10.3.4 Gums

Gums are complex polysaccharides obtained from various sources such as seaweeds, plants, fungi, and microbes. Gums are of three categories based on their synthesis: natural gums (from natural sources such as seeds or leaves of plants), modified gums (by modifying natural gums), and synthetic gums (synthesized chemically) (Taheri & Jafari, 2019a). Plant-based gums, from different parts of plants (seeds, bark, and leaves), are used in medicine, the food industry, and drug delivery applications due to their superior structural and functional properties. They contain xylan, glucose, galactose, arabinose, and rhamnose. They are biodegradable, biocompatible, flexible, safe, easy to use, and sustainable (Amiri et al., 2021; Taheri & Jafari, 2019a,b). Multiple available reactive sites make them suitable for various applications (Taheri & Jafari, 2019a). The most significant properties of plant-based gums are their high bioavailability, nontoxic nature, and stability (Amiri et al., 2021). They are hydrophilic due to the hydroxyl groups in their structure (Taheri & Jafari, 2019a). Their interaction with water results in the formation of emulsions and gels (Tanheri et al., 2019). These properties help in the encapsulation and delivery of various nutraceuticals, thus contributing to their increased bioavailability and stability (Amiri et al., 2021). However, they are known to have uncontrolled hydration rates, reduction in viscosity on storage, and susceptibility to microbial contamination. Additionally, their solubility is pH dependent. Nanosystems are known to help overcome these limitations (Amiri et al., 2021). Plant-based gums are used as green stabilizing agents for synthesizing nanoparticles incorporated with flavonoids instead of using metal ions for the purpose. Various plant-based gums have been examined for the preparation of nanocarriers for diverse nutraceuticals. (Anwar et al., 2019).

Gum Arabic, derived from the bark of *Acacia senegal*, is widely used alone and in combination with other polysaccharides and proteins for nutraceutical and pharmacological applications due to its film-forming capacity and surface activity. This is the result of the formation of an open and expanded structure resulting from the dissociation of carboxyl groups at neutral pH in addition to other favorable properties such as biodegradability, biocompatibility, emulsification, viscoelasticity, and suspension potential (Tan et al., 2016). Such systems include gum Arabic for the nanoencapsulation and delivery of hydrophobic bioactive compounds such as phenolics and carotenoids (Ilyasoglu & El, 2014), a gum Arabic–chitosan system for efficient encapsulation of curcumin (Tan et al., 2016), a gum Arabic–β-lactoglobulin colloidal system of oil and water emulsions for the encapsulation and delivery of lutein (Su et al., 2020), zein protein and gum Arabic nanoparticles in a Pickering emulsion for the encapsulation of thymol (J. Li et al., 2018), and zein/NaCS/gum Arabic nanoparticles for the encapsulation of microbial carotenoids (Wang et al., 2021). Gum Arabic–chitosan nanoparticles were utilized for the absorption and delivery of quercetin (Kim et al., 2019), saffron with high efficiency of encapsulation (>50%) (Mehrnia et al., 2017), and functional lipids such as eicosapentaenoic acid (EPA) and docosahexaenoic acid (DPA) from fish oil for fruit juice enrichment (Ilyasoglu & El, 2014). Whey protein and gum Arabic improved the colloidal stability of nanoemulsions incorporating fat-soluble vitamin E (Ozturk et al., 2015). Gum Arabic incorporation into a liposome delivery vehicle for curcumin helped reduce the size of the liposome and its polydispersity index. However, it provided lower retention and encapsulation of curcumin in comparison with sodium alginate (Li et al., 2021).

The two types of almond gums derived from the trees of wild or mountain almonds – Angum gum, also called Persian gum (composed of rhamnose and galactose units) and Almond gum (composed of arabinose, galactose, uronic acid, and galactose) – are known to have several advantages over other biopolymers, such as a straightforward, sustainable and low-cost method of production (Sedaghat Doost et al., 2018). Double emulsions (water-in-oil-in-water) of Angum gum and Arabic gum were observed to overcome crocin encapsulation limitations such as low stability under heat, oxygen, additives, acids, and light (Mehrnia et al., 2017). An Almond gum–shellac nanodelivery system was used to encapsulate quercetin with an efficiency of 98% using the antisolvent precipitation method (an aqueous phase containing the hydrophilic biopolymer and the hydrophobic compound [pre-dissolved in organic solvent] for the preparation of nanoparticles) (Sedaghat Doost et al., 2018).

Some of the other gum nanoparticle systems used for nutraceutical delivery include: Balangu seed gum nanocapsules for the encapsulation of essential oil from the extracts of *Mentha longifolia* L. (popular for its use as a flavoring and aromatic agent) as a fast flavor release system (Rezaeinia et al., 2019); Tara gum (with a peculiar 3:1 mannose:galactose ratio), obtained from the seeds of the plant *Caesalpinia spinosa*, and lactoferrin complex for encapsulation of fat-soluble vitamin D3 (Santos et al., 2021); Azivash gum electrospun nanofibers (from the leaves of the azivash plant, recognized for its viscosity suitable for electrospinning) with polyvinyl alcohol for the encapsulation of highly oxidation-sensitive compounds such as catechin (Hoseyni et al., 2021); *Alyssum homolocarpum* seed gum (AHSG) nanoparticles for the encapsulation of flaxseed gum (made up of rhamnogalacturonan and arabinoxylan) combined with flaxseed protein for the delivery of curcumin and lycopene *in vitro* (Nikbakht Nasrabadi et al., 2020); D-limonene (Ghasemi et al., 2018); corn fiber gum (made up of arabinoxylan) for the preparation of a core-shell nanocomposite as a nanodelivery system for hydrophobic curcumin (Wei et al., 2020); guar gum [a galactomannan composed of β-d-(1-4) mannopyranosyl units and α-D-galactopyranosyl units attached by (1-6) linkages] nanoparticles combined with starch nanoparticles for the encapsulation of salvianolic acid B (Sal B) and controlled intestinal and blood fluid release (Li et al., 2017); and acetylated cashew gum nanoparticles combined with fucan for the encapsulation of lycopene-rich extract (antioxidant, anti-inflammatory, antimicrobial, effective in treating neurological and cardiovascular disorders) (de Andrades et al., 2021).

10.3.5 Dextran

Dextran has been extensively studied for fabricating delivery systems with different designs of structure in the form of nanoparticles, nanoemulsions, nanocomplexes, nanocrystals, and even self-assembled micelles (di Martino et al., 2018; Fan et al., 2017; Hao et al., 2018; Mukwaya et al., 2019; Qin et al., 2018). Dextran-based delivery systems have been known for their potential applications in various areas such as nutraceuticals, pharmaceuticals, food science, and biomedicine (Deng et al., 2010; Duan et al., 2005; Semyonov et al., 2014; Shen et al., 2008b). Dextran is known for its biocompatibility, biodegradability, and solubility, which adds to the advantage of using it for synthesizing nanocarriers for drug delivery (Anirudhan & Binusreejayan, 2016; Shingel, 2004). The existence of six α-1glycosidic bonds in dextran increases chain mobility. It is also responsible for the solubility of dextran in a wide variety of solvents. Dextran-based nanocarriers could facilitate the easy dissolution of drugs and smooth permeation through the gastrointestinal membrane. They can also be easily metabolized by the digestive enzymes, leading to the biocompatibility of dextran-based nanocarriers (Cadée et al., 2000; Hu et al., 2021b). Further, the neutral charge of dextran facilitates the efficacy of the delivery of nutraceuticals. The positively charged nanocarriers often get trapped by the negatively charged intestinal mucus layer (Lai et al., 2009). In comparison, nanocarriers with a negative charge find it challenging to cross the mucus layer due to strong electric repulsion (Shan et al., 2015).

10.3.6 Inulin

Inulin is widely present in the roots and rhizomes of onion, garlic, wheat, leek, asparagus, chicory, banana, and dahlia (Abed et al., 2016a). Chicory root, the primary commercial source of inulin, contains 16–18% dry mass. It belongs to the class of dietary fibers known as fructans. Structurally, inulin consists of a combination of polysaccharides and oligosaccharides. Inulin is resistant to hydrolysis by human saliva or pancreatic or intestinal enzymes of the gastrointestinal tract. Inulin is selectively fermented or metabolized by the colonic bacteria into carbon dioxide and methane (Roberfroid, 2007). The degree of polymerization is known to have a significant effect on the characteristics of inulin. For example, inulin with a degree of polymerization of more than 10 is high-performance inulin.

Further, the water solubility of inulin also depends upon its chain length distribution (Giri et al., 2021). Similarly, to other polysaccharides, inulin also finds potential application in the delivery of drugs and nutraceuticals in the form of nanocarriers. These delivery systems could achieve site-specific delivery and hence, enhanced bioavailability (Giri et al., 2016). Their biodegradable nature and sensitivity towards colonic pH have attracted the applicability of inulin-based nanocarriers for colon targeting. These nanocarriers were found to exhibit good colon-specific release of the payload along with the minimum amount of premature release of the drug (Giri et al., 2021).

10.3.7 Future Prospects of Starch-Based Nutraceutical Nanocarriers

Plant-based polysaccharides offer a variety of advantages such as abundance, renewability, biodegradability, sustainability, low cost, and biocompatibility. They have been used for the encapsulation and delivery of vitamins, minerals, essential oils, polyphenols, and various other bioactive agents (Dhali et al., 2021). Polysaccharide nanoparticles such as nanocelluloses, in the form of cellulose nanocrystals and cellulose nanofibers, gum, starch, inulin, pectin, and dextran have been widely used for the production of nutraceutical nanocarriers with improved mechanical and chemical properties and increased solubility, bioavailability, and stability. The major limitation in using nanocelluloses lies in the extraction process, with multiple steps using harsh chemical conditions, limiting their commercial exploitation (Dhali et al., 2021). Few studies have demonstrated the improved bioavailability and stability of starch-based nanosystems over protein/gum Arabic–based delivery systems (Zhu, 2017). Pectin can be modified to produce various structural alternatives suitable for specific applications (Naqash et al., 2017). The stability of pectin nanoemulsions under techniques such as microwave, ultrasonication, and high-pressure techniques needs to be evaluated (Mungure et al., 2018). The existing knowledge in drug encapsulation and delivery using starch-based nanocarriers could be utilized for designing nutraceutical carriers and controlled release. The tendency of nanoparticles to aggregate could hinder their application (Kaur et al., 2018). Apart from this, the impact of various modifications on the properties and efficiency of nanocarriers needs to be studied in detail. Only limited studies are available on the *in vivo* mechanism of uptake, release, and target delivery of nutraceuticals (Zhu, 2017). Most studies do not evaluate the safety of nanoparticles and their derivatives. More research is needed into the safety and toxicology of nanocarriers (Qiu et al., 2019) and to develop adequate toxicity testing methods (Dhali et al., 2021). Future research should focus on the large-scale production of nanoparticles suitable for nutraceutical nanocarrier applications (Taheri & Jafari, 2019a) (Table 10.1).

TABLE 10.1
Plant-Based Polysaccharide Nanocarriers for Nutraceuticals

Polysaccharide nanocarrier type	Method	Particle size (nm)	Nutraceutical	References
Cellulose nanofiber	Oil-in-water emulsion encapsulation	60	Vitamin D3	(Winuprasith et al., 2018)
Cellulose acetate fiber	Electrospinning	300 ± 64	Curcumin	(Suwantong et al., 2007)
Cellulose nanofibers	Immobilization to form nanoformulation	500	Quercetin	(Li et al., 2019)
Cellulose nanofibers	Periodate oxidation	38.9	Cu(II)	(Lei et al., 2020)
Hydrophobin–cellulose nanofiber	Emulsion formulations	300–600	BCS class II drugs	(Paukkonen et al., 2017)
Cellulose acetate nanofiber mats	Electrospinning	247 ± 31	Vitamin A and Vitamin E	(Taepaiboon et al., 2007)
Cellulose acetate phthalate nanofiber	Electrospinning	300 ± 20	Curcumin	(Ravikumar et al., 2017)
Cellulose nanofiber	Emulsion formulation followed by spray drying	19–22	Ginger essential oil	(de Souza et al., 2021)
Cellulose acetate fibers	Electrospinning	375 ± 107	Gingerol	(Chantarodsakun et al., 2014)
Cellulose acetate fibers	Electrospinning followed by solvent casting	460 ± 84	*Acanthus ebracteatus* Vahl. extract	(Vongsetskul et al., 2016)
Cellulose acetate nanofibrous mats	Electrospinning	326 ± 134	Tannic acid–Fe^{3+} complex	(Yang et al., 2017)
Polyethylene glycol–cellulose nanocrystal	Liposome complex	150–200	Phycobiliprotein	(Patel, Lakshmibalasubramaniam, & Nayak, 2020a)
Cellulose nanocrystal	Dispersion	3–8	Spirooxazine dye	(Sun et al., 2014)
Cetyltrimethylammonium bromide–nanocrystalline cellulose (CTAB-NCC)	Microemulsions	12–110	Curcumin	(Zainuddin et al., 2021)
Phosphorylated cellulose nanocrystal–chitosan	Nanocapsules	428 ± 6–450 ± 8	Vitamin C	(Baek et al., 2021)
Zein/cellulose nanocrystals	Core-shell microparticles	1017.3 ± 17.6	Curcumin	(Wei et al., 2021)
Cellulose nanogels	Lysozyme–sodium carboxymethyl cellulose	241	5-fluorouracil	(Zhu et al., 2013)
β-Chitosan/cellulose nanocrystal	Loaded suspension	71.77 and 143.20	β-galactosidase	(Deng et al., 2020)
Alginate/cellulose nanocrystal	Ionotropic gelation technique	70 ± 15–100 ± 24	Rifampicin	(Thomas et al., 2018)
Chitosan–TPP/cellulose nanocrystal	Ionic gelation method	710	Folic acid	(Okamoto-Schalch et al., 2020)

(Continued)

TABLE 10.1 (CONTINUED)
Plant-Based Polysaccharide Nanocarriers for Nutraceuticals

Polysaccharide nanocarrier type	Method	Particle size (nm)	Nutraceutical	References
Dialdehyde cellulose nanocrystals	Pickering emulsions	263.33–113.24	Dihydromyricetin	(Xu et al., 2021)
Pectin/whey protein nanocapsules	Nanocomplexation technique	100	D-limonene	(Ghasemi et al., 2018)
Ovalbumin/pectin nanocomplex	Nanocomplexation technique	307.71	Vitamin D3	(Xiang et al., 2020)
Pectin/whey protein nanocapsules	Double emulsion technique	NA	Folic acid	(Assadpour et al., 2017)
Zein/pectin nanoparticles	Electrostatic deposition method	250	Curcumin	(Hu et al., 2015)
Whey protein/pectin nano-complex	Nanoencapsulation	500–800	Lactoferrin	(Raei et al., 2017)
Pectin-coated nanolipo-some	Pectonanolipo-somes	432 ± 19	Phloridzin (polyphenol)	(Haghighi et al., 2018)
Pectin/whey protein nanocomplex	Complexation technique	200	Orange peel oil	(Ghasemi et al., 2017)
Pectin–whey protein	Nanoencapsulation	<100	Saffron extract	(Esfanjani et al., 2015)
Quaternized curdlan/pectin polyelectrolyte nanovehicle	Complexation method	68	Curcumin	(Wu et al., 2019)
High methoxy pectin nanoemulsions	Microfluidization	11 ± 1	Essential oils	(Guerra-Rosas et al., 2017)
Alginate pectin nanoparticles	Nanoencapsulation	200–400	Folic acid	(Pamunuwa et al., 2020)
Lactoferrin/pectin polyelectrolyte complex nanoparticles	Complexation method	208	Curcumin	(Yan et al., 2017)
Zein/pectin nanoparticles	Complexation method	100–200	Resveratrol	(Contado et al., 2020)
Pectin-based nanoemul-sions	Nanoemulsion	300–500	β-Carotene	(Teixé-Roig et al., 2020)
Pectin nanoparticles	Nanoencapsulation	271.5 ± 5.3	Citrus peel extract flavonoids	(Hu et al., 2017)
Starch nanoparticles	Encapsulation	264.6	Methyl gallate	(Prakashkumar et al., 2021)
Starch acetate nanopar-ticles	Controlled precipitation	110	Piperine	(Chin et al., 2021)
Octenyl succinic anhydride (OSA)–modi-fied starch, and xanthan gum (XG)–modified starch emulsion nanoparticles	Emulsion technique	375.13 ± 6.37–503.63 ± 4.16	Conjugated linoleic acid	(Yang et al., 2020)

(*Continued*)

TABLE 10.1 (CONTINUED)
Plant-Based Polysaccharide Nanocarriers for Nutraceuticals

Polysaccharide nanocarrier type	Method	Particle size (nm)	Nutraceutical	References
β-Cyclodextrin/ debranched starch nanoparticles	Ultrasonication and recrystallization	351.12 ± 6.99–467.53 ± 6.13	Tangeretin	(Hu et al., 2020)
Debranched starch nanoparticles	Ionic gelation	50–100	Epigallocatechin gallate	(Liu et al., 2020)
Starch nanoparticles	Nanoencapsulation	322.7–615.6	Catechin	(Ahmad et al., 2019)
Banana starch nanoparticles (citric acid modified)	Nanoprecipitation and cross-linking	<250	β-Carotene	(Santoyo-Aleman et al., 2019)
Debranched starch nanoparticles (cationic modified) with κ-carrageenan and low-methoxyl pectin	Encapsulation	500	Epigallocatechin gallate	(Liu et al., 2019)
Starch nanoparticles	Nanoprecipitation	500	Quercetin	(Y et al., 2018)
Starch nanospheres	Nanoprecipitation	255–396	*Ginkgo biloba* extracts	(Wang et al., 2018)
Banana starch nanoparticles (acetylated)	Nanoprecipitation	135.1–190.2	Curcumin	(Acevedo-Guevara et al., 2018)
Taro starch nanoparticles	Pickering emulsions	467.93	Tea polyphenols	(Shao et al., 2018)
Starch nanoparticles	Electrospinning	44.7	Vitamin E–TPGS	(Kheradvar et al., 2018)
Hydroxyethyl starch nanoparticles	HES–DOCA conjugates	197	Flavonoid morin	(J. Li, Yang, et al., 2018)
Starch nanoparticles	Dialysis method and encapsulation technique	3–15	Benzo[a]pyrene	(Delsarte et al., 2018)
Starch nanoparticles	Nanoencapsulation	32.04–99.2	Vitamin D3	(Hasanvand et al., 2018)
TEMPO oxidation modified starch nanoparticles	Ionic gelation	30–50	Calcium ions	(Liu et al., 2018)
Aminated starch/ ZnO-coated iron oxide nanoparticles (folic acid tagged)	Nanocoprecipitation technique	50–90	Curcumin	(Saikia et al., 2017)
Anionic carboxymethyl starch (CMS); cationic quaternary ammonium starch (QAS)	Layer-by-layer assembly	NA	Protein	(Zhang et al., 2017)
Starch nanoparticles	Adsorption	100–400	Polyphenols (catechin, epicatechin, epigallocatechin-3-gallate, and proanthocyanins)	(Liu et al., 2016)

(*Continued*)

TABLE 10.1 (CONTINUED)
Plant-Based Polysaccharide Nanocarriers for Nutraceuticals

Polysaccharide nanocarrier type	Method	Particle size (nm)	Nutraceutical	References
Starch nanoparticles	Nanoencapsulation	20–30	Vitamin D3	(Hasanvand et al., 2015)
Octenyl succinic anhydride–modified starch	Nanoencapsulation	200–300	Coenzyme Q10	(Cheuk et al., 2015)
Corn starch	High-pressure treatment	50–60	Yerba mate (*Ilex paraguariensis*) antioxidants	(Deladino et al., 2015)
Starch nanoparticles	Emulsion method	351–990	Insulin	(Jain et al., 2008)
Starch nanoparticles	Microemulsion technique	48	Date palm pit extract	
Gum Arabic/chitosan nanoparticles	Polyelectrolyte complexation	250–290	Curcumin	(Tan et al., 2016)
Whey protein and gum (Angum gum and Arabic gum)	Double nanoemulsion	10	Crocin	(Mehrnia et al., 2017)
Pea protein isolate and mesquite gum	Encapsulation	100–300	Quercetin	(Cuevas-Bernardino et al., 2018)
Almond gum/shellac nanoparticles	Antisolvent precipitation	90–116	Quercetin	(Sedaghat Doost et al., 2018)
Balangu gum nanocapsules	Electrohydrodynamic encapsulation	96.53 ± 3.41	Essential oil	(Rezaeinia et al., 2019)
β-Lactoglobulin/gum Arabic nanoparticles	Pickering emulsions gels	244.6 ± 2.7	Lutein	(Su et al., 2020)
Azivash gum/polyvinylalcohol nanofibers	Electrospinning	130–148	Catechin	(Hoseyni et al., 2021)
Alyssum homolocarpum seed gum nanoparticles	Electrospraying	65–119	D-limonene	(Khoshakhlagh et al., 2017)
Sodium caseinate/gum Arabic nanocomplex	Complexation technique	232.2	Fish oil eicosapentaenoic acid (EPA)/docosahexaenoic acid (DHA)	(Ilyasoglu & El, 2014)
Persian gum/chitosan	Nanoencapsulation	65	Fish oil and garlic essential oil	(Raeisi et al., 2019)
Gum Arabic incorporated liposomes	Liposome delivery	100–300	Curcumin	(Li et al., 2021)
Zein/gum Arabic nanoparticle	Pickering emulsion	130.4	Thymol	(J. Li, Xu, et al., 2018)
Pea protein/carboxymethylated corn fiber	Nanocomposite	211–270	Curcumin	(Wei et al., 2020)
Whey protein/gum Arabic	Nanoemulsion	200	Vitamin E	(Ozturk et al., 2015)
Gelatin/gum Arabic nanoparticles	Ionic gelation	260–490	Quercetin	(Kim et al., 2019)
Dextran nanoparticles		100–450	Genistein	(Semyonov et al., 2014)

(Continued)

TABLE 10.1 (CONTINUED)
Plant-Based Polysaccharide Nanocarriers for Nutraceuticals

Polysaccharide nanocarrier type	Method	Particle size (nm)	Nutraceutical	References
Zein/dextran sulphate composite nanoparticles		~135	Curcumin	(Yuan et al., 2020)
Ovalbumin/dextran nanogels		–	Curcumin	(Feng et al., 2016)
Dextran-based nanosized carrier		40–50	Curcumin	(Anirudhan & Binusreejayan, 2016)

Abbreviations: BCS: Biopharmaceutics Classification system, TPP: tripolyphosphate, NA: not applicable, NPs: nanoparticles, TPGS: tocopherol polyethylene glycol succinate, HES-DOCA: hydroxyethyl starch-deoxycholic acid, TEMPO: (2,2,6,6-tetramethylpiperidine-1-yl)oxyl.

10.4 ANIMAL-DERIVED POLYSACCHARIDES

Animal-derived polysaccharides offer various advantages, such as biodegradability, renewability, and anti-inflammatory, antioxidant, antimicrobial, and other medical properties. Nanodelivery systems prepared from these polysaccharides are known to improve nutraceuticals' bioavailability, solubility, and stability (Wang et al., 2020). They are also used for target delivery and controlled release of nutraceuticals. Along with the various polysaccharides that can be obtained from animal sources, chitosan and hyaluronic acid have been studied for their application in nutraceutical delivery (Carrion et al., 2021; Pateiro et al., 2021).

10.4.1 CHITOSAN

Chitosan is obtained from the alkaline deacetylation of chitin, found in the exoskeleton of crustaceans (Estakhr et al., 2020). Chemically, it is a β(1-4) glycosidic bond–linked linear polymer of N-acetyl glucosamine and glucosamine units, where deacetylation leads to higher primary amine content and a corresponding increase in charge density (Levengood & Zhang, 2014). It is a cationic polymer with antimicrobial and antioxidant activities as well as biodegradability and biocompatibility. Functional characteristics of the polymer can be improved with ease along with enhancement in the sustained release by appropriate chemical and physical modifications (Pateiro et al., 2021). The ratio of N-acetylglucosamine and glucosamine determines the degree of deacetylation. Depending on the percentage of N-acetyl glucosamine, the polymer is denoted as chitin (higher degree) or chitosan (lower degree). The degree of deacetylation and molecular weight influence various properties such as degradation potential, solubility, and crystallinity (Levengood & Zhang, 2014).

Chitosan has been studied recently as a promising candidate for nanoparticle development due to its hydrophilicity, mucoadhesive properties, high surface-to-volume ratio, and small particle size (Pateiro et al., 2021). Due to its biodegradability, biocompatibility, enhanced solubility, and controlled release capabilities, chitosan was found to be suitable for application as a highly efficient drug carrier (Jeong et al., 2020).

Various nutraceutical-loaded chitosan nanoparticles have been prepared for food and therapeutic applications using various techniques. An ionic gelation technique was used for the preparation of chitosan nanoparticles loaded with lycopene with approximately 90% entrapment efficiency and >80% release rate at 6 hours (Dhiman & Bhalla, 2018). Encapsulation of catechin, a natural phenol and antioxidant, was studied in chitosan–alginate and chitosan–carrageenan delivery systems using ionic gelation and polyelectrolyte complexation with up to 97% encapsulation efficiency and up to

95% stability; the chitosan–alginate nanoparticles showed better performance (Manikkam et al., 2014). Chitosan–CPP (casein phosphopeptide) nanoparticles were utilized for the encapsulation of epigallocatechin gallate (EGCG) for the enhancement of free radical scavenging activity in biological systems with >80% encapsulation efficiency and sustained release (B. Hu et al., 2013).

Chitosan nanoparticles loaded with polyphenols have been studied for the treatment of cancer. The ionic cross-linking technique was used for the preparation of chitosan–TPP (sodium tripolyphosphate) nanoparticles (172–217 nm) for the encapsulation of resveratrol (polyphenol with anti-inflammatory and antioxidant activity). This was found to improve the stability of resveratrol under UV light and during storage as well as showing benefits during chemotherapy, such as lower cytotoxicity to hepatocyte and liver carcinoma cells (J et al., 2017). Chitosan nanoparticles loaded with curcumin (170–200 nm) were used to study carcinogenesis and were found beneficial in decreasing tumor incidence and multiplicity, with a sustained-release period of up to 180 h (Vijayakurup et al., 2019).

In another study, chitosan–TPP nanoparticles loaded with quercetin (a plant flavanol and dietary supplement) were prepared using an ionic gelation technique to enhance the solubility and bioavailability of quercetin and improve free radical scavenging potential (Zhang et al., 2008).

10.4.2 Hyaluronic Acid

Hyaluronic acid (HA), composed of N-acetyl glucosamine and glucuronic acid, can be obtained from bacteria, bovine sources, or rooster combs. HA used for commercial purposes is usually extracted from various animal and bacterial sources (Zhu et al., 2020). HA has recently been widely studied for targeted delivery of drugs and nutraceuticals due to various advantages, such as biodegradability, biocompatibility, nontoxic nature, hydrophilic nature, and the variety of chemicals it can interact with. It is widely used for the encapsulation of hydrophobic compounds within nanostructures. It is also suitable for the co-delivery of compounds with different chemical natures. Due to nontoxicity and safety, it is suitable for clinical trials (Carrion et al., 2021). HA has been used for various applications, such as cosmetics for skin hydration, and in the biomedical and nutrition industries. It has been widely studied for use in receptor-mediated transdermal delivery systems and enhanced permeability due to the presence of specific receptors for HA on target cells such as leukocytes, epithelial cells, and keratinocytes (Zhu et al., 2020). It has a role in immunomodulation and antioxidant scavenging. Oligo-hyaluronic acid molecules with molecular weight between 3000 and 10,000 Da were found to have better bioavailability, stability, and antioxidant potential and hence, are considered superior for utilization as a nanodelivery system for food encapsulation (Guo et al., 2018). It has mucoadhesive properties, a role in the mucosal diffusion of nanoparticles, and binds various biomolecules in the epithelial and connective tissues. It has been widely used for drug delivery in cancer cells, as it can bind to tumor and cancer cell receptors (Aguilera-Garrido et al., 2021). The suitability of HA for particular delivery systems is assessed on the basis of its molecular weight. HA with low molecular weight is considered suitable for transdermal delivery. HA with higher molecular weight is considered more appropriate for sustained release and localization due to its advantages, such as high viscosity. It is also known for its bioadhesive properties due to the few hydrophobic regions along the carbon chain of the hydrophilic hyaluronic acid molecule, which help in sustained release and also in enhancing its permeability across membranes. By conjugation with other molecules through groups such as alkylamine, its properties, such as hydrophilicity, can be altered to be used in hydrogels, nanoemulsions, and other delivery systems (Zhu et al., 2020).

Various types of nanocarriers prepared using HA have been used for the encapsulation and delivery of a variety of nutraceuticals. Some of these include an HA–containing liquid lipid nanodelivery system for improved bioaccessibility of curcumin (Aguilera-Garrido et al., 2021); oligo-hyaluronic acid–curcumin (oligohyalurosomes) for the delivery of polyphenols such as curcumin and resveratrol with enhanced stability, antioxidant activity, and bioavailability (Guo et al., 2018); HA nanoparticles for the immobilization of N-acetyl-d-galacturonic acid to stimulate the proliferation of chondrosarcoma cells (Şahin et al., 2019); hyaluronic acid nanoemulsions for the encapsulation of *Pterodon pubescens* fruit oil with improved stability (Kleinubing et al., 2021); and HA-based nanocapsules for encapsulation of hydrophobic anticancer compounds of garlic oil, such as diallyl disulfide (DADS) and diallyl trisulfide (DATS) (Janik-hazuka et al., 2021). Potential use of HA as a co-delivery system was explored in the development of a co-delivery system based on HA and hydrophobic zein protein prepared using a layer-by-layer method for long-term storage and improved stability of curcumin (encapsulated in the zein hydrophobic core) and quercetagetin (adsorbed on the hydrophilic surface) (Chen et al., 2019) and HA mucoadhesive nanoemulsions for controlled release to improve the stability, antioxidant ability, and nose-to-brain targetability of the polyphenols resveratrol and curcumin (Nasr, 2016).

10.4.3 Future Prospects of Animal-Derived Polysaccharide Nanocarriers

Animal-based polysaccharides such as chitosan and HA have been utilized for the encapsulation and delivery of various nutraceuticals. Future studies need to explore more animal-derived polysaccharides, such as heparin (Gupta et al., 2019). Due to the modification ability of chitosan, it is an ideal nanocarrier for nutraceuticals. Although animal-based nutraceuticals have been widely used for drug delivery, more research needs to be conducted on their application in nutraceutical delivery (Akbari-Alavijeh et al., 2020). More studies need to be conducted on the intestinal delivery and stability of the nanocarriers (Song & Chen, 2015). Apart from this, extensive research is required, including safety and toxicity studies and the economically viable large-scale production of animal-based polysaccharide nanoparticles (*Advances in Animal Biotechnology and Its Applications*, n.d.) (Table 10.2).

10.5 SEAWEED-DERIVED POLYSACCHARIDES

Seaweeds, or algae, have been identified as crucial sources of valuable bioactive constituents and metabolites, which possess a broad range of biological and physiological activities (Tanna & Mishra, 2019). Among all bioactive compounds, seaweed-derived polysaccharides have been recognized to have several valuable properties, such as antioxidant, anti-inflammatory, antiviral, anticoagulant, and

TABLE 10.2
Animal-Based Polysaccharide Nanocarriers for Nutraceuticals

Polysaccharide nanocarrier type	Method	Size (nm)	Nutraceutical	Reference
Chitosan	Ionic gelation	286 ± 36	Red ginseng extract	(Roshanpour et al., 2021)
Chitosan/gum Arabic	Ionic gelation	273–327	Cabbage and blueberry concentrates	(Jeong et al., 2020)
Chitosan	Complexation	332.4 ± 9.4	Curcumin	(Ng et al., 2020a)
Chitosan/gum Arabic	Ionic gelation	183–295	Saffron bioactives	(Rajabi et al., 2020)
Chitosan	Spray drier	217.7	Java tea	(Naibaho et al., 2019)
Chitosan/phosphatidyl choline	Sonication	144	Vitamin E	(Shwetha et al., 2020)
Hyaluronic acid/ Bovine Serum Albumin	Liquid lipid nanocapsules	136–143	Curcumin	(Aguilera-Garrido et al., 2021)
Oligo-hyaluronic acid	Co-encapsulation	134.5 ± 5.1	Curcumin and resveratrol	(Guo et al., 2018)
Hyaluronic acid nanoparticles	In vitro cell culture	175	Glucosamine	(Şahin et al., 2019)
Hyaluronic acid	Nanoemulsions	16–27	Fruit oil	(Kleinubing et al., 2021)

anticarcinogenic (Hamed et al., 2015; C. L & A, 2016). Seaweed-based polysaccharides have a variety of applications in numerous industries, such as food, nutraceuticals, cosmetics, and pharmaceuticals (N et al., 2016). The composition and diversity of seaweed-based polysaccharides play an essential role in their biological activities. For example, sulfated polysaccharides are a key component of the seaweed cell wall and are frequently utilized in the pharmaceutical industries (KKA et al., 2017; N et al., 2016). Moreover, fucoidan, ulvan, and carrageenan are achieved from brown, green, and red seaweed, respectively. They are used in the pharmaceutical industry as a drug delivery system (Table 10.3) due to their gelling properties (C. L & A, 2016). These natural seaweed-based polysaccharides can also be transformed into nanoparticles (NPs) using several approaches, like ionic gelation, polyelectrolyte complexing, and emulsion (Table 10.3). Polyelectrolyte complexing and ionic gelation are frequently employed techniques to develop NPs of the appropriate size, shape, and charge (Zhang et al., 2013). Seaweed polysaccharide–based nanoparticles (SPBNPs) exhibit high drug encapsulation, suitable particle size, and constant drug release with high biocompatibility (Efthimiadou et al., 2015; H & J, 2015). SPBNPs have exhibited favorable outcomes as delivery agents for proteins, anticancer drugs, and other drugs with improved bioavailability and drug release properties (W et al., 2014) (Table 10.3). However, the literature related to SPBNPs for nutraceutical delivery is limited.

10.5.1 Polysaccharides of Red Algae

Carrageenan and agar are two of the main red algae–based polysaccharides and reveal almost comparable structural and functional characteristics (Kraan, 2012). Carrageenan is the main ingredient of the cell wall of red algae, including *Eucheuma denticulatum*, *Chondrus crispus*, *Gigartina* species, and *Kappaphycus alvarezii* (Kraan, 2012; Usov & Zelinsky, 1992). A few red algae also contain xylan and porphyrin. However, the biological and physiological activity of xylan is not well known. It is found in large quantities, e.g., *Palmaria palmata* contains nearly 35% (cell dry weight basis) of xylan in its biomass (Kraan, 2012).

10.5.1.1 Carrageenans

Carrageenans from red algae possess various biological activities; e.g., *Stenogramme interrupta*–based carrageenans act as anti-hyperlipidemics (Panlasigui et al., 2003) and anti-coagulants (Wijesekara et al., 2011). Nowadays, carrageenans have been recognized as viscosity-enhancing agents and also possess the gelling property for controlled drug delivery, with huge applications in cancer treatment and tissue rejuvenation (Table 1.3) (Guan et al., 2017; L. L et al., 2014). A complex of chitosan–carrageenan-based-polyelectrolyte is accessible in the market. It can be utilized as a carrier system for numerous drugs (Mustafa et al., 2016).

The carrageenan-based delivery system (CBDS) has also been associated with the delivery of various bioactive compounds such as curcumin, quercetagetin, piperine, egg yolk immunoglobulin, and DHA-rich oil (Table 1.3). Nowadays, the CBDS for nutraceuticals are widely explored and primarily involve NP-based delivery aerogels, carriers, microcapsules, emulsions, microbeads, hydrogels, composites/complexes, and nanotubes (Dong et al., 2021). The CBDS exhibits an edge over other encapsulation materials (RC et al., 2020). The CBDS shows higher stability and electronegativity, and a better retention rate for bioactive ingredients and bioavailability of bioactive compounds, and improves loading efficiency and release amount of bioactive ingredients (Chen et al., 2020; Dong et al., 2021; Gu et al., 2021; RC et al., 2020). The thermo-reversible property of carrageenan-derived gels is an exceptional feature (Soukoulis et al., 2017). Though carrageenan helps in delivering biologically active ingredients in several ways, it still has a few unsolved issues in view of realistic employment in the nutraceutical and food industry (D. S et al., 2018).

10.5.1.2 Agar

Agar is derived from units of agaropectin and agarose and abundantly obtained from the *Gracilariaceae* and *Gelidiaceae* genera (Kraan, 2012). Agar has previously been investigated for

TABLE 10.3
Seaweed-Based Polysaccharide Nanocarriers for Nutraceuticals

Polysaccharide nanocarrier type	Methods	Particle size (nm)	Nutraceuticals	References
Alginate/chitosan	Ionotropic and polyelectrolyte complex	800	Insulin	(Sarmento et al., 2007)
κ-Carrageenan	Alginate/κ-carrageenan hydrogel beads	–	Egg yolk immunoglobulin	(Gu et al., 2021)
Alginate	Nanoparticles of oleoyl alginate ester	–	Vitamin D3	(F et al., 2012)
Alginate aldehyde/gelatin	Inverse miniemulsion technique	–	Curcumin	(Sarika & James, 2015)
Sodium alginate/gum tragacanth	Ionic gelation method	–	Phenolic compound	(A et al., 2020)
Alginate/whey protein	Ultrasonic-assisted technique	–	Omega-3 fatty acids	(Abbasi et al., 2019)
Alginate/starch/gelatin	Ultrasonication process	–	Beta-carotene	(Guedes Silva et al., 2021)
Gliadin/alginate	Centrifugation, homogenization	–	Curcumin	(F et al., 2012)
Alginate/chitosan	Ultrasonic emulsification, ionic gelation, freeze drying	–	Naringenin	(M. S et al., 2017)
Sodium alginate/whey protein isolate	–	–	Flaxseed oil	(Zamani Ghaleshahi & Rajabzadeh, 2020)
Alginate/pectin	Freeze-drying process	–	Proanthocyanidin	(K & H, 2019)
Alginate	Supercritical drying	–	Resveratrol	(MC et al., 2011)
Sodium alginate/whey protein isolate	Ultrasonication process	–	Olive oil	(Urbánková et al., 2021)
κ-Carrageenan	Gelled emulsions	–	DHA-rich algae oil	(Gayoso et al., 2019)
κ-Carrageenan	Soy protein isolate–κ-carrageenan complexes	–	Quercetagetin	(Wang et al., 2016)
Alginate/chitosan	Ionotropic pre-gelation	100–200	Insulin	(P et al., 2015)
κ-Carrageenan	Microbeads	–	β-carotene	(Perrechil et al., 2020)
Alginate–chitosan	Gelification	750	Insulin	(Sarmento et al., 2007)
Alginate/chitosan/TPP	Ionic gelation	260–525	Insulin	(FM et al., 2009)
Alginate NPs	Microemulsion	350	BSA	(N. J et al., 2012)
Alginate/chitosan	Ionic gelation	329–505	5-Fluorouracil	(Nagarwal et al. 2012)
Chitosan/carrageenan/TPP	Ionic gelation	150–300	BSA	(Budtova et al., 2020)
Carboxymethyl chitosan and carrageenan	–	–	Riboflavin	(GR et al., 2015)
κ-Carrageenan/chitosan	Complexation	–	Ascorbic acid, caffeine	(Yew & Misran, 2016)
κ-Carrageenan/lysozyme	Complexation	–	Curcumin	(Huang & Kuo, 2016)
κ-Carrageenan	Core-shell surface engineering	–	Coenzyme Q10 and piperine	(Chen et al., 2020)
κ-Carrageenan	Carrageenan hydrogels	–	β-galactosidase	(RC et al., 2020)
Chitosan–fucoidan	Coacervation process	154 and 453	–	(LC et al., 2012)

(Continued)

TABLE 10.3 (CONTINUED)
Seaweed-Based Polysaccharide Nanocarriers for Nutraceuticals

Polysaccharide nanocarrier type	Methods	Particle size (nm)	Nutraceuticals	References
Alginate/chitosan	Ionotropic and polyelectrolyte complex	800	Insulin	(Sarmento et al., 2007)
Fucoidan-coated poly(isobutylcyanoacrylate) NPs	Anionic emulsion polymerization	193 ± 4 to 399 ± 0.7	–	(MC et al., 2011)
Chitosan/fucoidan nanoparticles	Polyelectrolyte complexation	200	bFGF	(YC & YT, 2016)
Chitosan/fucoidan nanoparticles	Self-assembly	Approximately 100	PLL	(AC et al., 2015)
O-carboxymethyl chitosan/ fucoidan	Ionic cross-linking	270	Curcumin	(Huang & Kuo, 2016)
Fucoidan NPs	Self-assembly	140	Doxorubicin	(KW et al., 2013)

Abbreviations: bFGF: Basic fibroblast growth factor, BSA: bovine serum albumin, NA: not applicable, NPs: nanoparticles, PLL: prolymphocytic leukemia.

its role in lowering blood glucose levels, exhibiting the ability to absorb UV radiation, and preventing red blood cell clustering (Hamed et al., 2015; Rupérez et al., 2013). Agaro-oligosaccharide effectively suppressed the production of nitric oxide (NO)–producing enzymes and the effects of pro-inflammatory cytokines (Holdt & Kraan, 2011; Wijesekara et al., 2011). Agar also expresses an active role against alpha-glucosidase, which accounts for its antitumor and antioxidant properties (Holdt & Kraan, 2011).

10.5.2 Polysaccharides from Green Algae

Green algae such as Ulvales contain ulvans, sulfated polysaccharides, which possess various biological activities (Rupérez et al., 2013). Ulvans are cell wall components and cannot be assimilated or metabolized by humans. Ulvans exhibit water-holding abilities, making them beneficial dietary supplements (Hamed et al., 2015; Rupérez et al., 2013; Venkatesan et al., 2015). Ulvan has been demonstrated to possess hepatotoxic ability in an albino rat model (H et al., 2006). *In vivo* studies have revealed that the ulvans help in lowering hyperlipidemia in rats by reducing the total blood serum cholesterol, low-density lipid (LDL), and triglyceride, and increasing blood-serum high-density lipids (HDL) (BF & EJ, 2003). Ulvan extract exhibits potential as an immune stimulator against *Colletotrichum trifolii* (a pathogenic fungus) in barrel medick (JM et al., 2007). The extract also reveals anti-proliferative activity toward cancer (GF et al., 2017).

10.5.3 Polysaccharides of Brown Algae

Brown algae are the second largest group of multicellular algae after red algae. Brown algae contain a diverse range of polysaccharides (such as alginate, fucoidans, and laminaran) in their cell wall structure.

10.5.3.1 Fucoidans

Fucoidans have been isolated from *Ascophyllum nodusum*, *Ecklonia cava*, and *Undaria pinnatifida* (Athukorala et al., 2006; M. S et al., 2002). Fucoidans from *E. cava* reveal high anticoagulant activity in rat models (Wijesinghe et al., 2011). (Cumashi et al., 2007) investigated the heparin-mediated anti-thrombin activity of fucoidans, which depends on its sulfate content. A diverse range of fucoidan NPs has been demonstrated for encapsulating anticancer drugs and growth factors for cancer therapies and tissue regeneration, respectively (Table 1.3).

10.5.3.2 Alginate

Alginate is extracted in sodium and calcium complexes from brown algae such as *Ecklonia, Durvillaea, Laminaria, Macrocystis, Lessonia, Turbinaria, Sargassum*, and *Ascophyllum* species (Kraan, 2012; Rupérez et al., 2013). Alginate has gel-forming, thickening, and stabilizing properties. Alginate is also considered biodegradable, biocompatible, and less toxic (KM et al., 2015). Therefore, it is widely applicable for tissue-restricted controlled drug and nutraceutical delivery of depots or hydrogels in drug and nutraceutical delivery systems (KY & DJ, 2012).

Alginate has been used to prepare colon-specific delivery systems for peptides and proteins (Joye & McClements, 2014). Alginate NPs with some composite compounds have been used to deliver numerous bioactive constituents, such as β-carotene, resveratrol, and curcumin (Table 1.3). A study has demonstrated that the negatively surface-charged iron-loaded alginate nanoparticles exhibited exceptional iron encapsulation of about 75% and sustained iron release properties. This system provided an attractive option for iron supplementation over conventional systems (Katuwavila et al., 2016). Similarly, the coating of curcumin in Na alginate or zein or caseinate NPs enhanced its photochemical stability, solubility in water, pH stability, antioxidant activity, and resistance to digestion by gastro-intestinal fluids in comparison to pure protein nanoparticles (Ng et al., 2020; Q et al., 2019). Alginate-based NPs have been demonstrated in numerous drug delivery systems (Table 1.3) (JP et al., 2014). The extraction procedures affect the alginates' structure. However, modifying the recovery techniques can achieve a diverse range of physiochemical properties. For example, the alcalase assisted-alginate method stimulates inflammatory cytokines and nitric oxide in RAW 264.7 cells. In contrast, the enzyme-coupled method yields alginate with free radical scavenging properties (NJ et al., 2017). Alginate-based NPs have been widely examined as a drug delivery system due to their better encapsulation efficiency (Table 1.3). Generally, alginate-derived NPs have been examined as a drug delivery agent for anticancer drugs. Alginate-based NPs typically do not form organ clusters during the drug delivery process (Sarei et al., 2013). Chemically modified alginate-derived NPs with coating materials may show sustained release of the drug for a prolonged period. The stability of NPs is one of the crucial parameters in the drug delivery system. Azevedo and colleagues (MA et al., 2014) formulated highly stable alginate chitosan-based NPs. Alginate has been used to prepare colon-specific delivery systems for peptides and proteins (Joye & McClements, 2014).

10.5.3.3 Laminaran

Laminaran or leucosin is found in brown algae like *Laminaria* and *Saccharina* species. Other species of algae, like *Fucus, Ascophyllum*, and *Undaria*, exhibit lower levels of these polysaccharides. Laminaran has been recognized to possess anti-inflammatory action with immunostimulatory activity (JY et al., 2012).

10.5.4 Future Prospects of Seaweed-Derived Polysaccharide Nanocarriers

Seaweed-derived polysaccharides remain a subject of exhaustive future work. Various seaweed polysaccharide (such as ulvan and laminarin)–based derivatives are still left to be synthesized. Such derivatizations should be attempted for value addition to the seaweed-based products and target applications in various industries. Different seaweed-derived polysaccharides have their advantages and uses. For example, ulvan (an anionic polysaccharide) easily forms NPs with chitosan (cationic polymers), which demonstrates its promising role as a nutraceutical or drug delivery agent (Graiff et al., 2015; H & J, 2015).

Moreover, there is an urge for advanced economically viable synthesis processes, and the use of green and recyclable solvents (such as green neoteric solvents) may be encouraged for the commercial syntheses of SPBNPs in industries. The interaction of polysaccharides with DNA and protein may lead to developing novel derivatives or NPs having better pH-responsive anionic/cationic drug binding and releasing properties, chiral-sensing, and separations. Hence, the derivatization of seaweed-derived polysaccharides with other macromolecules may be focused to develop economical and sustainable derivatization approaches and products. Future research should also focus on the action of seaweed-based polysaccharides *in vivo* for the delivery of nutraceuticals.

10.6 SAFETY OF NANOCARRIERS FOR NUTRACEUTICALS

Increased use of nanotechnology for industrial applications has led to concerns regarding safety and toxicity, leading to potential health risks. The European Food Safety Authority has recommended a detailed risk evaluation of nanomaterials before consumption (Rashidi, 2021). However, studies on the clearance and toxicity of nutraceutical nanocarriers are limited. Detailed studies are needed on their mechanism of action to understand the interactions of the nanosystems with the nutraceuticals and the target locations. The use of surfactants in nanodelivery systems for the blood–brain barrier (BBB) could lead to toxicity. Detailed toxicity studies (*in vitro* and *in vivo*), cell viability studies, and studies on oxidative stress and DNA damage need to be conducted. At present, no specific guidelines and standard policies are available on safety and toxicity standards for nutraceutical nanocarriers (Meenambal & Srinivas Bharath, 2020). Parameters such as acceptable limits for daily intake of nutraceuticals loaded on nanodelivery systems needs to be defined (Rashidi, 2021). Besides, relevant parameters for toxicity studies need to be identified, for which complete characterization of nanostructural properties is important (Gonçalves et al., 2018).

10.7 CONCLUSION

Due to their size and surface area, nanocarriers help improve the bioavailability, solubility, and target delivery of nutraceuticals. The increasing demand for functional food has led to increased research on nutraceutical nanocarriers (Rashidi, 2021). Since polysaccharide nanocarriers have been extensively used in pharmaceutical applications in drug delivery, the available knowledge in the area can be utilized to enhance the properties of nutraceutical nanocarriers. More research needs to be conducted on improving the shelf life, permeability, and controlled release of nutraceuticals. There is immense potential in using nanocarriers for nutraceutical delivery for health applications. Future research in the field will involve exploring more polysaccharides for the delivery of nutraceuticals with improved stability and bioavailability. As yet, not enough animal and clinical studies have been conducted on nutraceutical nanocarriers. A significant limitation to developing nutraceutical nanocarriers is the high cost involved in the production process of controlled release techniques. Apart from this, further research should be conducted on improving safety and developing standard manufacturing policies for nanocarriers.

REFERENCES

Abbasi, F., Samadi, F., Jafari, S. M., Ramezanpour, S., & Shams Shargh, M. (2019). Ultrasound-assisted preparation of flaxseed oil nanoemulsions coated with alginate-whey protein for targeted delivery of omega-3 fatty acids into the lower sections of gastrointestinal tract to enrich broiler meat. *Ultrasonics Sonochemistry, 50*, 208–217. https://doi.org/10.1016/j.ultsonch.2018.09.014

Abd-Ellatef, G. E. F., Ahmed, O. M., Abdel-Reheim, E. S., & Abdel-Hamid, A. H. Z. (2017). *Ulva lactuca* polysaccharides prevent Wistar rat breast carcinogenesis through the augmentation of apoptosis, enhancement of antioxidant defense system, and suppression of inflammation. *Breast Cancer (Dove Medical Press), 9*, 67–83. https://doi.org/10.2147/BCTT.S125165

Abed, S. M., Ali, A. H., Noman, A., & Bakry, A. M. (2016). Inulin as prebiotics and its applications in food industry and human health: A review. *International Journal of Agriculture Innovations and Research, 5*(1), 2319–1473.

Acevedo-Guevara, L., Nieto-Suaza, L., Sanchez, L. T., Pinzon, M. I., & Villa, C. C. (2018). Development of native and modified banana starch nanoparticles as vehicles for curcumin. *International Journal of Biological Macromolecules*, *111*, 498–504. https://doi.org/10.1016/j.ijbiomac.2018.01.063

Adetunji, L. R., Adekunle, A., Orsat, V., & Raghavan, V. (2017). Advances in the pectin production process using novel extraction techniques: A review. *Food Hydrocolloids*, *62*, 239–250. https://doi.org/10.1016/j.foodhyd.2016.08.015

Advances in Animal Biotechnology and its Applications. (n.d.). https://doi.org/10.1007/978-981-10-4702-2

Aguilera-Garrido, A., del Castillo-Santaella, T., Galisteo-González, F., José Gálvez-Ruiz, M., & Maldonado-Valderrama, J. (2021). Investigating the role of hyaluronic acid in improving curcumin bioaccessibility from nanoemulsions. *Food Chemistry*, *351*(May 2020). https://doi.org/10.1016/j.foodchem.2021.129301

Ahmad, M., Mudgil, P., Gani, A., Hamed, F., Masoodi, F. A., & Maqsood, S. (2019). Nano-encapsulation of catechin in starch nanoparticles: Characterization, release behavior and bioactivity retention during simulated in-vitro digestion. *Food Chemistry*, *270*(July 2018), 95–104. https://doi.org/10.1016/j.foodchem.2018.07.024

Akbari-Alavijeh, S., Shaddel, R., & Jafari, S. M. (2020). Encapsulation of food bioactives and nutraceuticals by various chitosan-based nanocarriers. *Food Hydrocolloids*, *105*, 105774. https://doi.org/10.1016/J.FOODHYD.2020.105774

Amiri, M. S., Mohammadzadeh, V., Yazdi, M. E. T., Barani, M., Rahdar, A., & Kyzas, G. Z. (2021). Plant-based gums and mucilages applications in pharmacology and nanomedicine: A review. *Molecules*, *26*(6), 1–23. https://doi.org/10.3390/molecules26061770

Anirudhan, T. S., & Binusreejayan. (2016). Dextran-based nanosized carrier for the controlled and targeted delivery of curcumin to liver cancer cells. *International Journal of Biological Macromolecules*, *88*, 222–235. https://doi.org/10.1016/j.ijbiomac.2016.03.040

Anwar, A., Masri, A., Rao, K., Rajendran, K., Khan, N. A., Shah, M. R., & Siddiqui, R. (2019). Antimicrobial activities of green synthesized gums-stabilized nanoparticles loaded with flavonoids. *Scientific Reports*, *9*(1), 1–12. https://doi.org/10.1038/s41598-019-39528-0

Apoorva, A., Rameshbabu, A. P., Dasgupta, S., Dhara, S., & Padmavati, M. (2020). Novel pH-sensitive alginate hydrogel delivery system reinforced with gum tragacanth for intestinal targeting of nutraceuticals. *International Journal of Biological Macromolecules*, *147*, 675–687. https://doi.org/10.1016/J.IJBIOMAC.2020.01.027

Assadpour, E., & Mahdi Jafari, S. (2019). A systematic review on nanoencapsulation of food bioactive ingredients and nutraceuticals by various nanocarriers. *Critical Reviews in Food Science and Nutrition*, *59*(19), 3129–3151. https://doi.org/10.1080/10408398.2018.1484687

Assadpour, E., Jafari, S. M., & Maghsoudlou, Y. (2017). Evaluation of folic acid release from spray dried powder particles of pectin-whey protein nano-capsules. *International Journal of Biological Macromolecules*, *95*, 238–247. https://doi.org/10.1016/j.ijbiomac.2016.11.023

Asztalos, B. F., & Schaefer, E. J. (2003). HDL in atherosclerosis: Actor or bystander? *Atherosclerosis.* Supplements, *4*(1), 21–29. https://doi.org/10.1016/S1567-5688(03)00006-0

Athukorala, Y., Jung, W.-K., Vasanthan, T., & Jeon, Y.-J. (2006). An anticoagulative polysaccharide from an enzymatic hydrolysate of *Ecklonia cava*. *Carbohydrate Polymers*, *2*(66), 184–191. https://doi.org/10.1016/J.CARBPOL.2006.03.002

Azevedo, M. A., Bourbon, A. I., Vicente, A. A., & Cerqueira, M. A. (2014). Alginate/chitosan nanoparticles for encapsulation and controlled release of vitamin B2. *International Journal of Biological Macromolecules*, *71*, 141–146. https://doi.org/10.1016/J.IJBIOMAC.2014.05.036

Badar, A., Pachera, S., Ansari, A. S., & Lohiya, N. K. (2019). Nano based drug delivery systems: Present and future prospects. *Nanomedicine and Nanotechnology Journal*, *2*(1), 121.

Baek, J., Ramasamy, M., Willis, N. C., Kim, D. S., Anderson, W. A., & Tam, K. C. (2021). Encapsulation and controlled release of vitamin C in modified cellulose nanocrystal/chitosan nanocapsules. *Current Research in Food Science*, *4*(January), 215–223. https://doi.org/10.1016/j.crfs.2021.03.010

Bhushani, A., Harish, U., & Anandharamakrishnan, C. (2017). Nanodelivery of nutrients for improved bioavailability. In *Nutrient Delivery*. Elsevier Inc. https://doi.org/10.1016/b978-0-12-804304-2.00010-x

Blanco-Padilla, A., Soto, K. M., Hernández Iturriaga, M., & Mendoza, S. (2014). Food antimicrobials nanocarriers. *Scientific World Journal*, *2014*. https://doi.org/10.1155/2014/837215

Borazjani, N. J., Tabarsa, M., You, S., & Rezaei, M. (2017). Effects of extraction methods on molecular characteristics, antioxidant properties and immunomodulation of alginates from *Sargassum angustifolium*.

International Journal of Biological Macromolecules, *101*, 703–711. https://doi.org/10.1016/J.IJBIOMAC.2017.03.128

Borel, T., & Sabliov, C. M. (2014). Nanodelivery of bioactive components for food applications: Types of delivery systems, properties, and their effect on ADME profiles and toxicity of nanoparticles. *Annual Review of Food Science and Technology*, *5*(1), 197–213. https://doi.org/10.1146/annurev-food-030713-092354

Brodeur, G., Yau, E., Badal, K., Collier, J., Ramachandran, K. B., & Ramakrishnan, S. (2011). Chemical and physicochemical pretreatment of lignocellulosic biomass: A review. *Enzyme Research*, *2011*(1). https://doi.org/10.4061/2011/787532

Budtova, T., Aguilera, D. A., Beluns, S., Berglund, L., Chartier, C., Espinosa, E., Gaidukovs, S., Klimek-kopyra, A., Kmita, A., Lachowicz, D., Liebner, F., Platnieks, O., Rodríguez, A., Navarro, L. K. T., Zou, F., & Buwalda, S. J. (2020). Biorefinery approach for aerogels. *Polymers*, *12*(12), 1–63. https://doi.org/10.3390/POLYM12122779

Cadée, J. A., van Luyn, M. J. A., Brouwer, L. A., Plantinga, J. A., van Wachem, P. B., de Groot, C. J., den Otter, W., & Hennink, W. E. (2000). *In vivo biocompatibility of dextran-based hydrogels*, 397–404. https://doi.org/10.1002/(SICI)1097-4636(20000605)50:3

Carrion, C. C., Nasrollahzadeh, M., Sajjadi, M., Jaleh, B., Soufi, G. J., & Iravani, S. (2021). Lignin, lipid, protein, hyaluronic acid, starch, cellulose, gum, pectin, alginate and chitosan-based nanomaterials for cancer nanotherapy: Challenges and opportunities. *International Journal of Biological Macromolecules*, *178*, 193–228. https://doi.org/10.1016/j.ijbiomac.2021.02.123

Castro-Rosas, J., Ferreira-Grosso, C. R., Gómez-Aldapa, C. A., Rangel-Vargas, E., Rodríguez-Marín, M. L., Guzmán-Ortiz, F. A., & Falfan-Cortes, R. N. (2017). Recent advances in microencapsulation of natural sources of antimicrobial compounds used in food: A review. *Food Research International*, *102*, 575–587. https://doi.org/10.1016/J.FOODRES.2017.09.054

Chantarodsakun, T., Vongsetskul, T., Jangpatarapongsa, K., Tuchinda, P., Uamsiri, S., Bamrungcharoen, C., Kumkate, S., Opaprakasit, P., & Tangboriboonrat, P. (2014). [6]-Gingerol-loaded cellulose acetate electrospun fibers as a topical carrier for controlled release. *Polymer Bulletin*, *71*(12), 3163–3176. https://doi.org/10.1007/s00289-014-1243-x

Chen, J., Liu, W., Liu, C. M., Li, T., Liang, R. H., & Luo, S. J. (2015). Pectin modifications: A review. *Critical Reviews in Food Science and Nutrition*, *55*(12), 1684–1698. https://doi.org/10.1080/10408398.2012.718722

Chen, K., & Zhang, H. (2019). Alginate/pectin aerogel microspheres for controlled release of proanthocyanidins. *International Journal of Biological Macromolecules*, *136*, 936–943. https://doi.org/10.1016/J.IJBIOMAC.2019.06.138

Chen, S., Han, Y., Huang, J., Dai, L., Du, J., McClements, D. J., Mao, L., Liu, J., & Gao, Y. (2019). Fabrication and characterization of layer-by-layer composite nanoparticles based on zein and hyaluronic acid for codelivery of curcumin and quercetagetin. *ACS Applied Materials and Interfaces*, *11*(18), 16922–16933. https://doi.org/10.1021/acsami.9b02529

Chen, S., Zhang, Y., Qing, J., Han, Y., McClements, D. J., & Gao, Y. (2020). Core-shell nanoparticles for co-encapsulation of coenzyme Q10 and piperine: Surface engineering of hydrogel shell around protein core. *Food Hydrocolloids*, *103*. https://doi.org/10.1016/J.FOODHYD.2020.105651

Cheuk, S. Y., Shih, F. F., Champagne, E. T., Daigle, K. W., Patindol, J. A., Mattison, C. P., & Boue, S. M. (2015). Nano-encapsulation of coenzyme Q10 using octenyl succinic anhydride modified starch. *Food Chemistry*, *174*, 585–590. https://doi.org/10.1016/j.foodchem.2014.11.031

Chin, S. F., Azman, A., & Pang, S. C. (2014). Size controlled synthesis of starch nanoparticles by a microemulsion method. *Journal of Nanomaterials*, *2014*, 1–7. https://doi.org/10.1155/2014/763736

Chin, S. F., Pang, S. C., & Tay, S. H. (2011). Size controlled synthesis of starch nanoparticles by a simple nanoprecipitation method. *Carbohydrate Polymers*, *86*(4), 1817–1819. https://doi.org/10.1016/j.carbpol.2011.07.012

Chin, S. F., Salim, A., & Pang, S. C. (2021). Starch Acetate Nanoparticles as Controlled Release Nanocarriers for Piperine. *Starch/Staerke*, 2100054, 1–6. https://doi.org/10.1002/star.202100054

Contado, C., Caselotto, L., Mello, P., Maietti, A., Marvelli, L., Marchetti, N., & Dalpiaz, A. (2020). Design and formulation of Eudragit-coated zein/pectin nanoparticles for the colon delivery of resveratrol. *European Food Research and Technology*, *246*(12), 2427–2441. https://doi.org/10.1007/s00217-020-03586-w

Cuevas-Bernardino, J. C., Leyva-Gutierrez, F. M. A., Vernon-Carter, E. J., Lobato-Calleros, C., Román-Guerrero, A., & Davidov-Pardo, G. (2018). Formation of biopolymer complexes composed of pea protein and mesquite gum – Impact of quercetin addition on their physical and chemical stability. *Food Hydrocolloids*, *77*, 736–745. https://doi.org/10.1016/j.foodhyd.2017.11.015

Cumashi, A., Ushakova, n. A., Preobrazhenskaya, M. E., D'Incecco, A., Piccoli, A., Totani, L., Tinari, N., Morozevich, G. E., Berman, A. E., Bilan, M. I., Usov, A. I., Ustyuzhanina, N. E., Grachev, A. A., Sanderson, C. J., Kelly, M., Rabinovich, G. A., Iacobelli, S., Nifantiev, N. E., & and on behalf of the Consorzio Interuniversitario Nazionale per la Bio-Oncologia (CINBO), I. (2007). A comparative study of the anti-inflammatory, anticoagulant, antiangiogenic, and antiadhesive activities of nine different fucoidans from brown seaweeds. *Glycobiology*, *17*(5), 541–552. https://doi.org/10.1093/GLYCOB/CWM014

Cunha, L., & Grenha, A. (2016). Sulfated seaweed polysaccharides as multifunctional materials in drug delivery applications. *Marine Drugs*, *14*(3). https://doi.org/10.3390/MD14030042

da Silva, L. C., Garcia, T., Mori, M., Sandri, G., Bonferoni, M. C., Finotelli, P. V., Cinelli, L. P., Caramella, C. and Cabral, L. M. (2012). Preparation and characterization of polysaccharide-based nanoparticles with anticoagulant activity. *International Journal of Nanomedicine*, *7*, 2975–2986. https://doi.org/10.2147/IJN.S31632

David, S., Levi, C. S., Fahoum, L., Ungar, Y., Meyron-Holtz, E. G., Shpigelman, A., & Lesmes, U. (2018). Revisiting the carrageenan controversy: Do we really understand the digestive fate and safety of carrageenan in our foods? *Food & Function*, *9*(3), 1344–1352. https://doi.org/10.1039/C7FO01721A

de Andrades, E. O., da Costa, J. M. A. R., de Lima Neto, F. E. M., de Araujo, A. R., de Oliveira Silva Ribeiro, F., Vasconcelos, A. G., de Jesus Oliveira, A. C., Sobrinho, J. L. S., de Almeida, M. P., Carvalho, A. P., Dias, J. N., Silva, I. G. M., Albuquerque, P., Pereira, I. S., do Amaral Rabello, D., das Graças Nascimento Amorim, A., de Souza de Almeida Leite, J. R., & da Silva, D. A. (2021). Acetylated cashew gum and fucan for incorporation of lycopene rich extract from red guava (*Psidium guajava* L.) in nanostructured systems: Antioxidant and antitumor capacity. *International Journal of Biological Macromolecules*, *191*(September), 1026–1037. https://doi.org/10.1016/j.ijbiomac.2021.09.116

de Souza, H. J. B., Dessimoni, A. L. de A., Ferreira, M. L. A., Botrel, D. A., Borges, S. V., Viana, L. C., Oliveira, C. R. de, Lago, A. M. T., & Fernandes, R. V. de B. (2021). Microparticles obtained by spray-drying technique containing ginger essential oil with the addition of cellulose nanofibrils extracted from the ginger vegetable fiber. *Drying Technology*, *39*(12), 1912–1926. https://doi.org/10.1080/07373937.2020.1851707

Deladino, L., Teixeira, A. S., Navarro, A. S., Alvarez, I., Molina-García, A. D., & Martino, M. (2015). Corn starch systems as carriers for yerba mate (Ilex paraguariensis) antioxidants. *Food and Bioproducts Processing*, *94*(July), 463–472. https://doi.org/10.1016/j.fbp.2014.07.001

Delsarte, I., Delattre, F., Rafin, C., & Veignie, E. (2018). Investigations of benzo[a]pyrene encapsulation and Fenton degradation by starch nanoparticles. *Carbohydrate Polymers*, *186*(January), 344–349. https://doi.org/10.1016/j.carbpol.2018.01.037

Deng, W., Li, J., Yao, P., He, F., & Huang, C. (2010). Green preparation process, characterization and antitumor effects of doxorubicin-BSA-dextran nanoparticles. *Macromolecular Bioscience*, *10*(10), 1224–1234. https://doi.org/10.1002/MABI.201000125

Deng, Z., Zhu, K., Li, R., Zhou, L., & Zhang, H. (2020). Cellulose nanocrystals incorporated β-chitosan nanoparticles to enhance the stability and in vitro release of β-galactosidase. *Food Research International*, *137*(April), 109380. https://doi.org/10.1016/j.foodres.2020.109380

Dhali, K., Ghasemlou, M., Daver, F., Cass, P., & Adhikari, B. (2021). A review of nanocellulose as a new material towards environmental sustainability. *Science of the Total Environment*, *775*, 145871. https://doi.org/10.1016/j.scitotenv.2021.145871

Dhiman, A., & Bhalla, D. (2018). Development and evaluation of lycopene loaded chitosan nanoparticles. *Current Nanomedicine*, *9*(1), 61–75. https://doi.org/10.2174/2468187308666180815145855

di Martino, A., Trusova, M. E., Postnikov, P. S., & Sedlarik, V. (2018). Folic acid-chitosan-alginate nanocomplexes for multiple delivery of chemotherapeutic agents. *Journal of Drug Delivery Science and Technology*, *47*, 67–76. https://doi.org/10.1016/J.JDDST.2018.06.020

Dima, C., Assadpour, E., Dima, S., & Jafari, S. M. (2020). Nutraceutical nanodelivery; an insight into the bioaccessibility/bioavailability of different bioactive compounds loaded within nanocarriers. *Critical Reviews in Food Science and Nutrition*, *61*(18), 1–35. https://doi.org/10.1080/10408398.2020.1792409

Dong, Y., Wei, Z., & Xue, C. (2021). Recent advances in carrageenan-based delivery systems for bioactive ingredients: A review. *Trends in Food Science and Technology*, *112*, 348–361. https://doi.org/10.1016/J.TIFS.2021.04.012

Duan, H. L., Shen, Z. Q., Wang, X. W., Chao, F. H., & Li, J. W. (2005). Preparation of immunomagnetic iron-dextran nanoparticles and application in rapid isolation of E.coli O157:H7 from foods. *World Journal of Gastroenterology : WJG*, *11*(24), 3660. https://doi.org/10.3748/WJG.V11.I24.3660

Efthimiadou, E. K., Metaxa, A. F., & Koras, G. K. (2015). Modified polysaccharides for drug delivery. *Polysaccharides: Bioactivity and Biotechnology*, 1805–1835. https://doi.org/10.1007/978-3-319-16298-0_23

Esfanjani, A. F., Jafari, S. M., Assadpoor, E., & Mohammadi, A. (2015). Nano-encapsulation of saffron extract through double-layered multiple emulsions of pectin and whey protein concentrate. *Journal of Food Engineering*, *165*, 149–155. https://doi.org/10.1016/j.jfoodeng.2015.06.022

Estakhr, P., Tavakoli, J., Beigmohammadi, F., Alaei, S., & Mousavi Khaneghah, A. (2020). Incorporation of the nanoencapsulated polyphenolic extract of *Ferula persica* into soybean oil: Assessment of oil oxidative stability. *Food Science and Nutrition*, *8*(6), 2817–2826. https://doi.org/10.1002/FSN3.1575

Fan, Y., Yi, J., Zhang, Y., Wen, Z., & Zhao, L. (2017). Physicochemical stability and in vitro bioaccessibility of β-carotene nanoemulsions stabilized with whey protein-dextran conjugates. *Food Hydrocolloids*, *63*, 256–264. https://doi.org/10.1016/J.FOODHYD.2016.09.008

Farrag, Y., Ide, W., Montero, B., Rico, M., Rodríguez-Llamazares, S., Barral, L., & Bouza, R. (2018). Preparation of starch nanoparticles loaded with quercetin using nanoprecipitation technique. *International Journal of Biological Macromolecules*, *114*, 426–433. https://doi.org/10.1016/j.ijbiomac.2018.03.134

Fathi, M., Martín, Á., & McClements, D. J. (2014). Nanoencapsulation of food ingredients using carbohydrate based delivery systems. *Trends in Food Science and Technology*, *39*(1), 18–39. https://doi.org/10.1016/j.tifs.2014.06.007

Feng, J., Wu, S., Wang, H., & Liu, S. (2016). Improved bioavailability of curcumin in ovalbumin-dextran nanogels prepared by Maillard reaction. *Journal of Functional Foods*, *27*, 55–68. https://doi.org/10.1016/j.jff.2016.09.002

Fortunati, E., & Balestra, G. M. (2019). Lignocellulosic materials as novel carriers, also at nanoscale, of organic active principles for agri-food applications. In *Biomass, Biopolymer-Based Materials, and Bioenergy: Construction, Biomedical, and other Industrial Applications*. Elsevier Ltd. https://doi.org/10.1016/B978-0-08-102426-3.00009-6

Fox, S. C., Li, B., Xu, D., & Edgar, K. J. (2011). Regioselective esterification and etherification of cellulose: A review. *Biomacromolecules*, *12*(6), 1956–1972. https://doi.org/10.1021/bm200260d

Gayoso, L., Ansorena, D., & Astiasarán, I. (2019). DHA rich algae oil delivered by O/W or gelled emulsions: Strategies to increase its bioaccessibility. *Journal of the Science of Food and Agriculture*, *99*(5), 2251–2258. https://doi.org/10.1002/JSFA.9420

Gharehbeglou, P., Jafari, S. M., Hamishekar, H., Homayouni, A., & Mirzaei, H. (2019). Pectin-whey protein complexes vs. small molecule surfactants for stabilization of double nano-emulsions as novel bioactive delivery systems. *Journal of Food Engineering*, *245*(June 2018), 139–148. https://doi.org/10.1016/j.jfoodeng.2018.10.016

Ghasemi, S., Jafari, S. M., Assadpour, E., & Khomeiri, M. (2017). Production of pectin-whey protein nano-complexes as carriers of orange peel oil. *Carbohydrate Polymers*, *177*(July), 369–377. https://doi.org/10.1016/j.carbpol.2017.09.009

Ghasemi, S., Jafari, S. M., Assadpour, E., & Khomeiri, M. (2018). Nanoencapsulation of D-limonene within nanocarriers produced by pectin-whey protein complexes. *Food Hydrocolloids*, *77*, 152–162. https://doi.org/10.1016/j.foodhyd.2017.09.030

Ghorani, B., & Tucker, N. (2015). Fundamentals of electrospinning as a novel delivery vehicle for bioactive compounds in food nanotechnology. *Food Hydrocolloids*, *51*, 227–240. https://doi.org/10.1016/j.foodhyd.2015.05.024

Giri, S., Dutta, P., & Giri, T. K. (2021). Inulin-based carriers for colon drug targeting. *Journal of Drug Delivery Science and Technology*, *64*. https://doi.org/10.1016/j.jddst.2021.102595

Giri, T. K., Mukherjee, P., Barman, T. K., & Maity, S. (2016). Nano-encapsulation of capsaicin on lipid vesicle and evaluation of their hepatocellular protective effect. *International Journal of Biological Macromolecules*, *88*, 236–243. https://doi.org/10.1016/j.ijbiomac.2016.03.056

Gonçalves, R. F. S., Martins, J. T., Duarte, C. M. M., Vicente, A. A., & Pinheiro, A. C. (2018). Advances in nutraceutical delivery systems: From formulation design for bioavailability enhancement to efficacy and safety evaluation. *Trends in Food Science & Technology*, *78*, 270–291. https://doi.org/10.1016/J.TIFS.2018.06.011

Goycoolea, F. M., Lollo, G., Remunán-López, C., Quaglia, F., & Alonso, M. J. (2009). Chitosan-alginate blended nanoparticles as carriers for the transmucosal delivery of macromolecules. *Biomacromolecules*, *10*(7), 1736–1743. https://doi.org/10.1021/BM9001377

Graiff, A., Ruth, W., Kragl, U., & Karsten, U. (2015). Chemical characterization and quantification of the brown algal storage compound laminarin: A new methodological approach. *Journal of Applied Phycology*, *28*(1), 533–543. https://doi.org/10.1007/S10811-015-0563-Z

Gu, L., McClements, D. J., Li, J., Su, Y., Yang, Y., & Li, J. (2021). Formulation of alginate/carrageenan microgels to encapsulate, protect and release immunoglobulins: Egg Yolk IgY. *Food Hydrocolloids, 112*. https://doi.org/10.1016/J.FOODHYD.2020.106349

Guan, J., Li, L., & Mao, S. (2017). Applications of Carrageenan in Advanced Drug Delivery. *Seaweed Polysaccharides: Isolation, Biological and Biomedical Applications*, 283–303. https://doi.org/10.1016/B978-0-12-809816-5.00015-3

Guedes Silva, K. C., Feltre, G., Dupas Hubinger, M., & Kawazoe Sato, A. C. (2021). Protection and targeted delivery of β-carotene by starch-alginate-gelatin emulsion-filled hydrogels. *Journal of Food Engineering, 290*. https://doi.org/10.1016/J.JFOODENG.2020.110205

Guerra-Rosas, M. I., Morales-Castro, J., Cubero-Márquez, M. A., Salvia-Trujillo, L., & Martín-Belloso, O. (2017). Antimicrobial activity of nanoemulsions containing essential oils and high methoxyl pectin during long-term storage. *Food Control, 77*, 131–138. https://doi.org/10.1016/j.foodcont.2017.02.008

Gunathilake, T. M. S. U., Ching, Y. C., & Chuah, C. H. (2017). Enhancement of curcumin bioavailability using nanocellulose reinforced chitosan hydrogel. *Polymers, 9*(2), 64. https://doi.org/10.3390/polym9020064

Guo, C., Yin, J., & Chen, D. (2018). Co-encapsulation of curcumin and resveratrol into novel nutraceutical hyalurosomes nano-food delivery system based on oligo-hyaluronic acid-curcumin polymer. *Carbohydrate Polymers, 181*(November 2017), 1033–1037. https://doi.org/10.1016/j.carbpol.2017.11.046

Gupta, R. C., Lall, R., Srivastava, A., & Sinha, A. (2019). Hyaluronic acid: Molecular mechanisms and therapeutic trajectory. *Frontiers in Veterinary Science*, *6*(JUN), 1–24. https://doi.org/10.3389/fvets.2019.00192

Haghighi, M., Yarmand, M. S., Emam-Djomeh, Z., McClements, D. J., Saboury, A. A., & Rafiee-Tehrani, M. (2018). Design and fabrication of pectin-coated nanoliposomal delivery systems for a bioactive polyphenolic: Phloridzin. *International Journal of Biological Macromolecules, 112*, 626–637. https://doi.org/10.1016/j.ijbiomac.2018.01.108

Hamed, I., Özogul, F., Özogul, Y., & Regenstein, J. M. (2015). Marine bioactive compounds and their health benefits: A review. *Comprehensive Reviews in Food Science and Food Safety, 14*(4), 446–465. https://doi.org/10.1111/1541-4337.12136

Hao, Y., Chen, Y., Li, Q., & Gao, Q. (2018). Preparation of starch nanocrystals through enzymatic pretreatment from waxy potato starch. *Carbohydrate Polymers, 184*, 171–177. https://doi.org/10.1016/J.CARBPOL.2017.12.042

Hasanvand, E., Fathi, M., & Bassiri, A. (2018). Production and characterization of vitamin D3 loaded starch nanoparticles: Effect of amylose to amylopectin ratio and sonication parameters. *Journal of Food Science and Technology, 55*(4), 1314–1324. https://doi.org/10.1007/s13197-018-3042-0

Hasanvand, E., Fathi, M., Bassiri, A., Javanmard, M., & Abbaszadeh, R. (2015). Novel starch based nanocarrier for Vitamin D fortification of milk: Production and characterization. *Food and Bioproducts Processing, 96*, 264–277. https://doi.org/10.1016/j.fbp.2015.09.007

Holdt, S. L., & Kraan, S. (2011). Bioactive compounds in seaweed: Functional food applications and legislation. *Journal of Applied Phycology, 23*(3), 543–597. https://doi.org/10.1007/S10811-010-9632-5

Hoseyni, S. Z., Jafari, S. M., Shahiri Tabarestani, H., Ghorbani, M., Assadpour, E., & Sabaghi, M. (2021). Release of catechin from Azivash gum-polyvinyl alcohol electrospun nanofibers in simulated food and digestion media. *Food Hydrocolloids, 112*(September 2020), 106366. https://doi.org/10.1016/j.foodhyd.2020.106366

Hu, B., Ting, Y., Zeng, X., & Huang, Q. (2013). Bioactive peptides/chitosan nanoparticles enhance cellular antioxidant activity of (-)-epigallocatechin-3-gallate. *Journal of Agricultural and Food Chemistry, 61*(4), 875–881. https://doi.org/10.1021/JF304821K

Hu, K., Huang, X., Gao, Y., Huang, X., Xiao, H., & McClements, D. J. (2015). Core-shell biopolymer nanoparticle delivery systems: Synthesis and characterization of curcumin fortified zein-pectin nanoparticles. *Food Chemistry, 182*, 275–281. https://doi.org/10.1016/j.foodchem.2015.03.009

Hu, Q., Lu, Y., & Luo, Y. (2021). Recent advances in dextran-based drug delivery systems: From fabrication strategies to applications. *Carbohydrate Polymers*, *264*, 117999. https://doi.org/10.1016/J.CARBPOL.2021.117999

Hu, Y., Qin, Y., Qiu, C., Xu, X., Jin, Z., & Wang, J. (2020). Ultrasound-assisted self-assembly of β-cyclodextrin/debranched starch nanoparticles as promising carriers of tangeretin. *Food Hydrocolloids, 108*(May), 106021. https://doi.org/10.1016/j.foodhyd.2020.106021

Hu, Y., Zhang, W., Ke, Z., Li, Y., & Zhou, Z. (2017). In vitro release and antioxidant activity of Satsuma mandarin (Citrus reticulata Blanco cv. unshiu) peel flavonoids encapsulated by pectin nanoparticles. *International Journal of Food Science and Technology, 52*(11), 2362–2373. https://doi.org/10.1111/ijfs.13520

Huang, Y. C., & Kuo, T. H. (2016). O-carboxymethyl chitosan/fucoidan nanoparticles increase cellular curcumin uptake. *Food Hydrocolloids*, *53*, 261–269. https://doi.org/10.1016/J.FOODHYD.2015.02.006

Huang, Y. C., & Yang, Y. T. (2016). Effect of basic fibroblast growth factor released from chitosan-fucoidan nanoparticles on neurite extension. *Journal of Tissue Engineering and Regenerative Medicine*, *10*(5), 418–427. https://doi.org/10.1002/TERM.1752

Ilyasoglu, H., & El, S. N. (2014). Nanoencapsulation of EPA/DHA with sodium caseinate-gum Arabic complex and its usage in the enrichment of fruit juice. *LWT - Food Science and Technology*, *56*(2), 461–468. https://doi.org/10.1016/j.lwt.2013.12.002

Jain, A. K., Khar, R. K., Ahmed, F. J., & Diwan, P. V. (2008). Effective insulin delivery using starch nanoparticles as a potential trans-nasal mucoadhesive carrier. *European Journal of Pharmaceutics and Biopharmaceutics*, *69*(2), 426–435. https://doi.org/10.1016/j.ejpb.2007.12.001

Janik-hazuka, M., Kamiński, K., Kaczor-kamińska, M., Szafraniec-szczęsny, J., Kmak, A., Kassassir, H., Watała, C., Wróbel, M., & Zapotoczny, S. (2021). Hyaluronic acid-based nanocapsules as efficient delivery systems of garlic oil active components with anticancer activity. *Nanomaterials*, *11*(5). https://doi.org/10.3390/nano11051354

Jeong, S. J., Lee, J. S., & Lee, H. G. (2020). Nanoencapsulation of synergistic antioxidant fruit and vegetable concentrates and their stability during in vitro digestion. *Journal of the Science of Food and Agriculture*, *100*(3), 1056–1063. https://doi.org/10.1002/JSFA.10110

Joye, I. J., & McClements, D. J. (2014). Biopolymer-based nanoparticles and microparticles: Fabrication, characterization, and application. *Current Opinion in Colloid & Interface Science*, *19*(5), 417–427. https://doi.org/10.1016/J.COCIS.2014.07.002

Jun-xia, X., Hai-yan, Y., & Jian, Y. (2011). Microencapsulation of sweet orange oil by complex coacervation with soybean protein isolate/gum Arabic. *Food Chemistry*, *125*(4), 1267–1272. https://doi.org/10.1016/J.FOODCHEM.2010.10.063

Katuwavila, N. P., Perera, A. D. L. C., Dahanayake, D., Karunaratne, V., Amaratunga, G. A. J., & Karunaratne, D. N. (2016). Alginate nanoparticles protect ferrous from oxidation: Potential iron delivery system. *International Journal of Pharmaceutics*, *513*(1–2), 404–409. https://doi.org/10.1016/J.IJPHARM.2016.09.053

Kaur, J., Kaur, G., Sharma, S., & Jeet, K. (2018). Cereal starch nanoparticles: A prospective food additive: A review. *Critical Reviews in Food Science and Nutrition*, *58*(7), 1097–1107. https://doi.org/10.1080/10408398.2016.1238339

Khan, A., Wen, Y., Huq, T., & Ni, Y. (2018). Cellulosic nanomaterials in food and nutraceutical applications: A review. *Journal of Agricultural and Food Chemistry*, *66*(1), 8–19. https://doi.org/10.1021/acs.jafc.7b04204

Khanvilkar, A. M., Ranveer, R. C., & Sahoo, A. K. (2016). Carrier materials for encapsulation of bioactive components of food. *International Journal of Pharmaceutical Sciences Review and Research*, *40*(1), 62–73.

Kheradvar, S. A., Nourmohammadi, J., Tabesh, H., & Bagheri, B. (2018). Starch nanoparticle as a vitamin E-TPGS carrier loaded in silk fibroin-poly(vinyl alcohol): Aloe vera nanofibrous dressing. *Colloids and Surfaces B: Biointerfaces*, *166*, 9–16. https://doi.org/10.1016/j.colsurfb.2018.03.004

Khoshakhlagh, K., Koocheki, A., Mohebbi, M., & Allafchian, A. (2017). Development and characterization of electrosprayed *Alyssum homolocarpum* seed gum nanoparticles for encapsulation of D-limonene. *Journal of Colloid and Interface Science*, *490*, 562–575. https://doi.org/10.1016/j.jcis.2016.11.067

Kim, E. S., Kim, D. Y., Lee, J. S., & Lee, H. G. (2019). Mucoadhesive chitosan-gum arabic nanoparticles enhance the absorption and antioxidant activity of quercetin in the intestinal cellular environment. *Journal of Agricultural and Food Chemistry*, *67*(31), 8609–8616. https://doi.org/10.1021/acs.jafc.9b00008

Kim, H. Y., Park, S. S., & Lim, S. T. (2015). Preparation, characterization and utilization of starch nanoparticles. *Colloids and Surfaces B: Biointerfaces*, *126*, 607–620. https://doi.org/10.1016/j.colsurfb.2014.11.011

Kim, J. H., Shim, B. S., Kim, H. S., Lee, Y. J., Min, S. K., Jang, D., Abas, Z., & Kim, J. (2015). Review of nanocellulose for sustainable future materials. *International Journal of Precision Engineering and Manufacturing: Green Technology*, *2*(2), 197–213. https://doi.org/10.1007/s40684-015-0024-9

Kleinubing, S. A., Outuki, P. M., Hoscheid, J., Pelegrini, B. L., Antonio da Silva, E., Renata de Almeida Canoff, J., Miriam de Souza Lima, M., & Carvalho Cardoso, M. L. (2021). Hyaluronic acid incorporation into nanoemulsions containing Pterodon pubescens Benth. Fruit oil for topical drug delivery. *Biocatalysis and Agricultural Biotechnology*, *32*(December 2020), 1–9. https://doi.org/10.1016/j.bcab.2021.101939

Kraan, S. (2012). Algal polysaccharides, novel applications and outlook. *Carbohydrates: Comprehensive Studies on Glycobiology and Glycotechnology*. https://doi.org/10.5772/51572

Kringel, D. H., Dias, A. R. G., Zavareze, E. da R., & Gandra, E. A. (2020). Fruit wastes as promising sources of starch: Extraction, properties, and applications. *Starch: Stärke*, *72*(3–4), 1900200. https://doi.org/10.1002/STAR.201900200

Lai, S. K., Wang, Y. Y., & Hanes, J. (2009). Mucus-penetrating nanoparticles for drug and gene delivery to mucosal tissues. *Advanced Drug Delivery Reviews*, *61*(2), 158–171. https://doi.org/10.1016/j.addr.2008.11.002

Layek, B., & Mandal, S. (2020). Natural polysaccharides for controlled delivery of oral therapeutics: A recent update. *Carbohydrate Polymers*, *230*(November 2019), 115617. https://doi.org/10.1016/j.carbpol.2019.115617

Le Corre, D., Bras, J., & Dufresne, A. (2010). Starch nanoparticles: A review. *Biomacromolecules*, *11*(5), 1139–1153. https://doi.org/10.1021/bm901428y

Lee, J. L., Kim, Y. J., Kim, H. J., Kim, Y. S., & Part, W. (2012). Immunostimulatory effect of laminarin on RAW 264.7 mouse macrophages. *Molecules*, *17*(5), 5404–5411. https://doi.org/10.3390/MOLECULES17055404

Lee, K. W., Jeong, D., & Na, K. (2013). Doxorubicin loading fucoidan acetate nanoparticles for immune and chemotherapy in cancer treatment. *Carbohydrate Polymers*, *94*(2), 850–856. https://doi.org/10.1016/J.CARBPOL.2013.02.018

Lee, K. Y., & Mooney, D. J. (2012). Alginate: Properties and biomedical applications. *Progress in Polymer Science*, *37*(1), 106–126. https://doi.org/10.1016/J.PROGPOLYMSCI.2011.06.003

Lei, Z., Gao, W., Zeng, J., Wang, B., & Xu, J. (2020). The mechanism of Cu (II) adsorption onto 2,3-dialdehyde nano-fibrillated celluloses. *Carbohydrate Polymers*, *230*(July 2019), 115631. https://doi.org/10.1016/j.carbpol.2019.115631

Leiro, J. M., Castro, R., Arranz, J. A., & Lamas, J. (2007). Immunomodulating activities of acidic sulphated polysaccharides obtained from the seaweed *Ulva rigida* C. Agardh. *International Immunopharmacology*, *7*(7), 879–888. https://doi.org/10.1016/J.INTIMP.2007.02.007

Levengood, S. K. L., & Zhang, M. (2014). Chitosan-based scaffolds for bone tissue engineering. *Journal of Materials Chemistry. B*, *2*(21), 3161–3184. https://doi.org/10.1039/C4TB00027G

Li, J., Xu, X., Chen, Z., Wang, T., Lu, Z., Hu, W., & Wang, L. (2018a). Zein/gum Arabic nanoparticle-stabilized Pickering emulsion with thymol as an antibacterial delivery system. *Carbohydrate Polymers*, *200*(June), 416–426. https://doi.org/10.1016/j.carbpol.2018.08.025

Li, J., Yang, Y., Lu, L., Ma, Q., & Zhang, J. (2018b). Preparation, characterization and systemic application of self-assembled hydroxyethyl starch nanoparticles-loaded flavonoid morin for hyperuricemia therapy. *International Journal of Nanomedicine*, *13*, 2129–2141. https://doi.org/10.2147/IJN.S158585

Li, J., Zhai, J., Dyett, B., Yang, Y., Drummond, C. J., & Conn, C. E. (2021). Effect of gum Arabic or sodium alginate incorporation on the physicochemical and curcumin retention properties of liposomes. *LWT*, *139*(October 2020), 110571. https://doi.org/10.1016/j.lwt.2020.110571

Li, L., Ni, R., Shao, Y., & Mao, S. (2014). Carrageenan and its applications in drug delivery. *Carbohydrate Polymers*, *103*(1), 1–11. https://doi.org/10.1016/J.CARBPOL.2013.12.008

Li, X., Ge, S., Yang, J., Chang, R., Liang, C., Xiong, L., Zhao, M., Li, M., & Sun, Q. (2017). Synthesis and study the properties of StNPs/gum nanoparticles for salvianolic acid B-oral delivery system. *Food Chemistry*, *229*, 111–119. https://doi.org/10.1016/j.foodchem.2017.02.059

Li, X., Liu, Y., Yu, Y., Chen, W., Liu, Y., & Yu, H. (2019). Nanoformulations of quercetin and cellulose nanofibers as healthcare supplements with sustained antioxidant activity. *Carbohydrate Polymers*, *207*(November 2018), 160–168. https://doi.org/10.1016/j.carbpol.2018.11.084

Lira, M. C. B., Santos-Magalhães, N. S., Nicolas, V., Marsaud, V., Silva, M. P. C., Ponchel, G., & Vauthier, C. (2011). Cytotoxicity and cellular uptake of newly synthesized fucoidan-coated nanoparticles. *European Journal of Pharmaceutics and Biopharmaceutics : Official Journal of Arbeitsgemeinschaft Fur Pharmazeutische Verfahrenstechnik e.V*, *79*(1), 162–170. https://doi.org/10.1016/J.EJPB.2011.02.013

Liu, C., Ge, S., Yang, J., Xu, Y., Zhao, M., Xiong, L., & Sun, Q. (2016a). Adsorption mechanism of polyphenols onto starch nanoparticles and enhanced antioxidant activity under adverse conditions. *Journal of Functional Foods*, *26*, 632–644. https://doi.org/10.1016/j.jff.2016.08.036

Liu, Q., Cai, W., Zhen, T., Ji, N., Dai, L., Xiong, L., & Sun, Q. (2020). Preparation of debranched starch nanoparticles by ionic gelation for encapsulation of epigallocatechin gallate. *International Journal of Biological Macromolecules*, *161*, 481–491. https://doi.org/10.1016/j.ijbiomac.2020.06.070

Liu, Q., Han, C., Zhang, H., Jing, Y., & Tian, Y. (2019a). Encapsulation of curcumin in zein/ caseinate/ sodium alginate nanoparticles with improved physicochemical and controlled release properties. *Food Hydrocolloids*, *93*, 432–442. https://doi.org/10.1016/J.FOODHYD.2019.02.003

Liu, Q., Li, M., Xiong, L., Qiu, L., Bian, X., Sun, C., & Sun, Q. (2018). Oxidation modification of debranched starch for the preparation of starch nanoparticles with calcium ions. *Food Hydrocolloids*, *85*(July), 86–92. https://doi.org/10.1016/j.foodhyd.2018.07.004

Liu, Q., Li, M., Xiong, L., Qiu, L., Bian, X., Sun, C., & Sun, Q. (2019b). Characterization of cationic modified debranched starch and formation of complex nanoparticles with κ-carrageenan and low methoxyl pectin. *Journal of Agricultural and Food Chemistry*, *67*(10), 2906–2915. https://doi.org/10.1021/acs.jafc.8b05045

Mahdavinia, G. R., Etemadi, H., & Soleymani, F. (2015). Magnetic/pH-responsive beads based on caboxymethyl chitosan and κ-carrageenan and controlled drug release. *Carbohydrate Polymers*, *128*, 112–121. https://doi.org/10.1016/J.CARBPOL.2015.04.022

Maity, S., Mukhopadhyay, P., Kundu, P. P., & Chakraborti, A. S. (2017). Alginate coated chitosan core-shell nanoparticles for efficient oral delivery of naringenin in diabetic animals-An in vitro and in vivo approach. *Carbohydrate Polymers*, *170*, 124–132. https://doi.org/10.1016/J.CARBPOL.2017.04.066

Manikkam, R., & Pitchai, D. (2014). Catechin loaded chitosan nanoparticles as a novel drug delivery system for cancer: Synthesis and in vitro and in vivo characterization, *World Journal of Pharmacy and Pharmaceutical Science*, *3*(2), 1553–1577.

Matou, S., Helley, D., Chabut, D., Bros, A., & Fischer, A. M. (2002). Effect of fucoidan on fibroblast growth factor-2-induced angiogenesis in vitro. *Thrombosis Research*, *106*(4–5), 213–221. https://doi.org/10.1016/S0049-3848(02)00136-6

Meenambal, R., & Srinivas Bharath, M. M. (2020). Nanocarriers for effective nutraceutical delivery to the brain. *Neurochemistry International*, *140*, 104851. https://doi.org/10.1016/J.NEUINT.2020.104851

Mehrnia, M. A., Jafari, S. M., Makhmal-Zadeh, B. S., & Maghsoudlou, Y. (2017). Rheological and release properties of double nano-emulsions containing crocin prepared with Angum gum, Arabic gum and whey protein. *Food Hydrocolloids*, *66*, 259–267. https://doi.org/10.1016/j.foodhyd.2016.11.033

Miremadi, F., & Shah *. (2012). Applications of inulin and probiotics in health and nutrition. *International Food Research Journal*, *19*(4), 1337–1350.

Mukhopadhyay, P., Chakraborty, S., Bhattacharya, S., Mishra, R., & Kundu, P. P. (2015). pH-sensitive chitosan/alginate core-shell nanoparticles for efficient and safe oral insulin delivery. *International Journal of Biological Macromolecules*, *72*, 640–648. https://doi.org/10.1016/J.IJBIOMAC.2014.08.040

Mukwaya, V., Wang, C., & Dou, H. (2019). Saccharide-based nanocarriers for targeted therapeutic and diagnostic applications. *Polymer International*, *68*(3), 306–319. https://doi.org/10.1002/PI.5702

Mungure, T. E., Roohinejad, S., Bekhit, A. E. D., Greiner, R., & Mallikarjunan, K. (2018). Potential application of pectin for the stabilization of nanoemulsions. *Current Opinion in Food Science*, *19*, 72–76. https://doi.org/10.1016/j.cofs.2018.01.011

Mustafa, A., Tomescu, A., Mustafa, E., Cherim, M., & Sîrbu, R. (2016). Polyelectrolyte complexes based on chitosan and natural polymers. *European Journal of Interdisciplinary Studies*, *4*(1), 100. https://doi.org/10.26417/EJIS.V4I1.P100-107

Naibaho, J., Safithri, M., & Wijaya, C. H. (2019). Anti-hyperglycemic activity of encapsulated Java tea-based drink on malondialdehyde formation. *Journal of Applied Pharmaceutical Science*, *9*(4), 88–95. https://doi.org/10.7324/JAPS.2019.90411

Naqash, F., Masoodi, F. A., Rather, S. A., Wani, S. M., & Gani, A. (2017). Emerging concepts in the nutraceutical and functional properties of pectin: A Review. *Carbohydrate Polymers*, *168*, 227–239. https://doi.org/10.1016/j.carbpol.2017.03.058

Nasr, M. (2016). Development of an optimized hyaluronic acid-based lipidic nanoemulsion co-encapsulating two polyphenols for nose to brain delivery. *Drug Delivery*, *23*(4), 1444–1452. https://doi.org/10.3109/10717544.2015.1092619

Nayak, A. K., & Hasnain, M. S. (2019). Plant polysaccharides in drug delivery applications. *Springer Briefs in Applied Sciences and Technology*, 19–23. https://doi.org/10.1007/978-981-10-6784-6_2

Nesamony, J., Singh, P. R., Nada, S. E., Shah, Z. A., & Kolling, W. M. (2012). Calcium alginate nanoparticles synthesized through a novel interfacial cross-linking method as a potential protein drug delivery system. *Journal of Pharmaceutical Sciences*, *101*(6), 2177–2184. https://doi.org/10.1002/JPS.23104

Ng, S. W., Selvarajah, G. T., Hussein, M. Z., Yeap, S. K., & Omar, A. R. (2020). In vitro evaluation of curcumin-encapsulated chitosan nanoparticles against feline infectious peritonitis virus and pharmacokinetics study in cats. *BioMed Research International*, *2020*. https://doi.org/10.1155/2020/3012198

Nikbakht Nasrabadi, M., Goli, S. A. H., Sedaghat Doost, A., & Van der Meeren, P. (2020). Characterization and enhanced functionality of nanoparticles based on linseed protein and linseed gum biocomplexes. *International Journal of Biological Macromolecules*, *151*, 116–123. https://doi.org/10.1016/j.ijbiomac.2020.02.149

Okamoto-Schalch, N. O., Pinho, S. G. B., de Barros-Alexandrino, T. T., Dacanal, G. C., Assis, O. B. G., & Martelli-Tosi, M. (2020). Production and characterization of chitosan-TPP/cellulose nanocrystal system for encapsulation: A case study using folic acid as active compound. *Cellulose*, *27*(10), 5855–5869. https://doi.org/10.1007/s10570-020-03173-y

Ozturk, B., Argin, S., Ozilgen, M., & McClements, D. J. (2015). Formation and stabilization of nanoemulsion-based vitamin E delivery systems using natural biopolymers: Whey protein isolate and gum Arabic. *Food Chemistry*, *188*, 256–263. https://doi.org/10.1016/j.foodchem.2015.05.005

Paques, J. P., van der Linden, E., van Rijn, C. J., & Sagis, L. M. (2014). Preparation methods of alginate nanoparticles. *Advances in Colloid and Interface Science*, *209*, 163–171. https://doi.org/10.1016/J.CIS.2014.03.009

Pamunuwa, G., Anjalee, N., Kukulewa, D., Edirisinghe, C., Shakoor, F., & Karunaratne, D. N. (2020). Tailoring of release properties of folic acid encapsulated nanoparticles via changing alginate and pectin composition in the matrix. *Carbohydrate Polymer Technologies and Applications*, *1*(September), 100008. https://doi.org/10.1016/j.carpta.2020.100008

Panlasigui, L. N., Baello, O. Q., Dimatangal, J. M., & Dumelod, B. D. (2003). Blood cholesterol and lipid-lowering effects of carrageenan on human volunteers. *Asia Pacific J Clin Nutr*, *12*(2), 209–214.

Pateiro, M., Gómez, B., Munekata, P. E. S., Barba, F. J., Putnik, P., Kovačević, D. B., & Lorenzo, J. M. (2021). Nanoencapsulation of promising bioactive compounds to improve their absorption, stability, functionality and the appearance of the final food products. *Molecules*, *26*(6). https://doi.org/10.3390/MOLECULES26061547

Patel, A. S., Lakshmibalasubramaniam, S. P., & Nayak, B. (2020a). Steric stabilization of phycobiliprotein loaded liposome through polyethylene glycol adsorbed cellulose nanocrystals and their impact on the gastrointestinal tract. *Food Hydrocolloids*, *98*(July 2019), 105252. https://doi.org/10.1016/j.foodhyd.2019.105252

Patel, A. S., Lakshmibalasubramaniam, S. P., Nayak, B., Tripp, C., Kar, A., & Sappati, P. K. (2020b). Improved stability of phycobiliprotein within liposome stabilized by polyethylene glycol adsorbed cellulose nanocrystals. *International Journal of Biological Macromolecules*, *163*, 209–218. https://doi.org/10.1016/j.ijbiomac.2020.06.262

Paukkonen, H., Ukkonen, A., Szilvay, G., Yliperttula, M., & Laaksonen, T. (2017). Hydrophobin-nanofibrillated cellulose stabilized emulsions for encapsulation and release of BCS class II drugs. *European Journal of Pharmaceutical Sciences*, *100*, 238–248. https://doi.org/10.1016/j.ejps.2017.01.029

Perrechil, F. A., Maximo, G. J., Sato, A. C. K., & Cunha, R. L. (2020). Microbeads of sodium caseinate and κ-carrageenan as a β-carotene carrier in aqueous systems. *Food and Bioprocess Technology*, *13*(4), 661–669. https://doi.org/10.1007/S11947-020-02426-9

Pinheiro, A. C., Bourbon, A. I., Cerqueira, M. A., Maricato, E., Nunes, C., Coimbra, M. A., & Vicente, A. A. (2015). Chitosan/fucoidan multilayer nanocapsules as a vehicle for controlled release of bioactive compounds. *Carbohydrate Polymers*, *115*, 1–9. https://doi.org/10.1016/J.CARBPOL.2014.07.016

Prakashkumar, N., Sivamaruthi, B. S., Chaiyasut, C., & Suganthy, N. (2021). Decoding the neuroprotective potential of methyl gallate-loaded starch nanoparticles against beta amyloid-induced oxidative stress-mediated apoptosis: An in vitro study. *Pharmaceutics*, *13*(3), 1–24. https://doi.org/10.3390/pharmaceutics13030299

Qi, H., & Sheng, J. (2015). The antihyperlipidemic mechanism of high sulfate content ulvan in rats. *Marine Drugs*, *13*(6), 3407–3421. https://doi.org/10.3390/MD13063407

Qi, H., Zhang, Q., Zhao, T., Hu, R., Zhang, K., & Li, Z. (2006). In vitro antioxidant activity of acetylated and benzoylated derivatives of polysaccharide extracted from Ulva pertusa (Chlorophyta). *Bioorganic & Medicinal Chemistry Letters*, *16*(9), 2441–2445. https://doi.org/10.1016/J.BMCL.2006.01.076

Qin, Y., Xiong, L., Li, M., Liu, J., Wu, H., Qiu, H., Mu, H., Xu, X., & Sun, Q. (2018). Preparation of bioactive polysaccharide nanoparticles with enhanced radical scavenging activity and antimicrobial activity. *Journal of Agricultural and Food Chemistry*, *66*(17), 4373–4383. https://doi.org/10.1021/ACS.JAFC.8B00388

Qiu, C., Chang, R., Yang, J., Ge, S., Xiong, L., Zhao, M., Li, M., & Sun, Q. (2017). Preparation and characterization of essential oil-loaded starch nanoparticles formed by short glucan chains. *Food Chemistry*, *221*, 1426–1433. https://doi.org/10.1016/j.foodchem.2016.11.009

Qiu, C., Hu, Y., Jin, Z., McClements, D. J., Qin, Y., Xu, X., & Wang, J. (2019). A review of green techniques for the synthesis of size-controlled starch-based nanoparticles and their applications as nanodelivery systems. *Trends in Food Science and Technology*, *92*(October 2018), 138–151. https://doi.org/10.1016/j.tifs.2019.08.007

Raei, M., Shahidi, F., Farhoodi, M., Jafari, S. M., & Rafe, A. (2017). Application of whey protein-pectin nano-complex carriers for loading of lactoferrin. *International Journal of Biological Macromolecules*, *105*, 281–291. https://doi.org/10.1016/j.ijbiomac.2017.07.037

Raeisi, S., Ojagh, S. M., Quek, S. Y., Pourashouri, P., & Salaün, F. (2019). Nano-encapsulation of fish oil and garlic essential oil by a novel composition of wall material: Persian gum-chitosan. *LWT, 116*(August), 108494. https://doi.org/10.1016/j.lwt.2019.108494

Rajabi, H., Jafari, S. M., Feizy, J., Ghorbani, M., & Mohajeri, S. A. (2020). Preparation and characterization of 3D graphene oxide nanostructures embedded with nanocomplexes of chitosan gum Arabic biopolymers. *International Journal of Biological Macromolecules, 162*, 163–174. https://doi.org/10.1016/J.IJBIOMAC.2020.06.076

Rao, W., Zhang, W., Poventud-Fuentes, I., Wang, Y., Lei, Y., Agarwal, P., Weekes, B., Li, C., Lu, X., Yu, J., & He, X. (2014). Thermally responsive nanoparticle-encapsulated curcumin and its combination with mild hyperthermia for enhanced cancer cell destruction. *Acta Biomaterialia, 10*(2), 831–842. https://doi.org/10.1016/J.ACTBIO.2013.10.020

Rapaka, R. S., & Coates, P. M. (2006). Dietary supplements and related products: A brief summary. *Life Sciences, 78*(18), 2026–2032. https://doi.org/10.1016/J.LFS.2005.12.017

Rashidi, L. (2021). Different nano-delivery systems for delivery of nutraceuticals. *Food Bioscience, 43*, 101258. https://doi.org/10.1016/J.FBIO.2021.101258

Ravikumar, R., Ganesh, M., Ubaidulla, U., Young Choi, E., & Tae Jang, H. (2017). Preparation, characterization, and in vitro diffusion study of nonwoven electrospun nanofiber of curcumin-loaded cellulose acetate phthalate polymer. *Saudi Pharmaceutical Journal, 25*(6), 921–926. https://doi.org/10.1016/j.jsps.2017.02.004

Rebouillat, S., & Pla, F. (2013). State of the art manufacturing and engineering of nanocellulose: A review of available data and industrial applications. *Journal of Biomaterials and Nanobiotechnology, 4*(2), 165–188. https://doi.org/10.4236/jbnb.2013.42022

Rezaeinia, H., Ghorani, B., Emadzadeh, B., & Tucker, N. (2019). Electrohydrodynamic atomization of Balangu (Lallemantia royleana) seed gum for the fast-release of *Mentha longifolia* L. essential oil: Characterization of nano-capsules and modeling the kinetics of release. *Food Hydrocolloids, 93*(February), 374–385. https://doi.org/10.1016/j.foodhyd.2019.02.018

Roberfroid, M. (2007). Prebiotics: The concept revisited. *Journal of Nutrition, 137*(3) Supplement 2. https://doi.org/10.1093/jn/137.3.830s

Roshanpour, S., Tavakoli, J., Beigmohammadi, F., & Alaei, S. (2021). Improving antioxidant effect of phenolic extract of Mentha piperita using nanoencapsulation process. *Journal of Food Measurement and Characterization, 15*(1), 23–32. https://doi.org/10.1007/S11694-020-00606-X

Rostamabadi, H., Falsafi, S. R., & Jafari, S. M. (2019a). Nanostructures of starch for encapsulation of food ingredients. In *Biopolymer Nanostructures for Food Encapsulation Purposes*. Elsevier Inc. https://doi.org/10.1016/B978-0-12-815663-6.00015-X

Rostamabadi, H., Falsafi, S. R., & Jafari, S. M. (2019b). Starch-based nanocarriers as cutting-edge natural cargos for nutraceutical delivery. *Trends in Food Science and Technology, 88*(December 2018), 397–415. https://doi.org/10.1016/j.tifs.2019.04.004

Ruocco, N., Costantini, S., Guariniello, S., & Costantini, M. (2016). Polysaccharides from the marine environment with pharmacological, cosmeceutical and nutraceutical potential. *Molecules, 21*(5). https://doi.org/10.3390/MOLECULES21050551

Rupérez, P., Gómez-Ordóñez, E., & Jiménez-Escrig, A. (2013). Biological activity of algal sulfated and nonsulfated polysaccharides. *Bioactive Compounds from Marine Foods: Plant and Animal Sources*, 219–247. https://doi.org/10.1002/9781118412893.CH11

Şahin, Ş., Bilgiç, E., Salimi, K., Tuncel, A., Karaosmanoğlu, B., Taşkıran, E. Z., Korkusuz, P., & Korkusuz, F. (2019). Development, characterization and research of efficacy on in vitro cell culture of glucosamine carrying hyaluronic acid nanoparticles. *Journal of Drug Delivery Science and Technology, 52*(May), 393–402. https://doi.org/10.1016/j.jddst.2019.05.007

Saikia, C., Das, M. K., Ramteke, A., & Maji, T. K. (2017). Evaluation of folic acid tagged aminated starch/ZnO coated iron oxide nanoparticles as targeted curcumin delivery system. *Carbohydrate Polymers, 157*, 391–399. https://doi.org/10.1016/j.carbpol.2016.09.087

Sanjeeva, K. K., Lee, J. S., Kim, W. S., & Jeon, Y. J. (2017). The potential of brown-algae polysaccharides for the development of anticancer agents: An update on anticancer effects reported for fucoidan and laminaran. *Carbohydrate Polymers, 177*, 451–459. https://doi.org/10.1016/J.CARBPOL.2017.09.005

Santos, M. B., Geraldo de Carvalho, M., & Garcia-Rojas, E. E. (2021). Carboxymethyl tara gum-lactoferrin complex coacervates as carriers for vitamin D3: Encapsulation and controlled release. *Food Hydrocolloids, 112*(September 2020). https://doi.org/10.1016/j.foodhyd.2020.106347

Santoyo-Aleman, D., Sanchez, L. T., & Villa, C. C. (2019). Citric-acid modified banana starch nanoparticles as a novel vehicle for β-carotene delivery. *Journal of the Science of Food and Agriculture, 99*(14), 6392–6399. https://doi.org/10.1002/jsfa.9918

Sarei, F., Dounighi, N. M., Zolfagharian, H., Khaki, P., & Bidhendi, S. M. (2013). Alginate nanoparticles as a promising adjuvant and vaccine delivery system. *Indian Journal of Pharmaceutical Sciences*, *75*(4), 442. https://doi.org/10.4103/0250-474X.119829

Sarika, P. R., & James, N. R. (2015). Preparation and characterisation of gelatin-gum Arabic aldehyde nanogels via inverse miniemulsion technique. *International Journal of Biological Macromolecules*, *76*, 181–187. https://doi.org/10.1016/j.ijbiomac.2015.02.038

Sarmento, B., Ribeiro, A., Veiga, F., Sampaio, P., Neufeld, R., & Ferreira, D. (2007). Alginate/chitosan nanoparticles are effective for oral insulin delivery. *Pharmaceutical Research*, *24*(12), 2198–2206. https://doi.org/10.1007/s11095-007-9367-4

Sedaghat Doost, A., Muhammad, D. R. A., Stevens, C. V., Dewettinck, K., & Van der Meeren, P. (2018). Fabrication and characterization of quercetin loaded almond gum-shellac nanoparticles prepared by antisolvent precipitation. *Food Hydrocolloids*, *83*, 190–201. https://doi.org/10.1016/j.foodhyd.2018.04.050

Semyonov, D., Ramon, O., Shoham, Y., & Shimoni, E. (2014). Enzymatically synthesized dextran nanoparticles and their use as carriers for nutraceuticals. *Food and Function*, *5*(10), 2463–2474. https://doi.org/10.1039/c4fo00103f

Shan, W., Zhu, X., Liu, M., Li, L., Zhong, J., Sun, W., Zhang, Z., & Huang, Y. (2015). Overcoming the diffusion barrier of mucus and absorption barrier of epithelium by self-assembled nanoparticles for oral delivery of insulin. *ACS Nano*, *9*(3), 2345–2356. https://doi.org/10.1021/acsnano.5b00028

Shao, P., Zhang, H., Niu, B., & Jin, W. (2018). Physical stabilities of taro starch nanoparticles stabilized Pickering emulsions and the potential application of encapsulated tea polyphenols. *International Journal of Biological Macromolecules*, *118*, 2032–2039. https://doi.org/10.1016/j.ijbiomac.2018.07.076

Shen, Y., Wang, X., Xie, A., Huang, L., Zhu, J., & Chen, L. (2008). Synthesis of dextran/Se nanocomposites for nanomedicine application. *Materials Chemistry and Physics*, *109*(2–3), 534–540. https://doi.org/10.1016/J.MATCHEMPHYS.2008.01.016

Shingel, K. I. (2004). Current knowledge on biosynthesis, biological activity, and chemical modification of the exopolysaccharide, pullulan. *Carbohydrate Research*, *339*(3), 447–460. https://doi.org/10.1016/J.CARRES.2003.10.034

Shwetha, H. J., Shilpa, S., Mukherjee, M. B., Ambedkar, R., Raichur, A. M., & Lakshminarayana, R. (2020). Fabrication of chitosan nanoparticles with phosphatidylcholine for improved sustain release, basolateral secretion, and transport of lutein in Caco-2 cells. *International Journal of Biological Macromolecules*, *163*, 2224–2235. https://doi.org/10.1016/J.IJBIOMAC.2020.09.040

Silva, R. C., Trevisan, M. G., & Garcia, J. S. (2020). β-galactosidase encapsulated in carrageenan, pectin and carrageenan/pectin: Comparative study, stability and controlled release. *Anais Da Academia Brasileira de Ciencias*, *92*(1). https://doi.org/10.1590/0001-3765202020180609

Singh, A. R., Desu, P. K., Nakkala, R. K., Kondi, V., Devi, S., Alam, M. S., Hamid, H., Athawale, R. B., & Kesharwani, P. (2021). Nanotechnology-based approaches applied to nutraceuticals. *Drug Delivery and Translational Research*, 0123456789, *12*, 485–499. https://doi.org/10.1007/s13346-021-00960-3

Song, Y., & Chen, L. (2015). Effect of net surface charge on physical properties of the cellulose nanoparticles and their efficacy for oral protein delivery. *Carbohydrate Polymers*, *121*, 10–17. https://doi.org/10.1016/j.carbpol.2014.12.019

Soukoulis, C., Tsevdou, M., Andre, C. M., Cambier, S., Yonekura, L., Taoukis, P. S., & Hoffmann, L. (2017). Modulation of chemical stability and in vitro bioaccessibility of beta-carotene loaded in kappa-carrageenan oil-in-gel emulsions. *Food Chemistry*, *220*, 208–218. https://doi.org/10.1016/J.FOODCHEM.2016.09.175

Spizzirri, U. G., Aiello, F., Carullo, G., Facente, A., & Restuccia, D. (2021). Nanotechnologies: An innovative tool to release natural extracts with antimicrobial properties. *Pharmaceutics*, *13*(2), 1–32. https://doi.org/10.3390/pharmaceutics13020230

Srivastava, P., & Malviya, R. (2011). Sources of pectin, extraction and its applications in pharmaceutical industry: An overview. *Indian Journal of Natural Products and Resources*, *2*(1), 10–18.

Su, J., Guo, Q., Chen, Y., Dong, W., Mao, L., Gao, Y., & Yuan, F. (2020). Characterization and formation mechanism of lutein pickering emulsion gels stabilized by β-lactoglobulin-gum Arabic composite colloidal nanoparticles. *Food Hydrocolloids*, *98*(17), 105276. https://doi.org/10.1016/j.foodhyd.2019.105276

Sun, B., Hou, Q., He, Z., Liu, Z., & Ni, Y. (2014). Cellulose nanocrystals (CNC) as carriers for a spirooxazine dye and its effect on photochromic efficiency. *Carbohydrate Polymers*, *111*, 419–424. https://doi.org/10.1016/j.carbpol.2014.03.051

Sun, F., Ju, C., Chen, J., Liu, S., Liu, N., Wang, K., & Liu, C. (2012). Nanoparticles based on hydrophobic alginate derivative as nutraceutical delivery vehicle: Vitamin D3 loading. *Artificial Cells, Blood Substitutes, and Immobilization Biotechnology*, *40*(1–2), 113–119. https://doi.org/10.3109/10731199.2011.597759

Suwantong, O., Opanasopit, P., Ruktanonchai, U., & Supaphol, P. (2007). Electrospun cellulose acetate fiber mats containing curcumin and release characteristic of the herbal substance. *Polymer, 48*(26), 7546–7557. https://doi.org/10.1016/j.polymer.2007.11.019

Taepaiboon, P., Rungsardthong, U., & Supaphol, P. (2007). Vitamin-loaded electrospun cellulose acetate nanofiber mats as transdermal and dermal therapeutic agents of vitamin A acid and vitamin E. *European Journal of Pharmaceutics and Biopharmaceutics, 67*(2), 387–397. https://doi.org/10.1016/j.ejpb.2007.03.018

Taheri, A., & Jafari, S. M. (2019a). Gum-based nanocarriers for the protection and delivery of food bioactive compounds. *Advances in Colloid and Interface Science, 269*, 277–295. https://doi.org/10.1016/j.cis.2019.04.009

Taheri, A., & Jafari, S. M. (2019b). Nanostructures of gums for encapsulation of food ingredients. In *Biopolymer Nanostructures for Food Encapsulation Purposes.* Elsevier Inc. https://doi.org/10.1016/B978-0-12-815663-6.00018-5

Tan, C., Xie, J., Zhang, X., Cai, J., & Xia, S. (2016). Polysaccharide-based nanoparticles by chitosan and gum Arabic polyelectrolyte complexation as carriers for curcumin. *Food Hydrocolloids, 57*, 236–245. https://doi.org/10.1016/j.foodhyd.2016.01.021

Tanna, B., & Mishra, A. (2019). Nutraceutical potential of seaweed polysaccharides: Structure, bioactivity, safety, and toxicity. *Comprehensive Reviews in Food Science and Food Safety, 18*(3), 817–831. https://doi.org/10.1111/1541-4337.12441

Teixé-Roig, J., Oms-Oliu, G., Ballesté-Muñoz, S., Odriozola-Serrano, I., & Martín-Belloso, O. (2020). Improving the in vitro bioaccessibility of β-carotene using pectin added nanoemulsions. *Foods, 9*(4), 447. https://doi.org/10.3390/foods9040447

Thomas, D., Latha, M. S., & Thomas, K. K. (2018). Synthesis and in vitro evaluation of alginate-cellulose nanocrystal hybrid nanoparticles for the controlled oral delivery of rifampicin. *Journal of Drug Delivery Science and Technology, 46*(October 2017), 392–399. https://doi.org/10.1016/j.jddst.2018.06.004

Urbánková, L., Sedláček, T., Kašpárková, V., & Bordes, R. (2021). Formation of oleogels based on emulsions stabilized with cellulose nanocrystals and sodium caseinate. *Journal of Colloid and Interface Science, 596*, 245–256. https://doi.org/10.1016/j.jcis.2021.02.104

Usov, A. I., & Zelinsky, N. D. (1992). *Sulfated polysaccharides of the red seaweeds. Food Hydrocolloids, 6*, 9–23. https://doi.org/10.1016/S0268-005X(09)80055-6

Venkatesan, J., Lowe, B., Anil, S., Manivasagan, P., Kheraif, A. A. Al, Kang, K.-H., & Kim, S.-K. (2015). Seaweed polysaccharides and their potential biomedical applications. *Starch: Stärke, 67*(5–6), 381–390. https://doi.org/10.1002/STAR.201400127

Vijayakurup, V., Thulasidasan, A. T., Shankar, M., Retnakumari, A. P., Devika Nandan, C., Somaraj, J., Antony, J., Alex, V. V, Vinod, B. S., Liju, V. B., Sundaram, S., Kumar, G. S. V., & Anto, R. J. (2019). Chitosan encapsulation enhances the bioavailability and tissue retention of curcumin and improves its efficacy in preventing B[a]P-induced lung carcinogenesis, *Cancer Prev Res (Philia), 12*(4), 225–236. https://doi.org/10.1158/1940-6207.CAPR-18-0437

Vongsetskul, T., Phurayar, P., Chutimasakul, T., Tuchinda, P., Uamsiri, S., Kumkate, S., Pearngam, P., Jitpibull, J., Samphaongern, C., & Tangboriboonrat, P. (2016). *Acanthus ebracteatus* Vahl. extract-loaded cellulose acetate ultrafine fibers as a topical carrier for controlled-release applications. *Polymer Bulletin, 73*(12), 3319–3331. https://doi.org/10.1007/s00289-016-1658-7

Wang, K., Liu, M., & Mo, R. (2020). Polysaccharide-based biomaterials for protein delivery. *Medicine in Drug Discovery, 7*, 100031. https://doi.org/10.1016/J.MEDIDD.2020.100031

Wang, R., Wang, Y., Guo, W., & Zeng, M. (2021). Stability and bioactivity of carotenoids from *Synechococcus* sp. PCC 7002 in zein/NaCas/gum Arabic composite nanoparticles fabricated by pH adjustment and heat treatment antisolvent precipitation. *Food Hydrocolloids, 117*(January), 106663. https://doi.org/10.1016/j.foodhyd.2021.106663

Wang, T., Wu, C., Fan, G., Li, T., Gong, H., & Cao, F. (2018). Ginkgo biloba extracts-loaded starch nanospheres: Preparation, characterization, and in vitro release kinetics. *International Journal of Biological Macromolecules, 106*, 148–157. https://doi.org/10.1016/j.ijbiomac.2017.08.012

Wang, W., Liu, F., & Gao, Y. (2016). Quercetagetin loaded in soy protein isolate–κ-carrageenan complex: Fabrication mechanism and protective effect. *Food Research International, C*(83), 31–40. https://doi.org/10.1016/J.FOODRES.2016.02.012

Wang, W., Xue, C., & Mao, X. (2020). Radioprotective effects and mechanisms of animal, plant and microbial polysaccharides. *International Journal of Biological Macromolecules, 153*, 373–384. https://doi.org/10.1016/j.ijbiomac.2020.02.203

Wei, Y., Cai, Z., Wu, M., Guo, Y., Wang, P., Li, R., Ma, A., & Zhang, H. (2020). Core-shell pea protein-carboxymethylated corn fiber gum composite nanoparticles as delivery vehicles for curcumin. *Carbohydrate Polymers*, *240*(March), 116273. https://doi.org/10.1016/j.carbpol.2020.116273

Wei, Y., Guo, A., Liu, Z., Mao, L., Yuan, F., Gao, Y., & Mackie, A. (2021). Structural design of zein-cellulose nanocrystals core–shell microparticles for delivery of curcumin. *Food Chemistry*, *357*(17), 129849. https://doi.org/10.1016/j.foodchem.2021.129849

Wen, Y., & Oh, J. K. (2014). Recent strategies to develop polysaccharide-based nanomaterials for biomedical applications. *Macromolecular Rapid Communications*, *35*(21), 1819–1832. https://doi.org/10.1002/MARC.201400406

Wijesekara, I., Pangestuti, R., & Kim, S. K. (2011). Biological activities and potential health benefits of sulfated polysaccharides derived from marine algae. *Carbohydrate Polymers*, *84*(1), 14–21. https://doi.org/10.1016/J.CARBPOL.2010.10.062

Wijesinghe, W. A. J. P., Athukorala, Y., & Jeon, Y. J. (2011). Effect of anticoagulative sulfated polysaccharide purified from enzyme-assistant extract of a brown seaweed Ecklonia cava on Wistar rats. *Carbohydrate Polymers*, *86*(2), 917–921. https://doi.org/10.1016/J.CARBPOL.2011.05.047

Winuprasith, T., Khomein, P., Mitbumrung, W., Suphantharika, M., Nitithamyong, A., & McClements, D. J. (2018). Encapsulation of vitamin D3 in pickering emulsions stabilized by nanofibrillated mangosteen cellulose: Impact on in vitro digestion and bioaccessibility. *Food Hydrocolloids*, *83*, 153–164. https://doi.org/10.1016/j.foodhyd.2018.04.047

Wu, J., Wang, Y., Yang, H., Liu, X., & Lu, Z. (2017). Preparation and biological activity studies of resveratrol loaded ionically cross-linked chitosan-TPP nanoparticles. *Carbohydrate Polymers*, *175*, 170–177. https://doi.org/10.1016/J.CARBPOL.2017.07.058

Wu, L. X., Qiao, Z. R., Cai, W. D., Qiu, W. Y., & Yan, J. K. (2019). Quaternized curdlan/pectin polyelectrolyte complexes as biocompatible nanovehicles for curcumin. *Food Chemistry*, *291*(April), 180–186. https://doi.org/10.1016/j.foodchem.2019.04.029

Xiang, C., Gao, J., Ye, H., Ren, G., Ma, X., Xie, H., Fang, S., Lei, Q., & Fang, W. (2020). Development of ovalbumin-pectin nanocomplexes for vitamin D3 encapsulation: Enhanced storage stability and sustained release in simulated gastrointestinal digestion. *Food Hydrocolloids*, *106*(March), 105926. https://doi.org/10.1016/j.foodhyd.2020.105926

Xu, J., Li, X., Xu, Y., Wang, A., Xu, Z., Wu, X., Li, D., Mu, C., & Ge, L. (2021). Dihydromyricetin-loaded pickering emulsions stabilized by dialdehyde cellulose nanocrystals for preparation of antioxidant gelatin–based edible films. *Food and Bioprocess Technology*, *14*(9), 1648–1661. https://doi.org/10.1007/s11947-021-02664-5

Yan, J. K., Qiu, W. Y., Wang, Y. Y., & Wu, J. Y. (2017). Biocompatible polyelectrolyte complex nanoparticles from lactoferrin and pectin as potential vehicles for antioxidative curcumin. *Journal of Agricultural and Food Chemistry*, *65*(28), 5720–5730. https://doi.org/10.1021/acs.jafc.7b01848

Yang, J., He, H., Gu, Z., Cheng, L., Li, C., Li, Z., & Hong, Y. (2020). Conjugated linoleic acid loaded starch-based emulsion nanoparticles: In vivo gastrointestinal controlled release. *Food Hydrocolloids*, *101*(November 2019), 105477. https://doi.org/10.1016/j.foodhyd.2019.105477

Yang, W., Sousa, A. M. M., Fan, X., Jin, T., Li, X., Tomasula, P. M., & Liu, L. S. (2017). Electrospun ultra-fine cellulose acetate fibrous mats containing tannic acid-Fe3+ complexes. *Carbohydrate Polymers*, *157*, 1173–1179. https://doi.org/10.1016/j.carbpol.2016.10.078

Yew, H.-C., & Misran, M. (2016). Preparation and characterization of pH dependent κ-carrageenan-chitosan nanoparticle as potential slow release delivery carrier. *Iranian Polymer Journal*, *12*(25), 1037–1046. https://doi.org/10.1007/S13726-016-0489-6

Yuan, Y., Li, H., Zhu, J., Liu, C., Sun, X., Wang, D., & Xu, Y. (2020). Fabrication and characterization of zein nanoparticles by dextran sulfate coating as vehicles for delivery of curcumin. *International Journal of Biological Macromolecules*, *151*, 1074–1083. https://doi.org/10.1016/j.ijbiomac.2019.10.149

Zainuddin, N., Ahmad, I., Zulfakar, M. H., Kargarzadeh, H., & Ramli, S. (2021). Cetyltrimethylammonium bromide-nanocrystalline cellulose (CTAB-NCC) based microemulsions for enhancement of topical delivery of curcumin. *Carbohydrate Polymers*, *254*(November 2020). https://doi.org/10.1016/j.carbpol.2020.117401

Zaki, N. M. (2014). Progress and problems in nutraceuticals delivery. *Journal of Bioequivalence and Bioavailability*, *6*(3), 075–077. https://doi.org/10.4172/jbb.1000183

Zamani Ghaleshahi, A., & Rajabzadeh, G. (2020). The influence of sodium alginate and genipin on physicochemical properties and stability of WPI coated liposomes. *Food Research International*, *130*. https://doi.org/10.1016/J.FOODRES.2019.108966

Zhang, N., Wardwell, P. R., & Bader, R. A. (2013). Polysaccharide-based micelles for drug delivery. *Pharmaceutics*, *5*(2), 329. https://doi.org/10.3390/PHARMACEUTICS5020329

Zhang, Y., Chi, C., Huang, X., Zou, Q., Li, X., & Chen, L. (2017). Starch-based nanocapsules fabricated through layer-by-layer assembly for oral delivery of protein to lower gastrointestinal tract. *Carbohydrate Polymers*, *171*, 242–251. https://doi.org/10.1016/j.carbpol.2017.04.090

Zhang, Y., Yang, Y., Tang, K., Hu, X., & Zou, G. (2008). Physicochemical characterization and antioxidant activity of quercetin-loaded chitosan nanoparticles. *Journal of Applied Polymer Science*, *107*(2), 891–897. https://doi.org/10.1002/APP.26402

Zhu, F. (2017). Encapsulation and delivery of food ingredients using starch based systems. *Food Chemistry*, *229*, 542–552. https://doi.org/10.1016/j.foodchem.2017.02.101

Zhu, J., Tang, X., Jia, Y., Ho, C. T., & Huang, Q. (2020). Applications and delivery mechanisms of hyaluronic acid used for topical/transdermal delivery: A review. *International Journal of Pharmaceutics*, *578*(September 2019), 119127. https://doi.org/10.1016/j.ijpharm.2020.119127

Zhu, K., Ye, T., Liu, J., Peng, Z., Xu, S., Lei, J., Deng, H., & Li, B. (2013). Nanogels fabricated by lysozyme and sodium carboxymethyl cellulose for 5-fluorouracil controlled release. *International Journal of Pharmaceutics*, *441*(1–2), 721–727. https://doi.org/10.1016/j.ijpharm.2012.10.022

Zia, K. M., Zia, F., Zuber, M., Rehman, S., & Ahmad, S. (2015). Alginate based polyurethanes: A review of recent advances and perspective. *International Journal of Biological Macromolecules*, *79*, 377–387. https://doi.org/10.1016/J.IJBIOMAC.2015.04.076

11 Metal Nanoparticles in Encapsulation and Delivery Systems of Food Ingredients and Nutraceuticals

*H.C. Ananda Murthy**, *Gezahegn Tadesse Ayanie, Tegene Desalegn Zeleke, Yilkal Dessie Sintayehu, and C.R. Ravikumar*

CONTENTS

11.1 Introduction ... 301
11.2 Nanoparticle Synthesis Methods ... 303
11.2.1 Chemical Methods ... 303
11.2.2 Green (Biogenic) Synthesis Methods ... 304
11.3 Nutraceuticals ... 305
11.4 Encapsulation of Food Ingredients ... 306
11.5 Nanotechnologies in Food Science ... 306
11.6 Nanoencapsulated Nutrients ... 308
11.7 Nanoencapsulation Techniques ... 309
11.8 Metal Nanoparticles in Food Encapsulation ... 311
11.8.1 Gold Nanoparticles (Au NPs) ... 311
11.8.2 Silver Nanoparticles (Ag NPs) ... 312
11.8.3 Titanium Dioxide Nanoparticles (TiO_2) ... 315
11.8.4 Copper/Copper Oxide Nanoparticles ... 316
11.8.5 Zinc Oxide Nanoparticles ... 316
11.9 Future Perspectives ... 319
11.10 Conclusion ... 319
References ... 319

11.1 INTRODUCTION

Nanoparticles are one of the basic elements of nanotechnology, having numerous applications in several fields such as physics, chemistry, biology, agriculture, food sciences, health science, and engineering (Uikey and Vishwakarma 2016). In nanoscience, nanoparticles have been broadly studied and rapidly developed since the century of Richard Feynman, who stated that "There's Plenty of Room at the Bottom". Since 1959, the nanonization of particles has attracted the interest of researchers. Nanoparticles that exist in the 1–100 nm size range are obviously known by their high surface area and modified features that enable them to have new physical and chemical properties (Fadeel *et al.* 2007; Durgadevi and Mani 2018; Yaqoob *et al.* 2020; Zhao *et al.* 2014). The properties of these

* Corresponding author: anandkps350@gmail.com

DOI: 10.1201/9781003244721-11

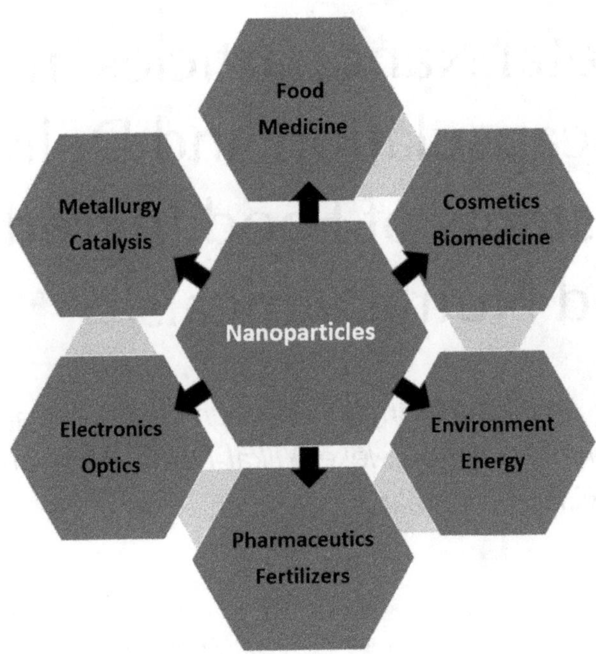

FIGURE 11.1 Different applications of nanoparticles.

particles are highly modified from those of their parent materials, which enables them to be used in a variety of applications, as shown in Figure 11.1. Due to these properties, nanoparticles are used in a variety of fields for several uses such as sensors, photocatalysis, antidiabetic, anticancer, antibacterial, antioxidant, antiviral, anti-inflammatory, fabrics, catalysis, antiseptic sprays, antiangiogenesis, electronics, photovoltaic devices, drug delivery, fuel cells, industrial lithography, optical devices, cosmetics, etc. (Naseer *et al.* 2018; Venkateasan, Prabakaran, and Sujatha 2017; Yoneyama and Katsumata 2006). In contrast to bulk materials, nanoparticles have high catalytic efficiency with effective viabilities in different applications. Due to this reason, current researchers are interested in fabricating numerous nanosized materials by using environmentally friendly based green techniques (Selim *et al.* 2020; Murthy *et al.* 2020).

Along with nanotechnology development, several forms of nanoparticles from polymers, inorganics, organics, carbon tubes, nanocrystals, and dendrimers have been effectively designed and used in different fields. Among these, inorganic nanoparticles are widely studied due to their unique mechanical, optical, electrical, chemical, magnetic, and catalytic properties. In addition, these nanoparticles have distinctive shape, crystal structure, band gap energy, plasmon excitation, and dielectric properties that offer them unlimited utilization in different areas as a platform for biomedical, drug delivery, and food encapsulation (Abdel-Fattah *et al.* 2017; Drahansky *et al.* 2016).

Nanoparticles can freely move in free state, exhibit particular surface areas, and have quantum confinement effects. Nowadays, a variety of nanoparticles are fabricated and manipulated within the size range of 1–100 nm. Based on the parent materials they are derived from, nanoparticles can be classified into several groups such as nanotubes, fullerenes, quantum dots, ceramic, liposomes, polymers, metallic, and metal oxide nanoparticles (Mauricio *et al.* 2018).

Among these metal nanoparticles (MNPs) are widely studied because of their unique physicochemical, electrical, and optical properties. Metallic nanoparticles such as gold (Au) (Zuo *et al.* 2017), silver (Ag) (Elemike *et al.* 2017), copper (Cu) (Ananda Murthy *et al.* 2018), iron (Fe) (Das *et al.* 2017), aluminum (Al) (Ansari *et al.* 2015), silica (Si) (Shi and Bedford 2018), cerium (Ce) (Das *et al.* 2013), manganese (Mn) (Shen *et al.* 2012), nickel (Ni) (Shafey 2020), platinum (Pt), and palladium (Pd) (Li and Tang 2014) are the most noble and widely used due to their distinctiveness. These nanoparticles can be easily synthesized by either top-down or bottom-up methods. The top-down approach involves the breaking of bulk materials into small pieces using techniques such as milling, grinding, sputtering, etching, and laser ablation, while the bottom-up or wet method involves the assembling of atoms or molecules from the basic pattern into nanosized particles. Of the two approaches, the bottom-up technique is preferred and more widely used due to the controlled morphology of nanoparticles, using fast and less expensive equipment during synthesis, which offers a high yield, having fewer defects, and having a more homogeneous chemical composition with short-range order (Murthy *et al.* 2020; Urabe and Aziz 2019). In bottom-up techniques, chemical techniques such as pyrolysis, hydrothermal, microwave, precipitation, microemulsion, sol-gel, solvothermal, electrochemical, and biogenic synthesis methods (such as plant extracts, microorganisms, algae, and fungi) are used for the production of different MNPs. In contrast to chemical techniques, researchers have been shifting their attention to the green methods route due to its environmentally benign properties, being simple with rapid, easy synthesis, using less expensive equipment, and using non-toxic chemicals (Agarwal, Venkat Kumar, and Rajeshkumar 2017). Figure 11.2 shows how MNPs were synthesized using both chemical and biogenic methods and were effectively designed for encapsulation and drug transport applications.

11.2 NANOPARTICLE SYNTHESIS METHODS

11.2.1 Chemical Methods

Several techniques have been developed for the production of nanoparticles. The best-known and most commonly used method is a chemical technique, which is widely used for synthesizing nanoparticles with at least one dimension. In this method, firstly, the template is assembled, forming a thermodynamically distinct new phase, which gradually grows to offer the desired nanoparticles (Karunagaran, Rajendran, and Sen 2017). Different classes of chemical methods, such as polyol, radiolytic, microemulsion, microwave-assisted, and electrochemical synthesis, are put forward for the fabrication of various nanoparticles. In this synthesis, the nucleation and agglomeration of the

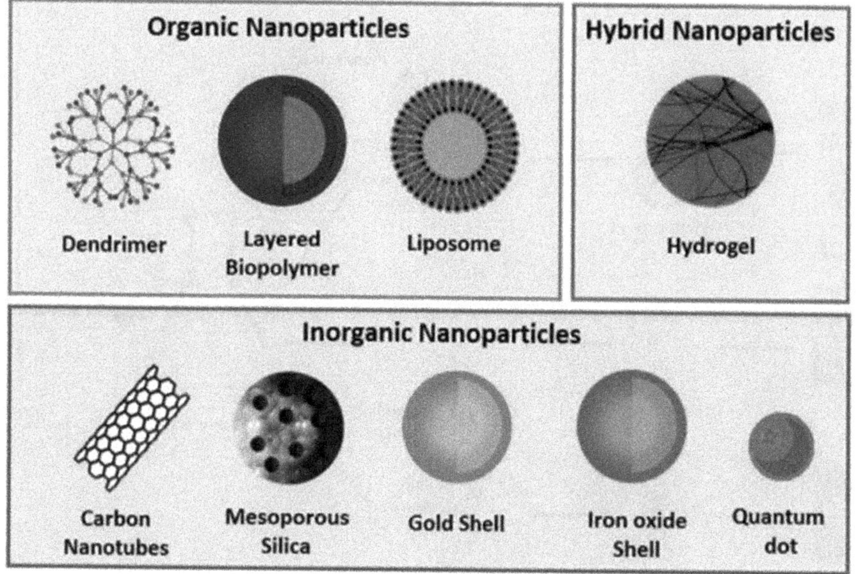

FIGURE 11.2 Different classes of organic and inorganic nanoparticles. (From Makvandi, P., *et al.*, *Environmental Chemistry Letters*, 19, 1, 583–611, 2021.)

nanoparticles are based on temperature, pH, concentration, and the physicochemical properties of the reducing agent (Liyanage *et al.* 2019).

$$m\mathrm{Me}^{n+} + n\mathrm{Red} \rightarrow m\mathrm{Me}^{0} + n\mathrm{O}_{x} \quad (11.1)$$

The MNPs formed by reaction (11.1) contain the reduced metal precursor salt and reducing agent, which prevent the over-agglomeration of nanoparticles. The chemical method is easy, low cost, high yield, and offers pure, thermally stable products. However, the chemicals used during the synthesis of nanoparticles are hazardous to health and toxic to the environment (Neha Desai *et al.* 2021; Nasrollahzadeh, Momeni, and Sajadi 2017).

11.2.2 Green (Biogenic) Synthesis Methods

Recently, biologically safe, low-cost methods with environmentally benign biosynthesis have been adopted and developed in the field of nanotechnology. In these methods, plant extracts, bacteria, fungi, and algae are used widely for the synthesis of metal nanoparticles. In contrast to chemical methods, biosynthesis methods are attracting the attention of many researchers due to their non-toxicity and use of low energy and temperature (Bouafia, Laouini, and Ouahrani 2020). In this biosynthesis method, extracts from different parts of plants, bacteria, algae, and fungi have been used as reducing and capping agents, as shown in Figure 11.3.

The bioactive molecules found in the extracts interact with the metal ions formed in the solution of precursor salt. Especially, the polyphenols of the extracts interact with the metal ions and form a complex with hydroxyl groups (-OH⁻) (Ghotekar *et al.* 2019). The complex undergoes oxidation by losing hydrogen and forming metal–oxygen nanoparticles. But, under suitable circumstances, the metal loses the oxygen atom and exists in the zerovalent oxidation state, which forms MNPs. Among the templates used in the biosynthesis of nanoparticles, plant extracts is preferred to microorganisms and fungi, which need proper isolation and suitable conditions and media for culturing, followed by a long reaction time (Tu 2019; Shivappa, Gangaiah, and Siddaramanna 2018).

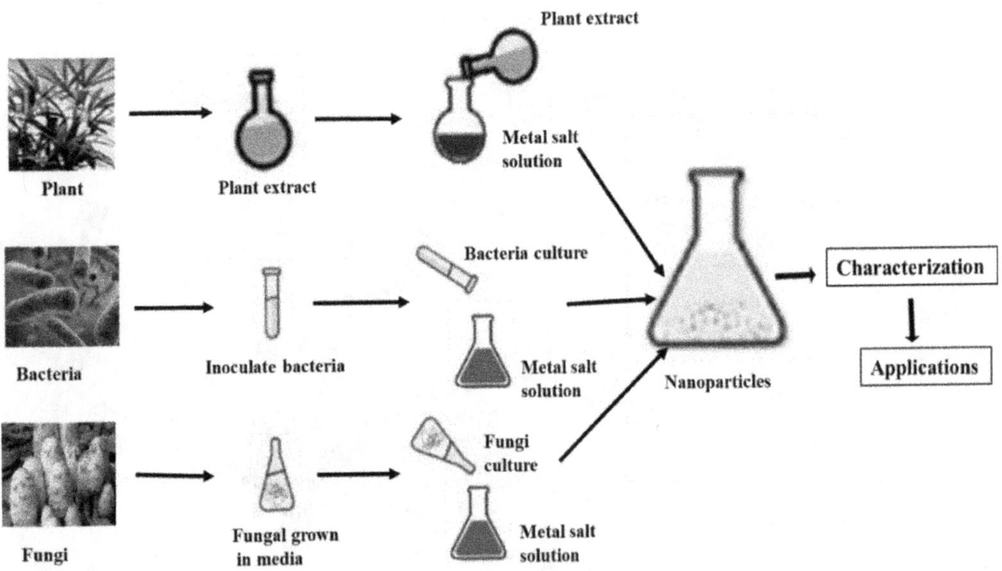

FIGURE 11.3 Plant, bacteria, and fungi extract–mediated biogenic synthesis of nanoparticles.

11.3 NUTRACEUTICALS

The term *nutraceuticals* is related to functional foods and was formed from two words, *nutrition* and *pharmaceuticals*, in 1989 by Stephen DeFelice, who is the founder and chairman of the Foundation for Innovation in Medicine. Due to their similar ingredients and ability to provide health benefits, the definition and subclassification of nutraceuticals are connected. According to the Institute of Food Technologists (IFT), functional foods are defined as "foods and food components that improves a health benefit and disease prevention beyond its basic nutrition" (Zhang and McClements 2016). Functional foods from rice, wheat, kidney beans, soybeans, lentils, carrots, chocolate, broccoli, citrus fruits, and nuts contain potentially disease-preventing chemical constituents (Jampilek, Kos, and Kralova 2019). These food items contain active ingredients such as collagen hydrolysate, carotenoids, and fatty acids that improve body immunity. Fundamentally, foods are determined by parameters such as standard of constituents, nutritional values, flavor, texture, taste, and appearance, which enable them to be effective in disease prevention and health improvement (Das *et al.* 2012; Arihara 2014).

Nutraceuticals contain a variety of metabolites, such as dietary fiber, vitamins, minerals, amino acids, peptides, proteins, carbohydrates, oils, and phytochemicals, as shown in Figure 11.4. These molecules are essential for life and beneficial in defense and provide desirable health. The desired functional food is achieved by inducing bioactive compounds that are used in the improvement of health and the treatment of different diseases, such as cancer, diabetes, oxidative stress, infectious diseases, obesity, heart disease, stroke, inflammation, etc. However, the efficacy in preventing diseases and improving health status depends on the appropriate utilization and preserving the bioavailability of those bioactive compounds (Al-Sheddi *et al.* 2018; Paolino *et al.* 2021).

For decades, nutraceuticals have attracted the interest of researchers concerning nutritional values, safety, therapeutic effects, and delivery systems in the body. The biological function of

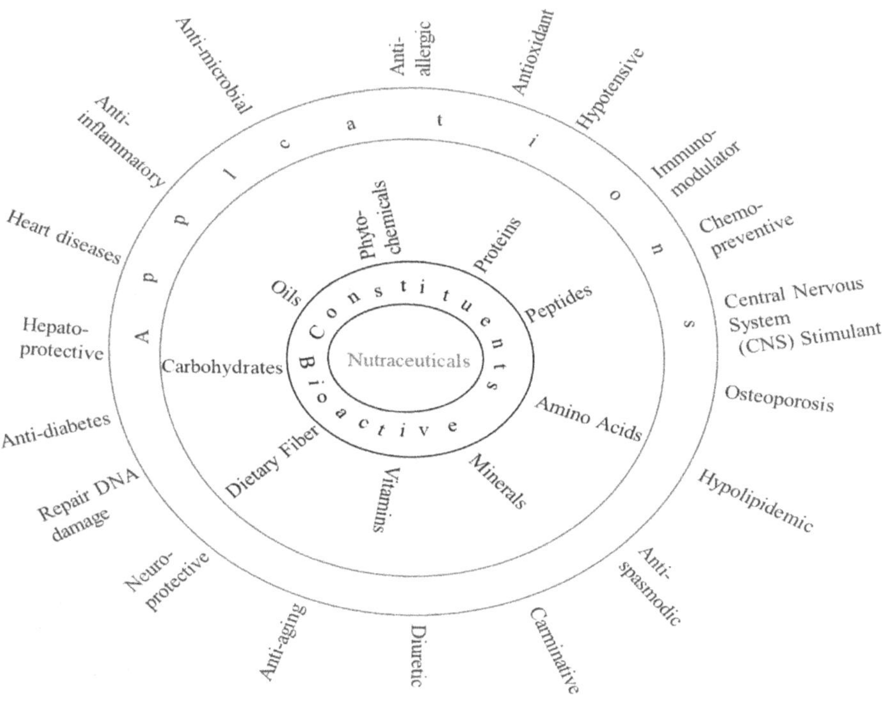

FIGURE 11.4 Some constituents of nutraceuticals and their applications. (From Nile, S.H. *et al.*, *Nanotechnologies in Food Science: Applications, Recent Trends, and Future Perspectives. Nano-Micro Letters*, 12, 2020.)

nutraceuticals is fundamentally based on solubility, bioavailability, target delivery, biocompatibility, and controlled release. But, the bioactive compounds are highly reactive and interact with any oxidizing agents, which affects their shelf life. Due to this justification, nutraceutical substances are enclosed within a variety of organic molecules, polymers, and nanomaterials. Of these, nanomaterials are preferred because of their bioavailability, small size, surface chemistry, durability, and adsorption characteristics (Rashidi 2021; Alfei, Schito, and Zuccari 2021; Ravichandran 2010).

11.4 ENCAPSULATION OF FOOD INGREDIENTS

In previous decades, food ingredients were encapsulated by different macro- and microencapsulates. In the food industry, encapsulation technology designed the macro- and microencapsulates for packaging and preservation (Alfei, Schito, and Zuccari 2021; Hoseinnejad, Jafari, and Katouzian 2018). Of the two, microencapsulation is the more accepted approach due to its appropriate coating, which enables it to protect against any external factors that affect the shelf life of food ingredients. The microsized material used for coating might exist in different forms, such as membrane, film, layer, shell, coat, and carrier, whereas the food ingredients resemble core particles, mobile particles, internal phase, and fill materials (Timilsena, Haque, and Adhikari 2020).

Micro-coating materials can enclose the core materials in all solid, liquid, and gaseous states. An example is a polymer substance used as film that helps to separate the core materials to protect them from external exposure (Henrique Rodrigues do Amaral, Lopes Andrade, and Costa de Conto 2019). The micro-coating is thus determined as a core-shell material structure, in which the food ingredients are enclosed within the shell part, which protects it from external deteriorating factors arising from bacteria, fungi, sunlight, and moisture, and minimizes the rate of loss due to evaporation of internal phase materials, pH, and oxidizing agents. The idea of microencapsulation was raised in the 1930s, and began with the fabrication of non-fading materials. Since then, microencapsulation has become a powerful food-preserving technique, ensuring food security (Desai and Park 2005). It offers a suitable concentration and similar distribution of food ingredients, which implies that the capsule materials should be carefully selected. As a benchmark, the capsule materials should be non-reactive and compatible with food ingredients (Chen *et al.* 2019).

Based on the type of food ingredients and their physicochemical properties, a variety of encapsulation approaches have been probed. Of these techniques, the encapsulating of bioactive ingredients using liquid film is most often selected in the food industry. The common recently developed encapsulation methods are coating, spray drying, air suspension, spray chilling and spray cooling, extrusion coating, liposome entrapment, coacervation, inclusion complexation, and ionotropic gelation. Further, encapsulation can also be classified into microencapsulation (1 µm–5 mm) and nanoencapsulation (<1 µm) based on size, and its applications are shown in Figure 11.5 (Walia *et al.* 2019; Primoži ̌ 2021).

Microencapsulated core materials are slowly absorbed and slightly soluble, with weakly controlled release and imperfect targeting of specific sites. Due to these limitations, researchers were motivated to develop nanoencapsulation to overcome the drawbacks determined in microencapsulation. Some reports indicate attempts to solve the limitations encountered in microencapsulation by using nanoencapsulation of food ingredients in different sectors (Ravichandran *et al.* 2019; Kashapov *et al.* 2021).

11.5 NANOTECHNOLOGIES IN FOOD SCIENCE

Nanotechnology is currently the leading area in almost all fields due to its numerous applications. Nanotechnology emerged with the manufacturing of nanosized materials that were used in several fields. Nowadays, developments and revolutions in food science are assisted by nanotechnology. However, the nanomaterials fabricated in nanotechnology are capable of providing safe, healthy, and high-quality foods. In contrast to conventional methods used for food processing in the food industry, nanotechnology is more advanced and safer. For the last two decades, nanomaterials have

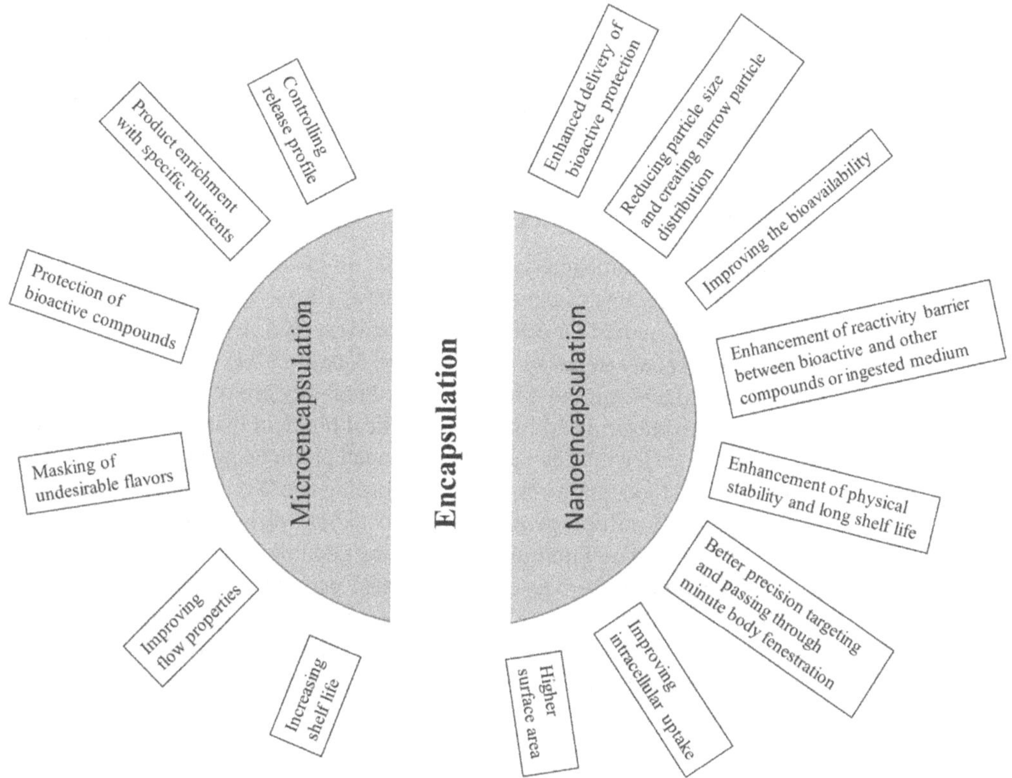

FIGURE 11.5 Some roles of micro- and nanoencapsulation. (From Timilsena, Y.P. *et al.*, *Food and Nutrition Sciences*, 11, 06, 481–508, 2020; Karunaratne, D.N. *et al.*, *Nutrient Delivery through Nanoencapsulation Nutrient Delivery*, Elsevier, 2017.)

been widely used for food encapsulation (Ubbink and Krüger 2006). In addition, nanomaterials are known in the field of biomedical applications, especially for drug delivery and pharmaceuticals. Nowadays, nanomaterials are used as food additives and preservation applications in the food industry. As shown in the literature, the bioavailability of flavors and bioactive compounds is well preserved in nanoencapsulated food (Jones and Jew 2007; Assadpour and Mahdi Jafari 2019).

Since 2000, the uses of nanomaterials in the food industry have become more familiar. Nanomaterials developed at the nanodimensions have unique features unlike those of their counterparts. Because of their large surface area to volume ratio, the materials have unique physicochemical properties, such as optical, magnetic, solubility, reactivity, strength, toxicity, bioavailability, etc. (Ravichandran 2010). Nanonization has been extensively expanded in both developed and developing countries, and provides a door for the development and application of nanoscale materials in different areas. Subsequently, in some sectors, such as microelectronics and pharmaceuticals, different nanomaterials have been used for several applications. Therefore, in many food industrial sectors, the fabricated nanomaterials are widely used for stable food encapsulation and delivery. The Acceptable Daily Intake (ADI) of nanoparticles from the food system is regulated by the Food and Drug Administration (FDA) (Jampilek, Kos, and Kralova 2019; Jafari 2017).

In the era of nanotechnology, researchers have designed different nanomaterials and concentrated on the encapsulation of drugs and food. This indicates that nanotechnology is constantly improving the novel properties of nanomaterials to enhance their performance. For example, a liposomal formulation for doxorubicin was ratified by the U.S. FDA for the treatment of Kaposi sarcoma. At the same time, albumin-based nano-optic protein-based paclitaxel was approved for the treatment of different cancers (Abdel-Fattah *et al.* 2017). Regardless of effective treatment and development of technology, the mechanistic interaction of nanoparticles with cells is still not well determined. Several potential nanomaterials have been investigation concerning protection of food from different contaminants. A review studied the production of nanomaterials such as silver, gold, copper, titanium, silicon, magnesium, iron, carbon nanotubes, zinc oxide, copper oxide, and titanium oxides for food encapsulation and delivery because of their potential disinfection activity (Jin and Maduraiveeran 2018).

11.6 NANOENCAPSULATED NUTRIENTS

Nanomaterial applications such as antibacterial, antifungal, antioxidant, and other significant health-related roles have been widely investigated. Due to their non-toxicity and large surface area, nanoparticles have been very well studied for drug and food delivery. But still, the accessibility and availability at the commercial level are in question (Khare and Vasisht 2014).

Manufacturing a balanced and high-quality food in the food industry is a serious problem worldwide. Additionally, microbes like bacteria and fungi are a critical problem that can cause food constituents to deteriorate, inducing toxicity. Thus, in the face of such problems, global food security is focusing on food preservation and food quality concerns. According to a World Health Organization (WHO) report, approximately 600 million people were infected by various diseases because of contaminated food they consumed. The Food and Agriculture Organization (FAO) also reported that 1.3 million tons of food produced for consumers is wasted per year (Brigger, Dubernet, and Couvreur 2002; Dhakal and He 2020; Teixeira 2018). Encapsulation is an important method to enhance shelf life and produce healthy food. In the field of food science, researchers are interested in developing a mechanism that can be used to manufacture high-quality, safe food that is stable, environmentally tolerable, and viable.

A few decades ago, food items were microencapsulated using different organic molecules and polymers. However, health problems due to low food quality and safety are still a serious problem worldwide. In order to solve such problems, nanotechnology finds an opportunity in the way it improves food quality and safety. The exposure of food to environmental contaminants such as temperature, odors, dust, moisture, physical exchange, light, microorganisms, and humidity affects its ingredients (Timilsena, Haque, and Adhikari 2020; Reis and Neufeld 2006).

In the food industry, manufacturing is intended to produce high-quality and healthy food. The food is mainly preserved by using different packaging materials to increase its shelf life after production. Among them, nanomaterials are good packaging candidates that improve the efficiency of food ingredients. The functionality, stability, and good bioavailability of bioactive compounds are obtained by a variety of nanoparticle encapsulates. Currently, several types of packaging nanomaterials – active packaging, smart packaging, and intelligent packaging – are used in the food industry, as shown in Figure 11.6 (Teixeira 2018; Ameta, Rai, and Hiran n.d.; Iqbal *et al.* 2019).

Nanoparticle encapsulation is widely studied by different researchers and has become preferable because of small surface area to volume ratio, small size, shape, biocompatibility, stability, selectivity, and non-toxicity (Shnoudeh *et al.* 2019). With nanotechnology development, the food preservation system has been modernized in processing, storage, and transportation. In the food industry, parameters such as taste, sensory, stability, durability, and colors are used to determine the standards of manufactured foods. In addition to this, water solubility, oral bioavailability of nutraceutical components, and thermal stability can be modified by nanomaterials. The bioactive ingredients may be lipid soluble or water soluble, which determines the rate of release. Nanomaterials utilized for encapsulation are developed using top-down and bottom-up methods, as shown in Figure 11.7 (Pateiro *et al.* 2021; Senthilkumar *et al.* 2017).

The clearly fabricated nanomaterials can be directly used with food ingredients such as vitamins, phenols, fatty acids, etc., as coloring agents and for antioxidant and fumigant activity. Nanoparticle-encapsulated food ingredients are released at targeted specific sites and at the right time. Several nanoencapsulation approaches are used to deliver both lipophilic and hydrophilic food ingredients, which determine the rate of release and erosion mechanism. The most common encapsulation

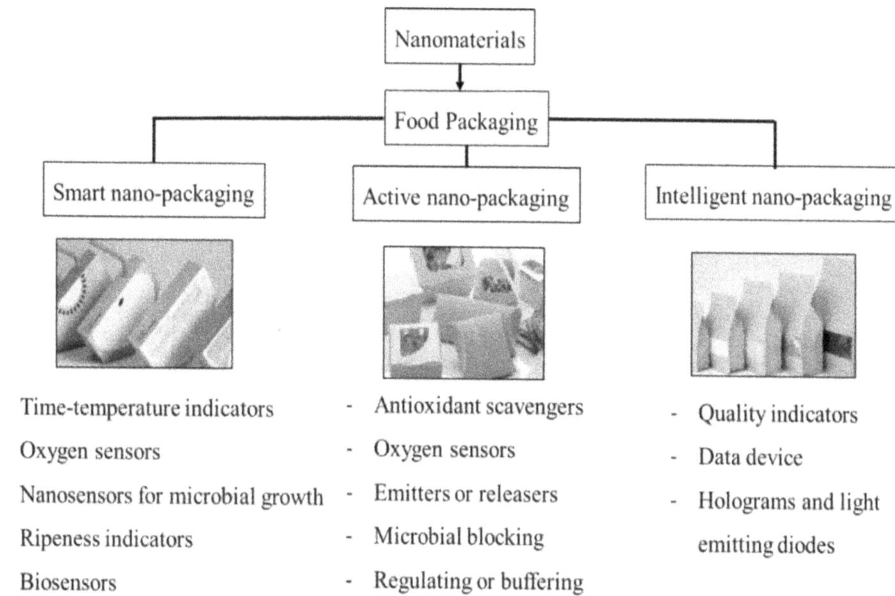

FIGURE 11.6 Different types of nanomaterials and their roles. (From He, X. and Hwang, H.M., *Journal of Food and Drug Analysis*, 24, 4, 671–681, 2016.)

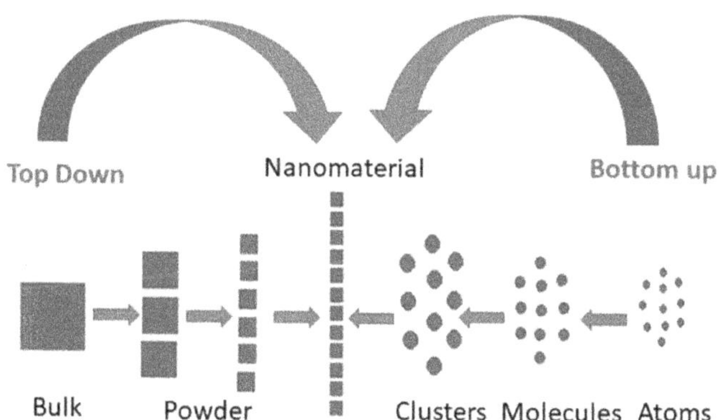

FIGURE 11.7 Nanoparticle fabrication using top-down (left to right) and bottom-up approaches (right to left).

techniques used in the food industry are coacervation, inclusion, complexation nanoprecipitation, emulsification, emulsification–solvent evaporation, and supercritical fluid (Ezhilarasi *et al.* 2013). In contrast to uncoated ingredients, the encapsulated food ingredients are well protected from environmental factors and are released at the right place and time. The rate of release is based on the binding force between core and shell nanomaterials, as shown in Figure 11.8. The physicochemical properties of the core particle are improved by nanosized encapsulate. Nanoencapsulates allow the food ingredients to enter the cell and increase their bioavailability (Patel and Mishra 2021).

11.7 NANOENCAPSULATION TECHNIQUES

The encapsulation method is used in the food industry to coat bioactive compounds and protect them from external factors, prolong the life time of food, and control release. Nanomaterials are

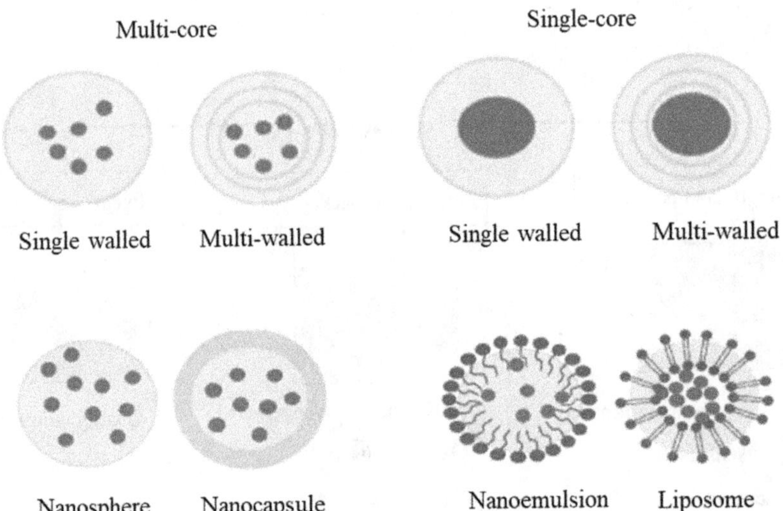

FIGURE 11.8 Different types of core and shell structures. (From Patel, M. and Mishra, S., *Asian Journal of Nanosciences and Materials*, 4, 3, 213–228, 2021.)

used to deliver and enhance the ability to reach the target specific site. The function of food ingredients is based on the encapsulation and delivery system (Assadpour and Mahdi Jafari 2019). The nanoencapsulation technique increases the efficiency of food ingredients by changing their size, distribution, bioavailability, dispersion, and solubility properties. A variety of nanoencapsulation techniques are used to coat and preserve food ingredients (Augustin and Hemar 2009). Among these, the best-known nanoencapsulation methods with their specific functions are shown in Table 11.1. In addition to food delivery using nanoencapsulation techniques, the non-toxic nanoencapsulates are

TABLE 11.1
Different Types of Nanoencapsulation Techniques and Their Function

Nanoencapsulation technique	Function	References
Emulsification	Oil-in-water used to transport lipophilic bioactive compounds Example: fish oil, like β-carotene, α-tocopherol, and dietary fats Water-in oil used to transport hydrophilic compounds such as polyphenols, water-soluble vitamins, and minerals	(Donsì *et al.* 2011)
Coacervation	Used for phase separation and differentiation of a single polyelectrolyte (protein and polysaccharide) from a solution and to form a coacervate phase that coated the core materials	(Iqbal *et al.* 2019)
Emulsification–solvent evaporation	Emulsify polymer into an aqueous phase and then, the solvent is evaporated	(Reis and Neufeld 2006)
Nanoprecipitation	Emulsify organic containing dissolved polymer, drug, and organic solvent into the aqueous external phase.The polymer is precipitated from the solution	(Patel and Mishra 2021)
Inclusion complexation	Encapsulate a supramolecular ligand and bind with some coating particles via hydrogen bonding and van der Waals force Encapsulate volatile organic molecules	(Ezhilarasi *et al.* 2013)
Freeze drying	Freeze the material to form a solid phase from the gas phase	(Yang *et al.* 2020)
Spray drying	Prepare a dry powder from a fluid material	(Sosnik and Seremeta 2015)

also added to the food matrix to improve food quality. So, the advantage of using nanoencapsulation over microencapsulation is that nanoparticles do not change the physicochemical properties of the food matrix (Ezhilarasi *et al.* 2013).

11.8 METAL NANOPARTICLES IN FOOD ENCAPSULATION

Uncoated food ingredients are easily exposed to and contaminated by different microorganisms and environmental conditions that have diverse health effects on consumers across the globe. As a result, disease is caused by contaminated food, and this can affect the economy too. Reports by the Centers for Disease Control and Prevention (CDC) in the United States indicate that around 46 million people are affected by disease due to the consumption of contaminated food (Pachaiappan *et al.* 2021). To manage these food-borne pathogens, the researchers were motivated to design sensors and encapsulate materials. Recently, in the food industry, nanoparticles have been used as a frequent food packaging system to protect from food-borne diseases, followed by enhancing the shelf life of foods (Kumar *et al.* 2020).

Researchers have used different nanoparticles for different applications because of their potential accessibility within a variety of fields. Among nanoparticles, metal- and metal oxide–based nanoparticles have attracted many researchers due to their flexible morphological structures, bioavailability, and biocompatibility. The presence of these properties enhances their performance in a range of applications. So, MNPs like silver, gold, zinc, copper, titanium oxide, zinc oxide, silicon oxide have been well studied due to their potential applications in the food industry (Galstyan, Id, and Sberveglieri 2018).

MNPs are also involved in gas sensing to protect food ingredients from contamination and monitor food quality. These particles can be used in two forms, as additives to improve the food quality and function of the food ingredients and to protect food from exposure to environmental conditions such as dry air, humidity, temperature, and microorganisms. The safety of packaged food has been ensured by modifying MNPs in the form of nanocoating, surface biocides, active packaging, nanocomposites, intelligent packaging, and bioplastics (Alfei, Schito, and Zuccari 2021; Galstyan, Id, and Sberveglieri 2018).

11.8.1 Gold Nanoparticles (Au NPs)

Gold (Au) is a transition metal. The first information on gold comes from articles written by researchers in India, Arabia, and China. Since then, Au has been used as a model gateway for the discovery of other transition metal–based nanoparticles. In the nanotechnology area, the bulky gold metal was reduced to nanoscale size and utilized in different fields (Długosz *et al.* 2020). Au NPs are ancient and widely used in different applications. This is due to their unique physicochemical properties. Au nanoparticles are widely used in biomedical applications in immunoassays, biosensors, genomics, optical bioimaging, target drug delivery, and monitoring of biological cells; as antibacterials, antifungals, antidiabetics, and antioxidants; and in food packaging (Venkateasan, Prabakaran, and Sujatha 2017; Paidari and Ibrahim 2021).

Nowadays, the processing, transportation, and quality of food are the major challenging issues in the food industry. In order to control or protect food ingredients from exposure to different external factors such as microorganisms and environmental conditions, researchers have designed nanoencapsulation techniques. In the food industry, highly stable and non-toxic Au NPs have been widely used for food coating and delivery. Au NPs have another unique feature; they interact with thiol groups to manage and operate intracellular release. This ensures that the Au NPs have the potential to prevent several contaminants from affecting food security and quality. The function of Au NPs is based on their synthesis method, size, and shape, which help in proper targeting of substrates (Huang and Pan 2020; De Oliveira 2014). Due to tunable size and novel functionality, Au NPs can deliver small as well as large biomolecules (Ghosh *et al.* 2008).

For example, Avella *et al.* reported that enzyme-assisted Au NPs detect bacteria developing in food ingredients. Au NPs on the surface of encapsulated food are released inside, then interact with

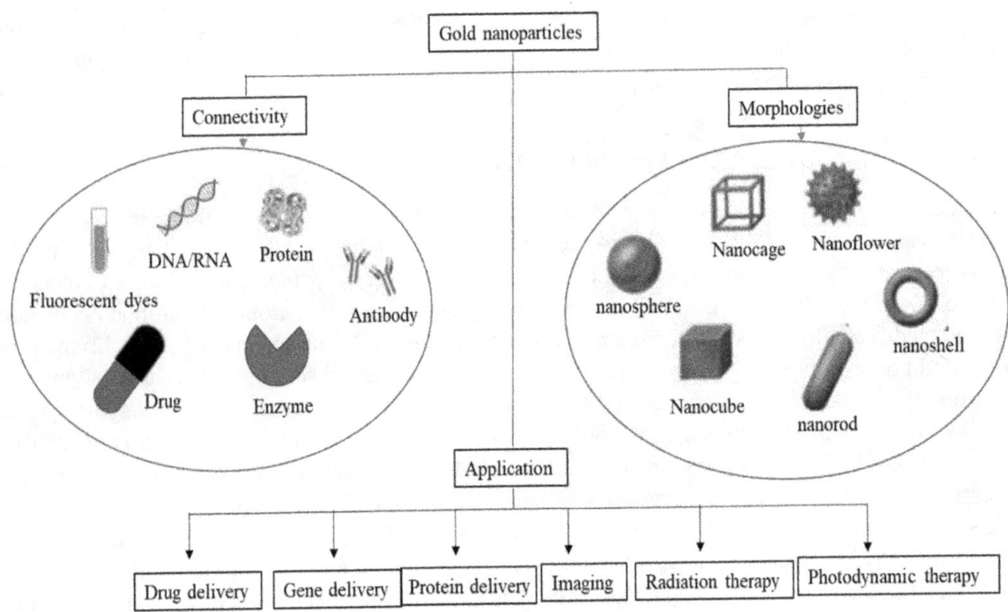

FIGURE 11.9 Connectivity, morphology, and some applications of gold nanoparticles. (From Hu, X. *et al.*, *Front. Bioeng. Biotechnol.*, 8, 990, 2020.)

the microorganisms to inhibit their growth, and finally, enhance the food's shelf life (Avella *et al.* 2005).

The unique features of Au NPs enable them to be biocompatible and easily functionalized by mimicking biological molecules such as drugs, proteins, enzymes, nucleic acids (DNA or RNA), and fluorescent dyes, as shown in Figure 11.9. The chemically and biologically synthesized Au NPs can exhibit nanocage, nanoflower, nanoshell, nanocube, and nanosphere morphology (Patil and Chandrasekaran 2020).

Au NPs can accommodate different compounds and can be used as a platform for several drugs and food ingredients. For example, antibiotics such as streptomycin, ciprofloxacin, neomycin, ampicillin, kanamycin, and vancomycin can easily conjugated to the surface of Au NPs to enhance bacterial activity against some drug-resistant bacteria. As shown in reports, Au NP–bound antibiotics have shown good growth inhibition of several bacteria (Elemike *et al.* 2019; Mittal, Chisti, and Banerjee 2013). Arief *et al.* have determined the antibacterial activity of Au NPs synthesized using *Uncaria gambir* Roxb leaf extract against bacteria. The particles showed better activity against *Escherichia coli* than *Staphylococcus aureus* because of the difference in the cell wall structure of bacteria (Arief, Nasution, and Labanni 2020). On the other hand, Arief *et al.* reported the antibacterial activity of Au NP–mediated *triethanolamine* plant leaf extract. The particles inhibit the growth of Gram-positive (*S. aureus*) and Gram-negative (*E. coli*) bacteria, as is clearly shown in Figure 11.10 (Arief, Nasution, and Labanni 2020). In detail, the inhibition zone of the green synthesized Au nanoparticles Au NPs-1 and Au NPs-2 against Gram-positive *S. aureus* were 6 and 7 mm, whereas they were 8 and 9 mm against Gram-negative *E. coli*. In the positive control, the inhibition zone for Gram-positive *S. aureus* and Gram-negative *E. coli* was 4 and 5 mm, respectively. The effective growth inhibition difference observed for Au NPs results from the components of the bacterial cell wall (Arief, Nasution, and Labanni 2020a).

11.8.2 Silver Nanoparticles (Ag NPs)

Silver (Ag) NPs have attracted the attention of researchers and have taken pole position among other nanoparticles. Ag NPs have been used in different applications for a long time, especially for

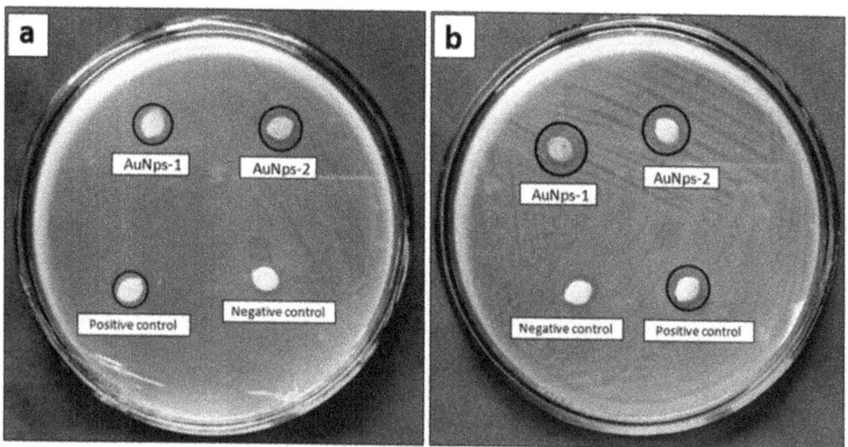

FIGURE 11.10 Antibacterial activity of gold NPs against (a) *S. aureus* and (b) *E. coli*. (From Arief, S. *et al.*, *Journal of Applied Pharmaceutical Science*, 10, 08, 124–130, 2020.)

the treatment of bacteria-borne diseases. In addition to this, Ag oxide NPs have been used for their antifungal and antiviral activity, even for HIV virus treatment (Duncan 2011). Ag NPs interact with different parts of microbial cell walls, leading to cell death. In bacteria, they bind to DNA, proteins, and enzymes, and attack membranes, causing bacteriostatic effects. Moreover, Ag NPs enter and disrupt the bacterial cytoplasmic membranes (Figure 11.11) and metabolic systems by inducing reactive oxygen species (ROS) (Sharma *et al.* 2017; Kumar *et al.* 2021).

Ag NPs induce adverse effects on bacteria by three main routes. Because of their tiny size, they disrupt the permeability and respiration of bacteria by binding to atoms of the cell membrane. Ag NPs have the ability to permeate bacterial cells and release the highly reactive Ag ion, which interacts with compounds containing phosphorus and sulfur heteroatoms and with the negatively charged cell membrane, finally resulting in the death of bacteria (Rutberg *et al.* 2008; Yoon *et al.* 2007).

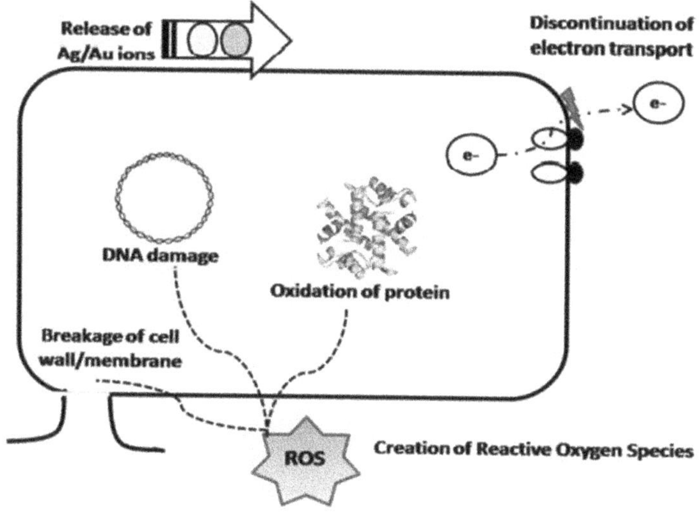

FIGURE 11.11 Cellular mechanism of interaction of Ag nanoparticles with bacterial cell. (From Sharma, C. *et al.*, *Front. Microbiol.*, 8, 1735, 2017.)

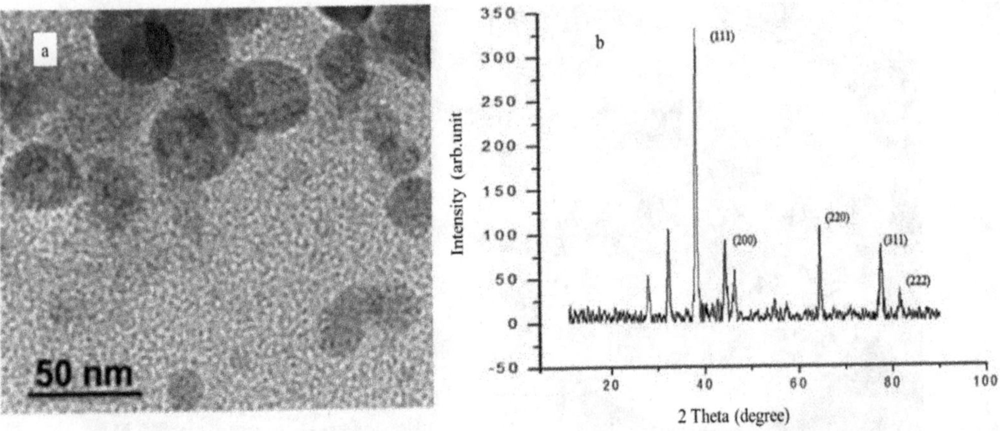

FIGURE 11.12 TEM image (a) and XRD pattern (b) of Ag nanoparticles. (Reprinted from *Results in Physics*, 15, Ravichandran, V., S. Vasanthi, S. Shalini, Syed Adnan Ali Shah, M. Tripathy, and Neeraj Paliwal, Green Synthesis, Characterization, Antibacterial, Antioxidant and Photocatalytic Activity of *Parkia speciosa* Leaves Extract Mediated Silver Nanoparticles, 102565, Copyright (2019), with permission from Elsevier.)

In the biomedical field, the biosynthesized Ag nanoparticles are utilized as antibacterial, antifungal, antioxidant, and anti-inflammatory agents. Fatimah (2016) has reported that Ag NPs synthesized using *Parkia speciosa* leaf extract were used to treat microbial-borne diseases. Phytochemicals extracted from *P. speciosa* interact with Ag ion in solution and reduce its size to the nanoscale. The extracts from these plant leaves contain several secondary bioactive molecules. Among these, biomolecules with phenolic and protein groups are commonly used as size reduction and capping agents, respectively. The electron released from the hydroxyl functional groups reduces the Ag ion to the zerovalent state. The zerovalent Ag NPs are capped and stabilized by proteins and other secondary bioactive molecules. The surface topography (transmission electron microscopy [TEM]) image of these particles indicated a spherical shape with an average size of 35 nm, as shown in Figure 11.12a. The crystallite size and its crystallinity behavior were also examined by X-ray diffraction (XRD), as shown in Figure 11.12b. Its potential activity against bacteria could also disrupt the bacterial metabolic system and facilitate cell death (Ravichandran *et al.* 2019). Manikandan *et al.* reported on the antibacterial activities of Ag NPs. Here, the NPs were synthesized using *Ficus benghalensis* prop root extract, and the inhibition activity was concentration dependent (Manikandan *et al.* 2017a).

Many years ago, people used Ag metal to preserve food for a long period of time. In addition to this, they used materials made of Ag metal for transportation over long distances. Nanoscale-sized Ag metal has been designed for sterilization and shows potent antimicrobial activity. Along with different polymers, Ag NPs have been used for food packaging purposes to prolong the shelf life of various food ingredients (Manikandan *et al.* 2017).

Some decades ago, Ag NPs were used in different food sectors as a preservative against different microorganisms, such as bacteria, fungi, yeasts, and viruses. Due to their large surface area, Ag NPs destroy the microorganism's cellular structure. Ag NPs can also be encapsulated with a variety of polymers and used for packaging purposes in different sectors (Murthy *et al.* 2020; Istiqola 2020).

Recently, Ag NPs and their hybrids have been used in food preservation because of their unique synergetic effects, like scavenging reactive molecules, inhibiting microorganisms, and in the treatment of different diseases (Aswathy *et al.* 2017). In the food industry, Ag NPs combined with low-density polyethylene are utilized for preservation and to enhance the shelf life of food ingredients. Ag NPs can also be synthesized by physical, chemical, and biological methods for food processing and food safety in the industry (Paredes *et al.* 2016).

It is noted that in the food industry, the food packaging section is the most challenging issue when food is exposed to different environmental factors. But, NPs such as Ag NPs have emerged as a primary solution due to their high efficacy in protection from foodborne diseases, as proved by different researchers (Al-Rubaye *et al.* 2020; Turrina, Berensmeier, and Schwaminger 2021; McClements and Xiao 2017).

Carbone reported that Ag NPs showed a greater antibacterial activity than natural antibacterial molecules. Additionally, the particle controls the release rate of Ag ion contained in the food as an additive. The report also indicated that Ag NPs showed high toxicity against *E. coli*, *S. aureus*, *Aspergillus niger*, and *Penicillium citrinum* at low concentration (Solomon 2015). Ag NP–packaged nuts stayed fresh for a long time in the presence of different pathogens. Also, Ag NPs combined with fluorescent carbon points showed effective antibacterial activity against *E. coli* and *S. aureus*. Similarly, the potential activity of Ag NPs conjugated with molecules having high–molecular weight electrolytes from proteins and polysaccharides against *E. coli* has been determined. Therefore, Ag NPs are used as a good food quality preservative, and therefore, they are considered as an effective food ingredient to prolong the shelf life of foods after packaging (Jayappa *et al.* 2020; S. Ahmad *et al.* 2019; Cushen *et al.* 2014).

11.8.3 Titanium Dioxide Nanoparticles (TiO_2)

TiO_2 is the most widely used inorganic material with white pigment and exists in three forms: rutile, anatase, and brookite. TiO_2 is used as a colorant in a variety of foods, such as candies, sweets, chewing gum, and coffee creamers. Additionally, the color of materials such as paint, plastic, toothpaste, and paper is enhanced by the addition of TiO_2. The American FDA also documented TiO_2 in the food list. This is due to its non-toxic properties, and it is used for drugs and food contact materials (Bekele *et al.* 2020).

Due to its multifunctional applications, researchers have aimed to fabricate TiO_2 NPs. It is well known to have a high surface area to volume ratio, highly structured electronic bands, and hygroscopic and stable characteristics. It is used in several fields, such as photocatalysis, green energy generation, cosmetics, antimicrobials, drug delivery, and food packaging, as shown in Figure 11.13. In addition to this, TiO_2 NPs could induce the development of vascular endothelial growth factor in humans (Chakra *et al.* 2017; Shi *et al.* 2013).

TiO_2 NPs are among the most widely utilized materials in food science., Properties such as lightness and brightness are induced in food ingredients by the TiO_2 NPs. In addition, TiO_2 NPs are

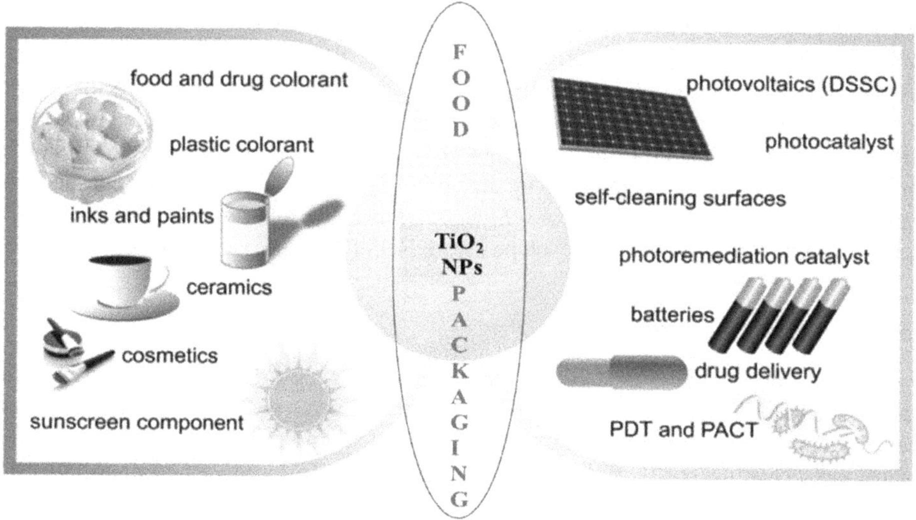

FIGURE 11.13 Different applications of TiO_2 nanoparticles. (From Ziental *et al.* 2020.)

utilized as an additive to improve food quality and for maintaining food preservation in the food industry (Jadoun *et al.* 2021).

Direct or indirect contact with TiO_2 NPs inhibits microbial growth through different mechanisms. These include blocking electron flows through the membrane, crushing the cell wall, and disordering metabolic systems by ROS. TiO_2 NPs encapsulate food ingredients and bind them through hydrogen bonding and covalent interactions to induce an antibacterial effect in packaging materials (Aravind, Amalanathan, and Mary 2021; Daneshniya, Maleki, and Nezhad 2021).

11.8.4 Copper/Copper Oxide Nanoparticles

Copper (Cu) is an inorganic transition metal that can exist in +1 and +2 oxidation states. It is reduced to its nanosize by using physical, chemical, and biogenic NP synthesis methods. Primarily, its hardness, ductility, flexibility, and rigidity enable its anticancer, antimicrobial, antidiabetic, antioxidant, and photocatalytic potential applications (Asemani and Anarjan 2019). NPs of Cu and its oxide are widely studied in the food industry because of their ability to inhibit different microorganisms. The massive surface area of Cu NPs enables them to interact with different parts of the bacterial cell structure and cause death (Ijaz *et al.* 2017).

Different physical and chemical techniques have been widely used to manufacture NPs of Cu and its oxide. Among these, laser ablation, microwave-assisted process, sol-gel, co-precipitation, vacuum vapor deposition, photochemical reduction, electrochemistry, electrospray synthesis, hydrothermal reaction, microemulsion, and chemical reduction are common. The biosynthesis of Cu NPs has been suggested as a valuable alternative to physical and chemical techniques due to its low cytotoxicity, economic prospects, environmental friendliness, enhanced biocompatibility, and high antioxidant and antimicrobial activities (Ghosh *et al.* 2020).

Recently, Cu metal and CuO NPs became the most operational agents to be utilized in the food industry for encapsulation. In addition to coating, both Cu and CuO nanoparticles are involved in the inhibition of different microorganisms (bacterial, fungal, yeast, and viral) (Kuswandi and Moradi 2019). However, Cu oxide NPs showed greater interaction with bacterial cells as compared with Cu metal NPs. Reports showed that CuO NPs inhibit the growth of both Gram-positive and Gram-negative bacteria in the food industry (Almasi, Jafarzadeh, and Mehryar 2018; Murthy *et al.* 2020). It is suggested that the mixed form of Cu oxide NPs inhibits both Gram-negative and Gram-positive bacteria by damaging cell wall components, followed by degradation of proteins, which leads to death. In the food industry, bacterial cell membranes are penetrated by an endocytosis mechanism during the Cu NP encapsulation of packaged food (Siddiqi *et al.* 2020).

Amiri *et al.* reported the antibacterial activity of CuO NPs against oral bacteria and candida species. In their discussions, the action of Cu NPs inhibited the bacterial growth that causes the deterioration of tooth enamel. As documented in the report, CuO NPs showed a greater efficiency than zinc oxide NPs against oral bacteria (Baghayeri *et al.* 2018). CuO NPs prevented biofilm formation at 0.0001–1 μg/mL concentration, as shown in Figure 11.14. Cu oxide nanoparticles inhibit oral bacteria such as *Streptococcus mutans*, *Lactobacillus casei*, and *Lactobacillus acidophilus* by hydrogen peroxide (H_2O_2) molecules developed from ROS (Baghayeri *et al.* 2018).

11.8.5 Zinc Oxide Nanoparticles

A few years ago, the FDA documented zinc oxide (ZnO) as generally recognized as safe (GRAS) and suitable for addition as a food improver. But, with the upsurge of nanonization, researchers fabricated materials at nanoscale to improve their physical and chemical properties, enhancing their activities (Jayappa *et al.* 2020; Hussain *et al.* 2019). ZnO reduced to nanoscale showed better activity in a variety of fields for different applications. In recent years, ZnO NPs have attracted the attention of many researchers because their physicochemical behaviors can be easily tuned by changing their surface morphology. ZnO NPs also have known applications for

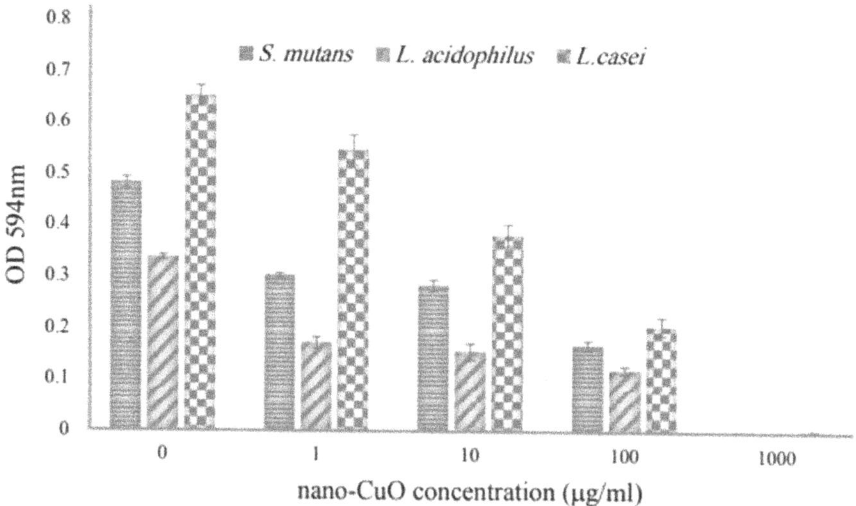

FIGURE 11.14 Antibacterial activity of CuO nanoparticles against oral bacteria (*S. mutans*, *L. casei*, and *L. acidophilus*). (From Siddiqi, K. S. *et al.*, *Agriculture and Food Security*, 9, 1, 1–15, 2020.)

antimicrobial, anti-inflammatory, antioxidant, anticancer, and antidiabetic activity (Sushma *et al.* 2016).

Furthermore, ZnO NPs are known for their semiconductor properties, which are determined by large band gap energy (3.37 eV) and excitation binding energy (60 meV). In contrast to other metal NPs, ZnO NPs are more stable and have a longer shelf life than even other organic molecules. Because of these properties, ZnO NPs are chosen for food packaging applications. To improve their packaging capacity, they are integrated with different polymer structures. In addition to their antibacterial activity, ZnO NPs are also used in food packaging materials due to having better optical properties in near-UV emission and optical transparency as well as effectiveness in electrical conductivity (Nagajyothi *et al.* 2015; Chaudhuri and Malodia 2017; Happy *et al.* 2019).

ZnO NPs can be synthesized by different physical and chemical methods. Nowadays, green (biogenic) synthesis methods are more common due to their non-toxicity, environmental friendliness, and cost-effectiveness (De Souza *et al.* 2019). For example, Sushma *et al.* manufactured ZnO NPs using zinc acetate dihydrate ($C_4H_6O_4$.Zn.$2H_2O$) as precursor salt and *Ocimum tenuflorum* as reducing and capping agent (Sushma *et al.* 2016). As another example of these biogenic synthesis methods, Tu *et al.* synthesized spherical ZnO NPs in the size range of 10–20 nm mediated by orange fruit peel extract as a reducing agent. Their antibacterial performance was tested using *S. aureus* and *E. coli* (Doan Thi *et al.* 2020). Furthermore, Ahmad and Kalra reported the potent antibacterial and antioxidant activities of ZnO NPs. The NPs were manufactured using *Euphorbia hirta* leaf extract as a reducing and capping agent, and the size of the NPs obtained was in the range of 20 to 25 nm. The crystal lattice of the synthesized ZnO NPs was effectively determined and demonstrated from the XRD pattern (Ahmad and Kalra 2020). Its cytotoxic effect was also further proved, and Azam *et al.* investigated the effect of ZnO nanoparticles against Gram-positive and Gram-negative bacteria that contaminate food. In addition, its antibacterial activity was explored by Padmavathy and Vijayaraghavan (Ahmadi and Ahmadi 2020).

The nanosized ZnO NPs protect food from bacterial contact due to their large surface reactive center. ZnO NPs release Zn^{+2}, which interacts with bacteria to hinder transport, disturb metabolism, and inhibit enzyme activity, leading to bacterial cell death. The ability of ZnO NPs to inhibit bacterial cells depends on their surface topography. Additionally, hydrogen peroxide (H_2O_2) is developed from ROS that occur on the surface of NPs. The produced H_2O_2 enters the cell membrane and destroys lipid, protein, and DNA of bacteria, which finally causes death. Due to their potential application against bacterial strains, ZnO NPs have been designed in industry for food packaging and delivery. The efficacy of ZnO NPs to penetrate the bacterial cell membrane and cause death is based on the concentration. ZnO NPs play a significant role in the food industry, helping to protect food from different bacterial infections. ZnO NPs can combine with several polymers and have indicated great antimicrobial activity in food packaging. For example, biodegradable polyvinyl alcohol (PVA)/Pluronic (PLUR)/ZnO films showed effective antibacterial activity against *E. coli* in the food industry (Sharmila *et al.* 2018; Sirelkhatim *et al.* 2015; Dobrucka and Długaszewska 2016). Due to their antioxidant and antimicrobial properties, ZnO NPs have been utilized as an active ingredient to enhance the quality of food. Ahmadi *et al.* reported good antibacterial and antioxidant activity of ZnO NPs modified with gelatin–cellulose-based nanofiber. In the report, gelatin–cellulose nanofiber is applied to keep the bacterial strain from growing. ZnO NPs added to gelatin–cellulose nanofiber showed good antibacterial activity, based on concentration, which was tested for 1–5% ZnO NPs. The bacterial inhibition zone of ZnO NPs incorporated into gelatin–cellulose nanofiber at a concentration of 5% is 10.44 ± 0.44 mm for *S. aureus* and 9.75 ± 0.11 mm for *Pseudomonas fluorescens* (Ahmadi, Ebrahimi, and Ahmadi 2021).

As stated in different reports, ZnO NPs induce ROS, which disrupt the metabolic activities of bacteria. Ngo *et al.* reported that a nanocomposite of ZnO and gelatin–cellulose nanofiber showed greater antibacterial activity against *S. aureus* than against the *P. fluorescens* strain. The difference in growth inhibition between those bacterial strains resulted from the types of bacterial cell wall structure (Jadoun *et al.* 2021) (Table 11.2).

TABLE 11.2
Different Nanoparticles and Their Applications in the Food Industry

Nanoparticle type	Target agent/ Action	Application/Function	References
Nanoclay	Toxic gases	Food packaging	(Dhapte *et al.* 2015)
Zinc oxide	Polyphenol oxidase	Antioxidants	(Siddiqi, Rahman, and Husen 2018)
Titanium dioxide	Ingredients	Coloring	(Musial *et al.* 2020)
Silver	Species of bacteria	Antimicrobials	(Rai *et al.* 2012)
Carbon nanotubes	Oxygen	Food packaging	(Arora and Padua 2010)
Magnesium oxide	Bacteria	Food packaging	(Swaroop and Shukla 2018)
Zerovalent iron	Oxygen	Food packaging	(Sreekanth and Sahu 2015)
Silver oxide	Bacteria	Food packaging	(Khodashenas and Ghorbani 2019)
Silver nanocomposite	Antimicrobial	Food packaging	(Hossen *et al.* 2019)
Silver	Bacterial cells	Food packaging	(Duncan 2011)
Zinc oxide	Bacteria	Food packaging	(Ahmadi and Ahmadi 2020)
Silver	Microbes	Food packaging	(Simbine *et al.* 2019)
Gold	Bacteria	Food packaging	(Yaqoob *et al.* 2020)
Zinc oxide	Bacteria	Active food packaging	(Amin *et al.* 2020)
Silver	Bacteria	Food packaging	(Wu *et al.* 2018)
Silver	Bacteria	Food packaging	(Naz *et al.* 2017)
Zinc oxide	Bacteria	Food packaging and antibacterial	(Wu *et al.* 2019)
Silver	Bacteria	Food packaging and antibacterial	(Shi *et al.* 2014)
Magnesium oxide	Bacteria	Food packaging	(Hoseinnejad, Jafari, and Katouzian 2018)
Zinc oxide	Bacteria	Antibacterial	(Mallikarjunaswamy *et al.* 2020)
Copper oxide	Bacteria	Antibacterial	(Kumar *et al.* 2021)
Copper oxide	Additive	Nutritional, dietary	(He, Deng, and Hwang 2018)

11.9 FUTURE PERSPECTIVES

Nanotechnology, which was raised by Richard Feynman in 1959, introduces the application of nanomaterials in food science. In food science, nanosized materials have played a significant role in the reduction of toxicity by packaging, preventing diseases, managing several microorganisms, transportation, and prolonging the shelf life of different food ingredients (Tabatabaei, Mahdavi-asl, and Khalili 2020). Among nanomaterials, metal and metal oxide NPs have attracted the attention of many researchers because of their limitless applications in food science and research areas. In the food industry, these NPs have potent applications in the protection of food ingredients from bacteria, the control of oxidation reactions, packaging, and the prevention of some effects of external exposure (Chauhan and Khan 2021). Additionally, carbon nanotubes are being used in renovating materials for food packaging and active packaging. Fifty years ago, nanoencapsulation materials were widely used in the packaging process in the food industry. In the pharmaceutical, nutraceutical, and functional food sectors, nanosized encapsulates have been used for drug delivery, food preservation, and improving the quality of food, respectively (Alimentaria 2020). A number of studies have reported that micro- and nanosized materials are used for food packaging in the food industry. However, nanomaterial coating is preferable to microencapsulation due to its target-specific selectivity, bioavailability, dispersibility, and lower toxicity. But, the development and fabrication of nanomaterials used in the food industry are too slow. The science of technology concerning the food industry is highly developed. However, the production, application, and availability of materials at market level are not yet solved effectively. In addition to this, the cytotoxicity of NPs is not well studied for various reasons, as discussed in the literature (Breast *et al.* 2019). Therefore, it is imperative to study the side effects, toxicity, and methods by which NPs can be manufactured in large amounts.

11.10 CONCLUSION

The most fundamental method for making food intelligently and extending its shelf life is by encapsulation of its constituents. Uncoated food is susceptible to several microorganisms, pH, temperature, sunlight, moisture, chemicals, and physical damage. Contact of food with those factors causes the development of microorganisms that result in foodborne diseases. Due to this, along with the advance of nanotechnology, researchers have developed nanosized materials that are used for food packaging. NPs like silver, gold, zinc, iron, copper, titanium oxide, magnesium oxide, zinc oxide, silicon oxide, nanoclay, and carbon nanotubes have been widely used to inhibit microorganisms and enhance the shelf life of food in the food industry and at market level. In addition to this, metal and metal oxide NPs are utilized from transportation until consumption. Metal and metal oxide NPs are directly involved in disinfection and ensure food quality in storage for a long time. So, this review has assessed the persuasive guarantee of metal NPs in the food industry in protecting food from contamination. Furthermore, food security has been obtained by safely delivering bioactive molecules to the required site in the cell. However, there are some problems with nanomaterials used in encapsulation and delivery systems. The vast usage of NPs for encapsulation and delivery leads to adverse effects on normal cells. But, metal and metal oxide NPs have shown great efficacy with few side effects relative to other materials.

REFERENCES

Abdel-Fattah, Wafa I., M. M. Eid, Sh I. Abd El-Moez, E. Mohamed, and Ghareib W. Ali. 2017. "Synthesis of Biogenic Ag@Pd Core-Shell Nanoparticles Having Anti-Cancer/Anti-Microbial Functions." *Life Sciences* 183: 28–36. https://doi.org/10.1016/j.lfs.2017.06.017.

Agarwal, Happy, S. Venkat Kumar, and S. Rajeshkumar. 2017. "A Review on Green Synthesis of Zinc Oxide Nanoparticles: An Eco-Friendly Approach." *Resource-Efficient Technologies* 3 (4): 406–13. https://doi.org/10.1016/j.reffit.2017.03.002.

Ahmad, Shabir, Sidra Munir, Nadia Zeb, Asad Ullah, Behramand Khan, Javed Ali, Muhammad Bilal, *et al.* 2019. "Green Nanotechnology: A Review on Green Synthesis of Silver Nanoparticles: An Ecofriendly Approach." *International Journal of Nanomedicine* 14: 5087–5107. https://doi.org/10.2147/IJN.S200254.

Ahmad, Waseem, and Divya Kalra. 2020. "Green Synthesis, Characterization and Anti Microbial Activities of ZnO Nanoparticles Using *Euphorbia hirta* Leaf Extract." *Journal of King Saud University - Science* 32 (4): 2358–64. https://doi.org/10.1016/j.jksus.2020.03.014.

Ahmadi, Azam, and Parisa Ahmadi. 2020. "Development of an Active Packaging System Containing Zinc Oxide Nanoparticles for the Extension of Chicken Fillet Shelf Life." *Food Science & Nutrition* 8 (10): 5461–73. https://doi.org/10.1002/fsn3.1812.

Ahmadi, Hanie, Alireza Ebrahimi, and Fatemeh Ahmadi. 2021. "Antibiotic Therapy in Dentistry." *International Journal of Dentistry* 2021. https://doi.org/10.1155/2021/6667624.

Al-Rubaye, Huda Ismail, Baidaa K. Al-Rubaye, Entisar E. Al-Abodi, and Enaam Ismail Yousif. 2020. "Green Chemistry Synthesis of Modified Silver Nanoparticles." *Journal of Physics: Conference Series* 1664 (1): 1–26. https://doi.org/10.1088/1742-6596/1664/1/012080.

Al-Sheddi, Ebtesam S., Nida N. Farshori, Mai M. Al-Oqail, Shaza M. Al-Massarani, Quaiser Saquib, Rizwan Wahab, Javed Musarrat, Abdulaziz A. Al-Khedhairy, and Maqsood A. Siddiqui. 2018. "Anticancer Potential of Green Synthesized Silver Nanoparticles Using Extract of *Nepeta deflersiana* against Human Cervical Cancer Cells (HeLA)." *Bioinorganic Chemistry and Applications* 2018. https://doi.org/10.1155/2018/9390784.

Alfei, Silvana, Anna Maria Schito, and Guendalina Zuccari. 2021. "Nanotechnological Manipulation of Nutraceuticals and Phytochemicals for Healthy Purposes: Established Advantages vs. Still Undefined Risks." *Polymers* 13 (14). https://doi.org/10.3390/polym13142262.

Alimentaria, Acta. 2020. "Nanotechnology In Food Systems: A Review" 49 (4): 460–74. https://doi.org/10.1556/066.2020.49.4.12.

Almasi, Hadi, Paria Jafarzadeh, and Laleh Mehryar. 2018. "Fabrication of Novel Nanohybrids by Impregnation of CuO Nanoparticles into Bacterial Cellulose and Chitosan Nanofibers: Characterization, Antimicrobial and Release Properties." *Carbohydrate Polymers* 186: 273–81. https://doi.org/10.1016/j.carbpol.2018.01.067.

Ameta, Satish Kumar, Avinash Kumar Rai, and Divya Hiran. n.d. *Use of Nanomaterials in Food Science.* https://doi.org/10.1007/978-981-15-2985-6.

Amin, Khaled M, Abir M Partila, Hassan A Abd El-rehim, and Noha M Deghiedy. 2020. "Antimicrobial ZnO Nanoparticle-Doped Polyvinyl Alcohol / Pluronic Blends as Active Food Packaging Films." https://doi.org/10.1002/ppsc.202000006.

Ananda Murthy, H C, Buzuayehu Abebe, Prakash C H, and Kumar Shantaveerayya. 2018. "A Review on Green Synthesis and Applications of Cu and CuO Nanoparticles." *Material Science Research India* 15 (3): 279–95. https://doi.org/10.13005/msri/150311.

Ansari, Mohammad A., Haris M. Khan, Mohammad A. Alzohairy, Mohammad Jalal, Syed G. Ali, Ruchita Pal, and Javed Musarrat. 2015. "Green Synthesis of Al2O3 Nanoparticles and Their Bactericidal Potential against Clinical Isolates of Multi-Drug Resistant *Pseudomonas aeruginosa*." *World Journal of Microbiology and Biotechnology* 31 (1): 153–64. https://doi.org/10.1007/s11274-014-1757-2.

Aravind, M., M. Amalanathan, and M. Sony Michael Mary. 2021. "Synthesis of TiO2 Nanoparticles by Chemical and Green Synthesis Methods and Their Multifaceted Properties." *SN Applied Sciences* 3 (4): 1–10. https://doi.org/10.1007/s42452-021-04281-5.

Arief, Syukri, Fri Wardana Nasution, and Arniati Labanni. 2020. "High Antibacterial Properties of Green Synthesized Gold Nanoparticles Using *Uncaria gambir* Roxb." *Leaf Extract and Triethanolamine* 10 (08): 124–30. https://doi.org/10.7324/JAPS.2020.10814.

Arihara, K. 2014. "Functional Foods." *Encyclopedia of Meat Sciences* 1: 32–36. https://doi.org/10.1016/B978-0-12-384731-7.00172-0.

Arora, Amit, and G. W. Padua. 2010. "Review: Nanocomposites in Food Packaging." *Journal of Food Science* 75 (1): 43–49. https://doi.org/10.1111/j.1750-3841.2009.01456.x.

Asemani, Marjan, and Navideh Anarjan. 2019. "Self-Dual Leonard Pairs Green Synthesis of Copper Oxide Nanoparticles Extract Assessment and Biological Properties." *Green Processing and Synthesis* 8: 557–67.

Assadpour, Elham, and Seid Mahdi Jafari. 2019. "A Systematic Review on Nanoencapsulation of Food Bioactive Ingredients and Nutraceuticals by Various Nanocarriers." *Critical Reviews in Food Science and Nutrition* 59 (19): 3129–51. https://doi.org/10.1080/10408398.2018.1484687.

Aswathy, R., B. Gabylis, S. Anwesha, and K. V. Bhaskara Rao. 2017. "Green Synthesis and Characterization of Marine Yeast-Mediated Silver and Zinc Oxide Nanoparticles and Assessment of Their Antioxidant Activity." *Asian Journal of Pharmaceutical and Clinical Research* 10 (10): 235–40. https://doi.org/10.22159/ajpcr.2017.v10i10.19979.

Augustin, Mary Ann, and Yacine Hemar. 2009. "Nano- and Micro-Structured Assemblies for Encapsulation of Food Ingredients." *Chemical Society Reviews* 38 (4): 902–12. https://doi.org/10.1039/b801739p.

Avella, Maurizio, Jan J De Vlieger, Maria Emanuela, Sabine Fischer, Paolo Vacca, and Maria Grazia. 2005. "Food Chemistry Biodegradable Starch / Clay Nanocomposite Films for Food Packaging Applications" *Food Chemistry* 93 (3): 467–74. https://doi.org/10.1016/j.foodchem.2004.10.024.

Baghayeri, Mehdi, Amirhassan Amiri, Behrooz Maleki, Zahra Alizadeh, and Oliver Reiser. 2018. "Sensors and Actuators B: Chemical A Simple Approach for Simultaneous Detection of Cadmium (II) and Lead (II) Based on Glutathione Coated Magnetic Nanoparticles as a Highly Selective Electrochemical Probe." *Sensors & Actuators: B. Chemical* 273 (July): 1442–50. https://doi.org/10.1016/j.snb.2018.07.063.

Bekele, Eneyew Tilahun, Bedasa Abdisa Gonfa, Osman Ahmed Zelekew, Hadgu Hailekiros Belay, and Fedlu Kedir Sabir. 2020. "Synthesis of Titanium Oxide Nanoparticles Using Root Extract of *Kniphofia foliosa* as a Template, Characterization, and Its Application on Drug Resistance Bacteria." *Journal of Nanomaterials* 2020: 1–10. https://doi.org/10.1155/2020/2817037.

Bouafia, Abderrhmane, Salah Eddine Laouini, and Mohammed Redha Ouahrani. 2020. "A Review on Green Synthesis of CuO Nanoparticles Using Plant Extract and Evaluation of Antimicrobial Activity." *Asian Journal of Research in Chemistry* 13 (1): 65. https://doi.org/10.5958/0974-4150.2020.00014.0.

Breast, Human, Cancer Cells, Zainab Jihad Taqi, and Mihailescu Dan Florin. 2019. "Nano Biomed Eng Zinc Oxide Nanoparticles Induces Apoptosis in Pathway" 11 (1): 35–43. https://doi.org/10.5101/nbe.v11i1.p35-43.Research.

Brigger, Irène, Catherine Dubernet, and Patrick Couvreur. 2002. "Nanoparticles in Cancer Therapy and Diagnosis." *Advanced Drug Delivery Reviews* 54 (5): 631–51. https://doi.org/10.1016/S0169-409X(02)00044-3.

Chakra, Ch Shilpa, V. Rajendar, K. Venkateswara Rao, and Mirgender Kumar. 2017. "Enhanced Antimicrobial and Anticancer Properties of ZnO and TiO_2 Nanocomposites." *3 Biotech* 7 (2): 1–8. https://doi.org/10.1007/s13205-017-0731-8.

Chaudhuri, Sadhan Kumar, and Lalit Malodia. 2017. "Biosynthesis of Zinc Oxide Nanoparticles Using Leaf Extract of *Calotropis gigantea*: Characterization and Its Evaluation on Tree Seedling Growth in Nursery Stage." *Applied Nanoscience* 7 (8): 501–12. https://doi.org/10.1007/s13204-017-0586-7.

Chauhan, Akshita, Tabassum Khan, and A. Omri. 2021. "Design and Encapsulation of Immunomodulators onto Gold Nanoparticles in Cancer Immunotherapy." *International Journal of Molecular Sciences* 22 (15): 8037. https://doi.org/10.3390/ijms22158037. PMID: 34360803; PMCID: PMC8347387.

Chen, Lei, Charles Gnanaraj, Palanisamy Arulselvan, Hesham El-Seedi, and Hui Teng. 2019. "A Review on Advanced Microencapsulation Technology to Enhance Bioavailability of Phenolic Compounds: Based on Its Activity in the Treatment of Type 2 Diabetes." *Trends in Food Science and Technology* 85 (June 2018): 149–62. https://doi.org/10.1016/j.tifs.2018.11.026.

Cushen, M, J Kerry, M Morris, and E Cummins. 2014. "Evaluation and Simulation of Silver and Copper Nanoparticle Migration from Polyethylene Nanocomposites to Food and an Associated Exposure Assessment." *Journal of Agricultural and Food Chemistry* 62 (6): 1403–11. https://doi.org/10.1021/jf404038y. Epub 2014 Jan 31. PMID: 24450547.

Daneshniya, Milad, Mohammad Hossein Maleki, and Hooman Jalilvand Nezhad. 2021. "Application of Titanium Dioxide (TiO 2) Nanoparticles." January.

Das, Lipi, Eshani Bhaumik, Utpal Raychaudhuri, and Runu Chakraborty. 2012. "Role of Nutraceuticals in Human Health." *Journal of Food Science and Technology* 49 (2): 173–83. https://doi.org/10.1007/s13197-011-0269-4.

Das, Ratul Kumar, Vinayak Laxman Pachapur, Linson Lonappan, Mitra Naghdi, Rama Pulicharla, Sampa Maiti, Maximiliano Cledon, Larios Martinez Araceli Dalila, Saurabh Jyoti Sarma, and Satinder Kaur Brar. 2017. "Biological Synthesis of Metallic Nanoparticles: Plants, Animals and Microbial Aspects." *Nanotechnology for Environmental Engineering* 2 (1): 1–21. https://doi.org/10.1007/s41204-017-0029-4.

Das, Soumen, Janet M. Dowding, Kathryn E. Klump, James F. Mcginnis, William Self, and Sudipta Seal. 2013. "Cerium Oxide Nanoparticles: Applications and Prospects in Nanomedicine." *Nanomedicine* 8 (9): 1483–1508. https://doi.org/10.2217/nnm.13.133.

Desai, Kashappa Goud H., and Hyun Jin Park. 2005. *Recent Developments in Microencapsulation of Food Ingredients*. Drying Technology. Vol. 23. https://doi.org/10.1081/DRT-200063478.

Desai, Neha, Munira Momin, Tabassum Khan, Sankalp Gharat, Raghumani Singh Ningthoujam, and Abdelwahab Omri. 2021. "Metallic Nanoparticles as Drug Delivery System for the Treatment of Cancer." *Expert Opinion on Drug Delivery* 18 (9): 1261–90. https://doi.org/10.1080/17425247.2021.1912008.

Dhakal, Shabana Praveen, and Jibin He. 2020. "Microencapsulation of Vitamins in Food Applications to Prevent Losses in Processing and Storage: A Review." *Food Research International* 137: 109326. https://doi.org/10.1016/j.foodres.2020.109326.

Dhapte, Vividha, Namrata Gaikwad, Priyesh V. More, Shaibal Banerjee, Vishwas V. Dhapte, Shivajirao Kadam, and Pawan K. Khanna. 2015. "Transparent ZnO/Polycarbonate Nanocomposite for Food Packaging Application." *Nanocomposites* 1 (2): 106–12. https://doi.org/10.1179/2055033215Y.0000000004.

Długosz, Olga, Krzysztof Szostak, Anita Staro, and Jolanta Pulit-prociak. 2020. "Methods for Reducing the Toxicity of Metal and Metal Oxide NPs as Biomedicine." *Materials* 13: 279. https://doi.org/10.3390/ma13020279.

Doan Thi, Tu Uyen, Trung Thoai Nguyen, Y. Dang Thi, Kieu Hanh Ta Thi, Bach Thang Phan, and Kim Ngoc Pham. 2020. "Green Synthesis of ZnO Nanoparticles Using Orange Fruit Peel Extract for Antibacterial Activities." *RSC Advances* 10 (40): 23899–907. https://doi.org/10.1039/d0ra04926c.

Dobrucka, Renata, and Jolanta Długaszewska. 2016. "Biosynthesis and Antibacterial Activity of ZnO Nanoparticles Using Trifolium Pratense Flower Extract." *Saudi Journal of Biological Sciences* 23 (4): 517–23. https://doi.org/10.1016/j.sjbs.2015.05.016.

Donsì, Francesco, Marianna Annunziata, Mariarenata Sessa, and Giovanna Ferrari. 2011. "LWT: Food Science and Technology Nanoencapsulation of Essential Oils to Enhance Their Antimicrobial Activity in Foods." *LWT: Food Science and Technology* 44 (9): 1908–14. https://doi.org/10.1016/j.lwt.2011.03.003.

Drahansky, Martin, M.t Paridah, Amin Moradbak, A.Z Mohamed, Folahan abdulwahab taiwo Owolabi, Mustapha Asniza, and Shawkataly H.P Abdul Khalid. 2016. "We Are IntechOpen, the World' s Leading Publisher of Open Access Books Built by Scientists, for Scientists TOP 1 %." *Intech* i (tourism): 13. https://doi.org/10.5772/57353.

Duncan, Timothy V. 2011. "Applications of Nanotechnology in Food Packaging and Food Safety: Barrier Materials, Antimicrobials and Sensors." *Journal of Colloid and Interface Science* 363 (1): 1–24. https://doi.org/10.1016/j.jcis.2011.07.017.

Durgadevi, S, and R Jothi Mani. 2018. "Green Synthesis of Copper Oxide Nano Particles Using Moringa Oleifera Leaf Extract." *International Journal of Trend in Research and Development* (IJTRD), ISSN: 2394-9333, Special Issue | REDEEMS-18, February 2018.

Elemike, Elias E., Damian C. Onwudiwe, Anthony C. Ekennia, Christopher U. Sonde, and Richard C. Ehiri. 2017. "Green Synthesis of Ag/Ag2O Nanoparticles Using Aqueous Leaf Extract of Eupatorium Odoratum and Its Antimicrobial and Mosquito Larvicidal Activities." *Molecules* 22 (5): 1–15. https://doi.org/10.3390/molecules22050674.

Elemike, Elias E., Damian C. Onwudiwe, Nirasha Nundkumar, Moganavelli Singh, and Osaro Iyekowa. 2019. "Green Synthesis of Ag, Au and Ag-Au Bimetallic Nanoparticles Using *Stigmaphyllon ovatum* Leaf Extract and Their In Vitro Anticancer Potential." *Materials Letters* 243: 148–52. https://doi.org/10.1016/j.matlet.2019.02.049.

Ezhilarasi, P. N., P. Karthik, N. Chhanwal, and C. Anandharamakrishnan. 2013. "Nanoencapsulation Techniques for Food Bioactive Components: A Review." *Food and Bioprocess Technology* 6 (3): 628–47. https://doi.org/10.1007/s11947-012-0944-0.

Fadeel, Bengt, Valerian Kagan, Harald Krug, Anna Shvedova, Magnus Svartengren, Lang Tran, and Lars Wiklund. 2007. "There's Plenty of Room at the Forum: Potential Risks and Safety Assessment of Engineered Nanomaterials." *Nanotoxicology* 1 (2): 73–84. https://doi.org/10.1080/17435390701565578.

Galstyan, Vardan, Manohar P Bhandari Id, and Veronica Sberveglieri. 2018. "Metal Oxide Nanostructures in Food Applications: Quality Control and Packaging." *Chemosensors* 6: 16. https://doi.org/10.3390/chemosensors6020016.

Ghosh, Mithun Kumar, Sanjay Sahu, Indersh Gupta, and Tanmay Kumar Ghorai. 2020. "Green Synthesis of Copper Nanoparticles from an Extract of *Jatropha curcas* leaves: Characterization, Optical Properties, CT-DNA Binding and Photocatalytic Activity." *RSC Advances* 10 (37): 22027–35. https://doi.org/10.1039/d0ra03186k.

Ghosh, Partha, Gang Han, Mrinmoy De, Chae Kyu Kim, and Vincent M Rotello. 2008. "Gold Nanoparticles in Delivery Applications ☆." 60: 1307–15. https://doi.org/10.1016/j.addr.2008.03.016.

Ghotekar, Suresh, Shreyas Pansambal, Sharad P. Pawar, Trupti Pagar, Rajeshwari Oza, and Sachin Bangale. 2019. "Biological Activities of Biogenically Synthesized Fluorescent Silver Nanoparticles Using *Acanthospermum hispidum* Leaves Extract." *SN Applied Sciences* 1 (11): 1342. https://doi.org/10.1007/s42452-019-1389-0.

Happy, Agarwal, Menon Soumya, S Venkat Kumar, S Rajeshkumar, R David Sheba, T Lakshmi, and V Deepak Nallaswamy. 2019. "Phyto-Assisted Synthesis of Zinc Oxide Nanoparticles Using *Cassia alata* and Its Antibacterial Activity against *Escherichia coli*." *Biochemistry and Biophysics Reports* 17 (September 2018): 208–11. https://doi.org/10.1016/j.bbrep.2019.01.002.

He, Xiaojia, and Huey Min Hwang. 2016. "Nanotechnology in Food Science: Functionality, Applicability, and Safety Assessment." *Journal of Food and Drug Analysis* 24 (4): 671–81. https://doi.org/10.1016/j.jfda.2016.06.001.

He, Xiaojia, Hua Deng, and Huey-min Hwang. 2018. "The Current Application of Nanotechnology in Food and Agriculture." *Journal of Food and Drug Analysis* 27 (1): 1–21. https://doi.org/10.1016/j.jfda.2018.12.002.

Henrique Rodrigues do Amaral, Pedro, Patrícia Lopes Andrade, and Leilane Costa de Conto. 2019. "Microencapsulation and Its Uses in Food Science and Technology: A Review." *Microencapsulation: Processes, Technologies and Industrial Applications*, 1–18. https://doi.org/10.5772/intechopen.81997.

Hoseinnejad, Mahmoud, Seid Mahdi Jafari, and Iman Katouzian. 2018. "Inorganic and Metal Nanoparticles and Their Antimicrobial Activity in Food Inorganic and Metal Nanoparticles and Their Antimicrobial Activity in Food Packaging Applications." *Critical Reviews in Microbiology* 44 (2): 161–81. https://doi.org/10.1080/1040841X.2017.1332001.

Hossen, Sarwar, M Khalid Hossain, M K Basher, M N H Mia, M T Rahman, and M Jalal Uddin. 2019. "Smart Nanocarrier-Based Drug Delivery Systems for Cancer Therapy and Toxicity Studies: A Review." *Journal of Advanced Research* 15: 1–18. https://doi.org/10.1016/j.jare.2018.06.005.

Hu, Xiaopei, Yuting Zhang, Tingting Ding, Jiang Liu, and Hang Zhao. 2020. "Multifunctional Gold Nanoparticles: A Novel Nanomaterial for Various Medical Applications and Biological Activities" 8 (August): 1–17. https://doi.org/10.3389/fbioe.2020.00990.

Huang, Gan, and Shu Ting Pan. 2020. "ROS-Mediated Therapeutic Strategy in Chemo-/Radiotherapy of Head and Neck Cancer." *Oxidative Medicine and Cellular Longevity* 2020. https://doi.org/10.1155/2020/5047987.

Hussain, Afzal, Mohammad Oves, Mohamed F. Alajmi, Iqbal Hussain, Samira Amir, Jahangeer Ahmed, Md Tabish Rehman, Hesham R. El-Seedi, and Imran Ali. 2019. "Biogenesis of ZnO Nanoparticles Using: *Pandanus odorifer* Leaf Extract: Anticancer and Antimicrobial Activities." *RSC Advances* 9 (27): 15357–69. https://doi.org/10.1039/c9ra01659g.

Ijaz, Faheem, Sammia Shahid, Shakeel Ahmad Khan, Waqar Ahmad, and Sabah Zaman. 2017. "Green Synthesis of Copper Oxide Nanoparticles Using *Abutilon indicum* Leaf Extract: Antimicrobial, Antioxidant and Photocatalytic Dye Degradation Activities." *Tropical Journal of Pharmaceutical Research* 16 (4): 743–53. https://doi.org/10.4314/tjpr.v16i4.2.

Iqbal, Muhammad Aamir, Abdul Hamid, Tanvir Ahmad, Muzammil Hussain Siddiqui, Imtiaz Hussain, Sajid Ali, Anser Ali, and Zahoor Ahmad. 2019. "Forage Sorghum-Legumes Intercropping: Effect on Growth, Yields, Nutritional Quality and Economic Returns." *SciELO* 78(1): 82–95.

Iqbal, Seemab, Muhammad Fakhar-e-Alam, Fozia Akbar, M. Shafiq, M. Atif, N. Amin, Muhammad Ismail, Atif Hanif, and W. Aslam Farooq. 2019. "Application of Silver Oxide Nanoparticles for the Treatment of Cancer." *Journal of Molecular Structure* 1189: 203–9. https://doi.org/10.1016/j.molstruc.2019.04.041.

Istiqola, Arsi. 2020. "A Review of Silver Nanoparticles in Food Packaging Technologies: Regulation, Methods, Properties, Migration, and Future Challenges." *Journal of the Chinese Chemical Society* 67 (April): 1942–1956. https://doi.org/10.1002/jccs.202000179.

Jadoun, Sapana, Rizwan Arif, Nirmala Kumari Jangid, and Rajesh Kumar Meena. 2021. "Green Synthesis of Nanoparticles Using Plant Extracts: A Review." *Environmental Chemistry Letters* 19 (1): 355–74. https://doi.org/10.1007/s10311-020-01074-x.

Jafari, Seid Mahdi. 2017. *An Introduction to Nanoencapsulation Techniques for the Food Bioactive Ingredients. Nanoencapsulation of Food Bioactive Ingredients*. Elsevier Inc. https://doi.org/10.1016/b978-0-12-809740-3.00001-5.

Jampilek, Josef, Jiri Kos, and Katarina Kralova. 2019. "Potential of Nanomaterial Applications in Dietary Supplements and Foods for Special Medical Purposes." *Nanomaterials* 9 (2). https://doi.org/10.3390/nano9020296.

Jayappa, Manasa Dogganal, Chandrashekar Konambi Ramaiah, Masineni Allapuramaiah Pavan Kumar, Doddavenkatanna Suresh, Ashwini Prabhu, Rekha Punchappady Devasya, and Sana Sheikh. 2020. "Green Synthesis of Zinc Oxide Nanoparticles from the Leaf, Stem and In Vitro Grown Callus of

Mussaenda Frondosa L.: Characterization and Their Applications." *Applied Nanoscience* 10 (8): 3057–74. https://doi.org/10.1007/s13204-020-01382-2.

Jin, Wei, and Govindhan Maduraiveeran. 2018. "Nanomaterial-Based Environmental Sensing Platforms Using State-of-the-Art Electroanalytical Strategies." *Journal of Analytical Science and Technology* 9: 18. https://doi.org/10.1186/s40543-018-0150-4.

Jones, Peter J., and Stephanie Jew. 2007. "Functional Food Development: Concept to Reality." *Trends in Food Science and Technology* 18 (7): 387–90. https://doi.org/10.1016/j.tifs.2007.03.008.

Karunagaran, Vithiya, Kumar Rajendran, and Shampa Sen. 2017. "Optimization of Biosynthesis of Silver Oxide Nanoparticles and Its Anticancer Activity." *International Journal of Nanoscience* 16 (5–6): 1–8. https://doi.org/10.1142/S0219581X17500181.

Karunaratne, Desiree Nedra, Dunusingha Asitha, Surandika Siriwardhana, Isuru Rangana Ariyarathna, Rajakaruna Mudiyanselage, Pradeepa Indunil, Frousnoon Thasneem Banu, and Veranja Karunaratne. 2017. *Nutrient Delivery through Nanoencapsulation. Nutrient Delivery*. Elsevier Inc. https://doi.org/10.1016/B978-0-12-804304-2/00017-2.

Kashapov, Ruslan, Anastasiya Lykova, Nadezda Kashapova, Albina Ziganshina, Tatiana Sergeeva, Anastasiia Sapunova, Alexandra Voloshina, and Lucia Zakharova. 2021. "Nanoencapsulation of Food Bioactives in Supramolecular Assemblies Based on Cyclodextrins and Surfactant." *Food Hydrocolloids* 113 (October): 106449. https://doi.org/10.1016/j.foodhyd.2020.106449.

Khare, Atul Ramesh, and Niraj Vasisht. 2014. *Nanoencapsulation in the Food Industry. Microencapsulation in the Food Industry*. Elsevier Inc. https://doi.org/10.1016/b978-0-12-404568-2.00014-5.

Khodashenas, Bahareh, and Hamid Reza Ghorbani. 2019. "Synthesis of Silver Nanoparticles with Different Shapes." *Arabian Journal of Chemistry* 12 (8): 1823–38. https://doi.org/10.1016/j.arabjc.2014.12.014.

Kumar, Antul, Anuj Choudhary, Harmanjot Kaur, Sahil Mehta, and Azamal Husen. 2021. "Metal-Based Nanoparticles, Sensors, and Their Multifaceted Application in Food Packaging." *Journal of Nanobiotechnology*. https://doi.org/10.1186/s12951-021-00996-0.

Kumar, P., P. Mahajan, R. Kaur, and S. Gautam. 2020. "Nanotechnology and Its Challenges in the Food Sector: A Review." *Materials Today Chemistry* 17: 100332. https://doi.org/10.1016/j.mtchem.2020.100332.

Kuswandi, Bambang, and Mehran Moradi. 2019. *Improvement of Food Packaging Based on Functional Nanomaterial*. Springer International Publishing.

Li, Guodong, and Zhiyong Tang. 2014. "Noble Metal Nanoparticle Metal Oxide Core/Yolk-Shell Nanostructures as Catalysts: Recent Progress and Perspective." *Nanoscale* 6 (8): 3995–4011. https://doi.org/10.1039/c3nr06787d.

Liyanage, Piumi Y., Sajini D. Hettiarachchi, Yiqun Zhou, Allal Ouhtit, Elif S. Seven, Cagri Y. Oztan, Emrah Celik, and Roger M. Leblanc. 2019. "Nanoparticle-Mediated Targeted Drug Delivery for Breast Cancer Treatment." *Biochimica et Biophysica Acta: Reviews on Cancer* 1871 (2): 419–33. https://doi.org/10.1016/j.bbcan.2019.04.006.

Makvandi, Pooyan, Sidra Iftekhar, Fabio Pizzetti, Atefeh Zarepour, Ehsan Nazarzadeh Zare, Milad Ashrafizadeh, Tarun Agarwal, *et al.* 2021. "Functionalization of Polymers and Nanomaterials for Water Treatment, Food Packaging, Textile and Biomedical Applications: A Review." *Environmental Chemistry Letters* 19 (1): 583–611. https://doi.org/10.1007/s10311-020-01089-4.

Mallikarjunaswamy, C., V. Lakshmi Ranganatha, Ramith Ramu, Udayabhanu, and G. Nagaraju. 2020. "Facile Microwave-Assisted Green Synthesis of ZnO Nanoparticles: Application to Photodegradation, Antibacterial and Antioxidant." *Journal of Materials Science: Materials in Electronics* 31 (2): 1004–21. https://doi.org/10.1007/s10854-019-02612-2.

Manikandan, Velu, Palanivel Velmurugan, Jung Hee Park, Woo Suk Chang, Yool Jin Park, Palaniyappan Jayanthi, Min Cho, and Byung Taek Oh. 2017. "Green Synthesis of Silver Oxide Nanoparticles and Its Antibacterial Activity against Dental Pathogens." *3 Biotech* 7 (1): 1–9. https://doi.org/10.1007/s13205-017-0670-4.

Mauricio, M. D., S. Guerra-Ojeda, P. Marchio, S. L. Valles, M. Aldasoro, I. Escribano-Lopez, J. R. Herance, M. Rocha, J. M. Vila, and V. M. Victor. 2018. "Nanoparticles in Medicine: A Focus on Vascular Oxidative Stress." *Oxidative Medicine and Cellular Longevity* 2018. https://doi.org/10.1155/2018/6231482.

McClements, David Julian, and Hang Xiao. 2017. "Is Nano Safe in Foods? Establishing the Factors Impacting the Gastrointestinal Fate and Toxicity of Organic and Inorganic Food-Grade Nanoparticles." *NPJ Science of Food* 1 (1). https://doi.org/10.1038/s41538-017-0005-1.

Mittal, Amit Kumar, Yusuf Chisti, and Uttam Chand Banerjee. 2013. "Synthesis of Metallic Nanoparticles Using Plant Extracts." *Biotechnology Advances* 31 (2): 346–56. https://doi.org/10.1016/j.biotechadv.2013.01.003.

Murthy, H. C. Ananda, Tegene Desalegn, Mebratu Kassa, Buzuayehu Abebe, and Temesgen Assefa. 2020. "Synthesis of Green Copper Nanoparticles Using Medicinal Plant *Hagenia abyssinica* (Brace) JF. Gmel. Leaf Extract: Antimicrobial Properties." *Journal of Nanomaterials* 2020. https://doi.org/10.1155/2020/3924081.

Musial, Joanna, Rafal Krakowiak, Dariusz T. Mlynarczyk, Tomasz Goslinski, and Beata J. Stanisz. 2020. "Titanium Dioxide Nanoparticles in Food and Personal Care Products: What Do We Know about Their Safety?" *Nanomaterials* 10 (6): 1–23. https://doi.org/10.3390/nano10061110.

Nagajyothi, P. C., Sang Ju Cha, In Jun Yang, T. V.M. Sreekanth, Kwang Joong Kim, and Heung Mook Shin. 2015. "Antioxidant and Anti-Inflammatory Activities of Zinc Oxide Nanoparticles Synthesized Using *Polygala tenuifolia* Root Extract." *Journal of Photochemistry and Photobiology B: Biology* 146: 10–17. https://doi.org/10.1016/j.jphotobiol.2015.02.008.

Naseer, Bazila, Gaurav Srivastava, Ovais Shafiq Qadri, Soban Ahmad Faridi, Rayees Ul Islam, and Kaiser Younis. 2018. "Importance and Health Hazards of Nanoparticles Used in the Food Industry." *Nanotechnology Reviews* 7 (6): 623–41. https://doi.org/10.1515/ntrev-2018-0076.

Nasrollahzadeh, Mahmoud, Seyedeh Samaneh Momeni, and S. Mohammad Sajadi. 2017. "Green Synthesis of Copper Nanoparticles Using *Plantago asiatica* Leaf Extract and Their Application for the Cyanation of Aldehydes Using K 4 Fe(CN) 6." *Journal of Colloid and Interface Science* 506: 471–77. https://doi.org/10.1016/j.jcis.2017.07.072.

Naz, M., N. Nasiri, M. Ikram, M. Nafees, M. Z. Qureshi, S. Ali, and A. Tricoli. 2017. "Eco-Friendly Biosynthesis, Anticancer Drug Loading and Cytotoxic Effect of Capped Ag-Nanoparticles against Breast Cancer." *Applied Nanoscience* 7 (8): 793–802. https://doi.org/10.1007/s13204-017-0615-6.

Nile, Shivraj Hariram, Venkidasamy Baskar, Dhivya Selvaraj, Arti Nile, Jianbo Xiao, and Guoyin Kai. 2020. *Nanotechnologies in Food Science: Applications, Recent Trends, and Future Perspectives. Nano-Micro Letters*. Vol. 12. Springer Singapore. https://doi.org/10.1007/s40820-020-0383-9.

Oliveira, Rachel De. 2014. "Development and Evaluation of Nanoparticles for Cancer Treatment." Université Paris Sud - Paris XI; Universidad de Estado de Rio de Janeiro (Brésil), 1–177. English. ffNNT: 2014PA114808ff. fftel-01310666f.

Pachaiappan, Rekha, Saravanan Rajendran, Pau Loke Show, Kovendhan Manavalan, and Mu Naushad. 2021. "Metal/Metal Oxide Nanocomposites for Bactericidal Effect: A Review." *Chemosphere* 272 (xxxx): 128607. https://doi.org/10.1016/j.chemosphere.2020.128607.

Paidari, Saeed, and Salam Adnan Ibrahim. 2021. "Potential Application of Gold Nanoparticles in Food Packaging: A Mini Review." *Gold Bull* 54: 31–36. https://doi.org/10.1007/s13404-021-00290-9.

Paolino, Donatella, Antonia Mancuso, Maria Chiara Cristiano, Francesca Froiio, Narimane Lammari, Christian Celia, and Massimo Fresta. 2021. "Nanonutraceuticals: The New Frontier of Supplementary Food." *Nanomaterials* 11 (3): 1–20. https://doi.org/10.3390/nano11030792.

Paredes, Alejandro Javier, Claudia Mariana Asencio, Llabot Juan Manuel, and Daniel Alberto Allemandi. 2016. "Nano Encapsulation in the Food Industry: Manufacture, Applications and Characterization" *Journal of Food Bioengineering and Nanoprocessing* 1 (1): 56–79.

Pateiro, Mirian, Belén Gómez, Paulo E.S. Munekata, Francisco J. Barba, Predrag Putnik, Danijela Bursać Kovačević, and José M. Lorenzo. 2021. "Nanoencapsulation of Promising Bioactive Compounds to Improve Their Absorption, Stability, Functionality and the Appearance of the Final Food Products." *Molecules* 26 (6). https://doi.org/10.3390/molecules26061547.

Patel, Monika, and Sunita Mishra. 2021. "A Review: Application and Production of Nanoencapsulation in the Food Sector." *Asian Journal of Nanosciences and Materials* 4 (3): 213–28. https://doi.org/10.26655/AJNANOMAT.2021.3.4.

Patil, Sunita, and Rajkuberan Chandrasekaran. 2020. "Biogenic Nanoparticles: A Comprehensive Perspective in Synthesis, Characterization, Application and Its Challenges." *Journal of Genetic Engineering and Biotechnology* 18 (1). https://doi.org/10.1186/s43141-020-00081-3.

Primoži, Mateja. 2021. "(Bio) Nanotechnology in Food Science:Food Packaging." *Sensors* 21: 2148.

Rai, M. K., S. D. Deshmukh, A. P. Ingle, and A. K. Gade. 2012. "Silver Nanoparticles: The Powerful Nanoweapon against Multidrug-Resistant Bacteria." *Journal of Applied Microbiology* 112 (5): 841–52. https://doi.org/10.1111/j.1365-2672.2012.05253.x.

Rashidi, Ladan. 2021. "Different Nano-Delivery Systems for Delivery of Nutraceuticals." *Food Bioscience* 43 (July): 101258. https://doi.org/10.1016/j.fbio.2021.101258.

Ravichandran, R. 2010. "Nanotechnology Applications in Food and Food Processing: Innovative Green Approaches, Opportunities and Uncertainties for Global Market." *International Journal of Green Nanotechnology: Physics and Chemistry* 1 (2): P72–96. https://doi.org/10.1080/19430871003684440.

Ravichandran, V., S. Vasanthi, S. Shalini, Syed Adnan Ali Shah, M. Tripathy, and Neeraj Paliwal. 2019. "Green Synthesis, Characterization, Antibacterial, Antioxidant and Photocatalytic Activity of *Parkia speciosa* Leaves Extract Mediated Silver Nanoparticles." *Results in Physics* 15 (December 2018): 102565. https://doi.org/10.1016/j.rinp.2019.102565.

Reis, Catarina Pinto, and Ronald J Neufeld. 2006. "Nanoencapsulation I. Methods for Preparation of Drug-Loaded Polymeric Nanoparticles" *Nanomedicine* 2: 8–21. https://doi.org/10.1016/j.nano.2005.12.003.

Rutberg, F. G., M. V. Dubina, V. A. Kolikov, F. V. Moiseenko, E. V. Ignat'eva, N. M. Volkov, V. N. Snetov, and A. Yu Stogov. 2008. "Effect of Silver Oxide Nanoparticles on Tumor Growth In Vivo." *Doklady Biochemistry and Biophysics* 421 (1): 191–93. https://doi.org/10.1134/S1607672908040078.

Selim, Yasser A., Maha A. Azb, Islam Ragab, and Mohamed H. M. Abd El-Azim. 2020. "Green Synthesis of Zinc Oxide Nanoparticles Using Aqueous Extract of *Deverra tortuosa* and Their Cytotoxic Activities." *Scientific Reports* 10 (1): 1–9. https://doi.org/10.1038/s41598-020-60541-1.

Senthilkumar, N., E. Nandhakumar, P. Priya, D. Soni, M. Vimalan, and I. Vetha Potheher. 2017. "Synthesis of ZnO Nanoparticles Using Leaf Extract of: *Tectona grandis* (L.) and Their Anti-Bacterial, Anti-Arthritic, Anti-Oxidant and In Vitro Cytotoxicity Activities." *New Journal of Chemistry* 41 (18): 10347–56. https://doi.org/10.1039/c7nj02664a.

Shafey, Asmaa Mohamed El. 2020. "Green Synthesis of Metal and Metal Oxide Nanoparticles from Plant Leaf Extracts and Their Applications: A Review." *Green Processing and Synthesis* 9 (1): 304–39. https://doi.org/10.1515/gps-2020-0031.

Sharma, Chetan, Romika Dhiman, Namita Rokana, and Harsh Panwar. 2017. "Nanotechnology: An Untapped Resource for Food Packaging" 8 (September). https://doi.org/10.3389/fmicb.2017.01735.

Sharmila, G., C. Muthukumaran, K. Sandiya, S. Santhiya, R. Sakthi Pradeep, N. Manoj Kumar, N. Suriyanarayanan, and M. Thirumarimurugan. 2018. "Biosynthesis, Characterization, and Antibacterial Activity of Zinc Oxide Nanoparticles Derived from *Bauhinia tomentosa* Leaf Extract." *Journal of Nanostructure in Chemistry* 8 (3): 293–99. https://doi.org/10.1007/s40097-018-0271-8.

Shen, Laifa, Hongsen Li, Evan Uchaker, Xiaogang Zhang, and Guozhong Cao. 2012. "General Strategy for Designing Core-Shell Nanostructured Materials for High-Power Lithium Ion Batteries." *Nano Letters* 12 (11): 5673–78. https://doi.org/10.1021/nl302854j.

Shi, Donglu, and Nicholas Bedford. 2018. "Core-Shell Nanocomposites Silicon Carbide Nanomaterials."

Shi, Hongbo, Ruth Magaye, Vincent Castranova, and Jinshun Zhao. 2013. "Titanium Dioxide Nanoparticles: A Review of Current Toxicological Data." *Particle and Fibre Toxicology* 10: 15. https://doi.org/10.1186/1743-8977-10-15.

Shi, Lu E., Zhen Hua Li, Wei Zheng, Yi Fan Zhao, Yong Fang Jin, and Zhen Xing Tang. 2014. "Synthesis, Antibacterial Activity, Antibacterial Mechanism and Food Applications of ZnO Nanoparticles: A Review." *Food Additives and Contaminants: Part A Chemistry, Analysis, Control, Exposure and Risk Assessment* 31 (2): 173–86. https://doi.org/10.1080/19440049.2013.865147.

Shivappa, Prashanth, Vijayakumar Gangaiah, and Ashoka Siddaramanna. 2018. "Materials Science in Semiconductor Processing Mesoporous CeO 2 Nanoparticles Modi Fi Ed Glassy Carbon Electrode for Individual and Simultaneous Determination of Cu (II) and Hg (II): Application to Environmental Samples." *Materials Science in Semiconductor Processing* 84 (January): 157–66. https://doi.org/10.1016/j.mssp.2018.05.010.

Shnoudeh, Abeer Jabra, Islam Hamad, Ruwaida W Abdo, Lana Qadumii, Abdulmutallab Yousef Jaber, Hiba Salim, and Shahd Z Alkelany. 2019. *Applications of Metal Nanoparticles. Biomaterials and Bionanotechnology*. Elsevier Inc. https://doi.org/10.1016/B978-0-12-814427-5.00015-9.

Siddiqi, Khwaja Salahuddin, Aziz Rahman, and Azamal Husen. 2018. "Properties of Zinc Oxide Nanoparticles and Their Activity Against Microbes." *Nanoscale Research Letters* 13: 141. https://doi.org/10.1186/s11671-018-2532-3.

Siddiqi, Khwaja Salahuddin, M. Rashid, A. Rahman, Tajuddin, Azamal Husen, and Sumbul Rehman. 2020. "Green Synthesis, Characterization, Antibacterial and Photocatalytic Activity of Black Cupric Oxide Nanoparticles." *Agriculture and Food Security* 9 (1): 1–15. https://doi.org/10.1186/s40066-020-00271-9.

Simbine, Emelda Orlando, Cunha Rodrigues, Judite Lapa-guimarães, Eliana Setsuko Kamimura, Carlos Humberto Corassin, Carlos Augusto, and Fernandes De Oliveira. 2019. "Application of Silver Nanoparticles in Food Packages: A Review." *Food Science and Technology* 2061 (4): 793–802.

Sirelkhatim, Amna, Shahrom Mahmud, Azman Seeni, Noor Haida Mohamad Kaus, Ling Chuo Ann, Siti Khadijah Mohd Bakhori, Habsah Hasan, and Dasmawati Mohamad. 2015. "Review on Zinc Oxide Nanoparticles: Antibacterial Activity and Toxicity Mechanism." *Nano-Micro Letters* 7 (3): 219–42. https://doi.org/10.1007/s40820-015-0040-x.

Solomon, Girmay. 2015. "Preliminary Phytochemical Screening and In Vitro Antimicrobial Activity of *Datura stramonium* Leaves Extracts Collected from Eastern Ethiopia." *International Research Journal of Biological Sciences* 4 (1): 55–59.

Sosnik, Alejandro, and Katia P Seremeta. 2015. "SC." *Advances in Colloid and Interface Science*. https://doi.org/10.1016/j.cis.2015.05.003.

Souza, Roberta C. De, Leticia U. Haberbeck, Humberto G. Riella, Deise H.B. Ribeiro, and Bruno A.M. Carciofi. 2019. "Antibacterial Activity of Zinc Oxide Nanoparticles Synthesized by Solochemical Process." *Brazilian Journal of Chemical Engineering* 36 (2): 885–93. https://doi.org/10.1590/0104-6632.20190362s20180027.

Sreekanth, K M, and Debjyoti, Sahu. 2015. "Effect of Iron Oxide Nanoparticle in Bio Digestion of a Portable Food-Waste Digester." *Synthesis* 7 (9): 353–59.

Sushma, N. John, B. Mahitha, K. Mallikarjuna, and B. Deva Prasad Raju. 2016. "Bio-Inspired ZnO Nanoparticles from *Ocimum tenuiflorum* and Their in Vitro Antioxidant Activity." *Applied Physics A: Materials Science and Processing* 122 (5): 1–10. https://doi.org/10.1007/s00339-016-0069-9.

Swaroop, Chetan, and Mukul Shukla. 2018. "Nano-Magnesium Oxide Reinforced Polylactic Acid Biofilms for Food Packaging Applications." *International Journal of Biological Macromolecules* 113 (2017): 729–36. https://doi.org/10.1016/j.ijbiomac.2018.02.156.

Tabatabaei, Leila, Negin Mahdavi-asl, and Ali Khalili. 2020. "Application of Copper Nano Particles in Antimicrobial Packaging: A Mini Review." *Frontiers in Sustainable Food Systems* 4 (5): 14–18.

Teixeira, José A. 2018. "Grand Challenges in Sustainable Food Processing." *Frontiers in Sustainable Food Systems* 2 (June): 2017–19. https://doi.org/10.3389/fsufs.2018.00019.

Timilsena, Yakindra Prasad, Md. Amdadul Haque, and Benu Adhikari. 2020. "Encapsulation in the Food Industry: A Brief Historical Overview to Recent Developments." *Food and Nutrition Sciences* 11 (06): 481–508. https://doi.org/10.4236/fns.2020.116035.

Tu, Hai Le. 2019. "Biosynthesis, Characterization and Photocatalytic Activity of Copper / Copper Oxide Nanoparticles Produced Using Aqueous Extract of Lemongrass Leaf." *Composite Materials* 3 (1): 30–35. https://doi.org/10.11648/j.cm.20190301.14.

Turrina, Chiara, Sonja Berensmeier, and Sebastian P. Schwaminger. 2021. "Bare Iron Oxide Nanoparticles as Drug Delivery Carrier for the Short Cationic Peptide Lasioglossin." *Pharmaceuticals* 14 (5). https://doi.org/10.3390/ph14050405.

Ubbink, Job, and Jessica Krüger. 2006. "Physical Approaches for the Delivery of Active Ingredients in Foods." *Trends in Food Science and Technology* 17 (5): 244–54. https://doi.org/10.1016/j.tifs.2006.01.007.

Uikey, Prateek, and Kirti Vishwakarma. 2016. "Review of Zinc Oxide (Zno) Nanoparticles Applications and Properties." *International Journal of Emerging Technology in Computer Science & Electronics (IJETCSE)* 21 (2): 976–1353.

Urabe, Aliyaa A., and Wisam J. Aziz. 2019. "Biosynthesis of Cobalt Oxide (Co_3O_4) Nanoparticles Using Plant Extract of *Camellia sinensis* (L.) Kuntze and *Apium graveolens* L. as the Antibacterial Application" *World News of Natural Sciences* 24 (April): 357–65.

Venkateasan, Alagesan, Rangasamy Prabakaran, and Venugopal Sujatha. 2017. "Phytoextract-Mediated Synthesis of Zinc Oxide Nanoparticles Using Aqueous Leaves Extract of *Ipomoea pes-caprae* (L.) R. Br Revealing Its Biological Properties and Photocatalytic Activity." *Nanotechnology for Environmental Engineering* 2 (1): 1–15. https://doi.org/10.1007/s41204-017-0018-7.

Walia, Niharika, Nandita Dasgupta, Shivendu Ranjan, Chidambaram Ramalingam, and Mansi Gandhi. 2019. "Methods for Nanoemulsion and Nanoencapsulation of Food Bioactives." *Environmental Chemistry Letters* 17 (4): 1471–83. https://doi.org/10.1007/s10311-019-00886-w.

Wu, Yongjun, Yan Gu, Ling Tong, Ronghua Chen, and Nina Xie. 2019. "Electrochemical Synthesis of ZnO Nanoparticles and Preparation of Pea Starch / ZnO Composite for Active Food Packaging Application." *International Journal of Electrochemical Science* 14: 10753. https://doi.org/10.20964/2019.12.38.

Wu, Zhengguo, Xiujie Huang, Yi-chen Li, Hanzhen Xiao, and Xiaoying Wang. 2018. "PT SC." *Carbohydrate Polymers*. https://doi.org/10.1016/j.carbpol.2018.07.030.

Yang, Mingyi, Ze Liang, Lei Wang, Ming Qi, Zisheng Luo, and Li Li. 2020. "Microencapsulation Delivery System in Food Industry: Challenge and the Way Forward." *Advances in Polymer Technology* 2020. https://doi.org/10.1155/2020/7531810.

Yaqoob, Asim Ali, Hilal Ahmad, Tabassum Parveen, Akil Ahmad, Mohammad Oves, Iqbal M.I. Ismail, Huda A. Qari, Khalid Umar, and Mohamad Nasir Mohamad Ibrahim. 2020. "Recent Advances in Metal Decorated Nanomaterials and Their Various Biological Applications: A Review." *Frontiers in Chemistry* 8 (May): 1–23. https://doi.org/10.3389/fchem.2020.00341.

Yaqoob, Sundas Bahar, Rohana Adnan, Raja Muhammad Rameez Khan, and Mohammad Rashid. 2020. "Gold, Silver, and Palladium Nanoparticles: A Chemical Tool for Biomedical Applications." *Frontiers in Chemistry* 8 (June): 1–15. https://doi.org/10.3389/fchem.2020.00376.

Yoneyama, Hiroshi, and Ryoichi Katsumata. 2006. "Antibiotic Resistance in Bacteria and Its Future for Novel Antibiotic Development." *Bioscience, Biotechnology and Biochemistry* 70 (5): 1060–75. https://doi.org/10.1271/bbb.70.1060.

Yoon, Ki Young, Jeong Hoon Byeon, Jae Hong Park, and Jungho Hwang. 2007. "Susceptibility Constants of *Escherichia coli* and *Bacillus subtilis* to Silver and Copper Nanoparticles." *Science of the Total Environment* 373 (2–3): 572–75. https://doi.org/10.1016/j.scitotenv.2006.11.007.

Zhang, Ruojie, and David Julian McClements. 2016. "Enhancing Nutraceutical Bioavailability by Controlling the Composition and Structure of Gastrointestinal Contents: Emulsion-Based Delivery and Excipient Systems." *Food Structure* 10: 21–36. https://doi.org/10.1016/j.foostr.2016.07.006.

Zhao, Daoli, Xuefei Guo, Tingting Wang, Noe Alvarez, and Vesselin N Shanov. 2014. "Simultaneous Detection of Heavy Metals by Anodic Stripping Voltammetry Using Carbon Nanotube Thread." *Electroanalysis* 26: 488–96. https://doi.org/10.1002/elan.201300511.

Ziental, D., B. Czarczynska-Goslinska, D.T. Mlynarczyk, A. Glowacka-Sobotta, B. Stanisz, T. Goslinski, and L. Sobotta. 2020. "Titanium Dioxide Nanoparticles: Prospects and Applications in Medicine." *Nanomaterials (Basel)* 10 (2): 387. https://doi.org/10.3390/nano10020387. PMID: 32102185; PMCID: PMC7075317.

Zuo, Yinxiu, Jingkun Xu, Fengxing Jiang, Xuemin Duan, Limin Lu, Guo Ye, Changcun Li, and Yongfang Yu. 2017. "Utilization of AuNPs Dotted S-Doped Carbon Nanoflakes as Electrochemical Sensing Platform for Simultaneous Determination of Cu (II) and Hg (II)." *Journal of Electroanalytical Chemistry* 794 (March): 71–77. https://doi.org/10.1016/j.jelechem.2017.04.002.

12 Applications of Nanotechnology-Based Approaches for Targeted Delivery of Nutraceuticals

*Hitesh Chopra, Shivani Sharma, Saba Yousaf, Rahat Naseer, Shakeel Ahmed, and Atif Amin Baig**

CONTENTS

12.1 Introduction 329
12.2 Present Status of Nutraceuticals 330
12.3 Existing Problems 330
12.4 Concept of Nano-Nutraceuticals 331
12.4.1 Lipids 331
12.4.2 Phenolics and Flavonoids 332
12.4.3 Vitamins 332
12.4.4 Minerals 333
12.5 Nanoformulations and Carriers for Nutraceuticals 333
12.6 Disadvantages 338
12.7 Future Directions and Conclusion 338
References 338

12.1 INTRODUCTION

Food, shelter, clothing, and medicine are the four essential components everyone needs to survive. People have known and utilized medications to manage diseases for thousands of years. An overall healthy diet is required in addition to medicines for the treatment of illness. Food is recognized as the primary source of nourishment for everyone. An essential biological activity, feeding, is responsible for sustaining normal physiological functions in the body (Panagiotou and Nielsen 2009). According to current research, food may be utilized as a medicine for illness prevention, treatment, and health promotion. This is where nutraceuticals originated (Subbiah 2007; Zelig and Rigassio Radler 2012). Eating as a therapy offers many potential benefits. Food is a safer option than medicines. Food is regarded as a source of sustenance in nutraceutical products to regulate and alter physiological processes inside the human body.

Nutraceuticals are extensively utilized in modern therapeutic practice (Kour et al. 2022). The focus is usually on the active component that may be useful in medicine. A significant issue in assessing nutraceutical product quality (Costello and Coates 2001). Medicines, dietary supplements, and food components are examples of nutraceuticals today. Many nutraceutical products are now accessible on the market due to the idea that nutraceuticals are "natural and safe, and may prevent illness, substitute prescription medications, and make up for a bad diet" (Pawar et al. 2013). The Dietary Supplements Office says ongoing nutraceutical research is essential to improving

* Corresponding author: atifamin@unisza.edu.my

DOI: 10.1201/9781003244721-12

supplement quality. The distribution of nutraceuticals for body usage is essential for successful use in internal medicine. Based on sophisticated nanotechnological developments, various novel methods for enhancing nutraceutical dispersion may be employed. This chapter provides current information about nutraceutical delivery.

12.2 PRESENT STATUS OF NUTRACEUTICALS

To profit from nutraceuticals, they must be brought into the human body. Nutraceutical products need to be correctly manufactured, like pharmaceuticals, to improve biokinetic and biodynamic responses in the human body. In contrast to medicines, nutraceutical compounds are typically taken orally rather than intravenously. Consequently, absorption of nutraceutical supplements via the gastrointestinal system is generally seen as a significant issue. In reality, significant advances in medicine delivery have been regularly implemented. However, the nutraceutical product is not generally regarded as a cause of worry.

As a consequence, the issue of identifying the intended treatment location may be seen (Shoji and Nakashima 2004). It may be stated that there are still many problems to be explored in the future with the dissemination of nutraceuticals. The bioavailability of nutritional supplements must be taken into account, as described in Shoji and Nakashima, to encourage the most efficient use of nutraceutical goods. The way this is carried out should thus be a significant future trend in nutraceutical research (Shoji and Nakashima 2004).

12.3 EXISTING PROBLEMS

It has previously been shown that the distribution of nutraceuticals is influenced by a wide range of factors. Several issues prevent nutraceuticals from reaching their target location consistently. Here are a few examples. Customer anxiety is centered on nutraceuticals, which are a cause of concern for many. There may be issues with some goods absorbing the active ingredients. Active chemicals that are both labile and insoluble in water may cause poor absorption in a variety of situations. The issue has previously been brought to light in the context of vitamin supplementation. It presents legitimate problems, since the vitamin is often labile, making it challenging to give the appropriate nutraceutical dose (Blair 1986). Another well-known example is the inclusion of calcium in food. Reduced calcium absorption is unquestionably a significant issue associated with the use of calcium supplements (Walden 1989).

Another possibility is that the issue is caused by a lack of communication with healthcare providers. The active ingredients in nutraceutical products may have undesirable side effects if they combine with other medicines or nutraceuticals, which is possible. Depending on the circumstances of interaction, the desired therapeutic impact may be increased or diminished, or it may result in new, unpleasant, and detrimental effects. When some meals and medications are eaten simultaneously, the body's capacity to utilize a specific food or drug may be hampered, and certain medications may produce severe adverse effects, according to Bushra and his collaborators (Bushra, Aslam, and Khan 2011). Medicines can have a detrimental impact on nutritional status in certain people, particularly those with abnormalities in nutrient absorption, metabolism, use, and excretion (Chan 2013). However, food or nutraceutical supplements may interfere with absorption and decrease the efficacy of pharmaceuticals and other medications.

Consequently, the issue of interaction has grown more significant in recent years. Indeed, when it comes to the use of nutraceutical supplements for the first time, as previously stated, there is a problem with interaction. This problem is often overlooked, and it may be very hazardous in actual practice (Markowitz and Zhu 2012). Assessing all nutraceutical goods is recommended to identify potential interactions between the products and the body's own systems (Santos and Boullata 2005). At the end of the day, it should be emphasized that cases with a bad family history should be given more consideration than others. In these circumstances, individuals with specialized knowledge of the correct dose and a combination of a nutraceutical product with pathophysiological conditions such as ageing, hepatic dysfunction, or renal impairment are required (Bhattacharya et al. 2022).

12.4 CONCEPT OF NANO-NUTRACEUTICALS

Nanoscience is the newest scientific field to develop on the "nano" level, and it deals with objects and processes that are extremely small compared with their surroundings. For instance, biomedical research can benefit in many ways from nanoscience (Chopra et al. 2021). It is possible to utilize nanotechnology in nutritional medicine (Helal et al. 2019). Nanotechnology has been developed recently in nutritional therapy, and it is likely to enhance the supply of nutraceutical components. The significance of quality control in nutraceuticals nowadays cannot be overestimated. Many nutraceutical products now face difficulties in providing nutrients to the body to be fully used *in vivo* (Nasri et al. 2014).

Nanotechnology can significantly increase the supply of drugs and the spread of nutraceuticals that have not yet been traced. Using nanotechnology may help resolve some of the fundamental problems behind the distribution of nutraceutical goods. The use of nanotechnologies in drug delivery may improve the precise delivery of medicines to their targeted locations. Nanodelivery methods have the potential to create molecules that, according to the national health institutes, can "combine single elements with the size, surface activity and charging of nanostructures." Advanced nanotechnology may reduce absorption and dissolution during the feeding process.

As a result, the possibility for enhancing existing nutraceutical products arises. Several beneficial properties may be achieved via the usage of nanotechnology. The discovered qualities may be useful in turning a troubled biological process into a good one for the future. If you reduce the surface area of the molecules, you speed up the required biological process in the first instance, and vice versa. If silymarin is synthesized into a nanoliposome and then exposed to light, the absorption is significantly increased (Javed, Kohli, and Ali 2011). In the second place, it is possible to help remove the unwanted biological effects of nanomolecular nutraceutical substances. For example, when it comes to vitamin and antioxidant nanoencapsulation, the level of gastrointestinal damage after consumption may decrease (Souto et al. 2013). As a result, it can be inferred that intravenous usage of the technique of nanoadministration may lead to effective barriers (Chrastina, Massey, and Schnitzer 2011).

12.4.1 LIPIDS

The demand for natural, non-animal industrial products has significantly grown across all sectors of society in recent years. The beauty industry is interested in vegetable oils as raw materials for producing natural, sustainable cosmetics because of the availability of renewable resources and ease of access, primarily because of seed wastes (Balboa et al. 2014). Triglycerides comprise different fatty acids enhanced by mono- and di-glycerides, and free fatty acids form a cohesive whole (Roncero et al. 2020). In the presence of natural antioxidants like triterpenes, carotenoids, and polyphenols, oils are anticarcinogenic and anti-inflammatory, protecting the skin from reactive oxygen (ROS) species (Câmara et al. 2021; Dhalaria et al. 2020; Dhavamani, Poorna Chandra Rao, and Lokesh 2014). Oils are used as moisturizers and emollients in cosmetic compositions because they assist in hydrating the skin (Alves et al. 2020). As a natural component of the lipid fraction, vegetable oils provide numerous advantages for the skin and therapeutic activities because they regulate oxidative stress (Badea et al. 2015). A range of natural sources, including grains, fruit, animal, oils, plants, and mushrooms contains lipids. Lipids are also present in several natural sources. While humans and other animals have a range of metabolic breakdown and lipid synthesis mechanisms, some lipids must be obtained via the diet.

A fatty acid is a carboxylic acid with an aliphatic tail, unsaturated or saturated (chain). For living organisms, fatty acids are essential nutrient components because they supply energy and fat. Long-chain polyunsaturated fatty acids (PUFAs) are vital components of human metabolism, especially in three forms: linolenic acid, acids such as docosahexaenoic acid (DHA), and carotenes. Linolenic acid helps to avoid heart conditions, inflammation, autoimmune disorders, hypertension, hypotriglyceridemia, and other cardiovascular issues (Bernal et al. 2011; Balić et al. 2020; Ruscica et al. 2020). Polyunsaturated fatty acids, such as DHA, found in nutraceuticals may protect against cardiovascular disease and other diseases (Zimet, Rosenberg, and Livney 2011). Carotenes are naturally generated by combining two geranyl-geranyl diphosphate molecules into a C40 tetraterpene. They are present in many species, including plants, algae, yeast, and fungi. They color a variety of fruits and vegetables yellow, orange, and red (Namitha and Negi 2010; Cömert, Mogol, and Gökmen 2020). The wide range, structural variation, and diversity of activities of these compounds are significant. Carotenoids are present in high concentrations in fruit and vegetables and are essential to human nutrition. In addition to this, a vitamin A carotenoid deficiency leads to visual impairment and higher mortality among humans due to decreased innate and adaptive immunity (Stephensen 2001). Lycopene lacks an effect on vitamin A, yet it is one of the most potent antioxidant carotenoids. Lycopene's strong antioxidant activity protects the biological system from reactive oxygen and nitrogen species and reduces the risk of cardiovascular diseases in humans (Saini, Nile, and Park 2015; Forni et al. 2019).

12.4.2 Phenolics and Flavonoids

A fatty acid is a carboxylic acid having an unsaturated or saturated aliphatic tail (chain). Fatty acids are important components of human metabolism and help avoid coronary heart conditions, inflammatory disorders, autoimmune disorders, high blood pressure, and cardiovascular issues (Grochowski et al. 2018; Yu et al. 2019; Yi, Ma, and Ren 2017). PUFAs, including DHA, present in nutraceuticals may contribute to protection from cardiovascular or other conditions (Zimet et al. 2011). The C40 tetraterpene's natural colors are produced by the conjunction of two C20 geranyl-geranyl diphosphate molecules. They exist in various organisms, such as plants, algae, yeast, and fungi. The broad distribution, morphological variety, and diversity of these compounds are important. Carotenoids are found in high quantities in fruit and vegetables and are essential to human nutrition. These chemicals have antioxidant and immunomodulatory properties that together, may prevent the occurrence of degenerative illnesses such as cardiovascular diseases, cataract, diabetes, and diverse malignancies, including prostate and stomach cancers (Chu, Chew, and Nyam 2021; Bhushan et al. 2018), causing further xerophthalmia, keratomalacia, and persistent blindness too. In addition to the aforementioned consequences, vitamin A deficiency in carotenoids reduces innate and adaptive immunity, causing vision loss and higher mortality. While lycopene has no effect on vitamin A, it is a powerful antioxidant (Huang et al. 2018). The strong antioxidant activity of lycopene often protects the cellular system from a range of reactive oxygen and nitrogen species and reduces the risk of cardiovascular illnesses in humans (Kurutas 2016).

12.4.3 Vitamins

Vitamins (vital amines) are a range of chemical compounds necessary to develop and preserve human life (Mora, Iwata, and Andrian 2008). They are sensitive and unstable and may be lost by incorrect exposure to high temperatures, oxygen, light, and moisture. Vitamins are integrated into the daily diet through high vitamins rich diet and various vitamin pills as nutraceutical or functional components. There are two kinds of vitamins: soluble in water and soluble in fat. There are 13 vitamins in the human body: four are fat-soluble and nine water-soluble.

Folic acid is a water-soluble vitamin that plays a vital role in human food. Green or natural food supplements contain this vitamin (folate). Some amounts of folic acid in the regular diet, such as in pregnancy and infancy, are needed for cell division and development (Argyridis 2019). Vitamin C is required for all animals and humans; vitamin C deficiency leads to a disease known as scurvy. It is used as a food additive because of the antioxidant characteristics of the molecule. Vitamin D is an approximately 40-molecule chemical complex referred to as vitamin D2 (Argyridis 2019). Vitamin D2 originates from plants and is present in tiny but readily synthesized amounts, making it a broad spectrum of chemical compounds that play an essential role in the development and preservation of life. Enough of the vitamin is taken in the daily diet via consuming foods high in the vitamin or as a nutraceutical or functional element in a wide range of tablets.

12.4.4 Minerals

Scientifically based mineral nutrients are considered necessary or vital for human health, since they promote the formation of bones and teeth and provide human muscular strength. Micronutrients and macronutrients are also needed in the daily diet. For instance, Fe is necessary to guarantee the correct workings of several human proteins, including hemoglobin, myoglobin, and cytochrome (Liu et al. 2014). Zinc is essential for healthy growth and development (Roohani et al. 2013). Ca is a crucial component of bone health, since it is a bone component (Vannucci et al. 2018). Increased usage of CaP produced calcium phosphates, denture, bone remodeling, pharmacological research, and osteoporosis prevention in food, drinks, and beverages. Selenium is an antioxidant vitamin necessary for thyroid iodine metabolism (Ventura, Melo, and Carrilho 2017). Chromium enhances insulin's action and raises glucose tolerance (Wang et al. 2014).

12.5 NANOFORMULATIONS AND CARRIERS FOR NUTRACEUTICALS

A wide variety of labile and sensitive compounds, from bioactive natural/nutraceutical chemicals to pharmaceutical medicines, may be delivered through new food delivery methods, increasing bioavailability. These bioactive compounds may be combined with solid lipid nanoparticles (NPs) to enhance stability and protect the gastrointestinal tract from damage. In general, the enormous potential of this technology has not been fully realized. The process/stages of formulating nano-nutraceuticals from nutraceuticals are described in Figure 12.1.

Besides the careful design of the biopolymer system, the functional components in the food matrix may be enhanced, labile components preserved, or functional ingredients selectively released. Proteins and polysaccharides may be utilized alone or in combination for many delivery strategies for nutraceuticals and dietary components (Mcclements et al. 2015).

The nanocapsules rupture only when the stomach is reached, avoiding the unpleasant smell of fish oil. Nanocapsules have been used to store and control the release of beneficial live probiotic species to promote good digestive function.

Dietary proteins, particularly globular proteins, have long been valued for their physical and industrial properties. Casein, a milk protein, has been utilized for vitamin D2 in nanosized micelles (Rehan, Ahemad, and Gupta 2019). To provide partial UV-induced protection against Vitamin D2 degradation, researchers demonstrated that nutraceutical compounds may be loaded into casein micelles through spontaneous self-assembly (Cohen et al. 2017). Penalva et al. developed casein NPs for oral delivery of folic acid using lysine or arginine as a stabilizing agent (Penalva et al. 2015). NPs measured about 150 nm and contained approximately 25 mg/mg of folic acid. In casein NPs, oral folic acid absorption was 52% higher than in regular aqueous solution. Casein is necessary to maintain and provide sensitive DHA through induction of Ucp1-receptor. DHA is a major PUFA, as shown by Zimet, Rosenberg, and Livney (2011). DHA may bind to it spontaneously because of its high affinity to casein. In addition, the results show that every protein molecule contains an average of three to four DHA sites. Thermal treatment had no obvious effect on the 50–60 nm DHA micelle casein particle size.

Carum copticum is an antibacterial and antioxidant herb. Its seeds contain thymol, g-terpinene, and r-cymene (Gokhale et al., 2019). To prevent degradation due to direct contact with environmental variables such as essential oils, Tripolyphosphate (TPP) and sodium hexametaphosphate have recently been incorporated into CS NPs (Gong et al., 2012). The optimum formula consisted of 1:1 CS/*C. copticum* oil with a TPP of 0.5% (w/v). Between 10.6 and 60.6% of scavenging activity was observed in *Staphylococcus aureus* and *Staphylococcus epidermidis*, *Bacillus cereus*, *Escherichia coli*, *Salmonella typhimurium*, and *Proteus vulgaris*.

Tripolis, a water-insoluble diterpenoid triepoxide, decreased angiogenesis and thyroid cancer cell growth in pancreatic carcinoma cells. Xu et al. synthesized methoxypoly(ethylene glycol)-poly(lactide) copolymer by a method of solvent evaporation loaded with triptolide (Xu et al. 2008).

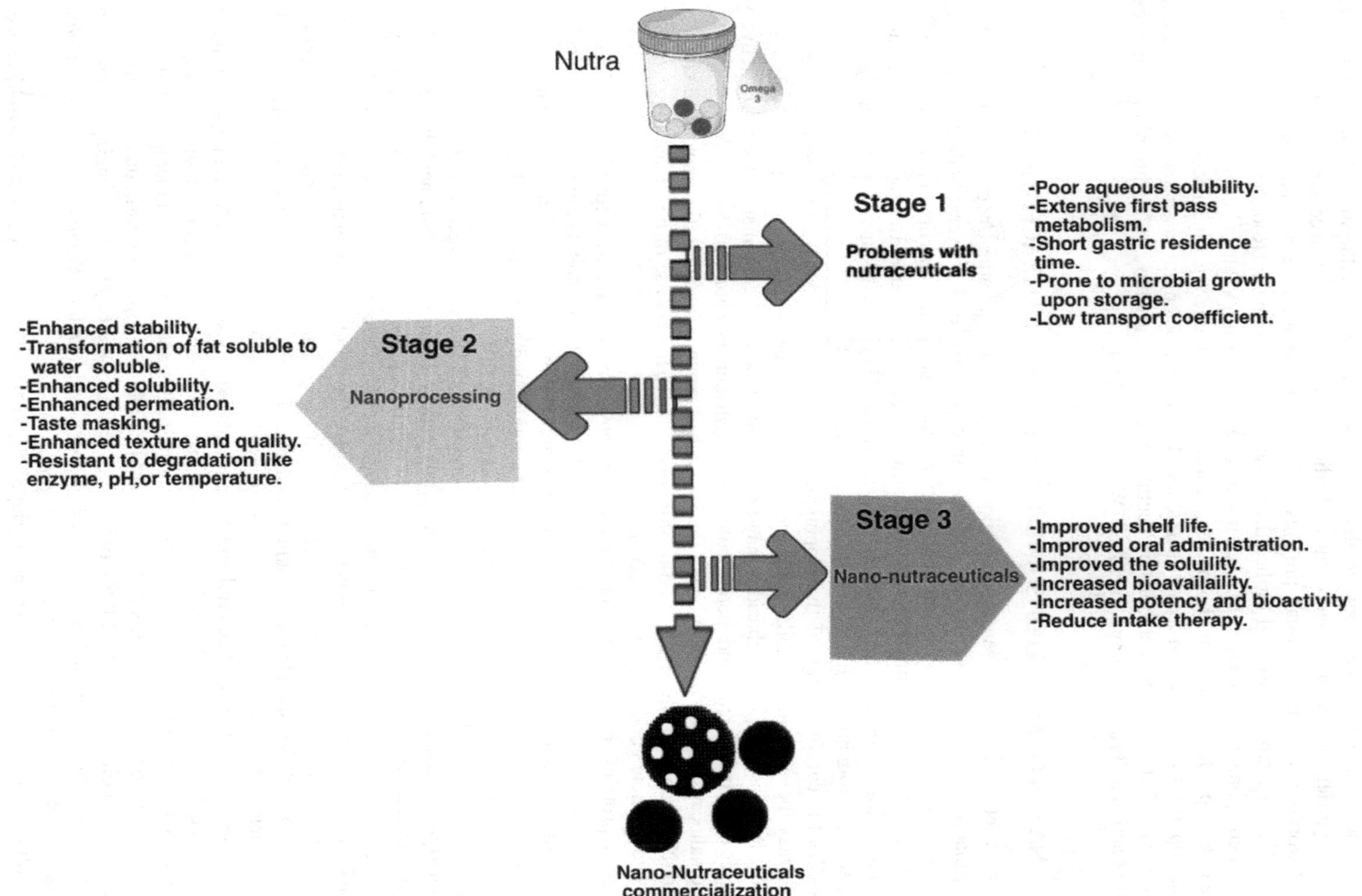
Nutra
Omega 3
Stage 1
Problems with nutraceuticals
-Poor aqueous solubility.
-Extensive first pass metabolism.
-Short gastric residence time.
-Prone to microbial growth upon storage.
-Low transport coefficient.
Stage 2
Nanoprocessing
-Enhanced stability.
-Transformation of fat soluble to water soluble.
-Enhanced solubility.
-Enhanced permeation.
-Taste masking.
-Enhanced texture and quality.
-Resistant to degradation like enzyme, pH,or temperature.
Stage 3
Nano-nutraceuticals
-Improved shelf life.
-Improved oral administration.
-Improved the soluility.
-Increased bioavailaility.
-Increased potency and bioactivity
-Reduce intake therapy.
Nano-Nutraceuticals commercialization

FIGURE 12.1 Schematic representation of various stages of nanoformulations along with their benefits and drawbacks.

The average particle size was 78.9 nm, and encapsulation efficiency was 66.7%. *In vitro* cytotoxicity to A2780 cancer cells was similar in micelles and pure triptolide; no immune-suppressive activity was observed, even though the development of the tumor was inhibited. Phosphatidylcholine and cholesterol interact to form liposomes that contain artemisinin via reverse evaporation (Dadgar et al. 2013). Artemisinin delivery was about 96% with artemisinin liposomes, and NPs with 500nm and 455 nm in diameter were 92% with polyethylene glycol (PEG). A cytotoxic test in MCF-7 breast cancer cells showed that both nanoformulations had a lower inhibitor concentration (IC50) than pure artemisinin. There are high levels of carvacrol and thymol, with well-known antioxidant and anti-bacterial properties, in oregano essential oil (OEO) produced from *Origanum vulgare* L. The OEO was incorporated successfully in CS NPs using an emulsion of O/W and sodium ionization TPP gelation (40–80 nm) (Hosseini et al. 2013). OEO-capable CS NPs at an initial OEO concentration of 0.1–0.8 g/g CS were encapsulated and loaded at about 21–47 and 3–8%, respectively. Research on *in vitro* release began with a breakthrough, and medication releases continued. Zerumbone was discovered to have a possible anticarcinogenic effect in colon, cervical, and breast cancer patients. According to Rahman et al. (2013), the encapsulation efficiency of Zerumbone-filled NLCs was 99.03%. The anticancer efficacy of Zerumbone in NLCs was not affected, and this was a possible therapy technique for leukemia.

Bala et al. investigated poly (lactic-co-glycolic acid) (PLGA) nanoparticles loaded with ellagic acid (EA) (Bala et al. 2006). For the purpose of didodecyl bromide (DMAB)/polyvinyl alcohol, PEG 400 was employed as a cosolvent, while the CS stabilizer was used. The release profiles show that because of the enhanced hydrophobicity of the DMAB, which restricts EA release from the polymer, the release profiles are more hydrophobic than PLGA. Berberine is an active component in the Chinese Coptis rhizome, which has been proven to inhibit the growth of cancer cells. Lin et al. have discovered a method to manufacture berberine–PEG liposomes (Lin et al. 2013). The average size was 121.6 nm, with 14% encapsulation. When tested *in vitro*, the cytotoxicity of HepG2 cells was 2.5 times that of liposomal berberine. Zhang et al. synthesized NPs with an average 150 nm diameter of gambogic acid and lactoferrin with 92.3/7.2% encapsulation effectiveness (Zhang et al. 2013). The natural resin gambogic acid inhibits the development of liver, breast, stomach, and lung cancer cells. Lactoferrin NPs were 1.39 times more efficacious than oral gambogic acid injections.

Noscapine is a decadal toxin suppressant that includes 10% alkaloid opium and is used in breast, cutaneous, ovarian, colon, and lung cancer treatment. Sebak et al. found a potential way of delivering medicines to cancer cells (Sebak et al. 2010). The application is based on noscapine–serum-loaded human NPs of 150–300 mm diameter and 85–96% encapsulation efficiency. The *in vitro* impact on breast cancer cells HER-2 and SK-BR-3 has been studied. The viability of cancer cells in noscapine NPs (40%) was lower than in drug-free NPs, while 93% of noscapine was at concentrations between 10 and 15 mg/mL. Merlin et al. used a dual emulsion method to synthesize ferulic acid–PGGA NPs (Merlin et al. 2012). The resulting NPs were around 483 nm in diameter and had 76% drug loading efficiency. NPs were more effective than pure ferulic acid in suppressing NCI-H460 cells (IC50 30 mM and 40 mM, respectively). Das et al. investigated apigenin–PLGA spherical nanoparticles (Das et al. 2013). The forms were around 101.3 m in diameter and showed about 87% encapsulation efficiency. The NPs improved medication and photoprotection. PLGA and PEG stabilized thymoquinone encapsulated in gastric, colorectal, and breast cancer cells (Ravindran et al. 2010). Researchers found NPs with a 150–200 nm diameter and a 97.5% loading efficiency. Cancer cells studied *in vitro* and *in vivo* had better outcomes than pure cancer cells.

Nanotechnology enables the creation of a range of labile and sensitive chemicals for novel food supply systems ranging from bioactive natural/nutraceutical substances to pharmaceutical compounds, thus improving the bioavailability of active molecules (Burdo 2005). These bioactive compounds may be combined with solid lipid nanoparticles to enhance stability and protect the gastrointestinal tract from damage. In general, the great potential of this technology has not been fully realized. Some of the nanosystems are described in Table 12.1.

TABLE 12.1
Various Nanosystems Studied for Nutraceuticals

Type of nanoformulation	Name of active ingredient	Notes	References
Nanoemulsion and microemulsion	Vitamin A, D, E, K	Improved absorption of liposoluble vitamins by nanoemulsion	(Shi et al. 2018; Choudhry et al. 2016; Il Kim et al. 2019; AL et al. 2019; Ozturk et al. 2015; Guttoff, Saberi, and Mcclements 2015)
	Gardenin B	Use of a nanoemulsion as a carrier system for 5-hydroxy-6,7,8,4-tetramethoxyflavone (Gardenin B)	(Abdallah et al. 2021)
	Tocopherol	Enhancement of tocopherol's oral bioavailability and pharmacological effects	(Gong et al. 2012; Kuo et al. 2008)
	β-Carotene	Determined the effect of carrier lipid concentration (MCT and LCT) on triglyceride digestibility	(Kuo et al. 2008)
		To improve the stability and bioaccessibility of β-carotene nanoemulsions	(Qian et al. 2012)
	Berberine	To increase berberine's oral bioavailability	(Xu et al. 2019; Li et al. 2017; Pund, Borade, and Rasve 2014; Floriano et al. 2021; Sharifi-Rad et al. 2021)
	Curcumin	To prevent lipophilic curcumin decomposition and escape	(Li et al. 2021)
		Curcumin bioavailability improved using organogel nanoemulsion	(Yu and Huang 2012)
		Prepare nonionic food-grade nanoemulsions using quercetin to stabilize curcumin	(Iqbal et al. 2020)
	Quercetin	The positive impact of nanoemulsion on the bioavailability of quercetin	(Pool et al. 2013; Gokhale, Mahajan, and Surana 2019)
	Coenzyme Q10	Effect of long-chain fatty acids on the bioavailability of coenzyme Q10	(Cho et al. 2014; Belhaj et al. 2012)
		Enhanced transdermal permeability	(El-Leithy et al. 2018)
	Capsaicin	Study the pharmacokinetics and process parameters of the formulated capsaicin nanoemulsions	(Choi et al. 2011; Akbas, Soyler, and Oztop 2018; J. H. Kim et al. 2014)
	Puerarin	Puerarin bioavailability following nasal nanoemulsion administration	(Yu et al. 2011)
	Fish oil	Improve intestinal absorption of fish oil by making nanoemulsions	(Santos et al. 2021; Alhakamy et al. 2020)
	Lycopene	Process optimization for lycopene nanoemulsion and beverage manufacturing	(S. O. Kim et al. 2014)
	Resveratrol	To improve resveratrol's oral bioavailability	(Sessa et al. 2014; Davidov-Pardo and McClements 2015)

(Continued)

TABLE 12.1 (CONTINUED)
Various Nanosystems Studied for Nutraceuticals

Type of nanoformulation	Name of active ingredient	Notes	References
	Silymarin	Enhance silymarin solubility and bioavailability	(Nagi et al. 2017; Ahmad et al. 2018; Parveen et al. 2015)
	(+)-Catechin	To increase oral and cerebral distribution of (+)-catechin	(Tsai and Chen 2016)
Liposomes	Curcumin	Liposomal curcumin uptake and antioxidant properties	(Chen et al. 2009; Bulboacă et al. 2019; Storka et al. 2015; Li et al. 2007; Basnet et al. 2012)
	Quercetin	Enhanced skin penetration	(Jeon, Yoo, and Park 2015; Liu et al. 2013; Hatahet et al. 2018)
	Coenzyme Q10	The antioxidant effects of liposomes and solid lipid nanoparticles were compared	(Gokce et al. 2012)
	Vitamin E acetate	To increase solubility, action onset, and bioavailability	(Muthu et al. 2011)
	Grape seed extract	Enhanced stability of extract	(Gibis et al. 2014; Gibis, Rahn, and Weiss 2013; Gibis, Vogt, and Weiss 2012)
	Carotenoids	To test a liposomal delivery system's ability to modify carotenoid bioaccessibility	(Tan et al. 2014; Xia et al. 2015)
	Melatonin	To increase melatonin bioavailability	(Dubey et al. 2006; Nguyen et al. 2015)
	Resveratrol	To produce a stable formulation with sustained release	(Alanazi et al. 2020; Alanazi et al. 2020; Wang et al. 2011; Narayanan et al. 2009)
Solid lipid nanoparticles	Camptothecin	To evaluate the biodistribution of camptothecin solid lipid nanoparticles after per se administration	(Du et al. 2018)
	β-Carotene	Improvement of encapsulated lipophilic component's chemoprotective capacity	(Jain et al. 2019)
	Melatonin	For sustained release in plasma	(He et al. 2021; Mirhoseini et al. 2019)
	Resveratrol	To increase resveratrol's oral bioavailability	(Mohseni et al. 2019; Ramalingam and Ko 2016)
Nanostructured lipid carriers (NLCs)	Lutein	To investigate the behavior of lutein NLCs in fish oil supplemented with ω-3 fatty acids	(Lacatusu et al. 2013)
	Coenzyme Q10	To develop a new and promising carrier technology for coenzyme Q10 delivery	(Nanjwade, Kadam, and Manvi 2013; Schwarz et al. 2013)
Nanocrystals	Curcumin	Curcumin solubility and bioavailability improved	(Vidlářová et al. 2016; Rachmawati et al. 2016; Oshi et al. 2020)
	Coenzyme Q10	To increase oral coenzyme Q10 bioavailability	(Song et al. 2016; Sun et al. 2012)
	Lutein	Lutein adsorption and dermal penetration	(Chang et al. 2018; Liu et al. 2017; Seto et al. 2019)

12.6 DISADVANTAGES

NPs are regularly produced in the nutraceutical and pharmaceutical sectors, but their environmental and ecological impacts should be carefully assessed. By exposing biological molecules to NPs, bio-coronas were produced. It is essential to interact with macromolecules or organelles for NPs since they may have less cellular influence (Lynch et al. 2013). In epigenetic processes, NPs have been shown to cause cell death, immunotoxicity, genotoxicity, cytotoxicity, oxidative stress, and aberrant genetic expression (Dusinska, Magdolenova, and Fjellsbø 2013; Dai et al. 2020). Although there may be no changes in the DNA sequence, epigenetic effects can induce heritable alterations to gene function. Mechanisms include histone modifications (methyl chromogenic remodeling, acetylation, phosphorylation, and ATP-ribosylation), regulatory DNA methylation, and post-transcription gene expression microRNA alterations. Environmental and lifestyle factors may induce epigenetic changes (Alegría-Torres, Baccarelli, and Bollati 2011; Jaenisch and Bird 2003). Transgenerational inheritance may occasionally pass epigenetic markers across generations (Jablonka and Raz 2009; Skinner et al. 2013). TiO_2 NPs are among the most often used NPs for antimicrobial packaging in the food industry. In most *in vitro* experiments using TiO_2 NPs, some research shows an increased degree of DNA disorder and oxidized DNA damage (Magdolenova et al. 2014).

12.7 FUTURE DIRECTIONS AND CONCLUSION

Nutraceutical products feature high dosing levels, low consistency and repeatability, challenges in characterization and solubility, problems with formulation and 2-year stability, and significant manufacturing problems. It is also essential to use nano-nutraceuticals from biotechnological goods that are genetically modified and tissue-culture products and compare their effectiveness with natural sources. Nanotechnology is widely used in the food industry. New foods based on nanotechnology may be tailored to tastes or health needs.

Nanotechnology will aid nutraceutical products in overcoming some of their current problems. It has the potential to reduce dosages and enhance nutraceutical stability. Nanostructured bioactive formulations will help to improve their characterization, patient acceptability, and more significantly, their therapeutic effectiveness. Many nutraceutical products are available in nano sizes. Nanosized nutraceuticals may also be more readily absorbed and more bioavailable. We thus conclude that nano-based carriers enhance nutraceutical efficiency and availability in the form of low solubility, stability, and bioavailability.

REFERENCES

Abdallah, Hossam M., Hany M. El-Bassossy, Ali M. El-Halawany, Tarek A. Ahmed, Gamal A. Mohamed, Azizah M. Malebari, and Noura A. Hassan. 2021. "Self-Nanoemulsifying Drug Delivery System Loaded with *Psiadia punctulata* Major Metabolites for Hypertensive Emergencies: Effect on Hemodynamics and Cardiac Conductance." *Frontiers in Pharmacology* 12. https://doi.org/10.3389/fphar.2021.681070.

Ahmad, Usama, Juber Akhtar, Satya Prakash Singh, B. Badruddeen, Farhan Jalees Ahmad, Sahabjada Siddiqui, and W. Wahajuddin. 2018. "Silymarin Nanoemulsion against Human Hepatocellular Carcinoma: Development and Optimization." *Artificial Cells, Nanomedicine and Biotechnology* 46 (2). https://doi.org/10.1080/21691401.2017.1324465.

Akbas, Elif, Betul Soyler, and Mecit Halil Oztop. 2018. "Formation of Capsaicin Loaded Nanoemulsions with High Pressure Homogenization and Ultrasonication." *LWT* 96. https://doi.org/10.1016/j.lwt.2018.05.043.

Alanazi, Abeer, Laila Fadda, Ahlam Alhusaini, and Rehab Ahmad. 2020. "Antioxidant, Antiapoptotic, and Antifibrotic Effects of the Combination of Liposomal Resveratrol and Carvedilol against Doxorubicin-Induced Cardiomyopathy in Rats." *Journal of Biochemical and Molecular Toxicology* 34 (7). https://doi.org/10.1002/jbt.22492.

Alanazi, Abeer M., Laila Fadda, Ahlam Alhusaini, Rehab Ahmad, Iman H. Hasan, and Ayman M. Mahmoud. 2020. "Liposomal Resveratrol and/or Carvedilol Attenuate Doxorubicin-Induced Cardiotoxicity by Modulating Inflammation, Oxidative Stress and S100A1 in Rats." *Antioxidants* 9 (2). https://doi.org/10.3390/antiox9020159.

Alegría-Torres, Jorge Alejandro, Andrea Baccarelli, and Valentina Bollati. 2011. "Epigenetics and Lifestyle." *Epigenomics*. https://doi.org/10.2217/epi.11.22.

Alhakamy, Nabil A., Hibah M. Aldawsari, Khaled M. Hosny, Javed Ahmad, Sohail Akhter, Ahmed K. Kammoun, Adel F. Alghaith, Hani Z. Asfour, Mohammed W. Al-Rabia, and Shadab Md. 2020. "Formulation Design and Pharmacokinetic Evaluation of Docosahexaenoic Acid Containing Self-Nanoemulsifying Drug Delivery System for Oral Administration." *Nanomaterials and Nanotechnology* 10. https://doi.org/10.1177/1847980420950988.

Alves, Thais F.R., Margreet Morsink, Fernando Batain, Marco V. Chaud, Taline Almeida, Dayane A. Fernandes, Classius F. da Silva, Eliana B. Souto, and Patricia Severino. 2020. "Applications of Natural, Semi-Synthetic, and Synthetic Polymers in Cosmetic Formulations." *Cosmetics*. https://doi.org/10.3390/COSMETICS7040075.

Argyridis, Savvas. 2019. "Folic Acid in Pregnancy." *Obstetrics, Gynaecology and Reproductive Medicine*. https://doi.org/10.1016/j.ogrm.2019.01.008.

Badea, Gabriela, Ioana Lăcătuşu, Nicoleta Badea, Cristina Ott, and Aurelia Meghea. 2015. "Use of Various Vegetable Oils in Designing Photoprotective Nanostructured Formulations for UV Protection and Antioxidant Activity." *Industrial Crops and Products* 67. https://doi.org/10.1016/j.indcrop.2014.12.049.

Bala, I., V. Bhardwaj, S. Hariharan, S. V. Kharade, N. Roy, and M.N.V. Ravi Kumar. 2006. "Sustained Release Nanoparticulate Formulation Containing Antioxidant-Ellagic Acid as Potential Prophylaxis System for Oral Administration." *Journal of Drug Targeting* 14 (1). https://doi.org/10.1080/10611860600565987.

Balboa, Elena M., Maria Luisa Soto, Daniele R. Nogueira, Noelia González-López, Enma Conde, Andrés Moure, María Pilar Vinardell, Montserrat Mitjans, and Herminia Domínguez. 2014. "Potential of Antioxidant Extracts Produced by Aqueous Processing of Renewable Resources for the Formulation of Cosmetics." *Industrial Crops and Products* 58. https://doi.org/10.1016/j.indcrop.2014.03.041.

Balić, Anamaria, Domagoj Vlašić, Kristina Žužul, Branka Marinović, and Zrinka Bukvić Mokos. 2020. "Omega-3 versus Omega-6 Polyunsaturated Fatty Acids in the Prevention and Treatment of Inflammatory Skin Diseases." *International Journal of Molecular Sciences*. https://doi.org/10.3390/ijms21030741.

Basnet, Purusotam, Haider Hussain, Ingunn Tho, and Natasa Skalko-Basnet. 2012. "Liposomal Delivery System Enhances Anti-Inflammatory Properties of Curcumin." *Journal of Pharmaceutical Sciences* 101 (2). https://doi.org/10.1002/jps.22785.

Belhaj, Nabila, Franois Dupuis, Elmira Arab-Tehrany, Frédéric M. Denis, Cédric Paris, Isabelle Lartaud, and Michel Linder. 2012. "Formulation, Characterization and Pharmacokinetic Studies of Coenzyme Q10 PUFA's Nanoemulsions." *European Journal of Pharmaceutical Sciences* 47 (2). https://doi.org/10.1016/j.ejps.2012.06.008.

Bernal, J., J. A. Mendiola, E. Ibáñez, and A. Cifuentes. 2011. "Advanced Analysis of Nutraceuticals." *Journal of Pharmaceutical and Biomedical Analysis*. https://doi.org/10.1016/j.jpba.2010.11.033.

Bhattacharya, Tanima, Giselle Amanda Borges E. Soares, Hitesh Chopra, Md Mominur Rahman, Ziaul Hasan, Shasank S. Swain, and Simona Cavalu. 2022. "Applications of Phyto-Nanotechnology for the Treatment of Neurodegenerative Disorders." *Materials* 15 (3): 804. https://doi.org/10.3390/MA15030804.

Bhushan, Brij, Dipendra Kumar Mahato, Deepak Kumar Verma, Mandira Kapri, and Prem Prakash, Srivastav. 2018. "Potential Health Benefits of Tea Polyphenols: A Review." *Engineering Interventions in Agricultural Processing*. https://doi.org/10.1201/9781315207377-10.

Blair, Kathryn A. 1986. "Vitamin Supplementation and Megadoses." *Nurse Practitioner* 11 (7). https://doi.org/10.1097/00006205-198607000-00003.

Bulboacă, Adriana Elena, Alina S. Porfire, Lucia R. Tefas, Paul Mihai Boarescu, Sorana D. Bolboacă, Ioana C. Stănescu, Angelo Corneliu Bulboacă, and Gabriela Dogaru. 2019. "Liposomal Curcumin Is Better than Curcumin to Alleviate Complications in Experimental Diabetic Mellitus." *Molecules* 24 (5). https://doi.org/10.3390/molecules24050846.

Burdo, O. G. 2005. "Nanoscale Effects in Food-Production Technologies." *Journal of Engineering Physics and Thermophysics* 78. https://doi.org/10.1007/s10891-005-0033-6.

Bushra, Rabia, Nousheen Aslam, and Arshad Yar Khan. 2011. "Food-Drug Interactions." *Oman Medical Journal*. https://doi.org/10.5001/omj.2011.21.

Câmara, José S., Bianca R. Albuquerque, Joselin Aguiar, Rúbia C.G. Corrêa, João L. Gonçalves, Daniel Granato, Jorge A.M. Pereira, Lillian Barros, and Isabel C.F.R. Ferreira. 2021. "Food Bioactive Compounds and Emerging Techniques for Their Extraction: Polyphenols as a Case Study." *Foods*. https://doi.org/10.3390/foods10010037.

Chan, Lingtak Neander. 2013. "Drug-Nutrient Interactions." *Journal of Parenteral and Enteral Nutrition* 37 (4). https://doi.org/10.1177/0148607113488799.

Chang, Daoxiao, Yanni Ma, Guoyu Cao, Jianhuan Wang, Xia Zhang, Jun Feng, and Wenping Wang. 2018. "Improved Oral Bioavailability for Lutein by Nanocrystal Technology: Formulation Development, In Vitro and In Vivo Evaluation." *Artificial Cells, Nanomedicine and Biotechnology* 46 (5). https://doi.org/10.1080/21691401.2017.1358732.

Chen, Changguo, Thomas D. Johnston, Hoonbae Jeon, Roberto Gedaly, Patrick P. McHugh, Thomas G. Burke, and Dinesh Ranjan. 2009. "An in Vitro Study of Liposomal Curcumin: Stability, Toxicity and Biological Activity in Human Lymphocytes and Epstein-Barr Virus-Transformed Human B-Cells." *International Journal of Pharmaceutics* 366 (1–2). https://doi.org/10.1016/j.ijpharm.2008.09.009.

Cho, H. T., L. Salvia-Trujillo, J. Kim, Y. Park, H. Xiao, and D. J. McClements. 2014. "Droplet Size and Composition of Nutraceutical Nanoemulsions Influences Bioavailability of Long Chain Fatty Acids and Coenzyme Q10." *Food Chemistry* 156. https://doi.org/10.1016/j.foodchem.2014.01.084.

Choi, Ae Jin, Chul Jin Kim, Yong Jin Cho, Jae Kwan Hwang, and Chong Tai Kim. 2011. "Characterization of Capsaicin-Loaded Nanoemulsions Stabilized with Alginate and Chitosan by Self-Assembly." *Food and Bioprocess Technology* 4 (6). https://doi.org/10.1007/s11947-011-0568-9.

Chopra, Hitesh, Protity Shuvra Dey, Debashrita Das, Tanima Bhattacharya, and Muddaser Shah. 2021. "Curcumin Nanoparticles as Promising Therapeutic Agents for Drug Targets."

Choudhry, Qaisra Naheed, Mi Jeong Kim, Tae Gyun Kim, Jeong Hoon Pan, Jun Ho Kim, Sung Jin Park, Jin Hyup Lee, and Young Jun Kim. 2016. "Saponin-Based Nanoemulsification Improves the Antioxidant Properties of Vitamin A and E in AML-12 Cells." *International Journal of Molecular Sciences* 17 (9). https://doi.org/10.3390/ijms17091406.

Chrastina, Adrian, Kerri A. Massey, and Jan E. Schnitzer. 2011. "Overcoming In Vivo Barriers to Targeted Nanodelivery." *Wiley Interdisciplinary Reviews: Nanomedicine and Nanobiotechnology* 3 (4). https://doi.org/10.1002/wnan.143.

Chu, Chee Chin, Sook Chin Chew, and Kar Lin Nyam. 2021. "Recent Advances in Encapsulation Technologies of Kenaf (*Hibiscus cannabinus*) Leaves and Seeds for Cosmeceutical Application." *Food and Bioproducts Processing*. https://doi.org/10.1016/j.fbp.2021.02.009.

Cohen, Yifat, Moran Levi, Uri Lesmes, Marielle Margier, Emmanuelle Reboul, and Yoav D. Livney. 2017. "Re-Assembled Casein Micelles Improve In Vitro Bioavailability of Vitamin D in a Caco-2 Cell Model." *Food and Function* 8 (6). https://doi.org/10.1039/c7fo00323d.

Cömert, Ezgi Doğan, Burçe Ataç Mogol, and Vural Gökmen. 2020. "Relationship between Color and Antioxidant Capacity of Fruits and Vegetables." *Current Research in Food Science* 2. https://doi.org/10.1016/j.crfs.2019.11.001.

Costello, Rebecca B., and Paul Coates. 2001. "In the Midst of Confusion Lies Opportunity: Fostering Quality Science in Dietary Supplement Research." *Journal of the American College of Nutrition* 20 (1). https://doi.org/10.1080/07315724.2001.10719010.

Dadgar, Neda, Dariush Norouzian, Mohsen Chiani, Hassan Ebrahimi Shamabadi, Mohammadreza Mehrabi Seyed, Ali Farhanghi, and Azim Akbarzadeh. 2013. "Effect of Artemisinin Liposome and Artemisinin Liposome Polyethyleneglycol on MCF-7 Cell Line." *International Journal of Life Sciences Biotechnology and Pharma Research* 2 (1).

Dai, Xiaofeng, Lihui Yu, Xijiang Zhao, and Kostya Ostrikov. 2020. "Nanomaterials for Oncotherapies Targeting the Hallmarks of Cancer." *Nanotechnology* 31 (39). https://doi.org/10.1088/1361-6528/ab99f1.

Das, Sreemanti, Jayeeta Das, Asmita Samadder, Avijit Paul, and Anisur Rahman Khuda-Bukhsh. 2013. "Strategic Formulation of Apigenin-Loaded PLGA Nanoparticles for Intracellular Trafficking, DNA Targeting and Improved Therapeutic Effects in Skin Melanoma in Vitro." *Toxicology Letters* 223 (2). https://doi.org/10.1016/j.toxlet.2013.09.012.

Davidov-Pardo, Gabriel, and David Julian McClements. 2015. "Nutraceutical Delivery Systems: Resveratrol Encapsulation in Grape Seed Oil Nanoemulsions Formed by Spontaneous Emulsification." *Food Chemistry* 167. https://doi.org/10.1016/j.foodchem.2014.06.082.

Dhalaria, Rajni, Rachna Verma, Dinesh Kumar, Sunil Puri, Ashwani Tapwal, Vinod Kumar, Eugenie Nepovimova, and Kamil Kuca. 2020. "Bioactive Compounds of Edible Fruits with Their Anti-Aging Properties: A Comprehensive Review to Prolong Human Life." *Antioxidants*. https://doi.org/10.3390/antiox9111123.

Dhavamani, Sugasini, Yalagala Poorna Chandra Rao, and Belur R. Lokesh. 2014. "Total Antioxidant Activity of Selected Vegetable Oils and Their Influence on Total Antioxidant Values in Vivo: A Photochemiluminescence Based Analysis." *Food Chemistry* 164. https://doi.org/10.1016/j.foodchem.2014.05.064.

Du, Yawei, Longbing Ling, Muhammad Ismail, Wei He, Qing Xia, Wenya Zhou, Chen Yao, and Xinsong Li. 2018. "Redox Sensitive Lipid-Camptothecin Conjugate Encapsulated Solid Lipid Nanoparticles for Oral Delivery." *International Journal of Pharmaceutics* 549 (1–2). https://doi.org/10.1016/j.ijpharm.2018.08.010.

Dubey, Vaibhav, Dinesh Mishra, Abhay Asthana, and Narendra Kumar Jain. 2006. "Transdermal Delivery of a Pineal Hormone: Melatonin via Elastic Liposomes." *Biomaterials* 27 (18). https://doi.org/10.1016/j.biomaterials.2006.01.060.

Dusinska, Maria, Zuzana Magdolenova, and Lise Marie Fjellsbø. 2013. "Toxicological Aspects for Nanomaterial in Humans." *Methods in Molecular Biology* 948. https://doi.org/10.1007/978-1-62703-140-0_1.

El-Leithy, Eman S., Amna M. Makky, Abeer M. Khattab, and Doaa G. Hussein. 2018. "Optimization of Nutraceutical Coenzyme Q10 Nanoemulsion with Improved Skin Permeability and Anti-Wrinkle Efficiency." *Drug Development and Industrial Pharmacy* 44 (2). https://doi.org/10.1080/03639045.2017.1391836.

Floriano, Barbara Freitas, Tamara Carvalho, Tairine Zara Lopes, Luandra Aparecida Unten Takahashi, Paula Rahal, Antonio Claudio Tedesco, and Marília Freitas Calmon. 2021. "Effect of Berberine Nanoemulsion Photodynamic Therapy on Cervical Carcinoma Cell Line." *Photodiagnosis and Photodynamic Therapy* 33. https://doi.org/10.1016/j.pdpdt.2020.102174.

Forni, Cinzia, Francesco Facchiano, Manuela Bartoli, Stefano Pieretti, Antonio Facchiano, Daniela D'Arcangelo, Sandro Norelli, et al. 2019. "Beneficial Role of Phytochemicals on Oxidative Stress and Age-Related Diseases." *BioMed Research International.* https://doi.org/10.1155/2019/8748253.

Gibis, Monika, Nina Rahn, and Jochen Weiss. 2013. "Physical and Oxidative Stability of Uncoated and Chitosan-Coated Liposomes Containing Grape Seed Extract." *Pharmaceutics* 5 (3). https://doi.org/10.3390/pharmaceutics5030421.

Gibis, Monika, Karina Thellmann, Chutima Thongkaew, and Jochen Weiss. 2014. "Interaction of Polyphenols and Multilayered Liposomal-Encapsulated Grape Seed Extract with Native and Heat-Treated Proteins." *Food Hydrocolloids* 41. https://doi.org/10.1016/j.foodhyd.2014.03.024.

Gibis, Monika, Effie Vogt, and Jochen Weiss. 2012. "Encapsulation of Polyphenolic Grape Seed Extract in Polymer-Coated Liposomes." In *Food and Function.* Vol. 3. https://doi.org/10.1039/c1fo10181a.

Gokce, Evren H., Emrah Korkmaz, Sakine Tuncay-Tanriverdi, Eleonora Dellera, Giuseppina Sandri, M. Cristina Bonferoni, and Ozgen Ozer. 2012. "A Comparative Evaluation of Coenzyme Q10-Loaded Liposomes and Solid Lipid Nanoparticles as Dermal Antioxidant Carriers." *International Journal of Nanomedicine* 7. https://doi.org/10.2147/IJN.S34921.

Gokhale, Jayanti P., Hitendra S. Mahajan, and Sanjay S. Surana. 2019. "Quercetin Loaded Nanoemulsion-Based Gel for Rheumatoid Arthritis: In Vivo and In Vitro Studies." *Biomedicine and Pharmacotherapy* 112. https://doi.org/10.1016/j.biopha.2019.108622.

Gong, Yinhua, Yunkai Wu, Chunli Zheng, Liya Fan, Fei Xiong, and Jiabi Zhu. 2012. "An Excellent Delivery System for Improving the Oral Bioavailability of Natural Vitamin E in Rats." *AAPS PharmSciTech* 13 (3). https://doi.org/10.1208/s12249-012-9819-y.

Grochowski, Daniel M., Marcello Locatelli, Sebastian Granica, Francesco Cacciagrano, and Michał Tomczyk. 2018. "A Review on the Dietary Flavonoid Tiliroside." *Comprehensive Reviews in Food Science and Food Safety.* https://doi.org/10.1111/1541-4337.12389.

Guttoff, Marrisa, Amir Hossein Saberi, and David Julian Mcclements. 2015. "Formation of Vitamin D Nanoemulsion-Based Delivery Systems by Spontaneous Emulsification: Factors Affecting Particle Size and Stability." *Food Chemistry* 171. https://doi.org/10.1016/j.foodchem.2014.08.087.

Hatahet, T., M. Morille, A. Hommoss, J. M. Devoisselle, R. H. Müller, and S. Bégu. 2018. "Liposomes, Lipid Nanocapsules and SmartCrystals®: A Comparative Study for an Effective Quercetin Delivery to the Skin." *International Journal of Pharmaceutics* 542 (1–2). https://doi.org/10.1016/j.ijpharm.2018.03.019.

He, Xiao Lie, Li Yang, Zhao Jie Wang, Rui Qi Huang, Rong Rong Zhu, and Li Ming Cheng. 2021. "Solid Lipid Nanoparticles Loading with Curcumin and Dexanabinol to Treat Major Depressive Disorder." *Neural Regeneration Research* 16 (3). https://doi.org/10.4103/1673-5374.293155.

Helal, Nada A., Heba A. Eassa, Ahmed M. Amer, Mohamed A. Eltokhy, Ivan Edafiogho, and Mohamed I. Nounou. 2019. "Nutraceuticals' Novel Formulations: The Good, the Bad, the Unknown and Patents Involved." *Recent Patents on Drug Delivery & Formulation* 13 (2). https://doi.org/10.2174/1872211313666190503112040.

Hosseini, Seyed Fakhreddin, Mojgan Zandi, Masoud Rezaei, and Farhid Farahmandghavi. 2013. "Two-Step Method for Encapsulation of Oregano Essential Oil in Chitosan Nanoparticles: Preparation, Characterization and In Vitro Release Study." *Carbohydrate Polymers* 95 (1). https://doi.org/10.1016/j.carbpol.2013.02.031.

Huang, Zhiyi, Yu Liu, Guangying Qi, David Brand, and Song Guo Zheng. 2018. "Role of Vitamin A in the Immune System." *Journal of Clinical Medicine* 7 (9): 258. https://doi.org/10.3390/JCM7090258.

Iqbal, Rashid, Zaffar Mehmood, Aisha Baig, and Nauman Khalid. 2020. "Formulation and Characterization of Food Grade O/W Nanoemulsions Encapsulating Quercetin and Curcumin: Insights on Enhancing Solubility Characteristics." *Food and Bioproducts Processing* 123. https://doi.org/10.1016/j.fbp.2020.07.013.

Jablonka, E. V.A., and G. A.L. Raz. 2009. "Transgenerational Epigenetic Inheritance: Prevalence, Mechanisms, and Implications for the Study of Heredity and Evolution." *Quarterly Review of Biology* 84 (2). https://doi.org/10.1086/598822.

Jaenisch, Rudolf, and Adrian Bird. 2003. "Epigenetic Regulation of Gene Expression: How the Genome Integrates Intrinsic and Environmental Signals." *Nature Genetics*. https://doi.org/10.1038/ng1089.

Jain, Ashay, Gajanand Sharma, Kanika Thakur, Kaisar Raza, U. S. Shivhare, Gargi Ghoshal, and Om Prakash Katare. 2019. "Beta-Carotene-Encapsulated Solid Lipid Nanoparticles (BC-SLNs) as Promising Vehicle for Cancer: An Investigative Assessment." *AAPS PharmSciTech* 20 (3). https://doi.org/10.1208/s12249-019-1301-7.

Javed, Shamama, Kanchan Kohli, and Mushir Ali. 2011. "Reassessing Bioavailability of Silymarin." *Alternative Medicine Review*.

Jeon, Soha, Cha Young Yoo, and Soo Nam Park. 2015. "Improved Stability and Skin Permeability of Sodium Hyaluronate-Chitosan Multilayered Liposomes by Layer-by-Layer Electrostatic Deposition for Quercetin Delivery." *Colloids and Surfaces B: Biointerfaces* 129. https://doi.org/10.1016/j.colsurfb.2015.03.018.

Kim, Jee Hye, Jung A. Ko, Jun Tae Kim, Dong Su Cha, Jin Hun Cho, Hyun Jin Park, and Gye Hwa Shin. 2014. "Preparation of a Capsaicin-Loaded Nanoemulsion for Improving Skin Penetration." *Journal of Agricultural and Food Chemistry* 62 (3). https://doi.org/10.1021/jf404220n.

Kim, Sang Oh, Thi Van Anh Ha, Young Jin Choi, and Sanghoon Ko. 2014. "Optimization of Homogenization-Evaporation Process for Lycopene Nanoemulsion Production and Its Beverage Applications." *Journal of Food Science* 79 (8). https://doi.org/10.1111/1750-3841.12472.

Kim, Tae Il, Tae Gyun Kim, Dong Hyun Lim, Sang Bum Kim, Seong Min Park, Tai Young Hur, Kwang Seok Ki, Eung Gi Kwon, Mayakrishnan Vijayakumar, and Young Jun Kim. 2019. "Preparation of Nanoemulsions of Vitamin A and C by Microfluidization: Efficacy on the Expression Pattern of Milk-Specific Proteins in MAC-T Cells." *Molecules* 24 (14). https://doi.org/10.3390/molecules24142566.

Kour, Jasmeet, Hitesh Chopra, Saba Bukhari, Renu Sharma, Rosy Bansal, Monika Hans, and Dharmesh Chandra Saxena. 2022. "Nutraceutical-A Deep and Profound Concept." *Nutraceuticals and Health Care*, January, 1–28. https://doi.org/10.1016/B978-0-323-89779-2.00021-1.

Kuo, Fonghsu, Balajikarthick Subramanian, Timothy Kotyla, Thomas A. Wilson, Subbiah Yoganathan, and Robert J. Nicolosi. 2008. "Nanoemulsions of an Antioxidant Synergy Formulation Containing Gamma Tocopherol Have Enhanced Bioavailability and Anti-Inflammatory Properties." *International Journal of Pharmaceutics* 363 (1–2). https://doi.org/10.1016/j.ijpharm.2008.07.022.

Kurutas, Ergul Belge. 2016. "The Importance of Antioxidants Which Play the Role in Cellular Response against Oxidative/Nitrosative Stress: Current State." *Nutrition Journal*. https://doi.org/10.1186/s12937-016-0186-5.

Lacatusu, Ioana, Elena Mitrea, Nicoleta Badea, Raluca Stan, Ovidiu Oprea, and Aurelia Meghea. 2013. "Lipid Nanoparticles Based on Omega-3 Fatty Acids as Effective Carriers for Lutein Delivery. Preparation and in Vitro Characterization Studies." *Journal of Functional Foods* 5 (3). https://doi.org/10.1016/j.jff.2013.04.010.

Li, Lan, Bilal Ahmed, Kapil Mehta, and Razelle Kurzrock. 2007. "Liposomal Curcumin with and without Oxaliplatin: Effects on Cell Growth, Apoptosis, and Angiogenesis in Colorectal Cancer." *Molecular Cancer Therapeutics* 6 (4). https://doi.org/10.1158/1535-7163.MCT-06-0556.

Li, Rui, Qiangsheng Fang, Peihong Li, Chunling Zhang, Yuan Yuan, and Hong Zhuang. 2021. "Effects of Emulsifier Type and Post-Treatment on Stability, Curcumin Protection, and Sterilization Ability of Nanoemulsions." *Foods* 10 (1). https://doi.org/10.3390/foods10010149.

Li, Yong Jiang, Xiong Bin Hu, Xiu Ling Lu, De Hua Liao, Tian Tian Tang, Jun Yong Wu, and Da Xiong Xiang. 2017. "Nanoemulsion-Based Delivery System for Enhanced Oral Bioavailability and Caco-2 Cell Monolayers Permeability of Berberine Hydrochloride." *Drug Delivery* 24 (1). https://doi.org/10.1080/10717544.2017.1410257.

Lin, Yung Chang, Jhan Yen Kuo, Chih Chieh Hsu, Wen Che Tsai, Wei Chu Li, Ming Chiang Yu, and Hsiao Wei Wen. 2013. "Optimizing Manufacture of Liposomal Berberine with Evaluation of Its Antihepatoma Effects in a Murine Xenograft Model." *International Journal of Pharmaceutics* 441 (1–2). https://doi.org/10.1016/j.ijpharm.2012.11.017.

Liu, Chen, Daoxiao Chang, Xinhui Zhang, Hong Sui, Yindi Kong, Rongyue Zhu, and Wenping Wang. 2017. "Oral Fast-Dissolving Films Containing Lutein Nanocrystals for Improved Bioavailability: Formulation Development, In Vitro and In Vivo Evaluation." *AAPS PharmSciTech* 18 (8). https://doi.org/10.1208/s12249-017-0777-2.

Liu, Dan, Haiyang Hu, Zhixiu Lin, Dawei Chen, Yongyuan Zhu, Shengtao Hou, and Xiaojun Shi. 2013. "Quercetin Deformable Liposome: Preparation and Efficacy against Ultraviolet B Induced Skin Damages In Vitro and In Vivo." *Journal of Photochemistry and Photobiology B: Biology* 127. https://doi.org/10.1016/j.jphotobiol.2013.07.014.

Liu, Jing, Saumen Chakraborty, Parisa Hosseinzadeh, Yang Yu, Shiliang Tian, Igor Petrik, Ambika Bhagi, and Yi Lu. 2014. "Metalloproteins Containing Cytochrome, Iron-Sulfur, or Copper Redox Centers." *Chemical Reviews*. https://doi.org/10.1021/cr400479b.

Lynch, Iseult, Arti Ahluwalia, Diana Boraschi, Hugh J. Byrne, Bengt Fadeel, Peter Gehr, Arno C. Gutleb, Michaela Kendall, and Manthos G. Papadopoulos. 2013. "The Bio-Nano-Interface in Predicting Nanoparticle Fate and Behaviour in Living Organisms: Towards Grouping and Categorising Nanomaterials and Ensuring Nanosafety by Design." *BioNanoMaterials* 14 (3–4): 195–216. https://doi.org/10.1515/BNM-2013-0011.

Magdolenova, Zuzana, Andrew Collins, Ashutosh Kumar, Alok Dhawan, Vicki Stone, and Maria Dusinska. 2014. "Mechanisms of Genotoxicity. A Review of In Vitro and In Vivo Studies with Engineered Nanoparticles." *Nanotoxicology*. https://doi.org/10.3109/17435390.2013.773464.

Markowitz, John S., and Hao Jie Zhu. 2012. "Limitations of In Vitro Assessments of the Drug Interaction Potential of Botanical Supplements." *Planta Medica*. https://doi.org/10.1055/s-0032-1315025.

Mcclements, David Julian, Liqiang Zou, Ruojie Zhang, Laura Salvia-Trujillo, Taha Kumosani, and Hang Xiao. 2015. "Enhancing Nutraceutical Performance Using Excipient Foods: Designing Food Structures and Compositions to Increase Bioavailability." *Comprehensive Reviews in Food Science and Food Safety* 14 (6). https://doi.org/10.1111/1541-4337.12170.

Merlin, Jose J.P., N. Rajendra Prasad, S. M.A. Shibli, and Mol Sebeela. 2012. "Ferulic Acid Loaded Poly- d,l-Lactide-Co-Glycolide Nanoparticles: Systematic Study of Particle Size, Drug Encapsulation Efficiency and Anticancer Effect in Non-Small Cell Lung Carcinoma Cell Line In Vitro." *Biomedicine and Preventive Nutrition* 2 (1). https://doi.org/10.1016/j.bionut.2011.12.007.

Mirhoseini, Mehri, Zahra Rezanejad Gatabi, Majid Saeedi, Katayoun Morteza-Semnani, Fereshteh Talebpour Amiri, Hamid Reza Kelidari, and Abbas Ali Karimpour Malekshah. 2019. "Protective Effects of Melatonin Solid Lipid Nanoparticles on Testis Histology after Testicular Trauma in Rats." *Research in Pharmaceutical Sciences* 14 (3). https://doi.org/10.4103/1735-5362.258486.

Mohseni, Roohollah, Zahra ArabSadeghabadi, Nasrin Ziamajidi, Roghayeh Abbasalipourkabir, and Azam RezaeiFarimani. 2019. "Oral Administration of Resveratrol-Loaded Solid Lipid Nanoparticle Improves Insulin Resistance Through Targeting Expression of SNARE Proteins in Adipose and Muscle Tissue in Rats with Type 2 Diabetes." *Nanoscale Research Letters* 14 (1). https://doi.org/10.1186/s11671-019-3042-7.

Mora, J. Rodrigo, Makoto Iwata, and Ulrich H. von Andrian. 2008. "Vitamin Effects on the Immune System: Vitamins A and D Take Centre Stage." *Nature Reviews. Immunology* 8 (9): 685. https://doi.org/10.1038/NRI2378.

Muthu, Madaswamy S., Sneha A. Kulkarni, Jiaqing Xiong, and Si Shen Feng. 2011. "Vitamin E TPGS Coated Liposomes Enhanced Cellular Uptake and Cytotoxicity of Docetaxel in Brain Cancer Cells." *International Journal of Pharmaceutics* 421 (2). https://doi.org/10.1016/j.ijpharm.2011.09.045.

Nagi, Amrita, Babar Iqbal, Shobhit Kumar, Shrestha Sharma, Javed Ali, and Sanjula Baboota. 2017. "Quality by Design Based Silymarin Nanoemulsion for Enhancement of Oral Bioavailability." *Journal of Drug Delivery Science and Technology* 40. https://doi.org/10.1016/j.jddst.2017.05.019.

Namitha, K. K., and P. S. Negi. 2010. "Chemistry and Biotechnology of Carotenoids." *Critical Reviews in Food Science and Nutrition* 50 (8). https://doi.org/10.1080/10408398.2010.499811.

Nanjwade, Basavaraj K., Vikrant T. Kadam, and F. V. Manvi. 2013. "Formulation and Characterization of Nanostructured Lipid Carrier of Ubiquinone (Coenzyme Q10)." *Journal of Biomedical Nanotechnology* 9 (3). https://doi.org/10.1166/jbn.2013.1560.

Narayanan, Narayanan K., Dominick Nargi, Carla Randolph, and Bhagavathi A. Narayanan. 2009. "Liposome Encapsulation of Curcumin and Resveratrol in Combination Reduces Prostate Cancer Incidence in PTEN Knockout Mice." *International Journal of Cancer* 125 (1). https://doi.org/10.1002/ijc.24336.

Nasri, Hamid, Azar Baradaran, Hedayatollah Shirzad, and Mahmoud Rafieian Kopaei. 2014. "New Concepts in Nutraceuticals as Alternative for Pharmaceuticals." *International Journal of Preventive Medicine* 5 (12).

Nguyen, Xuan Khanh Thi, Jaehwi Lee, Eun Joo Shin, Duy Khanh Dang, Ji Hoon Jeong, Thuy Ty Lan Nguyen, Yunsung Nam, et al. 2015. "Liposomal Melatonin Rescues Methamphetamine-Elicited Mitochondrial Burdens, pro-Apoptosis, and Dopaminergic Degeneration through the Inhibition PKCδ Gene." *Journal of Pineal Research* 58 (1). https://doi.org/10.1111/jpi.12195.

Oshi, Murtada A., Juho Lee, Muhammad Naeem, Nurhasni Hasan, Jihyun Kim, Hak Jin Kim, Eun Hee Lee, Yunjin Jung, and Jin Wook Yoo. 2020. "Curcumin Nanocrystal/pH-Responsive Polyelectrolyte Multilayer Core-Shell Nanoparticles for Inflammation-Targeted Alleviation of Ulcerative Colitis." *Biomacromolecules* 21 (9). https://doi.org/10.1021/acs.biomac.0c00589.

Ozturk, Bengu, Sanem Argin, Mustafa Ozilgen, and David Julian McClements. 2015. "Formation and Stabilization of Nanoemulsion-Based Vitamin E Delivery Systems Using Natural Biopolymers: Whey Protein Isolate and Gum Arabic." *Food Chemistry* 188. https://doi.org/10.1016/j.foodchem.2015.05.005.

Panagiotou, Gianni, and Jens Nielsen. 2009. "Nutritional Systems Biology: Definitions and Approaches." *Annual Review of Nutrition*. https://doi.org/10.1146/annurev-nutr-080508-141138.

Parveen, Rabea, Sanjula Baboota, Javed Ali, Alka Ahuja, and Sayeed Ahmad. 2015. "Stability Studies of Silymarin Nanoemulsion Containing Tween 80 as a Surfactant." *Journal of Pharmacy and Bioallied Sciences* 7. https://doi.org/10.4103/0975-7406.168037.

Pawar, Rahul S., Hemlata Tamta, Jun Ma, Alexander J. Krynitsky, Erich Grundel, Wayne G. Wamer, and Jeanne I. Rader. 2013. "Updates on Chemical and Biological Research on Botanical Ingredients in Dietary Supplements." *Analytical and Bioanalytical Chemistry*. https://doi.org/10.1007/s00216-012-6691-2.

Penalva, Rebeca, Irene Esparza, Maite Agüeros, Carlos J. Gonzalez-Navarro, Carolina Gonzalez-Ferrero, and Juan M. Irache. 2015. "Casein Nanoparticles as Carriers for the Oral Delivery of Folic Acid." *Food Hydrocolloids* 44. https://doi.org/10.1016/j.foodhyd.2014.10.004.

Pool, Hector, Sandra Mendoza, Hang Xiao, and David Julian McClements. 2013. "Encapsulation and Release of Hydrophobic Bioactive Components in Nanoemulsion-Based Delivery Systems: Impact of Physical Form on Quercetin Bioaccessibility." *Food and Function* 4 (1). https://doi.org/10.1039/c2fo30042g.

Pund, Swati, Ganesh Borade, and Ganesh Rasve. 2014. "Improvement of Anti-Inflammatory and Anti-Angiogenic Activity of Berberine by Novel Rapid Dissolving Nanoemulsifying Technique." *Phytomedicine* 21 (3). https://doi.org/10.1016/j.phymed.2013.09.013.

Qian, Cheng, Eric Andrew Decker, Hang Xiao, and David Julian McClements. 2012. "Nanoemulsion Delivery Systems: Influence of Carrier Oil on β-Carotene Bioaccessibility." *Food Chemistry* 135 (3). https://doi.org/10.1016/j.foodchem.2012.06.047.

Rachmawati, Heni, Annisa Rahma, Loaye Al Shaal, Rainer H. Müller, and Cornelia M. Keck. 2016. "Destabilization Mechanism of Ionic Surfactant on Curcumin Nanocrystal against Electrolytes." *Scientia Pharmaceutica* 84 (4). https://doi.org/10.3390/scipharm84040685.

Rahman, Heshu Sulaiman, Abdullah Rasedee, Chee Wun How, Ahmad Bustamam Abdul, Nazariah Allaudin Zeenathul, Hemn Hassan Othman, Mohamed Ibrahim Saeed, and Swee Keong Yeap. 2013. "Zerumbone-Loaded Nanostructured Lipid Carriers: Preparation, Characterization, and Antileukemic Effect." *International Journal of Nanomedicine* 8. https://doi.org/10.2147/IJN.S45313.

Ramalingam, Prakash, and Young Tag Ko. 2016. "Improved Oral Delivery of Resveratrol from N-Trimethyl Chitosan-g-Palmitic Acid Surface-Modified Solid Lipid Nanoparticles." *Colloids and Surfaces B: Biointerfaces* 139. https://doi.org/10.1016/j.colsurfb.2015.11.050.

Ravindran, Jayaraj, Hareesh B. Nair, Bokyung Sung, Sahdeo Prasad, Rajeshwar R. Tekmal, and Bharat B. Aggarwal. 2010. "Thymoquinone Poly (Lactide-Co-Glycolide) Nanoparticles Exhibit Enhanced Anti-Proliferative, Anti-Inflammatory, and Chemosensitization Potential." *Biochemical Pharmacology* 79 (11). https://doi.org/10.1016/j.bcp.2010.01.023.

Rehan, Farah, Nafees Ahemad, and Manish Gupta. 2019. "Casein Nanomicelle as an Emerging Biomaterial: A Comprehensive Review." *Colloids and Surfaces B: Biointerfaces*. https://doi.org/10.1016/j.colsurfb.2019.03.051.

Roncero, José M., Manuel Álvarez-Ortí, Arturo Pardo-Giménez, Adrián Rabadán, and José E. Pardo. 2020. "Review about Non-Lipid Components and Minor Fat-Soluble Bioactive Compounds of Almond Kernel." *Foods*. https://doi.org/10.3390/foods9111646.

Roohani, Nazanin, Richard Hurrell, Roya Kelishadi, and Rainer Schulin. 2013. "Zinc and Its Importance for Human Health: An Integrative Review." *Journal of Research in Medical Sciences*. https://doi.org/10.1016/j.foodpol.2013.06.008.

Ruscica, Massimiliano, Alberto Corsini, Nicola Ferri, Maciej Banach, and Cesare R. Sirtori. 2020. "Clinical Approach to the Inflammatory Etiology of Cardiovascular Diseases." *Pharmacological Research*. https://doi.org/10.1016/j.phrs.2020.104916.

Saini, Ramesh Kumar, Shivraj Hariram Nile, and Se Won Park. 2015. "Carotenoids from Fruits and Vegetables: Chemistry, Analysis, Occurrence, Bioavailability and Biological Activities." *Food Research International*. https://doi.org/10.1016/j.foodres.2015.07.047.

Santos, Cristina A., and Joseph I. Boullata. 2005. "An Approach to Evaluating Drug-Nutrient Interactions." *Pharmacotherapy*. https://doi.org/10.1592/phco.2005.25.12.1789.

Santos, Débora S., José Athayde V. Morais, Ísis A.C. Vanderlei, Alexandre S. Santos, Ricardo B. Azevedo, Luís A. Muehlmann, Osmindo R.P. Júnior, et al. 2021. "Oral Delivery of Fish Oil in Oil-in-Water Nanoemulsion: Development, Colloidal Stability and Modulatory Effect on in Vivo Inflammatory Induction in Mice." *Biomedicine and Pharmacotherapy* 133. https://doi.org/10.1016/j.biopha.2020.110980.

Schoener, AL, R Zhang, S Lv, J Weiss, and DJ McClements. 2019. "Fabrication of Plant-Based Vitamin D 3-Fortified Nanoemulsions: Influence of Carrier Oil Type on Vitamin Bioaccessibility." *Food & Function* 10 (4): 1826–35. https://doi.org/10.1039/C9FO00116F.

Schwarz, Julia C., Nuttakorn Baisaeng, Magdalena Hoppel, Monika Löw, Cornelia M. Keck, and Claudia Valenta. 2013. "Ultra-Small NLC for Improved Dermal Delivery of Coenyzme Q10." *International Journal of Pharmaceutics* 447 (1–2). https://doi.org/10.1016/j.ijpharm.2013.02.037.

Sebak, Safaa, Maryam Mirzaei, Meenakshi Malhotra, Arun Kulamarva, and Satya Prakash. 2010. "Human Serum Albumin Nanoparticles as an Efficient Noscapine Drug Delivery System for Potential Use in Breast Cancer: Preparation and In Vitro Analysis." *International Journal of Nanomedicine* 5 (1). https://doi.org/10.2147/IJN.S10443.

Sessa, Mariarenata, Maria Luisa Balestrieri, Giovanna Ferrari, Luigi Servillo, Domenico Castaldo, Nunzia D'Onofrio, Francesco Donsì, and Rong Tsao. 2014. "Bioavailability of Encapsulated Resveratrol into Nanoemulsion-Based Delivery Systems." *Food Chemistry* 147. https://doi.org/10.1016/j.foodchem.2013.09.088.

Seto, Yoshiki, Kodai Ueno, Hiroki Suzuki, Hideyuki Sato, and Satomi Onoue. 2019. "Development of Novel Lutein Nanocrystal Formulation with Improved Oral Bioavailability and Ocular Distribution." *Journal of Functional Foods* 61. https://doi.org/10.1016/j.jff.2019.103499.

Sharifi-Rad, Atena, Jamshid Mehrzad, Majid Darroudi, Mohammad Reza Saberi, and Jamshidkhan Chamani. 2021. "Oil-in-Water Nanoemulsions Comprising Berberine in Olive Oil: Biological Activities, Binding Mechanisms to Human Serum Albumin or Holo-Transferrin and QMMD Simulations." *Journal of Biomolecular Structure and Dynamics* 39 (3). https://doi.org/10.1080/07391102.2020.1724568.

Shi, Jia, Songlei Zhou, Le Kang, Hu Ling, Jiepeng Chen, Lili Duan, Yanzhi Song, and Yihui Deng. 2018. "Evaluation of the Antitumor Effects of Vitamin K2 (Menaquinone-7) Nanoemulsions Modified with Sialic Acid-Cholesterol Conjugate." *Drug Delivery and Translational Research* 8 (1). https://doi.org/10.1007/s13346-017-0424-1.

Shoji, Yoko, and Hideki Nakashima. 2004. "Nutraceutics and Delivery Systems." *Journal of Drug Targeting* 12 (6). https://doi.org/10.1080/10611860400003817.

Skinner, Michael K., Carlos Guerrero Bosagna M. Haque, Eric Nilsson, Ramji Bhandari, and John R. McCarrey. 2013. "Environmentally Induced Transgenerational Epigenetic Reprogramming of Primordial Germ Cells and the Subsequent Germ Line." *PLoS ONE* 8 (7). https://doi.org/10.1371/journal.pone.0066318.

Song, Yanzhi, Jie Han, Rui Feng, Mengjing Wang, Qingjing Tian, Ting Zhang, Xinrong Liu, Xiaobo Cheng, and Yihui Deng. 2016. "The 12-3-12 Cationic Gemini Surfactant as a Novel Gastrointestinal Bioadhesive Material for Improving the Oral Bioavailability of Coenzyme Q10 Naked Nanocrystals." *Drug Development and Industrial Pharmacy* 42 (12). https://doi.org/10.1080/03639045.2016.1195399.

Souto, Eliana B., Patrícia Severino, Rafael Basso, and Maria Helena A. Santana. 2013. "Encapsulation of Antioxidants in Gastrointestinal-Resistant Nanoparticulate Carriers." *Methods in Molecular Biology*. https://doi.org/10.1007/978-1-62703-475-3_3.

Stephensen, C. B. 2001. "Vitamin A, Infection, and Immune Function." *Annual Review of Nutrition*. https://doi.org/10.1146/annurev.nutr.21.1.167.

Storka, Angela, Brigitta Vcelar, Uros Klickovic, Ghazaleh Gouya, Stefan Weisshaar, Stefan Aschauer, Gordon Bolger, Lawrence Helson, and Michael Wolzt. 2015. "Safety, Tolerability and Pharmacokinetics of Liposomal Curcumin in Healthy Humans." *International Journal of Clinical Pharmacology and Therapeutics* 53 (1). https://doi.org/10.5414/CP202076.

Subbiah, M. T.Ravi. 2007. "Nutrigenetics and Nutraceuticals: The Next Wave Riding on Personalized Medicine." *Translational Research*. https://doi.org/10.1016/j.trsl.2006.09.003.

Sun, Jiao, Fan Wang, Yue Sui, Zhennan She, Wenjun Zhai, Chunling Wang, and Yihui Deng. 2012. "Effect of Particle Size on Solubility, Dissolution Rate, and Oral Bioavailability: Evaluation Using Coenzyme Q10 as Naked Nanocrystals." *International Journal of Nanomedicine* 7. https://doi.org/10.2147/IJN.S34365.

Tan, Chen, Yating Zhang, Shabbar Abbas, Biao Feng, Xiaoming Zhang, and Shuqin Xia. 2014. "Modulation of the Carotenoid Bioaccessibility through Liposomal Encapsulation." *Colloids and Surfaces B: Biointerfaces* 123. https://doi.org/10.1016/j.colsurfb.2014.10.011.

Tsai, Yin Jieh, and Bing Huei Chen. 2016. "Preparation of Catechin Extracts and Nanoemulsions from Green Tea Leaf Waste and Their Inhibition Effect on Prostate Cancer Cell PC-3." *International Journal of Nanomedicine* 11. https://doi.org/10.2147/IJN.S103759.

Vannucci, Letizia, Caterina Fossi, Sara Quattrini, Leonardo Guasti, Barbara Pampaloni, Giorgio Gronchi, Francesca Giusti, et al. 2018. "Calcium Intake in Bone Health: A Focus on Calcium-Rich Mineral Waters." *Nutrients*. https://doi.org/10.3390/nu10121930.

Ventura, Mara, Miguel Melo, and Francisco Carrilho. 2017. "Selenium and Thyroid Disease: From Pathophysiology to Treatment." *International Journal of Endocrinology*. https://doi.org/10.1155/2017/1297658.

Vidlářová, Lucie, Gregori B. Romero, Jaroslav Hanuš, František Štěpánek, and Rainer H. Müller. 2016. "Nanocrystals for Dermal Penetration Enhancement – Effect of Concentration and Underlying Mechanisms Using Curcumin as Model." *European Journal of Pharmaceutics and Biopharmaceutics* 104. https://doi.org/10.1016/j.ejpb.2016.05.004.

Walden, O. 1989. "The Relationship of Dietary and Supplemental Calcium Intake to Bone Loss and Osteoporosis." *Journal of the American Dietetic Association*.

Wang, Yanchun, Hanlin Xu, Qin Fu, Rong Ma, and Jizhou Xiang. 2011. "Protective Effect of Resveratrol Derived from Polygonum Cuspidatum and Its Liposomal Form on Nigral Cells in Parkinsonian Rats." *Journal of the Neurological Sciences* 304 (1–2). https://doi.org/10.1016/j.jns.2011.02.025.

Wang, Zhong Q., Yongmei Yu, Xian H. Zhang, and James Komorowski. 2014. "Chromium-Insulin Reduces Insulin Clearance and Enhances Insulin Signaling by Suppressing Hepatic Insulin-Degrading Enzyme and Proteasome Protein Expression in KKAy Mice." *Frontiers in Endocrinology* 5 (Jul). https://doi.org/10.3389/fendo.2014.00099.

Xia, Shuqin, Chen Tan, Yating Zhang, Shabbar Abbas, Biao Feng, Xiaoming Zhang, and Fang Qin. 2015. "Modulating Effect of Lipid Bilayer-Carotenoid Interactions on the Property of Liposome Encapsulation." *Colloids and Surfaces B: Biointerfaces* 128. https://doi.org/10.1016/j.colsurfb.2015.02.004.

Xu, Hong Yu, Chang Shun Liu, Chuan Li Huang, Li Chen, Yu Rong Zheng, Si Hang Huang, and Xiao Ying Long. 2019. "Nanoemulsion Improves Hypoglycemic Efficacy of Berberine by Overcoming Its Gastrointestinal Challenge." *Colloids and Surfaces B: Biointerfaces* 181. https://doi.org/10.1016/j.colsurfb.2019.06.006.

Xu, Lingyun, Huabing Chen, Huibi Xu, and Xiangliang Yang. 2008. "Anti-Tumour and Immuno-Modulation Effects of Triptolide-Loaded Polymeric Micelles." *European Journal of Pharmaceutics and Biopharmaceutics* 70 (3). https://doi.org/10.1016/j.ejpb.2008.07.017.

Yi, Lunzhao, Shasha Ma, and Dabing Ren. 2017. "Phytochemistry and Bioactivity of Citrus Flavonoids: A Focus on Antioxidant, Anti-Inflammatory, Anticancer and Cardiovascular Protection Activities." *Phytochemistry Reviews* 16 (3). https://doi.org/10.1007/s11101-017-9497-1.

Yu, Aihua, Haigang Wang, Jiali Wang, Fengliang Cao, Yan Gao, Jing Cui, and Guangxi Zhai. 2011. "Formulation Optimization and Bioavailability after Oral and Nasal Administration in Rabbits of Puerarin-Loaded Microemulsion." *Journal of Pharmaceutical Sciences* 100 (3). https://doi.org/10.1002/jps.22333.

Yu, Guohua, Zhiqiang Luo, Wubin Wang, Yihao Li, Yating Zhou, and Yuanyuan Shi. 2019. "*Rubus chingii* Hu: A Review of the Phytochemistry and Pharmacology." *Frontiers in Pharmacology*. https://doi.org/10.3389/fphar.2019.00799.

Yu, Hailong, and Qingrong Huang. 2012. "Improving the Oral Bioavailability of Curcumin Using Novel Organogel-Based Nanoemulsions." *Journal of Agricultural and Food Chemistry* 60 (21). https://doi.org/10.1021/jf300609p.

Zelig, Rena, and Diane Rigassio Radler. 2012. "Understanding the Properties of Common Dietary Supplements: Clinical Implications for Healthcare Practitioners." *Nutrition in Clinical Practice*. https://doi.org/10.1177/0884533612446198.

Zhang, Zhen Hai, Xiao Pan Wang, Waddad Y. Ayman, Were L.L. Munyendo, Hui Xia Lv, and Jian Ping Zhou. 2013. "Studies on Lactoferrin Nanoparticles of Gambogic Acid for Oral Delivery." *Drug Delivery* 20 (2). https://doi.org/10.3109/10717544.2013.766781.

Zimet, Patricia, Dina Rosenberg, and Yoav D. Livney. 2011. "Re-Assembled Casein Micelles and Casein Nanoparticles as Nano-Vehicles for ω-3 Polyunsaturated Fatty Acids." *Food Hydrocolloids* 25 (5): 1270–76. https://doi.org/10.1016/J.FOODHYD.2010.11.025.

13 Application of Nano-Nutraceuticals in Medicines

Kanika Sharma, Ahmed Salim, Shreyas R. Murthy, Gunjan Sharma, and Rupesh K. Gautam[*]

CONTENTS

13.1 Introduction 347
13.2 Classification on the Basis of Use in Various Diseases 350
  13.2.1 Cardiovascular Disorders 350
  13.2.2 Neurological Disorders 352
  13.2.3 Metabolic Disorders 356
  13.2.4 Gastrointestinal Disorders 358
  13.2.5 Treatment of Cancer 359
13.3 Targeted Delivery of Nano-Nutraceuticals for the Treatment of Cancer and Neurological Disorders 362
13.4 Concluding Remarks and Future Perspectives 362
References 366

13.1 INTRODUCTION

Nano-nutraceuticals are a sophisticated attempt to achieve the aim of the great Greek scholar Hippocrates, who said, "Let food be thy medicine, and let medicine be thy food". But even before Hippocrates, Ayurveda understood the ability of nutrition to prevent and treat illness. The application of turmeric and various other herbal preparations is well documented in Ayurveda history (Bradamante et al., 2003). The term *nutraceutical* was coined by Stephen L. DeFelice and derives from two terms – *nutrition* and *pharmaceutical*. He defines a nutraceutical as "Any substance that is a food or part of a food and provides medical or health benefits, including the prevention and treatment of disease" (DeFelice, 1995). In the last two decades, nutraceuticals have gained immense popularity with consumers as well as researchers due to their abundant physiological activities as well as their relatively low toxicity profile. These are chemically diverse phytochemicals with distinguishable mechanisms targeting a diverse range of ailments of the human body (Arora and Jaglan, 2016). They generally consist of varied bioactive compounds, including, but not limited to, carotenoids, phenolics, probiotics, dietary fibers, fatty acids and vitamins (Kalra, 2003). A true definition of nutraceuticals has not yet been coined. Definitions have been explored by the U.S. Food and Drug Administration (USFDA), the Canadian Ministry of Health and the European Food Safety Authority (EFSA) (Aryee and Boye, 2015). The Canadian Ministry of Health defines a nano-nutraceutical as "Product isolated or purified from the food, generally sold in medicinal form not associated with food and demonstrated to have a physiological benefit. It also provides benefit against chronic disease" (Esther et al., 2000).

When we discuss nutraceuticals, other terms, such as *functional foods* and *dietary supplements*, also demand explanation. A functional food, as defined by (Kalra, 2003), "Is any food which is

[*] Corresponding author: drrupeshgautam@gmail.com

DOI: 10.1201/9781003244721-13

prepared using scientific knowledge with or without knowledge of how or why it is being used". Vitamins, fats, proteins, carbohydrates, and so on, are provided by these functional foods (Kalra, 2003). The U.K. Ministry of Agriculture, Fisheries and Food defined a functional food as "a food that has a component incorporated into it to give it a specific medical or physiological benefit, other than purely nutritional benefit" (Cockbill, 1994). A dietary supplement, as defined by the Dietary Supplement Health and Education Act of 1994, United States, is

> a product (other than tobacco) that is intended to supplement the diet that bears or contains one or more of the following dietary ingredients: a vitamin, a mineral, an herb or other botanical, an amino acid, a dietary substance for use by man to supplement the diet by increasing the total daily intake, or a concentrate, metabolite, constituent, extract, or combinations of these ingredients.

Figure 13.1 illustrates different classifications of nutraceuticals. They are broadly categorized into two types:

i) Natural or traditional
ii) Non-traditional

These two are further classified as follows:

i) Traditional
    a. Chemical constituents
        i. Nutrients – Includes amino acids and vitamins that have nutritional functions
        ii. Herbals – They are either parts of a plant or whole plants that have pharmacological and nutritional value. Examples include cinnamon bark, leaves of senna
        iii. Phytochemicals – The extracted constituents from the plant source, usually the secondary metabolites of the plant, that have biological or pharmacological activity. Examples include alkaloids, glycoside, flavonoids, etc.
    b. Probiotic microorganisms – Probiotics are living bacteria that when taken in sufficient quantities, impart a health benefit to the host. They are beneficial bacteria that support proper digestion and nutrient absorption. They drive out pathogens such as yeasts, other bacteria and viruses that may otherwise cause disease and form a

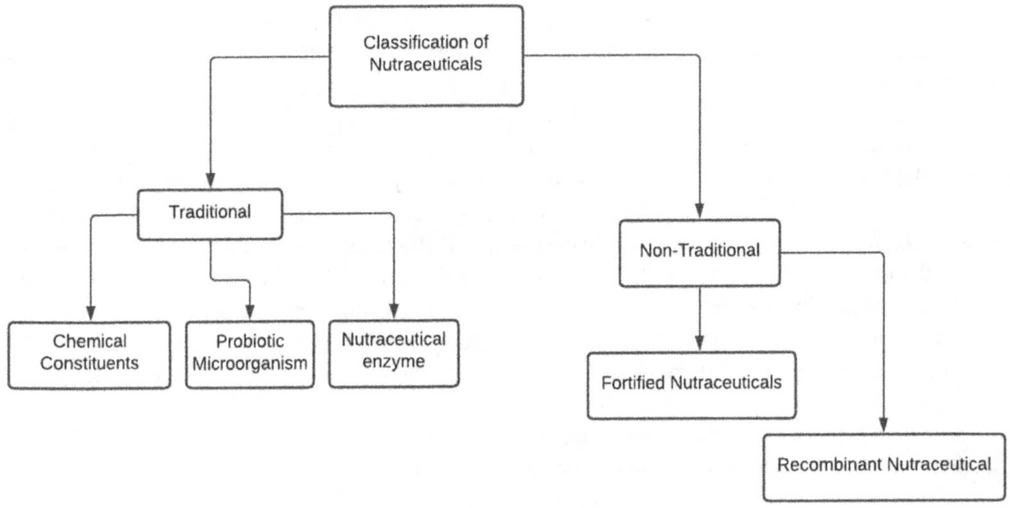

FIGURE 13.1 Classification of nutraceuticals based on the method of obtaining the compounds.

mutually beneficial symbiotic relationship with the human gastrointestinal system. They have an antimicrobial effect by altering the microflora, preventing pathogen adhesion to the intestinal epithelium, competing for nutrients required for pathogen survival, producing an antitoxin effect and reversing some of the effects of infection on the intestinal epithelium, such as secretory changes and neutrophil migration. Lactose intolerance can be treated with probiotics by increasing the synthesis of a particular enzyme (β-galactosidase). Examples of probiotic microorganisms that are obtained from natural sources:

i. Milk – *Lactobacillus lactis*
ii. Fermented products – *L. cellobiosus, L. curvatus, Bifidobacterium animalis*
iii. Vegetables/grains – *Saccharomyces cerevisiae, S. boulardii*

c. Nutraceutical enzymes – Enzymes are a necessary aspect of life; without them, our bodies would not operate. Those suffering from illnesses such as hypoglycemia, blood sugar imbalances, digestive issues and obesity might alleviate their symptoms by including enzyme supplements in their diet. These enzymes are obtained from microorganisms, plants and animals. Examples of nutraceutical enzymes include:

i. Microbial enzymes – Invertase, sucrase (obtained from yeast), glucoamylase (obtained from *Aspergillus niger, Saccharomycopsis fibuligera)*
ii. Plant enzymes – β-Amylase, bromelain (obtained from pineapple)
iii. Animal enzymes – Trypsin (pancreatic juice), α-amylase (saliva)

ii) Non-traditional nutraceuticals – These are compounds that are made with the help of biotechnology. They are made in such a way that the food has particular products that are enriched in the food or plant parts for better human wellness.

a. Fortified nutraceuticals – These consist of fortified food derived via agricultural breeding or from the addition of nutrients and/or additives: for example, orange juice fortified with calcium, cereals supplemented with vitamins or minerals, and wheat fortified with folic acid. Milk supplemented with cholecalciferol, for example, is used to treat vitamin D insufficiency. Prebiotic- and probiotic-enriched milk with *Bifidobacterium lactis* HN019 is utilized in children with diarrhea, respiratory infections, and severe diseases. In iron deficiency, banana is reinforced with the soybean ferritin gene.

b. Recombinant nutraceuticals – Biotechnology is used to make energy-producing foods such as bread, wine, fermented starch, yogurt, cheese, vinegar and others. Biotechnology enables the manufacture of probiotics and the extraction of bioactive components using enzyme/fermentation technologies as well as genetic engineering technologies. Examples include *Acetobacter xylinum* – β-glucuronidase and *Spirulina pacifica* – indoleamine 2,3-dioxygenase, which help to increase hemoglobin.

The ongoing studies will result in a new generation of food, resulting in a more permeable interface between food and medicine. The current state of nutraceutical knowledge offers a significant quest for nutritionists, doctors, food technologists and food chemists. Nutraceutical prevention and treatment is viewed by public health authorities as a strong tool for preserving health and combating nutritionally caused acute and chronic illnesses, therefore aiding optimal health, longevity and quality of life.

Nutraceuticals have shown promise in a wide range of disorders, from being used in metabolic disorders to being used as anticancer agents (Bergamin et al., 2019). Their broad array of effects is attributed to various molecular mechanisms, which range from reliving oxidative stress by quenching free radicals to inhibiting certain growth factors. These widely varying mechanisms are due to the vast structural differences found in nutraceuticals, from essential fatty acids (omega 3) to carotenoids (lycopene) to flavonoids and phytosterols (Jafari and McClements, 2017). For example, *Syzygium cumini* is known to decrease the level of reactive oxygen species (ROS), thereby preventing progression to infarct development. *Paeonia emodi* shows antithrombosis, free radical scavenging, and antihyperlipidemic properties. Resveratrol shows anti-atherosclerotic activity, imparting anti-inflammatory properties, but also could depolymerize the β amyloid peptide (Aβ) protein in Alzheimer's disease (AD) through a mechanism involving proteasomes. Resveratrol is not only neuroprotective but also has the marvelous ability to inhibit all three stages (initiation, promotion and progression) of tumor growth. Resveratrol is a stilbenoid obtained from skin of grapes, blueberries, raspberries, mulberries and peanuts. Furthermore, it also increases NO bioavailability and decreases tumor necrosis factor (TNF)α; both actions allow it to provide protection in cardiovascular (CV) diseases.

Curcumin is a nutraceutical obtained from *Curcuma longa* of the family Zingiberaceae. Curcumin shows antioxidant as well as angiotensin-converting enzyme (ACE) inhibitory activities. Curcumin is known to be beneficial in several disorders, ranging from CV diseases to metabolic disease, gastrointestinal diseases, and even cancer. In CV disease, curcumin enhances upregulation of calcium sensing receptor and calmodulin both which act as saviors in myocardial destruction. It is also known to decrease pulmonary hypertension.

Although nutraceuticals have a superior beneficial impact on the overall health of human beings, their stability, permeability, water solubility and hence, their bioavailability in the human body is below par. Poor bioavailability can result from other reasons as well, such as interaction with other components in the gastrointestinal tract, biotransformation and limited release from food (Arora and Jaglan, 2016). As well as these issues, problems such as crystallization and chemical instability must also be answered before nutraceuticals can be used as pharmaceuticals (Augustin and Sanguansri, 2012). Nano-nutraceuticals in all their grandeur can be envisioned as the critical bridge between pharmaceuticals and nutraceuticals. They allow us to deliver nutraceuticals with the help of nanotechnology. Although the structure of the compound delivered remains the same, its properties are vastly altered (Huang et al., 2010). Nanotechnology is a technique/technology conducted at the scale of 1–1000 nm (Javeri, 2016) The advantage of the nanoformulation of nutraceuticals is the superior pharmacokinetic profile that can be achieved. As most nutraceuticals are orally ingested, bioavailability across the intestine is fairly poor. Various techniques have since been tested to solve this issue. Nanoencapsulation techniques enhance bioavailability and also allow site-specific distribution, which is advantageous in the diseased state (Shibamoto et al., 2008; Shegokar and Müller, 2010). Polymeric nanoparticles, solid lipid nanoparticles, metal and inorganic nanoparticles, polymeric micelles, phospholipid micelles, colloidal nanoliposomes, polymeric nanoemulsions and hybrid nanocarriers have been prepared to deliver these nutraceuticals. In preparing these delivery forms, techniques like high-pressure homogenization, complex coacervation, co-precipitation, the salting out technique, the solvent displacement method, solvent emulsification–diffusion, the supercritical fluid method and the self-assembly method have been applied (Dasgupta et al., 2016; Gunasekaran et al., 2014; Yan Wang et al., 2016). Also, it provides thermal stability, which in certain functional foods, is the limiting factor (Kakkar et al., 2016). This chapter will give a detailed review of various disease conditions and how nano-nutraceuticals intervene in the pathology. A comprehensive mechanism of action of these nano-nutraceuticals will be discussed. Furthermore, various pharmaceutical formulations will be mentioned in this chapter.

13.2 CLASSIFICATION ON THE BASIS OF USE IN VARIOUS DISEASES

13.2.1 Cardiovascular Disorders

CV disease leads the infamous race in causes of death around the globe by taking nearly 17.9 million lives each year (WHO, 2019). CV diseases include coronary heart disease, cerebrovascular disease, rheumatic heart disease and many other conditions. Nearly 80% of the above-mentioned fatalities have occurred because of heart attack and stroke (WHO, 2019). Millions have been spent in the development and research of traditional pharmacological intervention, yet the problem still

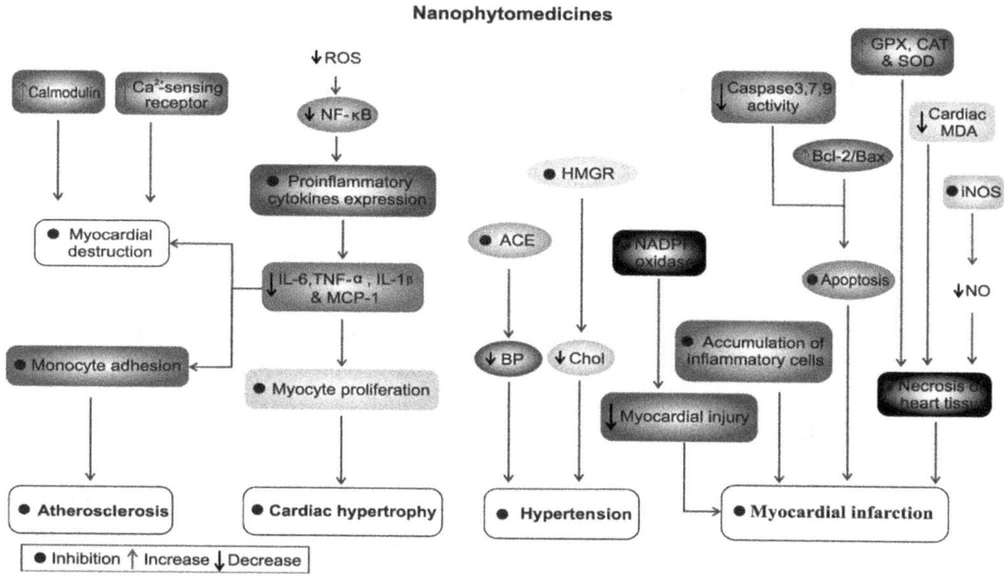

FIGURE 13.2 Mechanism of action of nano-nutraceuticals in cardiovascular disorders. (From Hesari, M. et al., *International Journal of Nanomedicine*, 16, 3293, 2021.)

remains (WHO, 2015). Psychosocial factors along with other factors are at the root of these diseases. The advent of lifestyle and dietary modification has been the mainstay, assisted by pharmacological interventions, in preventing cardiovascular events. Nano-nutraceuticals are a bridging intervention, one that uses the facets of dietary modification and pharmacological approach. Many phytochemicals have shown evidence of providing positive impacts in CV diseases, but their limiting stability, solubility, dissolution rate and other pharmacokinetic parameters restrict their groundbreaking impact. Nanoformulations allow us to fill this gap and provide a convincing solution (Pillai et al., 2021). A generalized mechanism of action of nano-nutraceuticals is depicted in Figure 13.2.

Syzygium cumini (*S. cumini*) has been part of traditional medicine for a very long time and has shown anti-inflammatory, antidiabetic and gastroprotective action (Ayyanar and Subash-Babu, 2012). Silver nanoparticles of *S. cumini* have been found to have enhanced delivery and increased therapeutic potential against cardiac diabetic myopathy; this ability of *S. cumini* is because of the ability to mitigate ROS (Hesari et al., 2021). ROS are known to increase nuclear factor κB (NFκB), which is a proinflammatory mediator and leads to vessel inflammation (Lingappan, 2018). This vessel inflammation, mediated via interleukin (IL)-6, TNFα and IL-1β, increases the likelihood of atherosclerotic plaque rupture, leading to an infarct. NFκB is also known to increase gene expression of adhesion molecules like selectin E and selectin P along with vascular cell adhesion molecule 1 (VCAM-1), all of which are associated with atherosclerosis (Sprague and Khalil, 2009). Also, NFκB has been implicated in cardiac cell proliferation, leading to cardiac hypertrophy, which in turn, is a significant factor in causing stroke and heart failure (Libby et al., 2016). Silver nanoparticles of *S. cumini* are also effective in restoring cell size and converting nuclear enlargement and DNA fragmentation into normal states in cell lines that were glucose stressed, thus showing its cardioprotective ability in the diabetic state (Atale et al., 2017).

Paeonia emodi, belonging to the Paeoniaceae family, possesses antithrombosis, free radical scavenging and antihyperlipidemic properties (Hesari et al., 2021). Gold nanoparticles of *Paeonia emodi* have been shown to control lipid profile in a pre-clinical model; these gold nanoparticles were found to decrease ALT (alanine transaminase), AST (aspartate aminotransferase), CPK (creatine phosphokinase), LDH (lactate dehydrogenase) and DNA damage, all of which are hallmark biomarkers of cardiovascular disorders (Ibrar et al., 2018; Ibrar et al., 2019).

Pinus plants, especially nanoparticles derived from *P. merkusii*, have been shown to decrease levels of LDH, MDA (malondialdehyde) and creatine kinase MB (CK-MB) in heart samples, all of which are biomarkers for cardiac tissue damage. Along with decreasing the levels of these biomarkers *P. merkusii* nanoparticles increase superoxide dismutase (SOD) and glutathione peroxidase (GPx), which curb the formation of ROS (Sudjarwo et al., 2019).

Curcumin is derived from turmeric and has long been known to have antioxidant properties. Curcumin also has ACE-inhibiting properties. A nanoemulsion of curcumin has been shown to have the edge over a normal curcumin formulation in ACE-inhibiting potential (Kocaadam and Şanlier, 2017). Encapsulated curcumin has been shown to have improved water solubility and biological activity (Rachmawati et al., 2016). It also enhances upregulation of calcium sensing receptor and calmodulin, both of which act as saviors in myocardial destruction. Curcumin encapsulated in chitosan polymer has shown ability to decrease pulmonary hypertension in pre-clinical studies; even protection against Q-T prolongation by curcumin nanoparticles hybridized with lipopolymer has been reported (Tong et al., 2018; Ranjan et al., 2013). Curcumin is known to operate synergistically with atorvastatin (a β-hydroxy β-methylglutaryl-CoA [HMG-CoA] reductase inhibitor); it not only increases the overall therapeutic effect but also decreases the side effect profile of the statin (Xiaoxia Li et al., 2019). The increased Bcl2/Bax ratio in cardiomyocytes inhibits apoptosis, thereby providing protection against myocardial infarction; this ratio is increased by curcumin nanoparticles. In a pre-clinical study of induced myocardial infarction in rats, curcumin nanoparticles prevented leakage of CK-MB, which is a marker for cardiac muscle injury. The serum concentrations of IL-6, TNF-α, monocyte chemoattractant protein (MCP)-1, IL-1α and IL-1β were the same following myocardial infarction, thereby affirming its cardioprotective potential and ability (Simion et al., 2016; Boarescu et al., 2019).

Quercetin-loaded phosphatidylcholine liposomes (PCLs) have been known to protect cardiomyocytes from myocardial injuries. Quercetin is a flavonoid isolated from citrus fruits. Quercetin acts as a scavenger and decomposer of peroxynitrite, which is known to cause myocardial injuries; furthermore, quercetin has the ability to restore normal contractility not only in isolated tissue preparations but also in whole anesthetized animals. The biphasic release of quercetin from nanoparticles allows it to prevent atherosclerosis and other CV disorders (Giannouli et al., 2018; Hesari et al., 2021).

Epigallocatechin ((−)-epigallocatechin-3-gallate), found abundantly in green tea, hinders atherosclerosis plaque formation by lowering inflammatory responses; it also decreases cholesterol accumulation in macrophages (Jhang et al., 2017). A nanoformulation of epigallocatechin was formulated, which targeted CD36, which is found on macrophages and platelets. This targeted accumulation of epigallocatechin by the nanoformulation in the intimal macrophages allows it to hinder plaque formation. Chitosan-derived nanoliposomes of epigallocatechin are also effective (Zhang et al., 2016).

Resveratrol exerts its anti-inflammatory actions through multiple pathways and targets. It increases NO bioavailability and decreases TNFα; both of these actions allow it to provide protection in CV diseases. It also decreases ROS activity, as it downgrades cyclooxygenase (COX)-2 activity. COX-2 is generally agreed to be the enzyme that is responsible for the generation of inflammatory cytokines like TNFα, IL-6, IL-2 and other prostaglandins (Yifan Wang et al., 2020; O'Leary et al., 2004; Bradamante et al., 2003).

Polyphenolic compounds are known to improve flow-mediated dilation (FMD), which is used to measure endothelial function. The dose dependency of polyphenols on FMD is well documented. Daily ingestion of coca, which is also a polyphenol, was found to increase FMD; consistently with this, the improved FMD was accompanied by a decrease in endothelin-1, which is a vasoconstriction mediator (de Oliveira et al., 2021).

13.2.2 Neurological Disorders

Disorders of the central nervous system range from proteinopathies such as Alzheimer's (AD) and Parkinson's disease (PD) to cerebral ischemia and traumatic brain injury. The commonalities shared by this wide range of conditions are nerve damage, oxidative stress, neural apoptosis and dendritic fragmentation, leading to synaptic loss (Ganguly et al., 2017). It has long been stated that the majority of this neural damage is due to high oxidative stress resulting from accumulation of abnormal protein or neural injury or some other secondary mechanism. Nutraceuticals, especially the polyphenolic groups and others, are known to provide decent neuroprotection; their nanoformulation allows enhanced delivery and therapeutic optimization (Costa et al., 2016).

Lycopene is a versatile carotenoid having pleiotropic pharmacological activity and is abundantly obtained from tomatoes as well as other red-colored fruits. It is a neuroprotective molecule that maintains cell viability, prevents fragmentation of mitochondria and lowers pro-apoptotic markers (Prakash and Kumar, 2013; Hwang et al., 2017). Lycopene's neuroprotective activity along with its ability to reduce the secretion of Aβ, the accumulation of which is a marker of AD, provides evidence of its usefulness in AD. Lycopene not only reduces Aβ secretion but also lowers tau hyperphosphorylation in the brain and even downregulates amyloid precursor proteins, all of which ultimately leads to less Aβ accumulation (Yu et al., 2017). Lycopene has exceptional antioxidant properties, which can be understood by its ability to improve SOD, catalase (CAT) and glutathione (GSH) concentrations in brain; this adds another weapon to its arsenal against AD (Sachdeva and Chopra, 2015). A detailed neuroprotective mechanism of lycopene is shown in Figure 13.3.

In pre-clinical models, lycopene is found to replenish concentrations of dopamine and its metabolic products in substantia nigra and striatum. This indicates its potential in PD, in which there is depletion of dopamine and diminution of the respective neurons (Paul et al., 2020). On a molecular level, lycopene is found to rescue cells from toxins that can induce PD. Also, it has the potential to reduce α-synuclein, which is believed to be the abnormal protein that accumulates in PD (Prema et al., 2015; Liu et al., 2013). Lycopene, as established earlier, is a potent antioxidant and therefore, confers neuroprotection in PD as well. In another pre-clinical model, lycopene was observed to decrease neuronal loss of tyrosine hydroxylase (TH)–positive neurons; these neurons have the ability to produce dopamine, and therefore, lycopene's ability to decrease TH-positive dopaminergic neuronal loss is a welcome sight (Paul et al., 2020). A schematic depiction of the neuroprotective mechanism of lycopene is shown in Figure 13.4.

Cerebral ischemia causes oxidative stress and inflammation in brain. Lycopene has the ability to safeguard brain tissue from both oxidative as well as nitrosative stress; the latter is done by decreasing the levels of NO, which is achieved by lycopene's ability to inhibit inducible neuronal nitric oxide synthase (Wei et al., 2010). In models of cerebral ischemia, it was found that iron and ferritin accumulation can lead to neurodegeneration. Lycopene diminishes the expression of neuronal iron transporters (hepcidin and ferroportin-1), thus providing a neuroprotective function. The microemulsion preparation of lycopene furthers its ability to act as a nano-nutraceutical (Guo et al., 2019).

Resveratrol is not only an anti-atherosclerotic agent imparting anti-inflammatory properties but also has the ability to depolymerize the Aβ protein in AD through a mechanism involving proteasomes. It also prevents Aβ aggregation by activating transthyretin, a protein that has the ability to sequester Aβ (Han et al., 2004). Further, it activates protein kinase C, which allows it to confer neuroprotective properties (Menard et al., 2013). Resveratrol inhibits calcium/calmodulin-dependent protein kinase II (CaMKII) and glycogen synthase kinase-3 (GSK3); it also activates protein phosphatase 2 (PP2A). This triad leads to protection against tau-phosphorylation and can even lead to dephosphorylation of tau proteins (Han et al., 2004). The development and progression of PD has been attributed to neuroinflammation and proinflammatory cytokines. Resveratrol reduces glial activation, and through NFκB, it reduces the levels and receptor expression of cytokines such as IL-1β, IL-6 and TNF-α. Resveratrol is thought to protect dopaminergic neurons through upregulation of decreased cytokine signaling-1 (SOCS-1) (Degan et al., 2018).

Interestingly, resveratrol increases the levels of serotonin and noradrenaline in brain, which in part is thought to be due to monoamine oxidase A (MOA-A) inhibition (Xu et al., 2010). Also, hippocampal brain derived neurotropic factor (BDNF) activation was found, which further validates its role as an antidepressant. In pre-clinical models, it was found to decrease serum cortisone levels,

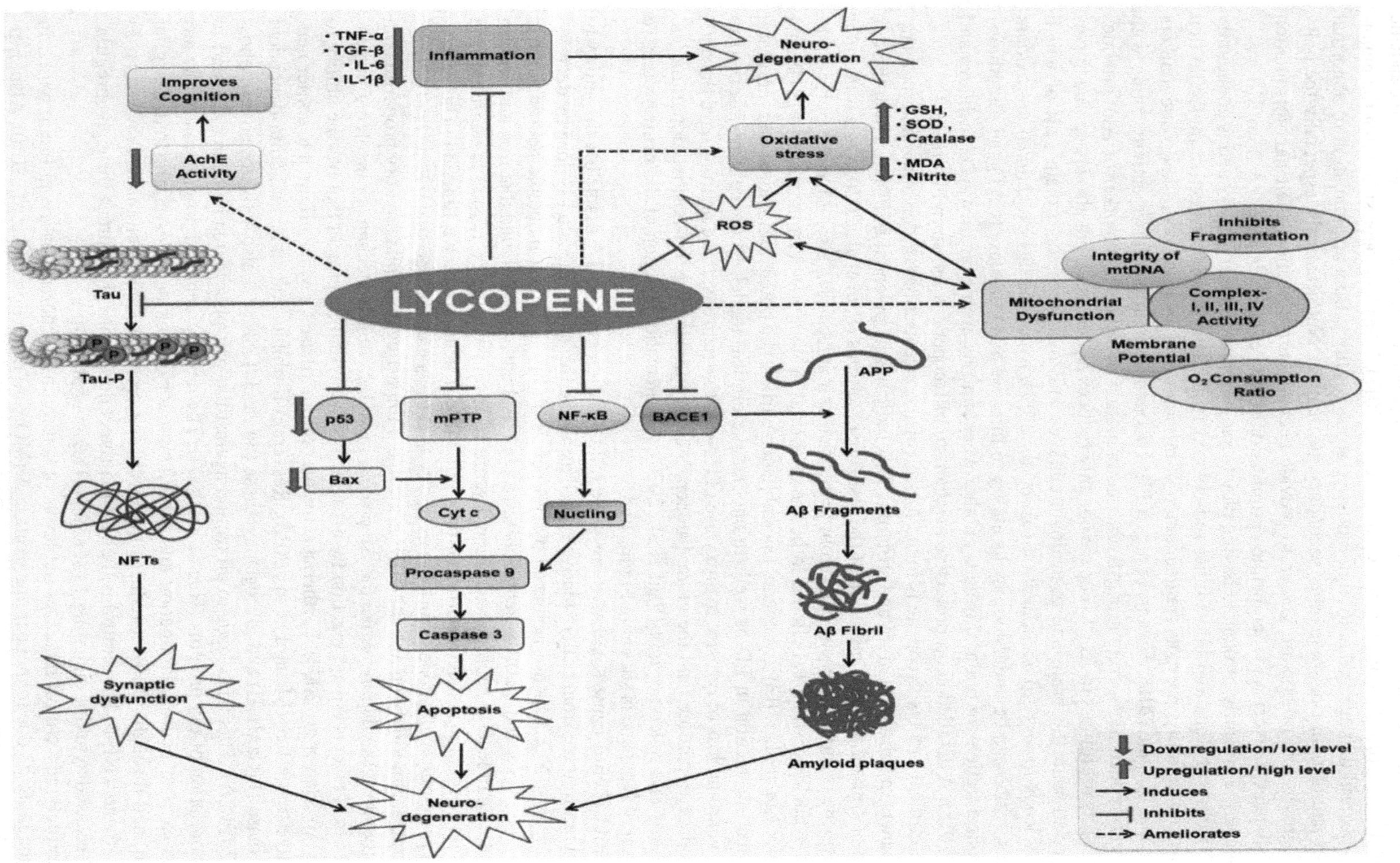

FIGURE 13.3 Neuroprotective mechanism of lycopene in Alzheimer's disease. (From Paul, R. et al., *Neurochemistry International*, 140, 104823, 2020.)

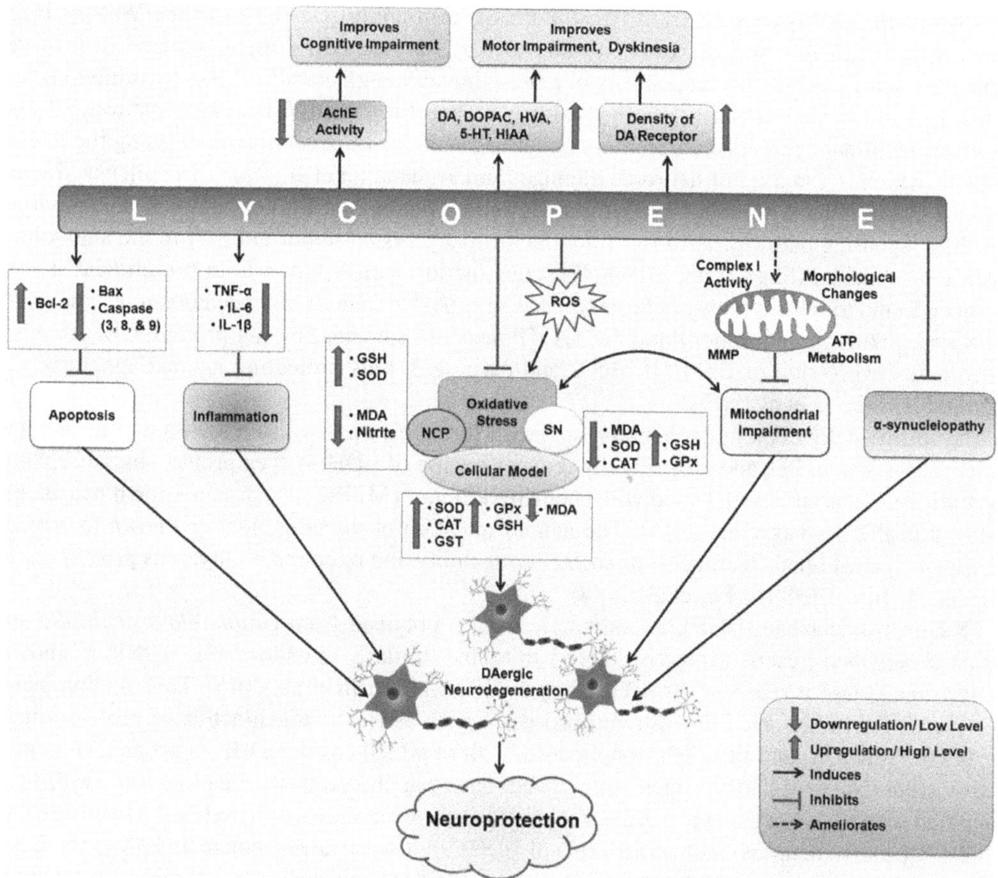

FIGURE 13.4 Schematic depiction of neuroprotective mechanism of lycopene in Parkinson's disease. (From Paul, R. et al., *Neurochemistry International*, 140, 104823, 2020.)

which are elevated during stress. It therefore appears to be a pleiotropic nano-nutraceutical agent (Hurley et al., 2014). Resveratrol confers protection against $FeCl_3$-induced posttraumatic seizures in rats and also confers protection in kainic acid–induced seizure models; both these models are used to establish antiseizure activity pre-clinically (Gupta et al., 2001; Gupta et al., 2002). The latter capability of resveratrol is attributed to its ability to decrease expression of kainite receptors. The overall ability of resveratrol to protect against epilepsy is because it can inhibit NFκB and the following inflammation; in addition, it plays a role in regulating the calcium-activated potassium channel of cortical neurons (Fonseca-Santos and Chorilli, 2020).

Another nutraceutical, ferulic acid, is abundantly found in cereal grains. Fluorescence electroscopic evaluation has shown that ferulic acid dose-dependently inhibits Aβ fibril formation; also, it has significant ability to destabilize preformed fibrils. Ferulic acid is also known to scavenge free radicals and therefore, provides additional artillery against AD (Ono et al., 2005). Chronic administration of ferulic acid has given insight into its ability to act as an antidepressant. This property is conferred on ferulic acid due its ability to modulate a variety of mediators and effectors, such as calcium/calmodulin-dependent protein kinase II (CaMKII), protein kinase C, protein kinase A and the PIK3 pathway. Modulation of oxidative stress and nitrosative stress is also implicated (Lenzi et al., 2015). Ferulic acid is also thought to be involved in neurogenesis in the hippocampal area. It also improves cognitive function by stimulating nerve growth factor (Antonella et al., 2015; Hassanzadeh et al., 2018).

Thymoquinone is extracted from *Nigella sativa*, commonly referred to in the West as black caraway seeds. In AD, the neurotoxicity due to Aβ is prevented by thymoquinone. It provides protection against Aβ-induced apoptosis in a pre-clinically engineered cell line by inhibiting caspases 3, 9 and 8, thus inhibiting both the intrinsic and the extrinsic apoptotic pathway. It also improves neuronal health and chromatin condensation and is implicated in decreasing the release of LDH, which is a marker of neuronal damage (Samarghandian et al., 2018). In epilepsy, thymoquinone is useful as it inhibits NFκB, reducing inflammation and excitotoxicity. It also modifies the NrF2 signaling pathway, a transcription factor that plays an important part in the antioxidant defense mechanism (Shao et al., 2017). The combination with Vitamin C in pre-clinical studies has been found to reduce high-grade seizures. The probable underlying mechanism is believed to be downregulation in phosphorylation of cAMP response element-binding protein (CREB). Also, it decreases expression of CaMKII, Bcl-2 and caspase 3, thus protecting against apoptosis and neuronal loss (Raza et al., 2008).

The ability of thymoquinone to inhibit apoptosis through caspases also allows it to protect dopaminergic neurons in PD, and it drastically reduces release of LDH. It even protects mesencephalic dopaminergic neurons from 1-methyl-4-phenylpyridinium (MPP+), which is a known neurotoxin involved in PD (Radad et al., 2015). The nanoformulation of thymoquinone is known to provide protection against renal clearance, enhanced permeability and retention ability, thus prolonging its pharmacological effect (El-Far et al., 2018).

7,8-Dihydroxyflavone (DHF), a monomeric flavone obtained from *Godmania aesculifolia* and *Tridax procumbens*, exerts protective effects in brain. Similarly to endogenous BDNF, it binds to the tyrosine kinase B receptor (TrkB) used by BDNF (Chitranshi et al., 2015). This binding therefore sparks the initiation of downstream processes such as Ras-mitogen-activated protein kinase (MAPK), PI3K/Akt, and PLC (phospholipase C), all of which enables DHF to act as a neuroprotective agent (Nie et al., 2019). Interestingly DHF has been observed to have prolonged half-life as compared with BDNF (Paul et al., 2021). PD patients have been shown to have lower levels of BDNF and TrkB-positive neurons, indicating a role of BDNF in disease advancement and progress. It has also been observed that consistent insufficiency of TrkB signaling causes loss of nigrostriatal dopaminergic neurons (Baydyuk et al., 2011). Chronic DHF treatment has been shown to prevent further loss of dopaminergic neurons in a pre-clinical model. It has also been shown to restore the levels of TH, which plays an important role in upregulating synthesis of dopamine synthesis. Also, in another pre-clinical model of the disease, it was found to decrease α-synuclein levels in substantia nigra. These results were further backed by similar results from primate studies (Nie et al., 2019; He et al., 2016). The dysregulation of TrkB is not limited to PD but is also found in AD. The dysfunction of TrkB causes synaptic debility. DHF acts a nerve growth factor in primarily cultured neurons; it facilitates synaptogenesis and also stimulates branching and lengthening of neurons (Aarons et al., 2019). DHF, as well as being neurotropic, has the ability to function as a robust antioxidant, as it increases the levels of SOD and GPx. Furthermore, it inhibits NFκB activation, which allows it to reduce and limit inflammatory cytokines (Wang et al., 2014).

DHF has been implicated in traumatic brain injury (TBI) as well. TBI causes loss of brain function, which is either acute or occurs after injury. This phase is called the primary injury. The loss of brain function can also be gradual along the lines of the primary injury site; this phase is referred to as secondary injury. DHF is implicated in the secondary injury state only. It acts by phosphorylation of TrkB, which ignites survival cascades involving PI3K/Akt-CREB. Nanoliposomal formulations of DHF are known to drastically increase the stability and antioxidant ability of DHF (Paul et al., 2021; Chen et al., 2020).

13.2.3 Metabolic Disorders

Metabolic disorders can be considered as a group of closely related disorders, which include diseases states such as central obesity, insulin resistance, hypertriglyceridemia, hypercholesterolemia and hypertension, as well as low high-density lipoprotein (HDL) concentration. These are often accompanied by other problematic co-morbidities like proinflammatory state, prothrombotic state and non-alcoholic fatty liver disease (NAFLD), all of which are a grave threat to public health (Wang et al., 2020). Nano-nutraceuticals have been indicated to combat them, some which are discussed further.

Hyperglycemic (diabetic) states are known to disrupt the normal homeostasis of the cellular environment and often cause accumulation of advanced glycation end products, which are referred to as AGEs. AGEs are known contributors to increasing oxidative stress, leading to further downstream disruptions (Fazel Nabavi et al., 2015). It is suggested that the formation of AGEs is suppressed by curcumin; the mechanism by which this is achieved involves the suppression of the receptor of AGEs, which is called RAGE. Curcumin suppresses RAGE by increasing the manifestation of peroxisome proliferator activated receptor gamma (PPARγ), and also, increasing the curcumin-mediated rise in glutathione levels has been considered (Stefanska, 2012; Jain et al., 2006).

When cells are exposed to hyperglycemic conditions, they are known to revert to a salvage pathway, which involves polyols. This pathway results in the synthesis of fructose and sorbitol, both of which are injurious to cells (Sai Varsha et al., 2015). Sorbitol damages the osmotic balance, and fructose, in turn, causes the generation of free radicals, which under hyperglycemic conditions are already quite prevalent in cells (Arun and Nalini, 2002). Curcumin is known to cause regulation of certain enzymes involved in the pathway; it causes both inhibition and potentiation of aldose reductase, the key enzyme in the pathway. The overall effect of this is protection in opposition to oxidative stress (Muthenna et al., 2009).

Curcumin is recognized for its anti-inflammatory and anti-oxidative characteristics, which are considered to be the reason behind its anti-hyperglycemic actions, but recent evidence has suggested that it also has the ability to stimulate beta cells (Best et al., 2007). Curcumin even affects membrane channels, and this is mediated by nuclear factor erythroid 2–related factor 2 (Nrf2) (Rashid and Sil, 2015; Fazel Nabavi et al., 2015).The fascinating array of mechanisms that curcumin has with regard to its antihyperglycemic activity is noteworthy, as it can also interact with M_1 cholinergic receptor. It increases glucose uptake by skeletal muscle, therefore improving insulin insensitivity (Na et al., 2011). Curcumin not only decreases NFκB levels but also acts by epigenetic mechanisms, regulating histone deacetylase and histone transferase, both of which are involved in gene expression (El-Azab et al., 2011). In a pre-clinical investigation, it was found that curcumin had the ability to modulate T-cell response, thus decreasing beta cell loss, which further provides insight into the fact that targeted nutraceuticals could be greatly beneficial in diabetic states (Castro et al., 2014).

Curcumin also shows beneficial effects in obesity and other metabolic disorders. Obesity is often linked with insulin resistance and other metabolic dysfunction. Adipocytes are known to be involved in cross talk between different organs such as liver, pancreas, brain and muscle; therefore, a local dysregulation in the functioning of adipocytes causes catastrophic central influence throughout the body (Romacho et al., 2014; Stern et al., 2016). In obesity, the local balance of pro- and anti-inflammatory molecules is often found to be dysregulated at the adipocyte level. The levels of leptin (proinflammatory) are found to be raised, and the levels of adiponectin (anti-inflammatory) are found to be low (Wellen and Hotamisligil, 2003). This observation is especially true for adipocytes of the abdominal area. The proinflammatory state that obesity causes is even more evident, as it is known that in obese individuals, the levels of TNF-α and IL-6 are high (Yamashita et al., 2018). Curcumin administration is known to decrease the levels of both of these notorious inflammatory molecules (Jain et al., 2009). Furthermore, chronic intake of curcumin has been demonstrated to curtail the levels of circulating free fatty acids (FFA), particularly in patients with diabetes mellitus. Even levels of leptin have been found to be decreased in patients treated with high-dose chronic curcumin therapy (Na et al., 2013). It has also been found that the dysfunction of endoplasmic reticulum is linked with release of misfolded proteins, leading to release of FFA. Curcumin prevents the increase of phospho-inositol-requiring kinase 1(p-IRE1) and phospho-eukaryotic initiation factor 2 (p-eIF2), both of which are known stress markers of endoplasmic reticulum (Varì et al., 2021). Also, in one pre-clinical model, it has been observed to reduce macrophage infiltration into adipocytes, which further supports the anti-inflammatory action of curcumin in obesity (Weisberg et al., 2008). Nonalcoholic fatty liver disease has been shown to be caused by insulin resistance and dyslipidemia. Curcumin has the ability to modulate LDL metabolism (Sahebkar, 2014). This is achieved by upregulating LDL receptor expression; this increased LDL expression increases the excretion of cholesterol and also inhibits a protein called Niemann–Pick C1-Like 1 (NPC1L1), which has a role in intestinal absorption of cholesterol (Rahmani et al., 2016).

Another nano-nutraceutical, resveratrol, which is a flavonoid obtained from various sources such as berries and soy, is an anti-inflammatory molecule, which not only acts as a cardioprotectant but also has a suggested role in diabetes (Appari et al., 2014). In pre-clinical studies, it has been established that resveratrol improves the biochemical attributes of diabetic neuropathy, diabetic nephropathy, diabetic retinopathy and obviously, diabetes-induced cardiopathies (Bastianetto et al., 2015; Mohammadshahi et al., 2014; Kubota et al., 2011). This protective action against a plethora of diabetes-induced morbidities is due to its capacity to stimulate adenosine monophosphate activated kinase (AMPK), which modulates critical intracellular processes including mitochondrial function, energy metabolism and cellular homeostasis (Rabinovitch et al., 2017). In type 2 diabetes, AMPK dysregulation has been found (Mohseni et al., 2019; Oyenihi et al., 2016). The protective effects of resveratrol in diabetic neuropathy are due to its ability to inhibit NFκB. Resveratrol has also been found to downregulate proprotein convertase subtilisin/kexin type 9 (PCSK9) along with repression of sphingosine kinase 1 (SphK1-S1P); these two mechanisms of action are considered to be involved in its ability to ameliorate the deleterious effects of diabetic nephropathy (Chen et al., 2016; Yan Wang et al., 2016). Insulin resistance is found to be ameliorated by resveratrol, as it is able to upregulate SIRT1, which is heavily involved in glucose homeostasis. The upregulation of SIRT1 also leads to an increase in GLUT-2 and GLUT-4 in pre-clinical models (Abbasi Oshaghi et al., 2017; Smoliga and Blanchard, 2014). Clinical trials of resveratrol have shown some promising signs; especially, chronic treatment has been shown to improve hemoglobin A1c, systolic blood pressure, total cholesterol and total protein in a type 2 diabetic population (Bhatt et al., 2012). It is also established that epigallocatechin gallate combined with resveratrol improves overall metabolic health (Most et al., 2014; Öztürk et al., 2017). Nanoparticles of resveratrol have shown superior performance in pharmacokinetic parameters. Also, in rat models, retinal expression of vascular endothelial growth factor (VEGF)-1, TNFα, MCP-1, intercellular adhesion molecule (ICAM)-1, IL-6 and IL-1β has been observed to be lower (Dong et al., 2019). Resveratrol can be contemplated as an acute lipogenic agent. In certain clinical trials, it has shown a decrease in total weight, body mass index (BMI), fat mass and waist circumference (WC), but its beneficial effects have still been not proven unequivocally (Zhou et al., 2019).

13.2.4 Gastrointestinal Disorders

Gastrointestinal disorders include diseases involving the esophagus, stomach, duodenum, jejunum, ileum, large intestine, sigmoid colon and rectum. The exacerbating factors range widely from bacterial infection with *Helicobacter pylori* to autoimmune disorders causing chronic inflammation. Gastrointestinal disorders cause great discomfort; sufferers cannot enjoy dietary pleasure due to the need to monitor their eating and lifestyle (Knutsson and Bøggild, 2010). Pharmaceutical interventions are often plagued with side effects; thus, nutraceuticals might provide a relieving solution.

Curcumin is one such nutraceutical, an anti-inflammatory molecule that is indicated in different gastrointestinal ailments (Rajasekaran, 2011). It has the ability to decrease levels of inflammatory regulators such as 5-lipoxygenase (5-LOX), COX-2) and inducible nitric oxide synthase (iNOS), which allows it to attenuate the discomfort in inflammatory bowel diseases (IBD) (Jurenka, 2009). Even in a pre-clinical model, it was able to exert mucosal protection in colitis (Lubbad et al., 2009). In IBD, stress is considered as a major aggravating factor, which is believed to increase proinflammatory factors such as NFκB (Kuebler et al., 2015). Curcumin, in addition to inhibition of NFκB, also has the ability to increase serotonin, BDNF and phosphorylation of cAMP response element-binding protein (pCREB), not only in colon but also in hypothalamus; this leads to regulation of a dysregulated hypothalamic-pituitary-adrenal (HPA) axis, which is also a hallmark of IBD (Yu et al., 2015; Pace et al., 2006). Curcumin is also observed to moderate immune response in chronic diarrhea (Vecchi Brumatti et al., 2014). Quercetin, like curcumin, is an anti-inflammatory molecule, and it too can modulate inflammatory modulators, especially NFκB, which allows it to reduce other inflammatory cytokines that cause aggravation in IBD. It can also inhibit phosphokinase (Rabiskova et al., 2009). Quercetin is indicated in *H. pylori* infection, as it is able to reduce gastric apoptosis, which is frequent in peptic ulcers. This is achieved by raising the ratio of BCL-2/BAX (Zhang et al., 2017).

In functional dyspepsia, gastric emptying time is impaired. The nutraceutical ginger is able to stimulate antral contraction (Hu et al., 2011), and in gastroenteritis, it is able to decrease gastrointestinal pain but shows no change in dyspepsia (Drozdov et al., 2012). Ginger is also indicated in dysphagia, as it aids in the improvement of swallowing function; gingerol and shogaol are responsible for this, as they act as TRPVI agonists (Abe et al., 2015). Ginger also, interestingly, has a spasmolytic effect, which is mediated by non-competitive inhibition of high potassium–induced contraction or due to some calcium antagonist effects (Talukder, 2019). A nanoformulation of ginger is superior to normal ginger preparations in therapeutic effect (Bakr et al., 2020).

Butyrate derivates are short-chain fatty acids, and direct dosing was found to decrease symptoms of IBD, such as abdominal pain, bloating and flatulence. Its beneficial effect was found to be more profound in the diarrheic subset of IBD; the possible mechanism is suggested to be via moderation of AMP-activated protein kinase (Nozu et al., 2019).

13.2.5 Treatment of Cancer

Cancer is known to attack virtually each and every organ of the human body. The global burden caused by cancer is enormous, leading to up to 9.6 million deaths, which can be visualized as one in six deaths, according to the data captured in 2018. In men, lung, prostate and colorectal cancer are more prevalent, whereas in women, the prevalence of breast cancer along with cervical cancer is high (WHO, 2018). Phytochemicals derived from food have been long indicated to possess anticancer activities, and nanonutraceuticals are a promising solution to this grave global problem (Salama et al., 2020).

Curcumin shows anticancer properties by interacting with a plethora of molecular targets. It upregulates death receptor DR4 and DR5, which upregulates the extrinsic apoptotic pathway, leading to cell death (Lee et al., 2012). Recently, it was found that curcumin had an unorthodox anticancer approach as it decreased glucose uptake and lactate production in cells. This decreased glucose uptake was achieved by downregulating pyruvate kinase M2 (PKM2) (Lee et al., 2012). Curcumin has been indicated in prostate cancer. It inhibits multiplication of cancer cells, and it also has the ability to induce apoptosis by interfering with NFκB, MAPK and EGFR, which has been specifically implicated in many carcinogenic pathways (Dorai et al., 2001; Mukhopadhyay et al., 2001). Curcumin also has the potential to trigger protein kinase D1 (PKD1) which diminishes oncogenic signaling, especially that involving beta catenin and MAPK (Sundram et al., 2011). The major molecular targets of curcumin are depicted in Figure 13.5.

In colorectal cancer, it is found that miR-21 gene is overexpressed; curcumin downregulates this gene by inhibiting its promoter AP-1 (Mudduluru et al., 2011). Curcumin also is able to inhibit cancer cells of head and neck by affecting NFκB as well as STAT3, both of which are abundantly overexpressed in head and neck tumor cells (Chun et al., 2003). In breast cancer cell lines, curcumin aids in the expression of p53 gene, which leads to apoptosis. (Moghtaderi et al., 2017). Also, curcumin has shown the ability to curb metastasis in breast cancer cells, as it inhibits certain inflammatory cytokines, such as CXCL1, which in turn, are known to suppress metastasis-promoting genes(Bachmeier et al., 2008).Curcumin even has the potential to arrest the cell cycle; the phase in which it generally arrests is G2/M. This is achieved because of its ability to increase protein kinase 1 (DAPK1). These effects were observed in brain cancer cell lines, especially in glioblastoma cells (Wu et al., 2013).

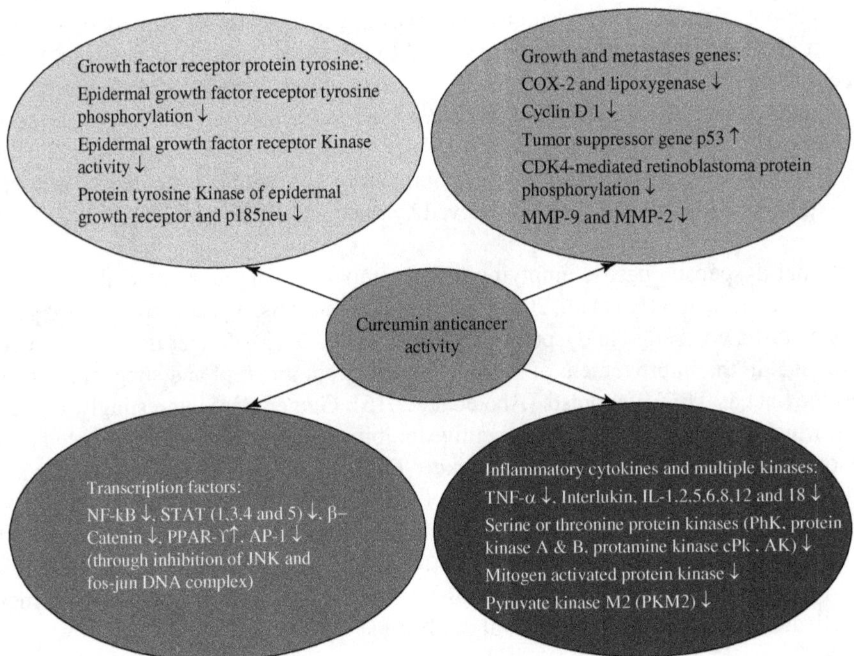

FIGURE 13.5 Molecular goals of curcumin. ↑: Increase; ↓: Decrease; MMP: Matrix metalloproteinase; AP-1: Activation protein 1. (From Tomeh, M.A. et al., *International Journal of Molecular Sciences*, 20, 5, 1033, 2019.)

Genistein is found in soyabean and other soy products; it is an isoflavone commonly consumed by Asian countries (Ronis, 2016). The anticancer ability of genistein is wide ranging, from anti-angiogenesis to antimetastatic effects. It is also known to arrest dividing cells in certain cancers (Tuli et al., 2019). In cancer of cervical origin, genistein causes cellular death by increasing the activity of caspases 9 and 3 (Dhandayuthapani et al., 2013). Genistein even has the ability to cause apoptosis through the endoplasmic reticulum stress mechanism. This is achieved by increasing the expression of glucose-regulated protein 78 (GRP78) and CCAAT/enhancer-binding protein homologous protein (Yang et al., 2016). It also increases the expression of proteins such as inositol-requiring enzyme 1α, calpain 1, growth arrest and DNA damage inducible gene 153, caspase 7 and caspase 4, all of which are implicated as endoplasmic stress proteins (Hsiao et al., 2019). The apoptotic ability of genistein is also thought to be due to its potential to increase the phosphorylation and stimulation of p53 and also by altering the balance of Bcl-2/Bax. Even a decrease in the phosphorylation of Akt is indicated by its ability to cause death in cancerous cells (Ouyang et al., 2009). It is also known to cause apoptosis through oxidative stress, particularly by inducing NO production. Interestingly, the inhibition of aerobic glycolysis by genistein also leads to apoptosis, as aerobic glycolysis leads to downregulation of hypoxia-inducible factor-1α, which disables glucose transporter 1 or/and hexokinase 2, thus disturbing the homeostatic balance in the cell (Li et al., 2017).

The cell cycle is a complicated and intricate process involving many regulatory mechanisms; checkpoints (Chk) are among these regulatory adapters. Genistein activates Chk-1 and Chk-2, which causes the inactivation of certain vital proteins, such as Cdc25C and Cdc25A, Cdc2. This inactivation ultimately leads to the arrest of cell cycle, thus inhibiting further proliferation of cancer cells (Ouyang et al., 2009). Another possible pathway of cell cycle arrest caused by genistein is through the ATM/p53 pathway (Zhang et al., 2013). Genistein's anti-angiogenic effects are mainly due to its capacity to downregulate VEGF, platelet-derived growth factor, urokinase plasminogen activator, matrix metalloprotease-2 (MMP-2) and MMP-9 expression, all of which are heavily involved in the formation of new blood vessels (Su et al., 2005). Genistein can inhibit metastasis of cancer cells by decreasing the amount and phosphorylation of focal adhesion kinase (FAK); it also inhibits COX-2, MMP-9, Ang-1 and vasodilator-stimulated phosphoprotein (VASP) (Kang et al., 2018)(Gu et al., 2009).

Furthermore, another nutraceutical, thymoquinone causes oxidative breakage in DNA, leading to antiproliferative effects. This oxidative breakage is due to its ability to interact with cellular copper present in chromatin. It can also inhibit telomerase function by stabilizing G-quadruplexes (Zubair et al., 2013; Salem et al., 2015). Thymoquinone can reduce the proliferation of cancer cells by decreasing F-actin polymerization and also by suppression of STAT3 phosphorylation (Badr et al., 2011). Additionally, thymoquinone can exert its antiproliferative activity by arresting cell proliferation and causing cell lysis by decreasing the Notch signaling pathway (Ke et al., 2015). Thymoquinone also targets the NLRP3 inflammasome, which leads to its anti-metastasis activity (Ahmad et al., 2013).

Quercetin causes cellular death in breast tumor cell lines of by suppressing Twist–p38MAPK, which is abundantly expressed in breast cancer cells (Liao et al., 2015). Silver nanoparticles of quercetin inhibit cellular proliferation by causing an escalation of proapoptotic proteins such as Bax and caspases 3; also, it downregulates Bcl-2, which is an anti-apoptotic protein (Balakrishnan et al., 2017). Triple negative breast cancer, which is considered to be more evasive as it lacks the normal receptor targets present on other cancer cells, is also arrested by quercetin interacting with Foxo3a (Nguyen et al., 2017). Quercetin in colon cancer cell lines is able to decrease metastasis by increasing E-cadherin protein; also, it decreases metastasis-related proteins such as MMP-2, MMP-9 and others in a dose-dependent fashion (Rauf et al., 2018). Also, recent research has indicated that suppression of NFκB by quercetin plays a significant role in colon cancer treatment. A twofold increase in caspases 3 is reported in certain colon cancer cell lines when treated with quercetin (Atashpour et al., 2015).

Quercetin and its glycosides are also indicated in pancreatic cancer, as they suppress the migration of cancer cell induced by transforming growth factor beta (TGF-β) and VEGF (Lee et al., 2013). Quercetin is even indicated as add-on therapy with gemcitabine, as the combination is able to decrease cellular proliferation, invasion and self-renewal capacity of cancer cells of pancreatic origin (Cao et al., 2015). Quercetin reduces viability of cells, reduces migration and triggers apoptosis in ductal adenocarcinoma of the pancreas (Appari et al., 2014). Interestingly, the glycoside of quercetin when administered in hepatic cancer shows less toxicity to normal hepatic cells than sorafenib, an approved drug (Sudan and Rupasinghe, 2015). Fatty acid synthase, an enzyme found in liver cells, is inhibited by quercetin, and this inhibition leads to apoptosis (Zhao et al., 2014). A nanoformulation of quercetin reduces histone deacetylase, which causes arrest of cellular division at sub-G stage (Bishayee et al., 2015). Nanoparticles of quercetin are also indicated in prostate cancer (Rauf et al., 2018).

Fascinatingly, the combination of curcumin and quercetin has shown superiority in suppressing cancer cell multiplication, especially in gastric cancer cell lines. It also disrupts the mitochondrial membrane potential, which eventually leads to apoptosis (Zhang et al., 2015).

Cancer metastasis is in part regulated by Toll like receptor 4. Epigallocatechin imparts its anticancer activity by downregulating this receptor (Byun et al., 2014). It even induces regulated cellular death in certain breast cancer cell lines via activation of proteins such as caspase 3, caspase 9 and PARP-1 (Zan et al., 2019). Additionally, it decreases mitochondrial membrane potential, which forces the cell to arrest proliferation (Zhang et al., 2015). Epigallocatechin is indicated in thyroid cancer, where it suppresses cell growth through an intricate signaling mechanism that involves EGFR/Ras/Raf/MAPK and extracellular signal-regulated kinase (ERK) (Wu et al., 2019). Epigallocatechin has been suggested also as an anti-angiogenic molecule because it hinders the binding of VEGF to its receptor, which suppresses the downstream phosphorylation of ERK/Akt (Jung and Ellis, 2001).

Resveratrol has the marvelous ability to inhibit or interfere with all the three stages of carcinogenesis: initiation, promotion and progression (Yousef et al., 2017). In colorectal cancer, resveratrol inhibited the cell cycle at the junction of S/G2. This is perhaps due to inhibition of ornithine decarboxylase. Also, the tumor growth is suppressed, as resveratrol induces Bax protein and inhibits bcl-2; even suppression of growth is postulated to be due to inhibition of NFκB (Buhrmann et al., 2016; Chen et al., 2009; Fuggetta et al., 2006). Resveratrol causes apoptosis of squamous epidermal cancer cells. This is brought into action by inactivation of Wnt-2 and its downstreaming genes (Liu et al., 2017). In melanoma cells, resveratrol causes apoptosis by suppression of survivin through STAT3/β-catenin (Habibie et al., 2014). Autophagy is also induced by resveratrol, as it stimulates proteins related to autophagy, such as Atg5, Atg12, Beclin-1 and LC3-II. Even cisplatin-resistant cancer cells undergo autophagy by resveratrol (Chang et al., 2017). Another mechanism by which it can induce autophagy is the SIRT1/AMPK-dependent inhibition of mTOR (Zhao et al., 2017). In nasopharyngeal carcinoma cells, resveratrol induces apoptosis but by a p53-independent mechanism, which is achieved by interaction with p63 (Chow et al., 2010). It also redistributes first apoptosis signal (FAS), which is a death receptor, into membrane rafts, especially in colon cancer cells. Resveratrol downregulates survivin, an antiapoptotic factor (Delmas, 2003). Moreover, resveratrol inhibits angiogenesis by suppressing VEGF (Kasiotis et al., 2013).

13.3 TARGETED DELIVERY OF NANO-NUTRACEUTICALS FOR THE TREATMENT OF CANCER AND NEUROLOGICAL DISORDERS

A number of nano-nutraceutical formulations have been formulated for targeted delivery in the treatment of tumors. The delivery mechanism of these drugs should be such that it provides increased bioavailability and efficacy. One study found that if a patient is administered 12,000 mg of curcumin, a meagre 51.2 ng/mL reaches the plasma after 4 hours (Lao et al., 2006). Furthermore, the formulation should also protect the drug from pharmacokinetic hurdles such as low solubility, interaction with other compounds present in the gastrointestinal tract, etc., and hence, low bioavailability. The formulation should have no influence on the texture and flavor of the drug, and should enable increased shelf life, protection against moisture and sustained release. Table 13.1 presents various thymoquinone formulations (El-Far et al., 2018). Table 13.2 lists various systems for delivery of curcuminoids in cancer (Tomeh et al., 2019), and Table 13.3 shows the different delivery systems of resveratrol used in neurological diseases (Fonseca-Santos and Chorilli, 2020).

13.4 CONCLUDING REMARKS AND FUTURE PERSPECTIVES

Nutraceuticals have been used in medicine for a long time. However, over the last two decades, a plethora of research has been done in this field. Researchers were able to establish the mechanism of action of a number of nutraceuticals along with their targets in various diseases. Furthermore, in-depth research has also been performed in delivery mechanisms. Poor water solubility, stability and interaction with other compounds have been the major obstacles to the successful delivery of these agents. A number of techniques have since been utilized to increase the bioavailability of these nutraceuticals. Nanotechnology is one of the relevant techniques that have been explored. Nanoencapsulation, nanoparticles, quantum dots, polymeric micelles, phospholipid micelles, colloidal nanoliposomes, nanoemulsions, dendrimers and polymerics have been utilized to produce bioavailable nutraceuticals. In the longer run, the cost of these food-based medicines as well as the cost involved in research could be reduced. Some of these nutraceuticals have reached the market, and for the rest, in-depth study involving the risk-to-benefit ratio could be conducted utilizing scientific studies.

TABLE 13.1
Thymoquinone Nanoformulations and Features for Use in Cancer

Method	Materials, stabilizers	Size (nm)	Animal, dose	Therapeutic effect	Reference
Single emulsion	PLGA	200–300		Enhanced anticancer	(Soni et al., 2015)
	Compritol ATO 888, gelucire	~200	Rats, 20 mg/kg, oral	Fourfold enchantment of oral bioavailability	(Elmowafy et al., 2016)
Nanoprecipitate	PVP, PEG_{200}, PEG_{400}, P123	20–40	Mice, 5 mg/kg, subcutaneous injection	Reduced tumor and increased life span	(Bhattacharya et al., 2015)
	Gum rosin, oleic acid, PVA, polysorbate 80	50–90	Wistar female albino rats, 20, 40, 80 mg/kg, oral	Significantly decreased blood glucose level and glycated hemoglobin	(Rani et al., 2018)
Double emulsion	Poly-N-acetyl glucosamine, PVA	185	Mice, ~150 mg/kg, subcutaneous injection	Inhibited tumor growth 43%, 31%	(El-Ashmawy et al., 2017)
Ionic gelation	Chitosan, TPP	150	Male Wistar rats, 2.52 mg/kg, intranasal	15-fold higher brain targeting efficiency	(Alam et al., 2012)
Film rehydration	Liposome	100		More potent anti-proliferative activity	(Odeh et al., 2012)
Cold wet milling	HPC-SSL	143	Male Sprague-Dawley rats, 2 mg/kg, oral	Sixfold enchantment of oral bioavailability	(Nihei et al., 2016)

Source: El-Far et al., 2018.

Abbreviations: PLGA, poly(D,L-lactide-co-glycolide); PVA, poly(vinyl alcohol); PVP, polyvinylpyrrolidone; PEG, poly(ethylene glycol); P123, poly(ethylene glycol)-b-poly(propyleneglycol)-b-poly(ethylene glycol); TPP, sodium triphosphate; HPC-SSL, hydroxypropyl cellulose grade SSL.

TABLE 13.2
Recent Curcumin Delivery Systems in Cancer

Nanoformulation	Particle size	Application	Outcome	Reference
Curcumin-loaded liposomal PMSA antibody	100–150 nm	Human prostate cancer (LNCa, C4-2B)	Enhanced antiproliferative efficacy and targeting	(Thangapazham et al., 2008)
Curcumin-loaded magnetic silk nanoparticles	100–350 nm	Human breast cancer (MDA-MB-231) cells	Enhanced cellular uptake and growth inhibition	(Song et al., 2017)
Curcumin/MPEG PCL micelles	27 ± 1.3 nm	Colon carcinoma (C-26) cells	Enhanced cancer growth inhibition	(Gou et al., 2011)
Curcumin nanoemulsion	<200 nm	Human ovarian adenocarcinoma cells (SKV3)	Increased cytotoxicity	(Ganta and Amiji, 2009)
Curcumin loaded liposomes coated with N-dodecyl chitosan–HPTMA chloride	73 nm	Murine fibroblasts (NIH3T3) and murine melanoma (B16F10) cells	Specific toxicity in murine melanoma (but not in fibroblasts)	(Karewicz et al., 2013)
Curcumin–PLGA nanoparticles	248 ± 1.6 nm	Erythroleukemia type 562 cells	Improved clinical management of leukemia	(Ganta and Amiji, 2009)
Curcumin-loaded lipo-PEGPEI [7] complexes	269 nm	Melanoma (B16F10) and colon carcinoma (CT-26) cells	Increased cytotoxicity	(Lin et al., 2012)
Curcumin–chitosan nanoparticles	100–250 nm	Melanomas	Enhanced antitumor effect	(Xingyi Li et al., 2012)
ApoE peptide–functionalized curcumin-loaded liposomes	132 nm	RBE4 cell monolayer	Increased accumulation in brain capillary endothelium	(Re et al., 2011)
Curcumin-crosslinked polymeric nanogels	10–200 nm	Breast and pancreatic cancers	Higher stability and enhanced antitumor effect	(Mangalathillam et al., 2012)
Curcumin-loaded chitin nanogels	70–80 nm	Human skin melanoma (A385) and human dermal fibroblasts (HDF)	Specific toxicity in skin melanoma (lower toxicity in HDF)	(Mangalathillam et al., 2012)
Curcumin-loaded lipid-core nanocapsules	196 ± 1.4 nm	Rat C6 and U251MG glioma cell lines	Decreased tumor size and prolonged survival	(Zanotto-Filho et al., 2013)
Liposome-encapsulated curcumin	Not reported	Head and neck squamous cell carcinoma (HNSCC) cell lines (CAL27 and UM-SCC1)	Cancer growth suppression both *in vitro* and *in vivo*	(Wang et al., 2008)

Source: Tomeh et al., 2019.
Abbreviations: PMSA: Prostate membrane specific antigen; MPEG: monomethoxy poly ethylene glycol; PCL: poly(ε-caprolactone); HPTMA: N-[(2-hydroxy-3-trimethylamine) propyl; PLGA: polylactic-co-glycolic acid; PEG: poly ethylene glycol; PEI: polyethyleneimine; ApoE: apolipoprotein E.

TABLE 13.3
Delivery System of Resveratrol in Neurological Diseases

Nanotechnology-based formulation	Formulation ingredients	Size and zeta potential	Drug entrapment	Disease or tissue	Route of administration	Major results	Model	Reference
Nanoparticles	PLGA, α-tocopheryl polyethylene glycol 1000, succinate	170 nm and −16 mV	60%	Glioma	Intravenous	The ability of resveratrol to kill glioma cells and pharmacokinetics shows the higher distribution in the brain compared with free drug	Cells and rats	(Vijayakumar et al., 2016)
Nanoparticles	PLA, polysorbate 80 and PVA	200 nm and −20 mV	70%	Parkinson's disease (PD)	Intraperitoneal	Nanoformulation shows neuroprotective effects in PD model	Mice	(Lindsay et al., 2010)
Nanoparticles	PLA–PEG and transferrin	150 nm and −10 mV	80%	Glioma	Intraperitoneal	Ability to destroy glioma cells, target the drug to the brain, decrease the volume of tumor in the brain and increase survival rate of rats	Rats	(Guo et al., 2013)
Nanoemulsions	Propylene glycol caprylate, vitamin E, polysorbate 80 and ethyldiglycol	100 nm and −35 mV	150 mg/mL	PD	Intranasal	Decreased degenerative changes and increased GSH and SOD levels in the brain; a higher concentration of the drug in the brain	Rats	(Pangeni et al., 2014)
Liposomes	Iron oxide, chitosan and oleanolic acid	150 nm	NA	PD	Intraperitoneal	Ability to cross BBB, and using a magnetic field above the head increased the transport of liposomes into the brain	Rats	(Wang et al., 2018)
Liposomes	Egg phosphatidylcholine, 1,2-dioleoyl-sn-glycero-3-phosphoethanolamine, cholesteryl hemisuccinate, DSPE-PEG and transferrin	200 nm and −35 mV	75%	Glioblastoma	Intraperitoneal	Evidence of internalisation into glioblastoma cells	Cells and mice	(Jhaveri et al., 2018)
Liposome	Phospholipid, cholesterol, dicetylphosphate and10 mM Tris buffer	200 nm and −30 mV	NA	Epilepsy	Intravenous	Reduced the spike frequency and antioxidant activity compared with the free form of resveratrol	Rats	(Ethemoglu et al., 2017)
Oil-core nano-capsules	PCL, caprylic/capric triglyceride mixture, polysorbate 80 and sorbitan monostearate	250 nm and −15 mV	1 mg/mL	Glioma	Intraperitoneal	Improvement of surviving glioma cells and ability to decrease brain volume of tumor	Cells and rats	(Figueiró et al., 2013)

Source: Fonseca-Santos and Chorilli, 2020.

REFERENCES

Aarons, T., Bradburn, S., Robinson, A., Payton, A., Pendleton, N., & Murgatroyd, C. (2019). Dysregulation of BDNF in prefrontal cortex in Alzheimer's disease. *Journal of Alzheimer's Disease*, *69*(4), 1089–1097.

Abbasi Oshaghi, E., Goodarzi, M. T., Higgins, V., & Adeli, K. (2017). Role of resveratrol in the management of insulin resistance and related conditions: Mechanism of action. *Critical Reviews in Clinical Laboratory Sciences*, *54*(4), 267–293.

Abe, N., Hirata, A., Funato, H., Nakai, M., Iizuka, M., Yagi, Y., Shiraishi, H., Jobu, K., Yokota, J., & Moriyama, H. (2015). Swallowing function improvement effect of Ginger (*Zingiber officinale*). *Food Science and Technology Research*, *21*(5), 705–714.

Ahmad, I., Muneer, K. M., Tamimi, I. A., Chang, M. E., Ata, M. O., & Yusuf, N. (2013). Thymoquinone suppresses metastasis of melanoma cells by inhibition of NLRP3 inflammasome. *Toxicology and Applied Pharmacology*, *270*(1), 70–76.

Alam, S., Khan, Z. I., Mustafa, G., Kumar, M., Islam, F., Bhatnagar, A., & Ahmad, F. J. (2012). Development and evaluation of thymoquinone-encapsulated chitosan nanoparticles for nose-to-brain targeting: A pharmacoscintigraphic study. *International Journal of Nanomedicine*, *7*, 5705.

Antonella, S., Daniela, G., & Martadi, C. (2015). Ferulic acid: A hope for Alzheimer's disease therapy from plants. *Nutrients*, *7*, 5764–5782.

Appari, M., Babu, K. R., Kaczorowski, A., Gross, W., & Herr, I. (2014). Sulforaphane, quercetin and catechins complement each other in elimination of advanced pancreatic cancer by miR-let-7 induction and K-ras inhibition. *International Journal of Oncology*, *45*(4), 1391–1400.

Arora, D., & Jaglan, S. (2016). Nanocarriers based delivery of nutraceuticals for cancer prevention and treatment: A review of recent research developments. *Trends in Food Science & Technology*, *54*, 114–126.

Arun, N., & Nalini, N. (2002). Efficacy of turmeric on blood sugar and polyol pathway in diabetic albino rats. *Plant Foods for Human Nutrition*, *57*(1), 41–52.

Aryee, A. N. A., & Boye, J. I. (2015). Current and emerging trends in the formulation and manufacture of nutraceuticals and functional food products. *Nutraceutical and Functional Food Processing Technology*, 1–52.

Atale, N., Saxena, S., Nirmala, J. G., Narendhirakannan, R. T., Mohanty, S., & Rani, V. (2017). Synthesis and characterization of *Sygyzium cumini* nanoparticles for its protective potential in high glucose-induced cardiac stress: A green approach. *Applied Biochemistry and Biotechnology*, *181*(3), 1140–1154.

Atashpour, S., Fouladdel, S., Movahhed, T. K., Barzegar, E., Ghahremani, M. H., Ostad, S. N., & Azizi, E. (2015). Quercetin induces cell cycle arrest and apoptosis in CD133+ cancer stem cells of human colorectal HT29 cancer cell line and enhances anticancer effects of doxorubicin. *Iranian Journal of Basic Medical Sciences*, *18*(7), 635.

Augustin, M. A., & Sanguansri, L. (2012). Challenges in developing delivery systems for food additives, nutraceuticals and dietary supplements. In *Encapsulation technologies and delivery systems for food ingredients and nutraceuticals* (pp. 19–48). Elsevier.

Ayyanar, M., & Subash-Babu, P. (2012). *Syzygium cumini* (L.) Skeels: A review of its phytochemical constituents and traditional uses. *Asian Pacific Journal of Tropical Biomedicine*, *2*(3), 240–246.

Bachmeier, B. E., Mohrenz, I. V, Mirisola, V., Schleicher, E., Romeo, F., Höhneke, C., Jochum, M., Nerlich, A. G., & Pfeffer, U. (2008). Curcumin downregulates the inflammatory cytokines CXCL1 and-2 in breast cancer cells via NFκB. *Carcinogenesis*, *29*(4), 779–789.

Badr, G., Mohany, M., & Abu-Tarboush, F. (2011). Thymoquinone decreases F-actin polymerization and the proliferation of human multiple myeloma cells by suppressing STAT3 phosphorylation and Bcl2/Bcl-XL expression. *Lipids in Health and Disease*, *10*(1), 1–8.

Bakr, A. F., Abdelgayed, S. S., EL-Tawil, O. S., & Bakeer, A. M. (2020). Ginger extract and ginger nanoparticles; characterization and applications. *Int J Vet Sci*, *9*, 203–209.

Balakrishnan, S., Mukherjee, S., Das, S., Bhat, F. A., Raja Singh, P., Patra, C. R., & Arunakaran, J. (2017). Gold nanoparticles–conjugated quercetin induces apoptosis via inhibition of EGFR/PI3K/Akt–mediated pathway in breast cancer cell lines (MCF-7 and MDA-MB-231). *Cell Biochemistry and Function*, *35*(4), 217–231.

Bastianetto, S., Ménard, C., & Quirion, R. (2015). Neuroprotective action of resveratrol. *Biochimica et Biophysica Acta (BBA)-Molecular Basis of Disease*, *1852*(6), 1195–1201.

Baydyuk, M., Nguyen, M. T., & Xu, B. (2011). Chronic deprivation of TrkB signaling leads to selective late-onset nigrostriatal dopaminergic degeneration. *Experimental Neurology*, *228*(1), 118–125.

Bergamin, A., Mantzioris, E., Cross, G., Deo, P., Garg, S., & Hill, A. M. (2019). Nutraceuticals: Reviewing their role in chronic disease prevention and management. *Pharmaceutical Medicine, 33*(4), 291–309.

Best, L., Elliott, A. C., & Brown, P. D. (2007). Curcumin induces electrical activity in rat pancreatic β-cells by activating the volume-regulated anion channel. *Biochemical Pharmacology, 73*(11), 1768–1775.

Bhatt, J. K., Thomas, S., & Nanjan, M. J. (2012). Resveratrol supplementation improves glycemic control in type 2 diabetes mellitus. *Nutrition Research, 32*(7), 537–541.

Bhattacharya, S., Ahir, M., Patra, P., Mukherjee, S., Ghosh, S., Mazumdar, M., Chattopadhyay, S., Das, T., Chattopadhyay, D., & Adhikary, A. (2015). PEGylated-thymoquinone-nanoparticle mediated retardation of breast cancer cell migration by deregulation of cytoskeletal actin polymerization through miR-34a. *Biomaterials, 51*, 91–107.

Bishayee, K., Khuda-Bukhsh, A. R., & Huh, S.-O. (2015). PLGA-loaded gold-nanoparticles precipitated with quercetin downregulate HDAC-Akt activities controlling proliferation and activate p53-ROS crosstalk to induce apoptosis in hepatocarcinoma cells. *Molecules and Cells, 38*(6), 518.

Boarescu, P.-M., Chirilă, I., Bulboacă, A. E., Bocşan, I. C., Pop, R. M., Gheban, D., & Bolboacă, S. D. (2019). Effects of curcumin nanoparticles in isoproterenol-induced myocardial infarction. *Oxidative Medicine and Cellular Longevity, 2019.*

Bradamante, S., Barenghi, L., Piccinini, F., Bertelli, A. A. E., De Jonge, R., Beemster, P., & De Jong, J. W. (2003). Resveratrol provides late-phase cardioprotection by means of a nitric oxide-and adenosine-mediated mechanism. *European Journal of Pharmacology, 465*(1–2), 115–123.

Buhrmann, C., Shayan, P., Popper, B., Goel, A., & Shakibaei, M. (2016). Sirt1 is required for resveratrol-mediated chemopreventive effects in colorectal cancer cells. *Nutrients, 8*(3), 145.

Byun, E.-B., Kim, J.-H., Song, D.-S., Lee, B.-S., Park, J.-N., Park, S.-H., Park, C., Jung, P.-M., Sung, N.-Y., & Byun, E.-H. (2014). Epigallocatechin-3-gallate-mediated Tollip induction through the 67-kDa laminin receptor negatively regulating TLR4 signaling in endothelial cells. *Immunobiology, 219*(11), 866–872.

Cao, C., Sun, L., Mo, W., Sun, L., Luo, J., Yang, Z., & Ran, Y. (2015). Quercetin mediates β-catenin in pancreatic cancer stem-like cells. *Pancreas, 44*(8), 1334–1339.

Castro, C. N., Barcala Tabarrozzi, A. E., Winnewisser, J., Gimeno, M. L., Antunica Noguerol, M., Liberman, A. C., Paz, D. A., Dewey, R. A., & Perone, M. J. (2014). Curcumin ameliorates autoimmune diabetes. Evidence in accelerated murine models of type 1 diabetes. *Clinical & Experimental Immunology, 177*(1), 149–160.

Chang, C.-H., Lee, C.-Y., Lu, C.-C., Tsai, F.-J., Hsu, Y.-M., Tsao, J.-W., Juan, Y.-N., Chiu, H.-Y., Yang, J.-S., & Wang, C.-C. (2017). Resveratrol-induced autophagy and apoptosis in cisplatin-resistant human oral cancer CAR cells: A key role of AMPK and Akt/mTOR signaling. *International Journal of Oncology, 50*(3), 873–882.

Chen, C., Huang, K., Hao, J., Huang, J., Yang, Z., Xiong, F., Liu, P., & Huang, H. (2016). Polydatin attenuates AGEs-induced upregulation of fibronectin and ICAM-1 in rat glomerular mesangial cells and db/db diabetic mice kidneys by inhibiting the activation of the SphK1-S1P signaling pathway. *Molecular and Cellular Endocrinology, 427*, 45–56.

Chen, J., Dong, X. S., & Guo, X. G. (2009). Inhibitory effect of resveratrol on the growth of human colon cancer ls174t cells and its subcutaneously transplanted tumor in nude mice and the mechanism of action. *Zhonghua Zhong Liu Za Zhi* [*Chinese Journal of Oncology*], *31*(1), 15–19.

Chen, Y., Xia, G., Zhao, Z., Xue, F., Gu, Y., Chen, C., & Zhang, Y. (2020). 7, 8-Dihydroxyflavone nano-liposomes decorated by crosslinked and glycosylated lactoferrin: Storage stability, antioxidant activity, in vitro release, gastrointestinal digestion and transport in Caco-2 cell monolayers. *Journal of Functional Foods, 65*, 103742.

Chitranshi, N., Gupta, V., Kumar, S., & Graham, S. L. (2015). Exploring the molecular interactions of 7, 8-dihydroxyflavone and its derivatives with TrkB and VEGFR2 proteins. *International Journal of Molecular Sciences, 16*(9), 21087–21108.

Chow, S. E., Wang, J. S., Chuang, S. F., Chang, Y. L., Chu, W. K., Chen, W. S., & Chen, Y. W. (2010). Resveratrol-induced p53-independent apoptosis of human nasopharyngeal carcinoma cells is correlated with the downregulation of ΔNp63. *Cancer Gene Therapy, 17*(12), 872–882.

Chun, K.-S., Keum, Y.-S., Han, S. S., Song, Y.-S., Kim, S.-H., & Surh, Y.-J. (2003). Curcumin inhibits phorbol ester-induced expression of cyclooxygenase-2 in mouse skin through suppression of extracellular signal-regulated kinase activity and NF-κB activation. *Carcinogenesis, 24*(9), 1515–1524.

Cockbill, C. A. (1994). Food law and functional foods. *British Food Journal, 96* (3), 3–4.

Costa, L. G., Garrick, J., Roque, P. J., & Pellacani, C. (2016). Nutraceuticals in CNS diseases: Potential mechanisms of neuroprotection. In *Nutraceuticals* (pp. 3–13). Elsevier.

Dasgupta, N., Ranjan, S., Mundra, S., Ramalingam, C., & Kumar, A. (2016). Fabrication of food grade vitamin E nanoemulsion by low energy approach, characterization and its application. *International Journal of Food Properties*, *19*(3), 700–708.

de Oliveira, G. V., Volino-Souza, M., Conte-Júnior, C. A., & Alvares, T. S. (2021). Food-derived polyphenol compounds and cardiovascular health: A nano-technological perspective. *Food Bioscience*, *41*, 101033.

DeFelice, S. L. (1995). The nutraceutical revolution: Its impact on food industry R&D. *Trends in Food Science & Technology*, *6*(2), 59–61.

Degan, D., Ornello, R., Tiseo, C., Carolei, A., Sacco, S., & Pistoia, F. (2018). The role of inflammation in neurological disorders. *Current Pharmaceutical Design*, *24*(14), 1485–1501.

Delmas, D., Rebe C, Lacour S, Filomenko R, Athias A, Gambert P, Cherkaoui-Malki M, Jannin B, Dubrez-Daloz L, Latruffe N, Solary E. (2003). Resveratrol-induced apoptosis is associated with fas redistribution in the rafts and the formation of a death-inducing signaling complex in colon cancer cells. *J Biol Chem*, *278*, 41482–41490.

Dhandayuthapani, S., Marimuthu, P., Hörmann, V., Kumi-Diaka, J., & Rathinavelu, A. (2013). Induction of apoptosis in HeLa cells via caspase activation by resveratrol and genistein. *Journal of Medicinal Food*, *16*(2), 139–146.

Dong, Y., Wan, G., Yan, P., Qian, C., Li, F., & Peng, G. (2019). Fabrication of resveratrol coated gold nanoparticles and investigation of their effect on diabetic retinopathy in streptozotocin induced diabetic rats. *Journal of Photochemistry and Photobiology B: Biology*, *195*, 51–57.

Dorai, T., Cao, Y., Dorai, B., Buttyan, R., & Katz, A. E. (2001). Therapeutic potential of curcumin in human prostate cancer. III. Curcumin inhibits proliferation, induces apoptosis, and inhibits angiogenesis of LNCaP prostate cancer cells in vivo. *The Prostate*, *47*(4), 293–303.

Drozdov, V. N., Kim, V. A., Tkachenko, E. V, & Varvanina, G. G. (2012). Influence of a specific ginger combination on gastropathy conditions in patients with osteoarthritis of the knee or hip. *Journal of Alternative and Complementary Medicine*, *18*(6), 583–588.

DSHEA. (2019). *Dietary supplements.* https://www.fda.gov/food/information-consumers-using-dietary-supplements/questions-and-answers-dietary-supplements

El-Ashmawy, N. E., Khedr, E. G., Ebeid, E.-Z. M., Salem, M. L., Zidan, A.-A. A., & Mosalam, E. M. (2017). Enhanced anticancer effect and reduced toxicity of doxorubicin in combination with thymoquinone released from poly-N-acetyl glucosamine nanomatrix in mice bearing solid Ehrlish carcinoma. *European Journal of Pharmaceutical Sciences*, *109*, 525–532.

El-Azab, M. F., Attia, F. M., & El-Mowafy, A. M. (2011). Novel role of curcumin combined with bone marrow transplantation in reversing experimental diabetes: Effects on pancreatic islet regeneration, oxidative stress, and inflammatory cytokines. *European Journal of Pharmacology*, *658*(1), 41–48.

El-Far, A. H., Al Jaouni, S. K., Li, W., & Mousa, S. A. (2018). Protective roles of thymoquinone nanoformulations: Potential nanonutraceuticals in human diseases. *Nutrients*, *10*(10), 1369.

Elmowafy, M., Samy, A., Raslan, M. A., Salama, A., Said, R. A., Abdelaziz, A. E., El-Eraky, W., El Awdan, S., & Viitala, T. (2016). Enhancement of bioavailability and pharmacodynamic effects of thymoquinone via nanostructured lipid carrier (NLC) formulation. *AAPS Pharmscitech*, *17*(3), 663–672.

Esther, B., Rapport, L., & Lockwood, B. (2000). What is nutraceutical. *Pharm. J*, *265*, 57–58.

Ethemoglu, M. S., Seker, F. B., Akkaya, H., Kilic, E., Aslan, I., Erdogan, C. S., & Yilmaz, B. (2017). Anticonvulsant activity of resveratrol-loaded liposomes in vivo. *Neuroscience*, *357*, 12–19.

Fazel Nabavi, S., Thiagarajan, R., Rastrelli, L., Daglia, M., Sobarzo-Sanchez, E., Alinezhad, H., & Mohammad Nabavi, S. (2015). Curcumin: A natural product for diabetes and its complications. *Current Topics in Medicinal Chemistry*, *15*(23), 2445–2455.

Figueiró, F., Bernardi, A., Frozza, R. L., Terroso, T., Zanotto-Filho, A., Jandrey, E. H. F., Moreira, J. C. F., Salbego, C. G., Edelweiss, M. I., & Pohlmann, A. R. (2013). Resveratrol-loaded lipid-core nanocapsules treatment reduces in vitro and in vivo glioma growth. *Journal of Biomedical Nanotechnology*, *9*(3), 516–526.

Fonseca-Santos, B., & Chorilli, M. (2020). The uses of resveratrol for neurological diseases treatment and insights for nanotechnology based-drug delivery systems. *International Journal of Pharmaceutics*, *589*, 119832.

Fuggetta, M. P., Lanzilli, G., Tricarico, M., Cottarelli, A., Falchetti, R., Ravagnan, G., & Bonmassar, E. (2006). Effect of resveratrol on proliferation and telomerase activity of human colon cancer cells in vitro. *Journal of Experimental and Clinical Cancer Research*, *25*(2), 189.

Ganguly, G., Chakrabarti, S., Chatterjee, U., & Saso, L. (2017). Proteinopathy, oxidative stress and mitochondrial dysfunction: Cross talk in Alzheimer's disease and Parkinson's disease. *Drug Design, Development and Therapy*, *11*, 797.

Ganta, S., & Amiji, M. (2009). Coadministration of paclitaxel and curcumin in nanoemulsion formulations to overcome multidrug resistance in tumor cells. *Molecular Pharmaceutics*, *6*(3), 928–939.

Giannouli, M., Karagkiozaki, V., Pappa, F., Moutsios, I., Gravalidis, C., & Logothetidis, S. (2018). Fabrication of quercetin-loaded PLGA nanoparticles via electrohydrodynamic atomization for cardiovascular disease. *Materials Today: Proceedings*, *5*(8), 15998–16005.

Gou, M., Men, K., Shi, H., Xiang, M., Zhang, J., Song, J., Long, J., Wan, Y., Luo, F., & Zhao, X. (2011). Curcumin-loaded biodegradable polymeric micelles for colon cancer therapy in vitro and in vivo. *Nanoscale*, *3*(4), 1558–1567.

Gu, Y., Zhu, C.-F., Dai, Y.-L., Zhong, Q., & Sun, B. (2009). Inhibitory effects of genistein on metastasis of human hepatocellular carcinoma. *World Journal of Gastroenterology: WJG*, *15*(39), 4952.

Gunasekaran, T., Haile, T., Nigusse, T., & Dhanaraju, M. D. (2014). Nanotechnology: An effective tool for enhancing bioavailability and bioactivity of phytomedicine. *Asian Pacific Journal of Tropical Biomedicine*, *4*, S1–S7.

Guo, W., Li, A., Jia, Z., Yuan, Y., Dai, H., & Li, H. (2013). Transferrin modified PEG-PLA-resveratrol conjugates: In vitro and in vivo studies for glioma. *European Journal of Pharmacology*, *718*(1–3), 41–47.

Guo, Y., Mao, X., Zhang, J., Sun, P., Wang, H., Zhang, Y., Ma, Y., Xu, S., Lv, R., & Liu, X. (2019). Oral delivery of lycopene-loaded microemulsion for brain-targeting: Preparation, characterization, pharmacokinetic evaluation and tissue distribution. *Drug Delivery*, *26*(1), 1191–1205.

Gupta, Y. K., Briyal, S., & Chaudhary, G. (2002). Protective effect of trans-resveratrol against kainic acid-induced seizures and oxidative stress in rats. *Pharmacology Biochemistry and Behavior*, *71*(1–2), 245–249.

Gupta, Y. K., Chaudhary, G., Sinha, K., & Srivastava, A. K. (2001). Protective Effect of Resveratrol Against Intracortical FeCl. *Methods Find Exp Clin Pharmacol*, *23*(5), 241–244.

Habibie, H., Yokoyama, S., Abdelhamed, S., Awale, S., Sakurai, H., Hayakawa, Y., & Saiki, I. (2014). Survivin suppression through STAT3/β-catenin is essential for resveratrol-induced melanoma apoptosis. *International Journal of Oncology*, *45*(2), 895–901.

Han, Y., Zheng, W., Bastianetto, S., Chabot, J., & Quirion, R. (2004). Neuroprotective effects of resveratrol against β -amyloid-induced neurotoxicity in rat hippocampal neurons: Involvement of protein kinase C. *British Journal of Pharmacology*, *141*(6), 997–1005.

Hassanzadeh, P., Atyabi, F., & Dinarvand, R. (2018). Nanoencapsulation: A promising strategy for biomedical applications of ferulic acid. *Biomedical Reviews*, *28*, 22–30.

He, J., Xiang, Z., Zhu, X., Ai, Z., Shen, J., Huang, T., Liu, L., Ji, W., & Li, T. (2016). Neuroprotective effects of 7, 8-dihydroxyflavone on midbrain dopaminergic neurons in MPP+-treated monkeys. *Scientific Reports*, *6*(1), 1–9.

Hesari, M., Mohammadi, P., Khademi, F., Shackebaei, D., Momtaz, S., Moasefi, N., Farzaei, M. H., & Abdollahi, M. (2021). Current advances in the use of nanophytomedicine therapies for human cardiovascular diseases. *International Journal of Nanomedicine*, *16*, 3293.

Hsiao, Y., Peng, S., Lai, K., Liao, C., Huang, Y., Lin, C., Lin, M., Liu, K., Tsai, C., & Ma, Y. (2019). Genistein induces apoptosis in vitro and has antitumor activity against human leukemia HL-60 cancer cell xenograft growth in vivo. *Environmental Toxicology*, *34*(4), 443–456.

Hu, M.-L., Rayner, C. K., Wu, K.-L., Chuah, S.-K., Tai, W.-C., Chou, Y.-P., Chiu, Y.-C., Chiu, K.-W., & Hu, T.-H. (2011). Effect of ginger on gastric motility and symptoms of functional dyspepsia. *World Journal of Gastroenterology: WJG*, *17*(1), 105.

Huang, Q., Yu, H., & Ru, Q. (2010). Bioavailability and delivery of nutraceuticals using nanotechnology. *Journal of Food Science*, *75*(1), R50–R57.

Hurley, L. L., Akinfiresoye, L., Kalejaiye, O., & Tizabi, Y. (2014). Antidepressant effects of resveratrol in an animal model of depression. *Behavioural Brain Research*, *268*, 1–7.

Hwang, S., Lim, J. W., & Kim, H. (2017). Inhibitory effect of lycopene on amyloid-β-induced apoptosis in neuronal cells. *Nutrients*, *9*(8), 883.

Ibrar, M., Khan, M. A., & Imran, M. (2018). Evaluation of *Paeonia emodi* and its gold nanoparticles for cardioprotective and antihyperlipidemic potentials. *Journal of Photochemistry and Photobiology B: Biology*, *189*, 5–13.

Ibrar, M., Khan, M. A., Nisar, M., & Khan, M. (2019). Evaluation of *Paeonia emodi* for its cardioprotective potentials: An investigative study towards possible mechanism. *Journal of Ethnopharmacology*, *231*, 57–65.

Jafari, S. M., & McClements, D. J. (2017). Nanotechnology approaches for increasing nutrient bioavailability. *Advances in Food and Nutrition Research*, *81*, 1–30.

Jain, S. K., Rains, J., Croad, J., Larson, B., & Jones, K. (2009). Curcumin supplementation lowers TNF-α, IL-6, IL-8, and MCP-1 secretion in high glucose-treated cultured monocytes and blood levels of TNF-α, IL-6, MCP-1, glucose, and glycosylated hemoglobin in diabetic rats. *Antioxidants & Redox Signaling, 11*(2), 241–249.

Jain, S. K., Rains, J., & Jones, K. (2006). Effect of curcumin on protein glycosylation, lipid peroxidation, and oxygen radical generation in human red blood cells exposed to high glucose levels. *Free Radical Biology and Medicine, 41*(1), 92–96.

Javeri, I. (2016). Application of "nano" nutraceuticals in medicine. In *Nutraceuticals* (pp. 189–192). Elsevier.

Jhang, K. A., Park, J.-S., Kim, H.-S., & Chong, Y. H. (2017). Resveratrol ameliorates tau hyperphosphorylation at Ser396 site and oxidative damage in rat hippocampal slices exposed to vanadate: Implication of ERK1/2 and GSK-3β signaling cascades. *Journal of Agricultural and Food Chemistry, 65*(44), 9626–9634.

Jhaveri, A., Deshpande, P., Pattni, B., & Torchilin, V. (2018). Transferrin-targeted, resveratrol-loaded liposomes for the treatment of glioblastoma. *Journal of Controlled Release, 277*, 89–101.

Jung, Y. D., & Ellis, L. M. (2001). Inhibition of tumour invasion and angiogenesis by epigallocatechin gallate (EGCG), a major component of green tea. *International Journal of Experimental Pathology, 82*(6), 309–316.

Jurenka, J. S. (2009). Anti-inflammatory properties of curcumin, a major constituent of Curcuma longa: A review of preclinical and clinical research. *Alternative Medicine Review, 14*(2), 141–153.

Kakkar, V., Modgill, N., & Kumar, M. (2016). From nutraceuticals to nanoceuticals. In *Nanoscience in Food and Agriculture* (vol. 3, pp. 183–198). Springer.

Kalra, E. K. (2003). Nutraceutical-definition and introduction. *AAPS Pharmsci, 5*(3), 27–28.

Kang, S., Kim, B. R., Kang, M.-H., Kim, D.-Y., Lee, D.-H., Oh, S. C., Min, B. W., & Um, J. W. (2018). Anti-metastatic effect of metformin via repression of interleukin 6-induced epithelial–mesenchymal transition in human colon cancer cells. *PloS One, 13*(10), e0205449.

Karewicz, A., Bielska, D., Loboda, A., Gzyl-Malcher, B., Bednar, J., Jozkowicz, A., Dulak, J., & Nowakowska, M. (2013). Curcumin-containing liposomes stabilized by thin layers of chitosan derivatives. *Colloids and Surfaces B: Biointerfaces, 109*, 307–316.

Kasiotis, K. M., Pratsinis, H., Kletsas, D., & Haroutounian, S. A. (2013). Resveratrol and related stilbenes: Their anti-aging and anti-angiogenic properties. *Food and Chemical Toxicology, 61*, 112–120.

Ke, X., Zhao, Y., Lu, X., Wang, Z., Liu, Y., Ren, M., Lu, G., Zhang, D., Sun, Z., & Xu, Z. (2015). TQ inhibits hepatocellular carcinoma growth in vitro and in vivo via repression of Notch signaling. *Oncotarget, 6*(32), 32610.

Knutsson, A., & Bøggild, H. (2010). Gastrointestinal disorders among shift workers. *Scandinavian Journal of Work, Environment & Health, 36*, 85–95.

Kocaadam, B., & Şanlier, N. (2017). Curcumin, an active component of turmeric (Curcuma longa), and its effects on health. *Critical Reviews in Food Science and Nutrition, 57*(13), 2889–2895.

Kubota, S., Ozawa, Y., Kurihara, T., Sasaki, M., Yuki, K., Miyake, S., Noda, K., Ishida, S., & Tsubota, K. (2011). Roles of AMP-activated protein kinase in diabetes-induced retinal inflammation. *Investigative Ophthalmology & Visual Science, 52*(12), 9142–9148.

Kuebler, U., Zuccarella-Hackl, C., Arpagaus, A., Wolf, J. M., Farahmand, F., von Känel, R., Ehlert, U., & Wirtz, P. H. (2015). Stress-induced modulation of NF-κB activation, inflammation-associated gene expression, and cytokine levels in blood of healthy men. *Brain, Behavior, and Immunity, 46*, 87–95.

Lao, C. D., Ruffin, M. T., Normolle, D., Heath, D. D., Murray, S. I., Bailey, J. M., Boggs, M. E., Crowell, J., Rock, C. L., & Brenner, D. E. (2006). Dose escalation of a curcuminoid formulation. *BMC Complementary and Alternative Medicine, 6*(1), 1–4.

Lee, H.-P., Li, T.-M., Tsao, J.-Y., Fong, Y.-C., & Tang, C.-H. (2012). Curcumin induces cell apoptosis in human chondrosarcoma through extrinsic death receptor pathway. *International Immunopharmacology, 13*(2), 163–169.

Lee, J. H., Lee, H.-B., Jung, G. O., Oh, J. T., Park, D. E., & Chae, K. M. (2013). Effect of quercetin on apoptosis of PANC-1 cells. *Journal of the Korean Surgical Society, 85*(6), 249–260.

Lenzi, J., Rodrigues, A. F., de Sousa Rós, A., de Castro, B. B., de Lima, D. D., Dal Magro, D. D., & Zeni, A. L. B. (2015). Ferulic acid chronic treatment exerts antidepressant-like effect: Role of antioxidant defense system. *Metabolic Brain Disease, 30*(6), 1453–1463.

Li, S., Li, J., Dai, W., Zhang, Q., Feng, J., Wu, L., Liu, T., Yu, Q., Xu, S., & Wang, W. (2017). Genistein suppresses aerobic glycolysis and induces hepatocellular carcinoma cell death. *British Journal of Cancer, 117*(10), 1518–1528.

Li, Xiaoxia, Xiao, H., Lin, C., Sun, W., Wu, T., Wang, J., Chen, B., Chen, X., & Cheng, D. (2019). Synergistic effects of liposomes encapsulating atorvastatin calcium and curcumin and targeting dysfunctional endothelial cells in reducing atherosclerosis. *International Journal of Nanomedicine, 14*, 649.

Li, Xingyi, Chen, S., Zhang, B., Li, M., Diao, K., Zhang, Z., Li, J., Xu, Y., Wang, X., & Chen, H. (2012). *In situ* injectable nano-composite hydrogel composed of curcumin, N, O-carboxymethyl chitosan and oxidized alginate for wound healing application. *International Journal of Pharmaceutics, 437*(1–2), 110–119.

Liao, H., Bao, X., Zhu, J., Qu, J., Sun, Y., Ma, X., Wang, E., Guo, X., Kang, Q., & Zhen, Y. (2015). O-Alkylated derivatives of quercetin induce apoptosis of MCF-7 cells via a caspase-independent mitochondrial pathway. *Chemico-Biological Interactions, 242*, 91–98.

Libby, P., Bornfeldt, K. E., & Tall, A. R. (2016). *Atherosclerosis: Successes, surprises, and future challenges. Am Heart Assoc, 118*, 531–534.

Lin, Y.-L., Liu, Y.-K., Tsai, N.-M., Hsieh, J.-H., Chen, C.-H., Lin, C.-M., & Liao, K.-W. (2012). A Lipo-PEG-PEI complex for encapsulating curcumin that enhances its antitumor effects on curcumin-sensitive and curcumin-resistance cells. *Nanomedicine: Nanotechnology, Biology and Medicine, 8*(3), 318–327.

Lindsay, K. W., Bone, I., & Fuller, G. (2010). *Neurology and neurosurgery illustrated e-book*. Elsevier Health Sciences.

Lingappan, K. (2018). NF-κB in oxidative stress. *Current Opinion in Toxicology, 7*, 81–86.

Liu, C.-B., Wang, R., Pan, H.-B., Ding, Q.-F., & Lu, F.-B. (2013). Effect of lycopene on oxidative stress and behavioral deficits in rotenone induced model of Parkinson's disease. *Zhongguo Ying Yong Sheng Li Xue Za Zhi= Zhongguo Yingyong Shenglixue Zazhi= Chinese Journal of Applied Physiology, 29*(4), 380–384.

Liu, Z., Li, H., Liu, J., Wu, M., Chen, X., Liu, L., & Wang, Q. (2017). Inactivated Wnt signaling in resveratrol-treated epidermal squamous cancer cells and its biological implication. *Oncology Letters, 14*(2), 2239–2243.

Lubbad, A., Oriowo, M. A., & Khan, I. (2009). Curcumin attenuates inflammation through inhibition of TLR-4 receptor in experimental colitis. *Molecular and Cellular Biochemistry, 322*(1), 127–135.

Mangalathillam, S., Rejinold, N. S., Nair, A., Lakshmanan, V.-K., Nair, S. V, & Jayakumar, R. (2012). Curcumin loaded chitin nanogels for skin cancer treatment via the transdermal route. *Nanoscale, 4*(1), 239–250.

Menard, C., Bastianetto, S., & Quirion, R. (2013). Neuroprotective effects of resveratrol and epigallocatechin gallate polyphenols are mediated by the activation of protein kinase C gamma. *Frontiers in Cellular Neuroscience, 7*, 281.

Moghtaderi, H., Sepehri, H., & Attari, F. (2017). Combination of arabinogalactan and curcumin induces apoptosis in breast cancer cells in vitro and inhibits tumor growth via overexpression of p53 level in vivo. *Biomedicine & Pharmacotherapy, 88*, 582–594.

Mohammadshahi, M., Haidari, F., & Soufi, F. G. (2014). Chronic resveratrol administration improves diabetic cardiomyopathy in part by reducing oxidative stress. *Cardiology Journal, 21*(1), 39–46.

Mohseni, R., ArabSadeghabadi, Z., Ziamajidi, N., Abbasalipourkabir, R., & RezaeiFarimani, A. (2019). Oral administration of resveratrol-loaded solid lipid nanoparticle improves insulin resistance through targeting expression of SNARE proteins in adipose and muscle tissue in rats with type 2 diabetes. *Nanoscale Research Letters, 14*(1), 1–9.

Most, J., Goossens, G. H., Jocken, J. W. E., & Blaak, E. E. (2014). Short-term supplementation with a specific combination of dietary polyphenols increases energy expenditure and alters substrate metabolism in overweight subjects. *International Journal of Obesity, 38*(5), 698–706.

Mudduluru, G., George-William, J. N., Muppala, S., Asangani, I. A., Kumarswamy, R., Nelson, L. D., & Allgayer, H. (2011). Curcumin regulates miR-21 expression and inhibits invasion and metastasis in colorectal cancer. *Bioscience Reports, 31*(3), 185–197.

Mukhopadhyay, A., Bueso-Ramos, C., Chatterjee, D., Pantazis, P., & Aggarwal, B. B. (2001). Curcumin downregulates cell survival mechanisms in human prostate cancer cell lines. *Oncogene, 20*(52), 7597–7609.

Muthenna, P., Suryanarayana, P., Gunda, S. K., Petrash, J. M., & Reddy, G. B. (2009). Inhibition of aldose reductase by dietary antioxidant curcumin: Mechanism of inhibition, specificity and significance. *FEBS Letters, 583*(22), 3637–3642.

Na, L.-X., Zhang, Y.-L., Li, Y., Liu, L.-Y., Li, R., Kong, T., & Sun, C.-H. (2011). Curcumin improves insulin resistance in skeletal muscle of rats. *Nutrition, Metabolism and Cardiovascular Diseases, 21*(7), 526–533.

Na, L., Li, Y., Pan, H., Zhou, X., Sun, D., Meng, M., Li, X., & Sun, C. (2013). Curcuminoids exert glucose-lowering effect in type 2 diabetes by decreasing serum free fatty acids: A double-blind, placebo-controlled trial. *Molecular Nutrition & Food Research, 57*(9), 1569–1577.

Nguyen, L. T., Lee, Y.-H., Sharma, A. R., Park, J.-B., Jagga, S., Sharma, G., Lee, S.-S., & Nam, J.-S. (2017). Quercetin induces apoptosis and cell cycle arrest in triple-negative breast cancer cells through modulation of Foxo3a activity. *The Korean Journal of Physiology & Pharmacology*, *21*(2), 205–213.

Nie, S., Ma, K., Sun, M., Lee, M., Tan, Y., Chen, G., Zhang, Z., Zhang, Z., & Cao, X. (2019). 7, 8-Dihydroxyflavone protects nigrostriatal dopaminergic neurons from rotenone-induced neurotoxicity in rodents. *Parkinson's Disease*, *2019*.

Nihei, T., Suzuki, H., Aoki, A., Yuminoki, K., Hashimoto, N., Sato, H., Seto, Y., & Onoue, S. (2016). Development of a novel nanoparticle formulation of thymoquinone with a cold wet-milling system and its pharmacokinetic analysis. *International Journal of Pharmaceutics*, *511*(1), 455–461.

Nozu, T., Miyagishi, S., Nozu, R., Takakusaki, K., & Okumura, T. (2019). Butyrate inhibits visceral allodynia and colonic hyperpermeability in rat models of irritable bowel syndrome. *Scientific Reports*, *9*(1), 1–13.

O'Leary, K. A., de Pascual-Tereasa, S., Needs, P. W., Bao, Y.-P., O'Brien, N. M., & Williamson, G. (2004). Effect of flavonoids and vitamin E on cyclooxygenase-2 (COX-2) transcription. *Mutation Research/Fundamental and Molecular Mechanisms of Mutagenesis*, *551*(1–2), 245–254.

Odeh, F., Ismail, S. I., Abu-Dahab, R., Mahmoud, I. S., & Al Bawab, A. (2012). Thymoquinone in liposomes: A study of loading efficiency and biological activity towards breast cancer. *Drug Delivery*, *19*(8), 371–377.

Ono, K., Hirohata, M., & Yamada, M. (2005). Ferulic acid destabilizes preformed β-amyloid fibrils in vitro. *Biochemical and Biophysical Research Communications*, *336*(2), 444–449.

Ouyang, G., Yao, L., Ruan, K., Song, G., Mao, Y., & Bao, S. (2009). Genistein induces G2/M cell cycle arrest and apoptosis of human ovarian cancer cells via activation of DNA damage checkpoint pathways. *Cell Biology International*, *33*(12), 1237–1244.

Oyenihi, O. R., Oyenihi, A. B., Adeyanju, A. A., & Oguntibeju, O. O. (2016). Antidiabetic effects of resveratrol: The way forward in its clinical utility. *Journal of Diabetes Research*, *2016*.

Öztürk, E., Arslan, A. K. K., Yerer, M. B., & Bishayee, A. (2017). Resveratrol and diabetes: A critical review of clinical studies. *Biomedicine & Pharmacotherapy*, *95*, 230–234.

Pace, T. W. W., Mletzko, T. C., Alagbe, O., Musselman, D. L., Nemeroff, C. B., Miller, A. H., & Heim, C. M. (2006). Increased stress-induced inflammatory responses in male patients with major depression and increased early life stress. *American Journal of Psychiatry*, *163*(9), 1630–1633.

Pangeni, R., Sharma, S., Mustafa, G., Ali, J., & Baboota, S. (2014). Vitamin E loaded resveratrol nanoemulsion for brain targeting for the treatment of Parkinson's disease by reducing oxidative stress. *Nanotechnology*, *25*(48), 485102.

Paul, R., Mazumder, M. K., Nath, J., Deb, S., Paul, S., Bhattacharya, P., & Borah, A. (2020). Lycopene-a pleiotropic neuroprotective nutraceutical: Deciphering its therapeutic potentials in broad spectrum neurological disorders. *Neurochemistry International*, *140*, 104823.

Paul, R., Nath, J., Paul, S., Mazumder, M. K., Phukan, B. C., Roy, R., Bhattacharya, P., & Borah, A. (2021). Suggesting 7, 8-dihydroxyflavone as a promising nutraceutical against CNS disorders. *Neurochemistry International*, 105068.

Pillai, S. C., Borah, A., Jacob, E. M., & Kumar, D. S. (2021). Nanotechnological approach to delivering nutraceuticals as promising drug candidates for the treatment of atherosclerosis. *Drug Delivery*, *28*(1), 550–568.

Prakash, A., & Kumar, A. (2013). Lycopene protects against memory impairment and mito-oxidative damage induced by colchicine in rats: An evidence of nitric oxide signaling. *European Journal of Pharmacology*, *721*(1–3), 373–381.

Prema, A., Janakiraman, U., Manivasagam, T., & Thenmozhi, A. J. (2015). Neuroprotective effect of lycopene against MPTP induced experimental Parkinson's disease in mice. *Neuroscience Letters*, *599*, 12–19.

Rabinovitch, R. C., Samborska, B., Faubert, B., Ma, E. H., Gravel, S.-P., Andrzejewski, S., Raissi, T. C., Pause, A., Pierre, J. S., & Jones, R. G. (2017). AMPK maintains cellular metabolic homeostasis through regulation of mitochondrial reactive oxygen species. *Cell Reports*, *21*(1), 1–9.

Rabiskova, M., Bautzova, T., Dvorackova, K., & Spilkova, J. (2009). Beneficial effects of rutin, quercitrin and quercetin on inflammatory bowel disease. *Ces. Slov. Farm*, *58*, 47–54.

Rachmawati, H., Soraya, I. S., Kurniati, N. F., & Rahma, A. (2016). In vitro study on antihypertensive and antihypercholesterolemic effects of a curcumin nanoemulsion. *Scientia Pharmaceutica*, *84*(1), 131–140.

Radad, K. S., Al-Shraim, M. M., Moustafa, M. F., & Rausch, W.-D. (2015). Neuroprotective role of thymoquinone against 1-methyl-4-phenylpyridinium-induced dopaminergic cell death in primary mesencephalic cell culture. *Neurosciences Journal*, *20*(1), 10–16.

Rahmani, S., Asgary, S., Askari, G., Keshvari, M., Hatamipour, M., Feizi, A., & Sahebkar, A. (2016). Treatment of non-alcoholic fatty liver disease with curcumin: A randomized placebo-controlled trial. *Phytotherapy Research*, *30*(9), 1540–1548.

Rajasekaran, S. A. (2011). Therapeutic potential of curcumin in gastrointestinal diseases. *World Journal of Gastrointestinal Pathophysiology*, *2*(1), 1.

Rani, R., Dahiya, S., Dhingra, D., Dilbaghi, N., Kim, K.-H., & Kumar, S. (2018). Improvement of antihyperglycemic activity of nano-thymoquinone in rat model of type-2 diabetes. *Chemico-Biological Interactions*, *295*, 119–132.

Ranjan, A. P., Mukerjee, A., Helson, L., & Vishwanatha, J. K. (2013). Mitigating prolonged QT interval in cancer nanodrug development for accelerated clinical translation. *Journal of Nanobiotechnology*, *11*(1), 1–8.

Rashid, K., & Sil, P. C. (2015). Curcumin enhances recovery of pancreatic islets from cellular stress induced inflammation and apoptosis in diabetic rats. *Toxicology and Applied Pharmacology*, *282*(3), 297–310.

Rauf, A., Imran, M., Khan, I. A., ur-Rehman, M., Gilani, S. A., Mehmood, Z., & Mubarak, M. S. (2018). Anticancer potential of quercetin: A comprehensive review. *Phytotherapy Research*, *32*(11), 2109–2130.

Raza, M., Alghasham, A. A., Alorainy, M. S., & El-Hadiyah, T. M. (2008). Potentiation of valproate-induced anticonvulsant response by *Nigella sativa* seed constituents: The role of GABA receptors. *International Journal of Health Sciences*, *2*(1), 15.

Re, F., Cambianica, I., Zona, C., Sesana, S., Gregori, M., Rigolio, R., La Ferla, B., Nicotra, F., Forloni, G., & Cagnotto, A. (2011). Functionalization of liposomes with ApoE-derived peptides at different density affects cellular uptake and drug transport across a blood-brain barrier model. *Nanomedicine: Nanotechnology, Biology and Medicine*, *7*(5), 551–559.

Romacho, T., Elsen, M., Röhrborn, D., & Eckel, J. (2014). Adipose tissue and its role in organ crosstalk. *Acta Physiologica*, *210*(4), 733–753.

Ronis, M. J. J. (2016). Effects of soy containing diet and isoflavones on cytochrome P450 enzyme expression and activity. *Drug Metabolism Reviews*, *48*(3), 331–341.

Sachdeva, A. K., & Chopra, K. (2015). Lycopene abrogates Aβ (1–42)-mediated neuroinflammatory cascade in an experimental model of Alzheimer's disease. *Journal of Nutritional Biochemistry*, *26*(7), 736–744.

Sahebkar, A. (2014). Low-density lipoprotein is a potential target for curcumin: Novel mechanistic insights. *Basic and Clinical Pharmacology and Toxicology*, *114*(6), 437–438.

Sai Varsha, M. K. N., Thiagarajan, R., Manikandan, R., & Dhanasekaran, G. (2015). Vitamin K1 alleviates streptozotocin-induced type 1 diabetes by mitigating free radical stress, as well as inhibiting NF-kB activation and iNOS expression in rat pancreas. *Nutrition*, *31*, 214–222.

Salama, L., Pastor, E. R., Stone, T., & Mousa, S. A. (2020). Emerging nanopharmaceuticals and nanonutraceuticals in cancer management. *Biomedicines*, *8*(9), 347.

Salem, A. A., El Haty, I. A., Abdou, I. M., & Mu, Y. (2015). Interaction of human telomeric G-quadruplex DNA with thymoquinone: A possible mechanism for thymoquinone anticancer effect. *Biochimica et Biophysica Acta (BBA)-General Subjects*, *1850*(2), 329–342.

Samarghandian, S., Farkhondeh, T., & Samini, F. (2018). A review on possible therapeutic effect of *Nigella sativa* and thymoquinone in neurodegenerative diseases. *CNS & Neurological Disorders-Drug Targets (Formerly Current Drug Targets-CNS &Neurological Disorders)*, *17*(6), 412–420.

Shao, Y., Li, B., Huang, Y., Luo, Q., Xie, Y., & Chen, Y. (2017). Thymoquinone attenuates brain injury via an antioxidative pathway in a status epilepticus rat model. *Translational Neuroscience*, *8*(1), 9–14.

Shegokar, R., & Müller, R. H. (2010). Nanocrystals: Industrially feasible multifunctional formulation technology for poorly soluble actives. *International Journal of Pharmaceutics*, *399*(1–2), 129–139.

Shibamoto, T., Kanazawa, K., Shahidi, F., & Ho, C.-T. (2008). *Functional food and health: An overview.* ACS publications.

Simion, V., Stan, D., Constantinescu, C. A., Deleanu, M., Dragan, E., Tucureanu, M. M., Gan, A.-M., Butoi, E., Constantin, A., & Manduteanu, I. (2016). Conjugation of curcumin-loaded lipid nanoemulsions with cell-penetrating peptides increases their cellular uptake and enhances the anti-inflammatory effects in endothelial cells. *Journal of Pharmacy and Pharmacology*, *68*(2), 195–207.

Smoliga, J. M., & Blanchard, O. (2014). Enhancing the delivery of resveratrol in humans: If low bioavailability is the problem, what is the solution? *Molecules*, *19*(11), 17154–17172.

Song, W., Muthana, M., Mukherjee, J., Falconer, R. J., Biggs, C. A., & Zhao, X. (2017). Magnetic-silk core–shell nanoparticles as potential carriers for targeted delivery of curcumin into human breast cancer cells. *ACS Biomaterials Science & Engineering*, *3*(6), 1027–1038.

Soni, P., Kaur, J., & Tikoo, K. (2015). Dual drug-loaded paclitaxel–thymoquinone nanoparticles for effective breast cancer therapy. *Journal of Nanoparticle Research*, *17*(1), 1–12.

Sprague, A. H., & Khalil, R. A. (2009). Inflammatory cytokines in vascular dysfunction and vascular disease. *Biochemical Pharmacology*, *78*(6), 539–552.

Stefanska, B. (2012). Curcumin ameliorates hepatic fibrosis in type 2 diabetes mellitus–insights into its mechanisms of action. *British Journal of Pharmacology, 166*(8), 2209–2211.

Stern, J. H., Rutkowski, J. M., & Scherer, P. E. (2016). Adiponectin, leptin, and fatty acids in the maintenance of metabolic homeostasis through adipose tissue crosstalk. *Cell Metabolism, 23*(5), 770–784.

Su, S.-J., Yeh, T.-M., Chuang, W.-J., Ho, C.-L., Chang, K.-L., Cheng, H.-L., Liu, H.-S., Cheng, H.-L., Hsu, P.-Y., & Chow, N.-H. (2005). The novel targets for anti-angiogenesis of genistein on human cancer cells. *Biochemical Pharmacology, 69*(2), 307–318.

Sudan, S., & Rupasinghe, H. P. V. (2015). Antiproliferative activity of long chain acylated esters of quercetin-3-O-glucoside in hepatocellular carcinoma HepG2 cells. *Experimental Biology and Medicine, 240*(11), 1452–1464.

Sudjarwo, S. A., Anwar, C., Eraiko, K., & Wardani, G. (2019). Cardioprotective activity of chitosan-Pinus merkusii extract nanoparticles against lead acetate induced cardiac cell damage in rat. *Rasāyan Journal of Chemistry, 12*(1), 184–191.

Sundram, V., Chauhan, S. C., & Jaggi, M. (2011). Emerging roles of protein kinase D1 in cancer. *Molecular Cancer Research, 9*(8), 985–996.

Talukder, J. (2019). Nutraceuticals in gastrointestinal conditions. In *Nutraceuticals in Veterinary Medicine* (pp. 467–479). Springer.

Thangapazham, R. L., Puri, A., Tele, S., Blumenthal, R., & Maheshwari, R. K. (2008). Evaluation of a nanotechnology-based carrier for delivery of curcumin in prostate cancer cells. *International Journal of Oncology, 32*(5), 1119–1123.

Tomeh, M. A., Hadianamrei, R., & Zhao, X. (2019). A review of curcumin and its derivatives as anticancer agents. *International Journal of Molecular Sciences, 20*(5), 1033.

Tong, F., Chai, R., Jiang, H., & Dong, B. (2018). In vitro/vivo drug release and anti-diabetic cardiomyopathy properties of curcumin/PBLG-PEG-PBLG nanoparticles. *International Journal of Nanomedicine, 13*, 1945.

Tuli, H. S., Tuorkey, M. J., Thakral, F., Sak, K., Kumar, M., Sharma, A. K., Sharma, U., Jain, A., Aggarwal, V., & Bishayee, A. (2019). Molecular mechanisms of action of genistein in cancer: Recent advances. *Frontiers in Pharmacology, 10*, 1336.

Varì, R., Scazzocchio, B., Silenzi, A., Giovannini, C., & Masella, R. (2021). Obesity-associated inflammation: does curcumin exert a beneficial role? *Nutrients, 13*(3), 1021.

Vecchi Brumatti, L., Marcuzzi, A., Tricarico, P. M., Zanin, V., Girardelli, M., & Bianco, A. M. (2014). Curcumin and inflammatory bowel disease: Potential and limits of innovative treatments. *Molecules, 19*(12), 21127–21153.

Vijayakumar, M. R., Kosuru, R., Singh, S. K., Prasad, C. B., Narayan, G., Muthu, M. S., & Singh, S. (2016). Resveratrol loaded PLGA: D-α-tocopheryl polyethylene glycol 1000 succinate blend nanoparticles for brain cancer therapy. *RSC Advances, 6*(78), 74254–74268.

Wang, B., Wu, N., Liang, F., Zhang, S., Ni, W., Cao, Y., Xia, D., & Xi, H. (2014). 7, 8-dihydroxyflavone, a small-molecule tropomyosin-related kinase B (TrkB) agonist, attenuates cerebral ischemia and reperfusion injury in rats. *Journal of Molecular Histology, 45*(2), 129–140.

Wang, D., Veena, M. S., Stevenson, K., Tang, C., Ho, B., Suh, J. D., Duarte, V. M., Faull, K. F., Mehta, K., & Srivatsan, E. S. (2008). Liposome-encapsulated curcumin suppresses growth of head and neck squamous cell carcinoma in vitro and in xenografts through the inhibition of nuclear factor κB by an AKT-independent pathway. *Clinical Cancer Research, 14*(19), 6228–6236.

Wang, H. H., Lee, D. K., Liu, M., Portincasa, P., & Wang, D. Q.-H. (2020). Novel insights into the pathogenesis and management of the metabolic syndrome. *Pediatric Gastroenterology, Hepatology & Nutrition, 23*(3), 189.

Wang, M., Li, L., Zhang, X., Liu, Y., Zhu, R., Liu, L., Fang, Y., Gao, Z., & Gao, D. (2018). Magnetic resveratrol liposomes as a new theranostic platform for magnetic resonance imaging guided Parkinson's disease targeting therapy. *ACS Sustainable Chemistry & Engineering, 6*(12), 17124–17133.

Wang, Yan, Zhang, C.-L., Liu, Y.-F., Chen, R.-Y., Wang, F.-Z., & Yu, D.-Q. (2016). Two new lupane saponins from *Schefflera kwangsiensis*. *Phytochemistry Letters, 18*, 19–22.

Wang, Yifan, Ocampo, M. F., Rodriguez, B., & Chen, J. (2020). Resveratrol and Spirulina: Nutraceuticals that potentially improving cardiovascular disease. *Journal of Cardiovascular Medicine and Cardiology, 7*(2), 138–145.

Wei, Y., Shen, X.-N., Mai, J.-Y., Shen, H., Wang, R.-Z., & Wu, M. (2010). The effects of lycopene on reactive oxygen species and anoxic damage in ischemia reperfusion injury in rats. *Zhonghua Yu Fang Yi Xue Za Zhi* [*Chinese Journal of Preventive Medicine*], *44*(1), 34–38.

Weisberg, S. P., Leibel, R., & Tortoriello, D. V. (2008). Dietary curcumin significantly improves obesity-associated inflammation and diabetes in mouse models of diabesity. *Endocrinology, 149*(7), 3549–3558.

Wellen, K. E., & Hotamisligil, G. S. (2003). Obesity-induced inflammatory changes in adipose tissue. *Journal of Clinical Investigation, 112*(12), 1785–1788.

WHO. (2015). *Cardiovascular and W. H. O. C. Diseases (CVDs).* https://www.who.int/health-topics/cardiovascular-diseases#tab=tab_1.

WHO. (2018). *Cancer.* https://www.who.int/health-topics/cancer#tab=tab_1

WHO. (2019). *Cardiovascular fact sheet.* https://www.who.int/en/news-room/fact-sheets/detail/cardiovascular-diseases-(cvds)

Wu, B., Yao, H., Wang, S., & Xu, R. (2013). DAPK1 modulates a curcumin-induced G2/M arrest and apoptosis by regulating STAT3, NF-κB, and caspase-3 activation. *Biochemical and Biophysical Research Communications, 434*(1), 75–80.

Wu, D., Liu, Z., Li, J., Zhang, Q., Zhong, P., Teng, T., Chen, M., Xie, Z., Ji, A., & Li, Y. (2019). Epigallocatechin-3-gallate inhibits the growth and increases the apoptosis of human thyroid carcinoma cells through suppression of EGFR/RAS/RAF/MEK/ERK signaling pathway. *Cancer Cell International, 19*(1), 1–17.

Xu, Y., Wang, Z., You, W., Zhang, X., Li, S., Barish, P. A., Vernon, M. M., Du, X., Li, G., & Pan, J. (2010). Antidepressant-like effect of trans-resveratrol: Involvement of serotonin and noradrenaline system. *European Neuropsychopharmacology, 20*(6), 405–413.

Yamashita, A. S., Belchior, T., Lira, F. S., Bishop, N. C., Wessner, B., Rosa, J. C., & Festuccia, W. T. (2018). Regulation of metabolic disease-associated inflammation by nutrient sensors. *Mediators of Inflammation, 2018.*

Yang, Y. M., Yang, Y., Dai, W. W., Li, X. M., Ma, J. Q., & Tang, L. P. (2016). Genistein-induced apoptosis is mediated by endoplasmic reticulum stress in cervical cancer cells. *Eur Rev Med Pharmacol Sci, 20*(15), 3292–3296.

Yousef, M., Vlachogiannis, I. A., & Tsiani, E. (2017). Effects of resveratrol against lung cancer: In vitro and in vivo studies. *Nutrients, 9*(11), 1231.

Yu, L., Wang, W., Pang, W., Xiao, Z., Jiang, Y., & Hong, Y. (2017). Dietary lycopene supplementation improves cognitive performances in tau transgenic mice expressing P301L mutation via inhibiting oxidative stress and tau hyperphosphorylation. *Journal of Alzheimer's Disease, 57*(2), 475–482.

Yu, Y., Wu, S., Li, J., Wang, R., Xie, X., Yu, X., Pan, J., Xu, Y., & Zheng, L. (2015). The effect of curcumin on the brain-gut axis in rat model of irritable bowel syndrome: Involvement of 5-HT-dependent signaling. *Metabolic Brain Disease, 30*(1), 47–55.

Zan, L., Chen, Q., Zhang, L., & Li, X. (2019). Epigallocatechin gallate (EGCG) suppresses growth and tumorigenicity in breast cancer cells by downregulation of miR-25. *Bioengineered, 10*(1), 374–382.

Zanotto-Filho, A., Coradini, K., Braganhol, E., Schröder, R., De Oliveira, C. M., Simões-Pires, A., Battastini, A. M. O., Pohlmann, A. R., Guterres, S. S., & Forcelini, C. M. (2013). Curcumin-loaded lipid-core nanocapsules as a strategy to improve pharmacological efficacy of curcumin in glioma treatment. *European Journal of Pharmaceutics and Biopharmaceutics, 83*(2), 156–167.

Zhang, J., Nie, S., Martinez-Zaguilan, R., Sennoune, S. R., & Wang, S. (2016). Formulation, characteristics and antiatherogenic bioactivities of CD36-targeted epigallocatechin gallate (EGCG)-loaded nanoparticles. *Journal of Nutritional Biochemistry, 30,* 14–23.

Zhang, J.-Y., Lin, M.-T., Zhou, M.-J., Yi, T., Tang, Y.-N., Tang, S.-L., Yang, Z.-J., Zhao, Z.-Z., & Chen, H.-B. (2015). Combinational treatment of curcumin and quercetin against gastric cancer MGC-803 cells in vitro. *Molecules, 20*(6), 11524–11534.

Zhang, S., Huang, J., Xie, X., He, Y., Mo, F., & Luo, Z. (2017). Quercetin from *Polygonum capitatum* protects against gastric inflammation and apoptosis associated with *Helicobacter pylori* infection by affecting the levels of p38MAPK, BCL-2 and BAX. *Molecules, 22*(5), 744.

Zhang, Y., Duan, W., Owusu, L., Wu, D., & Xin, Y. (2015). Epigallocatechin3gallate induces the apoptosis of hepatocellular carcinoma LM6 cells but not noncancerous liver cells. *International Journal of Molecular Medicine, 35*(1), 117–124.

Zhang, Z., Wang, C.-Z., Du, G.-J., Qi, L.-W., Calway, T., He, T.-C., Du, W., & Yuan, C.-S. (2013). Genistein induces G2/M cell cycle arrest and apoptosis via ATM/p53-dependent pathway in human colon cancer cells. *International Journal of Oncology, 43*(1), 289–296.

Zhao, H., Chen, S., Gao, K., Zhou, Z., Wang, C., Shen, Z., Guo, Y., Li, Z., Wan, Z., & Liu, C. (2017). Resveratrol protects against spinal cord injury by activating autophagy and inhibiting apoptosis mediated by the SIRT1/AMPK signaling pathway. *Neuroscience, 348,* 241–251.

Zhao, P., Mao, J., Zhang, S., Zhou, Z., Tan, Y., & Zhang, Y. (2014). Quercetin induces HepG2 cell apoptosis by inhibiting fatty acid biosynthesis. *Oncology Letters, 8*(2), 765–769.

Zhou, L., Xiao, X., Zhang, Q., Zheng, J., & Deng, M. (2019). Deciphering the anti-obesity benefits of resveratrol: The "gut microbiota-adipose tissue" axis. *Frontiers in Endocrinology, 10,* 413.

Zubair, H., Khan, H. Y., Sohail, A., Azim, S., Ullah, M. F., Ahmad, A., Sarkar, F. H., & Hadi, S. M. (2013). Redox cycling of endogenous copper by thymoquinone leads to ROS-mediated DNA breakage and consequent cell death: Putative anticancer mechanism of antioxidants. *Cell Death & Disease, 4*(6), e660.

14 Nano-Nutraceuticals and Oxidative Stress

*Rahat Naseer, Sadia Nawaz, Muzna Munir, Hitesh Chopra, Uday Younis Hussein Abdullah, Shakeel Ahmed, and Atif Amin Baig**

CONTENTS

14.1 Nutraceuticals 377
14.1.1 Definition 377
14.1.2 Classification 378
14.1.3 Examples 378
14.2 Nanoparticles 378
14.2.1 Structure of Nanoparticles 378
14.2.2 Mesoporous Nanoparticles 383
14.2.3 Classification of Nanoparticles 383
14.2.3.1 Composition-Based Classification 383
14.2.3.2 Structure-Based Classification 386
14.2.4 Synthesis of Nanoparticles 386
14.3 Delivery Systems and Challenges 387
14.3.1 Current Delivery Systems 387
14.3.2 Lipid-Based Nanoparticle System 387
14.3.3 Nanoliposomal Formulation 387
14.3.3.1 Solid-Core Micelles 388
14.3.3.2 Associated Colloids 388
14.3.3.3 Cholecalciferol (Vitamin D3) Mini-Tablets 388
14.3.3.4 Protein-Based Nanocarriers 389
14.4 Nanoparticles for Oral Delivery 389
14.4.1 Gastrointestinal Effect on Nanoparticles 389
14.4.2 Antioxidant Potential of Nano-Nutraceuticals 389
14.4.3 Mode of Action 390
14.4.4 Application in Medicine 391
14.5 Recommendations and Future Directions 391
14.5.1 Current Research Gaps 391
References 392

14.1 NUTRACEUTICALS

14.1.1 Definition

Nutra means nutrient, and *ceutical* is a refined healing item or therapy suffix. Although this concept dates back to the prehistoric period, this proposed name was formally expressed by Stephen

* Corresponding author: atifamin@unisza.edu.my

DOI: 10.1201/9781003244721-14

DeFelice in 1989, and probably the first certified and noted product is iodized table salt, introduced on a massive scale for the prevention of goiter (Benvenga et al. 2019).

Nutraceutical is a market-based term, and many variations are associated with this term, which varies from region to region. Although no regulatory definition exists so far, we can define a nutraceutical as 'Any food or component of food having health benefits' or 'Any non-toxic, palatable substance used for health benefits.' Research in the 1920s targeted the use of different plants and organic wastes as next-generation nutraceuticals (Varzakas, Zakynthinos, and Verpoort 2016). Hence, *nutraceuticals* is a broad term interchangeably used for functional foods or dietary supplements under certain conditions.

Keeping in view the diversity of the class, one can sum up the term as follows: 'Any food component having benefits beyond its nutritional role is a nutraceutical.'

14.1.2 Classification

Primarily, nutraceuticals have been classified into two main categories, natural and synthetic.

Naturally synthesized nutraceuticals are, as the name suggests, basically food ingredients known for their benefits, likely due to their chemical configuration and composition, like phytochemicals or enzymes, or even microbes themselves as probiotics (Tovar-Ramírez et al. 2010). Finding new agents is a continuous process, so no single classification system covers all the substances. A common way to classify nutraceuticals is by either nutritional sources or mode of action. Food sources used as nutraceuticals are classified as natural. Synthetic nutraceuticals are also available in the market under different brand names (Table 14.2). However, their therapeutic potential has its own challenges. Some common herbs used as nutraceuticals are mentioned in Table 14.1 (Pandey and Kumar 2011).

14.1.3 Examples

Some common composition-based classes are mentioned in Figure 14.1. Regardless of their variation in therapeutic potential, the health benefits of nutraceuticals are immense and proven. They are natural dietary supplements that can enhance health and facilitate humans' medical condition by preventing undesirable side effects. Intake of nutraceuticals improves the deficiencies of vitamins and minerals and restores a healthy gastrointestinal tract (GIT) and dietary routines. They are easily available and cost-effective.

In complex metabolic reactions, they can also act as substrates for essential biochemical reactions, cofactors, or even enzyme inhibitors, resulting in enhanced absorption and GIT scavenging.

14.2 NANOPARTICLES

Nanoparticles (NPs) are related to various materials having a size range between 1 and 100 nm (Laurent et al. 2008). NPs have been broadly used for their physicochemical properties. They possess a high surface area to volume ratio. In 1959, Richard P. Feynman was the first to forecast the idea of nanotechnology (Mody et al. 2010).

14.2.1 Structure of Nanoparticles

NPs are a form of aggregated molecules that comprise three major layers: the basal layer, responsible for the interaction of different molecules, is called the surface layer; the second layer, referred to as the shell layer, differs in composition from the other layers; and the third layer is made up of core, which is basically the central portion of the nanostructure (Shin et al. 2016).

TABLE 14.1
Common Natural Nutraceuticals

Common herb	Biological name	Components	Health benefits	References
Echinacea	Dried herb *Echinacea purpurea*	Alkylamide and echinacoside	Anti-inflammatory, immunomodulatory, antiviral	(Manayi, Vazirian, and Saeidnia 2015)
Ginger	Rhizomes of *Zingiber officinale*	Zingiberene and gingerols	Stimulant, chronic bronchitis, hyperglycemia, and sore throat	(Çifci et al. 2018)
Garlic	Dried bulbs of *Allium sativum*	Alliin and allicin	Anti-inflammatory, antibacterial, anti-gout, nervine tonic	(Radha Singh and Singh 2019)
Liquorice	Dried roots of *Glycyrrhiza glabra*	Glycyrrhizin and liquiritin	Anti-inflammatory and anti-allergic, expectorant	(Størmer, Reistad, and Alexander 1993)
Turmeric	Rhizome of *Curcuma longa*	Curcumin	Anti-inflammatory, antiarthritic, anticancer, and antiseptic	(Hewlings and Kalman 2017)
Onion	Dried bulb of *Allium cepa* Linn.	Allicin and alliin	Hypoglycemic activity, antibiotic and anti-atherosclerosis	(Upadhyay 2016)
Valerian	Dried root of *Valeriana officinalis* Linn.	Valerenic acid and valerate	Tranquilizer, migraine and menstrual pain, intestinal cramps, bronchial spasm	(Nandhini, Narayanan, and Ilango 2018)
Maiden hair tree	Leaves of *Ginkgo biloba*	Ginkgolide and bilobalide	PAF antagonist, memory enhancer, antioxidant	(Mahadevan and Park 2008)
St. John's wort	Dried aerial part of *Hypericum perforatum*	Hypericin and hyperforin	Antidepressant, against HIV and hepatitis C virus	(Shrivastava and Dwivedi 2015)
Aloes	Dried juice of leaves of *Aloe barbadensis*	Aloins and aloesin	Dilates capillaries, anti-inflammatory, emollient, wound healing properties	(Hashemi, Madani, and Abediankenari 2015)
Goldenseal	Dried root of *Hydrastis canadensis*	Hydrastine and berberine	Antimicrobial, astringent, antihemorrhagic, treatment of mucosal inflammation	(Ettefagh et al. 2011)
Brahmi	Herbs of *Centella asiatica*	Asiaticoside and madecassoside	Nervine tonic, spasmolytic, anti-anxiety	(Shinomol, Muralidhara, and Bharath 2011)
Bael	Unripe fruits of *Aegle marmelos* Corr.	Marmelosin	Digestive, appetizer, treatment of diarrhea and dysentery	(Benni, Jayanthi, and Suresha 2011)
Senna	Dried leaves of *Cassia angustifolia*	Sennosides	Purgative	(Blandizzi and Scarpignato 2011)
Asafetida	Oleo gum resin of *Ferula assafoetida*	Ferulic acid and umbellic acid	Stimulant, carminative, expectorant	(Mahendra and Bisht 2012)
Rambans	Leaf sap of *Agave americana*	Kaempferol, quercetin, isorhamnetin	Antiseptic, diuretics	(Misra, Varma, and Kumar 2018)
Oat straw	Dried herb *Avena sativa*	Avenanthramides	Diuresis, cholesterol control, anti-inflammatory, anti-itching	(Rajinder Singh, De, and Belkheir 2013)

(Continued)

TABLE 14.1 (CONTINUED)
Common Natural Nutraceuticals

Common herb	Biological name	Components	Health benefits	References
Chaulai	Oil from the seed of *Amaranthus* spp.	Alkaloids, flavonoids, saponin, betaine	Cardiovascular disease	(Hope 2013)
Kalmegha	Shoot powder of *Andrographis paniculata*	Andrographolide, apigenin, luteolin	Bacillary dysentery, respiratory tract infection	(Jayakumar et al. 2013)
Artemisia	Shoot decoction of *Artemisia annua*	Artemisinin, flavonoids	Fever, upper respiratory tract infection	(Nigam et al. 2019)
Shatavari	Roots of *Asparagus* spp.	Gallic acid, tannins, saponins, indole-3-carbinols	Tonic, astringent	(Gaonkar and Hullatti 2020)
Bugloss	Leaves and flowers of *Borago officinalis*	Anthocyanins, oleuropein	Skin care, anti-inflammatory, blood purifier	(Asadi-Samani, Bahmani, and Rafieian-Kopaei 2014)
Salai guggal	Gum resin of *Boswellia serrata*	Beta boswellic acid, acetyl-11-betaboswellic acid	Asthma, antiarthritis	(Siddiqui 2011)
RaktaKanchan	Roots and leaves of *Bauhinia purpurea*	5,6-dihydroxy-7-methoxyflavone, dibenzooxepins, lutein, astragalin	Catarrh, boil, glandular swelling	(Zakaria et al. 2011)
Barberry	Roots and barriers of *Berberis asiatica*	Berberine, tannins	Roots are used in treating ulcer, urethral discharge, ophthalmia, jaundice, fever, fruit is cooling and laxative	(Srivastava, Singh Rawat, and Mehrotra 2004)
Pot marigold	Floweral decoction *Calendula* spp.	Alpha amyrin, beta amyrin, calendu-ladiol	Anti-inflammatory, inhibit HIV, antibacterial, antitumor	(Muley, Khadabadi, and Banarase 2009)
Red pepper	Fruit powder of *Capsicum annuum*	Capsaicin, capsanthin, capsorubin, cryptoxanthin, lutein	Antiarthritic, antioxidant, stimulant, rubefacient	(Arimboor et al. 2015)
Gotu kola	Herbs powder of *Centella asiatica*	Centelloside, thankuniside, brahmoside	Improve memory, sedative, stress reduction, immune stimulant, wound healing, venous tonic	(Gohil, Patel, and Gajjar 2010)
Guggulu	Gum resin of *Commiphora wightii*	Guggulsterones	Cardioprotective, anti-inflammatory, rheumatic disease	(Sarup, Bala, and Kamboj 2015)
Lemon-grass	Dried leaves of *Cymbopogon citratus*	Luteolin, isoorientin 2O- rhamnoside, quercetin, kaempferol, apigenin	Stomachache	(Fernandes et al. 2012)
Ephedra	Dried stems of *Ephedra sinica*	Flavonoids, lignans, proanthocyanidins	Mild anti-asthmatic, nasal congestion, fluid retention, obesity	(Haller, Jacob, and Benowitz 2002)

(*Continued*)

TABLE 14.1 (CONTINUED)
Common Natural Nutraceuticals

Common herb	Biological name	Components	Health benefits	References
Cone flower	The whole plant of *Echinacea angustifolia*	Saponins, coumarins, tannins, caffeic acid, anthraquinones	Antibiotic, antiviral, anti-allergic, used in common cold	(Sharifi-Rad et al. 2018)
Fennel	Whole seed of *Foeniculum vulgare*	Anethole, estragole	Stomach bloating, stimulant, digestive spasm, galactagogue	(Badgujar, Patel, and Bandivdekar 2014)
Garcinia	Extract of the fruit of *Garcinia cambogia*	Hydroxycitric acid	Weight loss	(Lee and Lee 2007)
Mother-wort	Calyx powder or decoction of *Hibiscus subdariffa*	Protocatechuic acid, hibiscus acid, anthocyanins	Central nervous system depressant	(Wojtyniak, Szymański, and Matławska 2013)
Flaxseed	Seed powder of *Linum usitatissimum*	Alpha-linolenic acid	Constipation, irritable bowel syndrome, source of omega 3 essential fatty acids, cholesterol control, antiarthritis	(Basch et al. 2007)
Chamomile	Dried flowers of *Matricaria chamomilla*	Chamazulene, levomenol	Seductive, indigestion, insomnia, nausea, anti-inflammation, wound healing	(Abdullahzadeh, Matourypour, and Naji 2017)
Alfalfa	The dried leaf of *Medicago sativa*	Asparagines, trigonelline, stachydrine, coumarins, flavonoids	Appetite stimulation, antiarthritis	(Bora and Sharma 2011)
Moringa	*Moringa oleifera*	Vitamins C and E, beta carotene, isothiocyanate	Antiviral, antimicrobial, hepatoprotective, anticancer, asthma, and venomous bites	(Younus et al. 2016)
Psyllium	Dried seed and husk of *Plantago ovata*	Hemicelluloses, uronic acid, arabinoxylan	Constipation, lowering of cholesterol, type 2 diabetes control	(Solà et al. 2010)
Rosemary	Tubers of *Rosmarinus tuberose*	Diterpenes, carnosol, rosmanol	Digestion, rheumatism, stimulating appetite, stimulating circulation	(Andrade et al. 2018)
Saw palmetto	Dried whole fruit of *Serenoa repens*	Beta sitosterols, campestrol, stigmasterol, rhoifolin	Benign prostatic hyperplasia, anti-inflammatory, impotence	(Wilt, Ishani, and Mac Donald 2002)
Milk thistle	Whole or powdered seeds of *Silybum marianum*	Silymarin, linoleic acids, silibinin	Liver disorder, lactation problem, antioxidant	(Flora et al. 1998)
Harar	Fruit pulps of *Terminalia chebula*	Chebulinic acid, chebulagicaci, tannins	Antioxidant	(Pfundstein et al. 2010)

(Continued)

TABLE 14.1 (CONTINUED)
Common Natural Nutraceuticals

Common herb	Biological name	Components	Health benefits	References
Ashwagandha	Root powder of *Wathania somnifera*	Isopelletierine, anaferine, withanolides, withaferins	Stress, insomnia, cataract prevention	(Thiagarajan, Venu, and Balasubramanian 2003)
Valerian	Roots, powder, tea of *Valeriana officinalis*	Valerenic acid, flavonoids, tannins	Anxiety, insomnia, hypertension	(Shinjyo, Waddell, and Green 2020)
Stinging nettle	The dried leaf of *Urtica dioica*	Quercetin, ursolic acid	Benign prostate hyperplasia, dieresis, anemia, osteoarthritis	(Easton et al. 2021)
Chirata	The whole dried plant of *Swertia chirata*	Chiratin, ophelic-acids, amarogentin, mangiferin, sweroside	Migraine headache	(Kumar and Van Staden 2016)
Fenugreek	Seed of *Trigonella foenum graecum*	Flavonoids, saponins, galactomannans	Gastritis, excess cholesterol, diabetes, skin inflammation	(Basu and Srichamroen 2010)
Amla	Fruit pulp of *Phyllanthus emblica*	Ascorbic acid, ellagic acid, gallic acid, chlorogenic acid	Stress, diuretics, liver function, anti-aging	(Kunchana et al. 2021)
Bilikand	Tubers of *Pueraria tuberosa*	Puerarin, genistein, quercetin	Eases bowel movement, constipation, skin disease	(Bharti et al. 2021)
Ginseng	Dried roots of *Panax quinquefolius*	Ginsenoside	Convalescence, fatigue, diabetes, cholesterol control, aphrodisiac	(Murata et al. 2012)
Pineapple	*Ananas comosus*	Bromelain	Heart disease, aging effect, improves immune system, reduces arthritis	(Rabelo, Tambourgi, and Pessoa Jr 2004)
Camphor	Leaves extract of *Cinnamomum camphora*	Camphor, camphene	Inhalant to treat cold and flu	(Chen, Vermaak, and Viljoen 2013)
Strawberries/ raspberries	Fruit of *Rubus idaeus*	Ellagic acid	Phytochemical, fights cancers in humans	(Skrovankova et al. 2015)
Castor bean	*Ricinus communis*	Ricinoleic acid	Constipation, upper respiratory problems, liver and kidney issues	(Abdul et al. 2018)
Vanilla	*Theobroma cacao*	Flavonoids, theobromine, serotonin	Blood pressure, heart issues, antidepressant, stimulating effect	(Nehlig 2013)
Kachur	*Curcuma zedoaria*	Curcumin	Wound healer, nasal congestion, hair loss, arthritis	(Lobo et al. 2009)
Kale	*Brassica oleracea*	Zeaxanthin	Eye health, age-related macular degeneration	(Mageney, Baldermann, and Albach 2016)

(Continued)

TABLE 14.1 (CONTINUED)
Common Natural Nutraceuticals

Common herb	Biological name	Components	Health benefits	References
Cranberries	Fruit of *Vaccinium oxycoccus*	Resveratrol	Anti-inflammatory, blocks adhesion of blood cells, reduces skin and breast cancer	(Nemzer et al. 2022)
Chokeberries	Fruits of *Aronia melanocarpa*	Proanthocyanins	Urinary tract infection, inhibits adhesion of microorganisms	(Handeland et al. 2014)
Germinated corn	*Zea mays*	Phytosterols	Lower cholesterol, absorption in the digestive tract	(White and Rivin 2000)
Olive	*Olea europaea*	Olive oil	Maintains cholesterol level	(Besnard et al. 2008)
Feverfew	Leaves of *Tanacetum parthenium*	Parthenolide	Migraine, digestive irritation, anti-inflammatory	(Pareek et al. 2011)
Goldenseal	Roots, rhizome of *Hydrastis canadensis*	Berberine	Diarrhea	(Mandal et al. 2020)
Dong quai	Roots of *Angelica sinensis*	Trans-ferulic acid	Premenstrual tension, lowers blood pressure	(Yue et al. 2019)
Baheda	Seeds of *Terminalia bellirica*	Beta sitosterol, ellagic acid, tannins	Stress, treats stomach ulcer, wound healing, liver health, immunity booster	(Deb and Das 2016)
Flame lily	*Gloriosa superba*	Colchicines, thiocolchicoside	Cancer treatment, anti-inflammatory, anti-arthritic	(Reuter et al. 2010)
Bitter yam	*Dioscorea dumetorum*	Dioscoretine	Hyperlipidemia, hypercholesterolemia	(McKoy et al. 2015)
Cogon grass	The root of *Imperata cylindrica*	Dimethylsulfoniopropionate	Skin treatment	(Jung and Shin 2021)
Cinchona	Bark of *Cinchona* L., *Raiatea*	Quinine, alkaloids	Heals fever	(Nugraha et al. 2020)
Java turmeric	*Curcuma xanthorrhiza*	Curcuminoids, xanthorrhizol, camphor, curcumin	Antimicrobial, anti-inflammatory, antioxidant, antihypertensive, anticancer	(Rusdiana et al. 2021)

14.2.2 Mesoporous Nanoparticles

Mesoporous NPs are porous structures with large surface areas to bind different functional groups at a specific site (Shchipunov et al. 2006). This structure confers special properties on NPs. Their structure is honeycomb-like, containing active sites. The active surface allows the modification of surface properties and binding of therapeutic molecules (Klichko et al. 2009).

14.2.3 Classification of Nanoparticles

14.2.3.1 Composition-Based Classification

- **Carbon-Based NPs**

  NPs made of carbon sources include allotropes of carbon in fullerenes and carbon nanotubes. Fullerenes consist of globe-shaped cage-like structures. They have gained

TABLE 14.2
Synthetic Nutraceuticals

Products	Category	Contents	Manufacturer
Calcitriol D-3	Calcium supplement	Calcium and vitamins	Cadilla Healthcare Limited, Ahmadabad, India
GRD (23-25)	Nutritional supplement	Proteins, vitamins, minerals, and carbohydrates	ZydusCadila Ltd, Ahmadabad, India
Proteinex®	Protein supplement	Predigested proteins, vitamins, minerals, and carbohydrates	Pfizer Ltd, Mumbai, India
Coral calcium	Calcium supplement	Calcium and trace minerals	Nature's Answer, Hauppauge, NY, USA
Chyawanprash	Immune booster	Amla, ashwagandha, pippali	Dabur India, Ltd
Omega woman	Immune supplement	Antioxidant, vitamins, and phytochemicals	Wassen, Surrey, United Kingdom
Celestial healthcare	Immune booster	Dry fruit extract	Celestial Biolabs Limited
Amiriprash (Gold)	Good immune modulator	ChywannprashAvalenha, swarnabhasma, and RasSindur	Uap Pharma Pvt. Ltd
WelLife®	Amino acid supplement	Granulated-L-glutamine	Daesang America Inc., Hackensack, NJ, United States
Daytime Restore & Nighttime Repose	Restful sleep	Ginseng, ginkgo, biloba	Xigo Health
CogniSure	Amino acid supplement	Proline-rich polypeptide complex	Metagenics Inc.
PNerplus™	Neuropathic pain supplement	Vitamins and other natural supplements	Neurohelp, San Antonio, TX, United States
Biovinca™	Neurotonic	Vinpocetine	Cyvex nutrition, Irvine, CA, United States

importance because of their wide commercial applications, i.e., as fillers (Saeed and Khan 2016), a supporting medium for catalysts (Mabena et al. 2011), and adsorbents used in remediation (Ngoy et al. 2014).

- **Metal NPs**
  Metal NPs are basically formed by metal particles. These NPs exhibit unique optical and electrical properties. Alkaline-based NPs and noble metal–based NPs like Au, Ag, and Cu have been used recently because of their facets, size, shape-based synthesis, and broad absorption band detected in the visible region of the electromagnetic spectrum (Dreaden et al. 2012).
- **Clay-Based NPs**
  These are solid, porcelain, and inorganic NPs. These are prepared by increasing temperature and consecutive decreases in temperature. These may be amorphous, crystalline, compact, or hollow-shaped structures (Sigmund et al. 2006). Clay-based NPs can be used in photocatalysis reactions, different detoxifying dyes, and imaging (C Thomas, Kumar Mishra, and Talegaonkar 2015).
- **Partial Conductor NPs**
  Semiconductor NPs exhibit the properties of semiconductors (Ali et al. 2017). They have diverse applications due to their wide band gaps and tuning. These are used in catalysis,

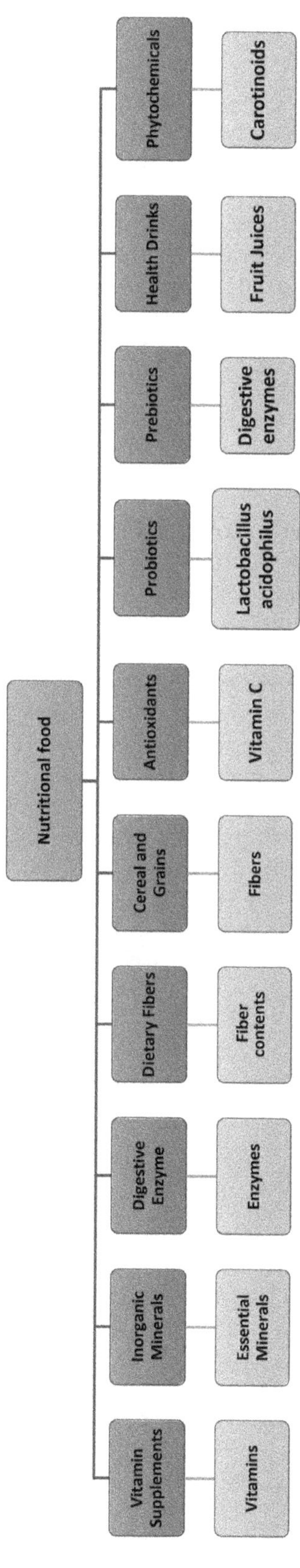

FIGURE 14.1 Compositional classes of dietary ingredients.

electronic equipment, photo-optics, and water processing applications (Hisatomi, Kubota, and Domen 2014).

- **Polymeric NPs**
  These NPs fall into the category of organic-based NPs. These are spherical and capsular in shape (Mansha et al. 2016). Nanocapsules are solids containing spherical surfaces for attaching other molecules (Rao and Geckeler 2011).
- **Lipid-Based NPs**
  Lipid-based NPs comprise lipid molecules that play a key role in different biomedical applications. These are spherical in shape and possess 10 to 1000 nm diameter. These nanoparticles are composed of a solid structure and a soluble lipotropic substance matrix. Surfactants are required to stabilize the outer core (Rawat, Jain, and Singh 2011). Lipid nanotechnology is a specialized field that emphasizes designing, synthesizing, and delivering target particles (Puri et al. 2009) and plays a role in cancer treatment (Gujrati et al. 2014).

14.2.3.2 Structure-Based Classification

- **0D NSMs**
  Zero-dimensional nanostructured materials (NSMs) include quantum dots, nanospheres, nanoarrays, core-shell nanoparticles, and hollow cubes. 0D NSMs have been used in light-emitting diodes, solar-based cells, single-electron transistors, and laser techniques (Tiwari, Tiwari, and Kim 2012).
- **1D NSMs**
  One-dimensional NSMs include nanowires, nanorods, nanotubes, and nanoribbons (Huang et al. 2002). 1D NSMs are used in nanoelectronics, nanosystems, nanocomposite materials, and national security due to their ideal characteristics (Kuchibhatla et al. 2007).
- **2D NSMs**
  Two-dimensional NSMs include junctions, branched-chain structures, nanodisks, nanosheets, nanowalls, and nanoplates. 2D NSMs have been recommended best for sensors, nanoreactors, and photocatalysts (Pradhan and Leung 2008).
- **3D NSMs**
  Three-dimensional nanostructured materials include nanoballs, nanocoils, nanoflowers, and nanopillars. 3D NSMs are widely used in catalysis, magnetic substances, and batteries (Tiwari, Tiwari, and Kim 2012).

14.2.4 Synthesis of Nanoparticles

Several methodological strategies have been used to synthesize NPs. The two main approaches to synthesizing nanoparticles are destructive and constructive.

- **Destructive Approach**
  This method is based on the disintegration of large molecules into smaller ones. These small molecules are processed further into NPs. Different destructive techniques like grinding, physical vapor deposition, laser ablation, sputtering, and electro-explosion are used (Iravani 2011).
- **Constructive Approach**
  This approach is entirely different from the destructive approach. In this approach, simpler molecules are used to synthesize relatively large molecules. Different constructive sedimentation and reduction techniques are used, e.g., sol-gel, greenway synthesis, biochemical and spin techniques (Iravani 2011).

A diagrammatic representation is shown in Figure 14.2.

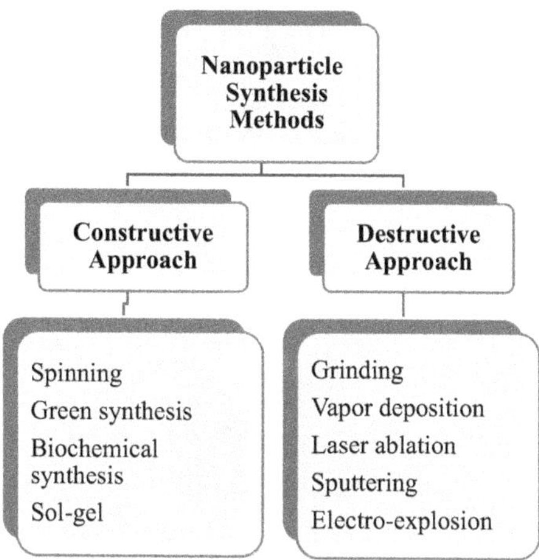

FIGURE 14.2 Nanoparticle synthesis methods.

14.3 DELIVERY SYSTEMS AND CHALLENGES

14.3.1 Current Delivery Systems

Nano-based delivery systems aim to deliver NPs to specific sites in a controlled manner. Different techniques are currently in practice. Either the molecule of interest is directly attached to the NP, or some systems use a peptide or antibody to attach themselves easily to a receptor. Encapsulation using substances tolerated by the human body, like dendrimers, alginate dextrin xanthan gum cellulose liposomes, and chitosan polymeric micelles, is also used. All have their own advantages and disadvantages. Nanoemulsions, dispersions, micelles, and associated colloids have the advantage of delivering the encapsulated ingredients directly to the active site, and controlled release is managed under optimum conditions. NPs such as lipid encapsulation-based NPs go directly to the site of action, mainly inducing the target cells to perform specific functions (Weiss, Takhistov, and McClements 2006).

Due to their large surface area and small size, nanotechnology-based emulsion formulations are more efficient in delivery than conventional formulations. A high-pressure homogenization process is used to prepare nanoemulsions. They are developed in the form of dispersed micelles at the interface of two liquids (Keservani and Sharma 2018). If these dispersed micelles have a diameter between 50 and 500 nm, they will qualify as a true emulsion (Debnath, Satayanarayana, and Kumar 2011). We will very briefly discuss a few systems.

14.3.2 Lipid-Based Nanoparticle System

The most widely used nanodelivery system is a lipid-based NP system. The lipid-based NP system gives rise to the formation of stable, circulating, tissue-targeted nanoparticles. Nanoliposomal formulations and solid-core micelles are the two widely used lipid-based NPs (Lu et al. 2021) (Figure 14.3 and Figure 14.4).

14.3.3 Nanoliposomal Formulation

Phosphatidylcholine nanoliposome complex, a nanoliposomal formulation, is one of the best examples of the nanoliposome formulation. This was specially designed for poorly absorbed silymarins. With the phosphatidylcholine nanoliposomes complex, the absorption increases threefold. The recently developed soya phosphatidylcholine andrographolides complex solid-lipid nanoparticles

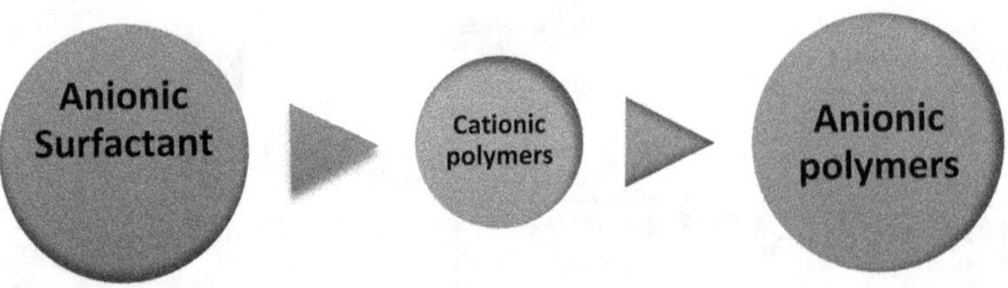

FIGURE 14.3 Lipid-based nanocarriers.

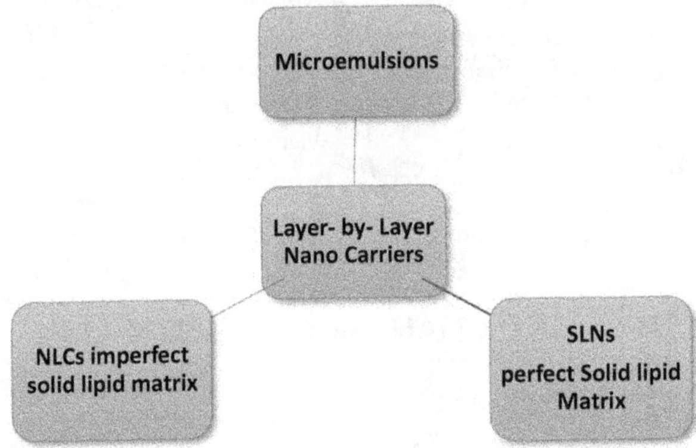

FIGURE 14.4 Layer-by-layer nanocarriers.

are a more advanced formulation. This formulation increases absorption. They are amphibolic in nature and hence, can entrap and deliver both hydrophilic and hydrophobic bioactive components like alpha tocopherols and glutathione. (From Mozafari et al. 2006; Khosravi-Darani et al. 2007.)

14.3.3.1 Solid-Core Micelles

Solid-core micelles mainly increase target cell proliferation and intracellular delivery. The assigned composition of lipids during nanomicelle design is a crucial factor determining the effectiveness of the solid-core micelles. The primary use of the nanodelivery system is for anticancer drug delivery. Particles of this design can also contain folate or vitamin B12. Micelles ranging from 5 to 100 nm in diameter are widely used for encapsulation of the bioactive components efficiently in release mechanisms (Rashidi and Khosravi-Darani 2011).

14.3.3.2 Associated Colloids

Associated colloids have been utilized in the encapsulation of nonpolar ingredients. These ingredients are trapped in the water-repelling core of surfactant vesicles. Lipid-based nanoencapsulation has greater bioavailability in the GIT and shows stability and resistance toward environmental stress as compared with other available biomaterial factors such as different proteins and polysaccharides. Hence antioxidants encapsulated in the lipid-based system due to increased activity are ideal for targeted delivery (Canani et al. 2011).

14.3.3.3 Cholecalciferol (Vitamin D3) Mini-Tablets

Cholecalciferol (vitamin D3) mini-tablets associated with either optimized bile salt or lipase alginate–glycerin deliver bioactive cholecalciferol much more efficiently *in vivo* and *in vitro.* The *in vivo* study showed that cholecalciferol's encapsulation with bioactive compounds enhances its bioavailability by 3.2 times compared with conventional products. Due to this, the serum level and concentration of 25-hydroxy vitamin D3 were also improved. Hence, bioactive association was a suitable vector for efficient delivery of cholecalciferol (Gupta et al. 2018).

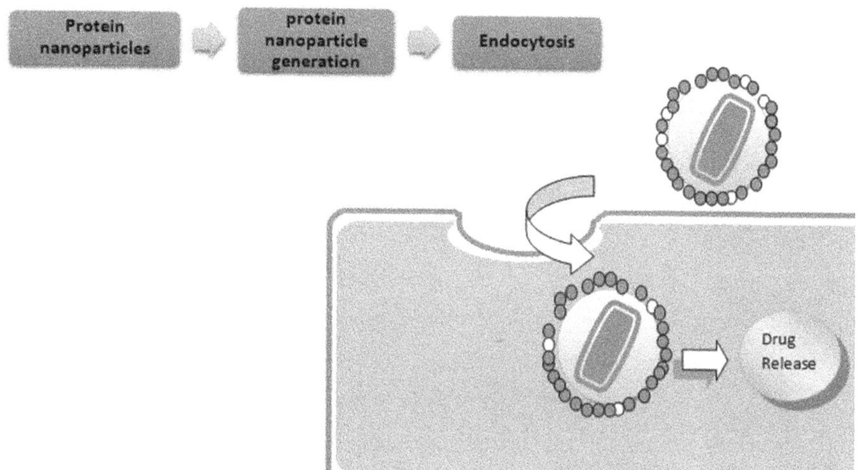

FIGURE 14.5 Protein-based nanocarriers to transport drugs through endocytosis process.

14.3.3.4 Protein-Based Nanocarriers

Protein-based nanocarriers or particles are also used. The properties of protein-based nanocarriers are associated with water behavior, such as water wetting, adsorption, and retention capacity. Other properties like solubility, dispensability, and viscosity are also important. Some properties are based on the interaction between proteins and ions, such as binding, self-assembly, the capacity to take shape, and complex polyelectrolytes formation. The proteins generally used for encapsulating nutraceuticals include whey, egg protein, casein, myofibrillar, bovine serum albumin, zein, gliadin, gelatin, etc. (Dima et al. 2020) (Figure 14.5).

14.4 NANOPARTICLES FOR ORAL DELIVERY

14.4.1 Gastrointestinal Effect on Nanoparticles

NP technology has increased the efficient intake of NPs into the GIT in different intestinal stages. Once in the GIT, the nutraceuticals will have to interact with the environmental components of the GIT. These components can be endogenous secretions like enzymes, hormones, membrane lipids, bile, and mucins, or exogenous, together with nutrients (Perez-Gregorio and Simal-Gandara 2017). These interactions may be positive or negative, in either case affecting the nutraceuticals' availability. For example, the interaction of anti-nutritional factors and protein negatively affects its bioavailability. The mechanism is also mentioned in Figure 14.6.

Due to the diversity of natural and synthetic substances, it is essential to understand how NPs are introduced into the body system. Hence, it is necessary to identify appropriate delivery systems and carrier molecules, depending upon the target. Endogenous NPs are found in the gut in a soluble form that is naturally occurring as a source of calcium and phosphate, and these minerals can recirculate as an ionic form in the gut environment (Dima et al. 2020).

14.4.2 Antioxidant Potential of Nano-Nutraceuticals

Reactive oxygen species (ROS) play a key role in oxidative stress in the body. The ROS family includes hydroxyl radicals, superoxide anion, singlet oxygen, and hydrogen peroxide. These are free radicals formed due to an external stimulus or toxin exposure. These species are generated through normal cell reactions, e.g., Fenton and non-Fenton reaction or myeloperoxidase activity with unpaired electrons. The organized structure of biomolecules, i.e., carbohydrates, lipids, and

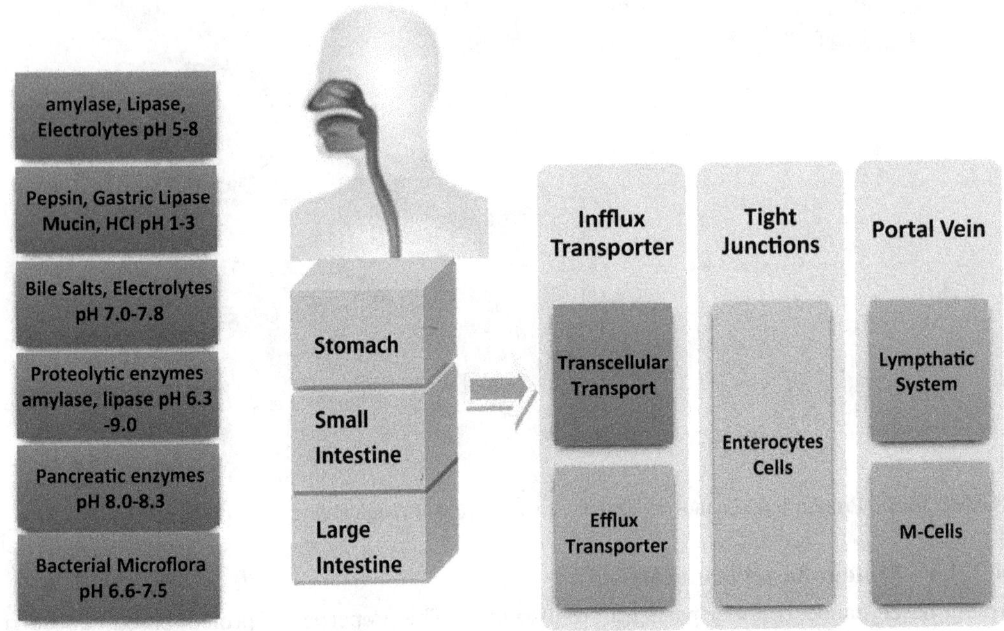

FIGURE 14.6 Gastrointestinal tract stages and transport of nutraceuticals through epithelial cell layers.

proteins, can be disrupted due to abnormal physiological and pathological conditions caused by the negative effects of radicals.

Oxidative stress results from an imbalance disproportion between ROS production and antioxidants in the living system. Mitochondria and NADPH oxidases are also among the reasons for oxidative stress. In an electron transport chain reaction, electrons will form an oxygen molecule. Superoxide dismutase helps to inactivate harmful oxygen molecules present in living tissues. It plays its role by converting O_2 to a less harmful compound, i.e., hydrogen peroxide. Inflammatory molecules produced after cellular damage activate ROS-generating enzymes, producing highly reactive oxygen, nitrogen, and halogens. Examples include peroxynitrite, hypochlorous acid, hypobromous acid, and nitric oxide.

Oxidative stress is linked with several autoimmune diseases. It activates the inflammatory pathway and results in cell death. Genetic and environmental influences are leading causes of such disorders. To determine oxidative stress, ROS level can be directly measured, or indirectly, the extent of damage to biomolecules caused by free radicals can be evaluated. Different markers of oxidative stress are freely employed. Protein carbonyl is the oxidative marker to determine the changes in protein induced by oxidative stress. In clinical samples, malondialdehyde is an essential oxidative stress marker to evaluate peroxidation in polyunsaturated fatty acids. Other markers are glutathione, catalase, and superoxide dismutase.

14.4.3 Mode of Action

The antioxidant potential of nano-nutraceuticals works on the principle of conversion of ROS to comparatively less toxic substances. This mechanism involves catalase, superoxide dismutase, glutathione peroxidase, etc. These antioxidants protect the body against harmful species. Any disorder in their activity leads to ROS formation and pathogenesis of many diseases.

Some non-enzymatic antioxidants that are basically present in plants are called phytochemicals. These phytochemicals can neutralize oxidative stress. Fruits and vegetables contain vitamins C and E, carotenoids, flavonoids, polyphenols, and terpenoids. These substances are present in diets that are helpful to treat chronic diseases. These phytochemicals can scavenge free radicals, induce beneficial effects on chronic diseases and aging, and exhibit cytotoxic and apoptotic activity. In clinical samples, antioxidant status is related to oxidative stress. Antioxidant mechanisms involve enzymatic and non-enzymatic components. Superoxide dismutase acts as an antioxidant enzyme that catalyzes the conversion of superoxide to H_2O_2 and O_2. Catalase also plays a role as an antioxidant enzyme responsible for converting hydrogen peroxide (H_2O_2) to water and oxygen. Glutathione involves converting hydrogen and lipid peroxides to alcohol and water through reduced glutathione (Katerji, Filippova, and Duerksen-Hughes 2019).

14.4.4 Application in Medicine

Nanomaterials that can therapeutically improve the comprehensive working of biological systems, from lower to high levels of organization, are known as nanomedicines. The GIT environment harbors many symbiotic microbes, so any disturbance can cause a wide range of diseases, and consequently, the introduction of NPs promises a positive balance (Quigley and Cone 2013).

NPs in cancer treatment improve conventional cancer management. This enhanced efficiency is attributed to the NPs targeting tumor cells and eventually accumulating in the tumor through vascular pores. Drugs like 5-fluorouracil and doxorubicin have been incorporated into the NP delivery system to enhance their efficacy against colorectal cancer (CRC) (Brar et al. 2021).

Nanotherapeutics and nanopharmaceuticals are effective in medicine for diagnostic and treatment purposes. Nano-nutraceuticals are novel products with multiple health benefits due to biosorption, stabilization, delivery, and encapsulation characteristics. Various documented applications include the use of *Lactobacillus rhamnosus GG* (LGG), a probiotic strain, nanoparticles (NPs) as an anticancer agent against HT-29 live cells; polylysine-induced polyglutamic acid showed antimicrobial activity; and the antitumor effect of *L. casei*–capped-Se NPs has been studied, which is linked with their cytotoxic and apoptotic activity.

14.5 RECOMMENDATIONS AND FUTURE DIRECTIONS

NPs lead to potential health risks, causing harmful effects on human health and the environment. Complicated and sometimes unpredicted interactions with biological systems may result from oral route drug administration (Homayun, Lin, and Choi 2019). So, there is a lot of apprehension concerning the safety of NPs. Diseases like Crohn's disease, ulcerative colitis, and cancer in the GIT are occasionally caused by NPs (Yang and Merlin 2019).

Currently, we are dealing with a double-edged sword. On the one hand, we need to develop good delivery systems for novel NPs, and on the other hand, a lot of research is required to check, monitor, and establish the toxicity profile of each novel NP.

14.5.1 Current Research Gaps

The scientific world is frantically pursuing herbal or natural ingredients as emerging therapeutic agents. Many substances are being used worldwide as part of traditional medicine. However, it is difficult to characterize these substances, which are generally used as a compound with an unknown composition and probably, an unknown list of side effects.

A second major challenge is establishing the therapeutic index, therapeutic dose, and therapeutic window of novel NPs. Considerable work is required, as even not so novel NPs lack this information for any treatment.

Appropriate delivery systems that are specific and targeted can be developed to ensure the maximum bioavailability, which is compromised in the case of oral delivery systems. The world is full of natural resources, and these vary from region to region and climate to climate. There is a need to develop a repository of natural ingredients. Scientists worldwide can generate a data bank of their indigenous nutraceuticals with comprehensive characterization. Nano-nutraceuticals offer a solution to many current problems related to living organisms, but an honest and joint collaboration is needed to explore and utilize their full potential.

REFERENCES

Abdul, Waseem Mohammed, Nahid H Hajrah, Jamal S M Sabir, Saleh M Al-Garni, Meshaal J Sabir, Saleh A Kabli, Kulvinder Singh Saini, and Roop Singh Bora. 2018. "Therapeutic Role of *Ricinus communis* L. and Its Bioactive Compounds in Disease Prevention and Treatment." *Asian Pacific Journal of Tropical Medicine* 11 (3): 177.

Abdullahzadeh M, Matourypour P, and Naji SA. 2017. Investigation Effect of Oral Chamomilla on Sleep Quality in Elderly People in Isfahan: A Randomized Control Trial. *Journal of Education and Health Promotion* 6: 53. https://doi.org/10.4103/jehp.jehp_109_15. PMID: 28616420; PMCID: PMC5470311.

Ali, Shahid, Ibrahim Khan, Safyan Akram Khan, Manzar Sohail, Riaz Ahmed, Ateeq ur Rehman, Muhammad Shahid Ansari, and Mohamed Ali Morsy. 2017. "Electrocatalytic Performance of Ni@ Pt Core–Shell Nanoparticles Supported on Carbon Nanotubes for Methanol Oxidation Reaction." *Journal of Electroanalytical Chemistry* 795: 17–25.

Andrade, Joana M, Célia Faustino, Catarina Garcia, Diogo Ladeiras, Catarina P Reis, and Patrícia Rijo. 2018. "*Rosmarinus officinalis* L.: An Update Review of Its Phytochemistry and Biological Activity." *Future Science OA* 4 (4): FSO283. https://doi.org/10.4155/fsoa-2017-0124.

Arimboor, Ranjith, Ramesh Babu Natarajan, K Ramakrishna Menon, Lekshmi P Chandrasekhar, and Vidya Moorkoth. 2015. "Red Pepper (*Capsicum annuum*) Carotenoids as a Source of Natural Food Colors: Analysis and Stability: A Review." *Journal of Food Science and Technology* 52 (3): 1258–71. https://doi.org/10.1007/s13197-014-1260-7.

Asadi-Samani, Majid, Mahmoud Bahmani, and Mahmoud Rafieian-Kopaei. 2014. "The Chemical Composition, Botanical Characteristic and Biological Activities of *Borago officinalis*: A Review." *Asian Pacific Journal of Tropical Medicine* 7: S22–28.

Badgujar, Shamkant B, Vainav V Patel, and Atmaram H Bandivdekar. 2014. "Foeniculum Vulgare Mill: A Review of Its Botany, Phytochemistry, Pharmacology, Contemporary Application, and Toxicology." *BioMed Research International* 2014: 842674. https://doi.org/10.1155/2014/842674.

Basch, Ethan, Steve Bent, Jeffrey Collins, Cynthia Dacey, Paul Hammerness, Michelle Harrison, Michael Smith, et al. 2007. "Flax and Flaxseed Oil (*Linum usitatissimum*): A Review by the Natural Standard Research Collaboration." *Journal of the Society for Integrative Oncology* 5 (3): 92–105. https://doi.org/10.2310/7200.2007.005.

Basu, Tapan K, and Anchalee Srichamroen. 2010. "Health Benefits of Fenugreek (*Trigonella foenum-graecum leguminosse*)." In *Bioactive Foods in Promoting Health*, 425–35. Elsevier.

Benni, Jyoti M, M K Jayanthi, and R N Suresha. 2011. "Evaluation of the Anti-Inflammatory Activity of *Aegle marmelos* (Bilwa) Root." *Indian Journal of Pharmacology* 43 (4): 393.

Benvenga, Salvatore, Ulla Feldt-Rasmussen, Daniela Bonofiglio, and Ernest Asamoah. 2019. "Nutraceutical Supplements in the Thyroid Setting: Health Benefits beyond Basic Nutrition." *Nutrients* 11 (9). https://doi.org/10.3390/nu11092214.

Besnard, G, C Garcia-Verdugo, R Rubio De Casas, U A Treier, N Galland, and P Vargas. 2008. "Polyploidy in the Olive Complex (*Olea europaea*): Evidence from Flow Cytometry and Nuclear Microsatellite Analyses." *Annals of Botany* 101 (1): 25–30. https://doi.org/10.1093/aob/mcm275.

Bharti, Ram, Bhupinder Singh Chopra, Sachin Raut, and Neeraj Khatri. 2021. "*Pueraria tuberosa*: A Review on Traditional Uses, Pharmacology, and Phytochemistry." *Frontiers in Pharmacology* 11 (January): 582506. https://doi.org/10.3389/fphar.2020.582506.

Blandizzi, Corrado, and Carmelo Scarpignato. 2011. "Gastrointestinal Drugs." In *A Worldwide Yearly Survey of New Data in Adverse Drug Reactions*, edited by J K B T—Side Effects of Drugs Annual Aronson, 33: 741–67. Elsevier. https://doi.org/10.1016/B978-0-444-53741-6.00036-2.

Bora, Kundan Singh, and Anupam Sharma. 2011. "Phytochemical and Pharmacological Potential of *Medicago sativa*: A Review." *Pharmaceutical Biology* 49 (2): 211–20. https://doi.org/10.3109/13880209.2010.504732.

Brar, Basanti, Koushlesh Ranjan, Ankur Palria, Rajesh Kumar, Mayukh Ghosh, Sweety Sihag, and Prasad Minakshi. 2021. "Nanotechnology in Colorectal Cancer for Precision Diagnosis and Therapy." *Frontiers in Nanotechnology* 3. https://doi.org/10.3389/fnano.2021.699266.

Canani, Roberto Berni, Margherita Di Costanzo, Ludovica Leone, Monica Pedata, Rosaria Meli, and Antonio Calignano. 2011. "Potential Beneficial Effects of Butyrate in Intestinal and Extraintestinal Diseases." *World Journal of Gastroenterology: WJG* 17 (12): 1519.

Chen, Weiyang, Ilze Vermaak, and Alvaro Viljoen. 2013. "Camphor—A Fumigant during the Black Death and a Coveted Fragrant Wood in Ancient Egypt and Babylon—A Review." *Molecules (Basel, Switzerland)* 18 (5): 5434–54. https://doi.org/10.3390/molecules18055434.

Çifci, Atilla, Cüneyt Tayman, Halil İbrahim Yakut, Halit Halil, Esra Cakir, Ufuk Cakir, and Salih Aydemir. 2018. "Ginger (*Zingiber officinale*) Prevents Severe Damage to the Lungs Due to Hyperoxia and Inflammation." *Turkish Journal of Medical Sciences* 48 (4): 892–900.

Deb, Anindita, and Biswajit Das. 2016. "Isolation, Purification and Characterization of Natural Polymer Obtained from Plant Baheda (*Terminelia bellarica*)." *Www.Wjpps.Com* 5 (6): 1118. https://doi.org/10.20959/wjpps20166-6868.

Debnath, SUBHASHIS, Kumar V G Satayanarayana, and G Vijay Kumar. 2011. "Nanoemulsion: A Method to Improve the Solubility of Lipophilic Drugs." *Pharmanest* 2 (2–3): 72–83.

Dima, Cristian, Elham Assadpour, Stefan Dima, and Seid Mahdi Jafari. 2020. "Bioavailability of Nutraceuticals: Role of the Food Matrix, Processing Conditions, the Gastrointestinal Tract, and Nanodelivery Systems." *Comprehensive Reviews in Food Science and Food Safety* 19 (3): 954–94. https://doi.org/10.1111/1541-4337.12547.

Dreaden, Erik C, Alaaldin M Alkilany, Xiaohua Huang, Catherine J Murphy, and Mostafa A El-Sayed. 2012. "The Golden Age: Gold Nanoparticles for Biomedicine." *Chemical Society Reviews* 41 (7): 2740–79.

Easton, Laura, Shalini Vaid, Angela K Nagel, Jineane V Venci, and Robert J Fortuna. 2021. "Stinging Nettle (*Urtica dioica*): An Unusual Case of Galactorrhea." *American Journal of Case Reports* 22 (December): e933999. https://doi.org/10.12659/AJCR.933999.

Ettefagh, Keivan A, Johnna T Burns, Hiyas A Junio, Glenn W Kaatz, and Nadja B Cech. 2011. "Goldenseal (*Hydrastis canadensis* L.) Extracts Synergistically Enhance the Antibacterial Activity of Berberine via Efflux Pump Inhibition." *Planta Medica* 77 (8): 835–40. https://doi.org/10.1055/s-0030-1250606.

Fernandes, CN, HF De Souza, G De Oliveria, JGM Costa, MR Kerntopf, and AR Campos. 2012. "Investigation of the Mechanisms Underlying the Gastroprotective Effect of Cymbopogon Citratus Essential Oil." *Journal of Young Pharmacists : JYP* 4 (1): 28–32. https://doi.org/10.4103/0975-1483.93578.

Flora, K, M Hahn, H Rosen, and K Benner. 1998. "Milk Thistle (*Silybum marianum*) for the Therapy of Liver Disease." *American Journal of Gastroenterology* 93 (2): 139–43. https://doi.org/10.1111/j.1572-0241.1998.00139.x.

Gaonkar, Vishakha Parab, and Kirankumar Hullatti. 2020. "Indian Traditional Medicinal Plants as a Source of Potent Anti-Diabetic Agents: A Review." *Journal of Diabetes and Metabolic Disorders* 19 (2): 1895–1908. https://doi.org/10.1007/s40200-020-00628-8.

Gohil, Kashmira J, Jagruti A Patel, and Anuradha K Gajjar. 2010. "Pharmacological Review on *Centella asiatica*: A Potential Herbal Cure-All." *Indian Journal of Pharmaceutical Sciences* 72 (5): 546–56. https://doi.org/10.4103/0250-474X.78519.

Gujrati, Maneesh, Anthony Malamas, Tesia Shin, Erlei Jin, Yunlu Sun, and Zheng-Rong Lu. 2014. "Multifunctional Cationic Lipid-Based Nanoparticles Facilitate Endosomal Escape and Reduction-Triggered Cytosolic SiRNA Release." *Molecular Pharmaceutics* 11 (8): 2734–44.

Gupta, Rahul, Chittaranjan Behera, Gourav Paudwal, Neha Rawat, Ashish Baldi, and Prem N Gupta. 2018. "Recent Advances in Formulation Strategies for Efficient Delivery of Vitamin D." *AAPS PharmSciTech* 20 (1): 11. https://doi.org/10.1208/s12249-018-1231-9.

Haller, Christine A, Peyton 3rd Jacob, and Neal L Benowitz. 2002. "Pharmacology of Ephedra Alkaloids and Caffeine after Single-Dose Dietary Supplement Use." *Clinical Pharmacology and Therapeutics* 71 (6): 421–32. https://doi.org/10.1067/mcp.2002.124523.

Handeland, Maria, Nils Grude, Torfinn Torp, and Rune Slimestad. 2014. "Black Chokeberry Juice (*Aronia melanocarpa*) Reduces Incidences of Urinary Tract Infection among Nursing Home Residents in the Long Term: A Pilot Study." *Nutrition Research* 34 (6): 518–25. https://doi.org/10.1016/j.nutres.2014.05.005.

Hashemi, Seyyed Abbas, Seyyed Abdollah Madani, and Saied Abediankenari. 2015. "The Review on Properties of *Aloe vera* in Healing of Cutaneous Wounds." *BioMed Research International* 2015.

Hewlings, Susan J, and Douglas S Kalman. 2017. "Curcumin: A Review of Its Effects on Human Health." *Foods* 6 (10): 92.

Hisatomi, Takashi, Jun Kubota, and Kazunari Domen. 2014. "Recent Advances in Semiconductors for Photocatalytic and Photoelectrochemical Water Splitting." *Chemical Society Reviews* 43 (22): 7520–35.

Homayun, Bahman, Xueting Lin, and Hyo-Jick Choi. 2019. "Challenges and Recent Progress in Oral Drug Delivery Systems for Biopharmaceuticals." *Pharmaceutics* 11 (3): 129. https://doi.org/10.3390/pharmaceutics11030129.

Hope, J. 2013. "A Review of the Mechanism of Injury and Treatment Approaches for Illness Resulting from Exposure to Water-Damaged Buildings, Mold, and Mycotoxins. Maček P, Aruoma O, Rocha JBT, editors. *Scientific World Journal* 2013: 767482. https://doi.org/10.1155/2013/767482.

Huang, Limin, Huanting Wang, Zhengbao Wang, Anupam Mitra, Dongyuan Zhao, and Yushan Yan. 2002. "Cuprite Nanowires by Electrodeposition from Lyotropic Reverse Hexagonal Liquid Crystalline Phase." *Chemistry of Materials* 14 (2): 876–80.

Iravani, Siavash. 2011. "Green Synthesis of Metal Nanoparticles Using Plants." *Green Chemistry* 13 (10): 2638–50.

Jayakumar, Thanasekaran, Cheng-Ying Hsieh, Jie-Jen Lee, and Joen-Rong Sheu. 2013. "Experimental and Clinical Pharmacology of *Andrographis paniculata* and Its Major Bioactive Phytoconstituent Andrographolide." *Evidence-Based Complementary and Alternative Medicine : ECAM* 2013: 846740. https://doi.org/10.1155/2013/846740.

Jung, Young-Kyung, and Dongyun Shin. 2021. "*Imperata cylindrica*: A Review of Phytochemistry, Pharmacology, and Industrial Applications." *Molecules* 26 (5): 1454.

Katerji, Meghri, Maria Filippova, and Penelope Duerksen-Hughes. 2019. "Approaches and Methods to Measure Oxidative Stress in Clinical Samples: Research Applications in the Cancer Field." Edited by Grzegorz Bartosz. *Oxidative Medicine and Cellular Longevity* 2019: 1279250. https://doi.org/10.1155/2019/1279250.

Keservani, Raj K, and Anil K Sharma. 2018. "Nanoemulsions: Formulation Insights, Applications, and Recent Advances." In *Nanodispersions for Drug Delivery*, 71–96. Apple Academic Press.

Khosravi-Darani, Kianoush, Abbas Pardakhty, Hamid Honarpisheh, V S N Malleswara Rao, and M Reza Mozafari. 2007. "The Role of High-Resolution Imaging in the Evaluation of Nanosystems for Bioactive Encapsulation and Targeted Nanotherapy." *Micron* (Oxford, England: 1993) 38 (8): 804–18. https://doi.org/10.1016/j.micron.2007.06.009.

Klichko, Yaroslav, Monty Liong, Eunshil Choi, Sarah Angelos, Andre E Nel, J Fraser Stoddart, Fuyuhiko Tamanoi, and Jeffrey I Zink. 2009. "Mesostructured Silica for Optical Functionality, Nanomachines, and Drug Delivery." *Journal of the American Ceramic Society*. American Ceramic Society 92 (s1): s2–10. https://doi.org/10.1111/j.1551-2916.2008.02722.x.

Kuchibhatla, Satyanarayana V N T, A S Karakoti, Debasis Bera, and S Seal. 2007. "One-Dimensional Nanostructured Materials." *Progress in Materials Science* 52 (5): 699–913.

Kumar, Vijay, and Johannes Van Staden. 2016. "A Review of *Swertia chirayita* (Gentianaceae) as a Traditional Medicinal Plant." *Frontiers in Pharmacology* 6. https://doi.org/10.3389/fphar.2015.00308.

Kunchana, Khwandow, Wattanased Jarisarapurin, Linda Chularojmontri, and Suvara K Wattanapitayakul. 2021. "Potential Use of Amla (*Phyllanthus emblica* L.) Fruit Extract to Protect Skin Keratinocytes from Inflammation and Apoptosis after UVB Irradiation." *Antioxidants* 10 (5): 703.

Laurent, Sophie, Delphine Forge, Marc Port, Alain Roch, Caroline Robic, Luce Vander Elst, and Robert N Muller. 2008. "Magnetic Iron Oxide Nanoparticles: Synthesis, Stabilization, Vectorization, Physicochemical Characterizations, and Biological Applications." *Chemical Reviews* 108 (6): 2064–2110. https://doi.org/10.1021/cr068445e.

Lee, Kyung Hwan, and Byung Mu Lee. 2007. "Evaluation of the Genotoxicity of (–)-Hydroxycitric Acid (HCA-SX) Isolated from *Garcinia cambogia*." *Journal of Toxicology and Environmental Health, Part A* 70 (5): 388–92. https://doi.org/10.1080/15287390600882192.

Lobo, Richard, Kirti S Prabhu, Annie Shirwaikar, and Arun Shirwaikar. 2009. "*Curcuma zedoaria* Rosc. (White Turmeric): A Review of Its Chemical, Pharmacological and Ethnomedicinal Properties." *Journal of Pharmacy and Pharmacology* 61 (1): 13–21. https://doi.org/10.1211/jpp.61.01.0003.

Lu, Hongyun, Shengliang Zhang, Jinling Wang, and Qihe Chen. 2021. "A Review on Polymer and Lipid-Based Nanocarriers and Its Application to Nano-Pharmaceutical and Food-Based Systems." *Frontiers in Nutrition* 8 (December): 783831. https://doi.org/10.3389/fnut.2021.783831.

Mabena, Letlhogonolo F, Suprakas Sinha Ray, Sabelo D Mhlanga, and Neil J Coville. 2011. "Nitrogen-Doped Carbon Nanotubes as a Metal Catalyst Support." *Applied Nanoscience* 1 (2): 67–77.

Mageney, Vera, Susanne Baldermann, and Dirk C Albach. 2016. "Intraspecific Variation in Carotenoids of *Brassica oleracea* var. sabellica." *Journal of Agricultural and Food Chemistry* 64 (16): 3251–57. https://doi.org/10.1021/acs.jafc.6b00268.

Mahadevan, S, and Y Park. 2008. "Multifaceted Therapeutic Benefits of *Ginkgo biloba* L.: Chemistry, Efficacy, Safety, and Uses." *Journal of Food Science* 73 (1): R14–19.

Mahendra, Poonam, and Shradha Bisht. 2012. "Ferula Asafoetida: Traditional Uses and Pharmacological Activity." *Pharmacognosy Reviews* 6 (12): 141–46. https://doi.org/10.4103/0973-7847.99948.

Manayi, Azadeh, Mahdi Vazirian, and Soodabeh Saeidnia. 2015. "*Echinacea purpurea*: Pharmacology, Phytochemistry and Analysis Methods." *Pharmacognosy Reviews* 9 (17): 63–72. https://doi.org/10.4103/0973-7847.156353.

Mandal, Sudip Kumar, Amal Kumar Maji, Siddhartha Kumar Mishra, Pir Mohammad Ishfaq, Hari Prasad Devkota, Ana Sanches Silva, and Niranjan Das. 2020. "Goldenseal (*Hydrastis canadensis* L.) and Its Active Constituents: A Critical Review of Their Efficacy and Toxicological Issues." *Pharmacological Research* 160 (October): 105085. https://doi.org/10.1016/j.phrs.2020.105085.

Mansha, Muhammad, Ahsanulhaq Qurashi, Nisar Ullah, Fatai Olawale Bakare, Ibrahim Khan, and Zain H Yamani. 2016. "Synthesis of In2O3/Graphene Heterostructure and Their Hydrogen Gas Sensing Properties." *Ceramics International* 42 (9): 11490–95.

McKoy, Marsha-Lyn, Kevin Grant, Helen Asemota, Oswald Simon, and Felix Omoruyi. 2015. "Renal and Hepatic Function in Hypercholesterolemic Rats Fed Jamaican Bitter Yam (*Dioscorea polygonoides*)." *Journal of Dietary Supplements* 12 (2): 173–83.

Misra, Arup Kumar, Sushil Kumar Varma, and Ranjeet Kumar. 2018. "Anti-Inflammatory Effect of an Extract of *Agave americana* on Experimental Animals." *Pharmacognosy Research* 10 (1): 104–8. https://doi.org/10.4103/pr.pr_64_17.

Mody, Vicky V, Rodney Siwale, Ajay Singh, and Hardik R Mody. 2010. "Introduction to Metallic Nanoparticles." *Journal of Pharmacy & Bioallied Sciences* 2 (4): 282–89. https://doi.org/10.4103/0975-7406.72127.

Mozafari, M Reza, John Flanagan, Lara Matia-Merino, Ajay Awati, Abdelwahab Omri, Zacharias E Suntres, and Harjinder Singh. 2006. "Recent Trends in the Lipid-Based Nanoencapsulation of Antioxidants and Their Role in Foods." *Journal of the Science of Food and Agriculture* 86 (13): 2038–45.

Muley, B P, S S Khadabadi, and N B Banarase. 2009. "Phytochemical Constituents and Pharmacological Activities of *Calendula officinalis* Linn (Asteraceae): A Review." *Tropical Journal of Pharmaceutical Research* 8 (5): 455–465.

Murata, Kazuya, Fumiaki Takeshita, Keiichi Samukawa, Tadato Tani, and Hideaki Matsuda. 2012. "Effects of *Ginseng rhizome* and *Ginsenoside ro* on Testosterone 5α-reductase and Hair Re-Growth in Testosterone-Treated Mice." *Phytotherapy Research* 26 (1): 48–53.

Nandhini, Sundaresan, Kasthuri Bai Narayanan, and Kaliappan Ilango. 2018. "*Valeriana officinalis*: A Review of Its Traditional Uses, Phytochemistry and Pharmacology." *Asian Journal of Pharmaceutical & Clinical Reearchs* 11 (1): 36–41.

Nehlig, Astrid. 2013. "The Neuroprotective Effects of Cocoa Flavanol and Its Influence on Cognitive Performance." *British Journal of Clinical Pharmacology* 75 (3): 716–27. https://doi.org/10.1111/j.1365-2125.2012.04378.x.

Nemzer, Boris V, Fadwa Al-Taher, Alexander Yashin, Igor Revelsky, and Yakov Yashin. 2022. "Cranberry: Chemical Composition, Antioxidant Activity and Impact on Human Health: Overview." *Molecules* 27 (5): 1503. https://doi.org/10.3390/molecules27051503.

Ngoy, Jacob M, Nicola Wagner, Luca Riboldi, and Olav Bolland. 2014. "A CO_2 Capture Technology Using Multi-Walled Carbon Nanotubes with Polyaspartamide Surfactant." *Energy Procedia* 63: 2230–48.

Nigam, Manisha, Maria Atanassova, Abhay P Mishra, Raffaele Pezzani, Hari Prasad Devkota, Sergey Plygun, Bahare Salehi, William N Setzer, and Javad Sharifi-Rad. 2019. "Bioactive Compounds and Health Benefits of *Artemisia* Species." *Natural Product Communications* 14 (7): 1934578X19850354.

Nugraha, Rhea Veda, Hastono Ridwansyah, Mohammad Ghozali, Astrid Feinisa Khairani, and Nur Atik. 2020. "Traditional Herbal Medicine Candidates as Complementary Treatments for COVID-19: A Review of Their Mechanisms, Pros and Cons." Edited by Yuan Xu. *Evidence-Based Complementary and Alternative Medicine* 2020: 2560645. https://doi.org/10.1155/2020/2560645.

Pandey, Manju, and V Kumar. 2011. "Nutraceutical Supplementation for Diabetes: A Review." *International Journal of Pharmacy and Pharmaceutical Sciences* 3 (supplement 4): 33–40.

Pareek, Anil, Manish Suthar, Garvendra S Rathore, and Vijay Bansal. 2011. "Feverfew (*Tanacetum parthenium* L.): A Systematic Review." *Pharmacognosy Reviews* 5 (9): 103–10. https://doi.org/10.4103/0973-7847.79105.

Perez-Gregorio, Rosa, and Jesus Simal-Gandara. 2017. "A Critical Review of Bioactive Food Components, and of Their Functional Mechanisms, Biological Effects and Health Outcomes." *Current Pharmaceutical Design* 23 (19): 2731–41.

Pfundstein, Beate, Samy K El Desouky, William E Hull, Roswitha Haubner, Gerhard Erben, and Robert W Owen. 2010. "Polyphenolic Compounds in the Fruits of Egyptian Medicinal Plants (*Terminalia bellerica, Terminalia chebula* and *Terminalia horrida*): Characterization, Quantitation and Determination of Antioxidant Capacities." *Phytochemistry* 71 (10): 1132–48. https://doi.org/10.1016/j.phytochem.2010.03.018.

Pradhan, D, and K T Leung. 2008. "Vertical Growth of Two-Dimensional Zinc Oxide Nanostructures on ITO-Coated Glass: Effects of Deposition Temperature and Deposition Time." *Journal of Physical Chemistry C* 112 (5): 1357–64.

Puri, Anu, Kristin Loomis, Brandon Smith, Jae-Ho Lee, Amichai Yavlovich, Eliahu Heldman, and Robert Blumenthal. 2009. "Lipid-Based Nanoparticles as Pharmaceutical Drug Carriers: From Concepts to Clinic." *Critical Reviews in Therapeutic Drug Carrier Systems* 26 (6): 523–580.

Quigley, Harry A, and Frances E Cone. 2013. "Development of Diagnostic and Treatment Strategies for Glaucoma through Understanding and Modification of Scleral and Lamina Cribrosa Connective Tissue." *Cell and Tissue Research* 353 (2): 231–44. https://doi.org/10.1007/s00441-013-1603-0.

Rabelo, Ana Paula Brescancini, Elias Basile Tambourgi, and Adalberto Pessoa Jr. 2004. "Bromelain Partitioning in Two-Phase Aqueous Systems Containing PEO–PPO–PEO Block Copolymers." *Journal of Chromatography B* 807 (1): 61–68.

Rao, J Prasad, and Kurt E Geckeler. 2011. "Polymer Nanoparticles: Preparation Techniques and Size-Control Parameters." *Progress in Polymer Science* 36 (7): 887–913.

Rashidi, Ladan, and Kianoush Khosravi-Darani. 2011. "The Applications of Nanotechnology in Food Industry." *Critical Reviews in Food Science and Nutrition* 51 (8): 723–30.

Rawat, Manoj K, Achint Jain, and Sanjay Singh. 2011. "Studies on Binary Lipid Matrix-Based Solid Lipid Nanoparticles of Repaglinide: In Vitro and In Vivo Evaluation." *Journal of Pharmaceutical Sciences* 100 (6): 2366–78.

Reuter, Simone, Sahdeo Prasad, Kanokkarn Phromnoi, Jayaraj Ravindran, Bokyung Sung, Vivek R Yadav, Ramaswamy Kannappan, Madan M Chaturvedi, and Bharat B Aggarwal. 2010. "Thiocolchicoside Exhibits Anticancer Effects through Downregulation of NF-KB Pathway and Its Regulated Gene Products Linked to Inflammation and Cancer." *Cancer Prevention Research (Philadelphia, Pa.)* 3 (11): 1462–72. https://doi.org/10.1158/1940-6207.CAPR-10-0037.

Rusdiana, Taofik, Yanni D Mardhiani, Norisca A Putriana, Dolih Gozali, Daisuke Nagano, Takuya Araki, and Koujirou Yamamoto. 2021. "The Influence of Javanese Turmeric (*Curcuma xanthorrhiza*) on the Pharmacokinetics of Warfarin in Rats with Single and Multiple-Dose Studies." *Pharmaceutical Biology* 59 (1): 639–46.

Saeed, Khalid, and Ibrahim Khan. 2016. "Preparation and Characterization of Single-Walled Carbon Nanotube/Nylon 6, 6 Nanocomposites." *Instrumentation Science & Technology* 44 (4): 435–44.

Sarup, Prerna, Suman Bala, and Sunil Kamboj. 2015. "Pharmacology and Phytochemistry of Oleo-Gum Resin of *Commiphora wightii* (Guggulu)." *Scientifica* 2015: 138039. https://doi.org/10.1155/2015/138039.

Sharifi-Rad, Mehdi, Dima Mnayer, Maria Flaviana Bezerra Morais-Braga, Joara Nályda Pereira Carneiro, Camila Fonseca Bezerra, Henrique Douglas Melo Coutinho, Bahare Salehi, Miquel Martorell, María del Mar Contreras, and Azam Soltani-Nejad. 2018. "Echinacea Plants as Antioxidant and Antibacterial Agents: From Traditional Medicine to Biotechnological Applications." *Phytotherapy Research* 32 (9): 1653–63.

Shchipunov, Yu.A., Yu.V. Burtseva, T.Yu. Karpenko, N M Shevchenko, and T N Zvyagintseva. 2006. "Highly Efficient Immobilization of Endo-1,3-β-d-Glucanases (Laminarinases) from Marine Mollusks in Novel Hybrid Polysaccharide-Silica Nanocomposites with Regulated Composition." *Journal of Molecular Catalysis B: Enzymatic* 40 (1): 16–23. https://doi.org/10.1016/j.molcatb.2006.02.002.

Shin, Won-Kyung, Jinhyun Cho, Aravindaraj G Kannan, Yoon-Sung Lee, and Dong-Won Kim. 2016. "Cross-Linked Composite Gel Polymer Electrolyte Using Mesoporous Methacrylate-Functionalized SiO2 Nanoparticles for Lithium-Ion Polymer Batteries." *Scientific Reports* 6 (1): 26332. https://doi.org/10.1038/srep26332.

Shindu, C Thomas, Pawan Kumar Mishra, and Sushama Talegaonkar. 2015. "Ceramic Nanoparticles: Fabrication Methods and Applications in Drug Delivery." *Current Pharmaceutical Design* 21 (42): 6165–88.

Shinjyo, Noriko, Guy Waddell, and Julia Green. 2020. "Valerian Root in Treating Sleep Problems and Associated Disorders-A Systematic Review and Meta-Analysis." *Journal of Evidence-Based Integrative Medicine* 25: 2515690X20967323. https://doi.org/10.1177/2515690X20967323.

Shinomol, G K, and Muchukunte M S Bharath. 2011. "Exploring the Role of 'Brahmi' (*Bacopa monnieri* and *Centella asiatica*) in Brain Function and Therapy." *Recent Patents on Endocrine, Metabolic & Immune Drug Discovery* 5 (1): 33–49. https://doi.org/10.2174/187221411794351833.

Shrivastava, Mansi, and L Dwivedi. 2015. "Therapeutic Potential of *Hypericum perforatum*: A Review." *International Journal of Pharmaceutical Sciences and Research* 6 (12): 4982–88.

Siddiqui, M Z. 2011. "*Boswellia serrata*, a Potential Antiinflammatory Agent: An Overview." *Indian Journal of Pharmaceutical Sciences* 73 (3): 255–61. https://doi.org/10.4103/0250-474X.93507.

Sigmund, Wolfgang, Junhan Yuh, Hyun Park, Vasana Maneeratana, Georgios Pyrgiotakis, Amit Daga, Joshua Taylor, and Juan C Nino. 2006. "Processing and Structure Relationships in Electrospinning of Ceramic Fiber Systems." *Journal of the American Ceramic Society* 89 (2): 395–407.

Singh, Radha, and Kusum Singh. 2019. "Garlic: A Spice with Wide Medicinal Actions." *Journal of Pharmacognosy and Phytochemistry* 8 (1): 1349–55.

Singh, Rajinder, Subrata De, and Asma Belkheir. 2013. "*Avena sativa* (Oat), A Potential Neutraceutical and Therapeutic Agent: An Overview." *Critical Reviews in Food Science and Nutrition* 53 (2): 126–44. https://doi.org/10.1080/10408398.2010.526725.

Skrovankova, Sona, Daniela Sumczynski, Jiri Mlcek, Tunde Jurikova, and Jiri Sochor. 2015. "Bioactive Compounds and Antioxidant Activity in Different Types of Berries." *International Journal of Molecular Sciences* 16 (10): 24673–706. https://doi.org/10.3390/ijms161024673.

Solà, Rosa, Eric Bruckert, Rosa-Maria Valls, Silvia Narejos, Xavier Luque, Manuel Castro-Cabezas, Gema Doménech, et al. 2010. "Soluble Fibre (*Plantago ovata* Husk) Reduces Plasma Low-Density Lipoprotein (LDL) Cholesterol, Triglycerides, Insulin, Oxidised LDL and Systolic Blood Pressure in Hypercholesterolaemic Patients: A Randomised Trial." *Atherosclerosis* 211 (2): 630–37. https://doi.org/10.1016/j.atherosclerosis.2010.03.010.

Srivastava, Sharad Kumar, Ajay Kumar Singh Rawat, and Shanta Mehrotra. 2004. "Pharmacognostic Evaluation of the Root of *Berberis asiatica*." *Pharmaceutical Biology* 42 (6): 467–73.

Størmer, F C, R Reistad, and J Alexander. 1993. "Glycyrrhizic Acid in Liquorice: Evaluation of Health Hazard." *Food and Chemical Toxicology* 31 (4): 303–12.

Thiagarajan, Geetha, Talla Venu, and Dorairajan Balasubramanian. 2003. "Approaches to Relieve the Burden of Cataract Blindness through Natural Antioxidants: Use of Ashwagandha (*Withania somnifera*)." *Current Science* 85 (7): 1065–71.

Tiwari, Jitendra N, Rajanish N Tiwari, and Kwang S Kim. 2012. "Zero-Dimensional, One-Dimensional, Two-Dimensional and Three-Dimensional Nanostructured Materials for Advanced Electrochemical Energy Devices." *Progress in Materials Science* 57 (4): 724–803.

Tovar-Ramírez, D, D Mazurais, J F Gatesoupe, P Quazuguel, C L Cahu, and J L Zambonino-Infante. 2010. "Dietary Probiotic Live Yeast Modulates Antioxidant Enzyme Activities and Gene Expression of Sea Bass (*Dicentrarchus labrax*) Larvae." *Aquaculture* 300 (1): 142–47. https://doi.org/10.1016/j.aquaculture.2009.12.015.

Upadhyay, Ravi Kant. 2016. "Nutraceutical, Pharmaceutical and Therapeutic Uses of *Allium cepa*: A Review." *International Journal of Green Pharmacy (IJGP)* 10 (1).

Varzakas, Theodoros, George Zakynthinos, and Francis Verpoort. 2016. "Plant Food Residues as a Source of Nutraceuticals and Functional Foods." *Foods* 5 (4): 88.

Weiss, Jochen, Paul Takhistov, and D Julian McClements. 2006. "Functional Materials in Food Nanotechnology." *Journal of Food Science* 71 (9): R107–16. https://doi.org/10.1111/j.1750-3841.2006.00195.x.

White, C N, and C J Rivin. 2000. "Gibberellins and Seed Development in Maize. II. Gibberellin Synthesis Inhibition Enhances Abscisic Acid Signaling in Cultured Embryos." *Plant Physiology* 122 (4): 1089–97. https://doi.org/10.1104/pp.122.4.1089.

Wilt, T, A Ishani, and R Mac Donald. 2002. "*Serenoa repens* for Benign Prostatic Hyperplasia." *Cochrane Database of Systematic Reviews* 3: CD001423. https://doi.org/10.1002/14651858.CD001423.

Wojtyniak, Katarzyna, Marcin Szymański, and Irena Matławska. 2013. "*Leonurus cardiaca* L. (Motherwort): A Review of Its Phytochemistry and Pharmacology." *Phytotherapy Research : PTR* 27 (8): 1115–20. https://doi.org/10.1002/ptr.4850.

Yang, Chunhua, and Didier Merlin. 2019. "Nanoparticle-Mediated Drug Delivery Systems for the Treatment of IBD: Current Perspectives." *International Journal of Nanomedicine* 14 (November): 8875–89. https://doi.org/10.2147/IJN.S210315.

Younus, Ishrat, Afshan Siddiq, Humera Ishaq, Laila Anwer, Sehrish Badar, and Muhammad Ashraf. 2016. "Evaluation of Antiviral Activity of Plant Extracts against Foot and Mouth Disease Virus In Vitro." *Pakistan Journal of. Pharmaceutical. Sciences* 29 (4): 1263–68.

Yue G, L-S Wong, H-W Leung, S Gao, J Tsang, Z Lin, et al. 2010. Is Danggui Safe to be Taken by Breast Cancer Patients? A Skepticism Finally Answered by Comprehensive Preclinical Evidence. *Frontiers in Pharmacology* 10. https://doi.org/10.3389/fphar.2019.00 article 706.

Zakaria, Z A, M S Rofiee, L K Teh, M Z Salleh, M R Sulaiman, and M N Somchit. 2011. "*Bauhinia purpurea* Leaves' Extracts Exhibited In Vitro Antiproliferative and Antioxidant Activities." *African Journal of Biotechnology* 10 (1): 65–74.

15 Nano-Nutraceuticals for the Treatment of Cancer

*Saba Yousaf, Fouzia Qamar, Zirwah Tahir, Zoya Faisal, Muniba Khaliq, Aiza Talat, Umar Bacha, Aaiza Naveed, Hitesh Chopra, Uday Younis Hussein Abdullah, Shakeel Ahmed, and Atif Amin Baig**

CONTENTS

15.1 Introduction 399
15.2 Entering the Era of Nanotechnology: The Better of Two Worlds—"Nutraceuticals and Nanotechnology" 401
15.3 What Precisely Does Nanotechnology Offer: Will It Prove a Stitch in Time? 401
15.4 Nanotechnology, Nano-Nutraceuticals: Empowering Health Benefits in Biomedical Sciences 402
15.5 Types of Nano-Nutraceutical Encapsulation 403
15.5.1 Nanoliposomes 403
15.5.2 Archaeosomes 403
15.5.3 Nanocochleates 403
15.5.4 Polymer-Based Nanoencapsulates 404
15.5.5 Micelles 404
15.5.6 Nanospheres 404
15.5.7 Nanoscale Shell-(Nanocapsules) and (Polymersomes) 404
15.5.8 Nanoemulsions 404
15.6 Nano-Nutraceuticals: Potential Candidates for Cancer Management 404
15.6.1 Limitations of Conventional Chemotherapy and How to Overcome Them by Nanotechnology 404
15.6.2 Nanotechnology in Cancer Targeting; Complementing Nano-Nutraceuticals with Current Strategies for Cancer Treatment 405
15.6.3 Advancement of Nanoproducts (Nanomedicine): Challenges and Promises 411
15.7 Conclusion 411
References 411

15.1 INTRODUCTION

Nutraceuticals are an emerging subject in the realm of the life sciences. The term was created in 1989 by Stephen L. DeFelice, founder of the Foundation of Innovation Medicine (Kalra Ekta 2003). It is a combination of two words: nutrition + medicines. According to Hippocrates (a Greek physician who was the founder of medicine), "let food be your medicine"; prevention is the prime focus behind the concept of nutraceuticals. One of the most appealing research areas is their application in human nutrition, which has far-reaching consequences for users, professionals of health

* Corresponding author: atifamin@unisza.edu.my

DOI: 10.1201/9781003244721-15

care departments, the people who make regulations (regulators), food production personnel, and wholesalers.

Thus, from food sources, nutraceuticals (products serving as medicine as well as nutrition) can provide extra medical health advantages and essential food nutrition. The definition of nutraceuticals and associated goods varies according to the source. These items can be categorized according to their natural origins, pharmacological conditions, and chemical composition. They are classified into four categories: dietary supplements, functional foods, medical foods, and pharmaceuticals. A product made up mainly of nutrients from food and often composited in medicinal forms like liquid, capsule, powder, or pill formulations is called a nutritional supplement. Although the Food and Drug Administration (FDA) regulates dietary supplements as food, their regulation differs from that of medicines and food items.

We know that functional foods comprise whole and fortified foods with enriched or enhanced dietary components that may lower the risk of chronic diseases, imparting a health benefit apart from the specific nutrients they contain. Therefore, the concept underlying this field of the life sciences is to allow food to act as a medication in addition to its nutritional benefits (Das et al. 2012). Nutraceuticals can be recognized as health promoters, retarding the ageing process, averting chronic illnesses, extending life span, or maintaining the body's structural or physiological well-being due to their potential nutritional, protective, and therapeutic benefits. Recent research has revealed that these chemicals have potential outcomes in various problems. Much effort has been directed towards the potential of nutraceuticals based on their disease-modifying manifestations (Nasri et al. 2014).

The need of the hour is to highlight the plant-based nutraceuticals that help treat challenging illnesses caused by oxidative stress, such as allergies, Alzheimer's disease/dementia, cardiovascular disease, malignancies, conditions related to diabetes (hyperglycemia), eye problems, immunological diseases, inflammation, and Parkinson's disease, as well as diseases associated with obesity. Recently published articles on various aspects of nutraceuticals as an alternative to pharmaceuticals were found by searching databases such as the scientific database of Medline, PubMed, and Google Scholar. The keywords for searching out the therapeutic importance of nutraceuticals were nutraceutical and allergy, Alzheimer's, cardiovascular, cancer, and diabetes (Nasri et al. 2014).

Nutraceuticals are classified as dietary supplements, functional foods, medical foods, and medicines based on their chemical makeup, natural sources, and pharmacological conditions. A dietary supplement is a product derived from high–nutritional value foods. A fascinating aspect regarding dietary supplements is that their regulation differs from that of pharmaceuticals and other foods, although they are classified as foods by the FDA. The functional foods category comprises multigrain, whole, fortified, enriched, or improved dietary components and foods that reduce the risk of chronic illnesses (Al-Obaidi et al. 2021; Oketch-Rabah et al. 2020). On the other hand, a medical food is a nutraceutical form taken under supervision and advice. Nutritional needs to counteract certain disorders are determined and given through medical examination. Last are medicines, essential medicinal components derived from genetically engineered crops or animals (Al-Obaidi et al. 2021).

Advocates for this type of nutraceutical believe it is considerably more cost-effective than traditional techniques and provides a more significant source of income for agricultural growers. Nutraceuticals promote the concept that prevention is preferable to treatment. This idea has gained a lot of attention because of its nutritional, safety, and medicinal benefits. Nutraceuticals, at first glance, can be utilized to keep healthy, manage symptoms, and prevent malignant processes. However, when it comes to the potential health advantages of nutraceuticals, the sky is the limit. Nutraceuticals are a one-stop shop for biological and cellular phenomena, like defense by antioxidants, cell proliferation, gene expression, and mitochondrial integrity, which are important to cell survival. A daily dosage of nutraceuticals can prevent Alzheimer's, cancer, eye problems, kidney illness, and many other life-threatening ailments (Al-Obaidi et al. 2021; Oketch-Rabah et al. 2020) due to their potential nutritional and medicinal impact coupled with safety.

15.2 ENTERING THE ERA OF NANOTECHNOLOGY: THE BETTER OF TWO WORLDS—"NUTRACEUTICALS AND NANOTECHNOLOGY"

New possibilities have arrived in the fields of science and technology, industry, economics, environment, and human lives by engagement with the initiative of taking nanotechnology on board with the beautiful aim to increase the quality of life, create job opportunities, enhance economic advancement, and last but not least, strengthen society's sustainability as well as wellbeing (Moraru et al. 2003; Baeumner 2004). This opened up a whole new universe for scientists and engineers, encouraging their curiosity and innovation. This ensured new scientific breakthroughs in nanoscale nutraceutical science and technology. Nanoscale phenomena have been used to formulate, produce, and process nutraceutical and functional foods. Nanotechnology-based concepts increased product efficacy and targeted transporting or delivery capability (DOA 2003; Shahidi 2006).

The grafting of selected biological and chemical ligands onto the surfaces of nanoparticles is done by nanotechnology. Because of the surface changes, nanoencapsulated particles can detect target cells. Nanoparticles exhibiting a built-in site-specific controlled-release mechanism may deliver functional chemicals to the target location, increasing efficacy and efficiency. This capacity may enhance the rate of delivery of active constituents. Specific molecular recognition, targeting a particular site, and attaching commands may eliminate unwanted components from human and animal systems (Patra et al. 2018).

The nanoscale in food networks will aid in developing inventive biomolecule-modulated protein and bioactive-compound shuttle networks (Chen, Weiss, and Shahidi 2006). Hence, nanoscale technology development may prove to be a doorway opened to multi-role nanoscale nutraceutical transporting systems that may identify the right location or site, evaluate contemporary and global demands, or certainly or positively determine to release the payload and check the reaction for feedback. This idea is analogous to "smart medication delivery" or "intelligent therapies," which are being investigated in the field of medical nanotechnology. Smart nanostructured materials can deliver drugs to the target sites in a controlled manner to minimize the side effects caused by therapeutic processes (Lombardo, Kiselev, and Caccamo 2019).

15.3 WHAT PRECISELY DOES NANOTECHNOLOGY OFFER: WILL IT PROVE A STITCH IN TIME?

At the molecular scale, nanotechnology refers to designing functioning systems that have dimensions from 1 to 100 nm (Kharat, Murthy, and Kamble 2017). One nanometer is equal to one-millionth of a millimeter. It is used to measure atoms and molecules. A single hair has a width of 1 nm. Materials show unique physical, chemical, and biological characteristics at the microscopic scale. These unique characteristics of nanoscale materials are expected to help solve several issues in the fields of physics, health, computers, electronics, energy, and aerospace. The technology is still in the early stages of development, but it is being utilized to create single-task products. Multitasking systems, multi-component systems, and integrated multi-level systems are some of the technology's functions as it develops (Roco 2011). The advent of nanotechnology is now referred to as the fifth technological revolution.

Nanotechnology definitions differ comprehensively; therefore, creating a uniform portrait/definition is critical. The United States National Nanotechnology Initiative defines "nanotechnology" as "the science, engineering, and technology linked to the understanding and control of matter at 1–100 nm." According to this, nanoscale technology is the research and development of materials, tools, and systems with the latest properties and novel functionalities that may be attributed to their nanoscale dimension and the capacity to quantify, control and control circumstances at a scale to change (Balogh 2010).

Nanobiotechnology is considered a unique combination of biotechnology and nanotechnology, through which the molecular biological approach is merged with the micro-technological approach

(Fakruddin, Hossain, and Afroz 2012). Nanotechnology projects use nanotechnological devices and ideas to investigate biological networks; to manipulate the architecture of biomolecules to perform roles distinctively varied from those they perform in nature; or to control biological systems in ways more precise or accurate than is possible by using the conventional molecular biological, synthetic chemical, or biochemical approaches that have been prevalent in research in the biological community for many years.

15.4 NANOTECHNOLOGY, NANO-NUTRACEUTICALS: EMPOWERING HEALTH BENEFITS IN BIOMEDICAL SCIENCES

Nanoscience and nanotechnology have a wide range of applications in science, including electronics, communication, energy generation, medical, and the food industry. In the medical sciences, drug and gene delivery, fluorescent biological labeling, detection of infections, tissue engineering, tumor detection, DNA structure analysis, magnetic resonance imaging (MRI) contrast enhancement, and phagokinetic investigations are all examples of scientific uses of nanotechnology (Wu and Li 2013; Doroudian et al. 2021; Mishra 2016).

The long-term research goals in nanomedicine include understanding nanomachines' structure and quantitative makeup at the molecular scale. Better knowledge of biological processes and the creation of more sophisticated technologies are possible if nanomachinery is precisely controlled and manipulated in cells. This discovery will pave the way for nanoscale imaging techniques that investigate chemical processes in live cells (Logothetidis 2012). Molecular imaging has developed as a valuable diagnostic technique to view molecular processes leading to the development of an underlying illness, particularly before the symptoms appear. Molecular imaging and nanotechnology provide a wide-ranging platform for developing new nanoprobes, which will aid in the detection and study of different biomarkers linked to human illness (Jones et al. 2008).

To get an understanding of the imaging network's sensitivity, resolution, and capacity to deliver information on biological systems at the molecular and cellular level, nanoparticle probes are used. Magnetic nanoparticles may be used as contrast enhancement probes for MRI. When fluorescent tags, radionuclides, and other biomolecules are added, the result is an imaging platform with the capacity to detect multiple biomarkers, transport DNA into cells, and carry out cellular trafficking. A hybrid MRI equipped with magnetic nanoparticle (MN) hybrid probes and adenovirus-based viral delivery probes may select cells and carry out delivery of genes with surveillance and expression of proteins in the form of green fluorescent protein (GFP) (Oh, Biswick, and Choy 2009). Positron emission tomography (PET) uses these nanoparticles in lower concentration. Positron emission tomography (PET) uses these nanoparticles in lower concentration.

Hybrid imaging using CT (computed tomography) with high PET sensitivity can map signals to atherosclerotic vascular regions (Nahrendorf et al. 2008). Driving the nanoparticles carrying the contrast agent to the target site helps increase the contrast agent's accumulation at the target location. This necessitates the employment of targeting groups, because access to target molecules concealed behind tissue barriers is required. A way to increase imaging sensitivity for modalities with poor sensitivity is by using nanoparticles that contain several contrast agents. While on the one hand, the contrast medium and the medication may both be given to the patient, on the other, it is possible to monitor the biodistribution and therapeutic action at the same time using nanoparticles (theranostics) (Debbage and Jaschke 2008). There is multiple pore dimension distributions, increased porosities, and huge surface-area-to-volume ratios for nanofiber-based scaffolds. Providing many properties that help cell attachment, development, and division while simultaneously proposing a framework for the future horizons of electrospun nanofibrous frames for the purpose of tissue engineering, this approach offers a wide range of benefits.

Nanotechnology has also proved its enormous potential in agricultural systems. Although the therapeutic effect of nutraceuticals against many illnesses has a long history, phytochemicals with potential benefits, such as carotenoids, curcumin, and resveratrol, have captured the attention of researchers due to the innate or intrinsic anti–free radical and antioxidant properties and immunity-promoting effects to treat a variety of health issues, such as cancer, heart issues, and neurodegenerative disorders like Alzheimer's disease (AlAli et al. 2021).

Nevertheless, the limited bioavailability and poor solubility associated with phytochemicals lead to insufficient absorption and as a result, decreased biological activity, which is a crucial concern for researchers. A portion of the nutraceutical dosage is absorbed in the conventional manner of nutraceutical administration and reaches the actual pharmacological site of action, while the remainder is excreted or causes nonspecific toxicity and severe side effects due to poor biodistribution. Nanotechnology concepts have been used to effectively deliver nutraceuticals to address these issues. The concept of the nano–delivery network has evolved to study different forms of administration to understand the linked issues along with their bioavailability/absorption. Nanotechnology has made great progress in nano-nutraceuticals (Patra et al. 2018). Nanotechnology has established a new frontier by imposing changed characteristics on nanoparticles and applying them to the production of nanoformulations, nutritional supplements, and the food business.

Based on nanoscience and technology, effective solubilization, encapsulation, and distribution as biocompatible systems are expected to confer excellent absorption at lower dosages, reduced frequency of administration, and improved therapeutic index. The nanoformulations also provide targeted distribution of encapsulated nutraceuticals and gradual and sustained release from the nanoformulation. Different *in vitro* and *in vivo* investigations and the development of various nanoformulations have demonstrated the remarkable improvement in the solubility and bioavailability of curcumin, a potential nutraceutical with numerous important characteristics (Patra et al. 2018).

15.5 TYPES OF NANO-NUTRACEUTICAL ENCAPSULATION

They are composed of a core consisting of more active bio-compounds surrounded by a wall acting as a barrier (García, Forbe, and Gonzalez 2010). They encapsulate a bioactive compound's incorporation, absorption, or dispersion within the nanosized vesicles. Hence, the bioactive compound is protected against degradation until it reaches the desired site of action. In this way, the nutraceutical leads to increased bioavailability and delivery to the cells and tissues (Letchford and Burt 2007). It offers opportunities for prolonged gastrointestinal retention time caused by bio-adhesive improvements in the mucus. Modulations of surface properties such as coating and the presence of biomolecular tags thereby ensure targeted delivery of compounds (Bouwmeester et al. 2009). Important lipid-based nanoencapsulates are used in food supplements; recent developments include nanoliposomes, archaeosomes, and nanocochleates. Several different types of nanoencapsulation are available (Mozafari et al. 2006). Some of them are listed in the following.

15.5.1 Nanoliposomes

These are vesicles possessing a lipid bilayer (less than 30 or 30–100 nm) in the nanometric size range. Due to their unique properties, they may encapsulate, transport, and dispatch water and lipid-soluble substances. They are pH and temperature sensitive (Mozafari et al. 2006).

15.5.2 Archaeosomes

These are liposomes made from Archaeobacteria. They are comparatively more thermostable and stress-resistant than normal liposomes. Therefore, they are ideal candidates to protect antioxidants (Mozafari et al. 2006).

15.5.3 Nanocochleates

These are defined by a solid lipid bilayer covering folded up in a spiral way, lacking any internal watery space (Reza Mozafari et al. 2008), that aims to deliver its contents to the target sites through the fusion of the cochleate.

15.5.4 Polymer-Based Nanoencapsulates

Nanoencapsulates are characterized by the polymerization of more than a single monomer that involves the assemblage of a hydrophilic and a hydrophobic unit. The resultant molecule has regions having different affinities for an aqueous solvent. Proteins like albumin, gelatin, collagen, and the milk protein α-lactalbumin, alongside the saccharides, e.g., alginate, chitosan, etc., are natural polymers.

15.5.5 Micelles

These are mainly comprised of the core-shell, in which the inner core is built from amphiphilic molecules' hydrophobic regions, constructing a cargo space that loads the lipophilic bioactive substance.

15.5.6 Nanospheres

These are solid colloidal particles entrapping and encapsulating the bioactive compounds that are dissolved and chemically bonded or absorbed into the polymer matrix. The co-polymeric composition may impart a solid appearance to the central core, making the structure indistinguishable from micellar or nanosphere particles (Chen, Remondetto, and Subirade 2006).

15.5.7 Nanoscale Shell-(Nanocapsules) and (Polymersomes)

The vesicular networks with the bioactive component are confined within a space surrounded by a coating (Letchford and Burt 2007).

15.5.8 Nanoemulsions

In the nutraceutical and food industries, the formulations based on nanoemulsions deliver lipophilic functional components such as vitamins, omega-3 fatty acids (DHA), oil-soluble flavors, preservatives, conjugated linoleic acid, etc. Nanoemulsions come with several potential advantages, such as clarity, stability, and enhanced bioavailability of nutrients. They are also referred to as "smart foods," which contain nanotech sensors that can detect pathogens, toxins, and other chemicals and help detect food degradation (Chandel et al. 2018).

15.6 NANO-NUTRACEUTICALS: POTENTIAL CANDIDATES FOR CANCER MANAGEMENT

15.6.1 Limitations of Conventional Chemotherapy and How to Overcome Them by Nanotechnology

Surgery, radiotherapy, chemotherapy, gene therapy, and immunotherapy are traditional approaches to treat cancer (Sutradhar and Amin 2014). However, side effects, recurrence after initial treatment, and resistance are the common issues that arise while using them (Frank et al. 2009). Abnormal blood vessel function and architecture, lack of lymphatics, interstitial hypertension, and dilated angiogenesis are the disease-relevant attributes of solid tumors that prevent chemotherapeutic drugs from penetrating the core of the tumor (Tannock et al. 2002). Multidrug resistance (MDR) is the second major challenge after penetration into the tumors (El-Readi and Althubiti 2019). MDR happens when cells of tumors develop protective resistance against structurally and functionally distinct classes of chemotherapeutic candidates (Moitra 2015). The potential benefits of nanoparticle formulations over standard chemotherapy dosage are well documented (Wang et al. 2017). However, particle size may be considered amongst the most important features of nanoparticles (Gindy and Prud'homme 2009).

Additionally, chemotherapy fails to target the tumor cells selectively, and its nonspecific action results in immune system damage, treatment-associated toxicities, and reduced treatment efficacy (Coates et al. 1983). On the contrary, nanoparticles recognize the cancerous cells and provide an accurate and selective drug delivery option, thus fending off any possible encounter with healthy cells.

Traditional chemotherapeutic agents are often engulfed by macrophages (Sutradhar and Amin 2014). Thus, chemotherapy is completely ineffective. The solubility of the intended medications/drugs is also an essential issue in a traditional chemotherapeutic process and may render them unfit to penetrate living membranes (Mousa and Bharali 2011). Nanoparticles can offer enhanced solubility. Drugs that repel water can be loaded in micelles to increase their solubility (Patri, Kukowska-Latallo, and Baker 2005). Dendrimers contain many binding sites for water-repelling and attracting molecules. Liposomes encapsulate hydrophobic drugs (Babu et al. 2014).

Furthermore, P-glycoprotein, a multi-resistance protein, is overexpressed on the surface of tumor cells, preventing the accumulation of a drug inside the tumor, thus acting as the efflux/outsider pump, leading to the conditions that promote the creation of resistance against anticancer medicines/drugs. Therefore, taking anticancer drugs cannot show the desired results (Davis, Chen, and Shin 2009). To address this issue, Verapamil or cyclosporine is administered simultaneously with the cytotoxic drug to limit the effect of P-glycoprotein.

Liposomes, dendrimers, nanoemulsions, polymeric micelles, nanocapsules, and nanotubes are the most extensively researched and evaluated drug delivery mechanisms for cancer therapy. They are administered orally or by topical and parenteral routes (Dong and Mumper 2010). DOXIL (liposomal doxorubicin) used for ovarian cancer and Abraxane (albumin-bound paclitaxel) are among the nanotechnology-generated formulations that have already been launched in the market (Nagahara et al. 2010).

15.6.2 Nanotechnology in Cancer Targeting: Complementing Nano-Nutraceuticals with Current Strategies for Cancer Treatment

We know the role of food in wellbeing, maintenance, and the prevention of diseases, especially in autoimmune disorders and degenerative and neoplastic health issues (Calvani, Pasha, and Favre 2020). The complementary role of nutraceuticals in enhancing the therapeutic effects of medication and health supplements is now widely acknowledged. Nutraceuticals are capturing the global market, estimated at US$165.62 billion in 2014 and expected to grow to US$278.96 billion by the end of 2021 (Shah et al. 2018). However, the factors restricting the proper usage of nutraceuticals include their inefficacy *in vivo*.

In recent years, functional or bioactive foods have shown multiple benefits as anticancer, antidiabetic, antimicrobial, gastroprotective, and immune modulators. Cancer still remains a challenging ailment to treat despite the significant breakthroughs in biomedical sciences and nanotechnology. According to the GLOBOCAN 2020 statistics (Nirmala et al. 2022) regarding the incidence and the mortality of a non-communicable multifactorial chronic disease, i.e., cancer, 19.3 million new cases of cancer and about 10 million deaths were reported in 2020 among men and women (Sung et al. 2021) with no discrimination between developed and developing parts of the globe. It is ranked as the leading cause of lowering life expectancy in every world region. The more terrifying fact that calls for immediate action is the predicted projection of cases to 28.4 million by 2040. Among prevailing therapies, surgery, chemotherapy, and radiation are the most suggested and available modalities effective for different cancer types (Zheng et al. 2016). A primary concern is to evaluate more possibilities for effective prevention and control of this life-threatening health issue. In addition to other contributing factors that contribute to cancer incidence, diet is considered a critical factor that directly impacts the patient's health and disease status. Nutraceuticals follow several cancer prevention and control mechanisms, e.g., through cell proliferation and differentiation inhibition, reduction in toxicity of chemotherapeutic drugs, arrest of cancer cell growth by inhibition of migration, and invasion of stem cells. An overview regarding the inhibition of metabolic activity of various cancer types induced by nutraceuticals is shown in Table 15.1.

Considering the gravity of the health issues due to cancer, more emphasis is needed to explore and understand the definitive role of these functional food components in treating and preventing cancer. The chemopreventive status of dietary or herbal chemicals was considered a turning point for purposing nutraceuticals as an excellent alternative strategy in cancer control (Salami, Seydi, and Pourahmad 2013). However, the factors restricting the proper usage of nutraceuticals include their inefficacy *in vivo*. Poor absorption and low solubility are the main problems that need to be overcome to benefit from these bioactive compounds. Low solubility and poor absorption translate into low stability, impeded oral bioavailability, and poor hepatic degradation (Greiner 2014). Nanotechnology has paved its way to fabricate or formulate novelty-based food products characterized by pronounced physiological, biochemical, and physical performances to address the issues of low aqueous solubility and bioavailability. In nanoencapsulation, the carrier system is comprised of cores that may be solid, liquid, or a mixture of solid and liquid components. Therefore, the role of nutraceuticals can be compounded manifold by using nanotechnology to produce engineered nanoscale formulations with an effective and efficient delivery mechanism that can prove to be a key to overcoming the limitations experienced so far in acquiring the full benefit of the nutraceuticals (Dutta, Moses, and Anandharamakrishnan 2018).

Nanotechnology encompasses a broad spectrum of manufacturing techniques, enlisting a wide variety of materials to design a product, especially its engagement in medicinal product preparation, which has captured the attention of scientists. In recent years, the nanoscale system has paved its way successfully in oncology (Salama, Hasanin, and Hesemann 2020). The capacity of chemotherapeutic agents to harm the nontarget healthy cells has been a matter of debate, because low-dose administration to avoid harm has caused cancer cells to acquire resistance, making the therapeutic approach more challenging. The delivery of drugs to specifically targeted sites through engineered natural and pharmaceutical products and the revolutionary achievement of nanotechnology-based imaging contrast agents for targeted tumor diagnosis has been considered a successful leap in the field of cancer prevention, diagnosis, and therapy. Having realized the potential of nanomedicine for its anticancer attributes, a rigorous and intensive approach is required to uncover the novel developments made in the fields of nanopharmaceuticals and nano-nutraceuticals.

In an attempt to screen, diagnose, treat, and prevent the conditions that cause cancer and related ailments, advanced nanomedicinal products like bio-containers for cell treatment and diagnostics are extensively studied. Some of the research ventures undertaken to highlight the potential role of nanomedicine in oncology are listed in Table 15.2.

The role that nutraceuticals, especially micronutrients, can play is debatable, as the therapeutic outcomes are mostly exaggerated. Therefore, discovering the true potential of nutraceuticals coupled with nanotechnology requires rigorous experimentation to control and treat malignancies. However, confronting cancer at the molecular level requires the personalized medicine approach to achieve formulations of nanotechnology-based nutraceuticals with active bio-compounds.

Nanoparticles can actively or passively target cancer cells. Considering the active attacking attribute, nanoparticles directly collaborate with neoplastic entities and cells either through ligand–receptor binding or through synergistic antibody–antigen interactions (Guo and Szoka 2003). The surface can be altered, and antibodies, folic acid, or peptides can be attached as targeting ligands to recognize the overexpressed receptor on tumor cells. The reticuloendothelial system (RES), also called "mononuclear phagocytes," is the prime immune system that rapidly removes nanoparticles from circulation (Li and Huang 2009). Therefore, it is considered a significant hurdle to circulating nanoparticles. Hydrophilic and flexible chains of polymers can cover the nanoparticles, which can hence, escape phagocytosis by the RES (Carrstensen, Müller, and Müller 1992). The nano–drug delivery networks exhibit the "enhanced permeation and retention" effect by passive targeting

TABLE 15.1
Metabolic Activity Inhibition of Various Cancers by Nutraceuticals

Nutraceuticals	Effects	Types of cancer	Factors/components	References
Factors of bovine milk fat globule membrane (MFGM)	Breast cancer cell growth inhibitors; agents (phospholipids) as colon cancer–fighting agents, signal transduction, anticancer factors; antioxidants; apoptosis in human adenocarcinoma cell lines	Breast cancer, colon carcinogenesis, ovarian cancer	BRCA 1; BRCA 2; fatty acid-binding protein (FABP); β-glucuronidase inhibitor; membrane-associated protein tyrosine kinase (cellular proto-oncogene c-src); FABP as selenium carrier; Vitamin E and carotenoids; sphingomyelin, sphingosine, and ceramide	(Spitsberg and Gorewit 1997; Fil et al. 2019; Bansal and Medina 1993; Whanger 2004; Lindmark-Månsson and Åkesson 2000; Parodi 2001)
Phytochemicals: Ginseng, spirulina, *Gingko biloba*, amino acids, *Aegle marmelos*, berberine, glucosamine, chondroitin, curcumin, citrus, gingerol, glycyrrhizin, lycopene, isoflavones, vanillin, apigenin, gingerols, resveratrol, etc.	Cell proliferation inhibitors via lowering angiogenic and metastatic rates; suppression of tumor cell lines through the process of apoptosis and G1 phase cell cycle arrest; reduction of necrosis; inhibition of Phase I and induction of phase II detoxification enzymes in the liver; scavengers of reactive or toxic chemicals or free radicals; antineoplastic cytotoxicity; antioxidation upregulation; reduce MMP2,9; cell proliferation; peroxisome proliferation; HMG CoA-LDH inhibitor; arrest carcinogenesis and metastasis through the effect on MAPK and NF-κB pathway; radical scavengers; pro-apoptotic properties, increase the cytotoxic tendency of antineoplastic agents	Breast cancer; osteosarcoma; arthritis cancer; ovarian tumors; colon cancer or colorectal adenocarcinoma; gastric cancer; pancreatic cancer; intestinal cancer; prostate cancer	Terpenoids (tocotrienols and tocopherol; carotenoids; limonoids; phytosterols); phenolic metabolites (phenolic acids, polyphenols; flavonoids; catechins and gallic acids; isoflavonoids; anthocyanidins); alkaloids and other nitrogen-containing plant constituents (glucosinolates, indoles, and fibers)	(Dillard and German 2000; Sharma 2009; Gupta et al. 2010)
Vitamins, enzymes, and minerals	cdk2, PKC, G1/S DNA breaks, delayed apoptosis, inhibit cell proliferation and angiogenesis, catalase inhibition, replication of DNA, supply of methyl groups, repairing of cells, high level of oxidative phosphorylation, pyruvate dehydrogenase activation, hormonally activate form 1,25-dihydroxycholecalciferol (Vit D) cancer suppression activity, antiproliferative activity of 5-methyl-tetrahydrofolate, arrest the cell cycle in S phase by reducing the cell number during G2/M phase, Na/K channels, Zn-activated MMP that inhibits ROS effect	All cancer types	Selenium, calcium, copper, zinc, manganese, copper, and potassium; vitamins A, B, C, D, E, K, folic acid; lipoic acid, co-enzyme Q-10, superoxide dismutase	(Calvani, Pasha, and Favre 2020; Iorio et al. 2019; Rabajdová et al. 2021; Akoglu et al. 2001; Mokbel and Mokbel 2019)

(Continued)

TABLE 15.1 (CONTINUED)
Metabolic Activity Inhibition of Various Cancers by Nutraceuticals

Nutraceuticals	Effects	Types of cancer	Factors/components	References
Oligosaccharides/ Polysaccharides	Generation of SCFAs that inhibit histone deacetylases and induce apoptosis of tumor cells; COS (chitosan oligosaccharides) inhibit the viability and proliferation of lung carcinoma cells and necrosis of tumor cells; XOS (xylooligosaccharides) trigger SCFAs that downregulate IFN-γ, interleukin 1β [IL-1β]; acidic oligosaccharides create defense barrier attributed to inulin mobility with anticancer property; antiangiogenic and anticancer activity increased by oversulfation of carrageenan oligosaccharides; fucoidan shows antitumor activity through NK cell activity enhancement, laminin, fibronectin conjugation	Colon cancer, lung cancer, hepatocellular carcinoma	COS, XOS, acidic oligosaccharides, carrageenan oligosaccharides, Ganoderma polysaccharides, Astragalus, lentinan, grifolan, and krestin (PSK), extracts from *Phellinus linteus*, *Lycium barbarum*, *Trametes versicolor*, *Poria cocos*, and *Atractylodes macrocephala*Mushroom, sporoderm-broken germinating spores (SBGS) of Reishi, *Pleurotus ostreatus* polysaccharides, and polysaccharide from sclerotia of *Pleurotus tuber-regium*	(Javier Fernández, Sául Redondo-Blanco, Elisa M. Miguélez, Claudio J. Villar, Alfonso Clemente 2015; Jiang et al. 2019; Fatima, Akhtar, and Sheikh 2017)
Polyunsaturated fatty acids (PUFA)	Cytostatic and antitumor activity, promote antioxidant defense systems, cell cycle disruption and suppression, stress induction triggered in reticuloendoplasmic set-up of cancer cells, reduction in MMPs expression, G1/S phase progression, S phase malignant cell arrest, cancer cell apoptosis, enhance the duration of G2/M cell cycle and hence impart cell proliferation inhibition, antiangiogenesis, accumulation of antitumor factors or suppressors (p21, p27, p53, pRb), cyclin E reduction, enhancement of neoplasm apoptosis and cell growth reduction, carcinogen scavengers, antioxidant properties	Gastric carcinoma, hepatocellular carcinoma, breast cancer, colorectal cancer	ω-3 fatty acids (Docosahexaenoic acid [DHA], eicosapentaenoic acid [EPA]); ω-6 fatty acids (linoleic acid)	(Asefy et al. 2021; Simopoulos 1991; Covington 2004; Jing, Wu, and Lim 2013; Lee Yee-Ki et al. 2010; Zurier 1991; Samson et al. 2020; Wei, Wang, and Pagliassotti 2007; Conklin 2002)
Dietary fibers (soluble and insoluble fibers), fruit juices	Antioxidant, cell proliferation inhibition, antineoplastic property, inhibit aromatase or estrogen biosynthesis	Colon cancer, breast carcinoma, ovarian carcinoma, endometrial malignancy, gastrointestinal tumor (GIST)	Broccoli, cabbage, cauliflower, blueberries, strawberries, orange, oatmeal, legumes, yellow and green leafy vegetables, dried beans, chicory as water-soluble fibers, etc.	(Pal, Banerjee, and Ghosh 2012)

TABLE 15.2
Research Ventures on Novel Applications of Nanotechnology in the Field of Oncology

Research ventures/ approaches/ nanoplatforms	Purpose	Forms	General advantages	Biological activity	Mechanism	References
Targeted drug delivery vehicles	Transportation of drug to the targeted site and monitoring of the outcomes	Nanoemulsions, nano/ microsphere capsulated formulations, liposomes, phytosomal formulations, transferases, vesicular formulations, etc.	Promote bioavailability and solubility; even macrophage distribution in the tissues, refuge from physical and chemical degradation; stability enhancement, etc.	Nutraceutical systemic antioxidant, anticancer, chemo-protective, anti-inflammatory, hepatoprotective, nutraceutical immunomodulation, analgesic	pH-sensitive liposomes release the contents inside the tumor or infected areas; interact with DNA; anticancer vaccine delivery to dendritic cells	(Ajazuddin and Saraf 2010)
Nanoshells, nanotubes, nanobubbles	Shell-to-core ratio (silica core to the metallic shell) is modifiable and adjustable for diagnostic, therapeutic, and immunological purposes in nanoshells; nanobubbles promote internalization of antitumor drug	Hexagonal network and geometrical characteristics render the nanotubes as optimal chemotherapy drug-carrying vehicles; nanobubbles generate structures similar to bubbles against hydrophobic formations in the liquids; nanoshells have silica core coated with metallic outer shell; nanotubes have tube-like structures	Nanobubbles loaded with antitumor drugs or medications can be easily tracked and visualized through ultrasound; they cause internalization of drugs to tumor cells	Antitumor; an *in vivo* diagnostic testing provision like tracking down the electrolyte and blood sugar levels; nanobubbles accumulate inside tumor interstitium through the capacity of extravasation through the defective tumor microvasculature; nanoshells act as contrast agents to image the cancer markers. They assist the targeting of the cancer cells through exposure to IgG PEGylated antibodies; nanotubes enhance the anticancer function of antitumor drugs	Nanotubes deliver the DNA into the cell for thermal ablation therapy. Near-infrared (NIR) light imbibed by nanoshells can produce extreme heat that is lethal for living cells. Energy can pass across the healthy cell/tissue, leaving the neighboring cells intact while showing a killing effect only on the tumor cells. Nanobubbles provide a substantial variation for ultrasound; once the image is taken, the drug is delivered from the nanobubble	(Dong and Ma 2019; Başpınar et al. 2019; Gao et al. 2008; Riley et al. 2018)

(*Continued*)

TABLE 15.2 (CONTINUED)
Research Ventures on Novel Applications of Nanotechnology in the Field of Oncology

Research ventures/ approaches/ nanoplatforms	Purpose	Forms	General advantages	Biological activity	Mechanism	References
Gold-coated nanoparticles (AuNPs)	For the transport of oligonucleotides/genes, drugs, and proteins to particular target sites and for the differential diagnosis of cancer type and selective phototherapy	Gold nanorods, nanocages, nanostars, nanospheres, nanoshells	Healthy cells remain unharmed, whereas cancerous or malignant cells are destroyed	The drug, when assembled onto AuNPs, accumulates in tumor cells more quickly than free drug	They manifest more effective, high photon-to-heat conversion that renders them a compelling and imperative therapeutic option in tumor cell decay	(Auría-Soro et al. 2019; Sztandera, Gorzkiewicz, and Klajnert-Maculewicz 2019)
Polymeric micelles: polylactic-co-glycolic acid (PLGA) nanoparticle is the most commonly used polymer	Used for the cell-targeted treatment of malignant cells	Self-aggregated nanoscale core-shell structures	Biocompatible and can carry both hydrophilic and hydrophobic components, low toxicity, and are cheaply and cost-effectively fabricated	PLGA nanoparticles with the peptide inhibit drug efflux pump, promote tumor suppression, and overpower drug resistance		(Deshantri et al. 2018; Huang and Zhang 2018; Kim et al. 2019; (Chen et al. 2019; Mukherjee et al. 2019; Widmer et al. 2018)
Nanosuspensions	Used for *in vivo* and *in vitro* studies to produce injectable anticancer drugs	Aqueous vehicle carrying suspended fine colloidal particles that are pharmaceutically active	Drug bioavailability and solubility are promoted;	Antitumor drug nanosuspensions exhibit superior tumor-suppressing ability compared with oral or injectable formulations	Sustained delivery of antitumor drugs, providing a viable pharmaceutical carrier candidate; prolong as well as enhance the antifibrotic and antiangiogenic effects of antitumor drugs	(Dong et al. 2019)

(Maeda 2001) due to the nature of blood vessels in cancer and endothelial cells. Moreover, the poor lymphatic drainage of tumors leads to enhanced drug retention in the tumor environment.

15.6.3 Advancement of Nanoproducts (Nanomedicine): Challenges and Promises

Formulations of nanopharmaceuticals have different challenges from traditional pharmaceutical products. One of the main challenges is the cost of production. Nanopharmaceuticals are very complicated molecules to meet therapeutic needs and proper storage demands. Large-scale production demands time, space, and high-efficiency tools. The cost of production is directly proportional to the complexity of the nonpharmaceutical (Salama et al. 2020). The cost-effectiveness of preparing nano-nutraceutical products should also be noted (Bosetti and Jones 2019). Therefore, a well-designed and carefully planned manufacturing process is vital (Navya et al. 2019) in order not to compromise the quality. For instance, many things impact the signaling of nanoparticle detection and thus, the specificity and sensitivity of cancer recognition (Lin, Huang, and Chang 2011). In active nanotargeting, it is necessary to chemically combine the targeting component with the nanocarrier and then deliver them to the tumor. Hence, it poses more challenges than passive nanotargeting. But, those difficulties can be overcome. The case of BIND-014 (docetaxel containing PSMA-targeted ACCURIN) is an example; in patients with prostatic cancer, 71% of patients accomplished relative progression–free survival for at least 6 months. Unfortunately, lack of interest resulted from the efficiency in phase III trials being a mere ~14% (Autio et al. 2018) 2018).

The characteristics of size, shape, surface charge, ligands, absorption, metabolization, dispersion, and elimination are among the properties of nanoparticles that have a vital role in possible toxicity in the body. Furthermore, from *in vivo* studies, it is impossible to thoroughly and quickly conclude the long-lasting toxicity effects from prolonged exposure to nanoparticles (primarily from nanobased imaging and treatment activities) (Navya et al. 2019).

FDA regulation of nano-nutraceuticals is another important challenge. Currently, the FDA's approach to reviewing nanopharmaceuticals is the same as scrutinizing products that do not encompass nanomaterials. The diversity of nanoformulations is increasing, and hence, the system of regulation devised by the FDA seems inadequate. Challenges involving safety measures, efficacy testing, and exact labeling and coding will likely arise (Navya et al. 2019).

Moreover, changes in the law sometimes occur during the preparation procedures of drugs in the laboratory (El-Readi et al. 2019). Therefore, groups like the Nanotechnology Characterization Laboratory must seek collaboration between scientists, government, laboratories, and pharmaceutical companies for the screening of nanomedicine legislation to ensure the therapeutic application of nanoparticles (Tinkle et al. 2014).

15.7 CONCLUSION

Thus, nanotechnology holds promise towards enhancing the efficiency of nutraceuticals and bioactive component delivery networks in functional foods for promoting human health. This may help enhance the solubility, regulation, discharge, and bioavailability of nutrients and bioactive chemicals during manufacturing, storage, and distribution. It increases food quality and functionality. Release under controlled circumstances may result in changes in physical and chemical qualities of the product, such as taste and color change, *in situ*. Finally, comprehending the procedures of targeted distribution would provide the groundwork for food producers to create smart and intelligent food networks capable of imparting optimal health benefits.

REFERENCES

Ajazuddin, and S Saraf. 2010. "Applications of Novel Drug Delivery System for Herbal Formulations." *Fitoterapia* 81 (7): 680–89. https://doi.org/10.1016/j.fitote.2010.05.001.

Akoglu, B, D Faust, V Milovic, and J Stein. 2001. "Folate and Chemoprevention of Colorectal Cancer: Is 5-Methyl-Tetrahydrofolate an Active Antiproliferative Agent in Folate-Treated Colon-Cancer Cells?" *Nutrition* 17 (7–8): 652–653. https://doi.org/10.1016/s0899-9007(01)00594-9.

Al-Obaidi, Jameel R, Khalid H Alobaidi, Bilal Salim Al-Taie, David Hong-Sheng Wee, Hasnain Hussain, Nuzul Noorahya Jambari, E I Ahmad-Kamil, and Nur Syamimi Ariffin. 2021. "Uncovering Prospective Role and Applications of Existing and New Nutraceuticals from Bacterial, Fungal, Algal and Cyanobacterial, and Plant Sources." *Sustainability* 13 (7): 3671.

AlAli, Mudhi, Maream Alqubaisy, Mariam Nasser Aljaafari, Asma Obaid AlAli, Laila Baqais, Aidin Molouki, Aisha Abushelaibi, Kok-Song Lai, and Swee-Hua Erin Lim. 2021. "Nutraceuticals: Transformation of Conventional Foods into Health Promoters/Disease Preventers and Safety Considerations." *Molecules* 26 (9): 2540.

Amir, Nirmala, Saleh Ariyanti, and Syahrul Said. 2022. "Guided Imagery to Improve Mental Health in Cancer Patients with Chemotherapy: Literature Review." *Jurnal Keperawatan Komprehensif* 8 (1): 108–18.

Asefy, Zahra, Asghar Tanomand, Sirus Hoseinnejhad, Zaker Ceferov, Ebrahim Abbasi Oshaghi, and Mohsen Rashidi. 2021. "Unsaturated Fatty Acids as a Co-Therapeutic Agents in Cancer Treatment." *Molecular Biology Reports* 48 (3): 2909–16. https://doi.org/10.1007/s11033-021-06319-8.

Auría-Soro, Carlota, Tabata Nesma, Pablo Juanes-Velasco, Alicia Landeira-Viñuela, Helena Fidalgo-Gomez, Vanessa Acebes-Fernandez, Rafael Gongora, María Jesus Almendral Parra, Raúl Manzano-Roman, and Manuel Fuentes. 2019. "Interactions of Nanoparticles and Biosystems: Microenvironment of Nanoparticles and Biomolecules in Nanomedicine." *Nanomaterials* 9 (10). https://doi.org/10.3390/nano9101365.

Autio, Karen A, Robert Dreicer, Justine Anderson, Jorge A Garcia, Ajjai Alva, Lowell L Hart, Matthew I Milowsky, et al. 2018. "Safety and Efficacy of BIND-014, a Docetaxel Nanoparticle Targeting Prostate-Specific Membrane Antigen for Patients With Metastatic Castration-Resistant Prostate Cancer: A Phase 2 Clinical Trial." *JAMA Oncology* 4 (10): 1344–51. https://doi.org/10.1001/jamaoncol.2018.2168.

Babu, Anish, Amanda K Templeton, Anupama Munshi, and Rajagopal Ramesh. 2014. "Nanodrug Delivery Systems: A Promising Technology for Detection, Diagnosis, and Treatment of Cancer." *AAPS PharmSciTech* 15 (3): 709–21. https://doi.org/10.1208/s12249-014-0089-8.

Baeumner, Antje. 2004. "Nanosensors Identify Pathogens in Food." *Food Technology* 58 (8): 51–56.

Balogh, Lajos P. 2010. "Why Do We Have so Many Definitions for Nanoscience and Nanotechnology?" *Nanomedicine: Nanotechnology, Biology, and Medicine* 3 (6): 397–98.

Bansal, M P, and D Medina. 1993. "Expression of Fatty Acid-Binding Proteins in the Developing Mouse Mammary-Gland." *Biochemical and Biophysical Research Communications* 191 (1): 61–69. https://doi.org/10.1006/bbrc.1993.1184.

Başpınar, Yücel, Gülşah Erel-Akbaba, Mustafa Kotmakçı, and Hasan Akbaba. 2019. "Development and Characterization of Nanobubbles Containing Paclitaxel and Survivin Inhibitor YM155 against Lung Cancer." *International Journal of Pharmaceutics* 566: 149–56. https://doi.org/10.1016/j.ijpharm.2019.05.039.

Bosetti, Rita, and Stephen L Jones. 2019. "Cost–Effectiveness of Nanomedicine: Estimating the Real Size of Nano-Costs." *Nanomedicine* 14 (11): 1367–70. https://doi.org/10.2217/nnm-2019-0130.

Bouwmeester, Hans, Susan Dekkers, Maryvon Y Noordam, Werner I Hagens, Astrid S Bulder, Cees De Heer, Sandra E C G Ten Voorde, Susan W P Wijnhoven, Hans J P Marvin, and Adriënne J A M Sips. 2009. "Review of Health Safety Aspects of Nanotechnologies in Food Production." *Regulatory Toxicology and Pharmacology* 53 (1): 52–62.

Calvani, M, A Pasha and C Favre. 2020. "Nutraceutical Boom in Cancer: Inside the Labyrinth of Reactive Oxygen Species." *International Journal of Molecular Sciences* 21 (6): 1936.

Carrstensen, H, R H Müller, and B W Müller. 1992. "Particle Size, Surface Hydrophobicity and Interaction with Serum of Parenteral Fat Emulsions and Model Drug Carriers as Parameters Related to RES Uptake." *Clinical Nutrition* 11 (5): 289–97. https://doi.org/10.1016/0261-5614(92)90006-C.

Chandel, S, P Jain, S Asati, and V Soni. 2018. Nanoemulsions: A New Application in Nutraceutical and Food Industry. In *NanoNutraceuticals* (pp. 121–146). CRC Press.

Chen, Hongda, Jochen Weiss, and Fereidoon Shahidi. 2006. "Nanotechnology in Nutraceuticals and Functional Foods." *Food Technology.*

Chen, Lingyun, Gabriel E Remondetto, and Muriel Subirade. 2006. "Food Protein-Based Materials as Nutraceutical Delivery Systems." *Trends in Food Science & Technology* 17 (5): 272–83.

Chen, Yan, NingXi Li, Bei Xu, Min Wu, XiaoYan Yan, LiJun Zhong, Hong Cai, et al. 2019. "Polymer-Based Nanoparticles for Chemo/Gene-Therapy: Evaluation Its Therapeutic Efficacy and Toxicity against Colorectal Carcinoma." *Biomedicine & Pharmacotherapy* 118: 109257. https://doi.org/10.1016/j.biopha.2019.109257.

Coates, Alan, Suzanne Abraham, S Betai Kaye, Timothy Sowerbutts, Cheryl Frewin, R M Fox, and M H N Tattersall. 1983. "On the Receiving End—Patient Perception of the Side-Effects of Cancer Chemotherapy." *European Journal of Cancer and Clinical Oncology* 19 (2): 203–8.

Conklin, Kenneth A B T. 2002. "Dietary Polyunsaturated Fatty Acids: Impact on Cancer Chemotherapy and Radiation. (Review: Essential Fatty Acids/Cancer)." *Alternative Medicine Review* 7 (1): 4–21. https://link.gale.com/apps/doc/A83582815/AONE?u=anon~90c05fe3&sid=googleScholar&xid=839a62d2.

Covington, M. 2004. "Omega-3 Fatty Acids." *American Family Physician* 70 (1): 133–40.

Das, Lipi, Eshani Bhaumik, Utpal Raychaudhuri, and Runu Chakraborty. 2012. "Role of Nutraceuticals in Human Health." *Journal of Food Science and Technology* 49 (2): 173–83. https://doi.org/10.1007/s13197-011-0269-4.

Davis, Mark E, Zhuo (Georgia) Chen, and Dong M Shin. 2009. "Nanoparticle Therapeutics: An Emerging Treatment Modality for Cancer." In *Nanoscience and Technology*, 239–50. Co-Published with Macmillan Publishers Ltd. https://doi.org/10.1142/9789814287005_0025.

Debbage, Paul, and Werner Jaschke. 2008. "Molecular Imaging with Nanoparticles: Giant Roles for Dwarf Actors." *Histochemistry and Cell Biology* 130 (5): 845–75.

Deshantri, Anil K, Aida Varela Moreira, Veronika Ecker, Sanjay N Mandhane, Raymond M Schiffelers, Maike Buchner, and Marcel H A M Fens. 2018. "Nanomedicines for the Treatment of Hematological Malignancies." *Journal of Controlled Release* 287: 194–215. https://doi.org/10.1016/j.jconrel.2018.08.034.

Dillard, Cora J, and J Bruce German. 2000. "Phytochemicals: Nutraceuticals and Human Health." *Journal of the Science of Food and Agriculture* 80 (12): 1744–56. https://doi.org/10.1002/1097-0010(20000915)80:12<1744::AID-JSFA725>3.0.CO;2-W.

DOA, U S. 2003. *Nanoscale Science and Engineering for Agriculture and Food Systems: A Report Submitted to Cooperative State Research, Education and Extension Service.* Department of Agriculture.

Dong, Dong, Cheng-Hui Hsiao, Beppino C Giovanella, Yifei Wang, Diana Sl Chow, and Zhijie Li. 2019. "Sustained Delivery of a Camptothecin Prodrug—CZ48 by Nanosuspensions with Improved Pharmacokinetics and Enhanced Anticancer Activity." *International Journal of Nanomedicine* 14 (May): 3799–3817. https://doi.org/10.2147/IJN.S196453.

Dong, Jie, and Qiang Ma. 2019. "Integration of Inflammation, Fibrosis, and Cancer Induced by Carbon Nanotubes." *Nanotoxicology* 13 (9): 1244–74. https://doi.org/10.1080/17435390.2019.1651920.

Dong, Xiaowei, and Russell J Mumper. 2010. "Nanomedicinal Strategies to Treat Multidrug-Resistant Tumors: Current Progress." *Nanomedicine* 5 (4): 597–615. https://doi.org/10.2217/nnm.10.35.

Doroudian, Mohammad, Andrew O'Neill, Ronan Mac Loughlin, Adriele Prina-Mello, Yuri Volkov, and Seamas C Donnelly. 2021. "Nanotechnology in Pulmonary Medicine." *Current Opinion in Pharmacology* 56: 85–92.

Dutta, Sayantani, Jeyan Arthur Moses, and C Anandharamakrishnan. 2018. "Encapsulation of Nutraceutical Ingredients in Liposomes and Their Potential for Cancer Treatment." *Nutrition and Cancer* 70 (8): 1184–98. https://doi.org/10.1080/01635581.2018.1557212.

El-Readi, Mahmoud Zaki, and Mohammad Ahmad Althubiti. 2019. "Cancer Nanomedicine: A New Era of Successful Targeted Therapy." *Journal of Nanomaterials* 2019.

Fakruddin, Md, Zakir Hossain, and Hafsa Afroz. 2012. "Prospects and Applications of Nanobiotechnology: A Medical Perspective." *Journal of Nanobiotechnology* 10 (1): 1–8.

Fatima, Naz, Tasleem Akhtar, and Nadeem Sheikh. 2017. "Prebiotics: A Novel Approach to Treat Hepatocellular Carcinoma." Edited by José L Mauriz. *Canadian Journal of Gastroenterology and Hepatology* 2017: 6238106. https://doi.org/10.1155/2017/6238106.

Fernández, Javier, Sául Redondo-Blanco, Elisa M. Miguélez, Claudio J. Villar, Alfonso Clemente, and Felipe Lombó. 2015. "Healthy Effects of Prebiotics and Their Metabolites against Intestinal Diseases and Colorectal Cancer." *AIMS Microbiology* 1 (1): 48–71. https://doi.org/10.3934/microbiol.2015.1.48.

Fil, Joanne E, Stephen A Fleming, Maciej Chichlowski, Gabriele Gross, Brian M Berg, and Ryan N Dilger. 2019. "Evaluation of Dietary Bovine Milk Fat Globule Membrane Supplementation on Growth, Serum Cholesterol and Lipoproteins, and Neurodevelopment in the Young Pig." *Frontiers in Pediatrics* 7. https://www.frontiersin.org/articles/10.3389/fped.2019.00417/full

Frank, Karen M, D Kyle Hogarth, Jonathan L Miller, Saptarshi Mandal, Philip J Mease, R Jude Samulski, Glen A Weisgerber, and John Hart. 2009. "Investigation of the Cause of Death in a Gene-Therapy Trial." *New England Journal of Medicine* 361 (2): 161–69.

Gao, Zhonggao, Anne M Kennedy, Douglas A Christensen, and Natalya Y Rapoport. 2008. "Drug-Loaded Nano/Microbubbles for Combining Ultrasonography and Targeted Chemotherapy." *Ultrasonics* 48 (4): 260–70. https://doi.org/10.1016/j.ultras.2007.11.002.

García, Mario, Tamara Forbe, and Eric Gonzalez. 2010. "Potential Applications of Nanotechnology in the Agro-Food Sector." *Food Science and Technology* 30 (3): 573–81.

Gindy, Marian E, and Robert K Prud'homme. 2009. "Multifunctional Nanoparticles for Imaging, Delivery and Targeting in Cancer Therapy." *Expert Opinion on Drug Delivery* 6 (8): 865–78.

Oehlke, Kathleen, Marta Adamiuk, Diana Behsnilian, Volker Gräf, Esther Mayer-Miebach, Elke Walz, and Ralf Greiner. 2014. "Potential Bioavailability Enhancement of Bioactive Compounds Using Food-Grade Engineered Nanomaterials: A Review of the Existing Evidence." *Food & Function* 5 (7): 1341–1359. https://doi.org/10.1039/c3fo60067j.

Guo, Xin, and Francis C Szoka. 2003. "Chemical Approaches to Triggerable Lipid Vesicles for Drug and Gene Delivery." *Accounts of Chemical Research* 36 (5): 335–41. https://doi.org/10.1021/ar9703241.

Gupta, Subash C, Ji Hye Kim, Sahdeo Prasad, and Bharat B Aggarwal. 2010. "Regulation of Survival, Proliferation, Invasion, Angiogenesis, and Metastasis of Tumor Cells through Modulation of Inflammatory Pathways by Nutraceuticals." *Cancer and Metastasis Reviews* 29 (3): 405–34. https://doi.org/10.1007/s10555-010-9235-2.

Huang, Wei, and Chenming Zhang. 2018. "Tuning the Size of Poly(Lactic-Co-Glycolic Acid) (PLGA) Nanoparticles Fabricated by Nanoprecipitation." *Biotechnology Journal* 13 (1): 1700203. https://doi.org/10.1002/biot.201700203.

Iorio, Jessica, Giulia Petroni, Claudia Duranti, and Elena Lastraioli. 2019. "Potassium and Sodium Channels and the Warburg Effect: Biophysical Regulation of Cancer Metabolism." *Bioelectricity* 1 (3): 188–200. https://doi.org/10.1089/bioe.2019.0017.

Jiang, Zhiwen, Hui Li, Jing Qiao, Yan Yang, Yanting Wang, Wanshun Liu, and Baoqin Han. 2019. "Potential Analysis and Preparation of Chitosan Oligosaccharides as Oral Nutritional Supplements of Cancer Adjuvant Therapy." *International Journal of Molecular Sciences* 20 (4). https://www.mdpi.com/1422-0067/20/4/920

Jing, Kaipeng, Wu, Tong, and Lim, Kyu. 2013. "Omega-3 Polyunsaturated Fatty Acids and Cancer." *Anti-Cancer Agents in Medicinal Chemistry (Formerly Current Medicinal Chemistry—Anti-Cancer Agents)*, 13 (8): 1162–77.

Jones, Ella F, Jiang He, Henry F Vanbrocklin, Benjamin L Franc, and Youngho Seo. 2008. "Nanoprobes for Medical Diagnosis: Current Status of Nanotechnology in Molecular Imaging." *Current Nanoscience* 4 (1): 17–29.

Kalra Ekta, K. 2003. Nutracetical-Definition and Introduction. *AAPS Pharm*: 1–2.

Kharat, Manoj Govind, Shankar Murthy, and Sheetal Jaisingh Kamble. 2017. "Environmental Applications of Nanotechnology: A Review." *ADBU Journal of Engineering Technology* 6 (3).

Kim, Ki-Taek, Jae-Young Lee, Dae-Duk Kim, In-Soo Yoon, and Hyun-Jong Cho. 2019. "Recent Progress in the Development of Poly(Lactic-Co-Glycolic Acid)-Based Nanostructures for Cancer Imaging and Therapy." *Pharmaceutics* 11 (6). https://doi.org/10.3390/pharmaceutics11060280.

Lee Yee-Ki, Carol, Wai-Hung Sit, Sheung-Tat Fan, Kwan Man, Irene Jor Wing-Yan, Leo Wong Lap-Yan, Murphy Wan Lam-Yim, Kian Tan-Un Cheng, and Jennifer Wan Man-Fan. 2010. "The Cell Cycle Effects of Docosahexaenoic Acid on Human Metastatic Hepatocellular Carcinoma Proliferation." *Int J Oncol* 36 (4): 991–98. https://doi.org/10.3892/ijo_00000579.

Letchford, Kevin, and Helen Burt. 2007. "A Review of the Formation and Classification of Amphiphilic Block Copolymer Nanoparticulate Structures: Micelles, Nanospheres, Nanocapsules and Polymersomes." *European Journal of Pharmaceutics and Biopharmaceutics* 65 (3): 259–69.

Li, Shyh-Dar, and Leaf Huang. 2009. "Nanoparticles Evading the Reticuloendothelial System: Role of the Supported Bilayer." *Biochimica et Biophysica Acta (BBA) - Biomembranes* 1788 (10): 2259–66. https://doi.org/10.1016/j.bbamem.2009.06.022.

Lin, Yang-Wei, Chih-Ching Huang, and Huan-Tsung Chang. 2011. "Gold Nanoparticle Probes for the Detection of Mercury{,} Lead and Copper Ions." *Analyst* 136 (5): 863–71. https://doi.org/10.1039/C0AN00652A.

Lindmark-Månsson, Helena, and B Åkesson. 2000. "Antioxidative Factors in Milk." *British Journal of Nutrition* 84 (S1): 103–10. https://doi.org/10.1017/S0007114500002324.

Logothetidis, S. 2012. "Nanomedicine: The Medicine of Tomorrow." In *Nanomedicine and Nanobiotechnology*, 1–26. Springer.

Lombardo, Domenico, Mikhail A Kiselev, and Maria Teresa Caccamo. 2019. "Smart Nanoparticles for Drug Delivery Application: Development of Versatile Nanocarrier Platforms in Biotechnology and Nanomedicine." *Journal of Nanomaterials* 2019.

Maeda, H. 2001. "The Enhanced Permeability and Retention (EPR) Effect in Tumor Vasculature: The Key Role of Tumor-Selective Macromolecular Drug Targeting." *Advances in Enzyme Regulation* 41: 189–207. https://doi.org/10.1016/s0065-2571(00)00013-3.

Mishra, Sundeep. 2016. "Nanotechnology in Medicine." *Indian Heart Journal* 68 (3): 437.

Moitra, Karobi. 2015. "Overcoming Multidrug Resistance in Cancer Stem Cells." *BioMed Research International* 2015: 1–8.

Mokbel, Kefah, and Kinan Mokbel. 2019. "Chemoprevention of Breast Cancer with Vitamins and Micronutrients: A Concise Review." *In Vivo* 33 (4): 983–97. https://doi.org/10.21873/invivo.11568.

Moraru, Carmen I, Chithra P Panchapakesan, Qingrong Huang, Paul Takhistov, Sean Liu, and Jozef L Kokini. 2003. "Nanotechnology: A New Frontier in Food Science Understanding the Special Properties of Materials of Nanometer Size Will Allow Food Scientists to Design New, Healthier, Tastier, and Safer Foods." *Nanotechnology* 57 (12): 24–29.

Mousa, Shaker A, and Dhruba J Bharali. 2011. "Nanotechnology-Based Detection and Targeted Therapy in Cancer: Nano-Bio Paradigms and Applications." *Cancers* 3 (3): 2888–2903.

Mozafari, M Reza, John Flanagan, Lara Matia-Merino, Ajay Awati, Abdelwahab Omri, Zacharias E Suntres, and Harjinder Singh. 2006. "Recent Trends in the Lipid-based Nanoencapsulation of Antioxidants and Their Role in Foods." *Journal of the Science of Food and Agriculture* 86 (13): 2038–45.

Mukherjee, Anubhab, Ariana K Waters, Pranav Kalyan, Achal Singh Achrol, Santosh Kesari, and Venkata Mahidhar Yenugonda. 2019. "Lipid-Polymer Hybrid Nanoparticles as a next-Generation Drug Delivery Platform: State of the Art, Emerging Technologies, and Perspectives." *International Journal of Nanomedicine* 14 (March): 1937–52. https://doi.org/10.2147/IJN.S198353.

Nagahara, Larry A, Jerry S H Lee, Linda K Molnar, Nicholas J Panaro, Dorothy Farrell, Krzysztof Ptak, Joseph Alper, and Piotr Grodzinski. 2010. "Strategic Workshops on Cancer Nanotechnology." *Cancer Research* 70 (11): 4265–68. https://doi.org/10.1158/0008-5472.CAN-09-3716.

Nahrendorf, Matthias, Hanwen Zhang, Sheena Hembrador, Peter Panizzi, David E Sosnovik, Elena Aikawa, Peter Libby, Filip K Swirski, and Ralph Weissleder. 2008. "Nanoparticle PET-CT Imaging of Macrophages in Inflammatory Atherosclerosis." *Circulation* 117 (3): 379–87.

Nasri, Hamid, Azar Baradaran, Hedayatollah Shirzad, and Mahmoud Rafieian-Kopaei. 2014. "New Concepts in Nutraceuticals as Alternative for Pharmaceuticals." *International Journal of Preventive Medicine* 5 (12): 1487.

Navya, P N, Anubhav Kaphle, S P Srinivas, Suresh Kumar Bhargava, Vincent M Rotello, and Hemant Kumar Daima. 2019. "Current Trends and Challenges in Cancer Management and Therapy Using Designer Nanomaterials." *Nano Convergence* 6 (1): 23. https://doi.org/10.1186/s40580-019-0193-2.

Oh, Jae-Min, Timothy T Biswick, and Jin-Ho Choy. 2009. "Layered Nanomaterials for Green Materials." *Journal of Materials Chemistry* 19 (17): 2553–63.

Oketch-Rabah, Hellen A, Amy L Roe, Cynthia V Rider, Herbert L Bonkovsky, Gabriel I Giancaspro, Victor Navarro, Mary F Paine, Joseph M Betz, Robin J Marles, and Steven Casper. 2020. "United States Pharmacopeia (USP) Comprehensive Review of the Hepatotoxicity of Green Tea Extracts." *Toxicology Reports* 7: 386–402.

Pal, Dilipkumar, Subham Banerjee, and Ashoke Kumar Ghosh. 2012. "Dietary-Induced Cancer Prevention: An Expanding Research Arena of Emerging Diet Related to Healthcare System." *Journal of Advanced Pharmaceutical Technology & Research* 3 (1): 16–24. https://doi.org/10.4103/2231-4040.93561.

Parodi, Peter W. 2001. "Cow's Milk Components with Anti-Cancer Potential." *Australian Journal of Dairy Technology* 56 (2): 65.

Patra, Jayanta Kumar, Gitishree Das, Leonardo Fernandes Fraceto, Estefania Vangelie Ramos Campos, Maria del Pilar Rodriguez-Torres, Laura Susana Acosta-Torres, Luis Armando Diaz-Torres, Renato Grillo, Mallappa Kumara Swamy, and Shivesh Sharma. 2018. "Nano Based Drug Delivery Systems: Recent Developments and Future Prospects." *Journal of Nanobiotechnology* 16 (1): 1–33.

Patri, Anil K, Jolanta F Kukowska-Latallo, and James R Baker. 2005. "Targeted Drug Delivery with Dendrimers: Comparison of the Release Kinetics of Covalently Conjugated Drug and Non-Covalent Drug Inclusion Complex." *Advanced Drug Delivery Reviews* 57 (15): 2203–14. https://doi.org/10.1016/j.addr.2005.09.014.

Rabajdová, Miroslava, Ivana Špaková, Zuzana Klepcová, Lukáš Smolko, Michaela Abrahamovská, Peter Urdzík, and Mária Mareková. 2021. "Zinc(II) Niflumato Complex Effects on MMP Activity and Gene Expression in Human Endometrial Cell Lines." *Scientific Reports* 11 (1): 19086. https://doi.org/10.1038/s41598-021-98512-9.

Reza Mozafari, M, Chad Johnson, Sophia Hatziantoniou, and Costas Demetzos. 2008. "Nanoliposomes and Their Applications in Food Nanotechnology." *Journal of Liposome Research* 18 (4): 309–27.

Riley, Rachel S, Rachel K O'Sullivan, Andrea M Potocny, Joel Rosenthal, and Emily S Day. 2018. "Evaluating Nanoshells and a Potent Biladiene Photosensitizer for Dual Photothermal and Photodynamic Therapy of Triple Negative Breast Cancer Cells." *Nanomaterials* 8 (9). https://doi.org/10.3390/nano8090658.

Roco, Mihail C. 2011. "The Long View of Nanotechnology Development: The National Nanotechnology Initiative at 10 Years." In *Nanotechnology Research Directions for Societal Needs in 2020*, 1–28. Springer.

Salama, Ahmed, Mohamed Hasanin, and Peter Hesemann. 2020. "Synthesis and Antimicrobial Properties of New Chitosan Derivatives Containing Guanidinium Groups." *Carbohydrate Polymers* 241: 116363. https://doi.org/10.1016/j.carbpol.2020.116363.

Salama, Lavinia, Elizabeth R Pastor, Tyler Stone, and Shaker A Mousa. 2020. "Emerging Nanopharmaceuticals and Nanonutraceuticals in Cancer Management." *Biomedicines* 8 (9). https://doi.org/10.3390/biomedicines8090347.

Salami, Ahmad, Enayatollah Seydi, and Jalal Pourahmad. 2013. "Use of Nutraceuticals for Prevention and Treatment of Cancer." *Iranian Journal of Pharmaceutical Research : IJPR* 12 (3): 219–20. https://pubmed.ncbi.nlm.nih.gov/24250626.

Samson, Faith Pwaniyibo, Ambrose Teru Patrick, Tosin Esther Fabunmi, Muhammad Falalu Yahaya, Joshua Madu, Weilue He, Srinivas R Sripathi, et al. 2020. "Oleic Acid, Cholesterol, and Linoleic Acid as Angiogenesis Initiators." *ACS Omega* 5 (32): 20575–85. https://doi.org/10.1021/acsomega.0c02850.

Shah, Ankita V, Heta H Desai, Prajwal Thool, Damon Dalrymple, and Abu T M Serajuddin. 2018. "Development of Self-Microemulsifying Drug Delivery System for Oral Delivery of Poorly Water-Soluble Nutraceuticals." *Drug Development and Industrial Pharmacy* 44 (6): 895–901. https://doi.org/10.1080/03639045.2017.1419365.

Shahidi, F. 2006. "Nanotechnology in Nutraceuticals and Functional Foods Emerging Technology Has Shown Great Potential for Delivering Bioactive Compounds in Functional Foods to Improve Human Health." *Nanotechnology* 60 (3).

Sharma, Rakesh. 2009. "Nutraceuticals and Nutraceutical Supplementation Criteria in Cancer: A Literature Survey." *The Open Nutraceuticals Journal* 2 (1): 92–106. https://doi.org/10.2174/1876396000902010092.

Simopoulos, A P. 1991. "Omega-3 Fatty Acids in Health and Disease and in Growth and Development." *American Journal of Clinical Nutrition* 54 (3): 438–63. https://doi.org/10.1093/ajcn/54.3.438.

Spitsberg, V L, and R C Gorewit. 1997. "In Vitro Phosphorylated Bovine Milk Fat Globule Membrane Proteins." *Journal of Nutritional Biochemistry* 8 (4): 181–89. https://doi.org/10.1016/S0955-2863(97)00001-6.

Sung, Hyuna, Jacques Ferlay, Rebecca L Siegel, Mathieu Laversanne, Isabelle Soerjomataram, Ahmedin Jemal, and Freddie Bray. 2021. "Global Cancer Statistics 2020: GLOBOCAN Estimates of Incidence and Mortality Worldwide for 36 Cancers in 185 Countries." *CA: A Cancer Journal for Clinicians* 71 (3): 209–49. https://doi.org/10.3322/caac.21660.

Sutradhar, Kumar Bishwajit, and Md Amin. 2014. "Nanotechnology in Cancer Drug Delivery and Selective Targeting." *International Scholarly Research Notices* 2014, Article ID 939378, 12 pages

Sztandera, Krzysztof, Michał Gorzkiewicz, and Barbara Klajnert-Maculewicz. 2019. "Gold Nanoparticles in Cancer Treatment." *Molecular Pharmaceutics* 16 (1): 1–23. https://doi.org/10.1021/acs.molpharmaceut.8b00810.

Tannock, Ian F, Carol M Lee, Jonathon K Tunggal, David S M Cowan, and Merrill J Egorin. 2002. "Limited Penetration of Anticancer Drugs through Tumor Tissue: A Potential Cause of Resistance of Solid Tumors to Chemotherapy." *Clinical Cancer Research* 8 (3): 878–84.

Tinkle, Sally, Scott E McNeil, Stefan Mühlebach, Raj Bawa, Gerrit Borchard, Yechezkel (Chezy) Barenholz, Lawrence Tamarkin, and Neil Desai. 2014. "Nanomedicines: Addressing the Scientific and Regulatory Gap." *Annals of the New York Academy of Sciences* 1313 (1): 35–56. https://doi.org/10.1111/nyas.12403.

Wang, Yiyue, Jing Li, Jing Jing Chen, Xuan Gao, Zun Huang, and Qi Shen. 2017. "Multifunctional Nanoparticles Loading with Docetaxel and GDC0941 for Reversing Multidrug Resistance Mediated by PI3K/Akt Signal Pathway." *Molecular Pharmaceutics* 14 (4): 1120–32.

Wei, Yuren, Dong Wang, and Michael J Pagliassotti. 2007. "Saturated Fatty Acid-Mediated Endoplasmic Reticulum Stress and Apoptosis Are Augmented by Trans-10, Cis-12-Conjugated Linoleic Acid in Liver Cells." *Molecular and Cellular Biochemistry* 303 (1): 105–13. https://doi.org/10.1007/s11010-007-9461-2.

Whanger, P D. 2004. "Selenium and Its Relationship to Cancer: An Update." *British Journal of Nutrition* 91 (1): 11–28. https://doi.org/10.1079/BJN20031015.

Widmer, Jérôme, Cédric Thauvin, Inès Mottas, Van Nga Nguyen, Florence Delie, Eric Allémann, and Carole Bourquin. 2018. "Polymer-Based Nanoparticles Loaded with a TLR7 Ligand to Target the Lymph Node for Immunostimulation." *International Journal of Pharmaceutics* 535 (1): 444–51. https://doi.org/10.1016/j.ijpharm.2017.11.031.

Wu, JiMin, and ZiJian Li. 2013. *Applications of Nanotechnology in Biomedicine*. Springer.

Zheng, Jie, Yue Zhou, Ya Li, Dong-Ping Xu, Sha Li, and Hua-Bin Li. 2016. "Spices for Prevention and Treatment of Cancers." *Nutrients* 8 (8). https://doi.org/10.3390/nu8080495.

Zurier, B. 1991. "Essential Fatty Acids and Inflammation." *Annals of the Rheumatic Diseases* 50 (11): 745–46. https://doi.org/10.1136/ard.50.11.745.

16 Nano-Nutraceuticals in Neurodegenerative Disorders

*Wardah Ali, Zirwah Tahir, Uday Younis Hussein Abdullah, Shakeel Ahmed, and Atif Amin Baig**

CONTENTS

16.1 Neurodegenerative Disorders 417
16.1.1 Molecular Basis of Neurodegenerative Disorders 418
16.1.1.1 Alzheimer's Disease 418
16.1.2 Huntington's Disease 420
16.1.2.1 Symptoms and Genetic Diagnosis 420
16.1.3 Parkinson's Disease 420
16.1.3.1 Symptoms and Diagnosis of Parkinson's Disease 420
16.1.4 Amyotrophic Lateral Sclerosis (ALS) 421
16.1.4.1 Diagnosis of ALS 421
16.2 Nano-Nutraceuticals 421
16.2.1 Types of Nano-Nutraceuticals for Treating Neurodegenerative Disorders 421
16.2.2 Main Nutraceuticals in the Treatment of Neurodegenerative Disorders 424
16.2.2.1 Nutraceuticals in Alzheimer's Disease 424
16.2.3 Significant Role of Nano-Nutraceuticals in Neurodegenerative Disorders 425
16.3 Multiple Approaches Utilizing Nano-Nutraceuticals 425
16.3.1 Therapeutic 425
16.3.2 Medicinal Approach (Natural Antioxidant Treatment Pathway) 425
16.3.3 Effectiveness of Nano-Nutraceutical Technology Utilizing Multiple Approaches in the Treatment of Neurodegenerative Disorders 426
16.3.4 Drawbacks in the Utilization of Nano-Nutraceuticals 426
16.3.5 Conclusion and Recommendations 427
References 427

16.1 NEURODEGENERATIVE DISORDERS

Neurodegenerative disorders are considered one of the significant healthcare issues worldwide. Multiple factors, such as current lifestyle changes in combination with a nutrient-deficient diet, have drastically increased the risk for these disorders. Globally, the healthcare and financial burden is deeply rooted in society. Specialized treatment with significantly fewer side effects against distinct disorders has been provided by multiple researchers. In Western countries, 3.1% of the population aged 70–79 years suffer from neurodegenerative disorders. Widespread neurodegenerative disorders exist today, such as Lewy body disease, Huntington's disease (HD), spinal muscular atrophy, Parkinson's disease (PD), amyotrophic lateral sclerosis (ALS), Alzheimer's disease (AD), etc. The detailed molecular basis of occurrence and detection of all these neurodegenerative disorders will be discussed in the following (Bungau and Popa, 2015; Rashita et al., 2020).

* Corresponding author: atifamin@unisza.edu.my

DOI: 10.1201/9781003244721-16

Historical data reports that 5 million U.S. citizens suffer from AD every year, and 1 million U.S. nationalists reportedly suffer from PD. A large number of people (400,000) suffer from multiple sclerosis. About 3000 U.S. nationals suffer from HD every year, and 30,000 people in the United States suffer from ALS annually. The risk rates of all these disorders increase according to the rate of increase in population age, as the onset of these neurodegenerative disorders appears in mid- to late life in infected individuals. One suggested way to curb these disorders is neuro-regeneration. One of the practically successful approaches is stem cell therapy, which has been evaluated in treating neurodegeneration and enhancement of the neuro-regeneration process (Chung et al., 2002; Rachakonda et al., 2004; Bungau and Popa, 2015; Rashita et al., 2020).

16.1.1 Molecular Basis of Neurodegenerative Disorders

A diet highly deficient in nutrients leads to the degeneration and deterioration of the central and peripheral central nervous system, resulting in various neurodegenerative disorders. The progressive loss of neural tissues results in abnormal neuron function in the whole degeneration process. Neurons of the brain lose their ability to regenerate, resulting in neuron cell death. On molecular levels, protein misfolding occurs in a specific diseased area. Early diagnosis of these disorders is necessary for the purpose of molecular diagnostic techniques comprising specific biomarkers used to detect and identify normal and diseased tissues and abnormal and normal biological processes in an infected person. Biomarkers identify the onset and presence of any neurodegenerative disorders. See Table 16.1 for understanding multiple disease identification on a molecular basis. On this basis, let us overview some neurodegenerative disorders to deeply understand the occurrence and biological influence of these disorders on the human body and their possible treatment. A promising biomarker has specifications in the identification of different diseases in combination with the quantitative imaging technique Magnetic Resonance Spectroscopy (MRS), which helps in the *in vivo* detection of neuronal metabolites as biomarkers that are further reported to investigate metabolic dysfunctions and irreversible neuron damage (Chung et al.,2002; Rachakonda et al., 2004; William et al., 2015; Megha and Abhijit, 2015).

16.1.1.1 Alzheimer's Disease

AD is a neurodegenerative disorder that attacks brain neurons, resulting in severe memory loss. Acetylcholine is present inside the brain's degenerated neurons, using notable neurotransmitters to communicate with other neurons and alter butyrylcholinesterase (BuChE) and acetylcholinesterase (AChE). Senile extracellular plaques and intracellular neurofibrillary tangles are two abnormal lesion types reported in the clogging of Alzheimer's patients' brains by APP (parent amyloid precursor) protein (cleavage occurs by the proteases α-, β-, and γ-secretases in the process of enzymatic cleavage and results in the formation of two types of abnormal lesions, called Aβ forms). In the degradation process, synaptic degradation occurs by NFTs, involving a nerve cell atrophy process that results in damaging synaptically connected axons (see Figure 16.1). NFTs are in the form of paired helical filaments. The tau protein hyperphosphorylated insoluble form forms PHF (el-Agnaf and Irvine, 2002; Hardy and Selkoe, 2002; Rachakonda et al., 2004).

16.1.1.1.1 Symptoms of Alzheimer's Disease

1—A major symptom that helped diagnose this disorder is the depletion of acetylcholine.
2—There is memory loss due to severe degradation of brain neurons.
3—Thinking and behavioral changes have been observed in patients.
4—This disorder mainly occurs in older people.

16.1.1.1.2 Diagnostic Biomarkers for Alzheimer's Disease

Cognitive testing such as neuropsychological testing is usually conducted, which helps in the detection and immediate treatment of AD. Demonstrating amyloid post-mortem eventually confirmed

TABLE 16.1
Summary of Identification, Mode of Action, and Specific Markers for Alzheimer's Disease, Huntington's Disease, Parkinson's Disease, and Amyotrophic Lateral Sclerosis

Neurodegenerative disorder	Identification	Mode of action	Diagnostic genetic biomarkers	Diagnostic biochemical biomarkers	References
1. Alzheimer's disease	Acetylcholine depletion	Attack on brain neurons forming abnormal plaques	1-ApoE isoforms2-Amyloid precursor protein mutation	1-Phospho-tau2-Plasma/CSF Aβ1–42 peptide	(Hardy and Selkoe, 2002; Rachakonda et al., 2004); Megha and Abhijit, 2015)
2. Huntington's disease	Huntingtin mutant protein (mHTT) production	Effect on motor neurons and cognitive ability	HTT gene mutations	Growth hormones	(Runne et al., 2008; Megha and Abhijit, 2015; Mastrokolias et al., 2015)
3. Parkinson's disease	Loss of dopamine-producing cells in substantia nigra	Dopaminergic neuron degradation between striatum and substantia nigra	Parkin mutated gene, α-synuclein mutated gene, PINK 1 mutated gene, UCH-L1 mutated gene used in diagnosis	Absence of dopamine transporter (DAT)	(Rachakonda et al., 2004; Molochnikov et al., 2012; Megha and Abhijit, 2015)
4. Amyotrophic lateral sclerosis	Stress in endoplasmic reticulum and excitotoxicity	Attack on voluntary muscle neurons	ALS2 gene mutations	mGLUR2	(Megha and Abhijit, 2015)

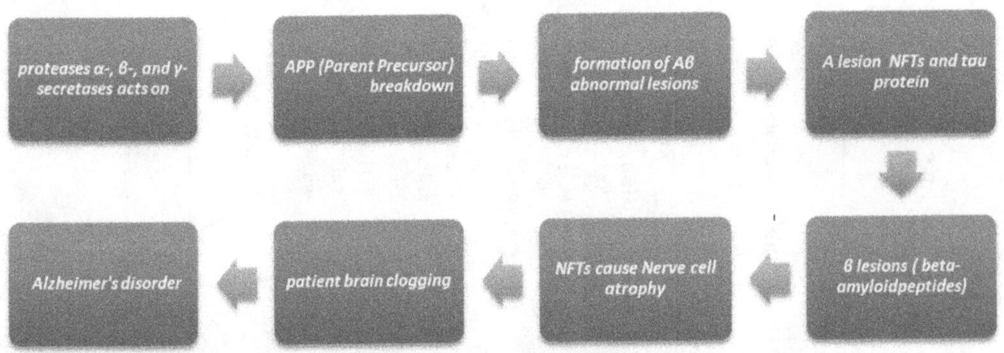

FIGURE 16.1 Flow chart of Alzheimer's disease production mechanism. (Adapted from Hardy, J. and Selkoe, D.J., *Science*, 297, 353–356, 2002; el-Agnaf, O.M. and Irvine, G.B., *Biochem. Soc. Trans.*, 30, 559–565, 2002.)

this disease. Before the clinical findings appear, the damage has already been done. Diagnostic biomarkers used to detect the onset and damage of AD are mentioned in Table 16.1 (Harvey, 2012; Megha and Abhijit, 2015).

16.1.2 Huntington's Disease

HD is a neurodegenerative heritable disorder that mainly affects a person's cognitive ability. A decline in cognitive ability shows that a person is symptomatic for HD; besides cognitive ability, it influences a person's motor neurons. In HD, changes occur in bases in the HTT gene. An increase in CAG repeats in the HTT gene results in mutant huntingtin protein (mHTT) production. In the process, aggregates of mutant proteins form, resulting in neural cell loss (Runne et al., 2008; Mastrokolias et al., 2015).

16.1.2.1 Symptoms and Genetic Diagnosis

An HD symptomatic person shows a change in personality, mentality, and psychiatric functioning. To diagnose HD, molecular diagnosis shows transcription deregulation in the brain tissue of the HD-affected person. The ultrasensitive immunoassay single-molecule counting mHTT technique is used for mHTT assessment in cerebrospinal fluid (CSF) samples. A disease progression biomarker is highly significant to identify the changes before the appearance of clinical symptoms of the disorder. In the bloodstream, changes in gene expression can be easily identified. The selection of biomarkers for diagnosing HD has always been challenging due to various symptoms and disease progression rates (Runne et al., 2008; Wild et al., 2015).

16.1.3 Parkinson's Disease

Dopaminergic neuron degradation between the striatum and substantia nigra (SN) results in PD. Patients diagnosed with PD, the majority of whom have lost dopamine-producing cells in their SN, have provided strong evidence for the cause of this disorder (Molochnikov et al., 2012).

16.1.3.1 Symptoms and Diagnosis of Parkinson's Disease

The most common reported symptoms of PD are slow movement, trembling in legs, face, arms, and hands, impaired coordination and balance, stiffness of the trunk and limbs, inability to complete daily tasks, and immense weakness in the body due to progressive destruction of neurons. Identifying this disease in the early stages of life; therefore, it's critical to treat it once it's too late. Most cases of PD are diagnosed in people over the age of 60. Effective neuroprotective therapies for PD are required once symptoms appear. Early-stage PD has been identified by specific molecular diagnostics biomarkers. Dopamine transporter (DAT), a biochemical, molecular diagnostic biomarker, Lewy body UCH-L1 gene Parkin gene mutations, α-synuclein gene mutations and PINK1 gene mutations are genetic biomarkers mentioned in Table 16.1, designed explicitly to diagnose the extent of disease in order to provide immediate treatment once the onset of the disease is identified. Currently, PD is only diagnosed on a clinical basis (Rachakonda et al., 2004; Molochnikov et al., 2012; Megha and Abhijit, 2015).

16.1.4 Amyotrophic Lateral Sclerosis (ALS)

ALS is a rapidly progressing and fatal neurodegenerative disorder that attacks neurons responsible for controlling the voluntary muscles. The degeneration mechanism occurs through messages transmitted from upper motor neurons (motor neurons in the brain) to the lower motor neurons (spinal cord motor neurons) and then to specific muscles. In most cases, both lower and upper motor neurons die, and being unable to send messages due to muscular weakness results in the outcome of muscle atrophy. The brain loses its control of voluntary movement. As reported in the research, the chemical basis for ALS occurrence is due to mitochondrial dysfunction, inflammation, dysregulated transcription, oxidative stress, RNA processing endoplasmic reticulum stress, excitotoxicity, dysregulated endosomal trafficking, apoptosis, and genetic susceptibility.

16.1.4.1 Diagnosis of ALS

Diagnosis of ALS is based on clinical examination, test series, and symptoms shown by patients. The research area for ALS has demonstrated that a combination of neurophysiological recording approaches such as the motor unit number estimation technique and electromyography, which detects muscles' electrical activity, has helped in the search for new biomarkers (DeJesus-Hernandez et al., 2011; Joyce and Carter, 2013).

16.2 NANO-NUTRACEUTICALS

In the past, diverse research on functional foods has been conducted to obtain functional foods having minimal side effects but possessing better therapeutic activity. For this purpose, extraction has been carried out using different extraction methods for specific bioactive compounds such as phytochemicals. Such bioactive compounds are known as vital foods, pharmafoods, functional foods, medicinal foods, and medifoods. Thus, the origin of functional foods has been intruded with other identical terms like svitafoods. The demand for nutraceuticals has been increasing because they possess features suitable for use as neurodegenerative therapeutic interventions. Nowadays, interest has moved from allopathic to Ayurveda to nutraceuticals. Traditional medicinal approaches, like pills and tablets in semi-solid form, lack efficacy and bioactive targeted delivery. Liposomes, niosomes, and ethosomes are among the approaches that have been utilized to deliver nutraceuticals (Rashita et al., 2020; Sahni, 2012).

16.2.1 Types of Nano-Nutraceuticals for Treating Neurodegenerative Disorders

Nutraceuticals have neuroprotective applications ranging from mild disorders to toxic malignant cancers and neurodegenerative disorders. Nano-nutraceuticals fall into different categories (Rashita and Tapan, et al., 2020):

1—Food-based nutraceuticals
2—Probiotic microorganisms
3—Nutrients
4—Recombinant nutraceuticals
5—Herbs or extracts and concentrates of botanical products

6—Nutraceutical enzymes
7—Based on the chemical nature of products
8—Mechanism of action
9—Based on neurodegeneration

1. **Food-Based Nutraceuticals**
   Those food ingredients that can be obtained naturally without change to their original contents and nutrients fall into this small category of nano-nutraceuticals. Examples include vegetables, fish, meat, grains, fruits, and dairy products like cheese that provide several health benefits (Bhat and Bhat, 2011; Bhaskarachary, 2016).
2. **Probiotic Microorganisms**
   Metchnikoff, a renowned scientist, suggested the term *probiotic microorganisms.* These microorganisms have antimicrobial characteristics, and they help in removal of toxic flora from the gut. Consumption of probiotics in adequate amounts has helped in the management of gastrointestinal disorders. Moreover, in many cases, probiotic therapy has been reported moderately effective in patients with weak immune systems (Gosálbez and Amón, 2015; Zucko et al., 2020).
3. **Nutrients**
   The primary metabolites of minerals possessing nutritional characteristics in metabolic pathways are termed *nutrients.* They have immense health benefits for neurodegenerative disorders when utilized with animal and plant products (Prakash and Boekel, 2010; Ramalingum and Mahomoodally, 2014).
4. **Recombinant Nutraceuticals**
   Nutraceuticals obtained from food products utilizing applications of biotechnology are termed *recombinant nutraceuticals.* These nutraceuticals extracted from food products such as bread and cheese have provided therapeutically beneficial results when used to treat neurodegenerative disorders (Andlauer and Fürst, 2002; Williams et al., 2015).
5. **Herbs or Extracts and Concentrates of Botanical Products**
   Herbs and concentrates extracted from botanical products and containing herbals and nutrients have helped treat multiple mental disorders (Dohrmann, et al., 2019; Barba, et al., 2020). Lavender (a tannin-containing compound) extracted from botanical products has helped lower blood pressure and played a role as a stress-reducing compound. Flavonoids isolated from concentrates of botanical products have enabled the prevention of heart diseases and diabetes. Another notable example of herbal botanical products, Aloe vera, contains anti-inflammatory properties and is used for wound healing and treatment of bronchospasms (Williams et al., 2016).
6. **Nutraceutical Enzymes**
   As nutraceutical enzymes have provided the most minor benefits in treating neurodegenerative disorders, but from the recent year's rare disorders such as Gaucher disease and Hunter syndrome have been cured (Punik et al., 2019).
7. **Based on the Chemical Nature of Products**
   Nutraceutical classification depends on the chemical nature of products called secondary metabolites. These include carbohydrates, amino acids, fatty acids, etc. The activity of compounds used to treat neurodegenerative disorders depends on the chemical nature of the secondary metabolite used in therapy. These secondary metabolites are used as adjunctive therapy to simultaneously treat different neurodegenerative disorders, such as AD and PD (see Figure 16.2). A research study reported that these bioactive compounds can react with host molecules with toxic results (Brower, 1998; Gupta, 2010; Rashita, 2020).
8. **Mechanism of Action**
   Nutraceuticals possess anti-obesity, antifungal, antioxidant, anti-inflammatory, and antibacterial characteristics for their utilization in therapeutic treatment for neurodegenerative

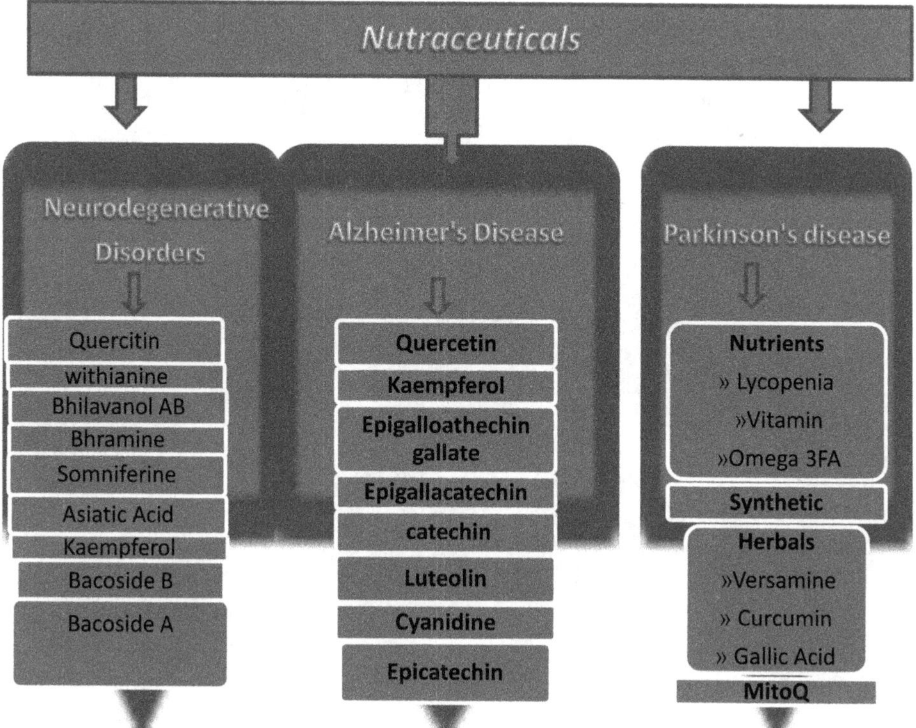

FIGURE 16.2 Chemical nature of nutraceuticals used in Parkinson's disease, Alzheimer's disease, and neurodegenerative disorders. (From Brower, V., *Nat. Biotechnol.*, 16, 728–731, 1998; Gupta, 2010; Rashita, 2020.)

disorders. As reported in a research study, a wide range of deaths globally are due to foodborne infections. Nutraceuticals contain major bioactive compounds, such as polyphenols and terpenoids, as an antimicrobial therapy in managing neurodegenerative disorders (Gutiérrez-Del-Río et al., 2018).

9. **Based on Neurodegeneration**
   Neurodegenerative disorders can develop due to protein misfolding. AD develops due to the abnormal misfolding of amyloid-β and tau proteins. Tau and TDP-43 malfunctioning proteins induce epilepsy and multiple tauopathies. Transactive response deoxyribonucleic acid (TAR DNA) and tau modifications result in trauma of the brain. Traumatic brain injury is also caused by alterations in Aβ proteins and binding protein-43 (TDP-43). Protein Aβ and another protein, α-synuclein, involved in detrimental degradation induce the cytotoxic cascade of cellular events in Down syndrome. All these misfolded proteins activate nuclear factor kappa-light-chain-enhancer. This kappa-light-chain enhancer further activates B cells (NF-κB). As a result, inflammatory cytokines such as interleukin-1β (IL-1β) proteins in combination with tumor necrosis factor-α (TNF-α) are produced, which further produce and activate destructive molecules like inducible nitric oxide synthase (iNOS) and cyclooxygenase (COX-2). Glutamate-induced oxidative damage occurs, causing mitochondrial toxicity, as shown in Figure 16.3. Furthermore, these misfolded proteins cause dysregulation of a major signaling pathway, the protein kinaseA/protein kinase B (PKB/PKA) pathway, ultimately resulting in synaptic process degradation and dysregulation in cognitive functions (Barber et al., 2006; Lin and Beal, 2006; Saldanha, Kelsey, et al., 2010; Asadi-Shekaari et al., 2012; Colin et al., 2015; Chauhan and Mehla, 2015).

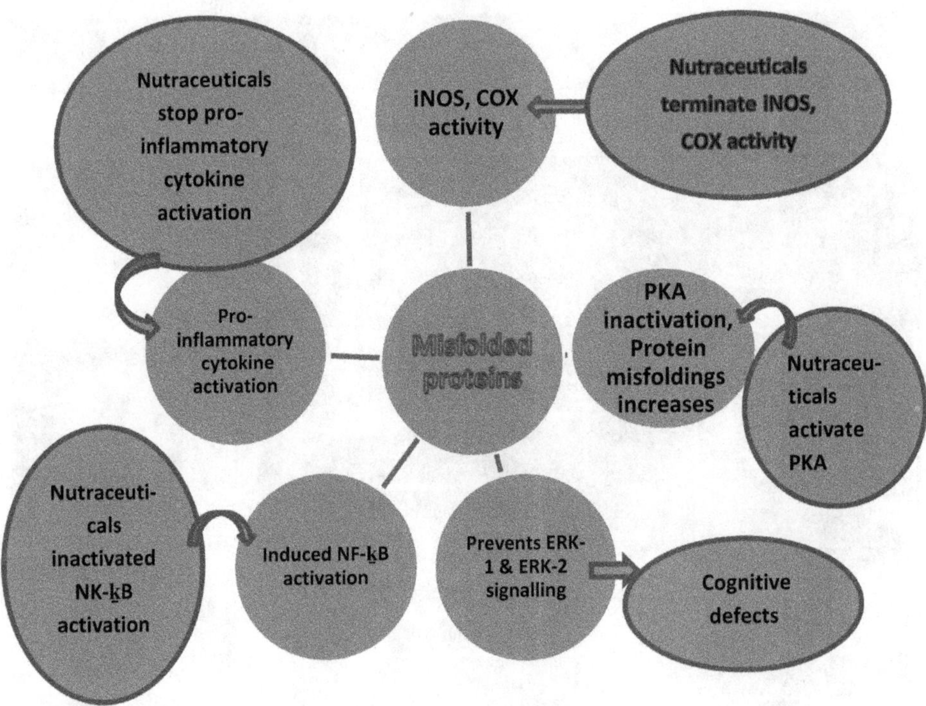

FIGURE 16.3 Neurodegeneration process activation due to misfolded proteins and other notable inhibitory proteins such as inflammatory proteins. Cytokines and interleukins, nuclear factor kappa-light-chain-enhancer, (iNOS) induction, and cyclooxygenase (COX) activation increases the neurodegeneration process and inflammation. Inhibiting cascade proteins by nutraceuticals in active form provides neuroprotective action with therapy. (From Rashita et al., 2020.)

16.2.2 Main Nutraceuticals in the Treatment of Neurodegenerative Disorders

Among all the nutraceuticals mentioned (Figure 16.2), the main nutraceuticals used to treat neurodegenerative disorders include bacoside B, brahmine, and bacoside A. Brahmine is an alkaloid extracted from brahmi (*Bacopa monnieri*), a nootropic plant. Brahmine possesses neurocognitive-enhancing properties; therefore, it has been utilized in Ayurveda. Bacoside B and bacoside possess saponin derivatives for therapeutic treatment of neurodegenerative disorders. Other nutraceuticals used to treat neurodegenerative disorders include bhilavanol A and B, quercetin, asiatic acid, somniferine, and kaempferol. Twithin. Former nutraceuticals have their targeted inhibitory protein misfolding specific properties against the activation and production of proinflammatory cytokines and their linked pathways. Quercetin and kaempferol have been reported as significantly decreasing the level of free radicals. Bhilavanol A and bhilavanol B, retrieved from bhallaatak, possess specific characteristics of inhibiting the inhibition of the acetylcholinesterase activation pathway and provide a beneficial anti-stress mechanism against stress-mediated neurodegeneration. The Mediterranean diet, including extra virgin olive oil, walnuts, and coffee, contains these nutraceuticals and is highly recommended due to their memory improvement mechanisms. Phenolic compounds in this category have provided notable medicinal benefits (González-Sarrías, et al, 2013; Preethi Pallavi and Sampath Kumar, 2018; Bungau et al., 2019; Barba et al, 2020).

16.2.2.1 Nutraceuticals in Alzheimer's Disease

The most commonly used nutraceuticals in AD are β-cryptoxanthin, including α- and β-carotenes, flavonoids (fruits, tea, coffee, and vegetables), anthocyanidins (cyanidin), and carotenoids (zeaxanthin, lycopene, lutein); flavones (apigenin and luteolin) have played a role as antioxidants during treatment of AD (Preethi Pallavi and Sampath Kumar, 2018; Ceskova and Silhan, 2018; Lama et al., 2020).

16.2.3 Significant Role of Nano-Nutraceuticals in Neurodegenerative Disorders

Notable and industrially approved phytochemicals such as rutin, resveratrol, blueberry, curcumin, herbal polyphenols, and carotenoids such as beta-carotene present in orange, green, and green-yellow vegetables and fruits possess specific anticancer characteristics to treat and prevent a huge number of neurodegenerative and cardiovascular disorders. Nano-nutraceuticals have exploited production and utilization as an anticancerous property to treat critical brain disorder cases. These substances have been reported to treat multiple disorders, including diabetes, cardiovascular diseases, cancer, and neurodegenerative disorders. They have been used for many years for numerous neurodegenerative disorders. Among them, various phytochemicals have provided physiological and significant health benefits. Nutraceuticals are products having medicinal importance as well as nutritional value. Different phytochemicals (nutraceuticals) such as rutin, resveratrol, curcumin, herbal polyphenols, and carotenoids (beta-carotene) that are present in orange, green, and yellow vegetables and fruits and possess anticancer properties have been used against various neurodegenerative disorders such as AD, HD, PD, and ALS. Nutraceuticals possess physiological benefits in providing a barrier against multiple acute to chronic diseases such as cardiovascular diseases, cancer, and neurodegenerative disorders and have the power to reduce the aging process. Multiple neurodegenerative disorders research studies by utilizing nano-nutraceuticals are enormous industrial and medical demands today. The role of nutraceuticals in modification of the molecular and cellular cascade is highlighted. These molecules act as supplementation therapy by facilitating the correction of misfolded proteins, preventing neurodegeneration. Therefore, nutraceuticals have proved an excellent alternative for the industry due to their decreased side effects and cost-effective and affordable price range (Brower, 1998; Lenaz, 2001; Ott et al., 2007; Chauhan et al., 2010; Sahni, 2012).

16.3 MULTIPLE APPROACHES UTILIZING NANO-NUTRACEUTICALS

Recently, multiple therapeutic and medicinal techniques have been reported for targeted delivery of nutraceuticals against AD. Their inherent antioxidant properties provided immune function improvement. In the therapeutic approach, curcumin used in different nanoformulations, like micelles, has provided bioactive targeted delivery.

16.3.1 Therapeutic

In the therapeutic approach, encapsulated nutraceuticals in the form of nanoformulations like micelles, nanocapsules, nanocrystals, nanocochleates, nanoparticles, and nanoemulsions have provided targeted delivery of encapsulated phytochemicals released from any selected nanoformulation, providing enhanced bioavailability and productive therapeutic results (Brower, 1998; Chauhan et al., 2010; Sahni, 2012). Most of the therapeutic interventions have been reported against AD by utilizing inherently antioxidant phytochemicals as nano-nutraceuticals to treat targeted neurodegenerative disorders. These inherent antioxidants provide beneficial health effects by improving immune system function. It has been reported that nutraceuticals have replaced many other synthetic drugs, such as when used in therapeutic treatment.

16.3.2 Medicinal Approach (Natural Antioxidant Treatment Pathway)

In the case of the medicinal approach, the major focus is on the bioavailability and solubility of the selected particles. One commonly used is curcumin, which possesses anticancer properties and is used as a natural antioxidant treatment pathway. The bioavailability and potential of curcumin have been approved by testing on different animals and cell culture models by making different nanoformulations like micelles, liposomes, and nanoparticles. These inherent antioxidants used in the therapeutic approach have provided beneficial health effects by improving immune system function. The different nano-nutraceutical particles introduced through therapeutic or medicinal approaches depend on the severity and the condition or disorder to be treated. Moreover, when used in adjunction therapy by augmentation of multiple pathways, such as increased uptake of inhibited monoamines, nutraceuticals help increase the effectiveness of various medications by providing enhanced neurobiological effects (Sahni, 2012; Van Der Burg et al., 2019).

16.3.3 Effectiveness of Nano-Nutraceutical Technology Utilizing Multiple Approaches in the Treatment of Neurodegenerative Disorders

As multiple neurodegenerative disorders have been treated utilizing therapeutic and medicinal approaches, the success of any of these techniques largely depends upon multiple points:

(a) Precise selection of type and properties of therapeutic and medicinal nano-nutraceutical particle.
(b) Large-scale availability of selected nano-nutraceuticals is of significant concern.
(c) Previous industrial value of selected therapeutic and medicinal nano-nutraceutical particle to be adopted.
(d) Bioactivity enhancement role of selected nano-nutraceuticals has been approved previously.
(e) Higher solubility is required, with the absorption of these nutraceutical chemicals inside the gastrointestinal tract.
(f) Selected nano-nutraceutical particles must be targeted against the disease to be treated, and they must possess anticancer properties.

Hence, research is ongoing into various newly introduced neurodegenerative disorders, but complications arise in their treatment by nano-nutraceuticals. The latest nano-nutraceutical technology has provided all-in-one effective treatment methods, but it massively depends on how successful those nutraceuticals are in the industry, suggesting desirable therapeutic results and benefits. In the case of the therapeutic approach, this depends on the utilization of selected pharmaceutical substances used already in the treatment and prevention of multiple neurodegenerative disorders. Of course, therapeutic interventions are highly targeted for many critical cases as compared with the therapeutic approach; however, in therapeutic treatment, more chances of complications in treating delicate areas of the brain could occur, but the utilization of therapeutic interventions in the later stages of disease has worked more efficiently than medicinal treatment. In both approaches, multiple treatment options and safety issues are equally overviewed and followed. Nano-nutraceuticals have exploited production and utilization as an anticancerous property to treat critical brain disorder cases. Most therapeutic interventions have been reported against Alzheimer's disease by utilizing inherent antioxidant phytochemicals as a nano-nutraceutical to treat targeted neurodegenerative disorders. In the medicinal approach, the major focus is on the bioavailability and solubility of the selected particle. One commonly used compound is curcumin, which possesses anticancer properties. The bioavailability and potential of curcumin have been proved by testing on different animals and cell culture models in different nanoformulations like micelles, liposomes, and nanoparticles (Brower, 1998; Chauhan et al., 2010; Sahni, 2012).

16.3.4 Drawbacks in the Utilization of Nano-Nutraceuticals

The major drawbacks are poor solubility and no biological activity. These issues have been addressed by introducing and using the latest nanotechnology, providing efficient and targeted delivery of nutraceuticals with the significant aim of increasing their biological activity. As in the case of phytochemicals (nutraceuticals), lower bioavailability and poor solubility in the gastrointestinal fluid have been observed due to their incomplete absorption in the gastrointestinal tract, resulting in diminished and poor biological activity, a major scientific concern today. In the future, due to the increasing demand for these nano-nutraceutical components, more research needs to be done to enhance their mass production to cope with these limitations. Therefore, to increase the availability and *in vitro* mass production of phytochemicals (nutraceuticals), more production strategies need to be implemented, so that fewer deficiency issues will be faced in utilizing nano-nutraceutical technology. However, improvements in the therapeutic potential and bioavailability of phytochemicals have been reported in various *in vivo* studies using nanoformulations, but more *in vivo* studies are still needed to improve the efficiency of results. Besides this, the safety aspects and potential toxicity of nanoencapsulated nutraceuticals to the human body need to be addressed in the future.

Moreover, previously approved and successful therapeutic and medicinal nano-nutraceutical particle selections have shown successful delivery results. The selected nano-nutraceutical must possess bioactivity enhancement properties, and higher solubility with the absorption of nutraceutical chemicals inside the gastrointestinal tract is an active research area, because in the case of phytochemicals (nutraceuticals), lower bioavailability and poor solubility in gastrointestinal fluid are noted due to their incomplete absorption in the gastrointestinal tract, resulting in diminished and poor biological activity, a major scientific concern today. Therapeutic and medicinal delivery of selective nano-nutraceutical particle selected against the disease to be treated must possess anticancerous andante oxidant properties. However, the major drawbacks are issues of poor solubility and lack of biological activity, which have been addressed by introducing and using the latest nanotechnology, providing efficient and targeted delivery of nutraceuticals to increase their biological activity. Therefore, to increase the availability and *in vitro* mass production of phytochemicals (nutraceuticals), more production strategies need to be implemented so that fewer deficiency issues will occur while utilizing nano-nutraceutical technology.

16.3.5 Conclusion and Recommendations

Nano-nutraceuticals have played a significant role in providing treatment against multiple neurodegenerative disorders. Due to innumerable characteristics, such as antioxidant, healing, anti-inflammatory, and hypolipidemic, possessed by nano-nutraceuticals as described in the chapter, nano-nutraceuticals are in high demand in both medicine and industry. The valuable herbal molecules used in medicines possess great potential to prevent and treat life-threatening disorders, notably neurodegeneration. Inhibition of protein folding leading to a neuroprotective response provided by therapy has helped protect against life-threatening diseases. The large-scale availability of these selected nano-nutraceuticals is of major concern. In the future, due to the increasing demand for these nano-nutraceutical components, more research needs to be done to enhance the production of nutraceutical components to cope with their limitations. Hence, due to the remarkable properties possessed by nutraceuticals, consuming these nano-nutraceuticals in appropriate dosages promotes good mental health and provides a cure for mental disorders (Brower, 1998; Chauhan et al., 2010; Sahni, 2012).

REFERENCES

Asadi-Shekaari, M., Kalantaripour, T. P., Nejad, F. A., Namazian, E., Eslami, A. (2012). The anticonvulsant and neuroprotective effects of walnuts on the neurons of rat brain cortex. *Avicenna J Med Biotechnol*, 4, 155.

Andlauer, W., Fürst, P. (2002). Nutraceuticals: A piece of history, present status and outlook. *Food Res Int*, 35, 171–176.

Barba, F. J., Putnik, P., Kovacevic, D. B. (2020). *Agri-Food Industry Strategies for Healthy Diets and Sustainability: New Challenges in Nutrition and Public Health*; Press, A., Ed.; MPS Limited Chennai India: Tamil Nadu, India.

Bhaskarachary, K. (2016). Traditional foods, functional foods and nutraceuticals. *Proc Indian Natl Sci Acad*, 82, 1565–1577.

Bhat, Z. F., Bhat, H. (2011). Milk and dairy products as functional foods: A review. *Int J Dairy Sci*, 6, 1–12.

Brower, V. (1998). Nutraceuticals: Poised for a healthy slice of the healthcare market? *Nat Biotechnol*, 16, 728–31.

Bungau, S., Abdel-Daim, M. M., Tit, D. M., Ghanem, E., Sato, S., Maruyama-Inoue, M., Yamane, S., Kadonosono, K. (2019). Health benefits of polyphenols and carotenoids in age-related eye diseases. *Oxidative Med Cell Longev.*

Bungau, S. G., & Popa, V. C. (2015). Between religion and science some aspects concerning illness and healing in antiquity. *Transylv Rev*, 24, 3–18.

Ceskova, E., Silhan, P. (2018). Novel treatment options in depression and psychosis. *Neuropsychiatr Dis Treat*, 14, 741.

Chauhan, D., Mehla, K., Sood, P., Nair, A. (2010). An overview of neutraceuticals: Current scenario. *J Basic Clin Pharm*, 1, 55–62.

Chauhan, N. B., Mehla, J. (2015). Ameliorative effects of nutraceuticals in neurological disorders. In *Bioactive Nutraceuticals and Dietary Supplements in Neurological and Brain Disease*. Elsevier: Amsterdam, the Netherlands, 245–260.

Chung, S., Sonntag, K. C., Andersson, T., Bjorklund, L. M., Park, J. J., Kim, D. W., et al. (2002). Genetic engineering of mouse embryonic stem cells by Nurrl enhances differentiation and maturation into dopaminergic neurons. *Eur J Neurosci*, 16, 1829–1838. doi: 10.1046/j.1460-9568.2002.02255.

Colín-González, A. L., Ali, S. F., Túnez, I., Santamaría, A. (2015). On the antioxidant, neuroprotective and anti-inflammatory properties of S-allyl cysteine: An update. *Neurochem Int*, 89, 83–91. [CrossRef] [PubMed].

Dohrmann, D. D., Putnik, P., Bursac´ Kovacˇevic´, D., Simal-Gandara, J., Lorenzo, J. M., Barba, F. J. (2019). Japanese, Mediterranean and Argentinean diets and their potential roles in neurodegenerative diseases. *Food Res Int*, 120, 464–477.

El-Agnaf, O. M., and Irvine, G. B. (2002). Aggregation and neurotoxicity of alpha-synuclein and related peptides. *Biochem Soc Trans*, 30, 559–565. doi: 10.1042/bst0300559.

González-Sarrías, A., Larrosa, M., García-Conesa, M. T., Tomás-Barberán, F. A., Espín, J. C. (2013). Nutraceuticals for older people: Facts, fictions and gaps in knowledge. *Maturitas*, 75, 313–334.

Gosálbez, L. R., Amón, D. (2015). Probiotics in transition: Novel strategies. *Trends Biotechnol*, 33, 195–196.

Gutiérrez-Del-Río, I., Fernández, J., Lombó, F. (2018). Plant nutraceuticals as antimicrobial agents in food preservation: Terpenoids, polyphenols and thiols. *Int J Antimicrob Agents*, 52, 309–315.

Hardy, J., & Selkoe, D. J. (2002). The amyloid hypothesis of Alzheimer's disease: progress and problems on the road to therapeutics. *Science*, 297, 353–356. doi:10.1126/science.1072994.

Kelsey, N. A., Wilkins, H. M., Linseman, D. A. (2010). Nutraceutical antioxidants as novel neuroprotective agents. *Molecules*, 15, 7792–7814.

Lenaz, G. (2001). The mitochondrial production of reactive oxygen species: Mechanisms and implications in human pathology. *IUBMB Life*, 52, 159–164.

Lama, A., Pirozzi, C., Avagliano, C., Annunziata, C., Mollica, M. P., Calignano, A., Meli, R., Mattace Raso, G. (2020). Nutraceuticals: An integrative approach to starve Parkinson's disease. *Brain Behav Immun Health*, 2, 100037.

Mastrokolias, A., Ariyurek, Y., Goeman, J. J., Van Duijn, E., Roos, R. A., van Der Mast, R. C., et al. (2015). Huntington's disease biomarker progression profile identified by transcriptome sequencing in peripheral blood. *Eur J Hum Genet.* doi: 10.1038/ejhg.2014.281.

Megha, A., Abhijit, B. (2015). Molecular diagnostics of neurodegenerative disorders. *Front Mol Biosci,* 2(54). doi: 10.3389/fmolb.2015.00054.

Molochnikov, L., Rabey, J. M., Dobronevsky, E., Bonucelli, U., Ceravolo, R., Frosini, D., et al. (2012). A molecular signature in blood identifies early Parkinson's disease. *Mol Neurodegener* 7, 26. doi: 10.1186/1750-13 26-7-26.

Ott, M., Gogvadze, V., Orrenius, S., Zhivotovsky, B. (2007). Mitochondria, oxidative stress and cell death. *Apoptosis*, 12, 913–922.

Prakash, V., Boekel. (2010). Nutraceuticals: Possible future ingredients and food safety aspects. In *Ensuring Global Food Safety*; Academic Press: Cambridge, MA, 333–338.

Preethi Pallavi, M. C., Sampath Kumar, H. M. (2018). Chapter 8—Nutraceuticals in prophylaxis and therapy of neurodegenerative diseases. In *Discovery and Development of Neuroprotective Agents from Natural Products*; Brahmachari, G., Ed.; Elsevier: Amsterdam, the Netherlands, 359–376.

Putnik, P., Gabric´, D., Roohinejad, S., Barba, F. J., Granato, D., Lorenzo, J. M., Bursac´ Kovacˇevic´, D. (2019). Bioavailability and food production of organosulfur compounds from edible Allium species. In *Innovative Thermal and Non-Thermal Processing, Bioaccessibility and Bioavailability of Nutrients and Bioactive Compounds*; Francisco, J., Barba, J. M. A. S., Giancarlo Cravotto, J., Lorenzo, M., Eds.; Woodhead Publishing: Cambridge, UK, 2019, 293–308.

Rachakonda, V., Pan, T. H., & LE, W. D. (2004). Biomarkers of neurodegenerative disorders: how good are they? *Cell Res*, 14, 347–358. doi: 10.1038/sj.cr.7290235.

Ramalingum, N., Mahomoodally, M. F. (2014). The therapeutic potential of medicinal foods. *Adv Pharmacol Sci*, 354264.

Rashita, M., Tapan, B., et al. (2020). Nutraceuticals in Neurological disorders. *Int J Mol Sci*, 21, 4424. doi:10.3390/ijms2112442.

Runne, H., Regulier, E., Kuhn, A., Zala, D., Gokce, O., Perrin, V., et al. (2008). Dysregulation of gene expression in primary neuron models of Huntington's disease shows that polyglutamine-related effects on the striatal transcriptome may not be dependent on brain circuitry. *J Neurosci*, 28, 9723–9731. doi: 10.1523/JNEUROSCI.3044-08.

Sahni, J. K. (2012). Exploring delivery of nutraceuticals using nanotechnology. *Int J Pharm Investig*, 2(2), 53. doi: 10.4103/2230-973X.100033.

Van der Burg, K. P., Cribb, L., Firth, J., Karmacoska, D., Sarris, J. (2019). Nutrient and genetic biomarkers of nutraceutical treatment response in mood and psychotic disorders: A systematic review. *Nutr Neurosci*, 1–17.

Wild, E. J., Boggio, R., Langbehn, D., Robertson, N., Haider, S., Miller, J. R., et al. (2015). Quantification of mutant huntingtin protein in cerebrospinal fluid from Huntington's disease patients. *J Clin Invest*, 125, 1979–1986. doi: 10.1172/JCI80743.

Williams, R. J., Mohanakumar, K. P., Beart, P. M. (2015). Neuro-nutraceuticals: The path to brain health via nourishment is not so distant. *Neurochem Int*, 89, 1–6.

Williams, R. J., Mohanakumar, K. P., Beart, P. M. (2016). Neuro-nutraceuticals: Further insights into their promise for brain health. *Neurochem Int*, 95, 1–3.

Zucko, J., Starcevic, A., Diminic, J., Oros, D., Mortazavian, A. M., Putnik, P. (2020). Probiotic: Friend or foe? *Curr Opin Food Sci*, 32, 45–49.

17 Use of Nano-Nutraceuticals as Anti-Inflammatory Tools in Cardiovascular Disease

*Rajat Goyal, Anjali Saharan, and Rupesh K. Gautam**

CONTENTS

17.1 Introduction 431
17.2 Studies of Relationships between Inflammation, Atherosclerosis, and CVDs 432
17.3 Role of Nanotechnology in Nutraceuticals 433
17.4 Nano-Nutraceuticals Used as Anti-Inflammatory Tools 433
17.5 Nano-Nutraceuticals Used in Cardiovascular Disorders 434
17.6 Conclusion 437
References 438

17.1 INTRODUCTION

Cardiovascular disorders (CVDs) such as peripheral arterial disease, cardiac heart failure, rheumatic arthritis, cerebrovascular disorders, pulmonary embolism, congenital heart defects, and deep vein thrombosis are thought to be caused by inflammation-induced endothelial aberrations, tobacco smoking, and nutritional habits. For decades, preventing CVDs with anti-inflammatory drugs has been a difficult undertaking. Nutraceuticals have numerous wellbeing benefits, and their usage in the prevention of CVDs and associated conditions such as hypertension, atherosclerosis, stroke, and heart attack can be quite beneficial (Jain et al., 2018).

The term "nutraceutical" originates from the combination of "nutrition" and "pharmaceutical." A nutraceutical is well defined as "a diet supplement that delivers a concentrated form of a biologically active component of food in a non-food matrix to enhance health" (Beconsini et al., 2020). The assimilation of nutraceuticals into food items is a simple and easy technique for the production of innovative and novel functional foods (FFs) (Gonclaves et al., 2018).

Evidence has shown that the mechanistic properties of natural products comprise a wide range of biological processes, involving antioxidant defense activation, cell survival–associated gene expression, signal transduction pathways, cell differentiation and proliferation, and mitochondrial integrity prevention. These activities appear to be important in protecting against the pathologies of a variety of chronic illnesses. Apart from that, nutraceutical derivatives in the form of dietary fibers, antioxidants, minerals, vitamins, and omega-3 polyunsaturated fatty acids, as well as physical activity, are indicated for the treatment therapy of CVDs (Rajasekaran et al., 2008). Despite their widespread usage, the therapeutic efficacies of nutraceuticals are frequently hampered by their limited oral bioavailability, which is caused by a variety of physical, chemical, physicochemical, and physiological variables. Researchers have used various delivery mechanisms to enhance the bioavailability of nutraceuticals in order to improve their efficacy (Ting et al., 2014).

* Corresponding author: drrupeshgautam@gmail.com

DOI: 10.1201/9781003244721-17

17.2 STUDIES OF RELATIONSHIPS BETWEEN INFLAMMATION, ATHEROSCLEROSIS, AND CVDs

In commercial countries, an "unhealthy" diet is alleged to be a major contributor to the rise in atherosclerotic cardiac disorders. Endothelial dysfunction is assumed to play a significant role in the expansion and progression of atherosclerosis, and portrayal of the endothelial effects of a variety of nutraceuticals could provide valuable information about their potential involvement in protection against CVDs (Zuchi et al., 2010).

Nutraceutical products are health-promoting foods or nutritional supplements that can be given to those who are on the verge of developing cardiovascular disease (Pillai et al., 2021). There are now 231 clinical trials with nutraceuticals. The bulk of them are concerned with the efficacy of nutraceuticals in the treatment of CVDs such as diabetes, hyperlipidemia, hypercholesterolemia, and hypertension (Morbidelli et al., 2018). Much experimental and clinical research has found that hypoxia and oxidative stress enhance arterial wall inflammation and play a role in cardiovascular disease (Pearson et al., 2003).

In the early stages of inflammation-induced atherosclerosis and CVDs, detecting the aberrant symptoms or biomarkers is very difficult. For example, it has been discovered that about 50% of strokes and heart attacks occur in people who have normal cholesterol and triglyceride levels. Tedious physical examinations are thus required to keep track of the dangers of cardiac diseases (Hughes et al., 2016). Endothelial inflammation is caused by a malfunctioning immune system, which allows extraneous pathogens to produce inflammation. The immune system's principal role is to protect the body from pathogenic attacks and maintain the integrity of physiological functioning. Because inflammation and the immune system are so intimately linked, immune system dysfunction can lead to the breakdown of cells/tissues, which can cause inflammatory or pathological alterations in the body organs (Scrivo et al., 2011; Russ et al., 2012). Various nutraceutical compounds used in inflammation and cardiovascular disorders are depicted in Figure 17.1.

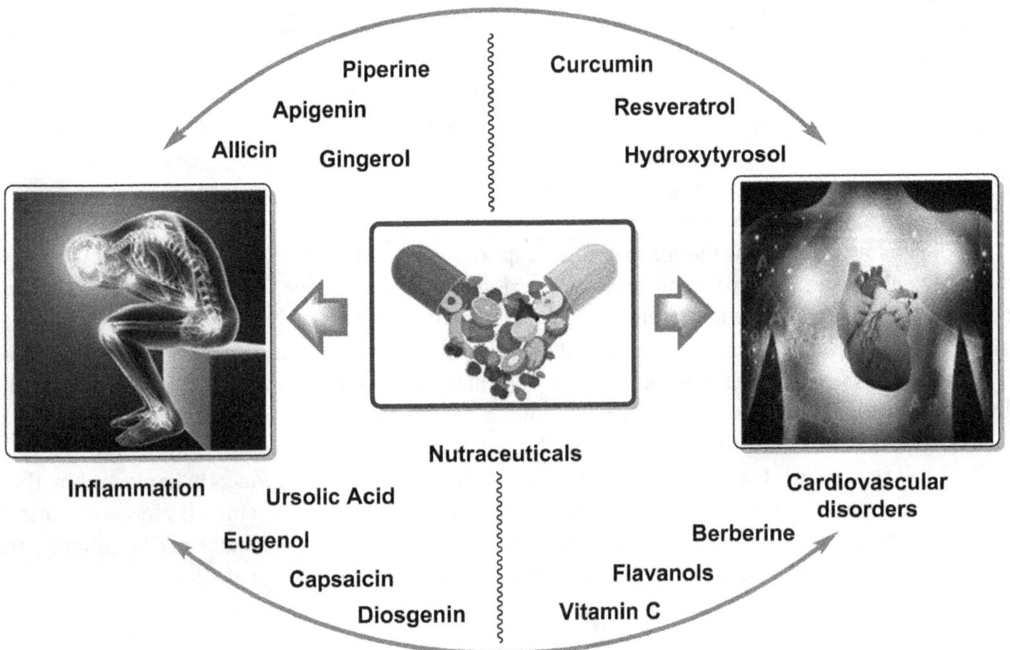

FIGURE 17.1 Nutraceutical compounds used in inflammation and cardiovascular disorders.

17.3 ROLE OF NANOTECHNOLOGY IN NUTRACEUTICALS

Several research organizations have been interested in the combination of nanotechnology and nutraceuticals in recent decades. Unfortunately, the usage of many nutraceutical products and their health advantages are limited by their inadequate physicochemical characteristics, including poor absorption, low stability, low water solubility, and probable chemical alterations after administration. Nanotechnology has been adopted as a breakthrough invention in activating the therapeutic characteristics of nutraceutical products for human wellbeing, thus boosting their effectiveness in a variety of ailments, based on their potential efficacy and limiting aspects. As a result, nanotechnology approaches might be a new frontier in supplementary nutrition (Paolini et al., 2021). The role of nanotechnology in the delivery of nutraceuticals is illustrated in Figure 17.2.

Fundamentally, nanoformulations of nutraceutical products follow the general principles of nanotechnology approaches. The therapeutic usage of nanotechnology comprises the delivery of small-molecular drugs, proteins, nucleic acids, and peptides. Nanoparticle therapy is a newer treatment option for heart diseases and other inflammatory conditions. When compared with medicinal agents, nanoparticles have more sophisticated pharmacological effects. Various nutraceutical products that have been formulated as nanoparticles (Nair et al., 2010) are described in Table 17.1.

17.4 NANO-NUTRACEUTICALS USED AS ANTI-INFLAMMATORY TOOLS

Inflammation is the biological response of the immune system, which can be triggered by numerous factors and shows as its primary characteristics the symptoms of redness, heat, swelling, and pain. It can be triggered by both infectious and non-infectious agents and causes cell damage by

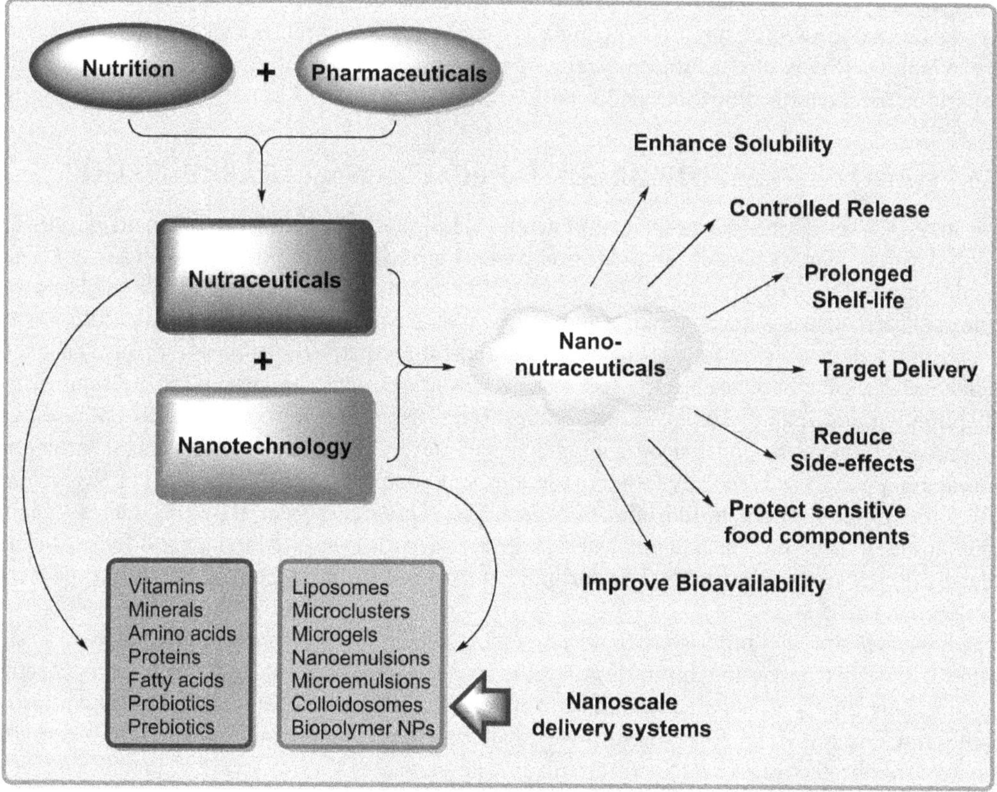

FIGURE 17.2 Role of nanotechnology approaches in nutraceutical delivery systems.

TABLE 17.1
List of Various Nutraceuticals Formulated as Nanoparticles and Their Properties

S. No.	Phytochemicals	Nanomaterials used	Properties
1.	Quercetin	Polylactide, polylactic-co-glycolic acid (PLGA)	Increased biocompatibility
2.	Eugenol	Chitosan	Increased hydrophilicity and biocompatibility
3.	Resveratrol	Poly-caprolactone-polyethylene glycol, polyethylene glycol-polycaprolactone	Increased bioavailability, biocompatibility, and circulation time
4.	Curcumin	Polylactic-co-glycolic acid, alginate-chitosan, poly(butyl)cyanoacrylate, casein	Increased biocompatibility
5	Daidzein	Polyethylene glycolated phospholipid	Increased stability
6.	Ellagic acid	Polylactic-co-glycolic acid, polycaprolactone	
7.	Ferulic acid	Bovine serum albumin	Increased bioavailability, good absorption
8.	Ursolic acid	Soybean phospholipid-poloxamer-188	Enhanced permeability
9.	Toxifolin	Polyvinyl pyrrolidone	Enhanced permeability
10.	Genistein	Egg lecithin MCT or ODD	Increased bioavailability
11.	Daidzein	PEGylated phospholipid	Enhanced stability

activating and triggering the inflammatory pathways, i.e., NF-κB, mitogen-activated protein kinase (MAPK) pathways, and cellular responses to cytokines (e.g., interleukin [IL]-6) and growth factors (e.g., epidermal growth factor [EGF]) (Aggarwal et al., 2009). Nutraceuticals alleviate inflammatory symptoms. Some of the nutraceuticals, such as allicin, apigenin, naringin, ursolic acid, and hesperidin, are mentioned in Table 17.2.

17.5 NANO-NUTRACEUTICALS USED IN CARDIOVASCULAR DISORDERS

Nowadays, CVDs are the foremost cause of death globally and are considered as the major growing health concern. Dietary aspects and nutrition play a significant part in the pathogenesis of CVDs. FFs or nutraceutical products are reported to have therapeutic properties and reduce the hazard of various chronic diseases beyond their rudimentary nutritional standards (Alissa et al., 2012).

The nutraceuticals mentioned earlier act as anti-inflammatory agents in the form of nanoemulsions that have been used to protect against cardiotoxicity. They mainly act by suppressing inflammation and lipid oxidative stress. From different sources, various chemical entities were isolated and found to be pharmacologically active as cardiotonics, such as lycopene, a carotenoid natural compound that is mainly found in tomatoes, pink guavas, and watermelons. It was reportedly found to reduce myocardial injury and ischemia (Quagliariello et al., 2018). Omega-3 fatty acids stored in food (such as fish oils and flax seeds) were also reported to be good for cardiovascular health by treating peripheral hyperlipidemia via reduction of blood triacylglycerol levels (Leslie et al., 2015).

Polyphenols are also reported to have numerous benefits by regulating several biological signaling pathways to accomplish the desired efficacy (Tangney et al., 2013). Among polyphenols, flavonoids are highly effective (Marzocchella et al., 2011). Phenolic compound such as resveratrol, mainly found in red wine), which contribute due to their anti-inflammatory properties, is revealed to have cardioprotective effects against peroxynitrite oxidation via protection of low-density lipoprotein (LDL) (Borriello et al., 2010). Some of the flavones, such as daidzein, cocoa flavanols, and green tea catechins, are nutraceuticals that work as anti-inflammatory and cardioprotective tools by

TABLE 17.2
Various Nutraceuticals Used as Anti-Inflammatory Agents

S. No.	Nutraceuticals with their chemical structures	Responses	References
1.	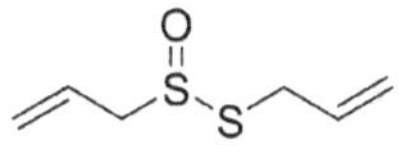 Allicin	It is used to suppress and inhibit tumor necrosis factor (TNF)-α-induced secretion, which causes suppression in the intestinal epithelial cells *in vitro* and the degradation of IkB.	(Lang et al., 2004)
2.	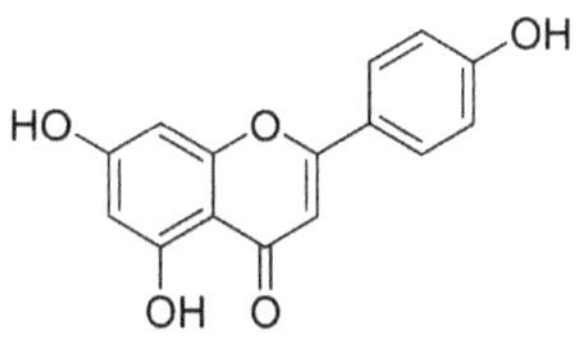 Apigenin	It was concluded that apigenin showed therapeutic effects on neuroinflammation in glial fibrillary acidic protein interleukin-6 (GFAPIL6).	(Liang et al., 2017)
3.	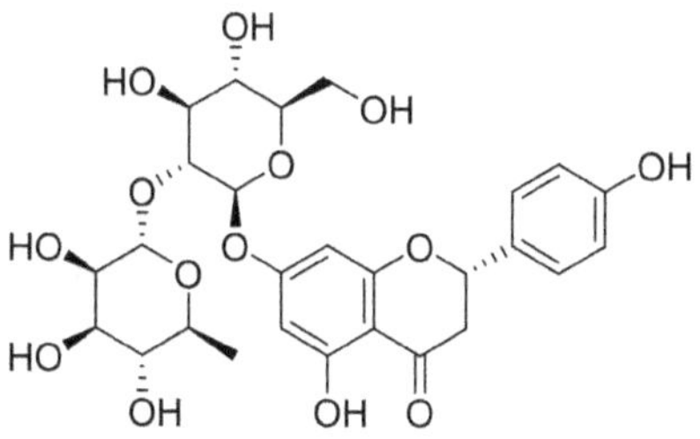 Naringin	It was concluded that naringin was used to decrease the inflammatory cell infiltration and fibrosis in the plasma or liver.	(Alam et al., 2014)
4.	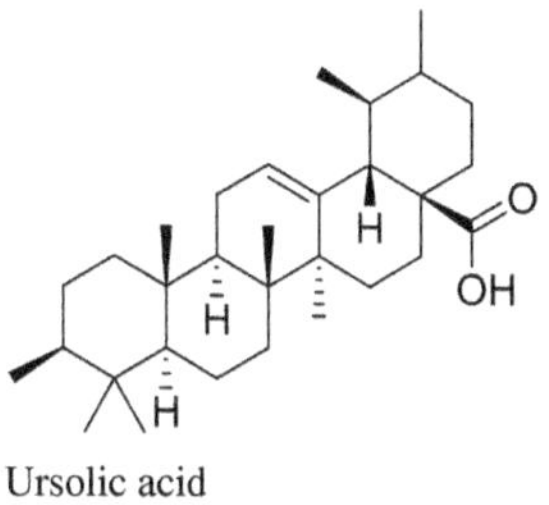 Ursolic acid	It has been reported that ursolic acid–containing piperazine, triazolone, and oxadiazole compounds have shown potent anti-inflammatory activities.	(Mlala et al., 2019)
5.	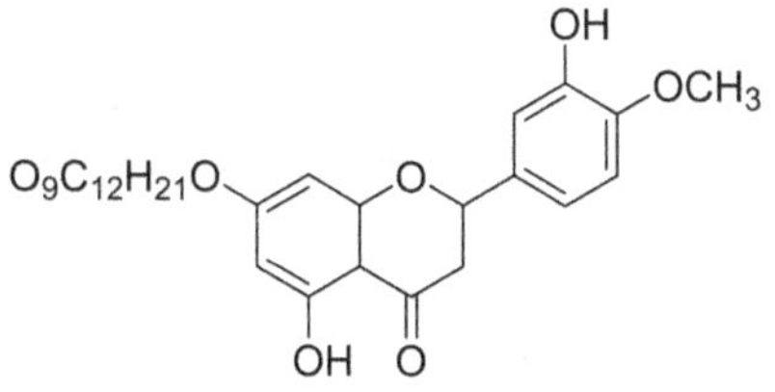 Hesperidin	It was reported that hesperidin was used to inhibit the receptor binding of angiotensin-converting enzyme (ACE)2, which results in suppressing the SARS-CoV-2 viral shedding and loading in nasal passages and treats inflamed lungs.	(Agrawal et al., 2021)
6.	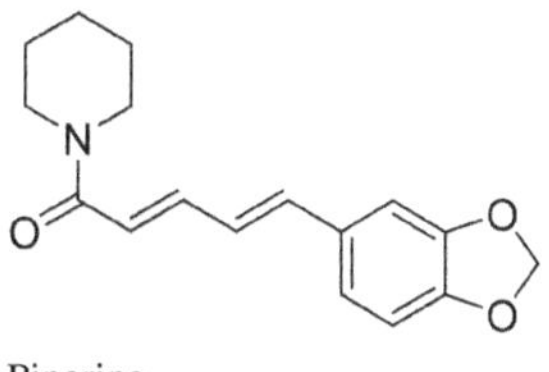 Piperine	Piperine was found to inhibit the expression of interleukins, which results in reducing the production of prostaglandin E2 with variation in dose frequency, and it reduced the inflammatory area in body joints.	(Bang et al., 2009)

(*Continued*)

TABLE 17.2 (CONTINUED)
Various Nutraceuticals Used as Anti-Inflammatory Agents

S. No.	Nutraceuticals with their chemical structures	Responses	References
7.	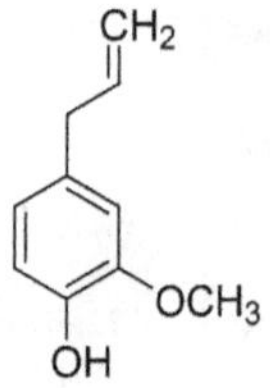 CH_2 OCH$_3$ OH Eugenol	Eugenol was found to inhibit anti-inflammatory leukocyte migration via different stimuli.	(Barboza et al., 2018)
8.	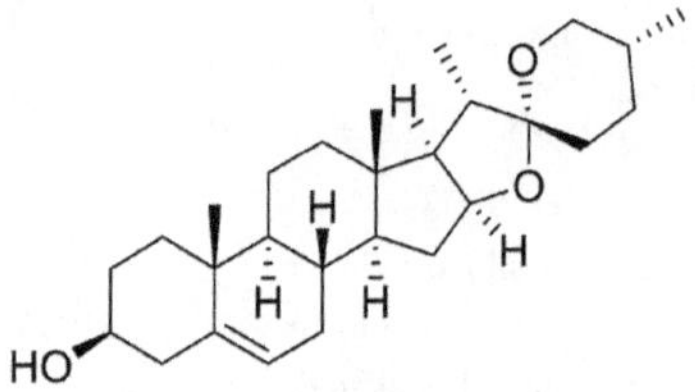 O H O H H H H HO Diosgenin	Diosgenin exhibited moderate efficacy in downregulation of cyclooxygenase (COX)-2 mediators of inflammation.	(Cai et al., 2020)
9.	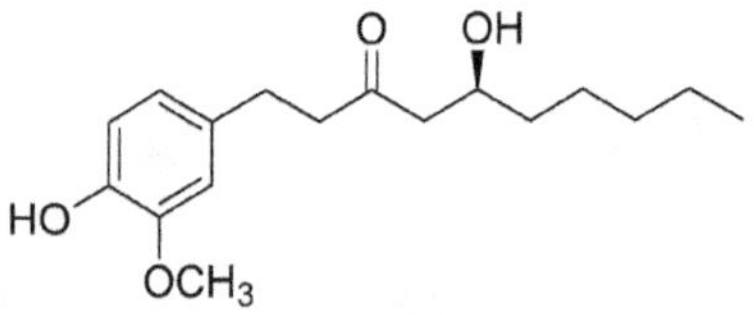 O OH HO OCH_3 Gingerol	Gingerol and its derivatives suppressed the biosynthesis of leukotrienes and prostaglandins and inhibited the synthesis of proinflammatory cytokines and cyclooxygenase inhibitors.	(Mashhadi et al., 2013)
10.	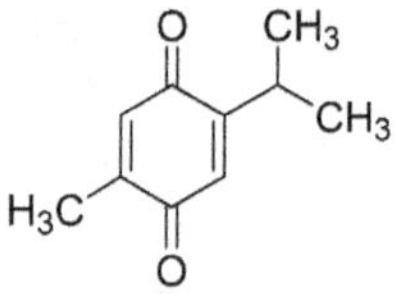 O CH_3 CH_3 H_3C O Thymoquinone	Thymoquinone inhibits the COX enzymes, which in turn, regulate the growth factor and cytokines, resulting in decreased inflammation, and acts as an anti-inflammatory mediator.	(Goyal et al., 2017)
11.	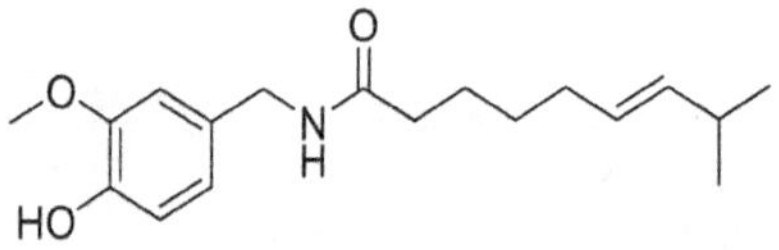 O O N H HO Capsaicin	Capsaicin acts as an anti-inflammatory agent by reducing the formation of inflammatory molecules in lipopolysaccharide (LPS)-stimulated murine peritoneal macrophages.	(Kim et al., 2003)
12.	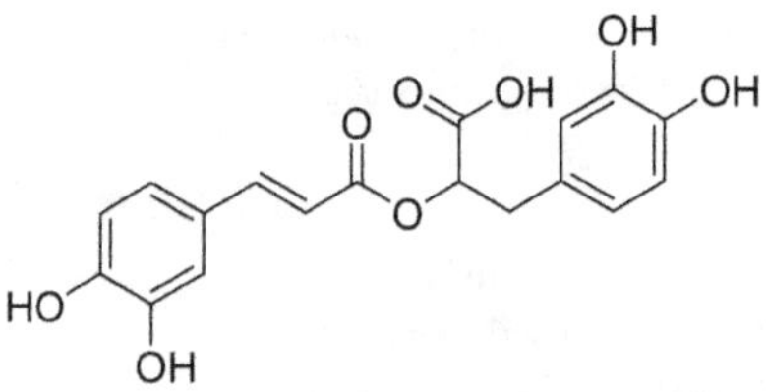 OH O OH OH O O HO OH Rosmarinic acid	It was concluded that rosmarinic acid inhibits LPS-induced nitric oxide and causes activation of receptors.	(Nadeem et al., 2019)

attenuating the cytokines, such as interleukin receptors, i.e., IL-6 and IL-8 (Hsu et al., 2007). The formulation of daidzein-loaded solid lipid nanoparticles with PEGylated phospholipid as an additive showed better efficacy towards the cardiovascular system via reduction of myocardial oxygen consumption and coronary resistance in the heart (Gao et al., 2008). A list of nutraceuticals used in the treatment and management of CVDs is provided in Table 17.3.

TABLE 17.3
Various Nutraceuticals Used in Cardiovascular Disorders

S. No.	Nutraceuticals with their chemical structures	Response	References
1	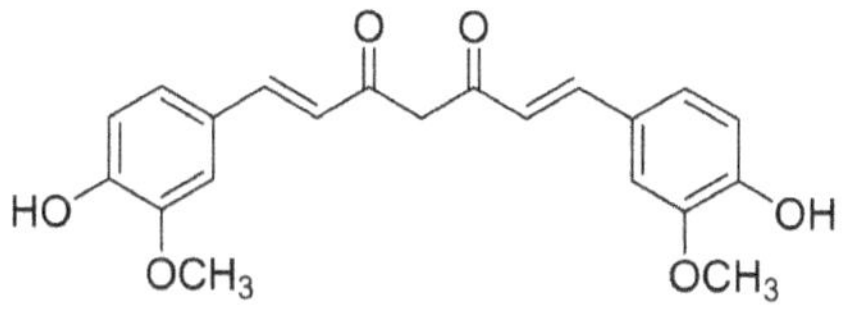 Curcumin	Curcumin was found to reduce the oxidative stress level and suppress arterial stiffness.	(Ramırez-Tortosa et al., 1999)
2	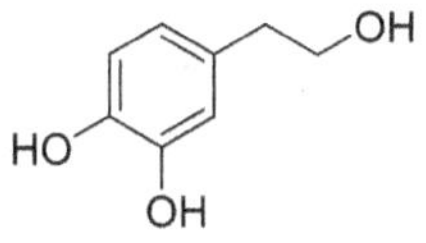 Hydroxytyrosol	Hydroxytyrosol was found in oily compounds having very-low-density lipoprotein, which increased the level of high-density lipoprotein in blood plasma.	(Mangas-Cruz et al., 2001)
3.	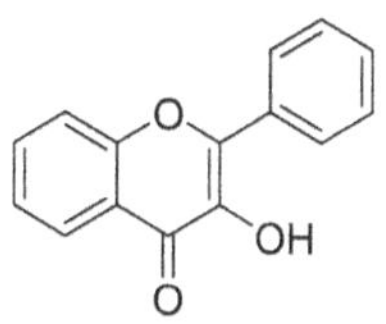 Flavanols	Flavanols were found to suppress arterial stiffness with endothelial functioning.	(Heiss et al., 2015)
4	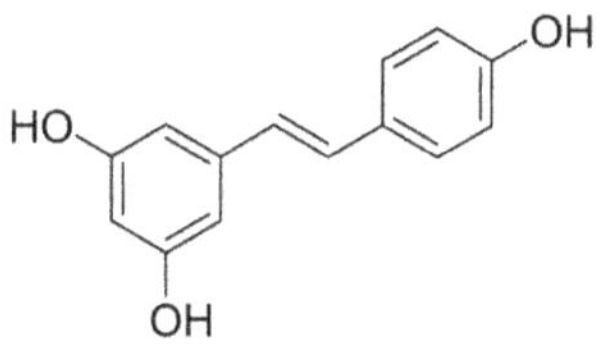 Resveratrol	Resveratrol was reported to suppress the formation of foam cells via Ox-LDL uptake inhibition and reduce the hazards of cardiovascular disorders.	(Voloshyna et al., 2013)
5	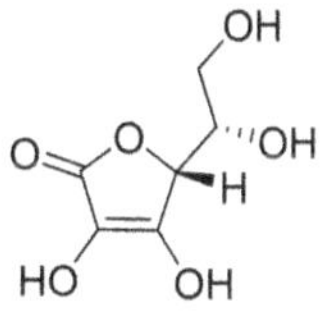 Vitamin C	It was reported to increase the endothelial functioning via increasing the activity of nitric oxide synthase (NOS) and suppress cardiovascular disorders.	(Khaw et al., 2001)
6	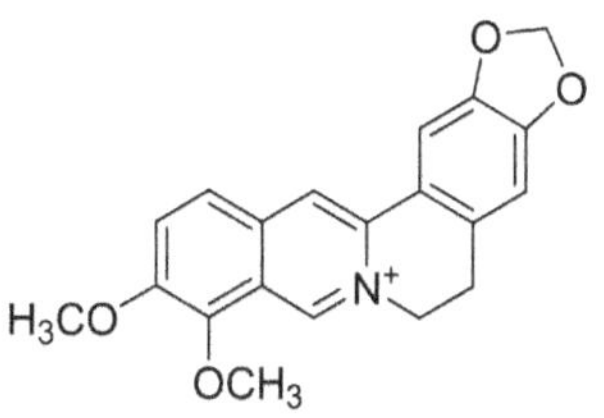 Berberine	It was found to enhance the cholesterol efflux gene expression and reduce the levels of cholesterol and triglycerides.	(Kong et al., 2004)

17.6 CONCLUSION

In this chapter, different nutraceutical compounds having promising anti-inflammatory and cardioprotective benefits were discussed in detail along with their reported outcomes. It was concluded that nutraceuticals act as a fusion in the health sector with multivariant benefits by providing a bridge between nutrition and medicines. They are predicted to have relatively lower toxicity and consequent adverse effects than medications used to cure identical symptoms, since they are derived from natural nutritional resources. Irrespective of this, various nutraceuticals are still not prominently used due to their lower bioavailability profile. Researchers have used various delivery mechanisms to enhance the bioavailability of nutraceuticals in order to improve their efficacy. From all these, it is evident that nanotechnology has promising results in providing its best potential for delivering nutraceuticals in inflammation and cardiovascular diseases.

REFERENCES

Aggarwal BB, Van Kuiken ME, Iyer LH, Harikumar KB, Sung B. Molecular targets of nutraceuticals derived from dietary spices: potential role in suppression of inflammation and tumorigenesis. *Experimental Biology and Medicine.* 2009 Aug;234(8):825–49.

Agrawal PK, Agrawal C, Blunden G. Pharmacological significance of hesperidin and hesperetin, two citrus flavonoids, as promising antiviral compounds for prophylaxis against and combating COVID-19. *Natural Product Communications.* 2021 Oct;16(10):1934578X211042540.

Alam MA, Subhan N, Rahman MM, Uddin SJ, Reza HM, Sarker SD. Effect of citrus flavonoids, naringin and naringenin, on metabolic syndrome and their mechanisms of action. *Advances in Nutrition.* 2014 Jul;5(4):404–17.

Alissa EM, Ferns GA. Functional foods and nutraceuticals in the primary prevention of cardiovascular diseases. *Journal of Nutrition and Metabolism.* 2012; 1–16.

Bang JS, Choi HM, Sur BJ, Lim SJ, Kim JY, Yang HI, Yoo MC, Hahm DH, Kim KS. Anti-inflammatory and antiarthritic effects of piperine in human interleukin 1β-stimulated fibroblast-like synoviocytes and in rat arthritis models. *Arthritis Research & Therapy.* 2009 Apr;11(2):1–9.

Barboza JN, da Silva Maia Bezerra Filho C, Silva RO, Medeiros JV, de Sousa DP. An overview on the anti-inflammatory potential and antioxidant profile of eugenol. *Oxidative Medicine and Cellular Longevity.* 2018 Jan 1;1–9.

Beconcini D, Felice F, Fabiano A, Sarmento B, Zambito Y, Di Stefano R. Antioxidant and anti-inflammatory properties of cherry extract: nanosystems-based strategies to improve endothelial function and intestinal absorption. *Foods.* 2020 Feb;9(2):207.

Borriello A, Cucciolla V, Della Ragione F, Galletti P. Dietary polyphenols: focus on resveratrol, a promising agent in the prevention of cardiovascular diseases and control of glucose homeostasis. *Nutrition, Metabolism and Cardiovascular Diseases.* 2010 Oct 1;20(8):618–25.

Cai B, Zhang Y, Wang Z, Xu D, Jia Y, Guan Y, Liao A, Liu G, Chun C, Li J. Therapeutic potential of diosgenin and its major derivatives against neurological diseases: recent advances. *Oxidative Medicine and Cellular Longevity.* 2020 Mar 6;1–16.

Gao Y, Gu W, Chen L, Xu Z, Li Y. The role of daidzein-loaded sterically stabilized solid lipid nanoparticles in therapy for cardio-cerebrovascular diseases. *Biomaterials.* 2008 Oct 1;29(30):4129–36.

Gonçalves RF, Martins JT, Duarte CM, Vicente AA, Pinheiro AC. Advances in nutraceutical delivery systems: from formulation design for bioavailability enhancement to efficacy and safety evaluation. *Trends in Food Science and Technology* 2018;78:270–91.

Goyal SN, Prajapati CP, Gore PR, Patil CR, Mahajan UB, Sharma C, Talla SP, Ojha SK. Therapeutic potential and pharmaceutical development of thymoquinone: a multitargeted molecule of natural origin. *Frontiers in Pharmacology.* 2017 Sep 21;8:656.

Heiss C, Sansone R, Karimi H, Krabbe M, Schuler D, Rodriguez-Mateos A, Kraemer T, Cortese-Krott MM, Kuhnle GG, Spencer JP, Schroeter H. Impact of cocoa flavanol intake on age-dependent vascular stiffness in healthy men: a randomized, controlled, double-masked trial. *Age.* 2015 Jun;37(3):1–2.

Hsu SP, Wu MS, Yang CC, Huang KC, Liou SY, Hsu SM, Chien CT. Chronic green tea extract supplementation reduces hemodialysis-enhanced production of hydrogen peroxide and hypochlorous acid, atherosclerotic factors, and proinflammatory cytokines. *American Journal of Clinical Nutrition.* 2007 Nov 1;86(5):1539–47.

Hughes MF, Patterson CC, Appleton KM, Blankenberg S, Woodside JV, Donnelly M, Linden G, Zeller T, Esquirol Y, Kee F. The predictive value of depressive symptoms for all-cause mortality: findings from the PRIME Belfast study examining the role of inflammation and cardiovascular risk markers. *Psychosomatic Medicine.* 2016 May 1;78(4):401–11.

Jain S, Buttar HS, Chintameneni M, Kaur G. Prevention of cardiovascular diseases with anti-inflammatory and anti-oxidant nutraceuticals and herbal products: an overview of pre-clinical and clinical studies. *Recent Patents on Inflammation & Allergy Drug Discovery.* 2018 Oct 1;12(2):145–57.

Khaw KT, Bingham S, Welch A, Luben R, Wareham N, Oakes S, Day N. Relation between plasma ascorbic acid and mortality in men and women in EPIC-Norfolk prospective study: a prospective population study. *Lancet.* 2001 Mar 3;357(9257):657–63.

Kim CS, Kawada T, Kim BS, Han IS, Choe SY, Kurata T, Yu R. Capsaicin exhibits anti-inflammatory property by inhibiting IkB-a degradation in LPS-stimulated peritoneal macrophages. *Cellular Signalling.* 2003 Mar 1;15(3):299–306.

Kong W, Wei J, Abidi P, Lin M, Inaba S, Li C, Wang Y, Wang Z, Si S, Pan H, Wang S. Berberine is a novel cholesterol-lowering drug working through a unique mechanism distinct from statins. *Nature Medicine.* 2004 Dec;10(12):1344–51.

Lang A, Lahav M, Sakhnini E, Barshack I, Fidder HH, Avidan B, Bardan E, Hershkoviz R, Bar-Meir S, Chowers Y. Allicin inhibits spontaneous and TNF-α induced secretion of proinflammatory cytokines and chemokines from intestinal epithelial cells. *Clinical Nutrition.* 2004 Oct 1;23(5):1199–208.

Leslie MA, Cohen DJ, Liddle DM, Robinson LE, Ma DW. A review of the effect of omega-3 polyunsaturated fatty acids on blood triacylglycerol levels in normolipidemic and borderline hyperlipidemic individuals. *Lipids in Health and Disease.* 2015 Dec;14(1):1–8.

Liang H, Sonego S, Gyengesi E, Rangel A, Niedermayer G, Karl T, Muench G. Anti-inflammatory and neuroprotective effect of apigenin: studies in the GFAP-IL6 mouse model of chronic neuroinflammation. *Free Radical Biology and Medicine.* 2017 Jul 1;108:S10.

Mangas-Cruz MA, Fernandez-Moyano A, Albi T, Guinda A, Relimpio F, Lanzon A, Pereira JL, Serrera JL, Montilla C, Astorga R, Garcia-Luna PP. Effects of minor constituents (non-glyceride compounds) of virgin olive oil on plasma lipid concentrations in male Wistar rats. *Clinical Nutrition.* 2001 Jun 1;20(3):211–5.

Marzocchella L, Fantini M, Benvenuto M, Masuelli L, Tresoldi I, Modesti A, Bei R. Dietary flavonoids: molecular mechanisms of action as anti-inflammatory agents. *Recent Patents on Inflammation & Allergy Drug Discovery.* 2011 Sep 1;5(3):200–20.

Mashhadi NS, Ghiasvand R, Askari G, Hariri M, Darvishi L, Mofid MR. Anti-oxidative and anti-inflammatory effects of ginger in health and physical activity: review of current evidence. *International Journal of Preventive Medicine.* 2013 Apr;4(Suppl 1):S36.

Mlala S, Oyedeji AO, Gondwe M, Oyedeji OO. Ursolic acid and its derivatives as bioactive agents. *Molecules.* 2019 Jan;24(15):2751.

Morbidelli L, Terzuoli E, Donnini S. Use of nutraceuticals in angiogenesis-dependent disorders. *Molecules.* 2018 Oct;23(10):2676.

Nadeem M, Imran M, Aslam Gondal T, Imran A, Shahbaz M, Muhammad Amir R, Wasim Sajid M, Batool Qaisrani T, Atif M, Hussain G, Salehi B. Therapeutic potential of rosmarinic acid: A comprehensive review. *Applied Sciences.* 2019 Jan;9(15):3139.

Nair HB, Sung B, Yadav VR, Kannappan R, Chaturvedi MM, Aggarwal BB. Delivery of antiinflammatory nutraceuticals by nanoparticles for the prevention and treatment of cancer. *Biochemical Pharmacology.* 2010 Dec 15;80(12):1833–43.

Paolino D, Mancuso A, Cristiano MC, Froiio F, Lammari N, Celia C, Fresta M. Nanonutraceuticals: The New Frontier of Supplementary Food. *Nanomaterials.* 2021 Mar;11(3):792.

Pearson TA, Mensah GA, Alexander RW, Anderson JL, Cannon III RO, Criqui M, Fadl YY, Fortmann SP, Hong Y, Myers GL, Rifai N. Markers of inflammation and cardiovascular disease: application to clinical and public health practice: a statement for healthcare professionals from the Centers for Disease Control and Prevention and the American Heart Association. *Circulation.* 2003 Jan 28;107(3):499–511.

Pillai SC, Borah A, Jacob EM, Kumar DS. Nanotechnological approach to delivering nutraceuticals as promising drug candidates for the treatment of atherosclerosis. *Drug Delivery.* 2021 Jan 1;28(1):550–68.

Quagliariello V, Vecchione R, Coppola C, Di Cicco C, De Capua A, Piscopo G, Paciello R, Narciso V, Formisano C, Taglialatela-Scafati O, Iaffaioli RV. Cardioprotective effects of nanoemulsions loaded with anti-inflammatory nutraceuticals against doxorubicin-induced cardiotoxicity. *Nutrients.* 2018 Sep;10(9):1304.

Rajasekaran A, Sivagnanam G, Xavier R. Nutraceuticals as therapeutic agents: a review. *Research Journal of Pharmacy and Technology.* 2008;1(4):328–40.

Ramırez-Tortosa MC, Mesa MD, Aguilera MC, Quiles JL, Baro L, Ramirez-Tortosa CL, Martinez-Victoria E, Gil A. Oral administration of a turmeric extract inhibits LDL oxidation and has hypocholesterolemic effects in rabbits with experimental atherosclerosis. *Atherosclerosis.* 1999 Dec 1;147(2):371–8.

Russ TC, Stamatakis E, Hamer M, Starr JM, Kivimäki M, Batty GD. Association between psychological distress and mortality: individual participant pooled analysis of 10 prospective cohort studies. *BMJ.* 2012 Jul 31;e4933.

Scrivo R, Vasile M, Bartosiewicz I, Valesini G. Inflammation as "common soil" of the multifactorial diseases. *Autoimmunity Reviews*. 2011 May 1;10(7):369–74.

Tangney CC, Rasmussen HE. Polyphenols, inflammation, and cardiovascular disease. *Current Atherosclerosis Reports*. 2013 May 1;15(5):324.

Ting Y, Jiang Y, Ho CT, Huang Q. Common delivery systems for enhancing in vivo bioavailability and biological efficacy of nutraceuticals. *Journal of Functional Foods*. 2014;7:112–28.

Voloshyna I, Hai O, Littlefield MJ, Carsons S, Reiss AB. Resveratrol mediates anti-atherogenic effects on cholesterol flux in human macrophages and endothelium via PPARγ and adenosine. *European Journal of Pharmacology*. 2013 Jan 5;698(1–3):299–309.

Zuchi C, Ambrosio G, Lüscher TF, Landmesser U. Nutraceuticals in cardiovascular prevention: lessons from studies on endothelial function. *Cardiovascular Therapeutics*. 2010 Aug;28(4):187–201.

18 Nanotechnology and Regulatory Issues

Abhijit Gupta, Saurabh Gupta, and Gunjan Mukherjee*

CONTENTS

18.1 Introduction 441
18.2 Background of Problems 442
18.3 Literature 442
18.4 Applications of Nanoparticles 443
18.4.1 In Medicine 443
18.4.2 In the Textile Industry 443
18.4.3 In the Cosmetic Industry 443
18.5 Preparation of Nanoparticles 445
18.5.1 Biological Synthesis of Nanoparticles 445
18.5.2 Antimicrobial Activity 447
18.5.3 Routes of Exposure of Nanoparticles for Dermal Absorption 447
18.6 Review of Nanoparticle Activity in Terms of Cytotoxicity 447
18.7 Regulatory Landscape of Nanotechnology 449
18.7.1 Legal Aspects 449
18.7.2 Principles for the Regulation of Nanotechnology 451
18.7.3 The Risk of Application of Nanomaterials 451
18.8 Ethical Considerations 452
18.8.1 Legislative Space 453
18.8.2 Analysis of Legislation 454
18.9 Conclusion 455
References 455

18.1 INTRODUCTION

You're probably thinking of the Greek term "nano," which means 10^{-9} nm, when you hear the word "nanotechnology." Nanotechnology is concerned with the modification of materials at the molecular level or lower (Horikoshi and Serphone, 2013). All other nanostructures are built on top of a nanoparticle. Nanoparticles, which have a diameter of less than 100 nm, are a common type of nanotechnology (Abhilash, 2010). Many features of proteins and polynucleic acid nanoparticles are similar to those of biomolecules, such as size (De *et al.*, 2008). The experimental circumstances and traditional techniques used to generate nanoparticles have a big impact on their size, shape, and stability (Marcato and Duran, 2008; Singh and Singh, 2011).

Traditional reduction procedures use a reducing chemical, but in biological approaches, the use of enzymes, such as nitrate reductase, plays an important role in the production of nanoparticles. Traditional nanoparticle manufacturing processes employ toxic chemicals, which could have a harmful influence on the environment (Shah *et al.*, 2015). Metallic catalysts for the production of nanoparticles can be found in the bio factories of bacteria, fungi, and plants. For the first time,

* Corresponding author: abhijitgupta.biotech@cumail.in

DOI: 10.1201/9781003244721-18

employing microorganisms and plants to make nanoparticles appears to be a viable option. The intriguing prospect of altering nanoparticle size distribution has piqued interest in biological synthesis utilizing plant extracts.

Novel aspects of nanoparticles are beneficial to biomedical device development and treatment. The cosmetics industry is another rapidly developing application for nanoparticles. Ag nanoparticles stand out from macroparticles in terms of physical, chemical, and biological properties, causing quite a stir. Colloidal silver is a great preservative for use in body lotions, bath gels, and other cosmetics due to its antibacterial properties. Colloidal silver helps in killing of germs which tends to grow in these compounds. Another commonly used type of nanoparticle in the industry is ZnO nanoparticles, which are used as a UV filter in cosmetics. Nanoparticles filter UV rays, deliver vitamins, and protect against contamination in skincare products like sunscreen and emulsion. Lipid nanoparticles and dendritic or highly branching polymers are possible delivery mechanisms for the incorporation of cosmetic components found in moisturizers, anti-ageing creams, and other skincare products. As a result, cosmetic companies are turning to nanotechnology to improve the value of their products and the satisfaction of their customers (Shokri, 2017).

Nanoparticles used topically are causing concern among scientists due to their smaller size and increased absorption and penetration into the body (Hett, 2004). Nanoparticles can freely go to the organs they are meant to reach after they enter the body, such as the liver or the brain. Nanoparticles outperform their bulk counterparts in many applications due to a size-dependent property.

Despite their superior performance, nanomaterials have yet to be widely adopted in many industrial applications due to concerns about potential negative consequences. Free radicals are produced by UV-absorbing TiO_2 and ZnO_2 nanoparticles, which harm cells and increase the risk of skin cancer. Metal oxide nanoparticles can be absorbed by the skin and circulated throughout the body. Because of their small size, nanoparticles are used in cosmetics, but their use is being scrutinized due to concerns about potential harm. According to current research, skin-borne ZnO_2 nanoparticles are unlikely to be a significant source of Zn ions released into the body from ZnO particles. Because of the ongoing debate over nanotoxicity, a solution must be found as soon as possible, despite the large number of products currently on the market. We are concerned about this because skincare products are applied directly to humans (Rasmussen *et al.*, 2010).

18.2 BACKGROUND OF PROBLEMS

People pay more attention to their health when their circumstances change. Cosmetics with all kinds of claims abound on the market. As a result, preventing microbiological contamination in cosmetic formulations is a primary priority throughout their shelf life. Nanoparticles have been added to the list of preservatives that can be used in the cosmetics industry. Due to their small size, skin cells can absorb and internalize nanoparticles, posing a risk. Nanoparticle safety is a contentious topic among scientists. Because of the wide range of nanoparticle-containing products currently on the market, research on nanoparticle toxicity is critical and important.

18.3 LITERATURE

Nanoparticles are distinguished from identical substances at the macroscopic level by their size dependence and quantum effects. Nanotechnology is all about changing the impact of small molecules to generate new characteristics and functionalities when it comes to structures and systems. Scientists have been working on a hybrid strategy that incorporates the best qualities of both systems in recent years. Pharmaceuticals and cosmetics are the only two industries that stand to gain a lot from nanotechnology advancements. In the pharmaceutical and cosmetic industries, nanotechnology has numerous applications, including the use of nanoparticles with atomic or molecularly specified constituent dimensions.

18.4 APPLICATIONS OF NANOPARTICLES

Nanoparticles have been studied for a variety of applications, ranging from medical devices to personal care goods. Medical nanotechnology holds special potential in areas such as disease detection, tailored medicine administration, and molecular imaging. Future nanotechnology is projected to aid in the construction of complex structures such as the retina and cochlea, while biotechnology will simplify the understanding of how complex biological systems are put together (Gutwein and Webster, 2003).

18.4.1 In Medicine

In oncology and neurology, the use of multifunctional nanoparticles as a medical delivery method and therapy is a potential innovation. According to a study by Giasuddin *et al.* (2012), gold nanoparticles have been employed in diagnostics, surgery, and medicine. Gold nanoparticles have low toxicity and attach quickly to physiologically relevant molecules. According to Holiday's (2008) research, gold nanoparticles are also used in microsurgical techniques to prevent bacterial infections, which demand extra care, and implant infections.

Innovative nanoscale materials are being developed, according to Sanvicens and Pilar (2008). The combination of nanoparticles with other nanotech-based materials, according to Agrawal *et al.* (2008), generates a lot of enthusiasm and opens up possibilities for cellular diagnosis.

Alexis and Rhee (2008) employed polymeric nanoparticles, the most effective chemotherapy carriers available, to treat cancer (Moghimi *et al.*, 2001). These surface-functionalized polymeric particles are better able to target cancer cells that have a higher systemic circulation. Pharmaceutical nanoparticles containing pharmaceuticals are a new type of treatment that has the potential to increase the efficacy and safety of presently available medications.

18.4.2 In the Textile Industry

Textile companies looking to improve tensile strength, elasticity, or fibre stiffness may find nanotechnology to be very valuable (Haggenmueller *et al.*, 2003; Poncharal *et al.*, 1999; Kalarikkal *et al.*, 2006). As a result, stronger, more elastic, and more damage-resistant fibres and materials may be developed. According to Siegfried (2007), fabrics can be stiffened by utilizing ZnO coating-based nanoparticles. Yadav *et al.* (2006) and Kalarikkal *et al.* (2006) employed Al_2O_3-coated nanoparticles to improve the fracture toughness of fibres.

Nanoscale ZnO_2 particles were synthesized and characterized before being employed as UV shields on cotton and wool fibres, according to Baglioni *et al.* (2008). These were made with a range of reaction media and temperatures (90 or 150°C). ZnO nanoparticles induced sunscreen action in cotton and wool. UV–Vis spectrophotometry was used to determine the efficiency, and the UV protection factor (UPF) was computed.

Studies by Daoud *et al.* (2005) and Parkin and Palgrave (2004) have shown that using nanoparticles as photocatalysts can eliminate organic dirt and stains. Huang et al. (2007) employed surface-modified carbon black molecules to coat nanoparticles in order to colour cotton, acrylic, and nylon fibres. Nanotechnology can help to make fabrics lighter, more flexible, and stain resistant. Nanotechnology, which is currently being developed, could help with a wide range of textile applications, including antimicrobial fabrics.

18.4.3 In the Cosmetic Industry

A two-phase system disperses micro-droplets 50–100 nm in size. Emulsions are nanoscale blends of commonly used cosmetic components such as water, oils, and surfactants. Emulsions, for example, are used to smooth out the finished product in conditioners and lotions. Nanosized components can now be found in a wide range of products. Raw nanomaterials, such as nanosized metals and carbon nanotubes, as well as metal oxides, will see increased market demand as a result. Nanoparticles or nanofibers are already found in over 800 consumer products, with the number projected to rise in the future. The most common mineral is silica, which is followed by titania dioxide (TiO_2), zinc oxide (ZnO), and cerium oxide (CeO) silver (Takhar and Mahant, 2011).

L'Oréal, a well-known cosmetics company, for example, has a huge number of nanotechnology patents and has been actively involved in this technology to improve its current goods (Wood *et al.*, 2003).

New cosmetics made by beauty industry innovators have boosted customers' expectations. Soaps and deodorants, as well as toothpaste. Food packaging and textiles are examples of other things that contain them. Among the many products provided are body firming lotions, bronzers, exfoliating scrubs, eye liners, and styling gels (Patel *et al.*, 2011). Gold nanoparticle products, which cost $1,398 a bottle, claim to have anti-inflammatory and antioxidant properties as well as to promote tissue regeneration, increase skin suppleness, and reduce the appearance of wrinkles and other signs of stress or ageing (Falaschetti, 2012).

According to Veronovski *et al.* (2014), TiO_2 nanoparticles can be employed as an inorganic UV absorber to create optically transparent coatings with a sufficient SPF. The sulphate synthesis approach is utilized as a starting point for making nanograde TiO_2. The technique is then fine-tuned to yield TiO_2 as the final nanomaterial. The surface of the TiO_2 nanoparticles was altered by using sodium silicate as a silica source to coat them. The synthesized silica-coated TiO_2 nanoparticles were examined using scanning electron microscopy (SEM)/transmission electronic microscopy (TEM). The sunscreen had an SPF of 28 thanks to 9.0% TiO_2 nanoparticles and 5.0% silica on the surface.

Gold nanoparticles have been used because of their catalytic, quantum-size-impact, and optical properties, according to a study by Taufikurohmah *et al.* (2014). Mercury steel has been linked to the use of gold nanoparticles in cosmetics. Mercury is employed in a variety of ways by the cosmetics industry, including in skin-whitening products such as soaps, lotions, and eye makeup removers. Glycerine was discovered to be an effective matrix for the production of gold nanoparticles, while sodium citrate was discovered to be an effective reduction agent. Gold nanoparticles are made up of clusters with sizes ranging from 10 to 40 nm. These clusters are useful for aesthetic applications, since they can enter skin pores.

Pant *et al.* (2013) claim that by using a plant extract from the *Solanum trilobatum* Linn. genus, as well as conventional and homogenization processes, silver nanoparticles can be made in an environmentally friendly manner. Silver nanoparticles were optimized in ideal conditions, but the best results came from exposing them to sunlight, which produced nanoparticles with a diameter of 15–20 nm. Researchers used a variety of bacterial and fungal species to test the toxicity of these particles. This approach to synthesizing silver nanoparticles is useful, since aqueous Ag^+ ions have been proven to have bio-reduction potential.

Allicin (from garlic) was found to be an effective antifungal agent when coupled with solid lipid nanoparticles (SLNs) in a herbal antidandruff shampoo, according to Abraham *et al.* (2013). Every single nanoparticle was tested for drug release and particle size. To improve appearance, viscosity, pH, spreadability, CMC and EDTA were employed in anti-dandruff shampoo in varied quantities. The pH was 3.7 to 7.2, while the viscosity was 1010 to 1700 centimetres per second (cps). It is more successful in treating dandruff on the scalp and hair while also being free of side effects, according to *in vitro* drug release values.

Wiechers and Musee (2010) published a review of reports on engineered nanoparticle (ENP)-containing cosmetics. When ENPs in cosmetic formulations are reduced in size, the compositions become more transparent, but the particles' defensive qualities are lost. The stratum corneum of the skin, according to their research, is impervious to cosmetic ENPs (TiO_2, ZnO, and Ag). Because cosmetic ENPs cannot be examined for skin penetration *in vitro*, they must be studied *in vivo* on people using a range of cosmetic formulations.

Muller *et al.* (2009) discovered that SLNs for cutaneous application are functionally separate from nanostructured lipoprotein carriers (NLCs). Liposomes and emulsions can be administered in different ways, including one or both of these methods. Well-established techniques were used to create lipid nanoparticles. SLNs and NLCs, for example, have been employed in cosmetics and pharmaceutical formulations to provide skin hydration, drug delivery control, controlled release, and occlusion, among other benefits. Biodegradable or physiologically relevant lipid nanoparticles are well tolerated. They can deliver a controlled amount of medicine, since they work as colloidal carriers. SLNs have a better possibility of contacting the stratum corneum directly, allowing the medicine to permeate deeper into the skin. The occlusive properties of lipid nanoparticles improve the skin's moisture.

18.5 PREPARATION OF NANOPARTICLES

Biological techniques, microwave-aided synthesis (MAS), and laser-mediated synthesis (LMS) are all strategies for regulating or altering nanoparticles. Nanoparticles have traditionally been synthesized using chemical and physical processes. Traditional treatment approaches are being replaced with green synthesis as environmental consciousness rises. It is quite usual to use naturally occurring microbes and plant extracts. Recent advancements have ushered in a new era, emphasizing the role of plants in nanoparticle synthesis, which appears to have piqued interest in the manufacturing of stable nanoparticles. Green synthesis, which has emerged as a viable alternative to conventional methods, relies heavily on plants and bacteria. Plant-derived biomolecules can act as capping and reducing agents, allowing nanoparticles to be reduced and stabilized more quickly by utilising bacteria (Kulkarni *et al.*, 2011).

Electronics has enhanced the efficiency of this technology for large-scale synthesis of additional inorganic materials, such as nanomaterials, by using nanoparticles in wound healing and medicine (Watal *et al.*, 2012).

In 2002, alfalfa plants were purportedly used to make gold nanoparticles with diameters ranging from 2 to 20 nm. Plants such as *Helianthus annuus*, *Oryza sativa*, *Zea mays*, and *Sorghum bicolor*, as well as gymnosperms such as *Artocarpus heterophyllus* (Thirumurugan *et al.*, 2010), were employed to create nanoparticles (Leela and Vivekanandan, 2008). Green synthesis nanoparticles produced from natural resources or extracts have been described by scientists. Another variety of tree found in the area was *Eucalyptus hybrida* (Bali *et al.*, 2006).

18.5.1 BIOLOGICAL SYNTHESIS OF NANOPARTICLES

According to the findings, processing ambient-temperature precursor Ag and Au to generate various nanoparticle forms from *Cinnamomum camphora* leaf biomass resulted in triangular or spherical Ag and Au nanoparticles after suitable incubation (Huang *et al.*, 2007).

Aqueous extracts from *Hybanthusennea spermus* leaves, stems, and roots were used to synthesize zinc oxide nanoparticles with zinc nitrate hexahydrate as an active ingredient ([Zn $(NO_3)_2.6H_2O$] solution) according to Shekhawat *et al.*, (2014). To generate the nanoparticles, herbal extracts were employed to reduce the volume of fluid. The colour shift of the reaction mixture was demonstrated using UV–Vis spectrophotometry. In reaction mixtures, leaf extracts exhibited a 300 nm absorption peak, stem extracts had a 290 nm absorption peak, and root extracts had a 288 nm absorption peak. Sarsar *et al.* (2014) used *Psidium guajava* leaf extract as a reductant in the synthesis of silver nanoparticles. Various characterization techniques were utilized to describe variations in conditions such as precursor concentration, reducing agent, pH, temperature, and synthesis time. Silver nanoparticles were discovered to exhibit antibacterial properties in studies on a variety of microorganisms.

Sindhura *et al.* (2014) used an extract from *Parthenium hysterophorus* leaves to make biogenic ZnO_2 nanoparticles. Zn nanoparticles possess enzymatic and antibacterial activity. To track the quantitative synthesis of zinc nanoparticles, sophisticated methods were used for characterization. The majority of nanoparticles were discovered to be spherical, with sizes ranging from 16 to 108.5 nm. Zinc nanoparticles are widely diffused, according to the findings, with zeta potential ranging from 100.4 to 117.20 mV.

Veerasamy *et al.* (2011) used a reducing agent, *Garcinia mangostana* leaf extract, to make silver nanoparticles quickly and easily. Silver ions in water were lowered using a reduction process before being exposed to leaf extract, where they expanded to a size of 35 nm on average. The materials were examined using a range of techniques, including UV–visible, Fourier transform infrared spectroscopy (FTIR), and TEM. These nanoparticles' antibacterial properties make them efficient against germs previously thought to be resistant to a variety of therapies. Despite the presence of harmful and lethal pollutants, bionanostructure creation could continue. It was also efficient against *Escherichia coli* and *Staphylococcus aureus.*

The silver nanoparticles, which have an average diameter of 33.6 nm and are meant to minimize pollution concerns in the environment, were reduced and capped using an onion (*Allium cepa*) extract. The final characterization was carried out using a UV–Vis spectrophotometer and dynamic light scattering (DLS). The morphology of the silver nanoparticles was validated using electron microscopy (TEM). The antibacterial activity of the nanoparticles was tested using *E. coli* and *Salmonella typhimurium* as test organisms. The antibacterial activity of *E. coli* and *S. typhimurium* nanoparticles generated at a dosage of 50g/mL was 100%.

Krishnaraj *et al.* (2010) used *Acalypha indica* to create green silver nanoparticles, and the nanoparticles' efficacy against water-borne bacterial infections was investigated. The reaction takes 30 minutes to complete. The researchers were able to confirm biosynthesis and characterize the chemical using UV–Vis spectrum, SEM, X-ray diffraction (XRD), and energy-dispersive spectroscopy (EDS). The nanoparticles were measured using high-resolution TEM (HRTEM) and determined to be 20–30 nm in diameter. The antibacterial activity of $AgNO_3$ nanoparticles was found to be ineffective against the aquatic pathogens *E. coli* and *Vibrio cholerae*, according to the researchers. The minimum inhibitory concentration (MIC) of silver nanoparticles at a concentration of 10 ng/mL proved efficient against *E. coli* and *V. cholerae.* The effect of silver nanoparticles on bacterial cell permeability and respiration was discovered in this experiment. Using a biological approach to make silver nanoparticles could save money while also helping the environment. These nanoparticles could help reduce microbial burdens in wastewater treatment.

Smitha *et al.* (2009) employed *Cinnamomum zeylanicum* leaf broth to make gold nanoparticles. The form of these nanoparticles was a mixed-gold nanoprism/sphere. There were fewer prism-shaped Au particles in the extract than spherical Au particles. The nanoparticles' good crystallinity is demonstrated by clear fringes in the fcc phase on XRD patterns and high-resolution TEM pictures, as well as selected area electron diffraction (SAED) patterns with vivid circular rings and HRTEM images. These gold nanoparticles are one of a kind because of their photoluminescence, and it was revealed that the power of their photoemission rose as the concentration of leaf broth increased. Because of their luminous properties, fluorescent Au nanoparticles can be employed for cell labelling and biological sensing studies.

Bar *et al.* (2009) employed *Jatropha curcas* latex as a reducing and capping agent in a green synthesis process to create silver nanoparticles. The final characterization was carried out using HRTEM, XRD, and UV–Vis absorption spectroscopy. XRD was used to determine the nanoparticles' fcc structure. When analysed with HRTEM pictures, nanoparticles with a radius of 10–20 nm were determined to be stabilized by the cyclic peptides, and their optimal cavity radius was found to be near to that of latex. Throughout the synthesis, the capping and reducing properties of *J. curcas*' renewable latex were demonstrated.

Honey was employed as a reduction and capping agent in aqueous solutions; hence, no harmful chemicals were used in the synthesis of gold nanoparticles. The particles have a diameter of roughly 15 nm according to the XRD pattern and TEM image. Proteins cap and stabilize honey-derived gold nanoparticles, according to infrared spectroscopy. Protein binding to Au surfaces was affected by the amide I and II bands, as well as carboxylic acid group vibrations.

Sadowski *et al.* (2008) used Neem (*Azadirachta indica*) leaf broth and aqueous $AgNO_3$ or $HAuCl_4$ solutions for the extracellular production of pure metallic silver and gold particles. When employing Ag^+ and Au^{3+} ions, there was a four-hour difference in reduction time between bacteria and fungi (24 hours vs. 120 hours). According to a new study, the surface-active components of leaf broth may have aided in the stabilization of a nanoparticle dispersion. The nanoparticles showed a negative zeta potential due to electrostatic repulsion and remained stable at pH values over 8.

18.5.2 Antimicrobial Activity

Azam and Ahmed (2012) discovered the antibacterial activities of ZnO, CuO, and Fe_2O_3 nanoparticles against Gram (+) and Gram (–) microorganisms in their investigation. Shameli *et al.* (2012) tested the antibacterial activity of different-sized silver nanoparticles against Gram (+) (*Staphylococcus aureus*) and Gram (–) (*Salmonella typhimurium* SL1344) bacteria using agar disc diffusion utilizing the Mueller–Hinton model (2012). Nanoparticles with a smaller size have better antibacterial capabilities. For the evaluation of growth parameters, protein content will be calculated subsequent to its analysis by performing sodium dodecyl sulphate–polyacrylamide gel electrophoresis (SDS PAGE). Experiments with zinc oxide nanoparticles revealed antibacterial efficacy against *Staphylococcus aureus*.

18.5.3 Routes of Exposure of Nanoparticles for Dermal Absorption

According to Nasir (2010), the three basic avenues for nanoparticles to enter the body are intercellular, transfollicular, and transcellular nanoparticle penetration. Because of the stratum corneum's lipid bilayers and the high protein levels below the stratum corneum's surface, passively transporting nanoparticles is conceivable but unlikely. Particles are more likely to enter if anatomical barriers or the skin are damaged. The particle size was also shown to be consistent with the size of follicular openings.

18.6 REVIEW OF NANOPARTICLE ACTIVITY IN TERMS OF CYTOTOXICITY

The impact of nanoparticles on the environment and its related cytotoxicity aspect has gained a lot of press in recent times, and this aspect has become one of the major concerns for the worldwide research community (Brayner, 2008; Panda *et al.*, 2016). Some of these research reports have made a huge impact in the research community. The next paragraphs go through a little of this research in further depth.

Incubation of various human culture cells with 10 mg/mL of encapsulated silver nanoparticles (AgNPs) (diameter 6–80 nm) had no discernible genotoxic effects (Hussain *et al.*, 2005; Asha Rani *et al.*, 2009; Lu *et al.*, 2010). Genotoxicity of capped AgNps was captured when human mesenchymal stem cells and human fibroblast cells were exposed to 0.1 mg/mL albumin (average size 46 nm; Hackenberg *et al.*, 2011) upon incubation with the nanoparticles, (sizes 6–20 nm).

In tests on human lung cells, Gliga *et al.* discovered that size and coating alter the toxicity of well-characterized AgNPs (2014). The lactate dehydrogenase (LDH) and Alamar Blue tests, as well as the 2'-7'dichlorofluorescin diacetate (DCFH-DA) assay for reactive oxygen species (ROS), were used to examine cell viability. The alkaline comet assay and the formation of H2AX foci were both employed to detect genotoxicity. Photon cross correlation spectroscopy (PCCS) was used to analyse the results of these studies. The study discovered that particle agglomeration in cell media produced cytotoxicity in human lung cells and enhanced overall DNA damage due to AgNPs (10 nm) generated during cellular uptake, intracellular localization, and Ag release after 24 hours.

According to Zhang *et al.*'s (2014) research, the AgNP form may have significant cytotoxicity and immunological repercussions. The human lung epithelial cell line A549 is significantly and greatly affected by silver nanowires bearing size range of (diameter 100–160 nm and 30 nm. When the same cells were exposed to AgNPs that were spherical in shape (30 nm), different reactions were found. A problem arose as a result of the wires being able to reach the cell's surface rather than being buried deep within it.

Liang *et al.* investigated the interactions of silica nanoparticles (50 nm SiO_2 NPs) in the cytoplasm, lysosomes, and autophagosomes of HaCaT cell lines (2013). Cell viability fell after reaching a threshold concentration of SiO_2 NPs (100 g/mL), and cell membrane damage resulted. The key mechanisms of SiO_2 NP cytotoxicity were glutathione depletion and the production of ROS. The samples were further characterized using an LDH assay and TEM.

Suker and Albadran (2013) discovered that TiO_2 NPs, a potential commercial material widely used in cosmetics, pharmaceuticals, and food colouring, have harmful effects. As a result, TiO_2 nanoparticles were determined to be the cause of these effects. Using an enzyme-linked immunosorbent assay (ELISA) reader with a 492 nm wavelength, cell proliferation was measured using tetrazolium bromide (MTT). The discovery was confirmed by cytotoxic inhibition, which was time and dose dependent. The authors investigated ZnO-NP-induced cutaneous inflammatory response signalling pathways and discovered some intriguing findings (2013). In HaCaT cells, ZnO-NPs increased the expression, promoter activity, and nuclear translocation of the early growth response-1 gene (Egr-1).

AgNPs have undeniably been used as an antibacterial and disinfectant, but their anticancer potential has yet to be discovered. Gurunathan *et al.* (2013) investigated the mechanism of cell death in MDA-MB-231 breast cancer cells exposed to AgNP cytotoxicity. Green synthesized AgNPs were made from *Bacillus funiculus* culture supernatant, and they were characterized using a number of analytical techniques. LDH, caspase-3, and ROS generation were activated in MDA-MB-231 cells, triggering apoptosis. AgNP may thus be a feasible alternative to chemotherapy in the treatment of human breast cancer. Various dosages of AgNP were given to these breast cancer cell lines.

Silver nanoparticles of various sizes, such as nano (20 nm) and submicron (200 nm), and titanium dioxide nanoparticles (TiO_2-NPs; 21 nm) were studied for their reproductive, cellular, and genotoxic effects by Asare (2012). Ntera 2 (a cell line derived from human testicular embryonic cancer) and cells from mice lacking the enzyme 8-oxoguanine DNA glycosylase (mOgg1) were used to synthesize nanoparticles from primary testicular cells from wild-type (WT) C57BL6 mice (KO). Using this paradigm, human testicular cells were compared with those of animals exposed to oxidative stress to assess how soon the latter recovered. With time and concentration, AgNPs were found to be more cytotoxic and cytostatic than TiO_2-NPs, resulting in increased apoptosis, necrosis, and decreased cell proliferation.

Mukherjee *et al.* (2012) employed human skin and cervical cancer cell lines to investigate the cytotoxic effects of Ag nanopowder. The tests used DLS, TEM, and atomic force microscopy (AFM) to examine different nanoparticle sizes (10, 50, 75, and 100 nm). To investigate how dose, exposure length, and the cell lines utilized influenced the findings, scientists examined cytotoxic endpoints.

Kokura *et al.* (2010) studied the effects of AgNPs on permeability of skin, bacteria, and cytotoxicity after exposing human keratinocytes to ultraviolet B irradiation. The stability of Ag nanoparticles was not demonstrated by sedimentation. When tested against bacterial and fungal cultures, they had no effect on normal human skin, as expected. AgNPs showed no effect on HaCaT lines and have no role in influencing the speeding up of UVB-induced cell death when employed at low concentrations (0.002–0.02 ppm), according to these findings. According to a new study, AgNPs could be used as a cosmetic preservative.

Only a few types of nanoparticles can penetrate the skin, according to Wiechers and Musee (2010), while ENPs (engineered nanoparticles) found in cosmetics do not penetrate past the stratum corneum and may cause irritation. Recent study of the qualities of ENPs that permeate human skin, on the other hand, has shown useful results.

Despite popular belief, nanoparticles may be more dangerous than bulk chemicals, according to some studies (Donaldson *et al.*, 1999). In 2009, Rani used a number of cellular models to investigate the toxicity of AgNPs. After being exposed to nanomaterials, Oberdorster (2004) discovered oxidative stress and lipid peroxidation in fish brain tissue. The formation of ROS by nanoparticles was thought to be one way for them to cause cytotoxicity (Foldbjerg and Autrup, 2013). Researchers are reviewing the idea of putting nanoparticles into daily consumer products due to concerns about nanoparticle safety.

18.7 REGULATORY LANDSCAPE OF NANOTECHNOLOGY

The most commonly debated issue in international literature is whether or not nanotechnology should be incorporated into the legal system. The European Union is now dealing with a number of binding and non-binding legal acts related to nanomaterials (regulations, directives, and the like), such as instructions on ethical scientific research or how to apply a uniform definition of a nanomaterial. For nearly two decades, European politicians have debated the legal concept that underpins nanotechnology. The strategy has altered when it comes to nanotechnology regulation. According to several authors, non-judicial approaches can help improve the safety of nanoparticle use (Matsuura, 2006). Due to the rapid proliferation of this subject, some consider that the application of soft legislation is urgently required (Brazell, 2012). A legislative act must include resolutions, directives, explanations, messages, programmes, and plans made by authorities participating in the legislative process to be referred to as "soft law." On top of these soft law tools, future hard legal legislation will be developed. Developing legislative processes necessitates the use of voluntary data collection programmes (Malloy, 2012).

Some experts advocate for bolstering present nanoscale collaboration programmes, while others advocate for governments to draft their own legal frameworks (Falkner *et al.*, 2010). Because EU regulations do not particularly cover nanotechnology, additional safeguards are needed to protect against unwanted effects of nanoparticles (Ponce del Castillo, 2010). Because we need to know everything there is to know about nanoparticles before we can establish a community division of law, this alternative is rarely discussed (Oud, 2007). Nanotechnology concerns have been recognized and must be addressed as soon as possible. Nanomaterials must be specified uniformly, metrology tools must be developed, and safety testing and risk assessments must be conducted.

The regulation of nanotechnology under the Future Framework Agreement is a difficult issue (Marchant and Doug, 2006). Because of the many different ways that nanoparticles might be employed in international trade and commerce, scholars believe that international regulation of nanotechnology is a serious challenge. Comprehensive nanotechnology regulation will undoubtedly be established in the near future. The development of global best practices for the management of nanomaterials could have a significant impact (Baran, 2015).

Nanotechnology's interdisciplinary nature allows a diverse range of applications. Nanotechnologies are now present in almost every facet of modern life.

According to published studies, intellectual property rights for nanotechnology are also required (Balcerzak, 2013; Ganguli and Jabade, 2012). As a result, concerns have arisen regarding the potential for nanoparticles to cause harm. Nanotechnology's availability and impact on human wellbeing are both problems. Concerns about nanotechnology splitting the world into prosperous and impoverished countries, as well as controlled and uncontrolled countries, are well founded (Bazela, 2008) (Figure 18.1).

18.7.1 Legal Aspects

The field of nanotechnology has already been part of EU strategic documents for the last two decades. *Nano Sciences and Nanotechnologies: An Action Plan for Europe, 2005–2009* was published by the European Commission (EC) in 2005. From 2007 to 2009, a new analysis revealed that nanotechnology could improve both quality of life and competitiveness in European businesses. A preliminary review of the legislative framework was conducted to see whether additional regulatory action was necessary to address the nanoparticle-related concerns. According to a preliminary study,

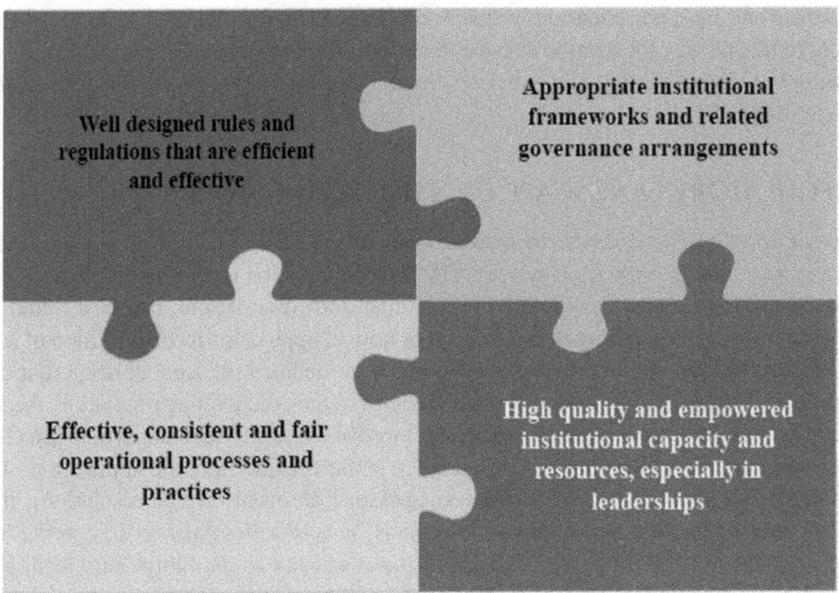

FIGURE 18.1 Necessary elements of better regulatory outcomes.

current rules address health and environmental concerns. State regulatory agencies were given the duty of assessing present national legislation and identifying any flaws, which they did. Legal methods such as thresholds and authorizations, hazardous waste qualification, strengthened conformity assessment procedures, and chemical introduction and usage limitations were also advocated. In 2009, the European Parliament accepted a study on nanomaterial regulatory issues, which included the EC's communication "Regulatory Aspects of Nanomaterials" (COM (2008, 366) dated 17 June 2008. The EC clearly sees both the potential benefits and the potential risks of nanotechnology development, as evidenced by this text. In general, there was a dearth of methodologies for appropriately assessing the dangers associated with concerns expressed about nanomaterials, as well as a thorough understanding of the numerous threats posed by nanomaterials and information on the risks posed by specific nanomaterials. The EC backed up these claims. Given the growing scepticism about nanomaterials, there appears to be a pressing need to incorporate them into the legal system. As a result, the EC began re-examining legislation restricting the use of nanomaterials in 2008, and issued recommendations for future action. A regulatory study commissioned by the EU indicated that EU law needed to be changed to ensure the safe use of nanomaterials, and this was done particularly for the EU (EC). The message highlights a wide spectrum of nanomaterials, from decades-old products like tyres or anticoagulants in food to cutting-edge items used in industry and cancer therapies. Nanoparticle hazards are becoming better known. They resist categorization, highlighting the importance of risk evaluations in specialized applications. According to the EC, risk assessments should be done on an individual basis, and solutions should be based on knowledge of potential exposure or danger threats. Many states have worked on legislation to limit the use of nanoparticles in recent years. CLP (Classification, Labelling, and Packaging) and REACH (Registration, Evaluation, and Authorization for Chemicals), which is EC No 1907/2006, are two current standards governing nanomaterial use. Nanoparticles are subject to sectoral rules as well. These regulations addressed biocides (EC 1223/2009), cosmetics (EC 1169/2011), and food additives (EC 1169/2011). The EU passed Directive 2001/83/EC on the Community Code for Medicinal Products for Human Use in 2001 in response to the expanding use of nanotechnology in medicine. Procedures for the approval of pharmaceuticals were also created (EC No 726/2004). For further protection, it's a good idea to keep track of nanoparticles and the goods that use them. It will be easier to maintain track of enterprises that are bringing nanomaterials to market using this register, and to ensure that product consumers have straightforward access to the information they require. This is currently a significant criterion in the cosmetics sector (Jurewicz, 2014).

While Europe continues to work on legislation, it is critical to remember the following: on a global basis, efforts to regulate nanotechnology are also underway. Many governments are concerned about the threats posed by this sector's continuous development. In a 2008 report, the National Research Council encouraged the U.S. government to regulate nanotechnology.

18.7.2 Principles for the Regulation of Nanotechnology

Nanomaterials, according to what we know so far, are similar to natural chemicals in that some are hazardous while others are not. Some nanoparticles, as well as their applications, may be hazardous. When dealing with nanoscale compounds, it is necessary to do a unique risk assessment for each nanomaterial using suitable data. Establishment of already-in-use methodology and techniques for the characterization and analysis of nanomaterials is a cumbersome task, and the risks associated with humans are more challenging. Despite all the study, the threats remain unquantifiable. Because of the vast amount of information available, it is difficult to estimate the toxicity of nanoparticles.

A phenomenon, product, or activity may pose a security risk that may be detected with sufficient assurance by scientific and objective evaluation. We must accept that we do not fully comprehend the threats that nanoparticles bring. This guiding idea underpins the legal systems of many countries. Article 191 paragraph 2 of the Treaty on the Functioning of the European Union (TFEU) requires Member States to incorporate this notion into their domestic legislation or risk consequences. Members of the EU must incorporate the concept into their national legislation. The application of this guideline makes it easier to notice potential problems before they become serious. To begin with, even if the presence of nanoparticles in the environment or waste products could be established, removing these particles would be technically difficult. Methods for reducing environmental and human health concerns at the end of the pollution chain have failed miserably. The European Parliament strongly advocated this phenomenon in 2009.

18.7.3 The Risk of Application of Nanomaterials

It's impossible to exaggerate the advantages of nanotechnology advancements. If this initiative is successful, it has the potential to minimize the amount of fossil fuels and energy needed, as well as the amount of waste generated. Cleaning and repair techniques based on nanotechnology, as well as pollutant removal, are now all possible. Small sensors, probes used for diagnostics, and complete testing systems are now being developed and implanted by medical researchers to aid in disease diagnosis. Thanks to nanotechnology, implant materials that are both bioactive and biocompatible have never been better (Maliszewska-Mazur, 2010).

Right now, the number of products based on nanomaterials is quickly expanding. During the 1990s, the number of patents awarded for nanotechnology surged substantially. Do we really worry about safety when it comes to the development, use, and disposal of nanotechnologies?

According to reports, nanomaterials represents their uniqueness, coz of their toxicity, their effective use in environmental protection and in the field of medicine which is appealing to be problematic" (Szponder, 2010). We don't know what will happen to nanomaterials and nanoparticles in the environment as a result of this. How essential is nano-risk if it is a result of chemical makeup? The focus of the investigation should be on the answer (Maliszewska-Mazur, 2010). Furthermore, most nanomaterials' toxicity cannot be assessed using currently known techniques or exposure scenarios. Even though many researchers are working with first-generation nanomaterials, due to rapid technical advancements, we are now in the third generation (Maynard, 2006).

Because nanotechnology is increasingly being used in a variety of sectors of the economy, despite the challenges it offers, regulating operations in this field is unquestionably a crucial step today. Before anything else, a consensus on what constitutes a nanomaterial is required. If there is a common definition, it will be easier to identify materials and apply legal laws to them.

This criterion will only be met by materials having a minimum average particle size of 100 nm. According to the 2011 European Commission Recommendation, at least 50% of the particles in the numerical particle size distribution of natural, random, or produced nanomaterials have one or more diameters between 1 and 100 nm. This term comprises both natural and manufactured nanoparticles. When protecting human health or the environment, numerical particle size distribution thresholds may be used instead of the numerical particle size distribution threshold.

Any value between 1% and 50% is likely to be right if the proportion equals 50%. Nanoparticles are a form of nanomaterial that can be employed in a variety of applications.

They, like other chemicals, are regulated by the EU. The REACH and CLP rules place restrictions on this. Early detection of an issue in the pollution control chain is provided by REACH and CLP. A new set of restrictions in many environmental law acts will be implemented for each CLP nanomaterial category to limit the amount of harmful compounds emitted into the biosphere.

Due to health considerations, customers must be informed if a product contains a nanoingredient. The following wording for the warning label was recommended at the Inter-Organization Programme for Chemical Sound Management (IOP) conference in May 2011: Sometimes nanoparticles have been processed purposefully but we are still having insufficient data to reach to any specific conclusions (Zapor, 2012). Regulation (1169/2011) of the EU requires the publication of information on the components in food items to include information on nanoparticles. The new restrictions apply to all of the product's ingredients, including half of the new ones.

The word "nano" is necessary for anything smaller than 100 nm. The "nano" symbol is required on cosmetics by the EU Regulation on Cosmetic Products (1223/2009). All compounds, including nanomaterials, must be listed in the ingredients list. The word "nano" must appear in brackets after the name of a component. Regulations will be systematically reviewed by EU organizations because of the changing nature of nanotechnology.

18.8 ETHICAL CONSIDERATIONS

We are in that phase of development where we are acknowledging significant progress in the development of new strategies. In the 21st century, biological engineering, genetic manipulation, and nanotechnology have all gained popularity. Although biotechnology has sparked a lot of debate and legislation, there is currently no common law covering genetic research, for example. As a result of comprehensive scientific inquiry and availability of novel solutions, the EU legislators implemented the concept of patents for providing protection to new, innovative ideas. Even if patent law in the field of nanotechnology is less complicated than in biotechnology, it stands to benefit from similar improvements in the future. According to studies, these two fields have less in common than the general public believes. Nanotechnology is based for the most part on methods and tactics for modifying nanoscale aspects of both living and non-living materials. As a result, nanobiotechnology has the potential to combine the most significant aspects of both domains into a single system (Balcerzak, 2013). To some extent, biotechnology legislation can be applied. The general public's acceptance of new technology is important to its development. In Europe, there is now a controversy about the use of nerve implants for Parkinson's and blindness patients. The usage of implants for non-rehabilitation purposes is being questioned. The EU Commission decided to take account of major public concerns. Nanotechnology products must be approved by customers as well as being secure and practical. Before people start adopting nanotechnology, they should be better educated on its benefits, risks, and safeguards. It is critical to encourage public debate in order to assist people in forming their own viewpoints. Scientists may help by teaching the general public about the concepts and applications of nanotechnology.

In 2008, the EC suggested code of conduct which is responsible for the field of nanotechnology as the foundation for future EU measures to ensure the safety, ethics, and long-term sustainability of N&N research. The EC came to an agreement and established the code in 2008.

Whether or not you follow the Code of Conduct is entirely up to you. According to the report, all N&N field participants must adhere to these general principles and norms. Current legislation will be made more user-friendly as a result of the Code's establishment, and contentious issues will be resolved once and for all. Remember that N&N research should follow the precautionary principle, which means analysing the possible impact of the research on the biosphere. A number of EU Member States have started establishing public education programmes about nanotechnology's rapid achievements. This is exactly what is happening in the United States, according to the DeEEP project in the United Kingdom. The Nanopodium initiative was one of the most important Dutch and EU social discourse initiatives. Belgium wanted to develop a platform for researchers, industry, and society to discuss nanotechnology with the Nanosoc programme.

Despite the fact that nanotechnology does not generate the same level of controversy as biotechnology, and societies are generally supportive of nanotechnology's diagnostic and therapeutic possibilities, they are increasingly calling for comprehensive information on nanoparticles' long-term effects on the human body. In Germany, a project called "Nanologo" was launched to educate people about the benefits and drawbacks of nanotechnology. It also attempted to educate the public about the ethical, social, and legal aspects of nanotechnology use, as well as to promote communication among the general public and other interested parties. The public's acceptance of nanoparticles and the eradication of their concerns will have a positive impact in the development phase of this field.

18.8.1 Legislative Space

Examining existing legislation can help you understand more about nanotechnology regulation. Despite fears that lawmakers in the United States are not paying enough attention to nanotechnology, several government leaders have acknowledged the need to deal with it. Since 1999, 185 pieces of nanotechnology legislation have been introduced in Congress, the majority of which never made it out of committee, much less to a vote. This section focuses on recently enacted legislation. This measure prioritizes research funding over imposing business regulations or norms.

1. **Law of 2000**

Under Section 221 of the Carbon Cycle Research, the Agriculture Risk Protection Act was established, whereby a group of scientists from various institutions received $15,000,000 to conduct research at a worldwide level, as well as to develop, analyse, and implement carbon cycle research (www.congress.gov/106/plaws/publ224/PLAW-106publ224.pdf). The funding will be used to promote soil carbon sequestration, among other things, by conducting research on further implementation of innovative technologies to boost effectiveness of carbon cycle.

2. **Law of 2001**

Under Section 314 of the Consolidated Appropriations Act 2001, the Marine Corps received $3 million in consolidated appropriations dollars for consequence management nanotechnology research. According to the *Air and Space Power Journal*, "organisational activities geared at stopping the progression of disease or the harm caused by injuries and lowering the long-term social incapacity instigated by any residual impairment. (Anthony *et al.*, 2009).

3. **Law of 2002**

The National Science Foundation (NSF) was established under the National Science Foundation Act of 2002. This act set aside $301 million from the NSF budget to encourage scientific and engineering research and teaching at nanoscale (www.congress.gov/107/plaws/publ368/PLAW-107publ368.pdf). For the first time, the NSF had set aside funds specifically for nanoscale research studies to uncover novel phenomena or processes, materials or methods to deal with key difficulties in the disciplines of materials science and engineering.

4. **Law of 2003**

The Defense Nanotechnology Research and Development Program was established under Section 246 of the Bob Stump National Defense Authorization Act of 2003. In order to solve national security challenges, the program's objectives are as follows: Reaffirming the United States' global leadership in nanotechnology. To ensure interagency cooperation and collaboration on this essential topic, the Department of Defense should coordinate its nanoscale research and development with that of other federal departments and agencies who are engaged in research and development. The more quickly nanoscale-derived technologies and concepts are transferred and applied within the armed forces, the more efficient it will be. By performing research, synthesizing it, and communicating it, new information about nanoscale development can be discovered. Fiscal year (FY) 2008, Section 240 of the National Defense Authorization Act improved the program's administration and reporting requirements, and for FY 2010, Section 242 of the National Defense Act changed those standards (www.gpo.gov/fdsys/pkg/CRPT-110hrpt477/pdf/CRPT-110hrpt477.pdf): National Defense Authorization Act for Fiscal Year 2010, Pub. Law No. 111-84 (2009), www.gpo.gov/fdsys/pkg/PLAW-111publ84/pdf/PLAW-111publ84.pdf.

5. **Law of 2003**

Representatives of the National Nanotechnology Initiative (NNI) began meeting on a regular basis in November 1996 to discuss their various organisations' nanotechnology strategy and initiatives. In September 1998, the Interagency Working Group on Nanotechnology (IWGN) was formed to formalize the ongoing interagency cooperation.

As a result of the suggestions provided by this working group, the Clinton administration formed this NNI organization in January 2000.

Just a few of the efforts being explored include research funding for single researchers and interdisciplinary teams, the development of an advanced technology user facility network, and the formation of merit-based and competitive multidisciplinary nanotechnology research centres increasing the United States' productivity and industrial competitiveness through long-term, stable, consistent, and coordinated scientific and engineering nanotechnology research increasing the private sector's deployment of nanotechnology research and development, including start-up businesses. In order for nanotechnology to foster a truly interdisciplinary research culture in these domains, scientists, engineers, computer scientists, and other professionals must have the necessary education and training in multidisciplinary perspectives.

Establishment of National Nanotechnology Coordination Office and a National Nanotechnology Advisory Panel, both of which will be made up of academics and private-sector executives. First, the panel must assess and make recommendations on nanotechnology trends and developments; second, progress has been made in implementing the Program; third, the Program must be revised; and fourth, the overall program balance, including funding levels for each component area, is required by law.

18.8.2 Analysis of Legislation

According to a review of important legislation on the subject, the U.S. government has backed nanotechnology research and development. Nanotechnology regulation has received very little attention from legislators. Legislation has had no impact on nanotechnology regulation thus far.

More people are coming into contact with nanomaterials as the use of nanotechnology in consumer products grows, even while society as a whole is still confused about the risks and the benefits. As shown earlier, legislative regulation is a reaction to a life-threatening or catastrophic incident. It was more difficult to discover flaws in the industry's monitoring and management when asbestos was widely used in new buildings. For example, legislation on regulation is not considered as a viable aspect of the regulatory system in this concept, because it is only passed after a tragic event or overwhelming evidence.

Legislators have done a great job in encouraging nanotechnology research. We feel that in light of this new information, the focus should stay on nanotechnology research. To fully comprehend the dangers of this technology and to maximize its potential for advancing American interests abroad, much more research is required. Without government funding, the majority of this study would be difficult to complete.

18.9 CONCLUSION

Currently, the field of nanotechnology is dealing with nanomaterials along with the help of the EU, where we have to deal with legal issues as well as other issues, which requires various recommendations regarding the fair conduct of scientific research in the field and in the production of nanomaterials. One of the major concerns is the regulation of activities in field of nanotechnology. Unfortunately, its implementation is not so easy to carry because it is going to effect the economic areas. The Food and Drug Administration (FDA) has a wide and rigorous regulatory framework but with shortcomings too. For the smooth working of regulatory agencies, currently, the FDA is facing many funding crises, which is not good for the future and is preventing the authority from meeting its expectations as efficiently and effectively as possible. Continued collaboration across agencies is greatly required and will ultimately help the industry in filling the gaps due to weak implementation of resources. Another important challenge, faced primarily by industry, is the establishment of innovative methods and validated techniques required for the detection, characterization, and analysis of nanomaterials, including their exposure and hazardous nature. The EU Commission states that collaborative states or agencies should strengthen the public debate on each and every aspect of this field, including its benefits, risks, and uncertainties. Finally, a collaborative approach by the regulatory authorities or neighbouring states has to start with the implementation of policies and new initiatives aimed at a better future, keeping public interest as the prime importance for this field of nanotechnology.

REFERENCES

Abhilash, M. (2010) Potential applications of Nanoparticles. *International Journal of Pharma and Biosciences*, 1 (1): 1–12.

Abraham, J, Logeswar, PI and Silambarasan, S. (2013) Eco-friendly synthesis of silver nanoparticles from commercially available plant powders and their antibacterial properties. *Scientia Iranica*, 20 (3): 1049–53.

Alexis, F and Rhee, J-W. (2008) New frontiers in nanotechnology for cancer treatment. *Urologic Oncology: Seminars and Original Investigations*, 26: 74–85.

Asare, N. (2012) Cytotoxic and genotoxic effects of silver nanoparticles in testicular cells. *Toxicology*, 291 (1–3): 27.

Agrawal, A, Deo, R, Wang, GD, Wang, MD and Nie, S. (2008) Nanometer-scale mapping and single-molecule detection with color-coded nanoparticle probes. *Proceedings of the National Academy of Sciences of the United States of America*, 105, 3298–3303.

Agriculture Risk Protection Act of 2000, Pub. Law. 106–224 (2000), https://www.congress.gov/106/plaws/publ224/PLAW-106publ224.pdf.

Asha Rani, PV, Mun, GLK, Hande, MP. (2009) Cytotoxicity and genotoxicity of silver nanoparticles in human cells. *ACS Nano*, 3: 279–290.

Azam, A and Ahmed, AS. (2012) Antimicrobial activity of metal oxide nanoparticles against Gram-positive and Gram-negative bacteria: A comparative study. *International Journal of Nanomedicine*, 7: 6003–6009.

Baglioni, P, Becher, A, Dürr, M and Nostro, PL. (2008) Synthesis and characterization of zinc oxide nanoparticles: Application to textiles as UV-absorbers. *Journal of Nanoparticle Research*, 10 (4): 679–689.

Balcerzak, M (2013) Zagadnienia nanotechnologii w prawie, Czy nanotechnologia może czerpać z doświadczeń biotechnologii? [The issues of nanotechnologies in law, Is nanotechnology can draw on the expertise of biotechnology?]. In DM Trzmielak (Ed.), *Innowacje i komercjalizacja w biotechnologii [Innovation and commercialization in biotechnology]*, Poznań–Łodź, Poland: Uniwerstytet Łodzki.

Bali, R, Razak, N, Lumb, A and Harris, AT. (2006) The synthesis of metallic nanoparticles inside live plants. In Proceedings of the 2006 International Conference on Nanoscience and Nanotechnology (ICONN'2006), 224–227.

Bar, H, Bhui, DK, Sahoo, GP, Sarkar, P, Sankar, P De and Misra, A. (2009) Green synthesis of silver nanoparticles using latex of *Jatropha curcas. Physicochemical and Engineering Aspects*, 339: 134–139.

Baran, A (2015) Prawne aspekty nanotechnologii w kontekście ochrony środowiska [Legal aspects of nanotechnology in environmental protection, Economics and Environment]. *Ekonomia i Środowisko*, 1(52): 28–40.

Bazela, M (2008) *Jakie są największe wyzwania w dzisiejszej bioetyce? [What are the biggest challenges in today's bioethics?]*. Retrieved from http://biotechnologia.pl/ bioetyka/aktualnosci/jakie-sa-najwieksze -wyzwania- w-dzisiejszej-bioetyce-maciej-bazela,11197

Bob Stump National Defense Authorization Act of 2003, Pub. Law No. 107-314 (2002), https://www.gpo.gov/ fdsys/pkg/PLAW-107publ314/pdf/PLAW-107publ314.pdf.

Brayner, R (2008) The toxicological impact of nanoparticles. *Nano Today*, 3: 48–55.

Brazell, L (2012) *Nanotechnology Law. Best Practices*. Alphen aan den Rijn: Wolters Kluwer, Law & Business.

Daoud, WA, Xin, JH and Zhang, YH (2005) Surface functionalization of cellulose fibers with titanium dioxide nanoparticles and their combined bactericidal activities. *Surface Science*, 599 (1–3): 69–75.

De, M, Gosh, PS and Rotello, VM. (2008) *Applications of Nanoparticles in Biology*. KGaA, Weinheim: Advanced Materials, Wiley- VCH Verlag Gmbh & Co., 1, 4225–4241.

Donaldson, K, Stone, V, MacNee, W. (1999) The toxicology of ultrafine particles. In: Maynard, LA, Howards, CA (eds) *Particulate Matter Properties and Effects upon Health* (pp 115–127)/ Oxford: Bios Scientific.

Directive 2001/83/EC of the European Parliament and of the Council of 6 November 2001 on the Community code relating to medicinal products for human use (*OJ L* 311, 28.11.2001).

EC/726/2004 Regulation (EC) No 726/2004 laying down Community procedures for the authorisation and supervision of medicinal products for human and veterinary use and establishing a European Medicines Agency (*OJ L* 136, 30.04.2004).

Falaschetti, C (2012) Nanotechnology and the science of beauty. Sep 19, https://helix.northwestern.edu/article /nanotechnology-and-science-beauty

Falkner, R, Breggin, L, Jaspers, N, Pendergrass, J, & Porter, RD (2010) International coordination and cooperation: The next agenda in nanomaterials regulation. In GA Hodge, DM Bowman and AD Maynard (Eds.), *International Handbook on Regulating Nanotechnologies*. Cheltenham and Northampton, Great Britain: Edward Elgar Publishing

Foldbjerg, R and Autrup, H. (2013) Mechanisms of Silver Nanoparticle Toxicity. *Archives of Basic and Applied Medicine*, 1(1): 5–15.

Ganguli, P, & Jabade, S (2012) *Nanotechnology, Intellectual Property Rights, Research, Design and Commercialization*. Boca Raton: CRC Press.

Giasuddin, ASM, Jhuma, KA and Mujibul Haq, AM. (2012) Use of gold nanoparticles in diagnostics, surgery and medicine: A review. *Bangladesh Journal of Medical Biochemistry* 5(2): 56–60.

Gurunathan, S, Han, JW, Eppakayala, V, Jeyaraj, M and Kim, JH. (2013) Cytotoxicity of biologically synthesized silver nanoparticles in MDA-MB-231 human breast cancer cells. *Hindawi Publishing Corporation, BioMed Research International*, Article ID 535796, 10 p.

Gutwein, LG and Webster, TJ. (2003) American Ceramic Society 26th Annual Meeting Conference Proceedings.

Gliga, AR, Skoglund, S, Wallinder, IO, Fadeel, B and Karlsson, HL. (2014) Size-dependent cytotoxicity of silver nanoparticles in human lung cells: The role of cellular uptake, agglomeration and Ag release. *Particle and Fibre Toxicology*, 11: 11.

Hackenberg, S, Scherzed, A, Kessler, M. (2011) Silver nanoparticles: Evaluation of DNA damage, toxicity and functional impairment in human mesenchymal stem cells. *Toxicology Letters*, 201: 27–33.

Haggenmueller, R, Zhou, W, Fischer, JE and Winey, KI. (2003) Production and characterization of polymer nanocomposites with highly aligned single-walled carbon nanotubes. *Journal of Nanoscience and Nanotechnology*, 3: 105–110.

Hett, A (2004) *Nanotechnology. Small Matter, Many Unknowns*. Zurich: Swiss Reinsurance Company.

Holiday, R (2008) *Use of Gold in Medicine and Surgery. Biomedical Scientist (The Official Gazette of the Institute of Biomedical Science, UK)*, 962–63.

Horikoshi, S and Serphone, N. (2013) *Microwaves in Nanoparticles Synthesis* (1st ed., pp. 1–24). Weinheim: Wiley- VCH Verlag GmbH & Co. KGaA.

Huang, J, Li Q, Sun D, Lu Y, Su Y, Yang, X, Wang, H, Wang, Y, Shao, W, He, W, Hong, J and Cuixue, C. (2007) Biosynthesis of silver and gold nanoparticles by novel sundried *Cinnamomum camphora* leaf. *Nanotechnology*, 18: 105104.

Hussain, SM, Hess, KL, Gearhart, JM. (2005) In vitro toxicity of nanoparticles in BRL 3A rat liver cells. *Toxicol In Vitro*, 19: 975–983.

Jurewicz, M (2014) *Nanotechnologia, Regulacje prawne, Legislacja Unii Europejskiej [Nanotechnology, Regulations, European Union Legislation]*. Warszawa, Poland: Difin.

Kalarikkal, SG, Sankar, BV and Ifju, PG (2006) Effect of cryogenic temperature on the fracture toughness of graphite/epoxy composites. *Journal of Engineering Materials and Technology*, 128(2): 151–157.

Kokura, S, Handa, O, Takagi, T, Ishikawa, T, Naito, Y and Yoshikawa, T. (2010) Silver nanoparticles as a safe preservative for use in cosmetics. *Nanomed*, 6: 570–574.

Krishnaraj, C, Jagan, EG, Rajasekar, S, Selvakumar, P, Kalaichelvan, PT and Mohan, N. (2010) Synthesis of silver nanoparticles using Acalypha indica leaf extracts and its antibacterial activity against water borne pathogens. *Biointerfaces* 76: 50–56.

Kulkarni, AP, Srivastava, AA, Harpale, PM and Zunjarrao, RS. (2011) Plant mediated synthesis of silver nanoparticles:tapping the unexploited sources. *Journal of Natural Product and Plant Resources*, 1 (4): 100–107.

Liang, H, Jin, C, Tang, Y, Wang, F, Ma, C and Yang, Y. (2013) Cytotoxicity of silica nanoparticles on HaCaT cells. *Journal of Applied Toxicology*, 34 (4): 367–372.

Leela, A and Vivekanandan, M. (2008) Tapping the unexploited plant resources for the synthesis of silver nanoparticles. *African Journal of Biotechnology*, 7 (17): 3162–3165.

Lu, W, Senapati D and Wang, S. (2010) Effect of surface coating on the toxicity of silver nanomaterials on human skin keratinocytes. *Chemical Physics Letters*, 487: 92–96.

Maliszewska-Mazur, M (2010) Nanotechnologia – nowe wyzwania, nowe możliwośći i nowe problem [Nanotechnology – new challenges, new possibilities and new problems]. *Ochrona Środowiska i Zasobów Naturalnych*, 45: 153–161.

Marcato, PD and Duran, N. (2008) New aspects of nanopharmaceutical delivery systems. *Journal of Nanoscience and Nanotechnology*, 8: 2216–2229.

Marchant, GE and Doug, JS. (2006) Transnational models for regulation of nanotechnology. *Journal of Law, Medicine & Ethics*, 714–725. Retrieved from http:// caat.jhsph.edu/programs/workshops/Transnational% 20models%20for%20reg.pdf

Matsuura, JH (2006) *Nanotechnology Regulation and Policy Worldwide*. Norwood, MA: Artech House, Inc.

Malloy, TF (2012) Soft law and nanotechnology: A functional perspective. *Jurimetrics: The Journal of Law, Sciences & Technology*, 52(3): 347–358.

Maynard, AD (2006) Nanotechnology: Assesing the risk, review feature. *Nanotoday*, 1(2): 22–33.

Moghimi, SM, Hunter, AC and Murray, JC. (2001) Long-circulating and target specific nanoparticles. Theory to practice. *Pharmacological Reviews*, 53: 283–318.

Mukherjee, SG, O'Claonadh, N, Casey, A and Gordon, C. (2012) Comparative in vitro cytotoxicity study of silver nanoparticles on two mammalian cell lines. *Toxicol In Vitro*, 26(2): 238–51. doi: 10.1016/j.tiv. 2011.12.004.

Muller, RH, Pardeike, J and Hommoss, A. (2009) Lipid nanoparticles (SLN, NLC) in cosmetic and pharmaceutical dermal products. *International Journal of Pharmaceutics*, 366: 170–184.

Nanotechnology Initiative. (2006) Washington, DC: National Academies Press, https://www.nap.edu/read/11752/chapter/3#17, 17.

Nasir, A. (2010) Nanotechnology and dermatology: Part II: risks of nanotechnology. *Clinics in Dermatology*, 28: 581–588.

National Defense Authorization Act for Fiscal Year. (2008) Pub. Law No. 110-477 (2008), https://www.gpo.gov/fdsys/pkg/CRPT-110hrpt477/pdf/CRPT-110hrpt477.pdf

National Defense Authorization Act for Fiscal Year. (2010) Pub. Law No. 111-84 (2009), https://www.gpo.gov/fdsys/pkg/PLAW-111publ84/pdf/PLAW-111publ84.pdf

National Research Council. (2001) *Preliminary Comments, Review of the National Nanotechnology Initiative*. Washington, DC: National Academies Press, https://doi.org/10.17226/10216, 1.

Oberdorster, E. (2004) Manufactured nanomaterials (fullerenes, C60) induce oxidative stress in the brain of juvenile largemouth bass. *Environ Health Perspect*, 112: 1058–1062.

Oud, M (2007) A European perspective. In GA Hodge, DM Bowman, & K Ludlow (Eds.), *New Global Frontiers in Regulation: The Age of Nanotechnology* (pp. 97–109). Cheltenham and Northampton, Great Britain: Edward Elgar Publishing.

Panda, KK, Achary, VM, Phaomie, G, Sahu, HK, Parinandi, NL and Panda, BB. (2016) polyvinyl polypyrrolidone attenuates genotoxicity of silver nanoparticles synthesized via green route, tested in *Lathyrus sativus* L. root bioassay, *Mutation Research/Genetic Toxicology and Environmental Mutagenesis*, 806 (11-23).

Pant, G, Nayak, N and Prasuna, RG. (2013) Enhancement of antidandruff activity of shampoo by biosynthesized silver nanoparticles from *Solanum trilobatum* plant leaf. *Applied Nanoscience* 3: 431–439.

Parkin, I and Palgrave, R. (2004) Self-cleaning clothes. *Journal of Materials Chemistry*, 15: 1689–1695.

Patel, A, Prajapati, P and Boghra, R. (2011) Overview on application of nanoaparticles in cosmetics. *Asian Journal of Pharmaceutical Sciences and Clinical Research*, 1 (2): 40–55.

Ponce del Castillo, AM (2010) *The UE Approach to Regulating Nanotechnology*. Brussels: ETUI.

Poncharal, P, Wang, ZL, Ugarte, D, Walt, A and Heer de. (1999) Electrostatic Deflections and Electromechanical Resonances of Carbon Nanotubes. *Science*, 283 (5407): 1513–1516.

Rasmussen, JW, Martinez, E, Louka, P and Wingett, DG. (2010) Zinc oxide nanoparticles for selective destruction of tumor cells and potential for drug delivery applications. *Expert Opinion on Drug Delivery*. 7(9): 1063–1077.

Regulation (EC) No 1169/2011 on the provision of food information to consumers (*OJ L* 304, 22.11.2011).

Regulation (EC) No 1223/2009 of the European Parliament and of the Council of 30 November 2009 on cosmetic products (Text with EEA relevance), (*OJ L* 342, 22.12.2009).

Regulation (EC) No 1907/2006 of the European Parliament and of the Council of 18 December 2006 concerning the registration, evaluation, authorization and restriction of chemicals (*OJ L* 396, 30.12.2006).

Sadowski, Z, Maliszewska, IH, Grochowalska, B, Polowczyk, I and Kozlecki, T. (2008) Synthesis of silver nanoparticles using microorganisms. *Materials Science-Poland*, 26 (2): 420–424

Sanvicens, N and Pilar, MM. (2008) Multifunctional Nanoparticles: Properties and prospects for their use in human medicine. *Trends in Biotechnology*, 26(8): 425–433. doi: 10.1016/j.tibtech.2008.04.005

Sarsar, V, Manjit, S and Kumar, K (2014) Significant parameters in the optimization of silver nanoparticles using *Psidium guajava* leaf. *An International Journal of Advances in Pharmaceutical Sciences*, 5 (1): 1769–1775.

Shah, M, Fawcett, D, Sharma, S, Tripathy, SK and Poinern, GEJ. (2015) Green synthesis of metallic nanoparticles via biological entities. *Materials*, 8: 7278–7308.

Shameli, K, Ahmad, MB, Jaffar Al-Mulla, EA, Ibrahim, NA, Shabanzadeh, P, Rustaiyan, A. (2012). Green biosynthesis of silver nanoparticles using *Callicarpa maingayi* stem bark extraction, *Molecules*, 17: 8506–8517.

Shekhawat, MS, Ravindran, CP and Manokari, M. (2014) A biomimetic approach towards synthesis of zinc oxide nanoparticles using *Hybanthus enneaspermus* (L.) F. Muell. *Tropical Plant Research*, 1 (2): 55–59.

Shokri, J (2017) Nanocosmetics: Benefits and risks. *BioImpacts*, 7(4): 207–208.

Sindhura, KS, Prasad, TNVKV, Selvam, PP and Hussain, OM. (2014) Synthesis, characterization and evaluation of effect of phytogenic zinc nanoparticles on soil exo-enzymes. *Applied Nanosciences*, 1: 1–9.

Singh, R and Singh, NH. (2011) Medical applications of nanoparticles in biological imaging, cell labeling, antimicrobial agents, and anticancer nanodrugs. *J Biomed Nanotechnol*, 7: 489–503.

Siegfried, B. (2007) *NanoTextiles: Functions, Nanoparticles and Commercial Applications, EMPA*.

Smitha, SL, Philip, D and Gopchandran, KG. (2009) Green synthesis of gold nanoparticles using *Cinnamomum zeylanicum* leaf broth. *Spectrochimica Acta Part A*, 74: 735–739.

Suker, DK and Albadran, RM. (2013) Cytotoxic effects of titanium dioxide nanoparticles on rat embryo fibroblast REF-3 cell line *in vitro*. *European Journal of Experimental Biology*, 3(1): 354–363.

Szponder, DK (2010) *Nanomateriały w środowisku – korzyści zagrożenia [Nanomaterials in the environment – benefits and risks]*. Krakow, Poland: V Krakowska Konferencja Młodych Uczonych

Takhar, P and Mahant, S. (2011) In vitro methods for nanotoxicity assessment: Advantages and applications. *Archives of Applied Science Research*, 3 (2): 389–403.

Taufikurohmah, T, Sanjaya, IGM, Baktir, A and Syahrani, A. (2014) TEM analysis of gold nanoparticles synthesisin glycerin: Novel safety materials in cosmetics to recovery mercury damage. *Research Journal of Pharmaceutical, Biological and Chemical Sciences*, 5(1): 397.

Thirumurugan, A, Neethu Anns, T, Jai Ganesh, R and Gobikrishnan, S. (2010) Biological reduction of silver nanoparticles using plant leaf extracts and its effect on increased antimicrobial activity against clinically isolated organism. *Scholars Research Library*, 2(6): 279–284.

Tvaryanas, AP, Brown, L and Miller, NL. (2009) Managing the human weapon system: A vision for an air force human-performance doctrine. *Air and Space Journal* 23 (2): 34–40, http://www.airuniversity.af.mil/Portals/10/ASPJ/journals/Volume-23_Issue-1-4/2009_Vol23_No2.pdf

Veerasamy, R, Xin, TZ, Gunasagaran, S, Xiang, TFW, Yang, EFC, Jeyakumar N and Dhanaraj SA. (2011) Biosynthesis of silver nanoparticles using mangosteen leaf extract and evaluation of their antimicrobial activities. *Journal of Saudi Chemical Society*, 15: 113–120.

Veronovski, N, Lešnik, M, Lubej, A and Verhovšek, D. (2014) Surface treated titanium dioxide nanoparticles as inorganic UV filters in sunscreen products. *Acta Chimica Slovenica*, 61 (3): 595–600.

Watal, G, Mubayi, A, Chatterji, S and Rai, PM. (2012) Evidence based green synthesis of nanoparticles. *Advanced Materials Letters*, 3(6): 519–525.

Wood, S, Jones, R and Geldart, A. (2003) *The Social and Economic Challenges of Nanotechnology (No. 0-86226-294-1)*. Swindon, UK: Economic and Social Research Council.

Wiechers, JW and Musee, N. (2010) Engineered inorganic nanoparticles and cosmetics: Facts, issues, knowledge gaps and challenges. *Journal of Biomedical Nanotechnology*, 6: 408–431.

Yadav, A, Prasad, V, Kathe, AA, Raj, S, Yadav, D, Sundramoorthy, C and Vigneshwaran, N. (2006) Functional finishing in cotton fabrics using zinc oxide nanoparticles. *Bulletin of Materials Science* 29 (6): 641–645. © Indian Academy of Sciences.

Zapor, L (2012) Bezpieczeństwo i higiena pracy a rozwoj nanotechnologii [Occupational safety and health and nanotechnology, Occupational Safety, Science and Practice]. *Bezpieczeństwo Pracy: nauka i praktyka*, 1: 4–7.

Zhang, T, Wang, L, Chen, Q and Chen, C. (2014) Cytotoxic potential of silver nanoparticles. *Yonsei Medical Journal* Mar 1, 55(2): 283–291.

Index

A

Absorption, 25–26, 53, 57, 60, 124, 129, 131, 137, 166, 187, 208, 209, 228, 403, 427, 447
Absorption, distribution, metabolism and elimination (ADME), 232, 248
Acceptable Daily Intake (ADI), 307
Acetylcholinesterase (AChE), 228, 236, 240, 418
Activity foods, 79
Additives, 6, 31, 115, 167, 191, 268, 272, 307, 450
Adenosine monophosphate activated kinase (AMPK), 358
Adenylyl cyclase (AC), 153
Alanine transaminase (ALT), 54, 351
Albumin, 18, 24, 72, 85, 404
Alginate, 88, 264, 272, 284, 404
Alkaline phosphatase (ALP), 54
Allergic contact dermatitis (ACD), 190
All-trans-retinoic acid (ATRA), 199
Almond gums, 272
Alyssum homolocarpum seed gum (AHSG), 272
Alzheimer's disease (AD), 4, 80, 89, 236, 354, 403, 418–420, 423, 424
Amide, 85–86
2-aminoethanesulfonic acid, 82
Amyotrophic lateral sclerosis (ALS), 417, 421
Analeptic agents, 51
Angiotensin-converting enzyme (ACE), 204, 350
Angum gum, 272
Anhydride, 13, 24, 85–86, 271
Animal-derived polysaccharides
  chitosan, 278–279
  hyaluronic acid, 279–280
  nutraceuticals, 280
Anthocyanin, 156, 158, 164
Anti-inflammatory nutraceuticals, 80
Antioxidants, 2, 4, 80–82
Antiviral drug acyclovir (ACV), 191
Arabic–chitosan system, 271
Archaeosomes, 403
Asiatic acid conjugates, 240
Aspartate transaminase (AST), 54, 351
Atomic absorption spectrometry (AAS), 11
Atomic force microscopy (AFM), 74–75, 113–114, 448
Auger electron spectroscopy (AES), 12

B

Beclomethasone dipropionate (BDP), 202
Bergamot, 52
β-Glucan, 123
  blood pressure, 129
  dyslipidemia, 127–129
  insulin resistance, 126–127
  mechanism of action
    biotransformation, 139
    cholesterol homeostasis, 135–138
    hypocholesterolemic capability, 133–134
    lipid metabolism/homeostasis, 139
    microbiome-mediated lipid assimilation, 139
    SCFA, 138–139
  obesity, 129–133
  therapeutic applications, 126
Bifidobacterium, 83
Bile acid hydroxylase (BSH), 133
Bioactive compounds, 3, 18
Bioavailability, 19, 24
  absorption, 25–26
  nanocarriers, 20–24
  solubilization, 24–25
  systemic circulation, 26
Biocompatible polyanhydrides, 86
Biodegradable polymers, 84
Biogenic methods, 304
Biogenic synthesis methods, 303
Biomedical sciences, 402–403
Bionanotubes (BNTs), 68
Blood–brain barrier (BBB), 53, 87, 99, 150, 228, 285
Blood pressure, 129
Bovine serum albumin (BSA), 24, 68, 240, 389
Brain-derived neurotropic factor (BDNF), 153, 353
Brown algae
  alginate, 284
  fucoidans, 283–284
  laminaran, 284
Bryostatin conjugates, 240
Butyrylcholinesterase (BuChE), 418

C

Calcium/calmodulin-dependent protein kinase II (CaMKII), 355
CAMP response element-binding (CREB), 153, 356
Camptothecin (CPT), 200
Cancer, 63
  anticancer therapeutics, 241
  antitumor therapeutics, 242
  dihydroartemisin conjugates, 242–343
  eugenol conjugates, 243
  methodology
    co-occurrences analysis, 65–67
    international co-authorship analysis, 67
    publication analysis, 64–65
  mortality rate, 64
  naringenin conjugates, 243
  research focus
    China, 70–71
    India, 71–75
    United States, 68–70
  resveratrol conjugates, 243
  thymoquinone conjugates, 242
Carbamazepine, 186
Carbohydrates, 124, 139, 348, 389
Carbon-based nanomaterials, 11
Carbon nanoparticles, 8
Carbon nanotubes (CNTs), 19, 150, 233, 319, 444
Cardiovascular disease (CVDs), 74, 80, 127, 226, 245, 332, 425, 431
β-Carotene, 3, 55, 269, 271, 284

Carotenoids, 16, 81
Carrageenan, 281
Carum copticum, 333
Casein, 71
Catechin (CAT), 74, 82
Catechol o-methyltransferase (COMT), 236
Cationic zein nanoparticles, 266, 269
Cavitation, 18
Cellulose, 268–269
Cellulose acetate (CA), 268
Cellulose nanocrystals (CNCs), 266–268, 273
Cellulose nanofibers (CNFs), 268, 273
Central nervous system (CNS), 82, 155, 204, 228, 352, 418
Centrifugation, 11
Cerebral ischemia, 353
Cetyl trimethylammonium bromide (CTAB), 268–269
Chemical force microscopy, 12
Chemotherapeutics, 241
Chitosan, 13, 16, 25, 165, 278–279
Chitosan-based nanoparticles, 165
Chitosan nanoparticles, 266
Cholecalciferol (vitamin D3) mini-tablets, 388
Cholecystokinin (CCK), 132
Cholesterylphosphonyl gemcitabine (CPNG), 211
Chondroitin sulfate, 82–83
Collagen, 72
Collagen-induced arthritis (CIA), 200
Colloidal silver, 442
Colorectal cancer (CRC), 67, 83, 359, 391
Commendamide, 139
Complexation techniques, 266
Composition-based classification
  carbon-based, 383–384
  clay-based, 384
  dietary ingredients, 385
  lipid-based, 386
  metal, 384
  partial conductor, 384, 386
  polymeric, 386
Confocal laser scanning microscopy (CLSM), 114
Conjugated linoleic acid (CLA), 187, 404
Conventional drug administration, 63
Copper/copper oxide nanoparticles, 316
Corn protein hydrolysate (CPH), 71
Cosmetics, 442
Cowpea mosaic virus (CPMV), 232
Cranberry, 82
Creatine phosphokinase (CPK), 351
Critical micellar concentration (CMC), 200
Curcumin, 54, 350, 352, 357, 364
Curcumin conjugates, 238–239
Curcumin-entrapped phospholipid-sodium deoxycholate-mixed micelles (CUR-PC-SDC-MMs), 202
Curcumin-loaded nanoemulsion, 187
Curcumin nanoparticles (Cur NPs), 71
Cutting-edge technology, 8
Cyclic monophosphate (cAMP) proteins, 153

D

Delivery system, 98
5-Demethylsuberosin (5DT), 68
Dendrimers, 232
Design of experiments (DoE), 74
Dextran, 272–273
Diabetic retinopathy (DR), 246
Diacerein-loaded SLNs (DC-SLNs), 198
Diallyl disulfide (DADS), 280
Diallyl trisulfide (DATS), 280
Didodecyl bromide (DMAB), 335
Dietary fibers, 123–129, 131, 135, 138, 139, 347
Dietary proteins, 333
Dietary supplements, 15, 31, 82–83, 347–348
Differential scanning calorimetry (DSC), 74, 111, 114, 188
Dihydroartemisin conjugates, 242–243
7,8-Dihydroxyflavone (DHF), 356
Dimyristoylphosphatidylglycerol (DMPG), 198
Dipalmitoylphosphatidylcholine (DPPC), 191, 198
Dipalmitoyl phosphatidylglycerol (DPPG), 190
Distearoyl phosphatidylethanolamine (DSPE), 190
Docosahexaenoic acid (DHA), 81
Dopamine transporter (DAT), 421
Double emulsions, 110–111, 266, 270, 272
Doxorubicin (DOX), 191
Drug encapsulation, 116
Drug-enriched core model, 232
Drug-enriched shell model, 232
Drug loading, 117
D-α-tocopherol polyethylene glycol 1000 succinate (TPGS), 202
Dual chlorambucil-tailed phospholipid (DCTP), 191
Dynamic light scattering (DLS), 73, 190, 446

E

Egg phosphatidylcholine, 186
Egg phospholipid-based nanoemulsion, 186
Eicosatetraenoic acid (EPA), 81, 272
Electrospinning, 270
Ellagic acid (EA), 335
Emulsions, 185
Encapsulation, 52
  archaeosomes, 403
  micelles, 404
  nanocapsules, 404
  nanocochleates, 403
  nanoemulsions, 404
  nanoliposomes, 403
  nanospheres, 404
  polymer-based, 404
  polymersomes, 404
Endothelial inflammation, 432
Energy dispersive X-ray spectroscopy (EDX), 11, 75
Engineered nanoparticle (ENP), 444
Enhanced permeation and retention (EPR), 71, 192
Epicatechin (ECAT), 74
Epigallocatechin, 352, 361
Epigallocatechin gallate (EGCG), 208
Essential amino acids, 82
Essential phospholipids, 185
Ester, 85–86
Ethosomes, 194–197
Eugenol conjugates, 243
European Food Safety Authority (EFSA), 9
Extracellular signal-regulated kinase (ERK), 361

F

Fabricated nanoparticles, 2
Farnesoid X receptor (FXR), 132
Field-flow fractionation (FFF), 11
Filtering, 11
First apoptosis signal (FAS), 362
Flavonoids, 151, 332
Flaxseed, 81
Flow-mediated dilation (FMD), 352
Food-based nutraceuticals, 422
Food ingredients, 306
Food nanotechnology, 3
Fortified nutraceuticals, 349
Fourier transform infrared spectroscopy (FTIR), 12, 73, 446
Free fatty acid (FFA), 187
Fructans, 273
Fruit juices, 4
Fucoidans, 283–284
Functional foods (FFs), 81, 347–348, 421, 431

G

Gamma-amino butyric acid (GABA), 169
Garlic, 81
Gastrointestinal (GI) microflora, 83
Gastrointestinal tract (GIT), 68, 131
Gelatin nanoparticles, 72
Generally recognized as safe (GRAS), 184, 317
Genistein, 360
Ghrelin, 132
Glucagon-like peptide 1 (GLP-1), 132, 200
Glucose transport type-4 (GLUT-4), 127
Glutathione peroxidase (GPx), 80
Gold nanoparticles (Au NPs), 73, 311–312
G protein-coupled receptor (GPCR), 153
Grape seed extract (GSE), 73
Greener methods, 267
Green fluorescent protein (GFP), 402
Green tea, 82
Gum Arabic (GA), 68
Gums, 271–272

H

Heat shock protein 90 (Hsp90), 200
Hesperidin (HRN), 73
High-density lipoprotein (HDL), 126, 357
High-dosage phospholipids, 185
High–methoxy pectin nanoemulsions, 267
High-performance liquid chromatographic (HPLC) analysis, 115
High-pressure homogenization (HPH), 105–107
High-pressure liquid chromatography (HPLC), 11
High-resolution TEM (HRTEM), 446
High-shear homogenization, 109–110
Homogeneous matrix model, 232
Human serum albumin (HSA), 72
Huntington's disease (HD), 420
Huperzine A conjugates, 240–241
Hyaluronic acid (HA), 279–280
Hydrocolloids, 72
Hydrogenated Soy PC (HSPC), 191
Hydrolyzable tannins, 151
Hydrophilic-lipophilic balance (HLB), 104, 188
Hydrophobic therapeutics, 186
Hydroxypropyl methylcellulose (HPMC), 68
Hypercholesterolemia, 129
Hypothalamic-pituitary-adrenal (HPA) axis, 359

I

Imperfect crystals, 100
Imprinted-like biopolymeric micelles (IBMs), 70
Inducible nitric oxide synthase (iNOS), 423
Inductively coupled plasma mass spectrometry (ICP-MS), 11
Inductively coupled plasma optical emission spectroscopy (ICP-OES), 11
Inflammation-induced atherosclerosis, 432
Inflammatory bowel disease (IBD), 358
Inorganic nanomaterials, 9
Insoluble fibers, 124
Institute of Food Technologists (IFT), 305
Inulin, 273

L

Lactate dehydrogenase (LDH), 351
Lactic acid–producing bacteria, 83
Lactobacillus, 83
Laminaran, 284
Laser diffraction (LD), 111
Laser Doppler electrophoresis technique, 113
Laser-mediated synthesis (LMS), 445
Lecithin-stabilized nanoemulsion, 186–187
Legislative space
  analysis of legislation, 454–455
  law of 2000, 453
  law of 2001, 453
  law of 2002, 453–454
  law of 2003, 454
Leucosin, 284
Lipid carriers, 99
Lipid matrix, 117
Lipid nanocarriers, 53
Lipid nanoparticles, 99
Lipids, 331–332
Lipoidal drug delivery systems, 99
Liposomal delivery mechanism, 53
Liposomes, 15–16, 87–88, 188–192, 231
  ethosomes, 194–197
  transferosomes, 192–194
Lipospheres, 204–205, 206
Liquid lipids, 99, 104
Locus coeruleus (LC), 153
Low-density lipoprotein (LDL), 81, 100, 127
Lower critical solution temperature (LCST), 86
Lutein, 56–57
Luteolin–phospholipid complex (LPC), 208
Lycopene, 54–55, 81, 353
Lycopene conjugates, 240
Lyophilization, 111

M

Magnetic nanoparticle (MN), 402
Maltodextrins, 124

Medium-chain triglycerides (MCT), 68
Medium spiny neurons (MSNs), 153
Melanoma, 53
Melting dispersion method, 109
Mesolimbic dopamine system, 152
Metabolic disorders, 124, 125
Metabolic syndrome, 123, 125, 134
Metallic catalysts, 441
Metallic nanoparticles, 302
Metal nanoparticles (MNPs), 302
  copper/copper oxide nanoparticles, 316
  food encapsulation, 311–318
  gold nanoparticles (Au NPs), 311–312
  silver nanoparticles (Ag NPs), 312–315
  titanium dioxide nanoparticles (TiO_2), 315–316
  zinc oxide nanoparticles, 316–318
Metastasis, 241
Micro-coating materials, 306
Microemulsion technique, 270
Microencapsulated core materials, 306
Microporous silica Nps (MSNs), 233
Microwave-aided synthesis (MAS), 445
Milk fat globule membrane (MFGM), 190
Milk proteins, 71–72
Minerals, 333
Mitogen-activated protein kinases (MAPK), 153
Monoacyl-phospholipids, 188
Mononuclear phagocytes, 406
Morphine, 149
Multifunctional micelles, 200–204
Multivariate analysis (MVA), 74
μ-opioid receptors (MORs), 149
Mutant huntingtin protein (mHTT), 420
Myricetins, 68

N

Nano-based nutraceutical technologies, 34
Nanocapsules, 18, 404
Nanocarrier-based systems, 241
Nanocarriers, 52, 63
  anthocyanin nanoparticles, 160–161
  chitosan, 165
  classification, 230
  CNT, 233
  dendrimers, 232
  hybrid, 233–234
  inorganic, structure of, 230
  liposomes, 163–164, 231
  metallic NPs, 233
  organic, structure of, 230
  PLGA, 159, 162–163
  PM, 232
  QDs, 233
  silica, 168–169
  SLN, 231–232
  TiO_2, 167–168
  uses, 231
  VNPs, 232–233
  ZnO_2 nanoparticles, 165–167
Nanocelluloses, 268, 273
Nanoceuticals, 6
  chemical nature, 423
  classification, 378, 400
  definition, 378
  disadvantages/pitfalls, 26–27
  examples, 378
  market appeal, 32–33
  nanoproducts, 27, 32
  natural, 379–383
  plant-based, 400
  synthetic, 384
  therapeutic efficiency, 28–31
Nanocochleates, 403
Nanocrystalline solid dispersions (NSDs), 73
Nanodelivery-based systems, 63
Nanodelivery mechanisms, 52
Nano-drug delivery systems, 150
Nanoemulsions, 17–18, 267, 404
Nanoencapsulated nutrients, 308–309
Nanoencapsulation, 6, 309–311
  β-carotene, 55
  challenges, 57, 59
  curcumin, 54
  lutein, 56–57
  lycopene, 54–55
  opportunities, 59–60
  resveratrol, 53
  technique and function, 58–59
  vitamin C, 55–56
Nanofibers, 16
Nano-foodstuffs, 34, 264
Nanoformulations, 57, 333–337
  carbon nanotubes, 19
  liposomes, 15–16
  nanocapsules, 18
  nanoemulsions, 17–18
  nanofibers, 16
  nanoparticles, 12–14
  nanospheres, 15
  niosomes, 14–15
  SLN, 14
Nanoliposome-based delivery systems, 267
Nanoliposomes, 403
Nanologo, 453
Nanomaterials
  conventional approaches, 11
  dispersions, 12
  labeling, 11
  physicochemical endpoints, 10
  physicochemical properties, 9
  physicochemical transitions, 11
  qualities, 9
  quantifying, 9
  scattering of light, 12
  therapeutic efficacy, 9
Nanomedicine, 226, 411
Nano-nutraceuticals
  anti-inflammatory tools, 433–434
  cancer management
    applications, 409–410
    cancer targeting, 405–406, 411
    conventional chemotherapy, 404–405
    metabolic activity inhibition, 407–408
    nanomedicine, 411
  classification, disease
    cancer, 359–362
    cardiovascular disorders, 350–352

gastrointestinal disorders, 358–359
metabolic disorders, 356–358
neurological disorders, 352–356
CVD, 434–437
drawbacks, 426–427
encapsulation
archaeosomes, 403
micelles, 404
nanocapsules, 404
nanocochleates, 403
nanoemulsions, 404
nanoliposomes, 403
nanospheres, 404
polymer-based, 404
polymersomes, 404
flavonoids, 332
lipids, 331–332
medicinal approach, 425–426
minerals, 333
neurodegenerative disorders, 426 (*see* Neurodegenerative disorders)
neurological disorders, 362
phenolics, 332
targeted delivery, cancer, 362–365
therapeutic approach, 425
thymoquinone, 363
vitamins, 332
Nanoparticles (NPs), 98
application in medicine, 391
applications, 302
cosmetics, 443–445
medicine, 443
textile, 443
composition-based classification
carbon-based, 383–384
clay-based, 384
dietary ingredients, 385
lipid-based, 386
metal, 384
partial conductor, 384, 386
polymeric, 386
cytotoxicity, 447–449
delivery systems and challenges
associated colloids, 388
cholecalciferol (vitamin D3) mini-tablets, 388
current delivery systems, 387
lipid-based systems, 387
nanoliposomal formulation, 387–388
protein-based nanocarriers, 389
solid-core micelles, 388
food industry, 318
literature, 442
mesoporous, 383
oral delivery
antioxidant potential, 389, 390
gastrointestinal effect, 389, 390
mode of action, 390–391
preparation
antimicrobial activity, 447
biological synthesis, 445–447
dermal absorption, 447
structure-based classification, 386
structure of, 378
synthesis, 386–387
synthesis methods, 265
biogenic methods, 304
chemical methods, 303–304
green (biogenic) synthesis methods, 304
Nanoparticulate delivery frameworks, 98
Nanoprecipitation technique, 235, 267
Nanoscience, 2, 6
Nanospheres, 15, 404
Nanostructured lipid carriers (NLCs), 99
advantages, 101
agents for preparation, 105
amorphous, 101, 102
applications, 117–118
characterization
AFM, 113–114
CLSM, 114
drug entrapment efficiency, 114–115
DSC, 114
HPLC, 115
laser Doppler electrophoresis technique, 113
NMR, 115–116
particle size analysis, 113
pH, 115
rheological study, 114
SEM, 113
TEM, 113
ultrafiltration, 115
XRD, 114
zeta potential measurement, 113
disadvantages, 101
drug encapsulation, 116
drug release, 116–117
emulsifiers, 104–105
lipids, 103
liquid, 104
solid, 104
methods of preparation
double emulsion technique, 110–111
high-shear homogenization, 109–110
HPH, 106, 107
melting dispersion method, 109
phase inversion, 106–107
solvent emulsification-diffusion technique, 109
solvent emulsification-evaporation technique, 108–109
solvent injection method, 106
ultrasonication technique, 109–110
process variables, 113
specifications, 112
stability, 111–113
structural model, 100
surfactants, 104–105
type I, 100
type II, 100–101
type III, 101
Nanostructured lipoprotein carriers (NLCs), 445
Nanotechnology, 2, 4, 150, 226
ethical considerations, 452–453
food science, 306–308
legislative space
analysis of legislation, 454–455
law of 2000, 453
law of 2001, 453
law of 2002, 453–454
law of 2003, 454

regulations
legal aspects, 449–451
principles, 451
risk of application, 451–452
Nanotechnology-based delivery systems
biodegradable polymers, 84
ester/anhydride/amide, 85–86
liposome, 87–88
micelles, 87
pH-sensitive polymers, 86–87
polymers, 84
polysaccharides, 85
proteins, 84–85
temperature-sensitive hydrogels, 86
Nanotechnology-based non-opioid delivery, 157–158
CNS protection, 158–159
nanocarriers
anthocyanin nanoparticles, 160–161
chitosan, 165
liposome-based, 163–164
PLGA, 159, 162–163
silica, 168–169
TiO_2, 167–168
ZnO_2 nanoparticles, 165–167
Nanotexturing, 6
Naringenin conjugates, 243
National Nanotechnology Initiative (NNI), 229
Natural bioactive compounds, 226
Natural chemotherapeutic agents, 53
Natural compounds, 226
Natural foods, 80
Natural killer (NK) cell activity, 83
Natural/organic nanomaterials, 8–9
Natural polysaccharides, 264
Natural/traditional nutraceuticals, 348–349
Neurodegeneration process, 424
Neurodegenerative disorders, 417–418
Alzheimer's disease, 236
asiatic acid conjugates, 240
bryostatin conjugates, 240
curcumin conjugates, 238–239
huperzine A conjugates, 240–241
lycopene conjugates, 240
molecular basis
AD, 418–420
ALS, 421
HD, 420
mode of action, 419
PD, 420–421
Parkinson's disease, 237–238
quercetin conjugates, 239
resveratrol conjugates, 239
role, 425
scylloinositol conjugates, 239–240
treatment, 424–425
types, 421–424
Neuropeptide Y (NPY), 200
New food constituents, 3
Niemann–Pick Cl-Like 1 (NPClL1), 358
Niosomes, 14–15
Nonalcoholic fatty liver disease (NAFLD), 357
Non-allergic natural phospholipids, 185
Non-digestible fibers, 124
Non-opioid delivery methods, 150
Non-steroidal anti-inflammatory drugs (NSAIDs), 86
Non-traditional nutraceuticals, 349
Noradrenaline, 153
Noscapine, 335
NP-nutraceutical (NP-NU) conjugation techniques
cancer management
anticancer therapeutics, 241
antitumor therapeutics, 242
dihydroartemisin conjugates, 242–343
eugenol conjugates, 243
naringenin conjugates, 243
resveratrol conjugates, 243
thymoquinone conjugates, 242
nanoprecipitation, 235
neurodegenerative diseases
Alzheimer's disease, 236
asiatic acid conjugates, 240
bryostatin conjugates, 240
curcumin conjugates, 238–239
huperzine A conjugates, 240–241
lycopene conjugates, 240
Parkinson's disease, 237–238
quercetin conjugates, 239
resveratrol conjugates, 239
scylloinositol conjugates, 239–240
oxidative stress
atherosclerosis, 246
biological consequences, 244
vs. cardiovascular disease, 245
curcumin conjugates, 246–247
diabetic retinopathy, 246
epigallocatechin gallate conjugates, 247
ischemic and reperfusion injuries, 245
lutein conjugates, 247
thromboembolic conditions, 245–246
spray-chilling, 235
spray-drying, 234
Nutraceutical delivery, 88–89
Nutraceutical enzymes, 349, 422
Nutraceuticals, 2, 264
animal-derived polysaccharides, 280
challenges, 227–229
classification of, 348
compensations, 4
constituents of, 305–306
disadvantages, 338
existing problems, 330
manufacturing to marketing, 5
nano (*see* Nano-nutraceuticals)
to nanoceutical, 6–8
nanoencapsulation (*see* Nanoencapsulation)
nanosystems, 336–337
natural/traditional, 348–349
non-traditional, 349
plant-based polysaccharide nanocarriers, 274–278
present status, 330
research gaps, 249–250
safety evaluation/toxicity concerns, 247–249
safety of nanocarriers, 285
seaweed-derived polysaccharides, 282–283
well-documented benefits, 228
Nutritional supplements, 400

O

Obesity, 123, 129–133
Oligofructose, 129
Oligo-hyaluronic acid, 279
Oligosaccharides, 124
One-dimensional nanostructured materials, 386
Opiates, 149
Opioids, 149, 152
Oral delivery
  antioxidant potential, 389, 390
  gastrointestinal effect, 389, 390
  mode of action, 390–391
Oral nutraceuticals, 24
Oregano essential oil (OEO), 335
Oxidative stress, 80
  atherosclerosis, 246
  biological consequences, 244
  vs. cardiovascular disease, 245
  curcumin conjugates, 246–247
  diabetic retinopathy, 246
  epigallocatechin gallate conjugates, 247
  ischemic and reperfusion injuries, 245
  lutein conjugates, 247
  thromboembolic conditions, 245–246

P

Paeonia emodi, 351
Paraventricular (PVT), 153
Parkinson's disease (PD), 237–238, 420–421
Particle-induced X-ray emission (PIXE), 11
Particle size analysis, 113
Pectin, 269–270
Pectin-based nanodelivery systems, 270
Pectin-based nanoemulsions, 269
Pectin-coated nanoliposomal delivery system, 267
Pectin–protein systems, 269
Peptide-loaded lipid nanoparticles (PLLNs), 111
Peptide YY (PYY), 131
Peroxisome proliferator-activated receptor (PPAR-γ), 127
Persian gum, 272
Phantraceuticals, 74
Phase inversion method, 106–107
Phenolics, 332
Phosphatidic acid (PA), 184
Phosphatidylcholine (PC), 184
Phosphatidylcholine liposomes (PCLs), 352
Phosphatidylethanolamine (PE), 184
Phosphatidylserine (PS), 184
Phospholipid-based nanoemulsion, 185–188, 189
Phospholipid–drug complexes, 205, 207
  pharmacosomes, 209–211
  phytosomes, 207–209
Phospholipids, 184
Phosphorylated cellulose nanocrystals (PCNs), 267
Phosphorylation of cAMP response element-binding protein (pCREB), 358
Photon correlation spectroscopy (PCS), 111
Physicochemical properties, 98
Phytochemicals, 390
Phytoconstituent nutraceuticals, 80
Pickering emulsion, 17, 271–272
Piperine–phospholipid complex (PPC), 208
Plant-based gums, 271–272
Plant-based polysaccharide nanocarriers
  cellulose, 268–269
  dextran, 272–273
  gums, 271–272
  inulin, 273
  nutraceuticals, 274–278
  pectin, 269–270
  starch, 270–271
Plymeric mcelles (PMs), 232
Polyacrylamide (PAAm), 86
Poly (acrylic acid) (PAA), 86
Poly (ε-caprolactone) (PCL) nanoparticles, 53
Poly(2-(dimethylamino) ethylmethacrylate) (PDMAEMA), 86
Poly-D-lysine (PDL), 68
Polyethylene glycol (PEG), 104, 112–113, 184, 231, 335
Polyethyleneimine (PEI), 233
Polyethylene oxide (PEO) polymers, 184
Polyglycolic acid (PGA), 85
Polylactic acid (PLA), 85
Poly (lactic-co-glycolic) acid (PLGA), 85, 159, 335
Poly lactide-co-glycolide-polyethylene glycol (PLGA-PEG), 238
Polymeric micelles, 200
Polymeric nanoparticles (PNPs), 73
Polymerization, 18
Polymers, 84
Polymersomes, 404
Polymethoxyflavones (PMFs), 68, 71
Poly(N-isopropylacrylamide) (PNIPAAm), 86
Poly(N-vinylcaprolactam) (PVCL), 86
Polyphenolic compounds, 82, 151, 352
Polysaccharide-based nanocarriers, 264–267
Polysaccharides, 85, 124, 264, 267
Polyunsaturated fatty acids (PUFAs), 2, 185
Polyurethane (PU) hydrogels, 86
Polyvinyl alcohol (PVA), 68, 318
Pomegranate-derived anthocyanin
  brain pathway, 152–153
  composition of, 151
  molecular changes in opioid, 153–154
  MOR proteins, 154–155
  non-opioid agents, 155–157
  opioids, 152
Pomegranate extract, 150
Positron emission tomography (PET), 402
Prefrontal cortex (PFC), 153
Probiotic microorganisms, 348, 422
Probiotics, 83
Proliferator activated receptor gamma (PPARγ), 357
Protein-based nanocarriers, 389
Protein kinase A/protein kinase B (PKB/PKA) pathway, 423
Proteins, 84–85

Q

Quantum dots (QDs), 233
Quercetin (QC), 68, 361
Quercetin conjugates, 239

R

Raffinose, 124
Rapid expansion of supercritical solutions (RESS), 191
Rapid reticuloendothelial system (RES), 232
Reactive oxygen species (ROS), 73, 80, 157, 243, 313, 349–350, 389, 447
Receptor acting glycation end-products (RAGEs), 246
Recombinant nutraceuticals, 349, 422
Red algae
- agar, 282, 283
- carrageenans, 281

Red wine, 82
Regulations
- legal aspects, 449–451
- principles, 451
- risk of application, 451–452

Resveratrol, 53, 68, 82, 352, 362, 365
Resveratrol conjugates, 239, 243
Reticulo-endothelial system (RES), 184, 192
R-lipoic acid, 52

S

Scanning electron microscopy (SEM), 113, 444
Scattering of light, 12
Scurvy, 332
Scylloinositol conjugates, 239–240
Seaweed-derived polysaccharides, 280–281, 284–285
- brown algae
  - alginate, 284
  - fucoidans, 283–284
  - laminaran, 284
- green algae, 283
- nutraceuticals, 282–283
- red algae
  - agar, 282, 283
  - carrageenans, 281

Seaweed polysaccharide–based nanoparticles (SPBNPs), 281
Secondary ion mass spectrometry (SIMS), 12
Secondary metabolites, 422
Secrete short-chain fatty acids (SCFAs), 71
Selected area electron diffraction (SAED), 446
Selective serotonin reuptake inhibitors (SSRIs), 236
Self-nanoemulsifying drug delivery system (SNEDDS), 208
Short-chain fatty acids (SCFAs), 127, 131
Short interfering RNA (siRNA), 233
Silica-based nanoparticles, 168–169
Silver nanoparticles (Ag NPs), 312–315
Single-walled and multi-walled nanotubes (SWNTs/MWNTs), 19
Size-controlled synthesis, 265
Small interfering RNA (siRNA), 74
Smart polymers, 86
Solid-core micelles, 388
Solid lipid nanoparticles (SLNs), 14, 57, 67, 68, 99, 201, 231–232
Solid lipids, 99, 104
Solid solution, 232
Solubilization, 24–25
Solvent emulsification-diffusion technique, 109
Solvent emulsification-evaporation technique, 108–109
Solvent injection method, 106
Soybean phosphatidylcholine (SPC), 198
Spray-chilling technique, 234, 235
Spray-dried emulsions, 266
Spray-drying, 111
Spray-drying technique, 234
Squalene, 185
Stabilized Centella extract (SCE), 208
Stabilizing agent
- poloxamer, 111–112
- polyethylene glycol, 112–113

Stachyose, 124
Starch nanocrystals, 270
Starch nanofibers, 270
Starch nanoparticles, 270
Sterically stabilized mixed micelles (SSMMs), 202
Structure–activity relationships (SAR), 226
Submicron emulsions, 53
Substantia nigra (SN), 420
Superoxide dismutase (SOD), 352
Superparamagnetic iron oxide Nps (SPIONs), 233
Surface plasmon resonance (SPR), 11
Surfactants, 117
Synergism, 74
Synthesis methods, 265
- biogenic methods, 304
- chemical methods, 303–304
- green (biogenic) synthesis methods, 304

Synthetic biodegradable polymers, 85
Synthetic chemotherapeutic agents, 53
Synthetic opioids, 152
Syzygium cumini, 351

T

Taurine, 82
Tetramethylammonium hydroxide (TMAH), 11
Three-dimensional nanostructured materials, 386
Thymoquinone (TQ), 227, 356
Thymoquinone conjugates, 242
Titanium dioxide nanoparticles (TiO2), 315–316
Tobacco mosaic virus (TMV), 232
Tomatoes, 81
Tomato phytochemicals, 81
Toxic Substances Control Act (TSCA), 31
Transactive response deoxyribonucleic acid (TAR DNA), 423
Transferosomes, 192–194
Transmission electron microscopy (TEM), 73, 113, 190, 444
Triacylglycerol (TG), 187
Trilaurin (TL), 198
Trimyristin (TM), 198
Tripolyphosphate (TPP), 266
Tumor necrosis factor (TNF), 350
Two-dimensional nanostructured materials, 386
Tyrosine hydroxylase (TH), 154
Tyrosine kinase B receptor (TrkB), 356

U

Ultrafiltration, 115
Ultrasonication technique, 109–110, 266
Ultrasound-aided emulsification, 18
United States Food and Drug Administration (USFDA), 2
Upper critical solution temperature (UCST), 86

V

Vanillyl alcohol (VA), 245
Vasoactive intestinal peptide (VIP), 200
Vasodilator-stimulated phosphoprotein (VASP), 361
Ventral tegmental area (VTA), 152
Virus-based Nps (VNps), 232–233
Vitamin C, 55–56
Vitamin E, 81
Vitamins, 332

W

Waist circumference (WC), 358
Water-insoluble ingredients, 6
Whey proteins, 71
Wide-angle X-ray diffraction (XRD), 114
Wine phytochemicals, 82

X

X-ray diffraction (XRD), 74
X-ray photoelectron spectroscopy (XPS), 12

Z

Zein-derived peptide nanoparticles, 57
Zero-dimensional nanostructured materials, 386
Zeta potential (ZP), 111
Zinc oxide nanoparticles, 316–318
Zwitterionic phospholipid emulsifiers, 185

For Product Safety Concerns and Information please contact our EU
representative GPSR@taylorandfrancis.com
Taylor & Francis Verlag GmbH, Kaufingerstraße 24, 80331 München, Germany

www.ingramcontent.com/pod-product-compliance
Lightning Source LLC
Chambersburg PA
CBHW080125220326
41598CB00032B/4961

* 9 7 8 1 0 3 2 1 5 5 6 7 8 *